World Scientific Lecture Notes in Physics Vol. 9

SPIN GLASS THEORY AND BEYOND

World Scientific Lecture Notes in Physics Vol. 9

SPIN GLASS THEORY AND BEYOND

Marc MEZARD

Laboratoire de Physique Theorique de l'Ecole Normale Superieure, 24 rue Lhomond, 75231 Paris Cedex 05

Giorgio PARISI

Dipartimento di Fisica, Universita di Roma II, Tor Vergata, via Orazio Raimondo, Roma 00179, and INFN, Sezione di Roma

Miguel Angel VIRASORO

Dipartimento di Fisica, Universita di Roma I, La Sapienza Piazzale Aldo Moro 2, Roma 00185, and INFN, Sezione di Roma

World Scientific

Singapore • New Jersey • Hong Kong

Published by

World Scientific Publishing Co. Pte. Ltd.
5 Toh Tuck Link, Singapore 569224
USA office: Suite 202, 1060 Main Street, River Edge, NJ 07661
UK office: 73 Lynton Mead, Totteridge, London N20 8DH

The editors and publisher would like to thank all authors and the following publishers for their permission to reproduce the reprinted papers found in this volume:
American Association for the Advancement of Science (*Science*);
American Physical Society (*Phys. Rev. & Phys. Rev. Lett.*); IOP Publishing Ltd. (*J. Phys.*) Les Editions de Physique (*J. de Phys., J. Phys. Lett. & Europhysics. Lett.*); North-Holland (*Nucl. Phys. & Phys. Lett.*); Taylor & Francis Ltd. (*Phil. Mag.*)

To those who have not granted us permission before publication, we have taken the liberty to reproduce their articles without consent. We shall however acknowledge them in future editions of this work.

First published 1987
Reprinted 1993, 2002, **2004**

SPIN GLASS THEORY AND BEYOND

ISBN 9971-50-115-5
ISBN 9971-50-116-3 (pbk)

Printed in Singapore.

PREFACE

Motivated initially by the will to understand the strange behavior of certain magnetic alloys (which it has not fully explained yet), the theory of spin glasses has acquired by now an independent existence. The theory developed at the mean field level demonstrates a very rich and complex structure; and yet we believe that the results so derived are applicable to a wide range of complex systems.

Our aim in this book is first to provide a self contained and coherent description of spin glass theory and second to point out some of its applications in other fields.

We are firmly convinced that the techniques developed for spin glasses: the replica theory, the TAP approach and the cavity method can be applied to a myriad of other problems that otherwise are very difficult to handle. On the other hand a newcomer, even convinced of the importance of the subject may feel rejected by the apparent difficulty of the complicated algebra. We think that the moment is ripe to try a presentation of these techniques that demonstrates their essential simplicity.

This is not a review on spin glasses. The interested reader is referred to the recent and complete one by Binder and Young (1986) and to the proceedings of the Heidelberg Conferences (Morgenstern and Van Hemmen eds, 1983, 1987).

In the first part on spin glass theory we present a rather complete account of our personal point of view on the subject. Emphasis is put on clarity and coherence, sacrificing somehow the historical development and giving up completely the idea of describing the various streams of the enormous literature.

The last two parts have a different objective. We try to give examples of the diffusion of the ideas and techniques of spin glass theory in two other domains: optimization and biology. These are new and very rapidly evolving research subjects. We give the highlights of the most active among them today. The understanding and even the focal points of attention may change a lot here, but perhaps some of the basic ideas and analogies inspired by spin glasses that we present will remain.

The space dedicated to the "Cavity Method" has grown all along while working on this book. In preparing a book one painfully realises how dispersive history is and one feels compelled to do something to unify the presentations. We were happy to find out that the cavity method originally introduced to replace the physically less explicit Replica Method could be used also to: 1) Derive TAP-like equations; 2) Analyse the fluctuations around the mean field solution and 3) Analyse the dynamical behavior.

Most of this work is new and presented here for the first time, though in general the same results are derivable by other methods.

Each part of the book is complemented by a collection of reprints. They will provide the reader with the details and the explicit calculations which are not given in our presentation. The collection is not a representative subset of the literature. Neither do they imply a judgement on historical relevance. They have been chosen simply because they complement the presentation.

It is a pleasure for us to thank our colleagues with whom the vision of this subject was shaped. In the first place our collaborators: C. De Dominicis, D. Gross, J. P. Nadal, G. Paladin, N. Parga, R. Rammal, N. Sourlas, G. Toulouse, J. Vannimenus but also P. W. Anderson, C. Bachas, E. Brézin, B. Derrida, C. Itzykson, S. Kirkpatrick. Special acknowledgement is due to D. Amati, Anderson, Kirkpatrick and Toulouse for carefully reading the first version of this book and helping us to improve it through useful comments.

This book has been written in Roma and Paris almost in equal parts. A grant from the CEE has facilitated this collaboration.

This book is dedicated to our children: Diego, Leonardo, Lorenza, Sabine, Vincent.

December 1986, Roma and Paris,
M. Mézard, G. Parisi, M. A. Virasoro

CONTENTS

Reprints SPIN GLASSES

Part 2 OPTIMIZATION

Reprints OPTIMIZATION

Part 3 BIOLOGICAL APPLICATIONS

Reprints BIOLOGICAL APPLICATIONS

Chapter 0
A KIND OF INTRODUCTION

Often in life we find out that our goals are mutually incompatible: we have to renounce some of them and we feel frustrated. For example, I may want to be a friend of both Mr. White and Mr. Smith. Unfortunately, they hate each other: it is then rather difficult to be a good friend of both of them (a very frustrating situation).

The situation is more complex when many individuals are present. In a classical tragedy the scenario may be the following: there is a fight between two groups and the various characters on the scene have to choose sides. In addition they all have strong personal feelings, positive or negative, towards each other (it is a tragedy!). Some of them are friends and some are enemies. For simplicity we will assume that all feelings are reciprocal; otherwise the system may never reach equilibrium (this more general case, though much more complicated can be studied. See Reprint 34 for one particular example). Let us consider three characters (A, B and C); if A and B, B and C, A and C do like each other, there is no problem: they will all choose the same side. In a similar way, if A and B are friends and C is an enemy of both, then A and B can be on one side and C will be on the other. Frustration follows, instead, if A, B and C hate each other because two personal enemies must then fight on the same side.

This analysis can be formalized by assigning to each pair a number J_{AB} which is $+1$ if A and B are friends and -1 if they hate each other; the relation among three characters is frustrated if (Ref. 1)

$$J_{AB} \cdot J_{BC} \cdot J_{CA} = -1 . \tag{0.1}$$

When many triples are frustrated, evidently the situation on the scene is unstable and many rearrangements of the two fields are possible.

At a given moment of the tragedy it is possible to define the "dramatic tension" as

$$\text{Number of frustrated triples/Total number of triples} . \tag{0.2}$$

Detailed studies[2] have shown that in many Shakespeare's plays the dramatic tension has a small value at the beginning of the tragedy, reaches a maximum in the middle and decreases by the end.

Mathematically we could say that we have N variables s_i, one for each character; s_i

takes values $+1$ or -1 depending on which side the ith character stays; for a given set of the J_{ik} we are interested in minimizing the "disconfort" function

$$H_J[s] = -\sum_{i>k} J_{ik} s_i s_k , \tag{0.3}$$

where the square parenthesis are used to stress that H is a function of all the s_i together.

The function H is often called (depending on the context) the cost function or the Hamiltonian (the energy).

It is a well-known, and hard, mathematical problem to find a fast algorithm which, for a given instance of the J's, computes the set of s_i that mimimize H_J. From the point of view of complexity theory this problem is NP complete, which means that very likely there is no algorithm that can find the minimum using a computer time increasing as a power of N for large N. In other words it is generally believed that the time needed by the best algorithm increases as $\exp(cN)$ with c constant when $N \to \infty$.

In this book we will describe an approach to this problem that makes essential use of the tools of statistical mechanics: we suppose that the J's are chosen at random (according to a probability law) and we are interested in finding analytical properties of both the absolute minimum of H and of other local minima. We will try to find out how much these minima are different and which are the relations among them. We are less concerned about constructing a fast algorithm for finding the absolute minimum, though it is clear that an analytical knowledge of the properties of these minima may be useful in this task.

A typical result that has been obtained is the following: if each J_{ik} is chosen $+1$ or -1 with equal probability and independently of the others, in the limit $N \to \infty$, the minimum of $H_J/N^{3/2}$ takes the same value for almost all realizations of the J's and this value is about $-.7633$. In other words, the best possible arrangement is that in average the number of friends minus the number of enemies among our allies be $(.7633/2)N^{1/2}$. For 10 000 persons, and to the best of everybody's effort, on my same side, I will have to bear with approximately 2462 enemies, thirty eight less than if the choice of side had been decided by tossing a coin! This sobering result is related to the large proportion of frustrated triples (1/2 of the total number), a consequence of the uncorrelated character of the J_{ik}. A different result, also discussed in this book follows if

$$J_{ik} = \sum_{p=1}^{P} \xi_i^p \xi_k^p , \tag{0.4}$$

where now the ξ_i^p are independent random ± 1 variables[3]. For $P = 1$ there are no frustrated triples while for $P \gg N$ it reduces to the previous case. There are interesting different regimes in P/N (see Chap. XIII).

The properties of those configurations $\{s_i\}$ which have low energy but are not minima are also interesting. They can be treated by assigning to each configuration s_i a probability

$$P_J[s] = \exp(-\beta H_J[s])/Z_J , \tag{0.5}$$

where the "partition function" Z_J is such that the sum of the probabilities is 1

$$Z_J = \sum_{\{s\}} \exp(-\beta H_J[s]). \tag{0.6}$$

In statistical mechanics H_J is the energy of the system and $P_J[s]$ is the Gibbs probability distribution ($\beta = 1/(kT)$, where k is Boltzmann constant and T is the absolute temperature).

We will denote by $\langle g \rangle$ the expectation value of a function $g[s]$ according to the probability distribution (0.5)

$$\langle g \rangle = \sum_{\{s\}} P_J[s] g[s] \tag{0.7}$$

(sometimes we will omit the J subindex to lighten the notation).

These are simple examples in which we want to find the minimum of a function in the presence of requirements that push towards different directions and it is difficult to establish which could be the best compromise. An essential ingredient, in addition to frustration is the so-called disorder. In this context this means that the structural parameters defining the problem (here the J_{ik}) are themselves the outcome of a random choice—or at least have large complexity. It is generally believed that the results obtained for one model are relevant for other disordered frustrated systems though to what degree they share the same properties is still unclear. Although this version of the model has its own interest, historically it was first studied in relation with spin glasses.

Spin glasses (also called amorphous magnets) are magnetic substances in which the interaction among the spins is sometimes ferromagnetic (it tends to align the spins; $J_{ik} > 0$), sometimes antiferromagnetic (it antialigns the spins; $J_{ik} < 0$). The sign of the interaction is supposed to be random.

In some spin glasses the spins can take only two values ± 1 (Ising spins) and the form of the energy is the same as in Eq. 0.3. The probability law for the J's distinguish various models. If all the J's have the same probability distribution, all spins interact with the others and the infinite range model so defined is called the Sherrington-Kirkpatrick (in short SK) model. Another possibility is to choose J_{ik} different from zero only for nearby pairs of spins (short range model: a spin interacts only with its neighbours).

In the following chapters we will discuss almost exclusively the SK model in the limit $N \to \infty$. We will show that the mean field theory can be considered to be satisfactorily understood. The replica theory with its strange $n \to 0$ limit will be discussed at length stressing the fact that though unusual, once one masters the different steps, it is a formidable method to derive the mean field approximation for disordered frustrated systems. The TAP approach (Chap. II) which was originally proposed as an alternative to replicas, when correctly complemented leads to the same result obtained by breaking replica symmetry (Chap. V). The main conclusion of this part concerns the complicated structure of the configuration space landscape of the

SK free energy which at low temperatures has an infinite number of valleys which are organized ultrametrically (Chap. IV). A chapter is dedicated to the dynamics of spin glasses (Chap. VI).

In the second part dedicated to Optimization, after two introductory chapters dedicated to the presentation of the subject, we demonstrate the power of the methods discussed in the first part by showing how they can be applied to the matching problem (Sec. IX.2), the travelling salesman problem (Sec. IX.3), the assignment problem (Sec. IX.4) and graph partitioning (Sec. IX.5).

In the third part we discuss applications to the problem of biological organization. The general strategy is presented in Chap. X while in Chap. XI we discuss why an infinite valley landscape can be useful to model some aspects of prebiotic evolution. In Chap. XII we introduce the ideas behind brain modelling while in Chap. XIII we show that one particular model, the Hopfield model, can again be analyzed as a spin glass.

References

1. Toulouse (1977).
2. Marcus, contribution to the *Seminaire de Semiotique Theatrale*, Urbino (1974), unpublished and Doreian (1971).
3. These examples are relevant to the so-called *voting problem* discussed in sociology (R. Axelrod, private communication).

Part 1
SPIN GLASSES

Chapter I
THE REPLICA APPROACH

I.1 Quenched Averages and Replicas

We suppose (as in the introduction) that we have a system whose Hamiltonian $H_J[s]$ depends on the configuration $[s]$ of the spins and on some control variables J, which are distributed according to a probability distribution $P[J]$. For each choice of the J's we can compute the partition function

$$Z_J = \sum_{\{s\}} \exp\{-\beta H_J[s]\} \tag{I.1}$$

and the free energy density

$$f_J = -1/(\beta N) \ln Z_J = -1/(\beta N) \ln \left\{ \sum_{\{s\}} \exp\{-\beta H_J[s]\} \right\}, \tag{I.2}$$

where N is the total number of variables s's.

Standard statistical mechanics deals with the problem of computing f at given J. Here we suppose that the J's are not known, but they are random variables and that their probability distribution is known. The J's are called quenched variables. Physically the necessity of computing the free energy f at fixed J's results from the fact that the changes of the J's happen on a time scale which is infinitely larger than the time scale characterizing the changes in the s's.

If the form of the interaction is not too pathological, standard thermodynamic arguments imply that in the limit N going to infinity (infinite volume limit) the free energy density assumes the same value (i.e. it is said to be a selfaveraging quantity) for each set of the control parameters J's which has a non vanishing probability. We naively expect that the fluctuations for different choice of the J's are proportional to $1/N^{1/2}$

$$\overline{f_J^2} - (\overline{f_J})^2 = O(1/N), \tag{I.3}$$

where we denote by a bar the average over the J's.

For models with short range interaction this result (Eq. (I.3)) can be established using a general argument[1]: we separate the system into many (let us say K) macroscopic subsystems and each of these subsystem is weakly interacting with the others; the fluctuations in the total free energy must be at most of order $1/K^{1/2}$: K is an arbitrary large number (although much smaller than N) so that the fluctuations of the free energy must vanish when $N \to \infty$, (although they may be proportional to $1/N^{\nu}$ with $\nu \neq 1/2$). For the infinite range model studied hereafter this property has not been rigorously demonstrated (even at high temperatures) although the selfaverageness of the free energy is well seen numerically[2]. (Of course this does not mean that other—non extensive—quantities are self averaging; we shall find counterexamples in Chap. IV).

We are interested in computing the average value of the free energy density

$$f = \sum_J P[J] f_J = \overline{f_J}. \tag{I.4}$$

The direct computation of Eq. (I.4) is not simple because it is not of the type which is usual in statistical mechanics; for example the computation of $\ln\{\sum_J P[J] \sum_{\{s\}} \cdot \exp\{-\beta H_J[s]\}\}$ (the so-called annealed case) would be more straightforward.

The replica method has been originally proposed as a trick to simplify the computation of Eq. (I.4) (Ref. 3); later on it has been found that the replica method (when the replica symmetry is broken) is very powerful in coding in a simple and compact way, complex properties of the system.

In a nutshell the idea at the basis of the replica method is very simple: it consists in computing the average of f (Eq. (I.4)) by some analytic continuation procedure from the average of the partition function of n uncoupled replicas of the initial system. We start with some preliminary definitions

$$Z_n = \sum_J P[J] \{Z_J\}^n \equiv \overline{(Z_J)^n},$$

$$f_n = -1/(\beta n N) \ln Z_n. \tag{I.5}$$

Using relations $\sum_J P[J] = 1$ and $A^n \approx 1 + n \ln(A)$ for $n \approx 0$, it is evident that

$$\lim_{n \to 0} f_n \equiv f_0 = \bar{f}. \tag{I.6}$$

For integer n we can write

$$(Z_J)^n = \sum_{\{s^1\}} \sum_{\{s^2\}} \cdots \sum_{\{s^n\}} \exp\left\{ -\sum_{a=1}^{n} \beta H_J[s^a] \right\}, \tag{I.7}$$

where we have introduced n replicas of the same system (with the same set of J's); indeed the partition function of n (non-interacting) replicas of the same system is the partition function of the original system to the power n; the variables $s^a{}_i$ carry two indices: the upper one denotes the replica and goes from 1 to n and the lower one denotes the site and goes from 1 to N. In the end we must perform the two limits $n \to 0$

and $N \to \infty$ (in principle one should take the limit $n \to 0$ first and afterwards $N \to \infty$; practically as we shall see the order of the limits is reversed, but it turns out that this should not be a source of troubles[4]).

The replica trick consists in the following steps: we use Eq. (I.7) to define the function f_n for integer n, we extend this function to an analytic function of n and finally we compute $f_0 = \bar{f}$. If the partition function is expanded in power of β (in this way we obtain the correct behaviour of the system in the high temperature region), f_n is a polynomial in n and the method is quite safe. Problems may arise and will arise in the low temperature region.

I.2 Replica Symmetric Solution of the *SK* Model

We can apply this method to the case of infinite range spin glasses; the Hamiltonian is

$$H_J[s] = -\sum_{1 \le i < k \le N} J_{i,k} s_i s_k - \sum_i h s_i , \tag{I.8}$$

where h is the magnetic field and the J's are independent random variables with zero mean and variance $1/N$

$$\overline{J_{i,k}} = 0, \qquad \overline{J_{i,k}{}^2} = 1/N, \qquad J_{i,k} = J_{k,i} . \tag{I.9}$$

It is usually supposed that the distribution of the J's is Gaussian, although the same results can also be obtained for non Gaussian distributions (such as $J_{i,k} = \pm 1/N^{1/2}$, with equal probability for the plus and the minus sign): as far as the J's are quantities of order $1/N^{1/2}$, the results depends only on the first two moments of the distribution[5]. The thermodynamic limit is obtained when N goes to infinity; the factor $1/N$ in Eq. (I.9) has been chosen in such a way that at fixed β the total energy is proportional to N and therefore the energy density is N-independent.

For fixed J's we expect that in the high temperature phase the local magnetization $m_i \equiv \langle s_i \rangle$ is different from zero only if a magnetic field is present and it vanishes when the magnetic field goes to zero; on the other hand we naively expect that in the low temperature region there should be (as in the case of ferromagnets) some freezing of the spins in the position which is mostly favoured energetically, hence the m_i should be different from zero also at $h = 0$. However the local magnetization m_i depends on the J's and it will sometimes be positive and sometimes negative (we will see afterwards that the global magnetization density $(1/N)\sum_i m_i$ is zero at $h = 0$) so it is convenient to characterize the system in terms of the quantity

$$q_{EA} \equiv (1/N) \sum_{i=1}^{N} m_i{}^2 \tag{I.10}$$

the so-called Edwards Anderson order parameter[3]. Using the definition given in Eq. (I.10), q_{EA} can depend on the J's; hereafter we shall compute the sample averaged value

$\bar{q}_{EA} \equiv \overline{m_i^2}$: we will make no distinction between $\bar{q}_{EA}$ and q_{EA}; in fact it is possible to show that (using a good definition of the m's as the averages within one pure state) q_{EA} does not depend on the J's in the limit $N \to \infty$ (see Chap. IV); we can thus conclude that q_{EA} is also equal to $\overline{m_i^2}$. Summarizing, q_{EA} should be equal to zero at $h = 0$ for a temperature greater than a certain critical temperature (T_c), while it should be different from zero at low temperature[6].

Let us now apply the replica method (Eq. I.5) to the Hamiltonian (I.8) (Ref. 5). We shall not work out all the details here but just explain the few basic steps and refer the reader to Kirkpatrick and Sherrington, Reprint 4, for a more extensive presentation. Z_n is expressed as

$$Z_n = \sum_J P(J) \sum_{\{s\}} \exp \sum_{a=1}^{n} \left[\beta \sum_{i<k} J_{ik} s_i^a s_k^a + \beta h \sum_i s_i^a \right] \tag{I.11}$$

where $\sum_{\{s\}}$ denotes the sum over all the configurations of the spins s_i^a ($a = 1, n$; $i = 1, N$). Eq. I.11 gives after sample averaging

$$\begin{aligned} Z_n &= \sum_s \exp\left[\beta^2/2N \sum_{i<k} \left(\sum_{a=1}^{n} s_i^a s_k^a \right)^2 + \beta h \sum_i \sum_{a=1}^{n} s_i^a \right] \\ &= \sum_s \exp\left[\beta^2 N n/4 + (\beta^2 N/2) \sum_{1\le a<b\le n} \left(\sum_i s_i^a s_i^b / N \right)^2 + \beta h \sum_i \sum_a s_i^a \right]. \end{aligned} \tag{I.12}$$

Using well-known properties of Gaussian integrals, this can be rewritten as

$$\begin{aligned} Z_n &= \int \prod_{a<b} (dQ_{ab} (N\beta^2/2\pi)^{1/2}) \exp\{-NA[Q]\}, \\ A[Q] &= -n\beta^2/4 + \beta^2/2 \sum_{1\le a<b\le n} (Q_{a,b})^2 - \ln Z[Q] \\ Z[Q] &= \sum_{\{S\}} \exp\{-\beta H[Q,S]\} \\ H[Q,S] &= -\beta \sum_{1\le a<b\le n} Q_{a,b} S_a S_b - h \sum_{a=1}^{n} S_a, \end{aligned} \tag{I.13}$$

where the matrix Q is an $n \times n$ symmetric matrix, zero on the diagonal, and the sum over $\{S\}$ goes over the 2^n configurations of the variables S_a, $a = 1, n$, ($S_a = \pm 1$).

Equation (I.13) suggest that Z_n can be computed through a saddle-point method which gives

$$f_n \equiv -1/(\beta N n) \log(Z_n) = 1/(\beta n) \min A[Q]. \tag{I.14}$$

Therefore we must find the solutions of the $n(n-1)/2$ saddle-point equations $\partial A/\partial Q_{a,b} =$

0. Once the saddle-point matrix Q_{sp} has been found, the free energy is obtained using (I.5, 6) as

$$f = \lim_{n\to 0} (1/\beta n) A[Q_{sp}]. \tag{I.15}$$

It is interesting to note that the equations $\partial A/\partial Q_{a,b} = 0$ can be written under the form of self consistency equations

$$Q_{a,b} = \langle S_a S_b \rangle_Q = \overline{\langle S_a S_b \rangle}, \qquad a < b \tag{I.16}$$

where the expectation value $\langle\ \rangle_Q$ is taken with respect to the "single site" Hamiltonian $H[Q, S]$.

The function $A[Q]$ is left invariant when we exchange some of the lines (or the rows) of the matrix Q and therefore the group of permutations of n elements (P_n or S_n) is a symmetry of the problem (all the replicas are equivalent!); this group is often called the replica group.

For positive integer n (different from zero) the minimum of A can be found and the matrix Q has the following form[4]

$$\forall a, b\ a \neq b : Q_{a,b} = q; \qquad \forall a : Q_{a,a} = 0. \tag{I.17}$$

This form of the matrix Q is the only one which is left invariant by the action of the replica group and it is therefore the natural solution, usually named the replica symmetric solution. Its properties are studied in details in Ref. 5 and we shall only sketch a small part of the results here.

If we analytically continue the solution of the equation $dA/dq = 0$, up to $n = 0$, we find the following equation for q

$$q = \int_{-\infty}^{+\infty} dz/(2\pi)^{1/2} \exp(-z^2/2)\, \mathrm{th}^2(\beta q^{1/2} z + \beta h). \tag{I.18}$$

The free energy (I.15) is

$$f = -(\beta/4)(1-q)^2 - \int_{-\infty}^{+\infty} dz/(2\pi)^{1/2} \exp(-z^2/2) \ln(2\cosh(\beta q^{1/2} z + \beta h)). \tag{I.19}$$

At zero magnetic field Eq. (I.18) has only the solution $q = 0$ for $1/\beta \equiv T > T_c = 1$, for $T < T_c$ there is another solution (the physical one) where q is different from zero. In conclusion there is a phase transition at $T = 1$, $h = 0$ while there is no transition at $h \neq 0$.

Everything seems perfect; unfortunately a detailed computation[5] shows that the entropy (which in a discrete system is non negative by definition: it is the logarithm of the number of configurations) becomes negative at small temperature. At zero temperature we get $S(0) = -1/(2\pi) \approx -.17$. In this case the replica method leads to a disaster!

I.3 Stability

After some reflection it became evident that the argument was too hasty: in particular the conditions $\partial A/\partial Q_{a,b} = 0$ do not imply that $A[Q]$ is a minimum function of Q. Moreover the number of independent components of Q is $n(n-1)/2$, which becomes negative for $0 < n < 1$, and it is not clear how to give a precise definition of the minimum of a function which depends on a negative number of variables. Indeed let us consider a simple example

$$Z = \int dQ \exp(-NTrQ^2) \equiv \int dQ \exp(-Ng[Q])$$

$$g[Q] \equiv TrQ^2 = n(n-1)q^2, \tag{I.20}$$

the last equality being correct only if the matrix Q has the form shown in Eq. (I.17).

Everybody would agree that $g[Q]$ has a minimum at $Q = 0$, in particular the eigenvalues of the Hessian matrix $(\partial^2 g/\partial Q \partial Q)$ are positive (they are identically equal to one), however it is also evident that for $0 < n < 1$, $g[Q]$ as a function of q has a maximum and not a minimum at $q = 0$! This surprising result is connected to the fact that for $n < 1$, the number of eigenvalues becomes negative.

Now, our aim is to use the saddle-point method to evaluate an integral. If one takes care of the corrections to the leading result one finds that

$$\int d[Q] \exp\{-NA[Q]\} \propto 1/\det^{1/2}[\partial^2 A/\partial Q \partial Q] \exp[-NA[Q_{sp}]]\{1 + O(1/N)\} \tag{I.21}$$

and both the square root of the determinant and the $O(1/N)$ corrections involve the square root of the products of the eigenvalues of the Hessian: a negative eigenvalue would imply an imaginary result. It is therefore absolutely necessary that the eigenvalues of the Hessian be non negative; if they are zero, one should look at higher order terms and the corrections to the leading term in Eq. (I.21) can then have a different behaviour, e.g. $O(1/N^{1/2})$.

This argument can be made more precise, and a careful computation[6] including the degeneracies of the various eigenvalues shows that for any value of the magnetic field, the Hessian, evaluated at the solution of Eq. (I.18), becomes negative at sufficiently low temperature when $n < 1$; the Ansatz (I.17) gives the correct saddle-point for evaluating the integral only above a certain temperature $T_c(h)$ (see Fig. I.1). At low temperatures we must look for a new solution of the equation

$$\partial A/\partial Q_{a,b} = 0 \tag{I.22}$$

whose Hessian $(\partial^2 A/\partial Q \partial Q)$ has no negative eigenvalues.

The naive replica approach fails because there is a phase transition in the region

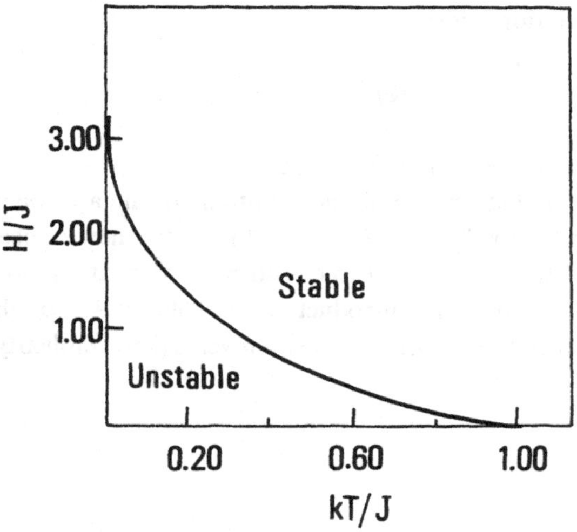

Fig. I.1. The phase diagram in the plane temperature-magnetic field. For $T > T_c(h)$—paramagnetic phase—the replica symmetric solution is correct. For $T < T_c(h)$—spin glass phase—the replica symmetric solution is unstable.

$0 < n < 1$ (Ref. 7) (in principle this statement can be checked directly because Z_n is well defined also for non integer n): we cannot do the analytic continuation for the free energy from its values on the integers down to its value at $n = 0$, but we can perform the analytic continuation of the equations which determine the matrix Q and look for other solutions of the equation at $n = 0$.

At this point it may be useful to summarize the order of the various steps of the replica approach. Starting from (I.13) one first takes the number of spins N to infinity which allows to compute Z_n with a saddle-point method. Then one writes down the saddle-point equations $\partial A/\partial Q = 0$ and the Hessian $\partial^2 A/\partial Q \partial Q$; then one exhibits an Ansatz for Q which is plugged into these equations, the $n \to 0$ limit being taken afterwards. Finally one solves the resulting $n \to 0$ limit of the saddle-point equations and works out the eigenvalues of the $n \to 0$ limit of the Hessian.

Unfortunately, if the matrix Q does not have the form shown in Eq. (I.17), it is not left invariant by the action of the replica group and we have to specify in details the value of the different elements of the matrix. As far as the matrix Q has no element in the limit n going to zero, the meaning of this operation is not clear *a priori*; moreover it is not clear at this stage why the saddle point should not be left invariant by the action of the replica group and which are the physical consequences of such non-invariance.

On the other hand the value of many thermodynamic functions computed within the above replica symmetric solution do not disagree too much with the numerical data, even at low temperatures where it is incorrect: for example the internal energy per spin at zero temperature (the ground state energy) comes out to be

$$U(0) = -(2/\pi)^{1/2} \approx -.798 \qquad \text{(I.23)}$$

while numerical simulations[8] give

$$U(0) = -.76 \pm .01 \tag{I.24}$$

which is a similar but definitely different value.

Superficially it seems that only small modifications of the approach are needed to get the correct results: we will see in Chap. IV that quite new concepts are needed. Before doing this it will be convenient to consider in the next chapter another more direct approach which avoids the introduction of replicas. Although this approach fails to solve the model, it is interesting because it reveals the complexity of the problem we face.

References

1. Brout, (1959).
2. This can be seen for instance from the results of Kirkpatrick and Sherrington (1978), Reprint 4, and Palmer and Pond (1978), but we do not know of any systematic study of this property. Analytic studies in the high temperature phase of the *SK* model have been done by Toulouse and Derrida (1981).
3. Edwards and Anderson, (1975) Reprint 1.
4. Van Hemmen and Palmer (1979, 1982).
5. Sherrington and Kirkpatrick (1975) Reprint 3. Kirkpatrick and Sherrington (1978) Reprint 4.
6. De Almeida and Thouless (1978) Reprint 5.
7. Kondor (1983).
8. The quoted result is from Kirkpatrick and Sherrington (1978) Reprint 4; Other results range from $-.755 \pm .01$ (Thouless, Anderson and Palmer 1978, Reprint 6) to $-.775 \pm .01$ (Palmer and Pond 1978).

Chapter II
THE TAP APPROACH

II.1 Mean Field Solution for a Given Sample

In the replica approach we first average out the disorder and later we construct a mean field theory with an effective order parameter, which is a very strange object: an n by n matrix with $n = 0$; one may feel that in this way we are putting the dirt under the carpet. The most natural choice seems to construct the mean field theory before averaging out the disorder.

The usual mean field equations in a magnet are

$$\langle s \rangle_i \equiv m_i = \text{th}(\beta \tilde{h}_i); \qquad \tilde{h}_i = h + \sum_{k=1}^{N} J_{i,k} m_k, \tag{II.1}$$

which can be obtained from the exact equations

$$\langle s \rangle_i = \left\langle \text{th}\left[\beta\left(h + \sum_{k=1}^{N} J_{i,k} s_k\right)\right]\right\rangle, \tag{II.2}$$

using the approximation $\langle f[s] \rangle \approx f[\langle s \rangle]$, i.e. by approximating the expectation value of a function of the s's with the function of the expectation values; this approximation amounts to neglect the fluctuations. It is usual to denote $\tilde{h}_i$ as the effective magnetic field on the point i.

In a normal ferromagnet the solution of Eq. (II.1) is very easy to find: the $J_{i,k}$ are known and they do not depend on the points i or k separately but only on the distance between the point i and k. In that case it is correct to assume that m_i does not depend on i. If for a given i a number z of coupling constants are different from zero and equal to $1/z$ (if the interaction is between nearest spins of a lattice, z is the coordination number of the lattice), Eq. (II.1) reduces to $m = \text{th}(\beta m + \beta h)$. In a spin glass the situation is much more complex because the J's are random quantities: they are not known but only their probability distribution is known; moreover the m_i must depend on i.

Before discussing the solution of Eq. (II.1) we should discuss whether it describes the infinite range model (defined in (I.8)) correctly. It is well known that for a ferromagnet when the range of the interaction (or the dimension of the space for a short range interaction) becomes infinite (generally speaking when z becomes infinite) Eq. (II.1) becomes correct and the mean field approach solves the model exactly.

In the case of the infinite range model for spin glasses it is possible to prove that in the limit $N \to \infty$, Eq. (II.1) is correct apart from a "small" modification: the magnetizations m_k, which appear in the definition of the effective magnetic field, are not the actual magnetizations of the spins, but they are the magnetizations that the spins would have in the absence of the spin i: we must consider a system with $N - 1$ spins, with the ith spin removed, and the magnetization of the ith spin can be computed from Eq. (II.1) (the cavity approximation).

In other words we must subtract from the effective field a "reaction term" which is the contribution due to the influence of the ith spin on the others. In this way we finally obtain

$$m_i = \mathrm{th}(\beta \tilde{h}_i) + O(1/N)$$

$$\tilde{h}_i \equiv h + \sum_{k=1}^{N} J_{i,k} \left\{ m_k - m_i J_{i,k} \chi_{k,k} \right\}$$

$$\approx \sum_{k=1}^{N} J_{i,k} m_k - \sum_{k=1}^{N} m_i J^2{}_{i,k} \beta (1 - m_k^2), \qquad \text{(II.3)}$$

where $\chi_{k,k}$ is the magnetic susceptibility of site k; it describes the answer of m_k to a small change of the effective field $\tilde{h}_k$

$$\chi_{k,k} = \partial m_k / \partial \tilde{h}_k = \partial [\mathrm{th}(\beta \tilde{h}_k)] / \partial \tilde{h}_k = \beta (1 - m_k{}^2). \qquad \text{(II.4)}$$

In the limit $N \to \infty$ the last expression in Eq. (II.3) reduces to

$$\tilde{h}_i = h + \sum_{k=1}^{N} J_{i,k} m_k - m_i (1 - q_{EA}), \qquad \text{(II.5)}$$

q_{EA} being defined in Eq. (I.10).

These equations have been studied in this context by Thouless, Anderson and Palmer[1] and they are often called the TAP equations. There are in the literature, many different proofs of this equation: the interested reader can find one in Ref. 2; the free energy corresponding to a solution of Eq. (II.3) can also be computed[1].

The TAP proposal was to compute directly the solution of Eq. (II.3) for a given choice of the J's: this should be possible because it is relatively simple to deduce analytically many properties of the matrix J when $N \to \infty$: for example the distribution of the eigenvalues of J is known and the maximum eigenvaue of J is equal to 2 (with probability one).

For instance let us study what happens in zero external field, starting from the high temperature (paramagnetic) regime. Assuming a continuous transition, one can see at what temperature the TAP equations can have a solution with non zero—but small—

values for the m_i's. Expanding the Eq. (II.3) to first order in the m's (with q of order m^2), we get

$$m_i = \beta \sum_j J_{ij} m_j - \beta m_i - O(m^3). \tag{II.6}$$

Let us call m_λ the projection of the magnetization vector onto an eigenvector $|\lambda\rangle$ of the matrix J, with corresponding eigenvalue J_λ. It satisfies

$$m_\lambda = \beta(J_\lambda - 1)m_\lambda - O(m^3). \tag{II.7}$$

As usual the terms in m^3 are saturation terms which tend to decrease the right-hand side of (II.7), so that a non zero solution for m_λ exists if and only if the slope $\beta(J_\lambda - 1)$ is larger than one. As the largest J_λ is equal to two, this means that, starting from high temperatures ($\beta \ll 1$) and increasing β, there is a phase transition at $\beta_c = 1/T_c = 1$. (It is interesting to notice that the same computation done without the reaction term would lead to $T_c = 2$, a completely wrong result.) At temperatures lower than one the local magnetizations acquire a spontaneous non zero expectation value; for small values of $\tau \equiv T_c - T$ one finds

$$m_i = \tau^{1/2}\phi_i + O(\tau), \tag{II.8}$$

where ϕ is the eigenvector of the matrix J corresponding to the eigenvalue 2.

In the original paper of TAP it was hoped that the computation of high orders in τ would be relatively simple: however it was later discovered that when one includes the order τ^3 the situation changes qualitatively: the solution m_i of the TAP equation has a zero projection on the vector ϕ.[3]

However this is only a minor complication which could be handled without difficulties: as we shall see in the next section, the serious problem is that as soon as $T < 1$, at $h = 0$, Eq. (II.3) has a number of solutions which increases exponentially with N: we face the embarassment of choosing.

A natural proposal is to assume that the statistical expectation value of a quantity is the average of the values that such a quantity takes in all possible solutions of the TAP equation, each solution contributing with a given weight. If for definiteness we consider the case of the local magnetization we should have

$$\langle s_i \rangle \equiv m_i = \sum_\alpha w_\alpha m^\alpha{}_i$$

$$\sum_\alpha w_\alpha = 1, \tag{II.9}$$

where the different solutions of the TAP equation are labeled by α ($m^\alpha{}_i$ is the magnetization at the point i of the αth solution of the TAP equation). The delicate point is the choice of the w_α; the first tentative was to take all w's equal to each other: one performs a white (unweighted) average out of all the solutions of the TAP equation[4]. This choice allows one to perform analytically all the computations needed to evaluate the thermodynamics of the model; unfortunately it leads to a terrible disaster, much worse than the replica approach: the value of the zero temperature energy is completely wrong ($U(0) \approx -.51$, the correct value being $\approx -.76$) and the specific heat becomes negative at low energy.

It soon became clear that the different solutions of the TAP equations have different free energy density. If we perform a weighted average (as we should do) and we set

$$w_\alpha \propto \exp(-\beta F_\alpha), \tag{II.10}$$

F_α being the total free energy of the αth solution, (which is simply expressed in terms of the $m^\alpha{}_i$), the only solutions which are relevant in the limit $N \to \infty$ are those which have a low value of the free energy.

II.2 About the Number of Solutions of the TAP Equations

In this framework the first step is to compute the number of solutions of the TAP equations as a function of the free energy density f (we recall that $f = F/N$, F being the total free energy). It has been argued[5] that

$$\overline{\text{Number of solutions}} \propto \exp\{N\omega_1(f)\} \tag{II.11}$$

the bar denoting the average over the J's, as usually. However in this case it is not evident that the average value of the number of solutions coincides with the most likely value of the number of solutions. At low temperature these numbers seem to be different; we still have

$$\text{Number of solutions} \propto \exp\{N\omega_0(f)\} \tag{II.12}$$

with probability 1 for large N (equivalently $\overline{\log(\text{Number of solutions})} \propto N\omega_0(f)$); ω_0 is smaller than ω_1 only in the low temperature (low free energy) region.

At a given temperature there exists a "critical" value of the free energy density f_0 such that for $f < f_0$ there are no solutions of the TAP equations ($\omega(f_0) < 0$), while at higher free energies there is an exponentially large number of such solutions ($\omega_0(f) > 0$ for $f > f_0$). This will be explained in Chap. V and we will find that $\omega_0(f)$ vanishes linearly at f_0: $\omega(f) = \lambda(f - f_0) + O((f - f_0)^2)$, with $\lambda < \beta$.

Because of the inequality $\lambda < \beta$, if one performs the weighted average (II.9), the only relevant solutions of the TAP equations are those which have the lowest value for the free energy; in other words in the limit $N \to \infty$ the number of relevant solutions does not increase exponentially with N: the multiplicity of the solutions does not contribute to the entropy density in this limit and the free energy density of the model is equal to the free energy density of each relevant solution (obviously the equality of the free energy densities ($f = F/N$) does not imply the equality of the total free energies (F); as we shall see in the next chapter, this fact has profound consequences).

Various people have attempted to compute the function $\omega_0(f)$ and more generally to evaluate the thermodynamics of the model in this framework: it finally turned out that it is very difficult to avoid the use of replicas and such a computation is at least as complex as the evaluation of the free energy using the formulae of the previous chapter. However introducing replicas one can show that the weighted average of the TAP equations gives the same thermodynamics as the replica approach.[6]

There is a very simple argument to understand the origin of this exponentially large number of solution. For simplicity we work at zero temperature ($\beta = \infty$) and at zero magnetic field; let us do the crucial hypothesis that

$$\beta(1 - q_{EA}) \to 0 \quad \text{for} \quad T \to 0. \tag{II.13}$$

As we shall see below, the condition (II.13) must hold for the solutions of the TAP equation having the smallest value of the free energy; Eq. (II.13) is also confirmed by a direct computation done using the replica approach in which one finds $q_{EA} = 1 - O(T^2)$ at low T. Using condition (II.13) Eq. (II.3) simplifies (at zero temperature) to

$$m_i = \text{sign}\left\{\sum_{k=1}^{N} J_{i,k} m_k\right\}. \tag{II.14}$$

This equation has an obvious meaning: at zero temperature the system should be in the ground state, only one configuration is allowed and the magnetizations m's coincide with the values of the s's in this configuration. Eq. (II.14) implies that each spin is oriented in the direction of the force acting on it and that the energy cannot be decreased by changing the value of a single spin. In the language of optimization theory the configuration of the s's is optimized with respect to one spin flip. We notice that if condition (II.13) were not satisfied for the ground state, we would not have obtained Eq. (II.14), so that Eq. (II.13) must hold at least for the ground state.

In other words Eq. (II.14) implies that the m's are at a *local* minimum of

$$H[m] \equiv - \sum_{1 \leq i < k \leq N} J_{i,k} m_i m_k, \qquad m_i = \pm 1, \tag{II.15}$$

but a local minimum is not necessarily a global minimum.

Numerically it is very easy to find solutions of Eq. (II.14) by using the following algorithm. We start from an arbitrary configuration of the m's ($m_i = \pm 1$), we scan all the sites (i) of the system and we change the sign of m_i if Eq. (II.14) is not satisfied; this operation is repeated an arbitrary number of times up to the moment where no spin is flipped during a scan of the system: at this point Eq. (II.14) is satisfied and any further application of the algorithm will not change the situation (this method is exactly the zero temperature Monte Carlo algorithm).

Each time we flip a spin the energy decreases: the energy cannot become arbitrarily small and the maximum number of sweeps through the whole lattice is clearly a polynomial in N (for example in the model where the J's are $\pm 1/N^{1/2}$, the maximum number of sweeps is proportional to $N^{3/2}$), numerically it converges rather quickly, the computer time being approximately proportional to N^3. On the other hand we also know that the problem of finding the ground state of a spin glass is an NP complete problem (see Chap. VII) which is unlikely to be solvable in a computer time growing like a power of N.

If the number of solutions of Eq. (II.14) would increase polynomially in N, it would be rather likely that, by applying the above algorithm a number of times which is much larger than the number of solutions of Eq. (II.14), starting each time from a different randomly chosen configuration, we should be able to find the ground state of the system in a time which is polynomial in N (we have done the tacit and natural assumption that the basin of attraction of the ground state with this algorithm is not exponentially smaller than the basin of attraction of the metastable states). Hence it is clear that the determination of the ground state is an exponentially hard problem because the ground state is one of the exponentially many solutions of Eq. (II.14).

II.3 Local Field Distribution

The TAP approach leads us to ask the following question: what is the probability distribution $P(\tilde{h})$ of the effective field $\tilde{h}$ on a single spin?

A detailed computation[7] shows that in the replica symmetric approach of the previous chapter the probability distribution of $\tilde{h}$ is given (at zero magnetic field h) by

$$P(\tilde{h}) = 1/(2\pi\tilde{q})^{1/2} \exp[-\tilde{h}^2/(2q)], \tag{II.16}$$

i.e. it is a Gaussian with variance q and zero mean.

This result can be easily understood from the cavity approach[8]: if in Eq. (II.1) the local magnetization m_k are computed in absence of the ith spin, the m_k cannot depend on the $J_{i,k}$. At $h = 0$, $\tilde{h}$ is the sum of N random numbers with zero mean and variance $1/N$ (the J's), multiplied by the local magnetization. The central limit theorem tells us that $\tilde{h}$ is a Gaussian distributed variable, with variance, the average variance of the m_i, i.e. q.

In this way Eq. (I.13) of the previous chapter can be identified as the self-consistency condition for q in the TAP approach (as we shall see in Chap. V)

$$q \equiv \overline{m^2}_i = \int d\tilde{h} P(\tilde{h}) [\mathrm{th}(\beta\tilde{h})]^2 . \tag{II.17}$$

Unfortunately this equation must be wrong as far as it leads to the disaster (negative entropy!) of the previous chapter. Moreover it was noticed numerically by TAP themselves that the true distribution of local fields $P(\tilde{h})$ is very far from being a Gaussian: at zero temperature and zero external magnetic field $P(\tilde{h})$ vanishes at $\tilde{h} = 0$ (later on a similar behaviour was observed at $T = 0$ also at non zero external magnetic field[9]). We also notice that the condition

$$P(0) = 0 \quad \text{at } T = 0 \tag{II.18}$$

is absolutely necessary for the validity of Eq. (II.13). If the probability distribution of

$\tilde{h}$ were Gaussian, Eq. (II.18) and consequently Eqs. (II.13–14) would be violated, leading to theoretical inconsistency.

These results are very interesting for two reasons:

a) While the replica approach is nearly correct for the internal energy (the error is only 5%), as far as the distribution of the local fields (and of the local magnetizations) is concerned, the error is 100% and the predictions are qualitatively wrong.

b) Equation (II.3) seems to be absolutely correct in the cavity approach for each particular solution; on the other hand if there exist many solutions the addition of a new spin may change the weight of the various solutions and this effect may modify the distribution of $\tilde{h}$.

Everything seems to suggest that we should modify the cavity approximation in order to take into account the existence of many solutions. This task is not easy and will be delayed to Chap. V. It turns out that the simplest way to obtain the correct spin glass theory is to modify the replica approach by adding to it the information on the existence of many relevant solutions of the TAP equations. This will be described in the next two chapters. Before leaving the TAP equations it is useful to see directly where this approach is internally inconsistent.

II.4 Stability and the Solutions of the TAP Equations

We will now describe how the system of TAP equations (II.3) makes sense only if the $\{m_i\}$ belongs to a certain part of the phase space.

Intuitively this restriction can be understood as a limit of the validity of the cavity approximation as follows: in Eq. (II.3) we have subtracted from $\tilde{h}_i \equiv \sum J_{i,k} m_k$ the contribution in m_k due to the influence of the ith spin. In fact there might be more indirect effects (e.g. s_i polarizes m_j which polarizes m_k which acts on s_i) that should be subtracted likewise. It just turns out that, because of the weakness of the couplings ($J_{i,k} \approx 1/N^{1/2}$), all these high order effects are damped by factors of order $1/N^{1/2}$ at least. Hence Eq. (II.3) is valid only provided that the sum of the series of all higher order effects is not divergent. The resulting constraint is[10]

$$x \equiv 1 - \beta^2(1 - 2q_{EA} + r) > 0, \tag{II.19}$$

where

$$q_{EA} = 1/N \sum_{i=1}^{N} m_i^2$$

$$r = 1/N \sum_{i=1}^{N} m_i^4. \tag{II.20}$$

Equation (II.19) can also be seen[11] as the condition for the Hessian matrix $\partial F/\partial m_i \partial m_j$ not to have negative eigenvalues; it also implies that the average of the square of the connected correlations functions

$$\chi_2 \equiv 1/N \sum_{i,j} (\langle s_i s_j \rangle - \langle s_i \rangle \langle s_j \rangle)^2 \tag{II.21}$$

does not become negative when it is computed within the replica approach (χ_2 is equal to C/x, C being a positive constant). In this last form the necessity of condition (II.19) is particularly obvious and indeed we shall return to it in this way in Chap. V where we shall derive it as a consistency condition of the cavity method.

In the approach of Sec. I.3 as well—the replica approach—condition (II.19) is violated at low temperature—it is nothing but the De Almeida Thouless stability condition—and this explains why wrong results are obtained. We will see later that using the correct approach at low temperature (where the naive approach predicts a negative x) x turns out to be identically zero: the stability of the phase is "marginal" and the above correlation functions diverge (see Chaps. V and VI).

The TAP equations are rather tricky: one can easily convince oneself of this statement by trying to solve them numerically for a given finite sample. It turns out that, apart from the region of very low temperatures in which one can decide to use Eq. (II.14), it is very difficult to find numerically any solution of Eq. (II.3) which satisfies the stability condition Eq. (II.19). This is due to the presence of unknown terms of order $1/N$ (or may be of order $1/N^{\delta}$) in the equations.

Nemoto and Takayama have invented a clever procedure to cope with this problem[12]: they look for the minima of

$$B \equiv 1/N \sum_i (m_i - \text{th}(\tilde{\beta} h_i))^2 . \tag{II.22}$$

In some cases (about 15% of the samples of system with a few hundreds of spins) the minimum lies at $B = 0$ and the TAP equations are satisfied: this is a "true" solution of the TAP equations and x is found to decrease to zero as $1/N^{2/3}$; very often one finds that B is not zero at the minima but vanishes as $1/N^{1/3}$ when $N \to \infty$; for these "quasi" solutions x is identically equal to zero. Both the 'true" solutions and the "quasi" solutions are relevant in the thermodynamic limit $N \to \infty$.

We shall return to the use which can be done of these solutions after developing the replica theory of the spin glass phase in the next two chapters.

References

1. Thouless *et al.* (1977) Reprint 6.
2. Sommers (1978); De Dominicis (1980); Plefka (1982); A derivation which is completely different from the diagrammatic analysis of these papers will be given in Chap. V.
3. Sompolinsky (1981).
4. De Dominicis *et al.* (1980); De Dominicis (1980).
5. Bray and Moore (1980) Reprint 7; De Dominicis *et al.* (1980); Tanaka and Edwards (1980).
6. De Dominicis and Young (1983).
7. De Almeida and Lage (1983); see also Kirkpatrick and Sherrington (1978) Reprint 4.
8. This approach will be developed in Chap. V. Let us mention that the quantity which is usually named "local field" in the literature is the hyperbolic arctangent of the local magnetization m_i. It is exactly what we shall call the cavity field in Chap. V.
9. Bantilan and Palmer (1981) Reprint 24.
10. Plefka (1982).
11. Bray and Moore (1979).
12. Nemoto and Takayama (1985).

Chapter III
BREAKING THE REPLICA SYMMETRY

III.1 Pure States

Before discussing how to break the replica symmetry let us summarize the main suggestions which have been obtained from the study of the TAP equations. Under the hypothesis that among the exponentially large number of solutions of the TAP equations only a few solutions (the one with the lowest free energies) contribute to the thermodynamic limit we can write

$$\langle s_1 s_2 \ldots s_k \rangle = \sum_{\alpha} w_\alpha \langle s_1 s_2 \ldots s_k \rangle_\alpha , \tag{III.1}$$

where α labels the relevant solutions of the TAP equation and some of the w's remain different from zero in the limit $N \to \infty$ (see Eq. (II.9, 10)).

In the TAP approach (as well as in any mean field approach where the fluctuations are neglected) the expectation value of the product of many spin can be computed as the product of the expectation values

$$\langle s_1 s_2 \ldots s_k \rangle_\alpha \equiv m^\alpha{}_1 m^\alpha{}_2 \ldots m^\alpha{}_k . \tag{III.2}$$

Indeed we have seen that the stability condition Eq. (II.19) implies that the connected correlation functions are not of O(1) when N goes to infinity: the vanishing of the connected correlation functions implies Eq. (III.2).

According to Eq. (III.1) the usual Gibbs equilibrium state, whose expectations values are denoted by $\langle \cdot \rangle$, can be decomposed as the sum of other equilibrium states which are labeled by α (the corresponding expectation values are denoted by $\langle \cdot \rangle_\alpha$): the Gibbs state is a mixture, while the states labeled by α cannot be decomposed as a sum of other equilibrium states and are consequently pure states.

The Gibbs state and the pure states (labeled by α) are intrinsically different because of the following property: the connected correlation functions are large in the Gibbs states and vanish in the pure states. Indeed in the case of the two point correlation we have

$$\langle s_1 s_2 \rangle = \sum_\alpha w_\alpha \langle s_1 s_2 \rangle_\alpha$$

$$\neq \langle s_1 \rangle \langle s_2 \rangle = \left\{ \sum_\alpha w_\alpha \langle s_1 \rangle_\alpha \right\} \left\{ \sum_\alpha w_\alpha \langle s_2 \rangle_\alpha \right\} . \qquad \text{(III.3)}$$

In the Gibbs state the connected correlation function

$$\langle s_1 s_2 \rangle_c \equiv \langle s_1 s_2 \rangle - \langle s_1 \rangle \langle s_2 \rangle \qquad \text{(III.4)}$$

has no obvious reasons to vanish, while we have already seen that in the TAP approach all the connected correlation functions vanish when $N \to \infty$.

Generally speaking we can characterize a pure state by the property that the connected correlation functions vanish at large distance (in the infinite range model, different points must be considered to be at a large distance); equivalently in a pure state intensive quantities like $1/N \sum_{i=1,N} A_i$ do not fluctuate when $N \to \infty$. It is possible to obtain Eq. (III.1) without making any reference to the TAP equations: the decomposition of an equilibrium state as the convex linear combination of other (pure) equilibrium states, in which intensive quantities do not fluctuate, is well known in statistical mechanics and it plays a central role in the precise definition of symmetry breaking.

Let us forget for the moment the spin glass and let us consider a usual ferromagnet in a finite dimensional space and zero external magnetic field at finite volume (the infinite volume limit will be taken only at the end). According to the conventional wisdom the expectation value of a spin (the magnetization density m) is zero at high temperatures and it becomes non zero at low temperature: as far as there is no external magnetic field to choose a preferred sign for the spontaneous magnetization we have two possibilities $m = \pm m_S$: the Z_2 symmetry which exchanges the sign of the spin is said to be spontaneously broken because the equilibrium states are not invariant under such a symmetry.

If we look at the problem more carefully we discover that we must be a little more precise: for a system of *finite volume* at zero magnetic field the symmetry of the problem implies that m is strictly zero (the integral of an odd function is always zero!): for each configuration of the spins $\{s_i\}$ there exists the reversed configuration $\tilde{s}$ ($\tilde{s}_i = -s_i$) which has the same energy (and consequently the same Boltzmann weight) and gives an opposite contribution to $\langle s_i \rangle$.

If we introduce a uniform external magnetic field (h), as soon as it becomes much larger than $1/N$ (it may be zero in the limit $N \to \infty$), we find that $m = \pm m_S$, the sign being the one of the external magnetic field. A possible definition of the spontaneous magnetization is

$$m_S = \lim_{h \to 0+} \left\{ \lim_{N \to \infty} \langle s \rangle \right\} \qquad \text{(III.5)}$$

or equivalently the limit of the magnetization in the limit $N \to \infty$, $h \to 0$ but $Nh \to \infty$.

Depending on the sign of the external magnetic field when it approaches zero, we can define two different equilibrium states at zero magnetic field such that

$$\langle s \rangle_+ = m_S \qquad \langle s \rangle_- = -m_S . \tag{III.6}$$

The expectations of products of an odd (even) number of spins have the opposite (the same) signs in the two states. From symmetry arguments it is also evident that the standard Boltzmann-Gibbs average is a mixture (with equal weights) of the above two equilibrium states

$$\langle \cdot \rangle = 1/2\langle \cdot \rangle_+ + 1/2\langle \cdot \rangle_- . \tag{III.7}$$

In order to use this approach for the definition of a pure state one must first know the direction of the spontaneous magnetization in order to be able to apply a magnetic field parallel to it: in the ferromagnetic case a staggered magnetic field which is positive on some spins and negative on some others (with zero average) would induce no spontaneous magnetization. For a non random system the choice of the magnetic field can be done *a priori* by studying the various possible schemes of spontaneous symmetry breaking.

In the case of spin glasses this approach is very unpleasant because one must know the local directions of the spontaneous magnetization in one state in order to add the correct point-dependent magnetic field which allows for the proper definition of the spontaneous magnetization. Briefly, the magnetic field must point in the direction of the spontaneous magnetization in order to define the spontaneous magnetization itself.

In order to avoid this problem, we should use some intrinsic definition of spontaneous magnetization which does not need an auxiliary magnetic field. As we have already remarked, in a "good" equilibrium state (i.e. a pure state) the connected correlation functions vanish at infinite distance (equivalently intensive quantities do not fluctuate). Using the fact that this property (the clustering property) holds for the two pure states which we have just introduced, it is easy to prove that the same property cannot hold for the Gibbs state. Indeed we have

$$\langle s \rangle = 1/2m_S - 1/2m_S = 0$$

$$\lim_{i-k\to\infty} [\langle s_i s_k \rangle] = 1/2m_S^2 + 1/2m_S^2 = m_S^2 \neq 0 . \tag{III.8}$$

On the other hand the decomposition of a non clustering equilibrium states as the sum of clustering (pure) equilibrium states is always possible. If the Gibbs state *is* invariant under the action of a group but it is not clustering and the pure states into which the Gibbs states is decomposed *are not* invariant under the action of the group, this group is said to be spontaneously broken.

It is possible to interpret Eq. (III.1) as the decomposition of the Gibbs states into pure clustering states; in this case Eq. (III.2) holds only when the distance between the various points is very large. In the infinite range model, where all points stay at the same distance, the connected correlation functions between any two points in the pure states must vanish when N goes to infinity.

The decomposition of the Gibbs state into pure equilibrium states is very important because, according to conventional wisdom, a system, which is not at equilibrium at initial time, evolves toward a pure equilibrium state and the time needed to go from one equilibrium state to another equilibrium state becomes exponentially large with the size of the system.

The pure states and not the Gibbs state correspond to our intuitive idea of an equilibrium state. For example if we consider H_2O at exactly zero Celsius, in the Gibbs state the system has 50% probability of being all water and 50% probability of all ice; if the system is in a pure state, the whole sample is water or ice. Usually the Gibbs state is a mixture (it can be decomposed into pure states) in the conditions where a first order phase transition is present and two (or more) phases may coexist.

III.2 Distribution of the Distance between the Pure States: the Order Parameter is a Function

If we suppose that many pure states exist, it is a natural question to ask how the pure states differ one from the other; to this end it is convenient to define a distance in the space of the states. A simple definition for the distance between the state α and the state β (here β is *not* $1/kT$) is the following

$$d^2{}_{\alpha\beta} = 1/N \sum_i (m_i^\alpha - m_i^\beta)^2 \tag{III.9}$$

where N is the total number of spin, the sum over i is from 1 to N and m_i^α is the average magnetization of the spin i in the state α, i.e. $m_i^\alpha = \langle s_i \rangle_\alpha$. In the same way we can define the overlap between two states α and β as

$$q_{\alpha\beta} = 1/N \sum_i m_i^\alpha m_i^\beta . \tag{III.10}$$

Let us assume that $q_{\alpha\alpha} = q_{\beta\beta} \equiv q_{EA}$, i.e. the overlap of a state with itself (the so-called self overlap) does not depend on the state (we will see later that this assumption is correct). In this case the distance is very simply related to the overlap

$$d^2{}_{\alpha,\beta} = 2(q_{EA} - q_{\alpha\beta}). \tag{III.11}$$

In order to characterize the distribution of distances between states, it is useful to define the probability distribution of overlaps $P(q)$:[1]

$$P(q) = \overline{P_J(q)} = \overline{\sum_{\alpha\beta} w_\alpha{}^J w_\beta{}^J \delta(q_{\alpha\beta} - q)}, \tag{III.12}$$

where (as usual) the bar denotes the average over different samples with different values of the couplings J. The w's are the weights defined in Eq. (III.1) for the decomposition of the Gibbs state into pure states. It will be useful to write the w's as

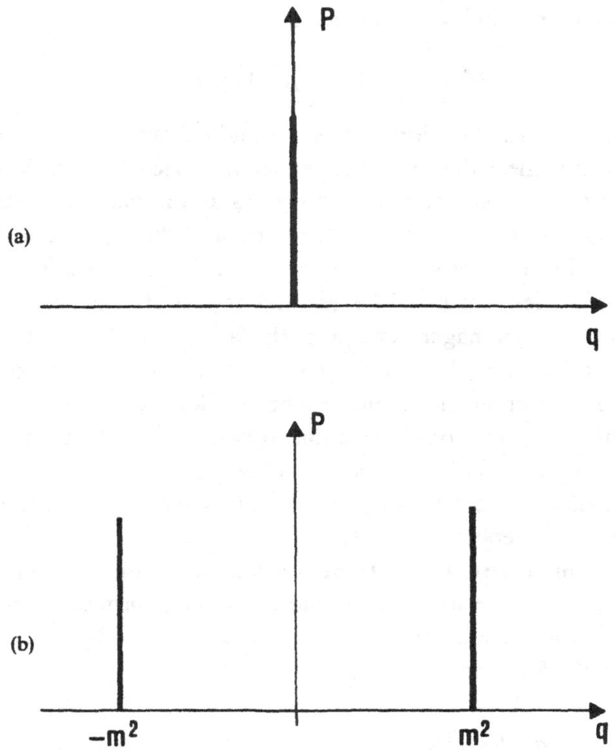

Fig. III.1. The $P(q)$ function in a ferromagnet a) above the Curie temperature b) at a temperature $T < T_c$ where the magnetization is M.

$$w_\alpha \propto \exp(-\beta F_\alpha), \tag{III.13}$$

where F_α can be interpreted as the total free energy of the state α: only states whose relative total free energy difference remains finite have a non vanishing value of the w's: all the relevant states have the same free energy density when $N \to \infty$.

In other words $P(q)$ is the probability of finding two states with overlap q, weighting each state with its probability of appearing in the ensemble. Only states which have a non zero probability ($w_\alpha \neq 0$) contribute to Eq. (III.12).

In the ferromagnetic Ising model at non zero magnetic field $P(q)$ is a delta function (Fig. III.1) because there is only one equilibrium state; when the temperature is sufficiently low, exactly at zero magnetic field the function $P(q)$ contains two delta functions, one at $q = m^2$, the other at $q = -m^2$, where m is the spontaneous magnetization.

In the case of spin glasses the previous arguments suggest that at low temperature there are many different states and the function $P(q)$ is not trivial; moreover we expect that the weights w's strongly fluctuate from one realization of the J's to another: the w's are sensitive to variations in the total free energy which are of order 1, i.e. of relative order $1/N$ (indeed we will see that the addition of a single spin strongly changes the values of the w's). If this happens the function $P_J(q)$ defined for one sample must also fluctuate when J changes and we should have

$$\overline{P_J(q_1)P_J(q_2)} \neq \overline{P_J(q_1)}\ \overline{P_J(q_2)}. \tag{III.14}$$

The non triviality of the function $P(q)$ is the main characteristic of the spin glass phase that has been neglected in the naive replica approach and which is likely to be responsible for the difference between spin glasses and normal magnets. Indeed for certain models there can be a freezing of spins in random directions, but with only two pure states, related by the reversal of all spins[2]—the $P(q)$ function is the sum of two delta functions, as in a ferromagnet. The essential physics of such a model is not much different from that of a ferromagnet, except for the fact that the "global magnetization" must be measured after having been projected on a non trivial direction of phase space. $P(q)$ is an order parameter function which can be used for characterizing "usual" phase transitions—bifurcation from one δ function to two at the critical temperature in a ferromagnet—but it is a much more general concept which is necessary for understanding the spin glass phase. Let us now see how this function $P(q)$ can be computed in the framework of the replica approach.

The decomposition of the Gibbs state into pure states is a rather complex operation so that it is more convenient to bypass it and to obtain information on the function $P_J(q)$ by looking at the correlation functions of the s's in the Gibbs state.

Let us consider the following quantity

$$\begin{aligned} q_J^{(1)} &\equiv 1/N \sum_{i=1}^{N} \langle s_i \rangle \langle s_i \rangle \\ &= 1/N \sum_{i=1}^{N} \left\{ \sum_\alpha w_\alpha \langle s_i \rangle_\alpha \right\} \left\{ \sum_\beta w_\beta \langle s_i \rangle_\beta \right\} \\ &= \sum_{\alpha,\beta} w_\alpha w_\beta q_{\alpha,\beta} = \int dq P_J(q) q. \end{aligned} \tag{III.15}$$

It is interesting to note that $q_J^{(1)}$, which can be computed directly in the Gibbs state, is equal to the average overlap among the different pure states. In a similar way we have

$$\begin{aligned} q_J^{(2)} &\equiv 1/N^2 \sum_{i_1=1}^{N} \sum_{i_2=1}^{N} \langle s_{i_1} s_{i_2} \rangle \langle s_{i_1} s_{i_2} \rangle \\ &= 1/N^2 \sum_{i_1=1}^{N} \sum_{i_2=1}^{N} \left\{ \sum_\alpha w_\alpha \langle s_{i_1} s_{i_2} \rangle_\alpha \right\} \left\{ \sum_\beta w_\beta \langle s_{i_1} s_{i_2} \rangle_\beta \right\} \\ &\approx \sum_{\alpha,\beta} w_\alpha w_\beta \left\{ 1/N \sum_{i_1=1}^{N} \langle s_{i_1} \rangle_\alpha \langle s_{i_1} \rangle_\beta \right\} \left\{ 1/N \sum_{i_2=1}^{N} \langle s_{i_2} \rangle_\alpha \langle s_{i_2} \rangle_\beta \right\} \\ &= \sum_{\alpha,\beta} w_\alpha w_\beta [q_{\alpha,\beta}]^2 = \int dq P_J(q) q^2, \end{aligned} \tag{III.16}$$

where we have used the property that in the pure states the connected correlation functions of two spins are generally negligible in the large N limit and therefore we can substitute $\langle s_{i_1}\rangle_\alpha \langle s_{i_2}\rangle_\alpha$ at the place of $\langle s_{i_1} s_{i_2}\rangle_\alpha$. More precisely, by "generally negligible", we mean that given two points i and k at random, the difference $\langle s_i s_k\rangle_\alpha - \langle s_i\rangle_\alpha \langle s_k\rangle_\alpha$ (i.e. the connected correlation function) is of order $1/N^\delta$, $\delta > 0$, apart may be for a vanishing fraction of spins. In other words we need

$$(1/N^2) \sum_{i,k=1}^{N} (\langle s_i s_k\rangle_\alpha - \langle s_i\rangle_\alpha \langle s_k\rangle_\alpha)^2 \to 0 \qquad \text{(III.17)}$$

when $N \to \infty$. Equation (III.17) is equivalent to the condition that the magnetization density (arbitrarily defined) does not fluctuate in the thermodynamic limit

$$\langle \mu^2\rangle \to \langle \mu\rangle^2 \qquad \text{(III.18)}$$

for $N \to \infty$, where

$$\mu = 1/N \sum_{i=1}^{N} \varepsilon_i s_i , \qquad \text{(III.19)}$$

where the ε's are arbitrary numbers equal to ± 1. Different choices of the ε's correspond to different definitions μ. Equation (III.18) must be valid for any choice of the ε's.

In the general case we have

$$q_J^{(k)} \equiv 1/N^k \sum_{i_1=1}^{N} \cdots \sum_{i_k=1}^{N} [\langle s_{i_1} \dots s_{i_k}\rangle]^2$$

$$\approx \sum_{\alpha,\beta} w_\alpha w_\beta [q_{\alpha,\beta}]^k = \int dq P_J(q) q^k . \qquad \text{(III.20)}$$

Equation (III.20) are approximates for a finite value of N but they become exact in the limit $N \to \infty$.

In a similar way we can define the sample averaged moments

$$q^{(k)} = \overline{q_J^{(k)}} = \int dq P(q) q^k . \qquad \text{(III.21)}$$

It is interesting to note that we could also write

$$q^{(1)} \equiv \overline{\langle s_{i_1}\rangle \langle s_{i_1}\rangle} ,$$

$$q^{(2)} \equiv \overline{\langle s_{i_1} s_{i_2}\rangle \langle s_{i_1} s_{i_2}\rangle} , \dots \qquad \text{(III.22)}$$

independently from the value of i_1 and i_2 provided that they are different ($i_1 \neq i_2$); in other words if we perform an average over the J's we do not need to do an average over the points. Equation (III.22) can be generalized to $q^{(k)}$ for any value of k without difficulties.

III.3 Computation of *P*(*q*) with Replicas: Existence of Many States and the Breaking of the Replica Symmetry

The main conclusion of the previous section is that the function $P(q)$ (which contains some information about the overlaps between various pure states) can be computed from the average value of the multiple correlations of the spins in the Gibbs state, bypassing in this way the need of the selection of one state by the application of an external magnetic field. We can summarize the discussion in the following identity[1]

$$P_J(q) = (1/Z^2) \sum_{\{s,\tau\}} \exp(-\beta H_J(s) - \beta H_J(\tau))\delta\left(\sum_i s_i\tau_i/N - q\right)$$

$$P(q) = \overline{P_J(q)}. \tag{III.23}$$

The right-hand side of Eq. (III.23) is a Gibbs Boltzmann average for two non interacting identical (i.e. with the same couplings) spin glass systems. In principle it can be computed with the usual methods of statistical mechanics. For instance one can use the Monte Carlo method to compute $P(q)$ numerically.[3] The two systems are then endowed with a dynamics that guarantees that the asymptotic occurrence frequency of each spin configuration be proportional to its Gibbs Boltzmann weight. At regular time intervals Δt the overlap of the spin configurations is measured and a histogram of the frequency of occurrence of the overlap gives $P(q)$

$$P_J(q) = \lim_{T\to\infty} (1/T) \sum_{k=1}^{T} \delta\left(\frac{1}{N}\sum_{i=1}^{N} s_i(k\Delta t)\tau_i(k\Delta t) - q\right). \tag{III.24}$$

If we compute the multiple correlations of the spins in the Gibbs state with the replica approach we can establish the link between the non triviality of the function $P(q)$ and the breaking of the replica symmetry.

In the replica formalism of Chap. I it is evident that

$$\begin{aligned} q^{(1)} &= \int dJP[J] \sum_{\{s,\tau\}} s_i\tau_i \exp(-\beta H_J(s) - \beta H_J(\tau))/Z_J^2 \\ &= \lim_{n\to 0} \int dJP[J]\{\langle s_i\rangle_J\}^2\{Z_J\}^n \\ &= \lim_{n\to 0} \overline{\langle s^a{}_i s^b{}_i\rangle}, \qquad (a \neq b), \end{aligned} \tag{III.25}$$

where a and b are two distinct replica indices (the sum over the spins in the $n-2$ other replicas gives a factor $Z_J{}^{n-2}$, which in the limit $n \to 0$ gives exactly the $Z_J{}^{-2}$ needed in

Eq. (III.25)). In proving Eq. (III.25) we have followed the strategy of first deriving some results for integer n, where everything is well defined: only at a later stage we analytically continue the equations to $n = 0$.

In a similar way we can derive the equations

$$q^{(2)} \equiv \lim_{n\to 0} \langle s^a{}_{i_1} s^a{}_{i_2} s^b{}_{i_1} s^b{}_{i_2} \rangle \qquad (i_1 \neq i_2), (a \neq b)$$

$$q^{(3)} \equiv \lim_{n\to 0} \langle s^a{}_{i_1} s^a{}_{i_2} s^a{}_{i_3} s^b{}_{i_1} s^b{}_{i_2} s^b{}_{i_3} \rangle \qquad (i_1 \neq i_2, i_1 \neq i_3, i_3 \neq i_2); (a \neq b). \quad \text{(III.26)}$$

In the replica formalism, with the saddle-point method explained in Sec. I.4, the various sites decouple in the limit $N \to \infty$, so that the averages on each site decouple in (III.26) and one gets

$$q^{(k)} = \lim_{n\to 0} \{\langle s^a{}_i s^b{}_i \rangle\}^k = \lim_{n\to 0} [Q_{a,b}]^k$$

$$Q_{a,b} \equiv \langle s^a{}_i s^b{}_i \rangle \qquad \text{(III.27)}$$

for an arbitrary point i, provided that $a \neq b$. $Q_{a,b}$ is the saddle point matrix solution of (I.16).

Equation (III.27) is not ambiguous if the replica symmetry is not broken and all $Q_{a,b}$ are equal ($a \neq b$). In the case where the replica symmetry is broken the various matrix elements Q_{ab} ($a \neq b$) are not all equal and one must be a bit more careful. The generalization of (III.27) is the following[1,4]:

$$q^{(k)} = \lim_{n\to 0} 1/[n(n-1)/2] \sum_{\{a,b\}} [Q_{a,b}]^k, \qquad \text{(III.28)}$$

where $\sum_{\{a,b\}}$ denotes (and it will denote in the rest of this chapter) the sum done over all the $n(n-1)/2$ pairs of replica indices a and b (with $a \neq b$ as usual) and $1/[n(n-1)/2]$ is a normalization factor (i.e. the number of pairs done with n objects) such that $q^{(0)} = 1$.

In order to understand Eq. (III.28) we note that, if the replica symmetry is broken in the sense that one exhibits a non trivial Q_{ab} matrix which is a stable solution of the saddle-point equations (I.15), then there are many other equivalent solutions: all the matrices which are obtained from this particular Q_{ab} through a permutation of the lines and the columns. In other words if there is a saddle point which breaks the symmetry of the permutation group s_n, then all the points obtained by applying one element of this permutation group to this saddle point are also saddle points with the same free energy, and the correct result will be found by averaging out all these saddle points[4]. This procedure is clearly equivalent to taking only one saddle point but averaging the observables one wants to compute over all the permutations of the replica indices, which is how we derived (III.28) from (III.27).

Let us notice that a fundamental assumption in deriving (III.28) is the use of the saddle-point method of Sec. (I.3): this is the property which enables us, after having introduced replicas, to neglect the correlations between the different sites. It might be that in systems different from spin glasses this assumption will turn out to be wrong, and this could reflect itself on different physical properties than the ones we will describe in the next chapter.

From Eqs. (III.20) and (III.28), we see that all the information on the structure of the states, their probability distribution and their mutual overlaps are encoded into the form of the matrix Q which is the solution of the saddle-point equations in the replica method; in this way it is possible (as we shall see later) to encode very complex properties of the states into relatively simple forms of the matrix Q. In fact if we compare these two equations we find that $P(q)$—the physical order parameter—has a simple interpretation in replica language: it is nothing else than the fraction of matrix elements $Q_{a,b}$ which take the value q

$$P(q) = \lim_{n \to 0} 1/[n(n-1)/2] \sum_{\{a,b\}} \delta(Q_{a,b} - q). \tag{III.29}$$

If the function $P(q)$ is not a delta function, $Q_{a,b}$ must depend on a and b in a non trivial way and the replica symmetry is broken. In this way the breaking of the replica symmetry (which in the infinite range spin glass is present at sufficiently low temperatures) is related to the existence of several equilibrium states.

In zero magnetic field there is an additional grain of salt which is the existence of an extra global symmetry: the simultaneous reversal of all the spins (sometimes called time reversal symmetry). Because of this symmetry, for each pure state α there exists the time reversed state "$-\alpha$" such that

$$\langle s_i \rangle_{-\alpha} = -\langle s_i \rangle_\alpha, \qquad i = 1, \ldots, N. \tag{III.30}$$

Of course the weights w_α and $w_{-\alpha}$ are equal.

This property implies that the distribution of overlaps $P(q)$ must be symmetric $P(q) = P(-q)$. In the replica computation we implicitly assume that this global symmetry is broken by an infinitesimal (but much larger than $1/N^{1/2}$) uniform magnetic field, so that only one of the two states α and $-\alpha$ are selected. If we call $\tilde{P}(q)$ the function in an infinitesimal magnetic field, which is the one we obtain from (III.29) in the replica approach, the true function $P(q)$ at exactly zero magnetic field is given by

$$P(q) = 1/2\,\tilde{P}(q) + 1/2\,\tilde{P}(-q) \tag{III.31}$$

the explicit computation of the function $\tilde{P}(q)$ in the replica approach shows that its support is concentrated in the region of positive q. Replica symmetry breaking thus corresponds to $\tilde{P}(q)$ containing more than one delta function.

The subtlety of the replica approach is due to the fact that we must work in the limit $n \to 0$ where the replica group is the group S_0, (i.e. the group of permutation of zero elements) and we must compute the off diagonal elements of $Q_{a,b}$ which is a 0×0 matrix.

The whole program seems to be completely crazy. Nevertheless we can follow the philosophy to always write formulae which are correct for integer n and make the analytic continuation at $n = 0$ only at the end. In other words we will consider some matrices $Q^{(n)}{}_{a,b}$ which depend on n in a simple way; compute all the quantities we need for integer n and only at the end will we perform the continuation of the results at $n = 0$.

From this point of view the 0×0 matrix $Q^{(0)}{}_{a,b}$ is defined in terms of the matrices $Q^{(n)}{}_{a,b}$ for all values of n and therefore the space of all 0×0 matrices is an infinite dimensional space; this result, strange as it may seem from the mathematical point of view, is very welcome from the physical point of view since we should be able to associate a 0×0 matrix $Q_{a,b}$ to an arbitrary function $P(q)$.

If we come back to the problem left unsolved in Chap. I, i.e. to compute the free energy of the infinite range model by using the replica approach and the method of the point of maximum to evaluate the integral in Eq. (I.13), we must proceed by trial and error: the space of 0×0 matrix is very large and it impossible at the present moment to write a workable expression for a generic element of such a space, such that $A[Q]$ in (I.13) can be evaluated explicitly.

For a given temperature and magnetic field we must find a matrix $Q_{a,b}$ (or equivalently a family of n-dependent matrices $Q^{(n)}{}_{a,b}$) which satisfies the following requirements:

a) The function $P(q)$ (and other similar functions) computed from Eq. (III.29) (and its generalizations) must never become negative because it is a probability.

b) It must be a solution of the equation $\partial A/\partial Q_{a,b} = 0$, where A is defined in (I.13).

c) The eigenvalues of the Hessian $\partial^2 A/\partial Q_{a,b}\partial Q_{c,d}$ must be non negative.

d) The entropy computed from this matrix must never become negative.

e) Last but not least the prediction of this approach should not contradict the existing numerical simulations of the model.

There have been several attempts to exhibit such an Ansatz[5]. At the present moment only one form of the matrix is known which satisfies conditions a), b) and c). Fortunately for this choice of the matrix the conditions d) and e) are automatically satisfied. There is a widespread agreement that this choice is the correct one because the results are very satisfactory. It is well possible that for systems which are different from spin glasses (e.g. for real glasses), different choices of the matrix $Q_{a,b}$ will be needed.

In order to present the final proposal for the matrix Q, it is convenient to go by stages: 1) we present a tentative solution, 2) we analyze the drawbacks and the advantages of this solution, 3) we modify the proposal for Q until conditions a–e) are satisfied.

III.4 A First Stage of Replica Symmetry Breaking

We come back now to the problem we left unsolved in Chap. I which is to exhibit a solution within the replica method (see (I.13)) which breaks the replica symmetry[6]. Let us try to find a first candidate for the matrix Q. A "natural" suggestion consists in dividing the n replicas into n/m groups of m replicas. (Of course n must be multiples of m, i.e. n/m must be an integer). We set $Q_{a,b} = q_1$, if a and b belongs to the same group, and $Q_{a,b} = q_0$, if a and b belongs to different groups (we do not consider the $Q_{a,a}$'s which are identically equal to 0). In other words

$$Q_{a,b} = q_1 \quad \text{if} \quad I(a/m) = I(b/m)$$

$$Q_{a,b} = q_0 \quad \text{if} \quad I(a/m) \neq I(b/m), \tag{III.32}$$

where $I(x)$ is an integer valued function: its value is the smallest integer which is greater than or equal to x.

Each line of the matrix has $m - 1$ off-diagonal elements which are equal to q_1 and $n - m$ which are equal to q_0 (the total number of off diagonal elements is $n - 1$, i.e. -1 in the limit $n \to 0$). According to Eq. (III.29) we have in the limit $n \to 0$

$$P(q) = m\delta(q - q_0) + (1 - m)\delta(q - q_1). \tag{III.33}$$

We see immediately that the function $P(q)$ is non-negative (as a probability should be) only if

$$0 \leq m \leq 1. \tag{III.34}$$

It is obvious that (if we exclude the two uninteresting cases $m = 0$ and $m = 1$) m cannot be an integer and satisfy the inequality (III.34). On the other hand the limit $n \to 0$ is obtained by doing an analytic continuation in m and nothing seems to forbid that in such a process m be non-integer.

The natural question "are we really allowed to take m non integer?" is very difficult to answer from a strict mathematical point of view as far as the mathematical foundations of the replica method are not yet established. We must proceed using common sense; it is rather satisfactory that at the end (as we will see in Chap. V) it is possible to reconstruct all the results of the present approach without using the replica method.

We must now evaluate the free energy (I.13, 15) using the Ansatz for Q_{ab} defined in Eq. (III.32). The only non trivial term is the "one site partition function" (I.13). This reads

$$Z[Q] = \sum_{\{S\}} \exp\{-\beta H[Q, S]\}$$

$$H[Q, S] = -\beta \sum_{1 \leq a < b \leq n} Q_{a,b} S_a S_b - h \sum_a S_a. \tag{III.35}$$

In order to evaluate $Z[Q]$ we write

$$\sum_{1 \leq a < b \leq n} Q_{a,b} S_a S_b = (1/2) \left[q_0 \left(\sum_a S_a \right)^2 + (q_1 - q_0) \sum_{\text{blocks}} \left(\sum_{a \subset \text{block}} S_a \right)^2 - n q_1 \right], \tag{III.36}$$

where there are n/m blocks labelled by $k = 0, \ldots, n/m$, the block k being defined as $\{a$

such that: $I(a/m) = k\}$. We now perform a Gaussian transform in $Z[Q]$ for each square in expression (III.36), giving

$$Z[Q] = \int dp_{q_0}(z) \prod_{k=0}^{n/m} \int dp_{(q_1-q_0)}(y_k)$$

$$\sum_{\{S\}} \exp \beta \left\{ (z+h) \sum_a S_a + \sum_{k=0}^{n/m} y_k \left(\sum_{a \subset \text{block}(k)} S_a \right) \right\}, \tag{III.37}$$

where we have used the short notation for the Gaussian probability distribution with variance q and zero average

$$dp_q(z) \equiv (2\pi q)^{-1/2} \exp(-z^2/2q)\, dz\,. \tag{III.38}$$

The sum over the spin configurations can now be done explicitly. The final result for the free energy is

$$A[Q] = A(q_0, q_1, m)$$

$$= -\beta/4[1 + mq_0^{\,2} + (1-m)q_1^{\,2} - 2q_1] - \ln 2$$

$$- \int dp_{q_0}(z) m^{-1} \ln \left\{ \int dp_{(q_1-q_0)}(y) \cosh^m [\beta(z+y+h)] \right\}. \tag{III.39}$$

A simple computation shows that when m is set to 0 or to 1 we recover (as we should) the free energy of the replica symmetric approach (see Eq. I.19) with $q = q_0$ or q_1 respectively.

We must now decide if we must minimize or maximize A with respect to the parameters q_0, q_1 and m. In principle we should look at the eigenvalues of the Hessian describing fluctuations around the solution of the equations $\partial A/\partial Q = 0$, however the approach becomes simpler if we can establish a variational computation.

Now as far as we have

$$\lim_{n \to 0} \left\{ 1/n \sum_{a,b} (Q_{a,b})^2 \right\} = -(1-m)q_1^{\,2} - mq_0^{\,2}, \tag{III.40}$$

it is clear that for $0 < m < 1$ the "correct extremum" (in the sense of a non negative Hessian) is obtained by finding the maximum with respect to q_0 and q_1.

The effective free energy Eq. (III.39) reduces to a quantity proportional to (III.40) in the limit of small β: we must therefore maximize the effective free energy with respect to q_0 and q_1; by analogy we can assume that the free energy should also be maximized with respect to m (partial justification for this prescription will be given later).

If we adopt this prescription the situation is rather satisfactory: we must maximize the free energy with respect to all the parameters of the theory. If a maximum is found in the region $0 < m < 1$, the results of this approach must be better than those obtained assuming unbroken replica symmetry: we have already remarked that the true free energy takes values greater than those obtained from the unbroken replica approach.

It is interesting to write down the saddle-point equations for q_0 and q_1

$$q_0 = \int dp_{q_0}(z-h)\left\{\int dp_{(q_1-q_0)}(y)\cosh^m[\beta(z+y)]\tanh(\beta(z+y))/D(z)\right\}^2$$

$$q_1 = \int dp_{q_0}(z-h)\int dp_{(q_1-q_0)}(y)\cosh^m[\beta(z+y)]\tanh^2(\beta(z+y))/D(z), \qquad \text{(III.41)}$$

where $D(z)$ is a short notation for

$$D(z) \equiv \int dp_{(q_1-q_0)}(y)\,\mathrm{ch}^m[\beta(z+y)]. \qquad \text{(III.42)}$$

The function A appearing in Eq. (III.39) can be computed numerically with high precision[6]. Maximizing A with respect to the q_0, q_1, and m one finds that, in the whole region where the replica symmetry should be broken (i.e. at sufficiently low temperatures), m is neither zero nor one ($m \to 0$ both for $T \to 0$ and $T \to 1$ at zero magnetic field).

The properties of the solution are satisfactory:

a) The zero temperature internal energy at zero magnetic field is $-.7652$ in good agreement with the numerical data.

b) The zero temperature entropy at zero magnetic field has collapsed from $S(0) \approx -.16$ to $S(0) \approx -.01$.

c) A detailed computation of the Hessian[7] shows that near the critical temperature at zero magnetic field the most negative eigenvalue is approximately equal to $-C\tau^2/9$ (τ being equal to $1 - T$ and C being a positive constant) where the same quantity in the replica symmetric case is equal to $-C\tau^2$. In other words the instability has decreased by a factor 9.

III.5 Replica Symmetry Breaking: the Final Formulation

We are clearly on the right track although we have not found the final answer yet. Hereafter we shall describe the correct solution without entering into the details of the computations which can be found in (Parisi, 1980, a, b, c, Reprints 9, 10 and 11). In order to generalize the solution it is convenient to use group theory to study the pattern of breaking of the replica symmetry. The group theory will be rather unusual due to the limit $n \to 0$.

In Eq. (III.32) we have divided the n replicas into n/m groups of replicas; no replica has been preferred: all lines of the matrix Q are a permutation one of the other and in this way we automatically impose that quantities like $n^{-1}\sum_{a,b=1}^{n}(Q_{a,b})^k$ are finite in the limit $n \to 0$.

The matrix Q described by Eq. (III.32) is left invariant by the following group of transformations

$$(S_m)^{\otimes n/m} \otimes S_{n/m}\,, \tag{III.43}$$

where $(S_m)^{\otimes n/m}$ is the direct product of the permutation group of m elements (S_m) with itself m times. Indeed we can permute both the replicas inside each group (and this leads to the product of S_m by itself n/m times) and the groups of replicas among themselves (this leads to $S_{n/m}$); (for experts in group theory the last product is a semi-direct product, not a direct product).

The group appearing in Eq. (III.43) is a subgroup of S_n as it should: when a symmetry is spontaneously broken the group of invariance is a subgroup of the original group of symmetry.

In the limit $n \to 0$ the group in Eq. (III.43) becomes

$$(S_m)^{\otimes 0} \otimes S_0\,. \tag{III.44}$$

In other words, S_0 contains itself as a subgroup! We have now the opportunity to break again the remaining group S_0 using the same pattern of symmetry breaking as before; in this way we remain with a symmetry group S_0 which is unbroken and the whole process can be done an infinite number of times in a hierarchical way. This generalization of the symmetry breaking pattern, which is quite natural in the replica formalism, will have far reaching consequences on the structure of states as we shall see in the next chapter.

The final form of the matrix Q is thus the following: we introduce a set of integer numbers m_i ($i = 0, \ldots, k+1$) such that $m_0 = n$ and $m_{k+1} = 1$ and m_i/m_{i+1} is an integer (for $i = 1, \ldots, k+1$). We can divide the n replicas in n/m_1 groups of m_1 replicas, each group of m_1 replicas is divided in m_1/m_2 groups of m_2 replicas and so on.... This iterative construction is schematized in Fig. III.2.

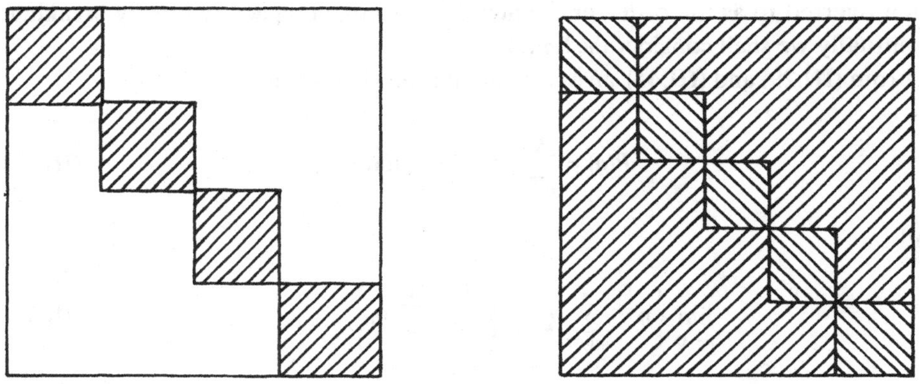

Fig. III.2. The iterative rule for breaking the replica symmetry: in the left half of the figure all the matrix elements have the same value in the small squares and have another value outside of them; after one iterative step each of the small squares is replaced by the square in the right half of the figure: outside the small squares the matrix elements take the same previous value while inside they take a new value, the same for all the small squares. This construction can be clearly repeated *ad infinitum*, provided that the dimensions of the starting matrix are sufficiently large.

$$\begin{matrix} q_2 & q_2 & q_1 & q_1 & q_0 & q_0 & q_0 & q_0 \\ q_2 & q_2 & q_1 & q_1 & q_0 & q_0 & q_0 & q_0 \\ q_1 & q_1 & q_2 & q_2 & q_0 & q_0 & q_0 & q_0 \\ q_1 & q_1 & q_2 & q_2 & q_0 & q_0 & q_0 & q_0 \\ q_0 & q_0 & q_0 & q_0 & q_2 & q_2 & q_1 & q_1 \\ q_0 & q_0 & q_0 & q_0 & q_2 & q_2 & q_1 & q_1 \\ q_0 & q_0 & q_0 & q_0 & q_1 & q_1 & q_2 & q_2 \\ q_0 & q_0 & q_0 & q_0 & q_1 & q_1 & q_2 & q_2 \end{matrix}$$

Fig. III.3. A typical form of the Q_{ab} matrix with two level of replica symmetry breaking ($n = 8$, $m_1 = 4$, $m_2 = 2$).

The off diagonal elements of the matrix Q are thus given by

$$Q_{a,b} = q_i \qquad \text{if } I(a/m_i) \neq I(b/m_i) \qquad \text{and}$$

$$I(a/m_{i+1}) = I(b/m_{i+1}), \qquad i = 0, \ldots, k, \tag{III.45}$$

where the q_i's are a set of $k + 1$ real parameters. For $k = 1$ we recover the previous attempt (Eq. (III.32)) and for $k = 0$ we recover the theory with unbroken replica symmetry theory; a typical form of the matrix with $k = 2$ is given in Fig. III.3. k is often referred to as the order, or the number of steps, of replica symmetry breaking. Eventually we shall consider the limit $k \to \infty$.

From (III.29) and (III.45), an easy computation shows that

$$P(q) = \sum_{i=0}^{k} (m_i - m_{i+1})\delta(q - q_i) \tag{III.46}$$

and

$$\lim_{n\to 0} \left\{ n^{-1} \sum_{a,b=1}^{n} Q^2{}_{a,b} \right\} = -\sum_{i=0}^{k} (m_{i+1} - m_i) q^2{}_i \,. \tag{III.47}$$

The $P(q)$ of Eq. (III.46) is positive definite only if the m's satisfy the conditions

$$0 \leq m_i \leq m_{i+1} \leq 1 \,. \tag{III.48}$$

If Eq. (III.48) is correct the same arguments as before suggest that we should look for the maximum of the free energy as a function of the q_i's and the m_i's (from now on we will assume that conditions (III.48) are satisfied).

If we assume that the q_i form an increasing sequence (as is suggested by a numerical study of the saddle-point equations), in order to keep track of the parameters q's and m's it is convenient to introduce the function $q(x)$ defined as

$$q(x) = q_i \qquad \text{if } m_i < x < m_{i+1} . \tag{III.49}$$

From Eq. (III.48), there is a one to one correspondence between the piecewise constant functions with k discontinuities and the parameters q's and m's. The relations between the functions $P(q)$ and $q(x)$ is the following

$$dx/dq = P(q), \tag{III.50}$$

where $q(x(q)) = q$; equivalently

$$x(q) = \int_0^q dq' P(q). \tag{III.51}$$

In the limit $k \to \infty$ the function $q(x)$ becomes arbitrary (any reasonable function can be approximated by a piecewice constant function) and

$$\lim_{n\to 0} \left\{ n^{-1} \sum_{a,b=1}^{n} Q^k{}_{a,b} \right\} = -\int_0^1 dx q^k(x). \tag{III.52}$$

In this formulation the free energy becomes a functional of $q(x)$ ($A[q]$) and it must be maximized with respect to $q(x)$. The approximations with finite k correspond to solving the variational problem in a space restricted to a given class of functions; in this way the procedure of maximizing with respect to the m's is fully justified. It is also clear that we would obtain the same results in the limit $k \to \infty$ by choosing $m_i = i/(k+1)$: the values of the m's which are taken at the intermediate steps (finite k) do not matter in the limit $k \to \infty$ (if they form a set which is dense in the interval $[0-1]$). However the best numerical results for k finite and small are obtained if we choose the m's in such a way that $\partial A/\partial m_i = 0$.

Although the exact solution requires $k = \infty$, good results are already obtained for small values of k.[8] (for $k = 2$ the zero temperature energy and entropy are respectively $-.7636$ and $-.003$, to be compared with the exact values of $k = \infty$: $-.7633$ and 0): near the critical temperature the difference between the exact value of the free energy and the approximated one is proportional to $(2k+1)^{-4}$.

The free energy can be explicitly computed near the critical temperature by expanding the functional $A[q]$ in powers of q; the details of such a study can be found in Ref. 9. The main results are that $q(x)$ is a continuous function, which is equal to

$$q(x) = q_m \qquad \text{for } x < x_m ,$$

$$q(x) = q_M \qquad \text{for } x > x_M\,,$$

$$q_m < q(x) < q_M\,, \qquad dq/dx > 0\,, \qquad \text{for } x_m < x < x_M\,, \tag{III.53}$$

where x_m and q_m are both proportional to $h^{2/3}$ (h being the external magnetic field); both $q(x)$ (in the region $x_m < x < x_M$) and x_M are weakly dependent on the magnetic field. When the magnetic field becomes zero the function $q(x)$ vanishes at $x = 0$ and when the magnetic field increases toward the critical value (for greater magnetic fields the replica symmetry becomes exact) $x_m \to x_M$ and $q_m \to q_M$.

As a consequence the function $P(q)$ has two delta functions: one at $q = q_m$, the other at $q = q_M$

$$P(q) = x_m\delta(q - q_m) + x_M\delta(q - q_M) + \tilde{P}(q), \tag{III.54}$$

where the function $\tilde{P}(q)$ is a smooth function with support in the interval $q_m < q < q_M$. In the limit $h \to 0$ the delta function at x_m disappears while when h increases and reaches the critical magnetic field (de Almeida Thouless line) the two delta functions collapse into a single one.

In other words if we choose two states α and β at random with their natural weights w_α and w_β, there is a probability x_M that these two states be the same one, in which case $q_{\alpha\beta}$ takes the maximal value q_M, there is a probability x_m that these two states have the minimum allowed overlap q_m, and there is a probability $1 - x_m - x_M$ of an intermediate situation. A detailed discussion of the structure of the states and their organization will be postponed to the next chapter.

The behaviour of $P(q)$ and $q(x)$ as a function of the magnetic field and of the temperature is illustrated in Fig. III.4.

A comparison of the prediction of this approach with the results of Monte Carlo simulations[3] are shown in Reprint 15. In Fig. III.5 we show the results for the function $q(x)$ obtained by looking at the numerical solutions of the TAP equation[10].

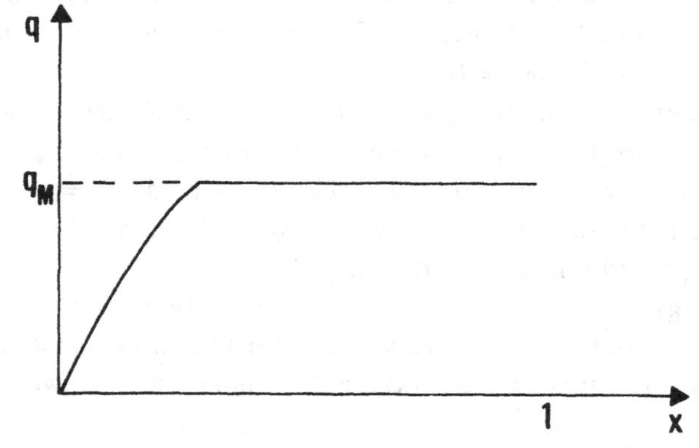

(a)

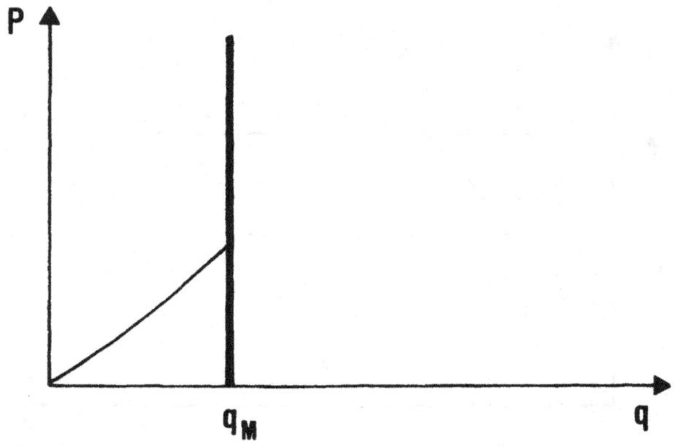

(b)

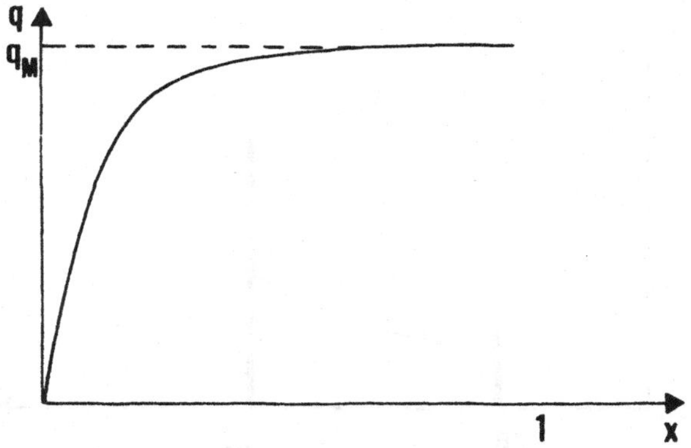

(c)

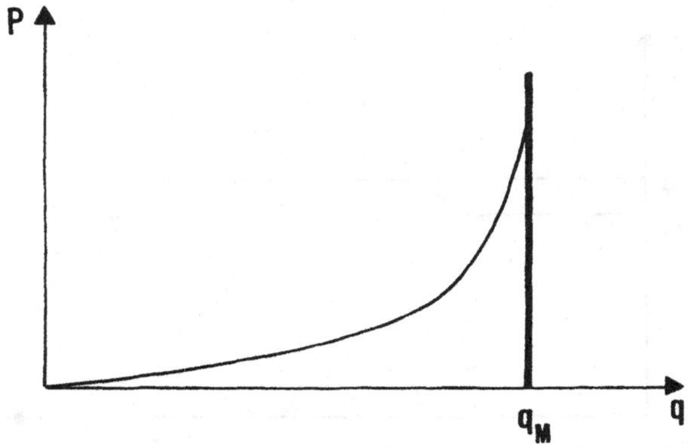

(d)

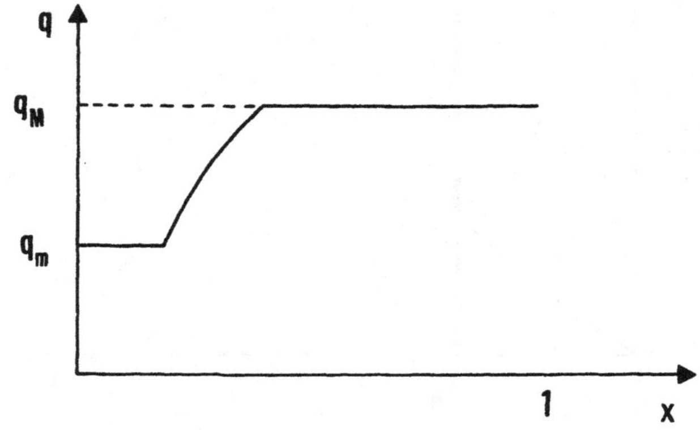

(e)

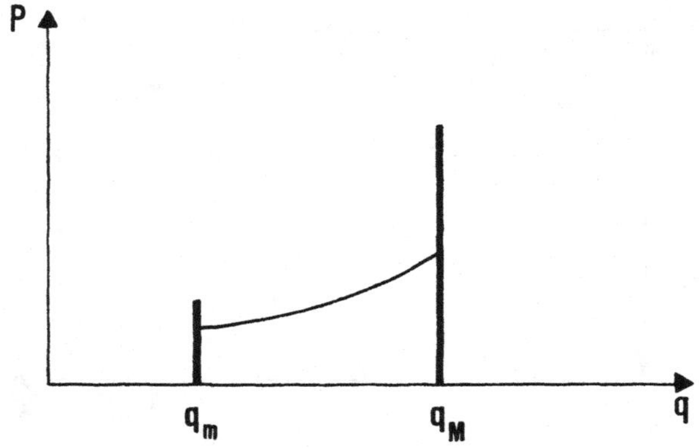

(f)

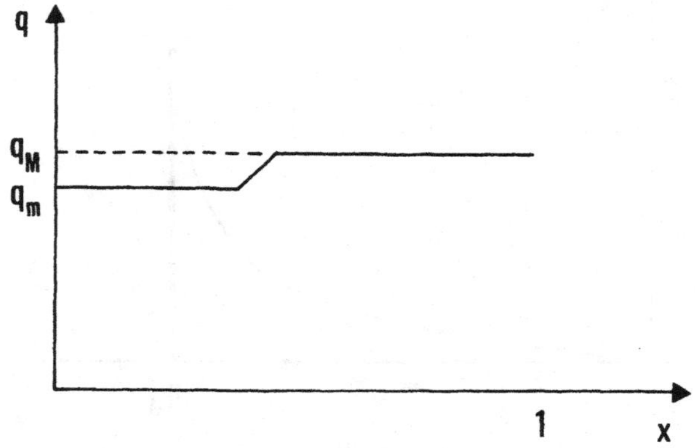

(g)

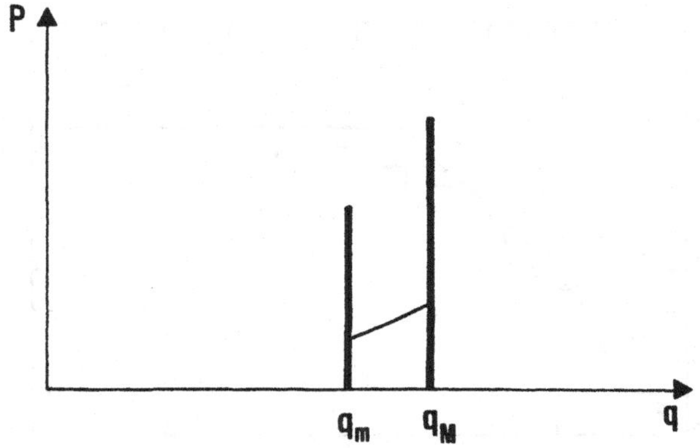

(h)

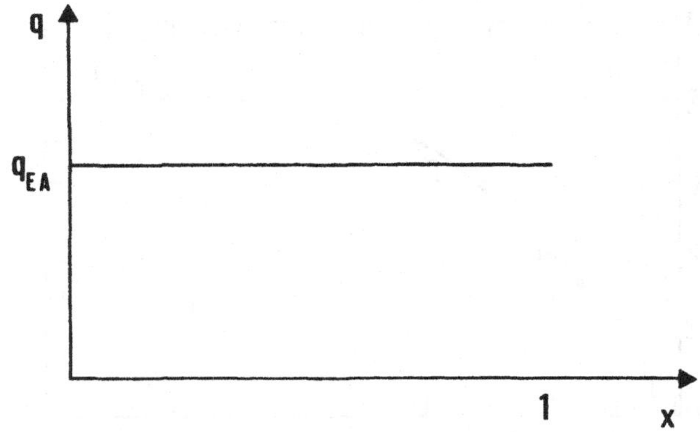

(i)

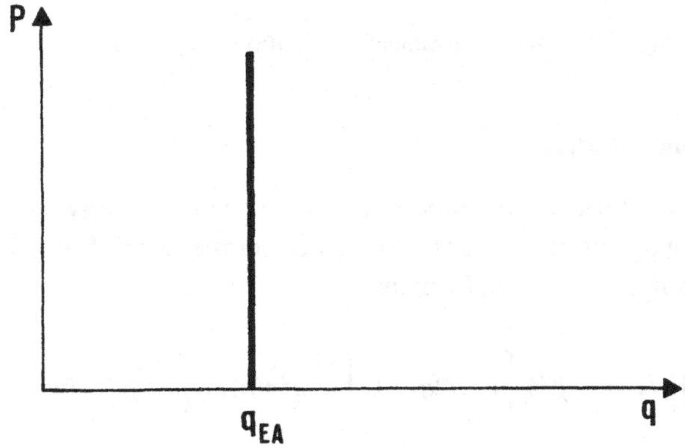

(l)

Fig. III.4. Behaviour of $q(x)$ and $P(q)$ for various values of the temperature and of the magnetic field (the delta functions in $P(q)$ are pictorially represented by spikes):
a, b) near the critical temperature at zero magnetic field: q_M is proportional to $T_c - T$, the slope of the function $q(x)$ is roughly temperature independent.
c, d) at small temperature at zero magnetic field: q_M is near to 1.
e, f) at small magnetic field: the value of q_m is proportional to $H^{4/3}$.
g, h) at larger magnetic field near the de Almeida Thouless line: the difference $q_M - q_m$ is proportional to the distance from this line.
i, l) in the replica symmetry region at non zero magnetic field: at $T > T_c$ the value of q_{EA} vanishes when H goes to zero.

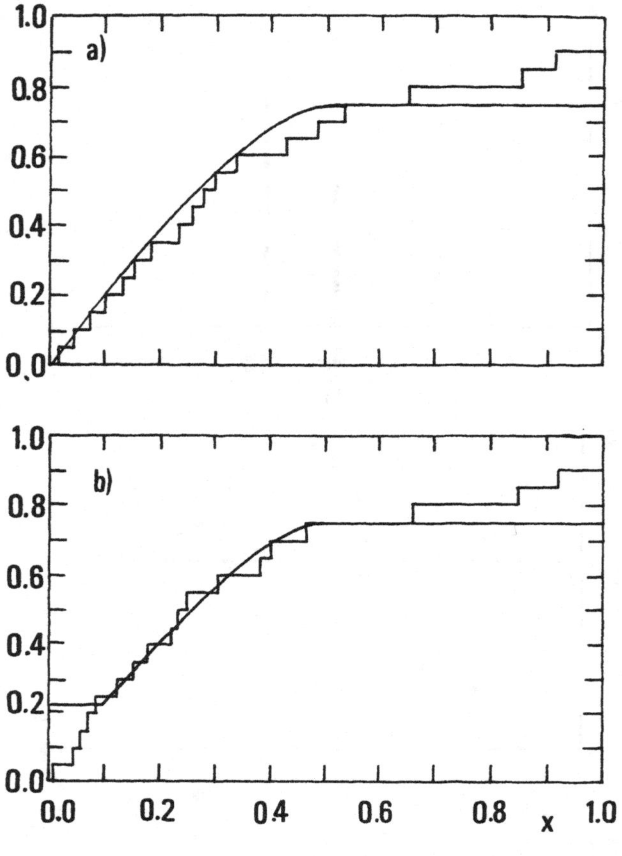

Fig. III.5. Numerical estimate of $q(x)$ (from Nemoto, 1986).

III.6 How to Compute $P(q)$

A numerical computation of the function $q(x)$ can be done at any temperature while analytic results can be obtained near the transition temperature. Indeed the functional $A[q]$ can be written in a compact form as[11]:

$$A[q] = -\beta/4\left\{1 - 2q_M + \int_{q_m}^{q_M} q^2 p(q)\, dq\right\} - \beta^{-1} a[q]$$

$$a[q] = \int dp_{q_m}(z) f(0, h + z), \tag{III.55}$$

where the function $f(q, y)$ satisfies the following differential equation in the interval $[q_m, q_M]$

$$\partial f/\partial q = -1/2\{\partial^2 f/\partial y^2 + x(q)(\partial f/\partial y)^2\} \tag{III.56}$$

with the boundary condition

$$f(q_M, y) = \ln[2\mathrm{ch}(\beta y)]. \qquad \text{(III.57)}$$

The stationarity equation $\delta A/\delta q(x) = 0$ becomes[12]

$$q = \int dy\, m^2(q, y), \qquad \text{(III.58)}$$

where $m(q, y) \equiv \partial f/\partial y$ satisfies the equation

$$\partial m/\partial q = -1/2\{\partial^2 m/\partial y^2 + 2x(q)\, m\, \partial m/\partial y\}. \qquad \text{(III.59)}$$

It is interesting to note that these equations can be written as stochastic differential equations[13]; to this end it is convenient to introduce an auxiliary function $w(q)$ which satisfies the following differential equation

$$dw/dq = \eta(q) - \beta x(q) m(q, w), \qquad \text{(III.60)}$$

where $\eta(q)$ is a white noise

$$\overline{\eta(q_1)\eta(q_2)} = \delta(q_1 - q_2). \qquad \text{(III.61)}$$

From now on in this chapter the bar will denote the average over the noise η (usually the bar denotes the average over the J's; we hope that this ambiguity in the use of the bar will not lead to confusion). The function $m(q, y)$ will finally coincide with the one introduced through Eq. (III.59). If we impose

$$m(q, y) = \overline{\mathrm{th}(\beta w(q_M))} \qquad \text{(III.62)}$$

(the average being done over all the trajectories such that $w(q) = y$), we find that m satisfies the differential equation (III.59). It is evident that in Eq. (III.60), after imposing the boundary condition $w(q) = y$, $w(q_M)$ depends only on the noise η in the interval $[q, q_M]$.

In this way Eqs. (III.60–62) determine $m(q, y)$ for a given $x(q)$. This function can be computed if we finally impose the selfconsistency condition Eq. (III.58) under the form

$$q = \overline{m^2(q, w(q))}, \qquad \text{(III.63)}$$

where the average is done over all the trajectories $w(q)$ which satisfy the boundary condition

$$w(0) = h. \qquad \text{(III.64)}$$

In a similar way we find that

$$\overline{\langle s_i \rangle^k} = \overline{m^k(q_M, w(q_M))} = \overline{\text{th}^k(w(q_M))}, \tag{III.65}$$

where w still satisfies the boundary condition Eq. (III.64). As a consequence $w(q_M)$ has the same probability distribution as the effective field ($\tilde{h}$) acting on a spin introduced in Chap. II (Ref. 14). A detailed computation[12,15] shows that if the replica symmetry is broken $\tilde{h}$ does not have a Gaussian distribution and its computed shape is in good agreement with the numerical data. In Fig. III.6 we show the results at zero temperature[16] (the ground state has been found using the Monte Carlo method) and in Fig. III.7 we show the results obtained using the numerical solutions of the TAP equations at temperature .2 and zero magnetic field[17]. Similar results in a field can be found in Ref. 18.

We conclude by observing that it is known that the following semiempirical rules are approximately true in the region where the replica symmetry is broken[19].

$$q_m(\beta, h) = q_m(h)$$

$$q_M(\beta, h) = q_M(\beta)$$

$$q(x, \beta, h) = q(x\beta) \qquad x_m < x < x_M\,. \tag{III.66}$$

These rules imply that at low temperature although $x(q)$ is proportional to T at small T, x_M at zero temperature is finite (1/2) and that

$$\lim_{\beta \to \infty} \beta x[q] \propto 1/(1-q)^{1/2} \tag{III.67}$$

for q near 1. This property is related to the vanishing of the probability distribution of the effective magnetic field $\tilde{h}$ at $\tilde{h} = 0$ when the temperature goes to zero.

References

1. Parisi (1983) Reprint 14.
2. Mattis (1976).
3. Young (1983) Reprint 15.
4. De Dominicis and Young (1983).
5. First attempts are by Blandin (1978), Blandin *et al.* (1980), Bray and Moore (1978); Another parametrization, equivalent to the one we explain here, has been introduced by De Dominicis *et al.* (1981), who find back with replicas the expressions of Sompolinsky (1981)—see more details in De Dominicis (1983).
6. Parisi (1979) Reprint 8.
7. Thouless (unpublished).
8. Parisi (1980) Reprint 9.
9. Parisi (1980) Reprint 10.
10. Nemoto (1986); see also Parga *et al.* (1984).
11. Parisi (1980) Reprint 9; Duplantier (1981).
12. De Almeida and Lage (1983).
13. Parisi (1980); Sommers and Dupont (1984).

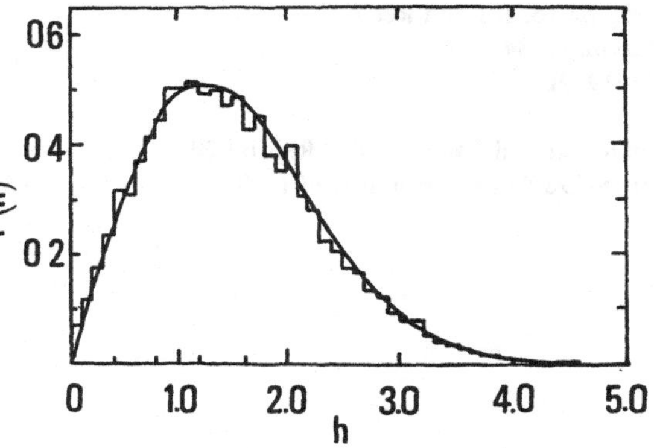

Fig. III.6. Distribution of local field. The histogram is from numerical simulations by Palmer and Pond (1979) while the curve is the theoretical prediction at $T = 0$ from Sommers and Dupont (1984).

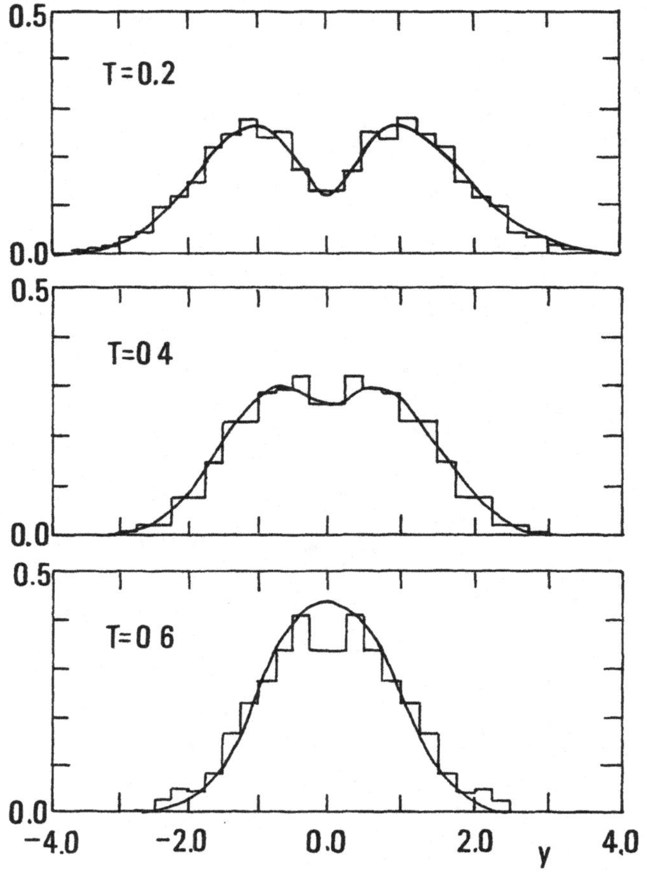

Fig. III.7. Distribution of local field (from Nemoto, 1986).

14. See, in Chap. II: Eq. (II.16, 17), and Ref. 8.
15. Sommers and Dupont (1984).
16. Palmer and Pond (1979).
17. Nemoto (1986).
18. Nemoto (1986); Bantilan and Palmer (1981) Reprint 24.
19. Parisi and Toulouse (1980); Vannimenus *et al.* (1981).

Chapter IV
THE NATURE OF THE SPIN GLASS PHASE

In the previous chapter we showed how the function $P(q)$, calculated in the somewhat mysterious replica formalism has a simple physical meaning in terms of quantities that are in principle observable—although in an ideal experiment because it assumes the possibility of creating two identical spin glass samples and then measuring the overlap between the two configurations.

Along this same line of reasoning, the next step would be to imagine a larger number of real identical replicas and check whether the correlations among the different configurations can be predicted on the basis of the analytical solution. This task was undertaken in Ref. 1 and the conclusions were so surprising that it reshaped our vision of the spin glass phase.

IV.1 The Discovery of Ultrametricity

To begin this analysis one can imagine three real replicas with three different overlaps

$$q_{12},\, q_{13},\, q_{23} \tag{IV.1}$$

and try to derive the joint probability distribution in the three variables

$$P_J(q_{12}, q_{13}, q_{23}). \tag{IV.2}$$

A short calculation[1] leads to

$$\overline{P_J(q, q', q'')} = \lim_{n\to 0} \frac{1}{n(n-1)(n-2)} \sum_{\substack{a,b,c \\ a\neq b\neq c \\ a\neq c}} \delta(Q_{ab} - q)\delta(Q_{ac} - q')\delta(Q_{bc} - q''). \tag{IV.3}$$

A surprising property follows from a direct inspection of the Ansatz for Q_{ab}: the probability is zero *unless* two of the overlaps are equal and not larger than the third one. In other words, in the $N \to \infty$ limit, the triangle constructed from any three configurations chosen according to their Gibbs-Boltzmann weight will always be equilateral or isosceles and in this case, the different side must be the smaller one. The same result will follow in the microcanonical ensemble, in which case the relevant configurations are those with energies lying in a particular interval. We will show later that this result also holds at the level of the states. Choosing three pure states weighted by their Boltzmann Gibbs probabilities, and defining their distances from the overlaps of their local magnetizations as in (IV.6), they generically form a triangle which is either equilateral or isosceles with a shorter third side. Furthermore this property is robust to changes in the definition of the distance[17].

A set in which all triangles are as described above is called *ultrametric*. In an ultrametric space the triangular inequality

$$d_{ab} \leq d_{ac} + d_{bc} \tag{IV.4}$$

is replaced by a stronger inequality

$$d_{ab} \leq \max(d_{ac}, d_{bc}). \tag{IV.5}$$

Therefore there are *no intermediate points* between *a and b*. In a metric space by successive repeated small steps in one direction one can get as far as one wants. In an ultrametric space after *any* number of fixed length steps one discovers oneself at the same initial distance from the starting point (the policy of small steps does not pay here!). In particular, brownian motion in an ultrametric space is non ergodic.

Any ultrametric ordering can be conveniently described if we put every point in the space in a one to one correspondence with the end tips of the branches (the leaves) of a tree. The distance between two points will be proportional to the height, at which the two corresponding branches converge (see Fig. IV.1).

Another way to visualize the ultrametric structure is by a partition of the space. One puts together in the same cluster all the points that are within a certain distance from each other. Then, by virtue of ultrametricity, one can easily prove that the partition exhausts all of the space with no overlappings among different clusters. In other words,

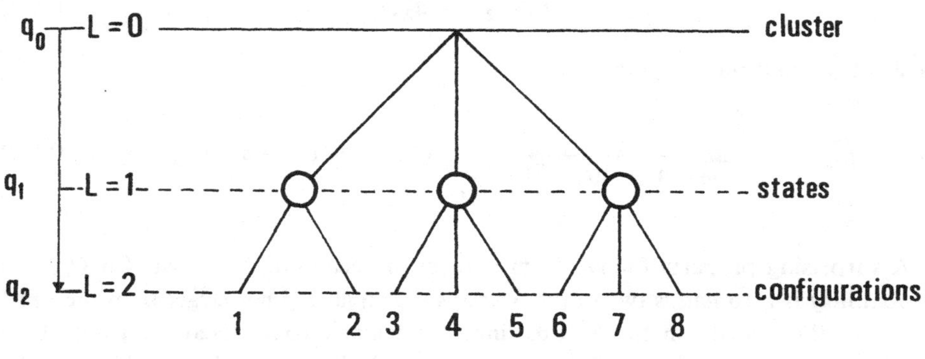

Fig. IV.1. An example of a tree with three levels. The overlaps $q_{12}, q_{34}, q_{45}, q_{35}, q_{67}, q_{78}, q_{68}$ are all equal to q_1 while the rest are q_0. In general $q_2 = 1, q_1 = q_{EA}$ and the vertices at level 1 correspond to three different pure states.

each point lies in just one cluster. The same procedure with a smaller distance will define a refinement of the partition. Any subcluster will be completely contained in one and only one cluster. A sequence of $L + 1$ decreasing distances or increasing overlaps

$$q_0, q_1, q_2, \ldots, q_L = 1 \qquad q_{i+1} > q_i$$

implies a structure of points contained in subclusters which are themselves contained in clusters which are contained in superclusters and so on (see Fig. IV.1 for $L = 2$).

If we choose $q_{L-1} = q_{\max} = q_{EA}$ (see Chaps. II and III) the $(L-1)$th level clusters are the *pure states*. To calculate the local magnetization at site i it is enough to average out a large representative set of configurations contained in the cluster. As $P(q)$ equals zero for $q > q_{\max}$, while the phase space obviously grows with decreasing mutual overlap, with probability one any two configurations chosen from the same cluster will have overlap $q_{\max}$. The overlap between two states defined as in II and III

$$q_{\alpha\beta} = (1/N) \sum_{i=1}^{N} m_i^{\alpha} m_i^{\beta} \tag{IV.6}$$

is identical to the overlap between configurations belonging to each of them. Therefore ultrametricity holds at the level of the states, as stated earlier.

The word ultrametricity was originally coined in mathematics[2] where the set is, generally speaking, finite dimensional and the ultradistance is somehow overimposed on it. On the other hand, in physics (as well as in other natural sciences) ultrametricity is a property of a sufficiently sparse subset of a metrical space and the ultradistance is just the distance restricted to the particular subset. In the limit of infinite dimensional space such a structure turns out to appear naturally. For finite N it is a probabilistic assertion. For instance in the infinite dimensional real vector space

$$\{\mathbf{v};\ \mathbf{v} = (v_1, v_2, \ldots),\ v_i = \text{real}\}, \tag{IV.7}$$

we define a sparse subset by simple random draw of each component according to independent, equally distributed statistics. The law of large numbers implies that the self overlap will be

$$(1/N) \sum_{i=1}^{N} v_i^{(a)} v_i^{(a)} = \overline{v^2} \tag{IV.8}$$

while the overlap between different vectors is

$$(1/N) \sum_{i=1}^{N} v_i^{(a)} v_i^{(b)} = (\overline{v})^2 \tag{IV.9}$$

and all triangles involving different states are equilateral.

A less trivial type of ultrametricity follows if the statistical draw becomes a stochastic branching process[2,3]. In the spin glass phase, as we saw in the previous chapter, the range of possible overlaps is continuous. The corresponding continuous branching process can then be defined by taking a suitable limit. We will come back to this analogy in Sec. 6. At this point we would like to comment on how ultrametricity *at finite N* appears in these examples. In this case it refers to probability distributions. For instance if we consider two of the sides as given, then the probability distribution of the third one will be peaked around the larger side with a width of the order of $N^{-1/2}$. For the *SK* model the situation is still more complicated, in part, because the tree becomes continuous when $N \to \infty$.

It is very hard to propose a non ultrametric Ansatz for Q_{ab}. In fact consistency with the saddle-point equations implies that the Ansatz must be the most general one invariant under a selected subgroup of replica permutation symmetry but then, in general, except for the examples discussed in Chap. III, one simply cannot write down an expression for arbitrary n such that the limit $n \to 0$ be feasible. Analytically, therefore, we cannot confront ultrametricity with an alternative hypothesis. The evidence in its favour is the evidence in favour of Parisi's solution, i.e. the stability test and the fact that it agrees well with the numerical simulations. There have been attempts to check directly (numerically) ultrametricity at finite N. There are two obstacles:

i) There are corrections at finite N that are not known analytically.

ii) The triangular inequality may imply constraints that are difficult to distinguish from the stronger ones imposed by ultrametricity[4]. For example, in a triangle with sides d_1, d_2, d_3 where d_3 is smaller than the other two the triangular inequality implies

$$|d_2 - d_3| \leq d_1 \leq d_2 + d_3 \tag{IV.10}$$

to be confronted with

$$d_1 = d_2 \tag{IV.11}$$

as required by ultrametricity. As finite N corrections smooth the constraints

$$d_1 = d_2 \pm \mathrm{O}(1/N^{\delta}) \tag{IV.12}$$

then for small d_3 the two hypotheses are impossible to distinguish.

The best numerical evidence for ultrametricity is depicted in Fig. IV.2 (from Bhatt and Young[5]). Though the data is convincing, larger samples and smaller temperatures should be considered before one can reach a definite conclusion.

The calculation of $\overline{P_J(q,q',q'')}$ from Eq. (IV.3) is straightforward. The details may be found in Ref. 1.The final formula is

$$\begin{aligned}\overline{P_J(q,q',q'')} = (1/2)P(q)x(q)\delta(q-q')\delta(q-q'') + 1/2[P(q)P(q')\theta(q-q')\delta(q'-q'') \\ + P(q')P(q'')\theta(q'-q'')\delta(q''-q) + P(q'')P(q)\theta(q''-q)\delta(q-q')]\,.\end{aligned} \tag{IV.13}$$

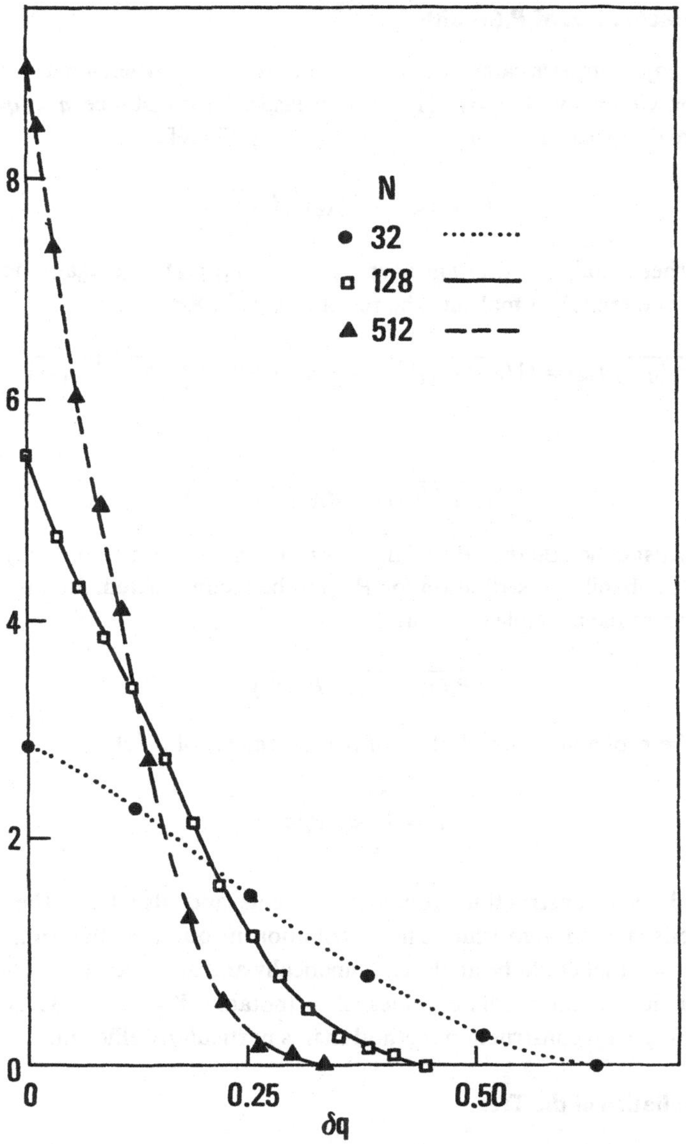

Fig. IV.2. Probability distribution of the difference between the two smaller overlaps ($\delta q = q_{mid} - q_{min}$) for a fixed larger overlap value $q = 0.5$. The temperature is $0.6T_{crit}$. In the $N \to \infty$ limit, the distribution is expected to become a δ function at the origin (from Ref. 5).

A new sign of the correlations among three different configurations appears in this formula: there is an abnormally large number of equilateral triangles as compared with isosceles ones. In fact, even without counting triangles with one side equal to the maximum possible value of q, i.e. the q_{EA}, where $P(q)$ has a δ-type contribution, the average number of equilateral triangles is 1/4 of the total.

IV.2 The Fluctuations of $P_J(q)$ with J_{ij}

A second major surprise came out in Ref. 1 (see also Ref. 6) when 4 real replicas were considered on which only two overlaps were measured, for instance: q_{12}, q_{34}.

The four real replicas are as usual non interacting. Therefore

$$P_J(q_{12}, q_{34}) = P_J(q_{12})P_J(q_{34}). \tag{IV.14}$$

On the other hand, the *average* over J_{ij} of $P_J(q_{12}, q_{34})$ can again be explicitly calculated in the replica formalism. The result, Eq. 20 in Ref. 1 is

$$\overline{P_J(q_{12}, q_{34})} = (1/3)\overline{P_J(q_{12})}\delta(q_{12} - q_{34}) + (2/3)\overline{P_J(q_{12})}\;\overline{P_J(q_{34})}. \tag{IV.15}$$

Therefore

$$\overline{P_J(q_{12})P_J(q_{34})} \neq \overline{P_J(q_{12})}\;\overline{P_J(q_{34})} \tag{IV.16}$$

so that one must conclude that $P_J(q)$ fluctuates with J_{ij} in a non trivial way. We must consider the probability distribution for $P_J(q)$ to be reconstructed, for instance, from the knowledge of the multiple correlations

$$\overline{P_J(q)P_J(q')\dots P_J(q^{(n)})}. \tag{IV.17}$$

In Ref. 1 the probability distribution of one parameter of $P_J(q)$

$$Y = \int_q^1 dq' P_J(q') \tag{IV.18}$$

was obtained by reconstruction from the moments and also from the maximum entropy reconstruction algorithm. The distribution in question has singularities at $Y = 1$ and $Y = 0$ that could be analysed numerically and analytically[7]. More recently new singularities at intermediate values of Y (notably $Y = 1/2$, $1/3\dots$) were discovered[9] through a reconstruction method that is particularly efficient.

IV.3 The Foliation of the Tree

Once we know that the set of configurations is organized ultrametrically, we have yet to analyze what type of tree one is dealing with. If at the end tips (the leaves of the tree) we locate the configurations, at the next level we will find the pure states. It is natural to define the thickness of a branch as the total probability of all configurations that ultimately stem from it (see Fig. IV.3).

With this definition the thickness of a branch corresponding to a pure state is the statistical probability of that state in the Gibbs-Boltzmann measure.

To characterize completely the tree it is enough to determine the probability distribution of the thickness of the branches. This is done in detail in Ref. 1. Again it follows by considering possible measurements in many different real replicas. Here we only

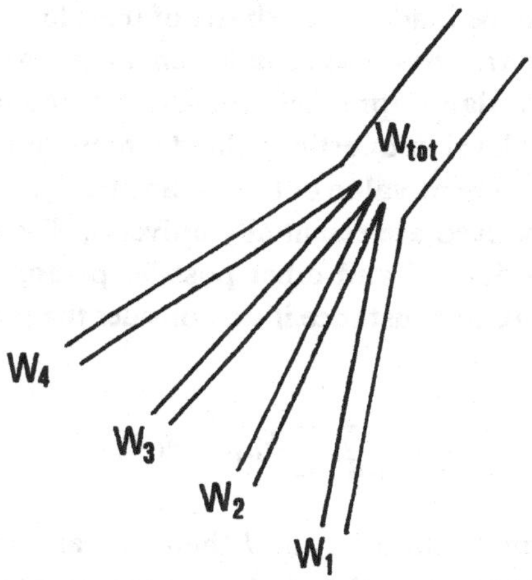

Fig. IV.3. Each branch is assigned a thickness equal to the sum of the statistical probabilities of all the states that descend from it. Consequently at a branching vertex the thickness is conserved i.e. $W_{tot} = W_1 + W_2 + W_3 + W_4$ in the figure. The replica formalism leads to the calculation of the W_i distributions.

quote the conclusions of such analysis:

i) At any level along the tree (labeled by the overlap q and the corresponding $x(q)$ which in Ref. 1 is sometimes written as $1 - y(q)$) the weights of the branches are random variables. The probability of having a branch thickness in $(W, W + dW)$ *disregarding* which are the values of the other branches is

$$f_J(W, y) = \frac{W^{y-2}(1-W)^{-y}}{\Gamma(y)\Gamma(1-y)} dW \tag{IV.19}$$

$$= \sum_{\text{branches}} \overline{\delta(W_{\text{branch}} - W)}.$$

ii) The sum of all the branch thicknesses is normalized to one. Therefore there exist correlations among the different branches. The probability of the combined occurrence of two branches with thicknesses in $(W, W + dW) \otimes (W', W' + dW')$ is

$$f_J(W, W')\,dW\,dW' = \sum_{\{\text{br}, \text{br}'; \text{br} \neq \text{br}'\}} \overline{\delta(W_{\text{br}} - W)\delta(W_{\text{br}'} - W')}$$

$$= (1-y)\frac{\theta(1 - W - W')(WW')^{y-2}(1 - W - W')^{1-2y}}{\Gamma(y)\Gamma(y)\Gamma(2-2y)}. \tag{IV.20}$$

iii) This result can be generalized (Eq. 43 in Ref. 1)

$$\sum_{\{\text{br}_1, \text{br}_2, \ldots; \text{br}_i \neq \text{br}_j\}} \overline{\prod_{i=1,k} \delta(W_{\text{br}_i} - W_i)}$$

$$= \frac{(1-y)^{k-1}\Gamma(k)}{(\Gamma(y))^k\Gamma(k-ky)}\theta\left(1 - \sum_i W_i\right)\left(\prod_i W_i\right)^{y-2}\left(1 - \sum_i W_i\right)^{k-ky-1}. \tag{IV.21}$$

Three observations can be made on the basis of these formulas:

a) The foliation of the tree has a self similar character. At different heights on the tree, measured by the overlap q (precisely the overlap among configurations whose branches converge at the level in question), the distribution in thicknesses is formally the same except for the different value of the parameter $y(q)$.

b) The distributions derived are essentially universal. The dependence on the temperature, external magnetic field and other possible parameters in the Hamiltonian appears only through $y(q)$. For instance if we consider the p-spin glass system[10] with the Hamiltonian

$$H_{(p)} = \sum_{i_1, i_2 \ldots i_p} J_{i_1 i_2 \ldots i_p} s_{i_1} s_{i_2} \ldots s_{i_p} \tag{IV.22}$$

with an appropriate normalization for the J then one can solve it with the same type of Ansatz for the Q_{ab}. The value of $P(q)$ that extremizes the free energy functional will be different and this will reflect itself in a different $y(q)$. However after having taken into account this change the distribution of the weights of the branches will be similar.

c) The ramification at each branching has a few dominating branches accounting for a large piece of the total thickness while an infinite number shares a small part of it. The tree is very far from being regular or symmetrical.

Observation b) suggests the following questions: Is there any Ansatz for Q_{ab} such that the W_I distributions be different from Eqs. (IV.19, 21)? To answer this question we have chosen a particular matrix Q_{ab} such that: i) ultrametricity still holds; ii) a row *is not* the permutation of any other one. We found, in a few examples, that in this case Eqs. (IV.19, 21) are not any more valid. From i) and ii) the original Ansatz for Q_{ab} follows necessarily. Therefore we conjecture that the proposal of this Ansatz is logically equivalent to assuming ultrametricity and the distributions of weights. The fact that the Ansatz solves the saddle-point equations in the replica method and the stability analysis constitute evidence of the self consistency of these hypotheses.

IV.4 The Energies and Free Energies as Independent Random Variables

IV.4.a The SK model

In spite of their universality the weight distributions are not very illuminating. It turns out that it is much better to focus the attention on a different set of variables more similar to energies or free energies. When considering a cluster I at a scale q, we could assign to it a "free energy" through

$$W_I = \exp(-\beta F_I) \Big/ \sum_K \exp(-\beta F_K). \tag{IV.23}$$

Notice that this equation cannot be used as a simple change of variables because it is impossible to solve F_I as a function of the W_I (there is one too many F's as compared with W's variables). The sheer information about the W_I distributions, Eqs (IV.19, 21) is not enough. We can obtain always through the replica formalism the joint distribution of the W_I at different cluster levels (an example with two levels was calculated in Ref. 8). This reduces the uncertainty on the F_I to one single overall scale which depends on the sample.

Of course we cannot say anything about the sample to sample fluctuations of this scale (Numerically it seems that there are fluctuations of order N^ε, $0 < \varepsilon < 1$, in absolute free energy—see Chap. I). However the *relative* fluctuations of all the other free energies F_I can be found. The simplest procedure (which is the original one) consists in guessing the distributions of the F_I's at the various levels and checking *a posteriori* that they give back through (IV.23) the correct W_I distributions found with the replica method.

The distribution of the F_I's can be defined completely if we assume that the possible overlap scales are discretized

$$0 < q_0 < q_1 < \cdots < q_k < 1 \tag{IV.24}$$

so that the tree of distances is well defined. (As always the final result should involve a continuous spectrum of possible overlaps, defined as the limit of (IV.24)).

To the top of the tree (the common root of all the configurations, sometimes denoted "grand ancestor"[3]) an overall scale $F^{(0)}$ is associated which depends on the sample and has an unknown distribution.

At the level q_1 each of the clusters I is assigned a free energy scale $F_I^{(1)}$. These are statistically independent, identically distributed random variables such that the average number in the interval $[F^{(1)}, F^{(1)} + dF^{(1)}]$ is

$$d\mathcal{N}^{(1)}(F^{(1)}) = \exp(\beta x(q_1)(F^{(1)} - F^{(0)}))\, dF^{(1)}. \tag{IV.25}$$

This process is iterated at the various scales: once the $F_I^{(l)}$ associated to the clusters at scale q_l have been chosen, the $F_J^{(l+1)}$ of the subclusters within *each* cluster I at scale q_l have again identically distributed independent random free energy scales, whose average number in the interval $[F^{(l+1)}, F^{(l+1)} + dF^{(l+1)}]$ is

$$d\mathcal{N}^{(l+1)}(F^{(l+1)}) = \exp(\beta x(q_{l+1})(F^{(l+1)} - F_I^{(l)}))\, dF^{(l+1)}. \tag{IV.26}$$

(This distribution depends on I through the scale $F_I^{(l)}$).

The F variables at the last level ($= k$) are the energies of the configurations while those at the level just above ($= k - 1$) are the free energies of the pure states (always up to an overall additive constant). Since the $x(q)$ at the last level is equal to one, the configuration energies have a distribution proportional to $\exp(\beta E)$, as they should from the definition of the temperature (see Sec. V.2; Eqs. (V.3–5)).

A complete description of the fluctuation of the free energies requires therefore the knowledge of the F_I variables at all the overlap levels. This is what enables one to derive for instance the correlations of the W_I's at various levels. However if one is concerned only with the spacings between the F_I at one given level, then one can forget about the values of the F_I at the other scales. For instance all the distributions of the weights of the states, P_α can be found from a choice of the f_α as identically distributed random free energies, with average number[8]

$$d\mathcal{N}(f) = \exp(\beta x(q_{k-1})f)\,df. \tag{IV.27}$$

The fact that from (IV.27) one effectively gets back the weight distributions of the states (IV.15–17) is demonstrated in Ref. 8. What might be less obvious to the reader is why the f_α's chosen through the whole process (IV.25, 26) also give this same weight distribution. In order to explain this let us imagine that there are only two clusters A and B. The density of states at free energy f is

$$d\mathcal{N}(f) = \{C_A \exp(\rho f) + C_B \exp(\rho f)\}\,df, \tag{IV.28}$$

where C_A and C_B are two constants related to the free energy scales of clusters A and B (constants which fluctuate with the clusters, the superclusters, ..., the sample...), and $\rho = \beta x(q_{k-1})$. Whatever these constants and their fluctuations, it is clear that the distribution of free energy spacings, which are the relevant quantities which determine the weights, are the same as those in (IV.27). An elegant mathematical reformulation of these free energy distributions, which avoids the cut-offs introduced in the computations in Ref. 8, can be found in Ref. 19.

IV.4.b Two instructive models: the REM and the GREM

It is interesting, particularly from a pedagogical point of view, to compare these distributions with the ones appearing in the Random Energy Model. The REM[10,11] is by definition a set of 2^N "configurations" whose energies are drawn at random with probability

$$e^{-E^2/N}\frac{dE}{\sqrt{\pi N}}. \tag{IV.29}$$

Multiplying this formula by the Gibbs-Boltzmann factor and by the total number of configurations we obtain

$$2^N e^{[-\beta E-(E^2/N)]}\frac{dE}{\sqrt{\pi N}}. \tag{IV.30}$$

The exponent is of order N so we can compute the internal energy E_0 by a saddle-point approximation to derive

$$E_0 = -\beta N/2. \tag{IV.31}$$

The total number of configurations at such an energy is

$$d\mathcal{N}(E) = 2^N e^{-E^2/N} \frac{dE}{\sqrt{\pi N}}. \tag{IV.32}$$

Of course the total number of configurations must be larger than one. (The entropy must be positive, old problem!)

$$2\sqrt{\ln 2}N > |E_0| = \beta N. \tag{IV.33}$$

The critical value $\beta_{crit} = 2\sqrt{\ln 2}$ marks a phase transition. For $\beta < \beta_{crit}$, $E_0 = \beta N/2$, while for $\beta > \beta_{crit}$ it is given by the boundary value $E_0 = \sqrt{\ln 2}N$.

The REM is the $p \to \infty$ limit of a spin model defined by the hamiltonian $H_{(p)}$ in Eq. (IV.22).[11] This fact is useful to unravel the similarity between the REM phase transition and the *SK* one. On a *p*-spin model the replica method can be used[10] and one checks that the onset of replica symmetry breaking coincides with β_{crit} in the limit $p \to \infty$ (as a bonus one verifies the reliability of the replica method). The solution for finite *p* can then be approximated by an expansion around $p = \infty$.[18]

In the low temperature region, $P(q)$ has two δ-type contributions

$$P(q) = (2\sqrt{\ln 2}/\beta)\delta(q) + (1 - 2\sqrt{\ln 2}/\beta)\delta(q-1). \tag{IV.34}$$

The distribution of the weights are given by Eq. (IV.21). On the other hand the energies are clearly independent random variables and for $E - E_0$ small of O(1) they become exponential distributions. In fact

$$d\mathcal{N}(E) = \begin{cases} dE \exp \beta(E - F(\beta)) & \beta < \beta_{crit} \\ dE \exp 2\sqrt{\ln 2}(E - F(\beta)) & \beta > \beta_{crit} \end{cases}, \tag{IV.35}$$

where

$$F(\beta) = N(-\beta/4 - \ln 2/\beta) \qquad \beta < \beta_{crit}$$

$$F(\beta) = -\sqrt{\ln 2}N \qquad \beta > \beta_{crit}. \tag{IV.36}$$

It follows, in this example, the equivalence of both kinds of distributions[12].

The ultrametric tree in the REM has only two levels: one root and the leaves. In Ref. 13 a generalization (the GREM) was defined that allows for any possible foliation. Again we begin with 2^N "configurations" but now located at the end tips of a regular tree. We define two sequences of L numbers where $L + 1$ is the number of levels in the tree

$$u_1, u_2, \ldots, u_L \qquad 0 < u_j < 1 \qquad \sum_j u_j = 1$$

$$\sigma_1, \sigma_2, \ldots, \sigma_L \tag{IV.37}$$

then any branch at the kth level will give rise to $2^{u_k N}$ $(k+1)$th level branches. The assignment of the energies is done in the following way:

1) At the first level we assign an energy to each one of the $2^{u_1 N}$ branches according to the probability distribution

$$dp_{\sigma_1 N}(E_1) = \exp(-E_1{}^2/(2\sigma_1 N))\, dE_1/(\sqrt{2\pi\sigma_1 N}). \tag{IV.38}$$

2) At the kth level, the $(k-1)$th branches already have an energy assigned. We then use the conditional probability distribution

$$dp_{\sigma_k N}(E_k - E_{k-1}) \tag{IV.39}$$

to determine the values of the E_k for the $2^{u_k N}$ branches.

The analysis now follows parallelly to the one of the REM. The total number of configurations with energy E_L will be

$$2^N \int_{E_1, E_2 \ldots E_{L-1}} dp_{\sigma_1 N}(E_1) \cdot dp_{\sigma_2 N}(E_2 - E_1) \ldots dp_{\sigma_L N}(E_L - E_{L-1}) \tag{IV.40}$$

This expression replaces the corresponding (IV.32) in the REM. We multiply by the Gibbs-Boltzmann factor to obtain the relevant configurations at a particular temperature $1/\beta$ by the saddle-point equations. There will now be L such equations but at the same time there will be L boundaries corresponding to the number of branchings at any level being necessarily greater than one. Therefore this system may have more than one freezing transition (see Ref. 13 for details).

IV.5 Self Averageness and Reproducibility

The leading contribution of order N to the energy of a relevant configuration is the same for each of them and does not fluctuate with J. This was checked by explicit calculation in Ref. 14. The non leading contributions, in particular the finite part of the energy when $N \to \infty$, are of course different. The lowest configuration energy fluctuates with J as N^{ε} with $0 < \varepsilon < 1$. A similar distinction holds for the states and their free energies. If an observable takes the same value for every possible state we call it *reproducible*, the word originating in the fact that if we were measuring in a sample but heating and cooling so as to allow a change of state, that quantity would not depend on the history but just on the temperature, external magnetic field and other thermodynamic parameters. A *non-reproducible* quantity, even if we average it over all the states weighted with the Gibbs-Boltzmann factors, will in general fluctuate with the sample. The reason can be seen in Eq. (IV.19–23), in particular observation c). There are in general an ∞ number of states (and also an ∞ number of clusters at any level) but when we consider the weights, the divergence comes from the very thin branches. Disregarding the thin branches up to a total weight equal δ there remains a finite number of states proportional to $\log \delta$. An average over a finite number of states cannot eliminate the fluctuations.

A powerful criterion to obtain reproducible, self averaging observables was derived in Ref. 3. Imagine any multilocal observable which is a function of the values of the spins in different sites $i_1, i_2, \dots i_k$. An extensive quantity is the sum over all sites $i_1, i_2, \dots i_k$. We will call it leading if it is non-zero when divided by N^k. If all spins refer to the same state, the thermal average of a leading extensive observable is reproducible and sample independent. If the spins refer to different states then the thermal average is exclusively a function of the mutual overlaps. Examples of reproducible, self averaging observables are

$$\mathcal{O}_1 = \left[\sum_i m_i^{(\alpha)} m_i^{(\alpha)} m_i^{(\alpha)}\right] \Big/ N$$

$$\mathcal{O}_2(q) = \left[\sum_i m_i^{(\alpha)} m_i^{(\alpha)} m_i^{(\beta)} m_i^{(\beta)}\right] \Big/ N \text{ for all } \alpha, \beta \text{ such that } q_{\alpha\beta} = q. \qquad \text{(IV.41)}$$

The criteria may be generalized to apply to observables where the coupling matrix appears[3].

An interesting example of an observable that is not self averaging is provided by the magnetic susceptibility inside one state

$$\chi = \frac{\beta}{N} \overline{\left\langle \sum_{i,j} (s_i - m_i^{(\alpha)})(s_j - m_j^{(\alpha)}) \right\rangle_\alpha} = \overline{\langle \chi[s] \rangle_\alpha} \qquad \text{(IV.42)}$$

In fact a simple calculation shows that at $h = 0$

$$\chi = \beta(1 - q_{EA}) \qquad \text{(IV.43)}$$

while the average value of the corresponding square is

$$\overline{\langle \chi[s]^2 \rangle_\alpha} = 3\beta^2(1 - q_{EA})^2 . \qquad \text{(IV.44)}$$

IV.6 A Branching Diffusion Process as a Generator of Ultrametricity

As we pointed out in Sec. 3 ultrametricity is a rather natural type of organization in infinite dimensional spaces. Consider the following stochastic process in L steps which must be carried out independently on each site i of the lattice, $1 \le i \le N$:

1. At the first step choose r^1 real number between -1 and $+1$ with a probability law

$$P_1(y^1)\, dy^1$$

Let us call these r^1 numbers $y_1{}^1, y_2{}^1, \dots y_{r^1}{}^1$.

2. In the second step choose *for each* $y_\iota{}^1$ a certain number $r^2{}_\iota$ of real numbers again between -1 and $+1$ with a probability law

$$P_2(y^2 | y_\iota{}^1) .$$

We will call these r^2_l numbers $y^2_{l1}, y^2_{l2}\dots$. Notice that the probability law is conditioned by the value of y_l^1. In this sense the $y_{lk}{}^2$ are descendants of y_l^1. It should be clear that we are drawing a tree with ramification r^1 followed by r^2_l.

3. Continue in this way up to the Lth step. At this point we will have a large number of y with L indices. We can now contract the y^L through

$$s_{\{L\ \mathrm{indices}\}} = \mathrm{sign}(y^L{}_{\{L\ \mathrm{indices}\}}). \tag{IV.45}$$

This is the value taken by the spin *at one site* in a particular configuration belonging to a state, belonging to a cluster, belonging to a hypercluster and continuing up the ladder of its genealogy as fixed by the family of L indices.

4. Repeat the whole procedure *independently* but with the same ramification (as fixed by the r numbers) at everyone of the N sites.

A simple probabilistic calculation shows that the overlap between two configurations having their closest common ancestor at the H-level has a well defined $N \to \infty$ limit, given by

$$q_H = \int\left[\prod_{k=1}^{H-1} (dy^k P_k(y^k|y^{k-1}\dots y^1))\right]\left[\int \mathrm{sign}(y^L) \prod_{k=H}^{L} (dy^k P_k(y^k|y^{k-1}\dots y^1))\right]^2 . \tag{IV.46}$$

Therefore for $N \to \infty$ such an ensemble of configurations is ultrametric.

In natural sciences there is a classical example of a hierarchical tree and ultrametric organization generates in this way. This is the evolutionary tree of the species. The idea of a natural distance has been introduced[15]. In one of the large proteins or a nucleic acid one compares the sequence of aminoacids or nucleotides. In this theory the stochastic branching process simply corresponds to neutral mutations (the neutral evolution of pseudogenes). In general, however, for *any* ultrametric organization that appears in an infinite dimensional space, under certain general conditions, it is possible to assign a stochastic process. We refer to Ref. 2 for the details on this converse theorem.

In the SK model, surprisingly enough, one can calculate explicitly such a process. Given the tree, with configurations at each end tip or leaf, we locate at each branching vertex an *ancestor* defined by a local value of the magnetization equal to an average, (without weighting with the Gibbs-Boltzmann factor), of the local spin value of all the descendants

$$m_i^{(A(k))} = \lim_{R\to\infty} (1/R) \sum_{c=1}^{R} s_i^{(c)}, \tag{IV.47}$$

where $m_i^{(A(k))}$ denotes the value of the magnetization at site i for the ancestor at the kth level on the tree; c denotes any configuration that lies at the end of any branch stemming from the branching vertex. As the Gibbs-Boltzmann weights are not included, the sum in (IV.47) is dominated by the configurations that are farthest away from each other (phase space grows as the overlap decreases). It follows that with probability one, any two configurations c and c' in the sum will have mutual overlap equal to the self-overlap $q(k)$.

It is convenient at this point to imagine a discretization of the generations. We then try and derive the conditional probability distribution of the local magnetization of the next descendant given the corresponding value for the ancestor.

$$\frac{\sum_{i=1}^{N} \delta(m_i^{(A(k))} - m)\delta(m_i^{(A(k-1))} - m')}{\sum_{i=1}^{N} \delta(m_i^{(A(k))} - m)}, \tag{IV.48}$$

where $A(k-1)$ is descendant of $A(k)$.

As in previous examples (IV.48) can be calculated in the replica formalism by computing the moments

$$(1/N) \sum_{i=1}^{N} (m_i^{A(k)})^r (m_i^{(A(k-1))})^s \tag{IV.49}$$

Taking into account Eq. (IV.47) we can calculate (IV.49) by considering $s + r$ real replicas such that s are at a distance $q(k)$ from each other while one of this s is surrounded by r at a larger overlap $q(k-1)$. Notice that (IV.49) fulfills our criteria and is self-averaging.

The details of the calculation are tedious[3] but the result is particularly simple. It can be written in terms of the function $m(q, y)$ introduced in Chap. III (Eqs. 56–59) and a new auxiliary function

$$\mathscr{P}_{q,q'}(y, y') \qquad q < q' \tag{IV.50}$$

introduced in Ref. 16 defined by the differential equation

$$-\frac{\partial \mathscr{P}}{\partial q} = \frac{1}{2}\frac{\partial^2 \mathscr{P}}{\partial y^2} + x(q) \cdot m(q, y)\frac{\partial \mathscr{P}}{\partial y}. \tag{IV.51}$$

This is a diffusion equation in a "time" q flowing from q' to q (i.e. decreasing q). The initial condition is

$$\lim_{q \to q'} \mathscr{P}_{q,q'}(y, y') = \delta(y - y') \tag{IV.52}$$

and the boundary conditions are

$$\lim_{y \to \pm\infty} \mathscr{P}_{q,q'}(y, y') = 0 \tag{IV.53}$$

and the auxiliary function $m(q, y)$ is the same function defined in (III.59).

In terms of these function expressions (IV.48) i.e. the probability distribution that the descendant $A(k-1)$ has magnetization m where the ancestor $A(k)$ has magnetization m' is equal to

$$|dy'/dm(q',y')| \cdot \mathscr{P}_{q,q'}(y,y') \qquad \text{(IV.54)}$$

where y, y', q and q' satisfy the conditions:

$$m(q',y') = m', \quad m(q,y) = m, \quad q(k) = q, \quad q(k-1) = q' \qquad \text{(IV.55)}$$

Therefore the knowledge of the function $\mathscr{P}_{q,q'}(y,y')$ amounts to the determination of the stochastic branching process, Markovian in this case.

We have proved that every relevant, low energy configuration is a particular outcome of the stochastic branching process. It is not true, however that every realization is a low energy configuration. Expectation values that include the J_{ij} have to be considered and they restrict the possible configurations. Amusingly enough, this situation resembles the one that appears in the theory of natural evolution. Every species must be the result of the stochastic process controlled by mutations but not every realization of mutations is a viable species. It is as if in the *SK* model there were place for mutations *and* natural selection, in spite of the fact that the primitive formulation only contained a hamiltonian whose minimization is rather reminiscent of natural selection.

References

1. Mézard *et al.* (1984a) and (1984b) Reprint 16.
2. Rammal *et al.* (1986).
3. Mézard and Virasoro (1985) Reprint 17.
4. Several proposals to deal with this obstacle can be found in Sourlas (1984); Parga, *et al.* (1984); Kirkpatrick and Toulouse (1985); Solla, *et al.* (1985).
5. Bhatt and Young (1986).
6. Young *et al.* (1984).
7. The singularity at $Y = 1$ was analysed in Ref. 1. The singularity at $Y = 0$ in Ref. 8.
8. Mézard *et al.* (1985) Reprint 18.
9. Derrida and Flyvjberg (1986).
10. Gross and Mézard (1984) Reprint 21.
11. Derrida (1981) Reprint 20.
12. Derrida and Toulouse (1985).
13. Derrida (1985); Derrida and Gardner (1985).
14. Mézard and Parisi (1984).
15. Zuckerlanl and Pauling (1985).
16. De Almeida and Lage (1982). See also Ref. 3.
17. Athanasiu *et al.* (1986).
18. Gardner (1985).
19. Ruelle (1986).

Chapter V
THE CAVITY METHOD

V.1 Introduction

The reason why the mean field approximation is so difficult in the SK model may be understood in the following way. In the infinite range model of a ferromagnet with N spins the instantaneous magnetic field on a generic site

$$h_i = \sum_{k=1}^{N} J_{ik} s_k \tag{V.1}$$

will be finite even if all the s_i are parallel because the couplings J_{ik} are of order $1/N$. Fluctuations of h_i will then be small because of the appearance of higher powers of J_{ik}.

In the SK model, on the contrary, each J_{ik} is much larger $(O(1/\sqrt{N}))$ and h_i is finite because of random cancellations. The fluctuations will therefore be relatively more important.

To control these fluctuations one may be tempted to consider the $\sum_j J_{ij} s_j$ as a random variable, sum of many uncorrelated variables and therefore controllable through the central limit theorem. This leads directly to the mean field approximation. Unfortunately this approach, though correct in unfrustrated systems, is essentially wrong in spin glasses and that because of the very features that make the latter so interesting.

The problem is that the different terms in the sum are correlated. In fact, when thermal fluctuations make $s_i = 1$, all terms $J_{ik} s_k$ are a bit larger than when $s_i = -1$ (s_k will feel the influence of s_i through the coupling $J_{ki} = J_{ik}$).

The "cavity field" introduced by Onsager[1] tries to take care of this particular problem. A new field h_i is defined as the field measured at site i once the spin i has been removed (thus the name cavity field). But of course one has to be careful not to throw the baby with the water in the sense that the reaction of the spins to the one in site i has to be included at the end. There is no doubt that the *instantaneous magnetic field acting at site i is not the cavity field.* In fact it is known that the local instantaneous magnetic field does not obey a Gaussian distribution[2].

We have already partially discussed this approach in previous sections (particularly in Chap. II). Here we would like to formalize it and show that it can become a powerful method to deal with amorphous frustrated systems. We will show how it can be used to derive TAP equations, the replica solution and in the next chapters to discuss the dynamical behavior of the *SK* model, neural models and optimization problems.

For pedagogical reasons we are going to discuss the method in the approximation equivalent to assuming one level of replica symmetry breaking as we did in Sec. III.4. Then $P(q)$ (see Eq. III.33) will have the form

$$P(q) = \delta(q - q_1)(1 - x) + \delta(q - q_0)x. \tag{V.2}$$

The formulae we are going to derive can be directly compared with those in Sec. III.4 and in Ref. 3. In the thermodynamic limit the space of configurations will be separated into an infinite number of pure states but the latter belong to a unique cluster.

The general idea of the method is to compare the behavior of a N spin system with a $N + 1$ spin system. The sheer existence of the thermodynamic limit implies that the differences are controllable. If the configurations inside a state are ordered according to their energies, when one adds a $N + 1$ spin (located conventionally at $i = 0$) they are reshuffled. This reshuffling is the reaction of the spins to the newcomer (up to this point our method parallels traditional derivations of TAP equations).

The reaction continues inter-valley. This is due to the fact that different valleys contribute to the equilibrium phase through their weights, and these change when adding the new spin. Notice that the high level of reactivity of the system is connected with the existence of infinitely many valleys with energies almost equal, i.e.

$$F_\alpha - F_\beta = O(1) \qquad \text{when } N \to \infty$$

so that adding a new spin completely upsets their relative importance.

The cavity method[8] allows us to discuss the two moments separately. In subsection 2a we will discuss the rearrangement inside a state. The TAP equations will easily follow. In Sec. 2b the rearrangements interstate are discussed. New TAP equations appear which apply to the ancestors as defined in Chap. IV. In Sec. 2c the average over the J_{ik} is performed. After this sample average, the stability of the thermodynamic limit leads to the computation of the free energy in equilibrium while self-consistency equations on the overlaps reproduce the saddle-point equations of the replica method. Finally in Sec. 3 we sketch how one can treat the fluctuations around the mean field approximation.

V.2 The Stability of the Thermodynamical Limit

V.2.a Inside a state—A new derivation of the TAP equations

The cavity method considers a system with N spins and adds a new one to it. Any configuration belonging to an equilibrium state of the $N + 1$ spins can be constructed by considering the configurations of the N spin system occurring in equilibrium combined with all possible values of the $(N + 1)$th spin.

The number of configurations with energy in the interval $(E, E + dE)$ is (definition of entropy!)

$$d\mathcal{N}(E) = e^{S(E)}\,dE\,. \tag{V.3}$$

At fixed temperature the relevant configurations have an energy near to E_α, the internal energy of the state[4], and

$$S(E) = S(E_\alpha) + \beta(E - E_\alpha), \tag{V.4}$$

where $|E - E_\alpha|$ is small. Therefore

$$d\mathcal{N}(E) = \exp(-\beta F_\alpha)e^{\beta E}\,dE\,. \tag{V.5}$$

By its derivation this expression, with the corresponding free energies $F_{\alpha(N)}$ and $F_{\alpha(N+1)}$, must be valid both for the N and the $N + 1$ systems. Notwithstanding this fact, it is useful, for our subsequent line of reasoning, to show how the exponential distribution of energies follows directly from the stability of the thermodynamic limit. This will become a crucial point in the next section where we will not be able to invoke Eq. (3).

When adding the spin at site 0 we choose randomly N couplings J_{0k}. The magnetic field produced by the configuration $\mathscr{C}$ of the spins acting on the new site will be

$$h^{\mathscr{C}} = \sum_{k=1}^{N} J_{0k} s^{\mathscr{C}}{}_k\,. \tag{V.6}$$

Though it may seem that we are back at Eq. (V.1), now by construction the J_{0k} are uncorrelated with the dynamical variables s_k, $k = 1, \ldots N$, which are in equilibrium among themselves but not with the newcomer. When we consider the ensemble of relevant configurations inside the N-system equilibrium state $h^{\mathscr{C}}$ becomes a random variable (to be denoted h). In other words we can define the probability that choosing at random any relevant configuration the value of the magnetic field at the new site 0 fall into a certain interval $(h, h + dh)$. This probability distribution is statistically independent of the value of the energy because h is a function of the new J_{0k} while the energy is a function of the old J_{ik}. Furthermore the central limit theorem can be applied to derive a Gaussian distribution with parameters

$$\langle h \rangle_{\alpha(N)} = \sum_{k=0}^{N} J_{0k} m_k{}^{\alpha(N)} = h_{\alpha(N)} \tag{V.7}$$

and

$$\langle (h - h_{\alpha(N)})^2 \rangle = \sum_{k,l=1}^{N} J_{0k} J_{0l} \langle (s_k - m_k{}^{\alpha(N)})(s_l - m_l{}^{\alpha(N)}) \rangle_{\alpha(N)}\,. \tag{V.8}$$

The suffix $\alpha(N)$ indicates that averages are taken on the state α of the N-system (at fixed energy so that the Gibbs-Boltzmann weights are irrelevant). So for instance $m_k{}^{\alpha(N)}$ is the average magnetization of the spin s_k in state α in the N ($\gg 1$) spin system. As we have already noticed the connected correlated function *inside a state* is such that

$$\lim_{N\to\infty} 1/N^2 \sum_{k,l=1}^{N} \langle (s_k - m_k^{\alpha(N)})(s_l - m_l^{\alpha(N)})\rangle^2 = 0 \tag{V.9}$$

so that the generic $k \neq 1$ terms can be assumed to go individually to 0 in the limit $N \to \infty$. As a consequence in (V.8) the sum over the $N(N-1)$ uncorrelated, randomly signed, non diagonal terms will also go to 0 and can be neglected in the thermodynamic limit. Therefore

$$\langle (h - h_{\alpha(N)})^2 \rangle = 1/N \sum_{k=1}^{N} \langle (s_k - m_k^{\alpha(N)})^2 \rangle = 1 - q_1 . \tag{V.10}$$

The number of configurations of the N spins with fixed energy and fixed magnetic field on site 0 is

$$d\mathcal{N}^{(N)}(E^{(N)}, h) = \exp(-\beta F_{\alpha(N)}) \cdot \exp(\beta E^{(N)})\, dE^{(N)}\, dp_{1-q_1}(h - h_{\alpha(N)}), \tag{V.11}$$

where again as in previous chapters we use a shorthand notation $dp_\sigma(x)$ to indicate a Gaussian measure of mean zero and width σ.

To each one of these configurations there correspond two configurations of the $N+1$ spin system, with energies

$$s_0 = 1, \qquad E^{(N+1)} = E^{(N)} - \beta h$$

$$s_0 = -1, \qquad E^{(N+1)} = E^{(N)} + \beta h \tag{V.12}$$

so that the joint distribution is

$$d\mathcal{N}^{(N+1)}(E^{(N+1)}, h, s_0) = \exp(-\beta F_{\alpha(N)} + \beta E^{(N+1)}\lambda + \beta h s_0)\, dE^{(N+1)}\, dp_{1-q_1}(h - h_{\alpha(N)}). \tag{V.13}$$

Integrating over h and summing over s_0 we get an exponential distribution of the energy $E^{(N+1)}$ as we should

$$d\mathcal{N}^{(N+1)}(E^{(N+1)}) = \exp(-\beta F_{\alpha(N+1)} + \beta E^{(N+1)})\, dE^{(N+1)}. \tag{V.14}$$

Only an exponential distribution in energies could have reproduced itself after the integration over h. The only change is in the normalization which in fact determines the free energy $F_{\alpha(N+1)}$ of the $N+1$ system in relation to the one of the N system

$$F_{\alpha(N+1)} - F_{\alpha(N)} = -1/\beta \ln \int 2\cosh \beta h\, dp_{1-q_1}(h - h_{\alpha(N)})$$

$$= -1/\beta \ln[2\cosh \beta h_{\alpha(N)}] - (\beta/2)(1 - q_1). \tag{V.15}$$

Equation (V.13) can be rewritten

$$d\mathcal{N}^{(N+1)}(E^{(N+1)}, h, s_0) = \exp(-\beta F_{\alpha(N+1)} + \beta E^{(N+1)})\mathcal{P}(h, s_0)\, dE^{(N+1)}, \qquad \text{(V.16)}$$

where $\mathcal{P}$ is the probability that the $N + 1$ *spin system* had a certain value of s_0 and of the field acting on s_0 (at a fixed value of $E^{(N+1)}$)

$$\mathcal{P}(h, s_0) = K \exp(-(h - h_{\alpha(N)})^2/2(1 - q_1) + \beta h s_0)\, dh, \qquad \text{(V.17)}$$

with K a normalisation constant. *The instantaneous magnetic field distribution is no more Gaussian and, as stressed in Sec. 1, it is correlated with the value of the spin* s_0.

In fact what happens is easy to understand: the configurations of the N-spin system which have low energy are few in number and the probability that one of them has a large value of the field is rather remote. On the other hand paying a certain amount of penalty in energy the number of configurations increases exponentially and we can choose one with a large magnetic field. These configurations will decrease considerably their energy when we add the new spin. Therefore the h distribution *a posteriori* will favour large values of h.

From this equation we derive the average magnetization and the average magnetic field

$$\langle s_0 \rangle_{\alpha(N+1)} = m_0{}^{\alpha(N+1)} = \sum_{s_0} \int dh \mathcal{P}(h, s_0) s_0 = \tanh \beta h_{\alpha(N)} \qquad \text{(V.18a)}$$

$$\langle h \rangle_{\alpha(N+1)} = h_{\alpha(N)} + \beta(1 - q_1) \tanh \beta h_{\alpha(N)}\, . \qquad \text{(V.18b)}$$

Equation (V.18a) is Onsager's cavity field equation. From (V.18b) we derive TAP equation because

$$m_0{}^{\alpha(N+1)} = \tanh \beta(\langle h \rangle_{\alpha(N+1)} - \beta(1 - q_1) m_0{}^{\alpha(N+1)}) \qquad \text{(V.19)}$$

with

$$\langle h \rangle_{\alpha(N+1)} = \sum_{k=1}^{N} J_{0k} \langle s_k \rangle_{\alpha(N+1)}, \qquad \text{(V.20)}$$

where now all averages are measured in the same $N + 1$ system.

If there was only one state, $h_{\alpha(N)}$ (which coincides with $\tilde{h}$ in Chap. II.2, see Eq. V.18a) would fluctuate solely because of the J_{ij}. Therefore it would be Gaussian (compare Sec. II.3). We will show in the next section that as there are many pure states, and as a consequence of the reshuffling of the free energies, this is not the case.

V.2.b Inside a cluster—Self-consistency of ultrametricity and the exponential distribution of the states

When the new spin is added the energy of a configuration can increase or decrease making it more or less important in the equilibrium measure. But also the free energy of the state changes so that its probability weight will be modified accordingly. This rearrangement will be described in this section. We shall also establish TAP equations for the ancestors. One must study the rearrangements at each level of the ultrametric tree before performing the average over the samples and regaining the results of the replica method, which will be the purpose of the next section.

In Chap. IV, end of Sec. 3, we argued that the Ansatz for the replica matrix Q_{ab} (Eqs. III.32 and III.45) encodes two basic physical properties: 1) the ultrametric topology, 2) the exponential free energy distribution of states inside a cluster and corresponding generalizations for clusters inside hyperclusters. The replica solution therefore could not be seen as a derivation of these properties but rather as a demonstration that they are self-consistent. The cavity method essentially does the same job but without invoking the troublesome mathematics of the $n \to 0$ trick. Furthermore these two physical hypotheses are formulated explicitly instead of being hidden in the form of an Ansatz for a 0×0 matrix.

We will thus assume (compare Sec. IV.4) that the distribution of free energies of the states which belong to the same cluster is given by

$$d\mathcal{N}(F) = C\exp(\beta x F)\,dF \qquad \text{with } x < 1 \tag{V.21}$$

and the constant, by analogy with (V.3), will be written

$$C = \exp(-\beta x F_\Gamma), \tag{V.22}$$

which is to be understood as a definition of the Γ cluster free energy. In addition we assume that the occurrence or not of a state inside the cluster at a particular free energy is a probabilistic event statistically independent of the presence of other states (see Sec. IV.4). In the region where (V.21) is small the equation is interpreted as giving the probability of occurrence; where the number predicted is large it also gives accurately the average number (with an error proportional to the square root of the number).

Comparing (V.21) and (V.5) we observe that we have to pay relatively more energy to obtain the same additional increase in the number of valleys. This reaction is then less effective (see discussion after Eq. (V.17)). The number of supervalleys is still flatter ($x' < x$) and in this way a hierarchy of responses appears without any reference to different time scales.

As in the previous section we want to analyse the stability of Eq. (V.21) under addition of the new spin. In the region of free energies where there are many states, an ensemble average over the states makes sense. We define it through the equation

$$\langle \mathcal{O} \rangle_\Gamma = (1/M) \sum_{\alpha=1}^{M} \langle \mathcal{O} \rangle_\alpha ,$$

where the sum is over a large number M of states, at a fixed value of the free energy, contained in cluster Γ.

Consistently with ultrametricity we assume that with probability one, any 2 states inside the cluster will have mutual overlap

$$q_0 < q_{EA} = q_1 . \tag{V.23}$$

Then the fact that the interstate overlap is always q_0 implies

$$(1/N^2) \sum_{i,j} |\langle m_i m_j \rangle_\Gamma - \langle m_i \rangle_\Gamma \langle m_j \rangle_\Gamma|^2 = 0. \tag{V.24}$$

Such an equation, reminiscent of the clustering condition that defines a pure state, would not be valid had we weighted the states with the Gibbs-Boltzmann weights. The weights of the states decrease like the $\exp(-\beta E)$ while their number increases like $\exp \beta x E$ with $x < 1$ so that they do not compensate. Consequently only a few low free energy states would contribute and in the derivation of (V.24) we would not have been able to neglect the intrastate overlap contribution q_1.

The magnetic field $h_{\alpha(N)}$ acting at the $(N+1)$th site and generated by the N spins in the state $\alpha(N)$ becomes a random variable (which we will denote h_1) in the ensemble of states previously defined. As a consequence of (V.24) it is Gaussian with

$$\langle h_1 \rangle_{\Gamma(N)} = \sum_{k=1}^{N} J_{0k} \langle m_k \rangle_{\Gamma(N)} = h_{\Gamma(N)} \tag{V.25}$$

$$\langle (h_1 - h_{\Gamma(N)})^2 \rangle_{\Gamma(N)} = q_1 - q_0 . \tag{V.26}$$

The distribution of F_N and of h_1, $d\mathcal{N}(F_N, h_1)$, is a product of the exponential distribution Eq. (V.21) and the Gaussian distribution of h_1 because as in the previous subsection, h_1 and F_N are independent random variables. Each state of the N spins system corresponds to one state of the $N+1$ spins system with a new free energy F_{N+1}. The increase in free energy due to the addition of the new spin is given by Eq. (V.15) so that if we want to find the distribution of states in the variables F_{N+1} and h_1 we simply compute

$$\begin{aligned} d\mathcal{N}(F_{N+1}, h_1) & \\ &= \int d\mathcal{N}(F_N, h_1) \delta(F_{N+1} - F_N - \Delta F(h_1)) \\ &= (2\cosh \beta h_1)^x \exp[\beta x (F_{N+1} - F_{\Gamma(N)}) + \beta^2 x (1 - q_1)/2]\, dF_{N+1}\, dp_{q_1 - q_0}(h_1 - h_{\Gamma(N)}) \end{aligned} \tag{V.27}$$

from which it follows, after averaging over h_1, that the change in the cluster free energy as defined in (V.20) is

$$F_{\Gamma(N+1)} - F_{\Gamma(N)} = -\beta(1 - q_1)/2 - (1/\beta x) \ln \left\{ \int (2\cosh \beta h_1)^x\, dp_{q_1 - q_0}(h_1 - h_{\Gamma(N)}) \right\} . \tag{V.28}$$

As in the previous section the assumption of an exponential distribution is *a posteriori* justified as the only one that could be stable.

The probability distribution of h_1 at fixed $h_{\Gamma(N)}$ and F_{N+1} becomes

$$\mathscr{P}(h_1)\,dh_1 = C\exp[-(h_1 - h_{\Gamma(N)})^2/(2(q_1 - q_0))](2\cosh\beta h_1)^x\,dh_1\,. \qquad \text{(V.29)}$$

Notice again that it is *not Gaussian* as anticipated in the previous section because of replica symmetry breaking (see also II.3).

Before going on with the average over the couplings, let us mention how Eq. (V.29) can be used to derive TAP-like equations for the ancestor. Multiplying (V.29) and (V.17) we derive the joint probability distribution of the random variables h, h_1, and s_0 and from it we obtain appropriate averages

$$\langle h\rangle_{\Gamma(N+1)} = \beta(1 - q_1)\langle\tanh\beta h_1\rangle_{\Gamma(N+1)} + \langle h_1\rangle_{\Gamma(N+1)} \qquad \text{(V.30)}$$

$$\langle h_1\rangle_{\Gamma(N+1)} = \beta x(q_1 - q_0)\langle\tanh\beta h_1\rangle_{\Gamma(N+1)} + h_{\Gamma(N)} \qquad \text{(V.31)}$$

$$\langle s_0\rangle_{\Gamma(N+1)} = \langle\tanh\beta h_1\rangle_{\Gamma(N+1)}\,, \qquad \text{(V.32)}$$

so that defining an auxiliary function

$$\tilde{m}(h_{\Gamma(N)}) \equiv \langle\tanh\beta h_1\rangle_{\Gamma(N+1)} \qquad \text{(V.33)}$$

the ancestor TAP equation follows

$$m_0 = \tilde{m}(h_{\Gamma(N)}) \qquad \text{(V.34)}$$

$$h_{\Gamma(N)} = \sum_k J_{0k}\langle s_k\rangle_{\Gamma(N+1)} - \beta[1 - q_1(1 - x) - xq_0]m_0\,. \qquad \text{(V.35)}$$

For future reference we mention that once the self-consistency equations for the q (Eq. (V.42)) are taken into account it is possible to prove that

$$\overline{\frac{\partial\tilde{m}(h)}{\partial h}} = \beta[1 - q_1(1 - x) - xq_0]\,. \qquad \text{(V.36)}$$

The function $\tilde{m}(h)$ is identical to the function $m(q_0, h)$ introduced in Chap. III, Eqs. 59–62 (see also IV, Eqs. 50–52) for the approximated $P(q)$ given in Eq. (V.1). The ancestor TAP equations have been derived with the replica method in Ref. 6.

V.2.c The average over the J_{ik}: self consistency equations and equivalence with the results of the replica method

There is a single cluster in the approximation we are considering. Therefore the field $h_{\Gamma(N)}$ can only fluctuate because of the J_{0k} fluctuations and this is the average which we shall study in this section. We shall first compute the sample averaged change of free energy when one adds a new spin. Then we note that after sample averaging the new site "0" has nothing special: the probability distribution of the local field on this site must be the same as on any of the other N sites. This gives two self consistency equations which determine the values of the overlaps q_1 and q_0. These equations turn out to be identical to those obtained with the replica method.

In the ensemble of samples the field $h_{\Gamma(N)}$ is a random variable which will be denoted by h_0. It is a sum of Gaussian random variables and therefore it is itself Gaussian (through the Central Limit theorem we know it will still be Gaussian for other choices of the distribution probability for the J_{0k}). The average value and the variance are

$$\overline{h_0} = 0$$

$$\overline{h_0^2} = q_0 \,. \tag{V.37}$$

Therefore (V.28) averaged over J_{0k} becomes

$$\overline{F_{\Gamma(N+1)} - F_{\Gamma(N)}} = -\beta(1-q_1)/2 - \int dp_{q_0}(h_0)(1/\beta x)\ln \int dp_{q_1-q_0}(h_1 - h_0)\cdot(2\cosh \beta h_1)^x \,. \tag{V.38}$$

In order to compare this average change of free energy with the result of the replica method, one must pay attention to the fact that adding a new spin the normalization of the couplings should change. Otherwise we are dealing with a $N+1$ system which is effectively at a slightly different temperature from the N system. In fact

$$\beta H_{N+1} = \beta \sum_{i>k=1}^{N+1} J_{ik} s_i s_k = \beta\sqrt{(1 + 1/N)} \sum_{i>k=1}^{N+1} J'_{ik} s_i s_k \,, \tag{V.39}$$

where the J'_{ik} have the correct normalization. Therefore introducing the partition function Z one has

$$\begin{aligned} \overline{F_{\Gamma(N+1)} - F_{\Gamma(N)}} &= -(1/\beta)\ln Z[N+1, \beta\sqrt{(1+1/N)}] + (1/\beta)\ln Z[N,\beta] \\ &= -(1/2N)(\partial(\ln Z)/\partial\beta) + F(N+1,\beta) - F(N,\beta) \end{aligned} \tag{V.40}$$

and in the thermodynamic limit $F(N,\beta) = Nf(\beta) = -(1/\beta)\ln Z[N,\beta]$:

$$\overline{F_{\Gamma(N+1)} - F_{\Gamma(N)}} = f + (1/2)\,\partial/\partial\beta(\beta f) \,. \tag{V.41}$$

It is a simple exercise to check that the result (V.38) for $\overline{F_{\Gamma(N+1)} - F_{\Gamma(N)}}$ is equal to $f + (1/2)\,\partial/\partial\beta(\beta f)$ where f is the free energy density found in (III.39). Therefore we have found the correct free energy we derived previously with the replica method.

In the same way the q_1 and the q_0 must be stable when one adds a new site. There follows two self-consistency equations

$$q_1 = \overline{\langle m_0{}^2 \rangle_{\Gamma(N+1)}}; \qquad q_0 = \overline{(\langle m_0 \rangle_{\Gamma(N+1)})^2} \tag{V.42}$$

which are the saddle-point equations of the replica method (see Eq. (III.41)). In that section there was another saddle-point equation to determine the best value of x. As mentioned there, such an equation, though it may result in an improved approximation when the number of replica symmetry breaking levels is small, is not logically required. We are approximating $q(x)$ by a piecewise constant function whose breakpoints can be fixed *a priori*. Clearly this makes no difference in the limit of a large number of breakings.

To summarize we have shown that if Eq. (V.42) holds the two hypotheses made on the N spin system (ultrametricity and the exponential distribution of free energies at each level of cluster) are also self consistently properties of the $N + 1$ spin system. We must also check whether the connected correlation functions inside one state and inside one cluster remain small. This is the object of the next section and will provide the range of stability of the present solution.

V.3 Fluctuations around the Mean Field

The calculation of the leading corrections to the saddle-point equations in the replica approach is by now a full fledged program[5].

The cavity method can be generalized to calculate the leading $O(1/\sqrt{N})$ terms in the correlation function of the spins. It has some advantages concerning particularly the physical interpretation of the mathematical steps involved. This may be useful for instance to understand the zero modes reported by the aforementioned authors.

As we are now trying to calculate expectation values of two correlated spins, we imagine a N-system, supposed to be in a given state $\alpha(N)$ to which we add 2 new spins (which we locate at points 0 and 0′). The magnetic fields at these locations, h and h', will be correlated to the same order, i.e.

$$\langle hh' \rangle - \langle h \rangle \langle h' \rangle = O(1/\sqrt{N}) \tag{V.43}$$

Because of the smallness of these correlations, in the absence of spins at sites 0 and 0′ we can assume the following distribution for h, h'

$$\exp[\delta(h - h_{\alpha(N)})(h' - h'_{\alpha(N)})]\, dp_{1-q_1}(h - h_{\alpha(N)})\, dp_{1-q_1}(h' - h'_{\alpha(N)}) \tag{V.44}$$

so that

$$\langle hh' \rangle_{c,\alpha(N)} = \langle hh' \rangle_{\alpha(N)} - \langle h \rangle \langle h' \rangle_{\alpha(N)} = \delta(1 - q_1)^2, \tag{V.45}$$

where the subscript c means that we are considering connected correlation functions. The probability distribution in the state $\alpha(N + 2)$ of h, h', s_0, $s_{0'}$ will be

$$C \exp[\delta(h - h_{\alpha(N)})(h' - h'_{\alpha(N)}) + \beta h s_0 + \beta h s_{0'} + \beta J_{00'} s_0 s_{0'}] \cdot$$

$$\cdot\, dp_{1-q_1}(h - h_{\alpha(N)})\, dp_{1-q_1}(h' - h'_{\alpha(N)}). \tag{V.46}$$

Integrating over h and h' leaves the following effective Hamiltonian for s_0, $s_{0'}$

$$-\beta H = \beta h_{\alpha(N)} s_0 + \beta h'_{\alpha(N)} s_{0'} + [\beta J_{00'} + \delta\beta^2(1-q_1)^2] s_0 s_{0'} \qquad \text{(V.47)}$$

so that

$$\langle s_0 s_{0'} \rangle_{c,\alpha(N+2)} = (1 - m_0{}^2)(1 - m_{0'}{}^2)[\beta J_{00'} + \delta\beta^2(1-q_1)^2]. \qquad \text{(V.48)}$$

On the other hand

$$\langle hh' \rangle_{c,\alpha(N)} = \sum_{kl} J_{0k} J_{0'l} \langle s_k s_l \rangle_{c,\alpha(N)}. \qquad \text{(V.49)}$$

Then, again because of the stability of the thermodynamic limit the correlation functions of the s in (V.48) and (V.49) must be equal in average. However the sample average of $\langle s_0 s_{0'} \rangle_c$ vanishes, and the relevant quantity we want to compute is rather the average of the square $\overline{(\langle s_0 s_{0'} \rangle_c)^2}$, from which we shall deduce the nonlinear susceptibility. We replace $\delta(1 - q_1{}^2)$ from (V.45) and (V.49) in (V.48) and then square the remaining equation to derive

$$(\langle s_0 s_{0'} \rangle_c)^2 = [(1 - m_0{}^2)(1 - m_{0'}{}^2)]^2 \left[\beta^2 J_{00'} + \beta^4 \sum_{klmn} J_{0k} J_{0'l} J_{0m} J_{0'n} \langle s_k s_l \rangle_c \langle s_m s_n \rangle_c \right]. \qquad \text{(V.50)}$$

(From now on we shall not write explicitly the indices $\alpha(N)$.)

In the multiple sum we separate the diagonal terms and perform the average

$$\overline{(\langle s_0 s_{0'} \rangle_c)^2} = \overline{[(1 - m_0{}^2)(1 - m_{0'}{}^2)]^2} [\beta^2/N + (\beta^4/N)\overline{(1 - m_k{}^2)^2} + \beta^4 \overline{(\langle s_k s_l \rangle_c)^2}] \qquad \text{(with } k \neq l) \quad \text{(V.51)}$$

There is a subtlety in the derivation of this result because for instance $1 - m_0^2$ depends on the couplings J_{0k} and is thus correlated with the last term in (V.50). However using the fact that in (V.50) the indices k, l, m, n have to be contracted by pairs in order not to have sign fluctuations in $\langle s_k s_l \rangle_c$, one finds the result (V.51). From this one gets

$$\overline{(\langle s_0 s_{0'} \rangle_c)^2} \equiv \frac{\beta^2/N}{1 - \beta^2 \overline{(1 - m_0^2)^2}} (\overline{(1 - m_0{}^2)^2})^2. \qquad \text{(V.52)}$$

The last equation determines directly the nonlinear susceptibility

$$(1/N) \sum_{k,l} \overline{(\langle s_k s_l \rangle_c)^2} = (\overline{(1 - m_0^2)^2})^2 / [1 - \beta^2 \overline{(1 - m_0{}^2)^2}]. \qquad \text{(V.53)}$$

Positivity implies the inequality

$$X = 1 - \beta^2 \overline{(1 - m_0{}^2)^2} \geq 0. \qquad \text{(V.54)}$$

We have recovered the stability criterion of the replica approach (Sec. I.4) and of the TAP approach (Sec. II.4). X is nothing but the lowest eigenvalue of the 2nd order derivative matrix on the saddle point. Eq. (V.54) coincides with the equation of the De Almeida-Thouless critical line.

When replica symmetry is broken one must check such inequalities for every level of symmetry breaking, and also inequalities involving product of connected correlation functions at various levels. The simplest case is of course the connected correlation function within one state $[\langle s_0 s_{0'} \rangle_{c,\alpha(N+2)}]^2$. The preceding derivation applies to this quantity also when replica symmetry is broken, the only difference being that X must be computed with the correct distribution of magnetic field, the one which takes into account the breaking. For the consistency of the cavity method one must also check the validity of the clustering property at the various levels of the clusters (Eq. (V.24)). Adding to these equations the constraints that other crossed correlation functions such as $\langle s_0 s_{0'} \rangle_{c,\alpha} \langle s_0 s_{0'} \rangle_{c,\beta}$ or $\langle s_0 s_{0'} \rangle_{c,\alpha} \langle s_0 \rangle_\beta \langle s_{0'} \rangle_\gamma$ must also remain finite, one should be able to regain on physical grounds the whole spectrum of fluctuations around the saddle point in the replica approach[5]; but this goes beyond the aim of the present book[7].

Turning to the simplest condition (V.54): $X > 0$, one finds that at any *finite* number of replica symmetry breakings this inequality is not satisfied. In the saddle point of the infinitely iterated breaking the inequality becomes an equality. This is one of the zero modes of the *SK* model[5]. In the approach of this chapter this result can be understood a bit more clearly. The pure state decomposition of the Gibbs-Boltzmann measure is necessitated by the breaking of the clustering property. A pure state is defined by a subset of the configuration space. Assume we have specified tentatively such a subset. When we add *any* new subset to it, the clustering equation must necessarily be violated. Otherwise we will have to conclude that the original definition of the pure state was too stingy. It follows that the clustering property must be satisfied *critically*. This is what happens if the inequalities (54) become equalities as shown by Eq. 52.

References

1. Onsager (1936).
2. The instantaneous magnetic field should not be confused with the effective field $\tilde{h}$ defined in chapter II through Eq. (II.3). The $\tilde{h}$ distribution is Gaussian if replica symmetry is not broken. Later in this chapter $\tilde{h}$ on site 0 is denoted $h_{\alpha(N)}$.
3. Parisi (1979) Reprint 8.
4. By definition the *relevant configurations* inside any state are those with energies: $|E - E_\alpha| < O(\sqrt{(2C/\beta^2)})$ and in that range

$$S(E) = S(E_\alpha) + \beta(E - E_\alpha) + \beta^2(E - E_\alpha)^2/2C.$$

   However the specific heat C is proportional to N so that the last term is negligible unless $E - E_\alpha = O(N)$ which pushes the configuration outside the state.
5. De Dominicis and Kondor (1983) Reprint 13. For a general recent review see De Dominicis and Kondor (1984). The zero modes are discussed in Goltsev (1984) and in De Dominicis and Kondor (1986).
6. Mézard and Virasoro (1985) Reprint 17.
7. Brunetti (1987) thesis, unpublished.
8. Mézard *et al.* (1986) Reprint 19.

Chapter VI
DYNAMICS

VI.1 Introduction

This chapter will be divided into two main sections in which we shall describe the dynamics on various time scales. In a problem in which ergodicity is broken, as is the case in the spin glass phase, one must distinguish between the various scales of time (and in particular their scaling with the size of the system) on which one wants to study the time-dependent effects. In a system of infinite volume the time evolution is confined to one of the available pure states from which the system will never be able to escape in a finite time. This relaxational dynamics inside one state has been studied in mean field theory by Sompolinsky and Zippelius (see Reprint 22) and shows very interesting behaviour, which constitutes another aspect of the "marginality" of the spin glass phase: The relaxation is not exponential and instead the spin correlations decay with time as a power law, with an exponent which depends on the temperature and the magnetic field. This will be treated in the first section.

On the other hand there have been many efforts to understand the dynamics on infinite time scales (i.e. time scales which diverge with the volume of the system), such that transitions between various states are allowed. This is an alternative approach to that of usual statistical mechanics which has been fully developed by Sompolinsky[8]. Various reasons suggest to study the dynamics in this regime of infinite times. Indeed many of the characteristic properties of spin glasses observed experimentally show up in time-dependent effects which can take place on really macroscopic time scales (which can be days or weeks...), and one might be tempted to get some kind of information on these effects by studying the transitions between various states in the spin glass phase. Practically the precise understanding of this regime requires a detailed knowledge of properties of spin glasses at finite (though large) size, such as the height of the free energy barriers between the states, which is not available at the moment. Instead various hypotheses have been put forward; we shall briefly describe and comment on some of them, and their predictions, in the second part of this chapter.

VI.2 Dynamics on Finite Time Scales

The *SK* model deals with Ising spins which have no natural intrinsic dynamics. Two approaches can be used to introduce time dependent effects in this model. One is

Glauber's dynamics which describes the time evolution of the probability of appearance of the various spin configurations[1,2,3]. The other, which we shall use hereafter, is to introduce a "soft spin" model in which s_i is a continuous variable submitted to a double well symmetric potential which strongly favours the values $s_i = \pm 1$, the Ising model being recovered in a suitable limit. The system at temperature $T = 1/\beta$ is described by the Hamiltonian

$$\beta \mathbf{H} = \sum_i [V(s_i) - \beta H_i s_i] - \beta \sum_{i<j} J_{ij} s_i s_j, \tag{VI.1}$$

where H_i is a site dependent magnetic field. Relaxational dynamics is introduced through the Langevin equation

$$\tau_0 \partial s_i/\partial t = -\partial(\beta \mathbf{H})/\partial s_i + \xi_i(t), \tag{VI.2}$$

where $\xi_i(t)$ is a Gaussian white noise with characteristics

$$\langle \xi_i(t) \rangle = 0; \qquad \langle \xi_i(t)\xi_j(t') \rangle = 2\tau_0 \delta_{ij} \delta(t - t'). \tag{VI.3}$$

The standard approach is to study a sample averaged dynamics. It is described in details in Ref. 4 and we shall only very briefly summarize the spirit of the method here. Using well-known functional methods[5] one can write the functional probability $P[\{s_i(t)\}]$ that the time evolution of the system be a set of functions $s_i(t)$, or its generating functional

$$Z[l] = \int DsP[s] \exp\left(\int dt \sum_i l_i(t) s_i(t)\right), \quad \text{with}$$

$$Ds = \prod_{i,t} ds_{i,t} \tag{VI.4}$$

as an action integral over the fields $s_i(t)$ and some auxiliary fields $\hat{s}_i(t)$ (which implement the constraints that $s_i(t)$ must satisfy (VI.2)), with a certain Lagrangean $L[s, \hat{s}]$. As was noticed by De Dominicis[6], since $Z[0] = 1$ identically, one can compute directly $\bar{Z}$, the average of Z over the distribution of the couplings, thus avoiding replicas. (In fact this is correct only if one does not insert a weight on initial conditions: we shall return to this point later on). After averaging, the various sites can be decoupled and one is reduced to the study of a single spin which evolves in time through the Langevin like equation[4]

$$\partial s/\partial t = -\partial V/\partial s + \beta \int^t G(t - t')s(t') + \beta H(t) + \varphi(t), \tag{VI.5}$$

where φ is a Gaussian variable with zero mean and width

$$\langle \varphi(t)\varphi(t') \rangle = 2\tau_0 \delta(t - t') + C(t - t'). \tag{VI.6}$$

$C(t)$ and $G(t)$ are respectively the correlation function and the response function of the spin s which must be determined self consistently from (VI.5–6)

$$C(t - t') = \langle s(t)s(t')\rangle; \qquad G(t - t') = \partial\langle s(t)\rangle/\partial H(t'). \tag{VI.7}$$

Hence the "average effect" of a spin onto the others is a memory of the noise $\varphi(t)$ whose correlations are proportional to the correlation function $C(t)$ and a time-dependent magnetic field $\beta \int^t G(t - t')s(t')$ proportional to the values that the spin had at previous times. Equations (VI.5–7) constitute a complicated set of coupled equations, and their solution gives the sampled average probability for one given spin s_i to have time behaviour $s(t)$.

In the spin glass phase where there exist long time correlations ($\lim_{t\to\infty} C(t) \neq 0$), there must be from (VI.6) a static component of the noise. This must be isolated from the time-dependent noise and averaged on afterwards, the distribution of this static noise being deduced from the static solution[4].

We shall not explain this separation here but rather present another approach to the problem, which avoids the necessity of sample averaging from the beginning and provides a clear physical explanation of the strange features of spin glass properties described above. In order to achieve this we resort once more to the cavity method.

We consider first a system of N spins s_i which evolve from a given initial condition at $t = -\infty$ according to the Langevin equations (VI.2–3) with a given realization of the noise $\xi_i(t)$. The trajectory of this system in phase space is described by the N functions $s_i(t)$.

Now we introduce a new spin s_0 and consider the system of $N + 1$ spins so obtained. The N spins s_i evolve in time from the *same* initial conditions as before and with the *same* realisation of the noise $\xi_i(t)$. The only difference is that they feel an additional magnetic field $\beta J_{i0}s_0(t)$ due to the presence of the new spin. Consequently their trajectories are new functions $\tilde{s}_i(t)$. As the additional field is very small (remember J_{i0} is of order $1/\sqrt{N}$) the perturbation due to it can be treated within the approximation of linear response and one has

$$\tilde{s}_i(t) = s_i(t) + \int^t dt'\partial s_i(t)/\partial H_i(t')J_{i0}s_0(t'). \tag{VI.8}$$

Consider now the evolution of the spin 0, given by the Langevin equation

$$\tau_0 \partial s_0/\partial t = -\partial V/\partial s_0 + \beta \sum_i J_{0i}\tilde{s}_i(t) + \beta H_0 + \xi_0(t). \tag{VI.9}$$

Using (VI.8) we find

$$\tau_0 \partial s_0/\partial t = -\partial V/\partial s_0 + \beta \sum_i J_{0i}s_i(t) + \beta H_0 + \xi_0(t)$$
$$+ \beta \int^t dt' s_0(t')\left[\sum_i \partial s_i(t)/\partial s_i(t')\right]\Big/ N. \tag{VI.10}$$

The quantity between brackets is nothing but the average response function of a spin which we denote by $\bar{G}(t-t')$. Equation (VI.10) gives the evolution of the new spin s_0 as a functional of the trajectories of the N other spins $s_i(t)$ *in the absence of* s_0. Instead of the whole set of $N+1$ coupled equations we started from, we are left with one equation for the spin $s_0(t)$ on which the other spins act as an external noise.

This external noise $\beta\sum_i J_{0i}s_i(t)$ can be decomposed into its average part $\beta h = \beta\sum_i J_{0i}\langle s_i(t)\rangle_N = \beta\sum_i J_{0i}m_i^c$ (here and in the following $\langle\ \rangle_N$ denotes average over the realisations of the noise $\xi_i(t)$, $i=1,\ldots,N$ and $\langle\ \rangle_{N+1}$ average over the ξ_is' and ξ_0) which depends on the cavity magnetizations of the N spins i.e. their average values before the adjunction of s_0, and a fluctuating part $\eta(t)=\beta\sum_i J_{0i}(s_i(t)-m_i^c)$. The average field h is the cavity field which we have studied before and its probability distribution is known from the static solution. Let us now study the fluctuating part $\eta(t)$. It satisfies

$$\langle\eta(t)\rangle_N = 0$$

$$\langle\eta(t)\eta(t')\rangle_N = \beta^2\sum_{i,k} J_{0i}J_{0k}\langle(s_i(t)-m_i^c)(s_k(t')-m_k^c)\rangle. \qquad \text{(VI.11)}$$

The connected correlation function which appears in the right side of this equation is generally very small, so that the above sum is dominated by the terms $i=k$ and one has

$$\langle\eta(t)\eta(t')\rangle_N = (\beta^2/N)\sum_i \langle(s_i(t)-m_i^c)(s_i(t')-m_i^c)\rangle$$

$$= (\beta^2/N)\sum_i C_{ii}(t-t') = \beta^2\bar{C}(t-t'). \qquad \text{(VI.12)}$$

Hence the correlation of the noise is proportional to the average correlation of all the other spins. The validity of the approximation which consists in keeping only the terms $i=k$ in (VI.11) is confirmed by the following equation (which is the mathematical formulation of the property used above saying that the connected correlation functions are "generally very small")

$$(1/N^2)\sum_{i,k}[\langle(s_i(t)-m_i^c)(s_k(t')-m_k^c)\rangle]^2 \to 0 \quad \text{when } N\to\infty. \qquad \text{(VI.13)}$$

Indeed the above correlation function should be decreasing when $t-t'$ increases and its value at $t=t'$ is the usual thermodynamic average $(1/N^2)\sum_{i,k}[\langle s_i s_k\rangle_c]^2$ which vanishes inside one pure state. Looking at correlation functions of η of higher order shows that it is a Gaussian variable.

Now the time evolution of s_0 is described by

$$\tau_0\,\partial s_0/\partial t = -\partial V/\partial s_0 + \beta h + \beta H_0 + \xi_0(t) + \eta(t) + \beta\int^t dt' s_0(t')G(t-t'), \qquad \text{(VI.14)}$$

where h is a static random field equal to $\sum_i J_{0i}m_i^c$, H_0 is the external field, ξ_0 is a white noise (see (VI.3)) and η is the noise described in (VI.12). As always in the cavity method one must impose the consistency equations: the correlation function $C_0(t-t') = \langle s_0(t)s_0(t')\rangle$, once averaged, must be equal to the average correlation function of all the other spins $\bar{C}(t-t')$, and the same must be true for the response function. Equations (VI.12–14) are identical to the average equations of motion (VI.5–6) once the time persistent part of the noise has been separated.

The physical reasons for the various new terms are clear: The fluctuating noise η represents the action on the spin s_0 of the fluctuating parts of the magnetizations of the N other spins, and its correlations must thus be proportional to the average autocorrelation of one spin. The field $\beta \int^t dt' s_0(t')\bar{G}(t-t')$ is an indirect effect of $s_0(t')$ on $s_0(t)(t' < t)$, through the polarization of the N other spins by s_0 at time t'. As we live in one given pure state $\bar{C}$ and $\bar{G}$ must be related by the fluctuation dissipation theorem: $\bar{G}(t) = -\theta(t)\partial\bar{C}/\partial t$. This is precisely what is needed in order to insure that the distribution of s_0 at thermal equilibrium be the Boltzmann distribution in the field $h + H_0$, so that

$$\langle s_0\rangle = \operatorname{th}\beta\left(H_0 + \sum_i J_{0i}m_i^c\right) \tag{VI.15}$$

as it should (see Chap. V).

As for the relaxation towards equilibrium, one must study the solution of (VI.14) at large times. This has been done by Sompolinsky and Zippelius[4,7] who find the following results for the average behaviour ($\bar{C}(t)$ and $\bar{G}(t)$)

— Above the critical temperature $T_c(H)$, solution of the equation

$$X(T,H) = 1 - (1/T^2)\int P(h)[1 - \operatorname{th}^2\beta(H+h)]^2 = 0, \tag{VI.16}$$

(which is the equation of the De Almeida Thouless line, recovered here in the dynamical approach), the relaxation is exponential, $\bar{C}(t)\exp(-t/\tau)$. There is critical slowing down as one approaches T_c from above, the relaxation time diverges as $\tau \propto 1/(T - T_c)$.

— Below T_c the behaviour depend on the value of $X(T,H)$. A necessary condition of dynamic stability (which means that correlations should not diverge at large time) is that X be nonnegative. This confirms the results found in the replica and in the cavity method. However it has been shown that within the hierarchical thermodynamic solution the equilibrium is marginal, i.e. X is in fact equal to 0 everywhere in the spin glass phase. Because of this property the relaxation can no longer be exponential. $\bar{C}(t)$ decays instead as a power law

$$\bar{C}(t) \propto t^{-\nu} \qquad \text{when } t \to \infty, \tag{VI.17}$$

where the exponent ν is the solution of the following equation

$$\overline{m^2(1-m^2)^2}/\overline{(1-m^2)^3} = \pi\operatorname{cotg}(\pi\nu)\Gamma(2\nu)/\Gamma(\nu)^2, \tag{VI.18}$$

the left side is computed at thermal equilibrium: m stands for th $\beta(H_0 + h)$ and $\bar{O}$ is $\int P(h)O(h)\,dh$. Near to T_c the distribution $P(h)$ can be approximated by the replica symmetric solution. The value $\nu \approx 1/2$ is found numerically in the range $.5 < T < T_c$.

Note that the power law behaviour (VI.17) has been derived for the sample averaged correlation function. One might be worried whether this behaviour could be the result of an average of exponential relaxations with relaxation times which fluctuate from sample to sample, but an examination of (VI.14) indicates that this is not the case.

VI.3 Dynamics on Infinite Time Scales and Statics

In the mean field theory the barriers between states diverge in the infinite volume limit. However it is interesting to study the behaviour of the *SK* model on infinite time scales, or equivalently what happens at large times in a system containing a large but finite number of spins, N.[19] The basic hypothesis, put forward by Sompolinsky[8], is that there exists a strong hierarchy of time scales for the evolution in this "infinite time" regime. This is an appealing structure around which all the approaches to this problem have been constructed so far. It means that all these time scales are divergent in the $N \to \infty$ limit, but also the ratio between two successive scales diverges in this limit. Consequently at a given time all the relaxation processes which occur on faster time scales are equilibrated while the relaxation processes with larger characteristic times are completely frozen: the solution one obtains in this way is really a static solution, the system on this given time scale is described by a certain "equilibrium" measure on phase space. The problem is to understand the evolution of this measure when one goes from one given time scale to another (infinitely longer) one.

As this may seem a bit abstract at this stage let us give a simple scenario of how this could happen, using fully the information on the nature of the spin glass phase derived in the previous chapters. In the absence of any further information about the barriers between the states, it is reasonable to assume that the time needed for the system to jump from one state to another is correlated with the distance between them or equivalently with their overlap, and the simplest assumption is that this time depends only on the overlap[9,10]. Then to each value of the overlap q one associates a typical time scale τ_q for hopping at a distance q. In the limit $N \to \infty$ the hypothesis of strong hierarchy reads

$$\lim_{N\to\infty} \tau_q = \infty, \qquad \lim_{N\to\infty} \tau_{q'}/\tau_q = \infty \qquad \text{if } q' < q. \tag{VI.19}$$

It is important to notice that this kind of hypothesis is selfconsistent only because of ultrametricity: a necessary consistency condition is that, given any three states α, β, γ, the transition time from α to β via γ should be larger or equal to the assumed transition time from α to β, that is

$$\tau_{q_{\alpha\gamma}} + \tau_{q_{\gamma\beta}} > \tau_{q_{\alpha\beta}} \tag{VI.20}$$

which reads in the limit of strong hierarchy

$$\mathrm{Max}(\tau_{q_{\alpha\gamma}}, \tau_{q_{\gamma\beta}}) > \tau_{q_{\alpha\beta}} \quad \text{or} \quad q_{\alpha\beta} > \mathrm{Min}(q_{\alpha\gamma}, q_{\gamma\beta}), \tag{VI.21}$$

this is precisely the ultrametric inequality (IV.5). (Remember that the larger the overlap between two states, the smaller their distance.)

Sompolinsky has studied directly the dynamical equations (VI.12–14) within the only hypothesis of a strong hierarchy, and independently of the above scenario. He has found a self consistent hierarchical solution of these equations which is presented in Refs. 8 and 11. Translated into the language of equilibrium states, this solution gives for the state at time τ_q a result which is nothing but the ancestor state at the level q (see Chap. IV, as well as Refs. 12 and 13)

$$\langle O \rangle_q = \lim_{M\to\infty} 1/M \sum_{\gamma \text{ in } S} \langle O \rangle_\gamma \tag{VI.22}$$

where S is any set of M states $\gamma_1, \ldots, \gamma_M$ which are all equidistant ($q_{\gamma\gamma} = q$) and in the same cluster as the state one starts from ($q_{\alpha\gamma} > q$). With this approach one can recover the evolution equations which lead from one ancestor to another and these are identical to the results of the replica method[20]; this approach provides a way of finding these results of the equilibrium theory without replicas.

In spite of this mathematical identity the two approaches have different physical contents. For instance the time persistent spin correlations

$$C(\tau_q) = \overline{\langle s_i(0)s_i(\tau_q)\rangle} = 1/N \sum_i \left\{ \lim_{T\to\infty} 1/T \int_0^T dt\; s_i(t)s_i(t+\tau_q) \right\} \tag{VI.23}$$

are equal to q in Sompolinsky's theory, where q is the typical overlap characteristic of the time τ_q. For large but finite times this gives q_{EA} as it should. The problem appears in the extreme other limit of "infinite infinite" times, $t_{\infty\infty}$ in which the system has been able to jump over all the barriers. From (VI.23) one gets $C(t_{\infty\infty}) = q(x = 0)$, the smallest possible overlap in the problem (and the self overlap of the "grand ancestor" state), while the statistical mechanics result is

$$C(t_{\infty\infty}) = \int_0^1 q' P(q')\, dq' = \int_0^1 q(x)\, dx. \tag{VI.24}$$

Going on with this scenario it is natural that, if the system started initially from a given state α, its situation at a given (infinitely large) time scale τ_q can be described by a restricted Gibbs measure which we now describe. The system has been able to visit all states β such that $q_{\alpha\beta} > q$, and since all the characteristic times for jumping from one of this state to the others are much smaller than τ_q it has thermalized among these states and it ignores completely the states which are further away than q. The state at time τ_q should then be such that the average value of any observable O equals[9]

$$\langle O\rangle_q = \left\{\sum_\gamma \exp(-\beta f_\gamma)\theta(q_{\alpha\gamma} - q)\langle O\rangle_\gamma\right\}\left\{\sum_\gamma \exp(-\beta f_\gamma)\theta(q_{\alpha\gamma} - q)\right\}^{-1}$$

$$= \left\{\sum_\gamma P_\gamma\theta(q_{\alpha\gamma} - q)\langle O\rangle_\gamma\right\}\left\{\sum_\gamma P_\gamma\theta(q_{\alpha\gamma} - q)\right\}^{-1} . \qquad \text{(VI.25)}$$

Note that the restricted Gibbs average (VI.25) would give

$$C(\tau_q) = \int_q^1 q'P(q')\,dq' \bigg/ \int_q^1 P(q')\,dq' \qquad \text{(VI.26)}$$

which disagrees with the previous result $C(\tau_q) = q$ on all the divergent time scales. (In fact the time is not well defined in either of these approaches. The best one can do is to take one of the correlation functions e.g. C to set up the scale, and then compute the other correlation functions: The "dynamic" predictions would be of the type[9] "at the moment where C is equal to .8, the other correlation functions are equal to . . .").

To our knowledge the questions of why the solution of dynamics equations give as quasi equilibrium state the ancestor state and above all why it does not recover the Boltzmann-Gibbs result in the limit $t_{\infty\infty}$ have not been answered so far.

On the other hand if one works in the limit of infinite volume then the ergodicity is broken and the choice of initial conditions for the dynamics matters. Houghton, Jain and Young[14] have discussed this problem and shown that, if one chooses the initial configurations with their Boltzmann weights, then the standard results from statistical mechanics can be recovered (at the price of a reintroduction of replicas in the game).

Finally let us mention that besides this specific approach to the dynamics of the *SK* model evocated above, there have been a number of works trying to extract the essential properties of the dynamics of systems with hierarchical distribution of time scales[15] by the study of hopping dynamics in ultrametric spaces[16] or self similar one dimensional structures[17]. A review of some of these results can be found in Ref. 18.

References

1. Glauber (1963).
2. Kirkpatrick and Sherrington (1978) Reprint 4.
3. Sommers (1986).
4. Sompolinsky and Zippelius (1981) Reprint 22.
5. Janssen (1976); Bausch *et al.* (1976); Martin *et al.* (1978); De Dominicis and Peliti (1978).
6. De Dominicis (1978).
7. Sompolinsky and Zippelius (1981).
8. Sompolinsky (1981).
9. Parisi (1983).
10. Dasgupta and Sompolinsky (1983).
11. Sommers (1983a, b).
12. Mezard and Virasoro (1984) Reprint 17.
13. Sommers and Dupont (1985).
14. Houghton *et al.* (1984) Reprint 23.

15. Palmer *et al.* (1984).
16. Ogielski and Stein (1985); Paladin *et al.* (1985); Schreckenberg (1985); see also Grossman *et al.* (1985).
17. Huberman and Kerszberg (1985); Teitel and Domany (1985); Maritan and Stella (1986).
18. Rammal *et al.* (1986).
19. Another point of view consists in saying that on infinitely large time scales the couplings can change and thus the free energy landscape is modified. Horner (1984).
20. Sompolinsky's solution[8] possesses a kind of "gauge invariance" and the formulas of the replica method are recovered in one special "gauge" in which we work here.

Reprints
SPIN GLASSES

J. Phys. F: Metal Phys., Vol. 5, May 1975. Printed in Great Britain. © 1975.

Theory of spin glasses

S F Edwards† and P W Anderson‡
Cavendish Laboratory, Cambridge, UK

Received 14 October 1974, in final form 13 February 1975

Abstract. A new theory of the class of dilute magnetic alloys, called the spin glasses, is proposed which offers a simple explanation of the cusp found experimentally in the susceptibility. The argument is that because the interaction between the spins dissolved in the matrix oscillates in sign according to distance, there will be no mean ferro- or antiferromagnetism, but there will be a ground state with the spins aligned in definite directions, even if these directions appear to be at random. At the critical temperature, the existence of these preferred directions affects the orientation of the spins, leading to a cusp in the susceptibility. This cusp is smoothed by an external field. If the potential between spins on sites i, j is $J_{ij}s_i \cdot s_j$ then it is shown that

$$kT_c = \left(\sum_{ij} \langle \tfrac{2}{3} J_{ij}^2 \varepsilon_{ij} \rangle\right)^{1/2}$$

where ε_{ij} is unity or zero according to whether sites i and j are occupied. Although the behaviour at low T needs a quantum mechanical treatment, it is interesting to complete the classical calculations down to $T = 0$. Classically the susceptibility tends to a constant value at $T \rightarrow 0$, and the specific heat to a constant value.

1. Introduction

A dilute solution of say Mn in Cu can be modelled by an array of spins on the Mn arranged at random in the matrix of Cu, interacting with a potential which oscillates as a function of the separation of the spins. To simplify our analysis we consider the spins as classical dipoles pointing in direction s_i, so the interaction energy is $J_{ij}s_i \cdot s_j$. Now if, when the probability of finding a pair at points i, j is ε_{ij}, it happens that $\Sigma J_{ij}\varepsilon_{ij} \neq 0$ the system can show residual ferromagnetism or antiferromagnetism at sufficiently low temperatures. If $\Sigma J_{ij}\varepsilon_{ij} = 0$, for the whole alloy, but still has domains in which it is nonzero, one may still construct a theory in which there are thermodynamic consequences, in particular in the susceptibility, a kind of macroscopic antiferromagnet (Adkins and Rivier 1974). In this paper however we argue that there is a much simpler and overriding model, in which it can be assumed that $\Sigma J_{ij}\ _{ij} = 0$ on any scale, and that the mere existence of a ground state is sufficient to cause a transition and a consequent cusp in the susceptibility, which is found experimentally (Canella and Mydosh 1972). There are many such states each of which is a local minimum and inaccessible from each other. This question is irrelevant to our argument.

† Present address: Science Research Council, State House, High Holborn, London WC1R 4TA.
‡ Present address: Bell Telephone Laboratories, Murray Hill, New Jersey, USA.

The argument is that there will be some orientation of the spins which gives the minimum of potential energy. This orientation is such that $\langle s_i \rangle = 0$ so the system is neither ferro- nor antiferromagnetic on any scale, nor need it be unique. Nevertheless there comes a critical temperature T_c at which the spins notice the existence of this state, and as $T \rightarrow 0$ the system settles into the state. This physical picture is simple enough, but it requires some new formalism to express. The problem has a resemblance to problems of gellation in polymer science. When a solution of very long molecules becomes dense there comes a density at which the mobility of a molecule falls essentially to zero and the system gels. Such a molecule will still appear as a random coil, but if viewed later will be the *same* random coil. Thus what we must argue is that if on one observation a particular spin is $s_i^{(1)}$ then if it is studied again a long time later, there is a nonvanishing probability that $s_i^{(2)}$ will point in the same direction, ie

$$q = \langle s_i^{(1)} . s_i^{(2)} \rangle \neq 0.$$

Recent observations by A T Fiory and co-workers using μ meson polarization have strikingly confirmed this qualitative change in behaviour. Above T_c there appears to be no mean magnetic field at the site of a stopped μ meson, below T_c there is. At $T = 0$ one expects $q = 1$, at $T \geqslant T_c$, $q = 0$. The parameter q then takes the role of the mean field of the Curie–Weiss theory and we now construct the theory at the level of accuracy of the Curie–Weiss theory.

2. The mean correlation theory

To illustrate the basis of the phase change we firstly consider a single spin.

The probability of finding orientation s_i is

$$P(s_i) = \exp\,[(F - \textstyle\sum J_{ij} s_i . s_j)/kT]. \tag{2.1}$$

The joint probability of finding $s_i^{(1)}$ at one time and $s_i^{(2)}$ at an infinitely remote time will be

$$P(s^{(1)}, s^{(2)}) = \exp\,\{[2F - \textstyle\sum J_{ij}(s_i^{(1)} . s_j^{(1)} + s_i^{(2)} . s_j^{(2)})]/kT\}. \tag{2.2}$$

The fields that spins $s_i^{(1)}$ and $s_i^{(2)}$ find themselves in are

$$\xi_i^{(1)} = \sum_j J_{ij} s_j^{(1)} \tag{2.3}$$

and

$$\xi_i^{(2)} = \sum_j J_{ij} s_j^{(2)}. \tag{2.4}$$

If a spin, as a function of position, is completely random,

$$\langle \xi \rangle = 0 \tag{2.5}$$

and

$$\langle \xi_i^{(1)} . \xi_i^{(1)} \rangle = \langle \textstyle\sum J_{i\alpha} s_\alpha^{(1)} J_{i\beta} s_\beta^{(1)} \rangle \tag{2.6}$$

$$= \sum_\alpha J_{i\alpha}^2 \varepsilon_{i\alpha} = J_0^2 \tag{2.7}$$

where ε is the probability of finding a spin at α given that there is one at i.

Also

$$\langle \xi_i^{(1)}.\xi_j^{(1)} \rangle = \langle \sum J_{i\alpha} s_\alpha^{(1)} J_{j\beta} s_\beta^{(1)} \rangle \tag{2.8}$$

$$= \sum J_{i\alpha} J_{j\alpha} \tag{2.9}$$

$$= 0 \qquad i \neq j. \tag{2.10}$$

However if

$$s_i^{(1)}.s_j^{(1)} = q \neq 0$$
$$\langle \xi_i^{(1)}.\xi_i^{(2)} \rangle = q \sum J_{i\alpha}^2 \varepsilon_{i\alpha} \tag{2.11}$$

$$= q J_0^2. \tag{2.12}$$

Now reconsider (2.2) from the point of view of one spin, s_1, say. Suppose that all the other spins are bundled into the fields ξ, so that

$$P(s_1^{(1)} s_1^{(2)} \xi^{(1)} \xi^{(2)}) = \mathcal{N} \exp\left(-\frac{1}{kT} \xi_1^{(1)} s_1^{(1)} - \frac{1}{kT} \xi_1^{(2)} s_1^{(2)} \right) \mathscr{P}(\xi_j^{(1)} \xi_j^{(2)}). \tag{2.13}$$

where $\mathscr{P}$ is the probability of finding the ξ_i's independently of any correlation caused by coupling to $s^{(1)}$, which we assume obeys equations (2.10)–(2.12), and $\mathcal{N}$ a normalization. If there are a large number of s's arranged at random the ξ variables can be expected to have a gaussian distribution so that

$$P(s^{(1)}, s^{(2)}) = \mathcal{N}^* \exp\left[-\frac{1}{(kT)^2} (s_1^{(1)^2} \langle \xi_1^{(1)^2} \rangle - s_1^{(1)^2} \langle \xi_1^{(2)^2} \rangle - 2 s_1^{(1)}.s_1^{(2)} \right.$$
$$\left. \times \langle \xi_1^{(1)}.\xi_1^{(2)} \rangle) \right]. \tag{2.14}$$

But $s_1^{(1)^2} = s_1^{(2)^2} = 1$. Hence

$$P(s_1^{(1)}, s_1^{(2)}) = \tilde{\mathcal{N}} \exp(-s_1^{(1)}.s_1^{(2)} \rho q) \tag{2.15}$$

where

$$\rho = \frac{2J_0^2}{3(kT)^2}. \tag{2.16}$$

Finally, therefore,

$$q = \langle s_1^{(1)}.s_1^{(2)} \rangle \tag{2.17}$$

$$= \int \mu e^{-\mu \rho q} \Big/ \int e^{-\mu \rho q} \tag{2.18}$$

$$= \coth \rho q - \frac{1}{\rho q} \tag{2.19}$$

as usual. As $T \to 0$, $\rho \to \infty$, $q \to 1$ correctly. Expanding near $q = 0$

$$q = \frac{1}{\rho q} - \frac{1}{\rho q} + \frac{1}{3} q \rho - \frac{1}{45} q^3 \rho^3 \tag{2.20}$$

so that $q = 0$, or, writing $\frac{1}{3}\rho^2 = (T_c/T)^2$ (2.21)

$$\frac{1}{5}q^2\left(\frac{T_c}{T}\right)^6 = \left(\frac{T_c}{T}\right)^2 - 1 \tag{2.22}$$

$$q^2 = 5\left[1 - \left(\frac{T}{T_c}\right)^2\right]\left(\frac{T}{T_c}\right)^4. \tag{2.23}$$

The structure is similar to the standard Curie–Weiss theory, with the proviso that, as $T \to 0$, $q \to +1$ not -1, whereas either root is permitted in ferromagnetism. So far we have considered the life of a single spin, and find an abrupt change in its behaviour at a T_c. Clearly if we considered the single spin on three separate occasions we would get correlation $s^{(1)}.s^{(2)}$, $s^{(2)}.s^{(3)}$, and more complex correlations. So far we have not related these functions to thermodynamics, and it is not clear that the q above is directly related to the free energy. In the next section a development of disordered thermodynamics will be given following the method used in rubber elasticity (Edwards 1970, 1971) which permits the calculation of the free energy. A new definition of q will be given which will be directly related to the thermodynamic functions.

3. The formulation of the thermodynamic functions

Consider a particular spin glass specified by a set of occupation numbers. We can absorb these into the definition of J_{ij} so that there is a probability of finding a particular interaction operative, ie let

$$\mathscr{J}_{ij} = J_{ij}\varepsilon_{ij}. \tag{3.1}$$

$P(\mathscr{J})$ is then the probability of finding a $\mathscr{J}$.

A particular spin glass will have a free energy $\mathscr{F}(\mathscr{J})$ defined by

$$\exp[-\mathscr{F}(\mathscr{J})/kT] = \int \exp[+\sum \mathscr{J}_{ij}s_i.s_j]\Pi(\mathrm{d}s_i) \tag{3.2}$$

where the integration allows for the probability of occupation. The ensemble free energy is then

$$F = \int \mathscr{F}(\mathscr{J})\mathscr{P}(\mathscr{J})[\mathrm{d}\mathscr{J}] \tag{3.3}$$

$$= -kT\int \mathscr{P}(\mathscr{J})\left[\log \int (\Pi\ \mathrm{d}s)\exp\left(\sum \mathscr{J}ss\right)\right]\mathrm{d}\mathscr{J}. \tag{3.4}$$

In order to be able to perform the integrals over $\mathscr{J}$ and s it appears to be essential to be able to alter the order of the integration. A way to do this is to consider m systems and define an $\tilde{F}(m)$ by

$$\exp[-\tilde{F}(m)/kT] = \int \prod_{\alpha=1}^{m}(\Pi\ \mathrm{d}s^{(\alpha)})\exp\left(\sum_{\alpha=1}^{m}\sum_{ij}\mathscr{J}_{ij}s_i^{(\alpha)}s_j^{(\alpha)}\right)P(\mathscr{J})\,\mathrm{d}\mathscr{J} \tag{3.5}$$

$$= \int \exp[-m\mathscr{F}(\mathscr{J})/kT]\,P(\mathscr{J})\,\mathrm{d}\mathscr{J}. \tag{3.6}$$

We consider that (3.6) can truly be evaluated for all m and continue the integral to a small m value and expand it

$$\exp[-\tilde{F}(m)/kT] = 1 - \frac{mF}{kT} + \mathrm{O}(m^2). \tag{3.7}$$

Hence

$$\tilde{F}(m) = mF + O(m^2). \tag{3.8}$$

If, then, one can evaluate the $3mN$ dimensional integral (3.5) one can obtain the free energy of the system. The integral can be evaluated by the method of the previous section. Firstly, we note that whereas in the $s^{(1)}$, $s^{(2)}$ discussion above one had $s_i^{(1)2} = s_i^{(2)2} = 1$, $\langle s_i^{(1)} . s_i^{(2)} \rangle = q$ and the argument works at the level of approximation in which $s_i . s_j = 0$. We now will have

$$s_i^{(\alpha)2} = 1 \qquad \langle s_i^{(\alpha)} . s_i^{(\beta)} \rangle = q \qquad \alpha \neq \beta$$

when we now employ the same symbol q, but it will follow a different definition. We take the simplest possible probability distribution for the $\mathscr{J}_{ij}^2$, ie

$$P(\mathscr{J}_{ij}) = \exp(-\mathscr{J}_{ij}^2/2J^2\rho_0^2) \tag{3.9}$$

where $J^2 = \sum_{ij} J_{ij}^2$ and ρ_0 is the density of occupation.

Then

$$\int \exp\left(\sum_{ij\alpha} \mathscr{J}_{ij} s_i^{(\alpha)} . s_j^{(\alpha)}\right) = \exp\left[+\frac{3}{2}\frac{\rho}{\rho_0} \sum_{\alpha\beta ij} (s_i^{(\alpha)} . s_j^{(\alpha)} s_i^{(\beta)} . s_j^{(\beta)})\right] \tag{3.10}$$

where

$$\rho = \tfrac{2}{3} J^2 \rho_0^2/(kT)^2. \tag{3.11}$$

To do the integral over the s we have to resort to a variational principle of the Feynman type, replacing the quartic form by a best quadratic. The replacement is then to write

$$\exp\left\{+\tfrac{1}{2} \sum_{i,\alpha\neq\beta} \eta s_i^\alpha . s_i^\beta + C - \left[\tfrac{1}{2}\eta \sum_{i,\alpha\neq\beta} s_i^\alpha . s_i^\beta + C - \tfrac{3}{2}(\rho/\rho_0) \sum s_i^\alpha s_j^\alpha s_i^\beta s_j^\beta\right]\right\} \tag{3.12}$$

$$> \exp\left\{+\tfrac{1}{2} \sum_{\alpha\neq\beta} \eta s_i^\alpha . s_i^\beta + C\right\}(1 + [\text{as above}]). \tag{3.13}$$

So one chooses C so that

$$\int \Pi\, ds \exp\left(+\tfrac{1}{2}\sum \eta ss + C\right)\left[\tfrac{1}{2}\eta \sum ss - C - \tfrac{3}{2}(\rho/\rho_0)\sum s_i^\alpha s_j^\alpha s_i^\beta s_j^\beta\right] = 0. \tag{3.14}$$

Thereupon one performs the final integral over s_i^α and minimizes

$$\int \Pi\, ds \exp\left(-\tfrac{1}{2}\sum \eta ss + C\right),$$

to determine η.

Note that a term in $s_i^\alpha . s_i^\alpha$ is not required since it is unity, and terms in $s_i . s_j$ need not be included to the order of accuracy of this paper and without, of course, violating the extremal property of the solution. To evaluate we note that

$$\sum_{\alpha\neq\beta} s_i^\alpha . s_i^\beta = \left(\sum s_i^\alpha\right) . \left(\sum s_i^\alpha\right) - m \tag{3.15}$$

and that

$$\exp\left[\left(\sum s_i^\alpha\right)^2 \eta/2\right] = \frac{1}{(2\pi)^{3/2}} \int \exp\left[\eta^{1/2}\left(\sum s_i^\alpha . r\right) - \tfrac{1}{2}r^2\right] d^2 r. \tag{3.16}$$

Thus

$$\int \Pi(\mathrm{d}s) \exp\left(\tfrac{1}{2}\eta \sum_{\alpha \neq \beta} s_i^\alpha s_i^\beta\right) = \mathrm{e}^{-m\eta/2}\left(\frac{1}{2\pi}\right)^{3/2} \int \left(\frac{\sinh r\eta^{1/2}}{r\eta^{1/2}}\right)^m \mathrm{e}^{-1/2r^2}\,\mathrm{d}^3 r. \tag{3.17}$$

Likewise

$$\int \exp\left(\tfrac{1}{2}\eta \sum s^\alpha s^\beta\right)\left(\sum_{\alpha,\beta} s_i^\alpha . s_j^\alpha s_i^\beta . s_j^\beta\right) = \frac{\mathrm{e}^{-m\eta}}{(2\pi)^{3/2}} \int \exp\left[\eta^{1/2}\left(\sum s_i^\alpha . r\right)\right]$$

$$\times \left(\sum s_i^\alpha . s_j^\alpha s_i^\beta . s_j^\beta\right) \mathrm{e}^{-r^2/2}\,\mathrm{d}^3 r\, \Pi(\mathrm{d}s). \tag{3.18}$$

Let

$$\mathscr{E}(m, n) = \int \exp\left(\tfrac{1}{2}\eta \sum_{i, \alpha \neq \beta} s_i^{(\alpha)} . s_i^{(\beta)}\right) \Pi\,(\mathrm{d}s) \tag{3.19}$$

$$= \mathrm{e}^{-m\eta/2} \frac{1}{(2\pi)^{3/2}} \int \left(\frac{\sinh r\eta^{1/2}}{r\eta^{1/2}}\right)^m \mathrm{e}^{-r^2/2}\,\mathrm{d}^3 r. \tag{3.20}$$

Then

$$\langle \sum s_i^\alpha . s_j^\alpha s_i^\beta . s_j^\beta \rangle = \tfrac{1}{3}\langle \sum s_i^\alpha . s_i^\beta s_j^\alpha . s_j^\beta \rangle \tag{3.21}$$

$$= \tfrac{1}{3}\langle \sum s_i^\alpha . s_i^\beta \times \sum s_j^\alpha s_j^\beta \rangle \tag{3.22}$$

$$= \tfrac{m}{3} + \tfrac{1}{3} \sum_{\alpha \neq \beta} \langle s_i^\alpha . s_i^\beta \rangle \sum \langle s_j^\alpha . s_j^\beta \rangle \tag{3.23}$$

$$= \frac{m}{3} + \frac{1}{3}\left(\frac{2}{\varepsilon}\frac{\partial \mathscr{E}}{\partial \eta}\right)^2 \frac{1}{m(m-1)}. \tag{3.24}$$

If q is now defined to be $\langle s_i^{(\alpha)} . s_i^{(\beta)} \rangle$ then

$$m(m-1)q = (2/\mathscr{E})(\partial \mathscr{E}/\partial \eta) \tag{3.25}$$

and

$$\sum \langle s_i . s_i s_j . s_j \rangle = \tfrac{m}{3}[1 - q^2(1-m)]. \tag{3.26}$$

Collecting the terms together

$$-\frac{\tilde{F}}{kT} = \log \mathscr{E} - \frac{\eta}{\varepsilon}\frac{\partial \mathscr{E}}{\partial \eta} + \tfrac{1}{2}\rho/\rho_0[1 - q^2(1-m)]. \tag{3.27}$$

$\tilde{F}$ will be a minimum with respect to variations in η and since q is defined solely in terms of η, it follows that

$$\frac{1}{m}\frac{\partial \tilde{F}}{\partial \eta} = 0 = 2\eta \frac{\partial q}{\partial \eta} - 2\left(\frac{\rho}{\rho_0}\right) q \frac{\partial q}{\partial \eta} \tag{3.28}$$

or

$$\eta = \left(\frac{\rho}{\rho_0}\right) q \tag{3.29}$$

where higher order terms in m have now been neglected. The definition of q in terms of η now yields (again keeping only terms in m)

$$1 - q = \int \frac{r}{[(\rho/\rho_0)q]^{1/2}} (\coth\{[\rho/\rho_0)q]^{1/2} r\} - [(\rho_0/\rho)q]^{1/2} r^{-1}) e^{-r^2/2} \frac{d^3r}{(2\pi)^{3/2}}. \tag{3.30}$$

$$-\frac{F}{kT} = -\frac{\tilde{F}}{mkT} = + \int \log\left\{\frac{\sinh r[(\rho/\rho_0)q]^{1/2}}{r[(\rho/\rho_0)q]^{1/2}}\right\} e^{-r^2/2} \frac{d^3r}{(2\pi)^{3/2}}. \tag{3.31}$$

$$-(\rho/\rho_0)q(1-q) + \tfrac{1}{2}(\rho/\rho_0)(1-q^2) = \int \log(\text{as above}) e^{-r^2/2} \frac{d^3r}{(2\pi)^{3/2}}$$

$$+ \frac{1}{2}\left(\frac{\rho}{\rho_0}\right)(1 - 2q + \cdots) \tag{3.32}$$

$$-\frac{F}{kT} = \frac{1}{2}\left(\frac{\rho}{\rho_0}\right)(1-q)^2 + \int \log(\text{as above}) e^{-r^2/2} \frac{d^3r}{(2\pi)^{3/2}}. \tag{3.33}$$

Note that $\partial F/\partial q = 0$ from the definition of q, and in accordance with the variational property of F.

The equation for q and hence for F is not simple but we can solve it in the limits $T \to T_c$ and $T \to 0$. For $T \to T_c$, one has since

$$\coth x = \frac{1}{x} - \frac{x}{3} + \frac{x^3}{45} - \frac{2x^5}{945} + \cdots \tag{3.34}$$

firstly the identity at $q = 0$ of

$$1 = \frac{1}{(2\pi)^{3/2}} \int \frac{r^2}{3} e^{-r^2/2} d^3r \tag{3.35}$$

and then

$$q = \frac{1}{(2\pi)^{3/2}} \int \frac{r^4}{45} e^{-r^2/2} \left(\frac{\rho}{\rho_0}\right) d^3r \tag{3.36}$$

ie T_c is where $\rho_0 = \frac{1}{3}\rho$, $T_c^2 = 2J_0^2/9k^2$ as before. Near $T = T_c$, for $T < T_c$

$$[1 - (T_c/T)^2] + 2q = 0. \tag{3.37}$$

Near $T = 0$, putting $q = 1 - \epsilon$

$$\left(\frac{\rho q}{\rho_0}\right)^{1/2} (1-q) = 2\left(\frac{2}{\pi}\right)^{1/2} \tag{3.38}$$

$$\epsilon = \left(\frac{8}{3\pi}\right)^{1/2} \left(\frac{T}{T_c}\right). \tag{3.39}$$

From these results we can now calculate the specific heat and susceptibility.

4. The specific heat

The free energy has the form $F = -Tf$, so the internal energy is

$$E = T^2 \frac{\partial f}{\partial T}. \tag{4.1}$$

Since F is stationary with respect to ξ ie q, only ρ_0 need be differentiated in f to yield

$$E = -T^2 \frac{2}{T}\frac{1}{2}\left(\frac{\rho}{\rho_0}\right)(1-q^2) \tag{4.2}$$

$$= -T(\rho/\rho_0)(1-q^2). \tag{4.3}$$

$$C_v = \frac{\partial E}{\partial T} = -\left(\frac{\rho}{\rho_0}\right)(1-q^2) - T\frac{\partial}{\partial T}\left[\left(\frac{\rho}{\rho_0}\right)(1-q^2)\right]. \tag{4.4}$$

Putting

$$(\rho/\rho_0) = \lambda \tag{4.5}$$

we have

$$C_v = -(1-q^2)\lambda - T\frac{\partial\lambda}{\partial T}\frac{\partial}{\partial\lambda}[\lambda(1-q^2)] \tag{4.6}$$

$$= -\lambda(1-q^2) + 2\lambda\frac{\partial}{\partial\lambda}[\lambda(1-q^2)] \tag{4.7}$$

$$= +\lambda(1-q^2) - 4\lambda^2 q\frac{\partial q}{\partial\lambda}. \tag{4.8}$$

This implies a cusp in C_v. There is a little experimental data which neither excludes nor entirely supports this (de Nobel and du Chatenier 1959, Zimmerman and Hoare 1960, Zimmerman and Crane 1961).

We are specially interested near $T \sim 0$ where

$$1 - q = \frac{\gamma}{(q\lambda)^{1/2}} \qquad \gamma = \int r\,\mathrm{e}^{-r^2/2}\frac{\mathrm{d}^3 r}{(2\pi)^{3/2}}. \tag{4.9}$$

Using this form one gets

$$C_v/k = \lambda\left[-\frac{(1-q)^3 + (1+3q^2)/2\lambda q}{3q - 1 + 1/2\lambda q}\right] \tag{4.10}$$

which tends to a constant as $T \to 0$. This is of course a classical result.

We are grateful to Dr K Fischer for pointing out an error in the first calculation of (4.10).

Dynamics near $T = 0$ always are of the utmost importance and this region cannot really be properly discussed in the present theory. A possible description is given by Anderson (1973).

5. The susceptibility

The argument is again straightforward, and if χ_c is the normal paramagnetic susceptibility, one finds by adding the magnetic field to the energy and differentiating F in the usual way:

$$\chi = \chi_c(1-q). \tag{5.1}$$

Thus $\chi = \chi_c$ above $T = T_c$. Below $T = T_c$ one finds, since

$$q = -\frac{1}{2}\left[1 - \left(\frac{T_c}{T}\right)^2\right] \tag{5.2}$$

and

$$\chi_c = \frac{a}{T} \qquad a = \text{usual Curie constant} \tag{5.3}$$

that

$$\chi = \frac{a}{T_c}\left[1 - \left(\frac{T_c - T}{T_c}\right)\right]\left\{1 + \frac{1}{2}\left[1 - 1 + 2\left(\frac{T_c - T}{T_c}\right)\right] + O(T_c - T)^2\right\} \tag{5.4}$$

$$\chi = \frac{a}{T_c} - O(T_c - T)^2. \tag{5.5}$$

The cusp is thus linear on one side but quadratic on the lower T side. From one's experience with the Bragg–Williams theory one can expect this lack of symmetry on either side of the cusp to be an artifact of the molecular field approximation employed here. The true structure will probably be more symmetric.

At low temperatures one has from (3.38)

$$q = 1 - \left(\frac{2}{3\pi}\right)^{1/2}\left(\frac{T}{T_c}\right) \tag{5.6}$$

so

$$\chi = \frac{a}{T}\left(\frac{2}{3\pi}\right)^{1/2}\left(\frac{T}{T_c}\right) \tag{5.7}$$

and is therefore independent of T as $T \to 0$.

Note that the cusp at $T = T_c$ is destroyed by an external magnetic field. This has the effect of altering r in $\sinh r[(\rho/\rho_0)q]^{1/2}\,[(\rho/\rho_0)q]^{1/2} r]^{-1}$ to $|r\eta + \mu B|$ where $\mu =$ dipole moment$/kT$. This ensures that $q > 0$ for all T, being of order B^2/T^2 as $T \to \infty$. However since the cusp has a strong theory dependent shape we do not pursue the algebra of the cusp form as B increases.

It will be noted that the cusp as calculated here is not symmetric unlike the experimental finding. This situation is analogous to the use of the mean field theory with thermodynamics of ordering in alloys (or the Ising model) after Bragg and Williams. The simple on or off theories give an asymmetry. Improvements like the Bethe–Peierls or Rushbrooke expansions redress the asymmetry to a certain extent, but exact theories find exact symmetry at the critical point. The purpose of this paper is simply to uncover the effect however and we do not attempt to apply the well known improvements to the mean field type of theory.

6. Conclusion

In this paper we have applied the simplest theory available to elucidate a new effect in disordered system physics. There are, apart from the obvious improvements required and possible in the present treatment of the phase change, several new avenues of study opened up. Firstly the methods should be made quantum mechanical in order to give a reliable treatment near $T = 0$. Also, just as the mean field theory has many applications in ordered physics, the present theory will have many other applications in disordered state physics. One has already been mentioned, that of rubber elasticity which antedates the present work. But more generally the present approach permits the use of second quantization methods in problems which have hitherto been studied only as first quantization problems. It is hoped to return to these questions in later papers.

Acknowledgments

The manuscript has been critically discussed with Dr Conyers Herring and Dr David Sherrington who made several helpful suggestions. A letter from Dr K H Fischer corrected a serious error in (4.10) and made helpful comments. P W Anderson thanks the Air Force Office of Scientific Research for support under grant AFOSR-73-2449.

References

Adkins K and Rivier N 1974 *J. Phys., Paris* **35** C4–237

Anderson P W 1973 *Amorphous Magnetism* ed H O Hooper and A M de Graaf (New York: Plenum Press) p 1

Canella V and Mydosh J A 1972 *Phys. Rev.* B **6** 4220

Edwards S F 1970 *Statistical Mechanics of Polymerised Materials* in *4th Int. Conf. on Amorphous Materials* ed R W Douglas and B Ellis (New York: Wiley)

—— 1971 *Statistical Mechanics of Rubber, in Polymer Networks* ed A J Chompff and S Newman (New York: Plenum Press)

de Nobel J and du Chatenier F J 1959 *Physica* **25** 969

Zimmerman J E and Hoare F E 1960 *J. Phys. Chem. Solids* **17** 52

Zimmerman J E and Crane L T 1961 *J. Phys. Chem. Solids* **21** 310

Communications on Physics 2 115–119 (1977)

Theory of the frustration effect in spin glasses : I

G. Toulouse
Laboratoire de Physique de l'Ecole Normale Supérieure,
24 rue Lhomond, 75231 Paris Cedex 05, France

Received 4 April 1977

An analysis of disorder is given, based on the concept of local invariance. The frustration effect is defined and several fundamental concepts are introduced. Spin glasses are considered here as representative of a new vast class of condensed matter phases.

The problem of spin glasses is at present receiving much attention [1]. There is no consensus on the interpretation of the results, and a general feeling of dissatisfaction reigns at the theoretical level.

We wish here to present an analysis of disorder, based on the concept of local invariance. This allows us to isolate the most genuine and interesting feature, the 'frustration effect', from the rest. The frustration effect is simple and fundamental: it is probably active in many physical systems (besides spin glasses). A theoretical investigation is needed, to develop all its consequences.

Let us consider a lattice, with sites (i) and bonds (ij), spins $\mathbf{S}_i$ on the sites interacting according to the Hamiltonian :

$$\mathscr{H} = -\sum_{ij} J_{ij} \mathbf{S}_i \cdot \mathbf{S}_j$$

For simplicity, we may have in mind a two-dimensional square lattice, with Ising spins ($S_i = \pm 1$), nearest-neighbour interactions, and $|J_{ij}| = 1$. If $J_{ij} = +1$, there is a tendency to align spins and this is a ferromagnetic model. If $J_{ij} = -1$, the tendency is to alternate spins and the model is antiferromagnetic. If $J_{ij} = \pm 1$ randomly, this is a model for a disordered magnet.

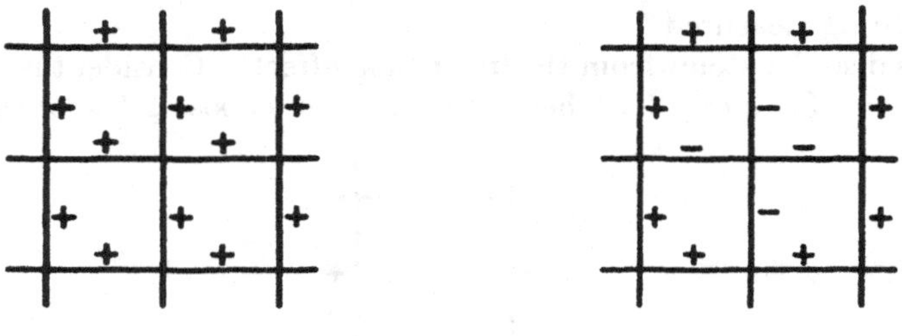

(*a*) (*b*)

Figure 1 (*a*) A ferromagnetic configuration of bonds. (*b*) All bonds around the central site have been made negative ; this has the same ground-state energy, with the central spin flipped.

Suppose that, starting from a ferromagnetic configuration with $J_{ij} = +1$ everywhere (*Figure* 1 (*a*)), the sign of J_{ij} is reversed on all bonds around one site (*Figure* 1 (*b*)) : this is not a serious disorder because, by flipping the spin on that site, one obtains a spin configuration which has the same energy as for the perfect ferromagnet. This is the manifestation of a local invariance ; the Hamiltonian $\mathscr{H}$ is invariant under the local transformation :

$$\left.\begin{array}{ll} S_i & \rightarrow -S_i \\ J_{ij}\ (j \text{ adjacent to } i) & \rightarrow -J_{ij} \end{array}\right\} \qquad (1)$$

First remark : a considerable amount of apparent disorder can be realized by flipping the bond signs around randomly chosen sites (this is the content of the Mattis model [2] : $J_{ij} = \epsilon_i \epsilon_j$, $|\epsilon| = 1$), but this sort of disorder has no effect on thermodynamic properties. Of course, it does have an effect on the magnetic susceptibility because this involves a coupling term with a homogeneous external field, which is not invariant under the local transformation (*Equation* 1). Such a model exhibits a phase transition, but this is just a hidden usual phase transition with all its attributes (long-range order, order parameter, ground state degeneracy $= 2, \ldots$) : the susceptibility gets a cusp as a function of temperature, but for a trivial reason.

Second remark : the local transformation (*Equation* 1) is quite similar to the local transformation :

$$\left.\begin{array}{l} \psi(\mathbf{x}) \rightarrow e^{i\alpha(\mathbf{x})} \cdot \psi(\mathbf{x}) \\ \mathbf{A}(\mathbf{x}) \rightarrow \mathbf{A}(\mathbf{x}) + \nabla\alpha \end{array}\right\} \qquad (2)$$

under which the Hamiltonian of electrodynamics is invariant. In the case of electrodynamics, this invariance expresses conservation of local charge. The local transformation (*Equation* 2) is called a gauge transformation. The gauge group is a one-dimensional Abelian continuous group : $G = U(1) = SO(2)$. On the other hand, the local transformation (*Equation* 1) corresponds to a simpler ' gauge ' group, namely $G = Z_2$, the discrete Abelian group with two elements. The bond interactions J_{ij} play a role somewhat similar to the electromagnetic potential **A**. We know that if we add to **A** a gradient term (' pure gauge ' term), the physics is unchanged. In this light, the Mattis model may be called a ' pure gauge ' model.

How does one distinguish serious disorder from non-serious disorder ? How is it to be measured ?

Serious disorder stems from the frustration effect. Consider the elementary configuration of *Figure* 2. There is no way of choosing the orientations of

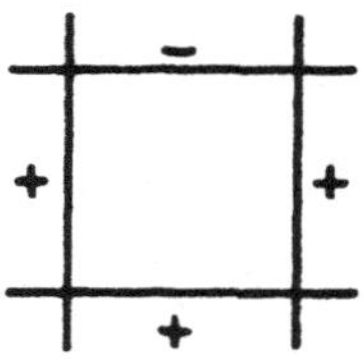

Figure 2 A frustrating configuration of bonds ; there is no fully satisfactory way of orienting Ising spins around this square.

the site spins around the square without frustrating at least one bond. This frustration effect is measured with the frustration function [3], defined on any closed contour (c) along the bonds of the lattice,

$$\Phi = \prod_{(c)} J_{ij}$$

If $\Phi = +1$, it is possible to orient the spins along (c) without frustration ; if $\Phi = -1$, it is not. Take two points A and B on the contour of *Figure* 3 ; it may be said that A sends to B contradictory orders along the two paths forming (c), if $\Phi(c) = -1$.

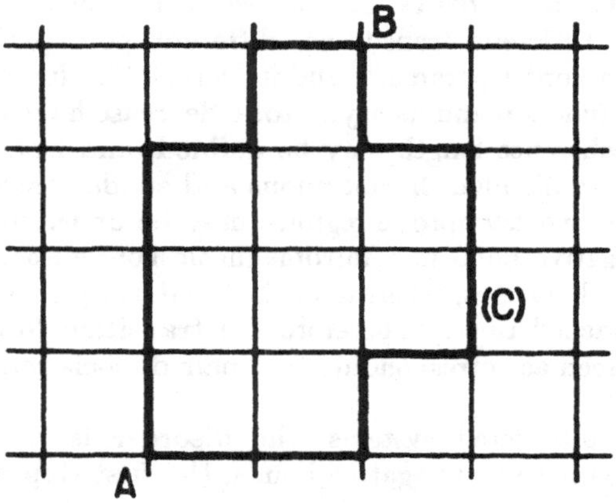

Figure 3 If the frustration function around contour (c) takes the value -1, then A sends to B contradictory orders along the two paths forming (c).

This is reminiscent of ' parallel transport ' of a tangent vector on a curved surface : the transport from P to Q depends on the path followed. In this case, the misfit angle in Q may be expressed as an integral of the curvature over the area bounded by the two paths going from P to Q.

So the frustration effect may be visualized as giving ' curvature ' to the lattice. Actually, the frustration function may be written also

$$\Phi = \prod_{k \in S} \Phi_k$$

where k is an index for the elementary squares of the lattice (more generally called ' plaquettes '), Φ_k is the value of Φ for the contour of plaquette k and S is the surface bounded by the chosen contour (c) [4]. If $\Phi_k = +1$, we shall say that plaquette k is a ' flat ' plaquette ; if $\Phi_k = -1$, it is a ' curved ' plaquette. Obviously, the Mattis model is a ' flat ' model.

We may wish to define an average of Φ. If some frustration is present, this average will tend asymptotically to zero, as the size of the contour increases. Two canonical asymptotic decreases may be distinguished :

(1) *Weak decrease*

$$\Phi \approx e^{-L}$$

where L is the perimeter of (c) ;

(2) *Strong decrease*

$$\Phi \approx e^{-S}$$

where S is the area of a surface bounded by (c).

In various models, as frustration or curvature is increased, a sharp transition from weak decrease to strong decrease is observed [5]. This transition is analogous to the usual phase transitions observed when the temperature is raised, but with a fundamental difference. For the usual transitions, the low-temperature phase corresponds to a breaking of a *global* symmetry ; for this transition, the low ' frustration ' phase corresponds to a breaking of a *local* symmetry. Different transitions, different concepts. Instead of the usual concepts of an order parameter and its correlation functions, the concepts of the frustration function and its asymptotic decrease have to be used. In an obvious way, a coherence length may be defined, and from the behaviour of that length one may distinguish continuous and sudden transitions.

Physically, the two asymptotic regimes may be understood by noting that to an isolated negative bond in a ferromagnetic configuration is associated a pair of ' curved ' plaquettes, whereas an isolated plaquette has a ' string ' of flipped bonds attached to it. Therefore the transition from weak to strong decrease may be seen as corresponding to a pair dissociation transition for the curved plaquettes.

In the usual disordered systems, the disorder is frozen. Given some configuration of positive and negative bonds, the first step must be to extract and analyse the ' frustration network '. For the two-dimensional Ising case, the elementary curved unit is one plaquette ; but, in three dimensions, a curved plaquette cannot be isolated because of a conservation law for curvature, which is the strict analogue of the conservation of magnetic flux in electromagnetism : the elementary unit is a line of curved plaquettes. Such a line may close on itself, i.e. form a ring, or go to infinity.

The frustration network acts as a system of sources—sinks for defects of the spin system. In the two-dimensional Ising system, the spin defects are linear. A ground-state configuration is obtained by associating the curved plaquettes in pairs, in such a way that the sum of the elongation of the pairs is minimal [6]. For each such pair of curved plaquettes, there is a linear segment of spin defects (frustrated bonds) the ends of which are on the curved plaquettes. The total number of different ways of associating the curved plaquettes in pairs with minimal summed elongation is the ground-state degeneracy of the spin system.

In the three-dimensional Ising system, the spin defects are planar and the frustration network is linear : in a ground-state configuration, the area of spin-wall defects, bounded by the frustration network, is minimal.

The next step is to complement this geometrical picture with a precise description of the ground-state degeneracy, of the metastable configurations and of the low-energy excited states. This is a vast programme, which will be discussed elsewhere. After this brief exposition of concepts, let us terminate with a few remarks :

(1) Many extensions have to be explored, in order to build for the phase transitions associated with a broken local symmetry the same general under-

standing that we have for the usual phase transitions (arbitrary dimensionalities, arbitrary symmetries, ...).

(2) The frustration effect may be induced by disorder but it may also exist on perfectly regular lattices (example : the two-dimensional triangular lattice with all bonds negative) ; spin-glass like (or ' liquid ') phases may therefore be expected on regular frustrating lattices.

(3) Besides the frustration effect in disordered magnets, there is a ' hole ' effect due to disconnected bonds ; this hole effect is to be studied with percolation concepts ; in itself, it does not lead to qualitatively new phases and from a theoretical point of view, it is desirable to separate the hole effect from the frustration effect ; this is not the case with the models where one assumes a continuous (for example, Gaussian) distribution for the bond values.

(4) It appears that numerical simulation techniques will help a great deal in this analysis of the various disorder effects.

Acknowledgments

I wish to thank Adrien Douady, of the Centre de Mathematiques de l'Ecole Normale Supérieure, for illuminating seminars on the geometrical content of gauge theories.

References

1 For a recent paper with a useful list of references, see for example, Young A P and Stinchcombe R B *J Phys C* **9** 4419 (1976).
2 Mattis D C *Phys Lett* **56A** 421 (1976) ; Bidaux R, Carton J P and Sarma G *Phys Lett* **58A** 467 (1976).
3 The word frustration is borrowed from P W Anderson (lecture at Aspen, August 1976), the function from K Wilson [5].
4 In dimensions higher than two, S is a surface made of plaquettes and bounded by (c).
5 Wilson K *Phys Rev* **D10** 2445 (1974) ; Balian R, Drouffe J M and Itzykson C *Phys Rev* **D10** 3376 (1974) ; **D11** 2098 2104 (1975).
6 The elongation of a pair is measured along a path going through adjacent plaquette centres.

Solvable Model of a Spin-Glass

David Sherrington* and Scott Kirkpatrick
IBM Thomas J. Watson Research Center, Yorktown Heights, New York 10598
(Received 16 October 1975)

We consider an Ising model in which the spins are coupled by infinite-ranged random interactions independently distributed with a Gaussian probability density. Both "spin-glass" and ferromagnetic phases occur. The competition between the phases and the type of order present in each are studied.

Compelling experimental[1,2] and theoretical[3-5] evidence has accumulated in recent years suggesting that a new magnetic phase may occur in spatially random systems with competing exchange interactions. In this "spin-glass" phase, moments are frozen into equilibrium orientations, but there is no long-range order. Edwards and Anderson (EA) have demonstrated[3] that such a phase occurs within a novel form of molecular-field theory, and they propose that spin correlations between Gibbs-like replicas of the random system play the role of a spin-glass order parameter.

A closely related replica formalism has been employed in several recent papers[6,7] applying renormalization-group methods to random magnetic systems. The possibility of an EA-type[4] order parameter was not considered in that work, although some of the models studied[7] appear likely to exhibit spin-glass phases.

VOLUME 35, NUMBER 26 PHYSICAL REVIEW LETTERS 29 DECEMBER 1975

It is well known that molecular-field theory for a pure ferromagnet becomes exact in the thermodynamic limit for a constant infinite-ranged exchange interaction provided that the interaction is appropriately scaled with the number of spins in the system.[8] In this Letter we define and solve the analogous infinite-ranged problem for a disordered system. We obtain a spin-glass solution characterized by the EA order parameter in the appropriate regime of temperature and the strength of the exchange fluctuations. A simple interpretation of this order parameter is given. The various thermodynamic quantities and the competition with ferromagnetic long-range order are explored in some detail.

We consider N Ising spins interacting through infinite-ranged exchange interactions which are independently distributed with a Gaussian probability density. The Hamiltonian is

$$\mathcal{H} = -\tfrac{1}{2}\sum_{i\neq i} J_{ij} S_i S_j, \quad S_i = \pm 1, \tag{1}$$

with the J_{ij} distributed according to

$$p(J_{ij}) = [(2\pi)^{1/2} J]^{-1} \exp[-(J_{ij} - J_0)^2/2J^2], \tag{2}$$

and J_0 and J scaled by

$$J_0 = \tilde{J}_0/N, \quad J = \tilde{J}/N^{1/2}, \tag{3}$$

so that $\tilde{J}_0$ and $\tilde{J}$ are both intensive. Following the usual procedure, we calculate the averaged free energy. (Averaging the free energy and not the partition function corresponds to treating a "quenched" rather than an "annealed" system.) With use of the identity

$$\ln x = \lim_{n\to 0}(x^n - 1)/n, \tag{4}$$

the averaged free energy F may be expressed as

$$F = -kT \lim_{n\to 0} n^{-1}\{\int \prod_{(ij)} [p(J_{ij}) dJ_{ij}] \operatorname{Tr}_n \exp(\sum_{\alpha=1,\dots,n} \sum_{i\neq j} J_{ij} S_i^\alpha S_j^\alpha / 2kT) - 1\}$$
$$= -kT \lim_{n\to 0} n^{-1}\{\operatorname{Tr}_n \exp(\sum_{i\neq j}[\sum_\alpha S_i^\alpha S_j^\alpha J_0/2kT + \sum_{\alpha,\beta} S_i^\alpha S_j^\alpha S_i^\beta S_j^\beta J^2/4(kT)^2]) - 1\}, \tag{5}$$

where α and β label n dummy replicas. Reordering and dropping terms which vanish in the thermodynamic limit yields

$$F = -kT \lim_{n\to 0} n^{-1}\{\exp[J^2N^2n/4(kT)^2] \operatorname{Tr}_n \exp[\sum_{(\alpha\beta)}(\sum_i S_i^\alpha S_i^\beta)^2 J^2/2(kT)^2 + \sum_\alpha(\sum_i S_i^\alpha)^2 J_0/2kT] - 1\}, \tag{6}$$

where $(\alpha\beta)$ refers to combinations of α and β with $\alpha \neq \beta$. Using the identity

$$\exp(\lambda a^2) = (2\pi)^{-1/2}\int dx \exp[-x^2/2 + (2\lambda)^{1/2} a x], \tag{7}$$

we rewrite (6) as

$$F = -kT \lim_{n\to 0} n^{-1}\{\exp[\tilde{J}^2 N n/4(kT)^2] \int [\prod_{(\alpha\beta)} (N/2\pi)^{1/2} dy^{(\alpha\beta)}][\prod_\alpha (N/2\pi)^{1/2} dx^\alpha]$$
$$\times \exp[-N\sum_{(\alpha\beta)} (y^{(\alpha\beta)})^2/2 - N\sum_\alpha (x^\alpha)^2/2$$
$$+ N \ln \operatorname{Tr} \exp((\tilde{J}/kT)\sum_{(\alpha\beta)} y^{(\alpha\beta)} S^\alpha S^\beta + (\tilde{J}_0/kT)^{1/2}\sum_\alpha x^\alpha S^\alpha)] - 1\}, \tag{8}$$

where the trace is now over n replicas at a single spin site.

It is assumed that the limit $n \to 0$ and the thermodynamic limit $N \to \infty$ can be interchanged. For integral $n \geq 2$, the integrals may be done by steepest descents. Since the replicas are indistinguishable, we consider only the extremum of the exponential for which all the $y^{(\alpha\beta)}$ are equal, as are all the x^α. We denote their values by y and x. This permits the replacement $\sum y^{(\alpha\beta)} S^\alpha S^\beta \to y[(\sum_\alpha S^\alpha)^2 - n]$, and $(\sum_\alpha S^\alpha)^2$ may be absorbed by the introduction of a random field.[3] Continuation to arbitrary n, extraction of the terms linear in n as $n \to 0$, and the substitutions $y \to q(\tilde{J}/kT)$ and $x \to m(\tilde{J}_0/kT)^{1/2}$ then yield

$$F = NkT\{-\tilde{J}^2(1-q)^2/(2kT)^2 + \tilde{J}_0 m^2/2kT - (2\pi)^{-1/2}\int dz \exp(-z^2/2) \ln[2\cosh(\tilde{J}q^{1/2}z/kT + \tilde{J}_0 m/kT)]\}, \tag{9}$$

where q and m satisfy the simultaneous equations

$$q = 1 - (2\pi)^{-1/2}\int dz \exp(-z^2/2)\, \operatorname{sech}^2[(\tilde{J}q^{1/2}/kT)z + \tilde{J}_0 m/kT], \tag{10a}$$

$$m = (2\pi)^{-1/2}\int dz \exp(-z^2/2) \tanh[(\tilde{J}q^{1/2}/kT)z + \tilde{J}_0 m/kT]. \tag{10b}$$

VOLUME 35, NUMBER 26 PHYSICAL REVIEW LETTERS 29 DECEMBER 1975

To show the physical significance of m and q we note that the thermal average of the spin at site i, $\langle S_i \rangle$, and its square may be written

$$\langle S_i \rangle = (\partial/\partial h) \ln \mathrm{Tr} \exp\Big(\sum_{i \neq j} J_{ij} S_i^\alpha S_j^\alpha / 2kT + h S_i^\alpha \Big)_{h=0}, \tag{11a}$$

$$\langle S_i \rangle^2 = (\partial/\partial h') \ln \mathrm{Tr} \exp\Big[\sum_{i \neq j} J_{ij} (S_i^\alpha S_j^\alpha + S_i^\beta S_j^\beta)/2kT + h' S_i^\alpha S_i^\beta \Big]_{h'=0}, \tag{11b}$$

where $\alpha \neq \beta$ are dummy labels. Averaging over the J_{ij} distribution, which we denote by $\langle \rangle_J$, we see that $\langle\langle S_i \rangle\rangle_J$ and $\langle\langle S_i \rangle^2\rangle_J$ are given by taking the $n \to 0$ limits respectively of $\langle S_i^\alpha \rangle$ and $\langle S_i^\alpha \times S_i^\beta \rangle_{\alpha \neq \beta}$ evaluated for a system characterized by the J-averaged n-ensemble partition function. This result is valid for finite-ranged interactions as well as infinite-ranged ones. Thus[9]

$$m \equiv \langle\langle S_i \rangle\rangle_J, \tag{12a}$$

$$q \equiv \langle\langle S_i \rangle^2\rangle_J, \tag{12b}$$

independent of i. A nonzero q indicates magnetic order, while nonzero m (in addition to q) indicates that that order is ferromagnetic. When $m = 0$ but q is nonzero, we shall call the state a "spin-glass."

Equations (10) indicate that magnetic order sets in as kT is reduced below the greater of $\tilde{J}_0$ or $\tilde{J}$. If $\tilde{J}_0 > \tilde{J}$, the phase that is reached is ferromagnetic, but when the converse is true, spin-glass order ensues, and m remains zero for $kT < \tilde{J}$. The full phase diagram is plotted in Fig. 1, in terms of the dimensionless combinations $\tilde{J}_0/\tilde{J}$ and $kT/\tilde{J}$, and may easily be rescaled to describe models in which $\tilde{J}_0$ and $\tilde{J}$ are known functions of external parameters (see, e.g., Ref. 4).

For $\tilde{J}_0/\tilde{J} \gg 1$ the effects of fluctuations are weak, and one can show from (10a) and (10b) that $q \sim m^2$, in accord with the physical interpretation (12b) of q as the square of the modulus of the frozen moment per site. The zero-temperature magnetization is diminished by weak fluctuations as

$$m \sim 1 - (2/\pi)^{1/2} (\tilde{J}/\tilde{J}_0) \exp(-\tilde{J}_0^2/2\tilde{J}^2), \tag{13a}$$

and vanishes continuously at the spin-glass phase boundary as

$$m \sim (18\pi)^{1/4} (\tilde{J}_0/\tilde{J})^2 [(2/\pi)^{1/2} - \tilde{J}/\tilde{J}_0]^{1/2}. \tag{13b}$$

Values of $m(T)$ and $q^{1/2}$ obtained by numerical solution of (10a) and (10b) are plotted in Fig. 2 for various values of $\tilde{J}_0/\tilde{J}$. We note that the effect of fluctuations is strongest at low temperatures, causing the decrease in the magnetization as $T \to 0$ in Fig. 2, and a line of second-order transitions from ferromagnet to spin-glass in the phase diagram of Fig. 1.

The frozen moment, $q^{1/2}(T)$, as is shown in Fig. 2 and as can be derived from (10a) and (10b), is proportional to $(T_c - T)^{1/2}$ just below T_c, tends to unity as $T \to 0$, and is always greater than $m(T)$ at the same temperature. The linear low-temperature dependence of q and $q^{1/2}$,

$$1 - q(T) \sim (2/\pi)^{1/2} (kT/\tilde{J}) \exp(-\tilde{J}_0^2 m^2/2\tilde{J}^2), \tag{14}$$

as $T \to 0$, contrasts with that of $m(T)$ in a uniform Ising magnet, for which all temperature derivatives vanish at $T = 0$, since excitations from the ferromagnetic ground state require a

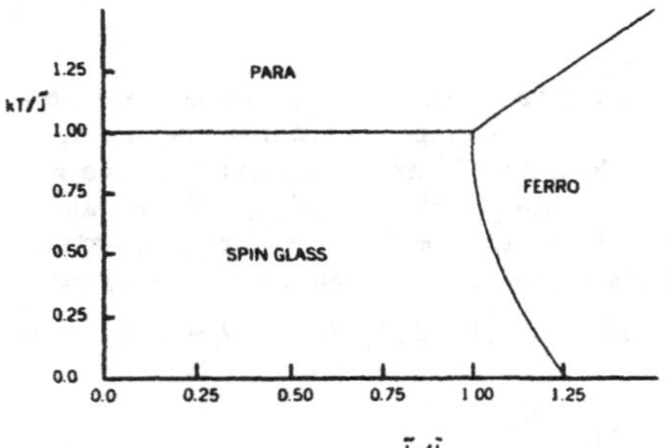

FIG. 1. Phase diagram of spin-glass ferromagnet.

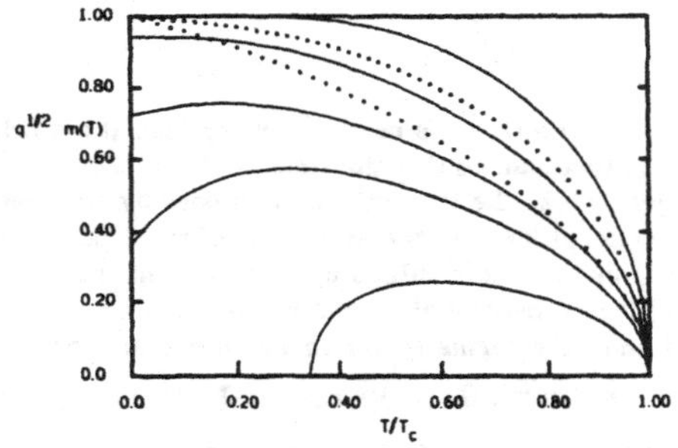

FIG. 2. Solid lines denote $m(T)$ for ratios $\tilde{J}_0/\tilde{J}$ of (top to bottom) ∞, 2.0, 1.5 1.3, and 1.1. Dotted lines show $q^{1/2}(T)$ for $\tilde{J}_0/\tilde{J} = 2.0$ (upper line) and 0.0 (lower line).

finite energy.

The differential susceptibility, χ, may be obtained by repeating the steps leading to (9) and (10) with an external field term, $\sum_i HS_i$, in the Hamiltonian (1). This simply adds an extra contribution, H/kT, to the arguments of sech and tanh in (10a) and (10b). Differentiating (10b) with respect to H, and taking the limit $h\to 0$, we then obtain

$$\chi(T)=[1-q(T)]/\{kT-\tilde{J}_0[1-q(T)]\}$$
$$=\chi^{(0)}/(1-\tilde{J}_0\chi^{(0)}), \qquad (15)$$

where $\chi^{(0)}$ is the result for $\tilde{J}_0=0$. Above the ordering temperature, where $q=0$, this is just a Curie-Weiss law. In the spin-glass phase, the fluctuations decrease χ, while $\tilde{J}_0$ enhances it. Two examples are plotted in Fig. 3. The dotted lines in Fig. 3 show the effect of a finite field, $H=0.1\tilde{J}$, on the differential susceptibility in each case.

From (9), we obtain the internal energy, U:

$$U=-N[m^2\tilde{J}_0/2+\tilde{J}^2(1-q^2)/2kT]. \qquad (16)$$

In the spin-glass phase the leading term in the specific heat at low temperatures is

$$C\sim Nk(kT/\tilde{J})(2/\pi)^{1/2}[(\pi^2/12)-1/2\pi]. \qquad (17)$$

At the spin-glass ordering temperature C has a cusp. For all $\tilde{J}_0$, C equals $Nk\tilde{J}^2/2(kT)^2$ above the ordering temperature,[10] in contrast to the corresponding pure systems for which it vanishes. The linear temperature dependences seen in C, m, and $q^{1/2}$ suggest that the system possesses excitations from the ground state whose density remains finite down to zero energy. The entropy S equals $Nk[\ln 2-\tilde{J}^2/(2kT)^2]$ above the spin-glass ordering temperature, but goes to a negative limit, $-Nk/2\pi$, at $T=0$. We speculate that this unphysical behavior has its origin in the interchange of limits $N\to\infty$ and $n\to 0$, but that the consequences are confined to low temperatures.

When an Ising system described by (1) and (2) with nearest-neighbor interactions is treated with mean-field theory (of the EA type), equations identical to (10) are obtained, with zJ_0 and $z^{1/2}J$ replacing $\tilde{J}_0$ and $\tilde{J}$, where z is the average number of neighbors.[11] For interactions which are on the average antiferromagnetic or which include second-neighbor terms, analogous equations result in which m is replaced by the appropriate sublattice magnetization.

For the finite-ranged interactions occurring in real systems, the results presented here have at best mean-field significance, and thus cannot treat critical phenomena correctly. However, it is evident that any more sophisticated treatment of critical properties in the presence of random competing exchange interactions should allow for the possibility of order parameters $\langle\langle S_i\rangle^n\rangle_J$ with $n>1$, such as q discussed above.

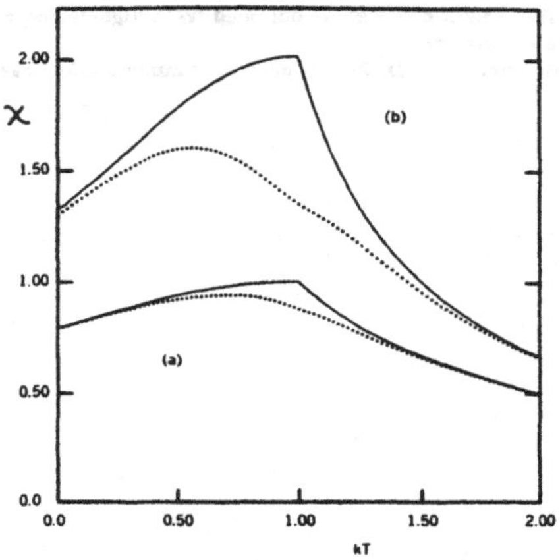

FIG. 3. Differential susceptibility without external field (solid lines) and with a field $H=0.1\tilde{J}$ (dotted lines) for $\tilde{J}_0/\tilde{J}=0$, curves a, and $\tilde{J}_0/\tilde{J}=0.5$, curves b.

*On sabbatical leave from Department of Physics, Imperial College, London SW7 2BZ, England.

[1]J. A. Mydosh, in *Magnetism and Magnetic Materials—1974*, edited by C. D. Graham, Jr., G. H. Lander, and J. J. Rhyne, AIP Conference Proceedings No. 24 (American Institute of Physics, New York, 1975), p. 131.

[2]D. E. Murnick, A. T. Fiory, and W. J. Kossler, to be published; J. M. D. Coey and R. W. Readman, Nature (London) 246, 476 (1973).

[3]S. F. Edwards and P. W. Anderson, J. Phys. F: Met. Phys. 5, 965 (1975).

[4]D. Sherrington and B. W. Southern, J. Phys. F: Met. Phys. 5, L49 (1975).

[5]K. H. Fischer, Phys. Rev. Lett. 34, 1438 (1975).

[6]V. J. Emery, Phys. Rev. B 11, 239 (1975); G. Grinstein, Ph.D. thesis, Harvard University, 1974 (unpublished); G. Grinstein and A. H. Luther, to be published.

[7]A. Aharony, Phys. Rev. Lett. 34, 590 (1975).

[8]See, e.g., H. E. Stanley, *Phase Transitions and Critical Phenomena* (Oxford Univ. Press, New York, 1971) p. 91.

[9]The identification (12b) of q has also been pointed out by Fischer in Ref. 5.

[10]This result can also be obtained from high-temperature expansions.

[11]R. Harris and D. Zobin (private communication) have investigated analogous mean-field equations for the Heisenberg spin-glass model of Ref. 4. They find a phase diagram similar to Fig. 1.

PHYSICAL REVIEW B VOLUME 17, NUMBER 11 1 JUNE 1978

Infinite-ranged models of spin-glasses

Scott Kirkpatrick
IBM Thomas J. Watson Research Center, Yorktown Heights, New York 10598

David Sherrington
IBM Thomas J. Watson Research Center, Yorktown Heights, New York 10598, Physics Department, Imperial College, London, England SW7 2BZ, * *and Institut Laue-Langevin, BP 156, 38042 Grenoble Cedex, France* †

(Received 8 December 1977)

A class of infinite-ranged random model Hamiltonians is defined as a limiting case in which the appropriate form of mean-field theory, order parameters and phase diagram to describe spin-glasses may be established. It is believed that these Hamiltonians may be exactly soluble, although a complete solution is not yet available. Thermodynamic properties of the model for Ising and XY spins are evaluated using a "many-replica" procedure. Results of the replica theory reproduce properties at and above the ordering temperature which are also predicted by high-temperature expansions, but are in error at low temperatures. Extensive computer simulations of infinite-ranged Ising spin-glasses are presented. They confirm the general details of the predicted phase diagram. The errors in the replica solution are found to be small, and confined to low temperatures. For this model, the extended mean-field theory of Thouless, Anderson, and Palmer gives physically sensible low-temperature predictions. These are in quantitative agreement with the Monte Carlo statics. The dynamics of the infinite-ranged Ising spin-glass are studied in a linearized mean-field theory. Critical slowing down is predicted and found, with correlations decaying as $e^{-[(T-T_c)/T]^2 t}$ for T greater than T_c, the spin-glass transition temperature. At and below T_c, spin-spin correlations are observed to decay to their long-time limit as $t^{-1/2}$.

I. INTRODUCTION

In recent years considerable interest has arisen in the possibility of the existence in suitably spatially disordered systems of a new type of magnetic order not found in pure systems. This is the so-called spin-glass phase[1,2] in which magnetic moments are believed to be frozen into thermal equilibrium orientations but with no average long-range order; one way of defining this statement mathematically is to say that $\langle|\langle\vec{S}_i\rangle|\rangle_d \neq 0$ but $\langle\langle\vec{S}_i\rangle\cdot\langle\vec{S}_j\rangle\rangle_d \rightarrow 0$ as $\vec{R}_i - \vec{R}_j \rightarrow \infty$, where $\langle\ \rangle$ refers to a thermal average and $\langle\ \rangle_d$ to an average over the spatial disorder (in the latter case with $\vec{R}_i - \vec{R}_j$ held fixed).

A number of physical examples appear to exist; the canonical cases[1] are metallic alloys with substitutional magnetic impurities, such as *Cu*Mn or *Au*Fe; other examples are found in amorphous systems and in compounds with inequivalent sites randomly available to magnetic ions. A necessary requirement seems to be a locally random competition between ferromagnetic and antiferromagnetic forces, although this competition may have a number of possible microscopic origins; for example, fixed positions but random exchange, fixed exchange interaction as a function of distance but random positions, topological disorder as in an amorphous system with antiferromagnetic exchange but no possibility of a sublattice, etc.

In order to demonstrate theoretically the possible existence of a spin-glass phase Edwards and Anderson[3] (EA) introduced a simple model with exchange disorder and were able within a novel mean-field theory to demonstrate the existence of the phase. Their model is of a set of classical spins $\vec{S}_i$ on a periodic lattice interacting via an exchange interaction

$$\mathcal{H} = -\sum_{(ij)} J_{ij}\vec{S}_i\cdot\vec{S}_j\ , \tag{1.1}$$

where the sum is over nearest-neighbor pairs and the J_{ij} are independently distributed with the Gaussian probability distribution

$$p(J_{ij}) = [(2\pi)^{1/2}J]^{-1}\exp(-J_{ij}^2/2J^2)\,. \tag{1.2}$$

The disorder is quenched; that is J_{ij}'s are chosen randomly but then fixed for all thermodynamic purposes. It is evident that none of the conventional types of order is possible. To study this problem, EA employed a novel replication procedure with spin correlation between Gibbs-like replicas playing the role of a spin-glass order parameter in a generalized mean-field solution. Since that work, extensions have been made[2] to include interactions beyond nearest neighbors,[4,5] exchange distributions offset from zero to allow for competition between spin-glass and ferro (or antiferro-) magnetism,[6] and some quantum effects.[4,6] These studies have all employed mean-field theories with the unconventional EA order parameter.

It is well known that molecular-field theory for a pure ferromagnet becomes exact in the thermodynamic limit for a constant infinite-range exchange interaction provided that the interaction is appropriately scaled with the number of spins in the system.[7] In this paper, we examine the analogous situation for systems which can exhibit spin-glass and ferromagnetic behavior. The Hamiltonians we employ are analogous to (1.1) but the summation $\sum_{(ij)}$ runs over all pairs of sites (ij) in the system and we concentrate primarily, but not exclusively, on an Ising interaction. We show that to lead to physical but nontrivial thermodynamic consequences a cumulant moment of J_{ij} (denoted $\langle J_{ij}^r\rangle_c$) must scale inversely as the number of spins N. Thus, for possible ferromagnetism we require

$$\langle J_{ij}\rangle_c = \tilde{J}_0 N^{-1} + O(N^{-r}) \qquad (r>1), \tag{1.3a}$$

and for potential spin-glass behavior

$$\langle J_{ij}^2\rangle_c = \tilde{J}^2 N^{-1} + O(N^{-r}) \qquad (r>1). \tag{1.3b}$$

The relative magnitudes of $\tilde{J}_0$ and $\tilde{J}$ determine whether ferromagnetism or spin-glass ordering occurs at low temperature. Higher cumulants will scale as higher inverse powers of N than the first two and thus do not affect the thermodynamic properties of an infinite-ranged model. We shall therefore employ a Gaussian distribution of interactions without loss of generality.

The plan of the paper is as follows. In Sec. II we employ the replication procedure[3] of EA to analyze an infinite-ranged Ising spin-glass–ferromagnet model with a Gaussian distribution of interactions the moments of which are as given in (1.3a) and (1.3b). A condensed version of Sec. II was originally presented in Ref. 5. In Sec. III, we present a high-temperature series expansion for the same model, which confirms the predictions of the replication procedure for the transition from paramagnet to spin-glass or ferromagnet with decreasing temperature. [This expansion has been described by Thouless *et al.*[8] (TAP) for the case of pure spin-glass ordering.] In Sec. IV, we investigate the corresponding classical planar spin model using the replication method. As stated earlier we believe the general classical m-vector model based on an infinite-ranged version of (1.1) with distributions scaled as in (1.3) is in principle exactly solvable. Section V reports Monte Carlo tests of the replica and other theories, necessary since no real systems with the assumed interactions are known. The simulations also give access and insight into some of the microscopic phenomena which are unique to spin glasses. Dynamics of spin-glasses (assuming Ising interactions and single spin-flip relaxation processes) are analyzed and then studied by Monte Carlo in Sec. V.

II. REPLICA THEORY

For purposes of calculation, it is usually convenient to average over any randomness in a physical system at the earliest possible stage. When the randomness is quenched (immobile) the averaging must however be carried out on a physical observable. In the present thermodynamic problem, we must therefore average the free energy F and not, for example, the partition function Z. Normally, this is a much more awkward procedure than that of averaging Z, as would be appropriate to a system with annealed disorder. However, at the cost of increasing the effective spin dimensionality and using limiting procedures, a free-energy average can be transformed into a partition function average using a procedure which appears to have been first used in statistical mechanics by Kac[9] but rediscovered independently by Edwards,[3,10] Grinstein and Luther,[11] and Emery.[12] This procedure is essentially the identity

$$\ln x = \lim_{n\to 0}\,(x^n - 1)/n, \tag{2.1}$$

with x taken as the partition function Z. For integral n, Z^n may be expressed as

$$Z^n = \prod_{\alpha=1}^{n} Z_\alpha, \tag{2.2}$$

where α is a dummy label. The set $\alpha = 1, \ldots, n$ may be interpreted as identical replicas of the real system. In a disordered system Z_α is a function of the disorder, all the replicas α having the same disorder but not interacting in any way with one another. Averaging the free energy over the disorder leads, via (2.1), to an averaging of Z^n over the disorder. For the case of integral n, averaging Z^n leads in turn to an effective interaction between the replicas α and thus to an effective pure system with an interaction of higher order than the real (impure) system. This effective system is analytically continued to small n to give the averaged free energy using (2.1). In this section, we apply this replica procedure to the infinite-ranged Ising model with Gaussian exchange distribution.

The system we consider here is characterized by the Hamiltonian

$$\mathcal{H} = -\sum_{(ij)} J_{ij} S_i S_j - H \sum_i S_i, \tag{2.3}$$

where the spin operators S_i take the values ± 1, the (ij) sum is over all bonds, the J_{ij} are distributed according to

$$p(J_{ij})=[(2\pi)^{1/2}J]^{-1}\exp\frac{-(J_{ij}-J_0)^2}{2J^2}\,, \tag{2.4}$$

and H is an external field. Using (2.1) and (2.2) the averaged free energy per spin, $f(\equiv F/N)$ in the thermodynamic limit may be expressed as

$$f=-kT\lim_{N\to 0}\lim_{n\to 0}(Nn)^{-1}\left\{\int\prod_{(ij)}[p(J_{ij})dJ_{ij}]\,\mathrm{Tr}_n\exp\left[\sum_{\alpha=1}^{n}\left(\sum_{(ij)}J_{ij}S_i^\alpha S_j^\alpha/kT+H\sum_i S_i^\alpha/kT\right)\right]-1\right\} \tag{2.5}$$

$$=-kT\lim_{N\to\infty}\lim_{n\to 0}(Nn)^{-1}\left\{\mathrm{Tr}_n\exp\left[\sum_{(ij)}\left(\sum_\alpha S_i^\alpha S_j^\alpha J_0/kT+\sum_{\alpha,\beta}S_i^\alpha S_j^\alpha S_i^\beta S_j^\beta J^2/2(kT)^2\right)+\frac{H}{kT}\sum_\alpha\sum_i S_i^\alpha\right]-1\right\}, \tag{2.6}$$

where α,β label n dummy replicas and Tr_n denotes the trace over spins in each of the n replicas. After some rearranging we obtain

$$f=-kT\lim_{N\to\infty}\lim_{n\to 0}(nN)^{-1}\left[\exp\{[J^2/4(kT)^2](N^2n-n^2N)-(J_0/2kT)Nn\}\,\mathrm{Tr}_n\exp\left(\frac{J_0}{2kT}\sum_\alpha\left(\sum_i S_i^\alpha\right)^2\right.\right.$$
$$\left.\left.+\frac{J^2}{2(kT)^2}\sum_{(\alpha\beta)}\left(\sum_i S_i^\alpha S_i^\beta\right)^2+\frac{H}{kT}\sum_\alpha\sum_i S_i^\alpha\right)-1\right], \tag{2.7}$$

where (α,β) refers to combinations of α and β with $\alpha\neq\beta$. Note that the exchange terms in the second exponent are now in the form $\lambda(\sum_i O_i)^2$ where O_i is a local intensive operator, which leads to physical but nontrivial thermodynamic consequences only if $\lambda\propto N^{-1}$. The physically sensible scaling of J_0, J is, thus,

$$J_0=\tilde{J}_0/N\,, \tag{2.8a}$$

$$J=\tilde{J}/N^{1/2}\,, \tag{2.8b}$$

with $\tilde{J}_0$, $\tilde{J}$ both intensive. More complicated distributions than (2.4) will, in general, give rise to terms of sixth and higher order in (2.6), but these will be of order N^{-1} or smaller, and thus are without thermodynamic consequences.

It also follows from the form $\lambda(\sum_i O_i)^2$ that a transformation may be made to a Gaussian-averaged single-site problem, using the identity

$$\exp(\lambda a^2)=(2\pi)^{-1/2}\int dx\,\exp[-\tfrac{1}{2}x^2+(2\lambda)^{1/2}ax]\,. \tag{2.9}$$

Dropping terms which vanish in the thermodynamic limit, we may rewrite (2.7) as

$$f=-kT\lim_{N\to\infty}\lim_{n\to 0}(nN)^{-1}\left(\exp[\tilde{J}^2Nn/4(kT)^2]\int\left[\prod_\alpha\left(\frac{N}{2\pi}\right)^{1/2}dx^\alpha\right]\left[\prod_{(\alpha\beta)}\left(\frac{N}{2\pi}\right)^{1/2}dy^{(\alpha\beta)}\right]\right.$$
$$\times\exp\left\{-N\left[\sum_\alpha\tfrac{1}{2}(x^\alpha)^2+\sum_{(\alpha\beta)}\tfrac{1}{2}(y^{(\alpha\beta)})^2\right.\right.$$
$$\left.\left.\left.-\ln\mathrm{Tr}\exp\left(\frac{H}{kT}\sum_\alpha S^\alpha+\left(\frac{\tilde{J}_0}{kT}\right)^{1/2}\sum_\alpha x^\alpha S^\alpha+\frac{\tilde{J}}{kT}\sum_{(\alpha\beta)}y^{(\alpha\beta)}S^\alpha S^\beta\right)\right]\right\}-1\right), \tag{2.10}$$

where the trace is now over n replicas at a single site.

For large N and integral $n\geq 2$ the integrations in (2.10) can be performed by the method of steepest descents, the integral being dominated by the region of maximum integrand. At the maximum all the x^α are equal as also are all the $y^{(\alpha\beta)}$; we denote their values by x_n, y_n. A convenient parametrization permitting analytic continuation to $n\to 0$ then follows from the substitution

$$\sum_{(\alpha\beta)}y_{\max}^{(\alpha\beta)}S^\alpha S^\beta-\tfrac{1}{2}y_n\left[\left(\sum_\alpha S^\alpha\right)^2-n\right], \tag{2.11}$$

together with the use of identity (2.9). With the further substitution

$$q_n=y_n(kT/\tilde{J})\,, \tag{2.12a}$$

$$m_n=x_n(kT/\tilde{J}_0)^{1/2}\,, \tag{2.12b}$$

the integral in (2.10) for $n\geq 2$ becomes

$$(\det\Lambda_n)^{-1/2}\exp\Big[-N\Big(n(\tilde{J}_0 m_n^2/2kT)+(\tilde{J}/2kT)^2[n(n-1)q_n^2+2nq_n]$$
$$-\ln\int dz(2\pi)^{-1/2}\exp(-\tfrac{1}{2}z^2)(2\cosh\Xi_n)^n\Big)\Big][1+O(N^{-1})], \tag{2.13}$$

where $\Xi_n \equiv (\tilde{J}_0 m_n+\tilde{J}q_n^{1/2}z+H)/kT$, and m_n, q_n satisfy the coupled equations

$$m_n=\frac{\int dz(2\pi)^{-1/2}\exp(-\frac{1}{2}z^2)(2\cosh\Xi_n)^n\tanh\Xi_n}{\int dz(2\pi)^{-1/2}\exp(-\frac{1}{2}z^2)(2\cosh\Xi_n)^n}, \tag{2.14}$$

$$1+(n-1)q_n=\frac{\int dz(2\pi)^{-1/2}(kT/\tilde{J})(z/q_n^{1/2})(2\cosh\Xi_n)^n\tanh\Xi_n}{\int dz(2\pi)^{-1/2}(2\cosh\Xi_n)^n}. \tag{2.15}$$

Λ_n is an $[\frac{1}{2}n(n+1)]\times[\frac{1}{2}n(n+1)]$ matrix whose elements are simple functions of m and q of order unity; it is given explicitly in Appendix A. Continuing these equations to small n and substituting (2.13) into (2.10) we find

$$f=kT\Big((\tilde{J}_0 m^2/2kT)-[\tilde{J}^2(1-q)^2/4(kT)^2]$$
$$-(2\pi)^{-1/2}\int dz\exp(-\tfrac{1}{2}z^2)\ln(2\cosh\Xi)\Big), \tag{2.16}$$

with

$$m=\int dz(2\pi)^{-1/2}\exp(-\tfrac{1}{2}z^2)\tanh\Xi, \tag{2.17}$$

$$q=\int dz(2\pi)^{-1/2}\exp(-\tfrac{1}{2}z^2)\tanh^2\Xi. \tag{2.18}$$

Solutions of (2.17) and (2.18) may be obtained by expansion near the points $T=0$ and $T=T_c$. Before exhibiting the solutions and their consequences for the thermodynamics let us note the physical significance of m and q. As shown in Appendix B,

$$m=\langle\langle S_i\rangle\rangle_c, \tag{2.19}$$

$$q=\langle\langle S_i\rangle^2\rangle_c. \tag{2.20}$$

A nonzero q thus indicates magnetic order, while nonzero m (in addition to q) indicates that the order is ferromagnetic. When $m=0$ but $q\neq 0$ we call the state a spin-glass. A uniform infinite range model does not permit periodic antiferromagnetic orderings although these are possible in related short-range interaction models.

The solution of (2.17) and (2.18) leads for $H=0$ to the phase diagram shown in Fig. 1. All the phase transition lines are second order. The phase transition from paramagnetic to ordered magnetic state occurs at a temperature equal to the larger of $\tilde{J}_0/k$, $\tilde{J}/k$, the ordered phase being ferromagnetic if $\tilde{J}_0>\tilde{J}$, spin-glass if the converse holds. A finite field H removes sharp phase transitions by allowing m and q to be nonzero at all temperatures.

For $\tilde{J}_0/\tilde{J}\gg 1$ the effect of exchange fluctuations are weak and $q\sim m^2$, in accord with the physical interpretation of q as the mean-square frozen moment per site. Although $q(T=0)=1$, $m(0)$ is diminished by weak fluctuations as

$$m(0)\sim 1-(2/\pi)^{1/2}(\tilde{J}/\tilde{J}_0)\exp(-\tilde{J}_0^2/2\tilde{J}^2), \tag{2.21}$$

and vanishes continuously at the spin-glass phase boundary as

$$m(0)\sim(18\pi)^{1/4}(\tilde{J}/\tilde{J}_0)^2[(2/\pi)^{1/2}-\tilde{J}/\tilde{J}_0]^{1/2}. \tag{2.22}$$

Values of $m(T)$ and $q^{1/2}(T)$ obtained by numerical solution of (2.17) and (2.18) are exhibited in Fig. 2 for four choices of $\tilde{J}_0/\tilde{J}$. We note that the effect of fluctuations is strongest at low temperatures, causing a decrease in magnetization as $T\to 0$, and

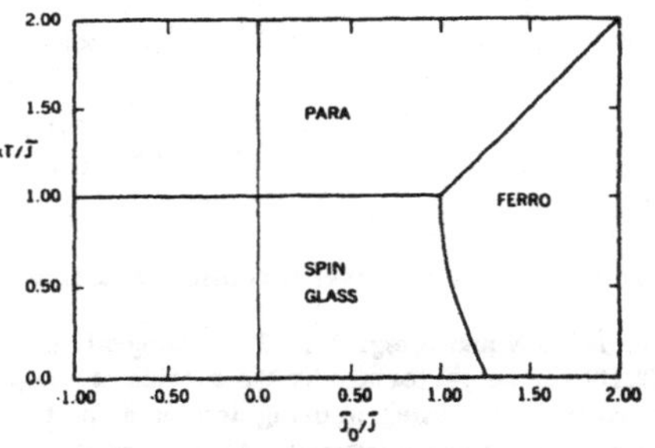

FIG. 1. Magnetic phase diagram of the random infinite-ranged Ising model defined by Eqs. (2.3) and (2.4), using the natural units (2.8). A spin-glass phase is obtained at low temperatures for all negative $\tilde{J}_0$, even though the corresponding uniform system has no phase transition.

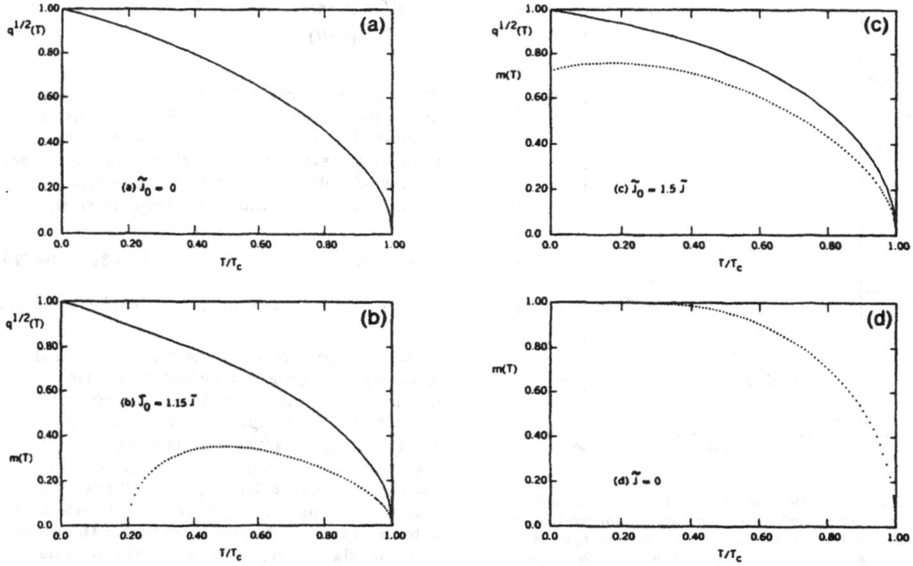

FIG. 2. Frozen moment, $q^{1/2}(T)$ (solid line), and magnetization, $m(T)$ (dotted line), for four magnetic systems described by (3), with parameters chosen to exhaust the possibilities of the phase diagram in Fig. 1. T is scaled to the transition temperature from the paramagnetic phase to whichever ordered phase forms first. In (a), $\tilde{J}_0=0$, and $m=0$ for all T. In (b), $\tilde{J}_0=1.15\,\tilde{J}$, and the ferromagnetic phase gives way to a spin-glass phase at lower T. In (c), $\tilde{J}_0=1.5\tilde{J}$, and only a ferromagnetic phase is found, with $q^{1/2}(T)$ significantly greater than $m(T)$ for all $T \leq T_c$. Case (d) is a pure ferromagnet ($\tilde{J}=0$), with $q^{1/2}(T)=m(T)$.

a line of second-order transitions from ferromagnet to spin-glass with decreasing temperature for $\tilde{J}_0/\tilde{J}$ between 1 and $(\frac{1}{2}\pi)^{1/2}$.

The frozen moment $q^{1/2}(T)$, as is shown in Fig. 2, is found from (2.17) and (2.18) to be proportional to $(T_c-T)^{1/2}$ just below T_c, tends to unity as $T\to 0$, and is always greater than $m(T)$ at the same temperature (for $\tilde{J}\neq 0$). The low temperature behavior of $q(T)$ is predicted to be linear

$$q(T)\sim 1-(2/\pi)^{1/2}(kT/\tilde{J})\exp(-\tilde{J}_0^2 m^2/2\tilde{J}^2). \quad (2.23)$$

This contrasts with the behavior of $m(T)$ in a uniform Ising magnet, for which all temperature derivatives vanish at $T=0$ since excitations from the ferromagnetic ground state require a finite energy.

The standard thermodynamic functions follow straightforwardly. For example, the internal energy may be obtained via the Gibbs-Helmholtz relation, yielding

$$U/N=-[\tfrac{1}{2}\tilde{J}_0 m^2+\tilde{J}^2(1-q^2)/2kT+Hm]. \quad (2.24)$$

The low-temperature limit of U/N is extracted by substituting (2.23) into (2.24). In particular,

$$U(0)/N=-\{\tfrac{1}{2}\tilde{J}_0 m(0)^2+Hm(0)+\tilde{J}(2/\pi)^{1/2}\exp[-\tilde{J}_0^2 m(0)^2/\tilde{J}^2]\}. \quad (2.25)$$

Higher-order terms in (2.23) were calculated in order to obtain analytic expressions for the heat capacity C at low temperature. Numerical results are displayed in Fig. 3. The following features should be noted. C is zero at $T=0$, as required by general thermodynamic considerations. At the ordering temperature, T_c, C has a singularity. If the transition is to a spin-glass, C displays a cusp [Fig. 3(a)]; if the ordered phase is ferromagnetic, there is a discontinuity [Figs. 3(b)–3(d)]. In both cases, a finite contribution to $C(T)$ is found above T_c:

$$C/N=\tilde{J}^2/2kT^2,\quad T>T_c. \quad (2.26)$$

This latter behavior should be contrasted with the situation for a pure infinite-ranged ferromagnet (or a finite-ranged ferromagnet treated in mean-field theory) for which C vanishes above the

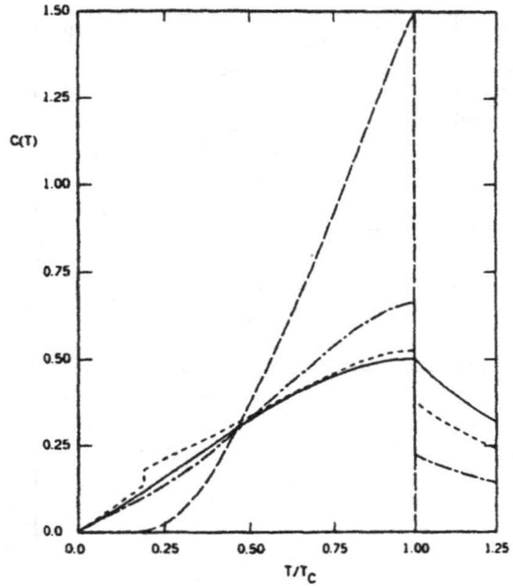

FIG. 3. Specific heat as a function of T (normalized to the relevant ordering temperature), for the four cases treated in Fig. 2: (a) $\tilde{J}_0=0$, solid line; (b) $\tilde{J}_0=1.15\tilde{J}$ short dashed line; (c) $\tilde{J}_0=1.5\tilde{J}$ long-short dashed line; (d) $\tilde{J}=0$, long-dashed line.

ordering temperature.

In the spin-glass phase the leading contribution to the specific heat at low temperatures is given by

$$C/N \sim (k^2T/\tilde{J})(2/\pi)^{1/2}[\tfrac{1}{12}\pi^2-(2\pi)^{-1}], \quad (m=0). \tag{2.27a}$$

In the ferromagnetic phase, fluctuations still give rise to a linear term in C, the amplitude of which decreases to zero as $m(0)$ tends to 1 with increasing ratio $\tilde{J}_0/\tilde{J}$:

$$\frac{C}{N}\sim\left(\frac{k^2T}{\tilde{J}}\right)\left(\frac{2}{\pi}\right)^{1/2}\{\tfrac{1}{12}\pi^2-(2\pi)^{-1}\exp[-(\tilde{J}_0m/\tilde{J})^2]\}$$
$$\times\exp[-\tfrac{1}{2}(\tilde{J}_0m/\tilde{J})^2]. \tag{2.27b}$$

When $\tilde{J}_0/\tilde{J}$ lies between 1.0 and $\pi^{1/2}\approx1.25$, the transition from spin-glass to ferromagnet with increasing temperature is indicated by a second discontinuity in $C(T)$. This case is shown in Fig. 3(b).

The susceptibility χ at general H may be obtained directly from (2.15) and has been illustrated for various cases within the spin-glass parameter range in Fig. 3 of Ref. 5. In the limit $H\to0$ the susceptibility may be simply expressed in terms of q as

$$\chi(T)=[1-q(T)]/\{kT-\tilde{J}_0[1-q(T)]\}$$
$$=\chi^{(0)}/(1-\tilde{J}_0\chi^{(0)}), \tag{2.28}$$

where $\chi^{(0)}$ is the result for $\tilde{J}_0=0$. Above the ordering temperature, where $q=0$, this is just a Curie-Weiss law. In the spin-glass phase, fluctuations decrease $\chi^{(0)}$ and χ, giving rise to a cusp. Positive $\tilde{J}_0$ enhances χ at all temperatures.

Within this formulation the entropy is given by

$$S/N=-[\tilde{J}_0m^2/2T+(\tilde{J}^2/4kT^2)(1-q)(1+3q)-Hm/T]$$
$$+k(2\pi)^{-1/2}\int dz\exp(-\tfrac{1}{2}z^2)\ln(2\cosh\Xi). \quad 2.29)$$

At high temperatures this yields physical results, which we shall see are in accord with series expansions, but it leads to unacceptable consequences as $T\to0$; for example if $H=0$, $m=0$, the $T\to0$ limit of (2.28) is $-k/2\pi$. This incorrect result is the most obvious indication that the procedure used above of continuing integral $n\geq2$ results to small continuous n is not correct in all details as T tends to zero. Other evidence lies in the computer simulations reported below. On the other hand, the computer simulations verify many of the qualitative features below the ordering temperature. Both the simulations and partial summations of high-temperature series indicate that the results for temperatures greater than the ordering temperature are correct, as also is the prediction of that temperature.

Another apparent difficulty is that f of (2.16) cannot be considered a Landau variational free-energy function. Although it has an extremum at (q,m) as given by (2.17), (2.18), this extremum is not a minimum in the spin-glass phase but a maximum.[13] This particular difficulty does not invalidate the solution since f is not such a variational function but rather is defined only at the physical values of q,m. In particular it is evident from series expansions (see Sec. III) that the spurious minimum of f at $q=0$ found, say, for $H=0$, $\tilde{J}_0<\tilde{J}$ does not represent a stable solution.

Finally, by way of further support for aspects of the above analysis, it should be noted that since the prelininary report of this work the spherical model analogue of (2.3), (2.4), and (2.8) has been solved at all temperatures by Kosterlitz *et al.*[14] They use two methods of analysis; the $n\to0$ procedure, and a direct method, not generalizable to the Ising model, which employs the exact eigenvalue spectrum of a Gaussian random infinite matrix. They find the same results by both procedures, provided care is taken in the order of certain multiple integrations. The spherical model also gives a negative entropy at $T=0$ but this is normal for classical continuum models. They do however also find that f is maximized for the spin-glass q in the $n\to 0$ limit. These results give us extra confidence in the general nature of the $n\to 0$ procedure.

III. HIGH-TEMPERATURE SERIES

In our preliminary letter it was noted that the specific heat above the ordering temperature as derived above is identical with that given by high-temperature series expansions. In fact, however, as noted by TAP,[8] the high-temperature series for the free energy for the model given by (2.3) and (2.4) can be summed completely at least to order N in F without using any $n\to 0$ tricks giving (i) the same results as above for $kT>\tilde{J}_0$ and $\tilde{J}$ and (ii) divergences at a temperature equal to the larger of $\tilde{J}_0/k$ and $\tilde{J}/k$. This demonstrates that the transition temperatures from paramagnetic to ordered state are correctly given in Sec. II at least with respect to transitions of greater than first order. In this section, we discuss these series for the Hamiltonian of (2.3) with $H=0$.

Using standard manipulations, the partition function may be expressed as

$$Z=\mathrm{Tr}\prod_{(ij)}\exp(-K_{ij}S_iS_j) \tag{3.1}$$

$$=2^N\Big(\prod_{(ij)}\cosh K_{ij}\Big)\mathrm{Tr}\,2^{-N}\prod_{(ij)}(1+S_iS_jt_{ij}), \tag{3.2}$$

where

$$K_{ij}=J_{ij}/kT, \tag{3.3}$$

$$t_{ij}=\tanh K_{ij}. \tag{3.4}$$

The averaged free energy is therefore given by

$$\begin{aligned} f=-kT\ln 2-\frac{kT}{N}\sum_{(ij)}\int dJ_{ij}p(J_{ij})\ln\cosh K_{ij}\\ -\frac{kT}{N}\int\prod_{(ij)}dJ_{ij}p(J_{ij})\\ \times\ln\Big(2^{-N}\,\mathrm{Tr}\prod_{(ij)}(1+S_iS_jt_{ij})\Big). \end{aligned} \tag{3.5}$$

Diagrammatically the argument of the ln in the last term of (3.5) is the sum of closed loop diagrams with an even number of bonds at each vertex and no repeated bonds. Repeated bonds can occur once the logarithm is expanded. Taking the average against $P(\tilde{J}_{ij})$ given by (2.4) with the scalings (2.8), one retains to order N^{-1} in the last term of (3.5) only single and double polygons.[15] One finds

$$\begin{aligned} f=-kT\ln 2-\tilde{J}^2/4kT+(kT/2N)\ln(1-\tilde{J}_0/kT)\\ -(kT/2N)\ln[1-(\tilde{J}/kT)^2]+\tilde{J}_0/2N+O(N^{-2}), \end{aligned} \tag{3.6}$$

where the ln strictly signifies the first $(N-1)$ terms of the expansion of the logarithm. From the (3.6) we can note the following: (i) for $kT>\max(\tilde{J}_0,\tilde{J})$ the free energy is given to leading order by

$$f=-kT\ln 2-\tilde{J}^2/4kT, \tag{3.7}$$

in agreement with (2.16) and (ii) the series diverges at $kT=\max(\tilde{J}_0,\tilde{J})$, signalling the breakdown of the paramagnetic phase at the same temperatures as obtained in Sec. II. Below this temperature, the paramagnetic phase is unstable. When the single-bond polygon series is divergent, one expects a transition to a phase in which $\langle\langle S_i\rangle\rangle_d$ is nonzero, while a divergence of the double-bond polygon series is consistent with nonvanishing $\langle\langle S_i\rangle^2\rangle_d$. These observations are in accord with the relevant part of the phase diagram (Fig. 1) predicted by the replica theory.

IV. OTHER INFINITE-RANGED SPIN-GLASS MODELS

Equation (2.3) can be generalized to give a Hamiltonian for general classical m-vector spin-glasses in the form

$$\mathcal{H}=-\sum_{(ij)}J_{ij}\vec{S}_i\cdot\vec{S}_j-\vec{H}\cdot\sum_i\vec{S}_i,\quad |S|=1, \tag{4.1}$$

with $P(J_{ij})$ a continuous distribution as before. We believe that this model is in principle solvable for arbitrary m and give below a solution for the planar model ($m=2$). We shall use the $n\to 0$ limiting procedure.

With the transformations as employed for the Ising problem the free energy for an arbitrary m-vector model may be expressed as

$$f=-kT\lim_{N\to\infty}\lim_{n\to\infty}(nN)^{-1}\Big\{\mathrm{Tr}_n\exp\Big[\sum_{(ij)}\Big(\frac{J_0}{kT}\sum_\alpha\vec{S}_i^\alpha\cdot\vec{S}_j^\alpha+\frac{J^2}{2(kT)^2}\sum_{\alpha\beta}\vec{S}_i^\alpha\cdot\vec{S}_j^\alpha\vec{S}_i^\beta\cdot\vec{S}_j^\beta\Big)\Big]-1\Big\}. \tag{4.2}$$

For the planar model this can be put into a form suitable for steepest descents analysis using

$$\cos(\varphi_i^\alpha - \varphi_j^\alpha)\cos(\varphi_i^\beta - \varphi_j^\beta) = \tfrac{1}{2}(\vec{S}_i^{\alpha\beta}\cdot\vec{S}_j^{\alpha\beta} + \vec{T}_i^{\alpha\beta}\cdot\vec{T}_j^{\alpha\beta}), \quad (4.3)$$

where $\vec{S}_i^{\alpha\beta}$ is a two-dimensional unit vector characterized by an angle

$$\varphi_i^{\alpha\beta} = \varphi_i^\alpha - \varphi_i^\beta, \quad (4.4)$$

and $\vec{T}_i^{\alpha\beta}$ is a similar unit vector with the angle

$$\psi_i^{\alpha\beta} = \varphi_i^\alpha + \varphi_i^\beta. \quad (4.5)$$

For $\alpha = \beta$, $\vec{T}^{\alpha\beta}$ reduces to $\vec{U}^{\alpha\beta}$, a unit vector oriented at an angle $2\varphi_i^\alpha$. With this notation, f may be rewritten

$$f = -kT \lim_{N\to\infty} \lim_{n\to 0} (nN)^{-1}\Big\{\mathrm{Tr}_n \exp\Big[J_0/2kT \sum_\alpha \Big|\sum_i \vec{S}_i^\alpha\Big|^2 + \frac{J^2}{8(kT)^2}\sum_{\alpha\beta}{}'\Big(\Big|\sum_i \vec{S}_i^{\alpha\beta}\Big|^2 + \Big|\sum_i \vec{T}_i^{\alpha\beta}\Big|^2\Big) + \frac{J^2}{8(kT)^2}\sum_\alpha\Big|\sum_i \vec{U}_i^\alpha\Big|^2 - \frac{J_0}{2kT}Nn^2 + \frac{J^2}{8(kT)^2}(nN^2 - 2Nn^2)\Big] - 1\Big\} \quad (4.6)$$

$$= -kT \lim_{N\to\infty}\lim_{n\to 0}(nN)^{-1}\Big(\exp(\tilde{J}^2 nN/8kT)^2 \int \prod_\alpha (N/2\pi)\, ds^\alpha\, du^\alpha \prod_{\alpha\beta}(N/2\pi)\, ds^{\alpha\beta}\, dt^{\alpha\beta} \times \exp\Big\{-N\Big[\sum_\alpha(\tfrac{1}{2}|s^\alpha|^2 + \tfrac{1}{2}|u^\alpha|^2) + \sum_{\alpha\beta}{}'(\tfrac{1}{2}|s^{\alpha\beta}|^2 + \tfrac{1}{2}|t^{\alpha\beta}|^2) - \ln \mathrm{Tr}\exp\Big(\sum_\alpha[(\tilde{J}_0/kT)^{1/2}\vec{s}^\alpha\cdot\vec{S}^\alpha + (\tilde{J}/2kT)\vec{u}^\alpha\cdot\vec{U}^\alpha] + (\tilde{J}/2kT)\sum_{\alpha\beta}{}'(\vec{s}^{\alpha\beta}\cdot\vec{S}^{\alpha\beta} + \vec{t}^{\alpha\beta}\cdot\vec{T}^{\alpha\beta})\Big)\Big]\Big\} - 1\Big). \quad (4.7)$$

The trace is taken over spins at a single site only. This integral can be performed by steepest descents as was done for the Ising model in Sec. II.

The general steepest descents treatment is complicated. We therefore make the physical Ansatz that for small enough $\tilde{J}_0$ there exists a spin-glass extremum with $s^\alpha = u^\alpha = t^{\alpha\beta} = 0$ and with all the $|s^{\alpha\beta}|$ equal. The angle of $\vec{s}_{\alpha\beta}$ in its reference plane is arbitrary. We set it to zero for convenience. Anticipating this identification of the order parameter in the $\lim_{n\to 0}$ we define

$$q_n = q^{\alpha\beta} = |s^{\alpha\beta}|\,2kT/\tilde{J} \quad (4.8)$$

at the extremum. The trace in (4.7) may now be expressed as

$$\mathrm{Tr}\exp\Big((\tilde{J}/2kT)^2 q_n \sum_{\alpha\beta}{}'\cos(\varphi^\alpha - \varphi^\beta)\Big) = \int \prod_{\alpha=1}^{n}\frac{d\varphi^\alpha}{2\pi}\int\frac{d\vec{r}}{2\pi}\exp(-\tfrac{1}{2}r^2)\exp\Big(\frac{\tilde{J}}{kT}(\tfrac{1}{2}q_n)^{1/2}\vec{r}\cdot\sum_\alpha\vec{S}^\alpha\Big)\exp\Big[-nq_n\Big(\frac{\tilde{J}}{2kT}\Big)^2\Big], \quad (4.10)$$

where the r integration is two dimensional and leads to

$$\int_0^\infty r\,dr\exp(-\tfrac{1}{2}r^2)[I_0(\tilde{J}r/kT)(\tfrac{1}{2}q_n)^{1/2}]^n \times \exp[-nq_n(\tilde{J}/2kT)^2], \quad (4.11)$$

where

$$I_n(\lambda) = \int_0^{2\pi}(d\varphi/2\pi)\cos(n\varphi)e^{\lambda\cos\varphi} \quad (4.12)$$

is a modified Bessel function of nth order.

Applying the extremal condition to (4.7) and taking the $\lim_{n\to 0}$ we obtain

$$q = 1 - (kT/\tilde{J})(2/q)^{1/2} \times \int_0^\infty r\, dr \exp(-\tfrac{1}{2}r^2) \frac{I_1[\tilde{J}r(\frac{1}{2}q)^{1/2}/kT]}{I_0[\tilde{J}r(\frac{1}{2}q)^{1/2}/kT]}. \quad (4.13)$$

This yields a spin-glass ordering temperature $T_g = \tilde{J}/2k$, as compared with $\tilde{J}/k$ for the Ising case and $\tilde{J}/3k$ found for the Heisenberg system by Edwards and Anderson. We speculate that for an m-vector classical model there will be a second-order spin-glass transition at $\tilde{J}/mk$. Below the ordering temperature, q increases as

$$q = \tfrac{2}{5}t + O(t^2), \quad (4.14)$$

where $t \equiv (T_g - T)/T_g$. At low temperature, q approaches unity linearly in T:

$$q = 1 - \pi^{1/2}(kT/\tilde{J}) + O(T^2). \quad (4.15)$$

The free energy is given by

$$f = -(\tilde{J}^2/8\,kT)(1-q)^2 - \int r\, dr \exp - \tfrac{1}{2} r^2 \ln I_0[\tilde{J}r(\tfrac{1}{2}q)^{1/2}/kT], \quad (4.16)$$

and the internal energy by

$$U = -(\tilde{J}^2/4\,kT)(1-q^2). \quad (4.17)$$

This leads to a specific heat which tends to a constant $\frac{1}{2}k$ per spin, at $T=0$, a consequence of the classical nature of the system. It has a cusp of the usual sort at T_g, and decreases as $\tilde{J}^2/4kT^2$ above T_g.

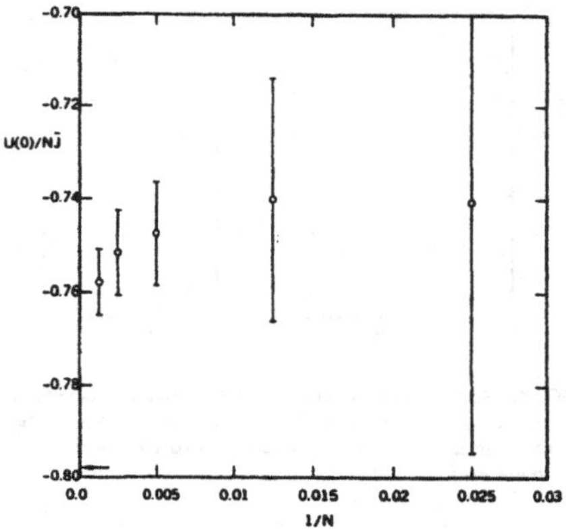

FIG. 4. Ground-state energies found for finite samples with interactions coupling all pairs of spins, distributed as in (1.2), with zero mean. For each size, N the number of samples studied was: 40 spins (200 cases); 80 (200); 200 (60); 400 (40); and 800 (20). The arrow indicates the ground-state energy predicted in (5.1) for an infinite sample.

The TAP analysis[8] is confined to the case $\tilde{J}_0 = 0$, and differs significantly from the present work only for $T \lesssim 0.5\, T_g$. We compare Monte Carlo results with the predictions of the two theories, emphasizing the low-temperature results in Figs. 7–11.

In order to obtain predictions for the low-tem-

V. MONTE CARLO TESTS OF THE THEORY

The infinite-ranged interactions that make it possible to demonstrate the existence of a spin-glass transition in the model Hamiltonian (1.1) do not occur in nature. In order to test the predictions of the present replica methods or the extended mean-field theory of TAP for the low-temperature phase, the necessary experiments were performed by computer simulation. Data obtained in samples of up to 800 spins are reported and compared with theory in this section. Since space and time constraints limited the sample size N to a relatively small number of spins ($N \leq 800$), some attention is given to determining the dependence of the results on N.

Since the replica method predicts an unphysical (slightly negative) entropy in the limit $T \to 0$, we have made a fairly extensive study of the ground-state properties of the infinite-ranged Ising spin-glass as a function of $\tilde{J}_0$, and compare these with the other predictions of the theory in Fig. 4–6.

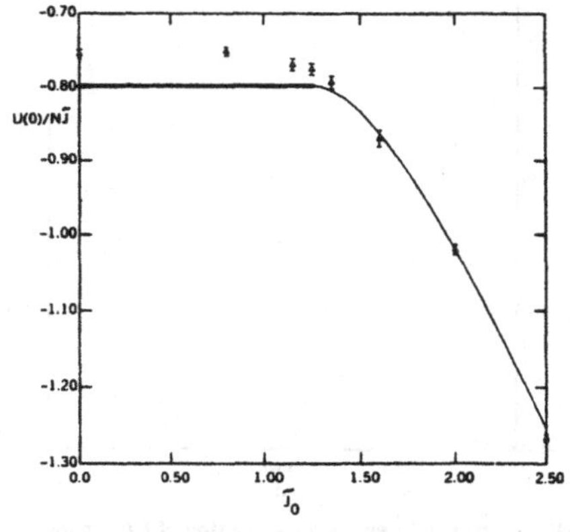

FIG. 5. Ground-state energies $U(0)/N\tilde{J}$, for 500-spin samples, as a function of $\tilde{J}_0$. Each point represents an average over 20 cases. The solid line (heavier in the spin-glass phase) is the predicted result (2.25).

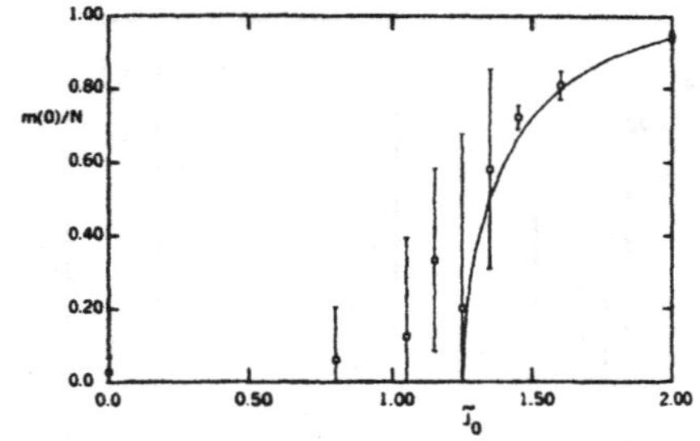

FIG. 6. Zero-temperature magnetization for 500-spin samples in an applied field $h = 0.01\tilde{J}$. Data points represent average and rms deviations found in 20 cases at each value of $\tilde{J}_0$.

perature thermodynamics, TAP characterize the low-energy excitations of a spin-glass by a distribution of single-spin molecular fields $p(h)$ neglecting possible excitations which might require the simultaneous reversal of more than one spin. They argue that such a distribution is only stable against further spin rearrangements if $p(h)$ increases from zero no faster than linearly in h at small fields. [A related argument has been used to show that $p(0) = 0$ for random exchange interactions which decay as $1/R^3$; see Ref. 16.]. This implies that $q(T) = 1 - \alpha(kT/\tilde{J})^2$, instead of the linear T dependence given by (2.23). TAP suggest that α should be the value which gives the maximum density of low-energy excitations consistent with their mean-field equations, $\alpha = 2(\ln 2)^{1/2} \approx 1.665$.

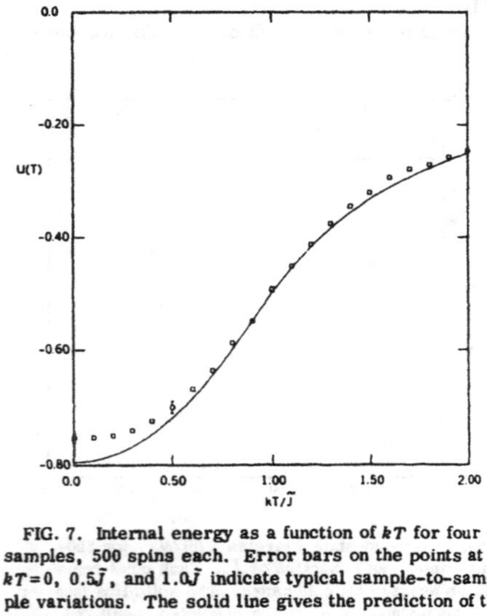

FIG. 7. Internal energy as a function of kT for four samples, 500 spins each. Error bars on the points at $kT = 0$, $0.5\tilde{J}$, and $1.0\tilde{J}$ indicate typical sample-to-sample variations. The solid line gives the prediction of the replica theory.

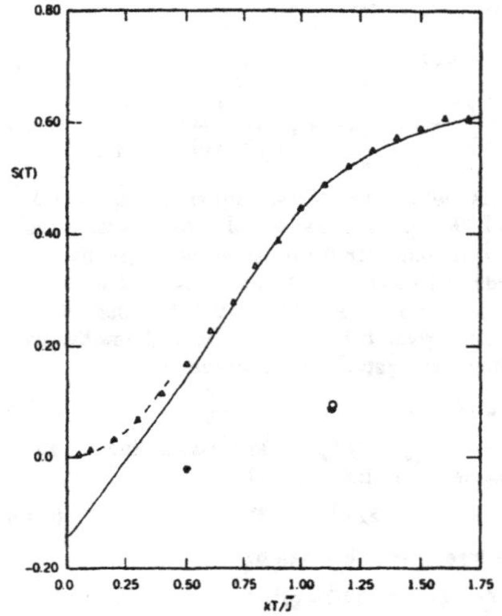

FIG. 8. Entropy as a function of temperature, obtained by integrating the Monte Carlo data of Fig. 7 (data points), and as predicted by the replica theory (solid line). The Monte Carlo results remain positive at all temperatures, and are in good agreement with the TAP prediction (dashed line) at low temperatures.

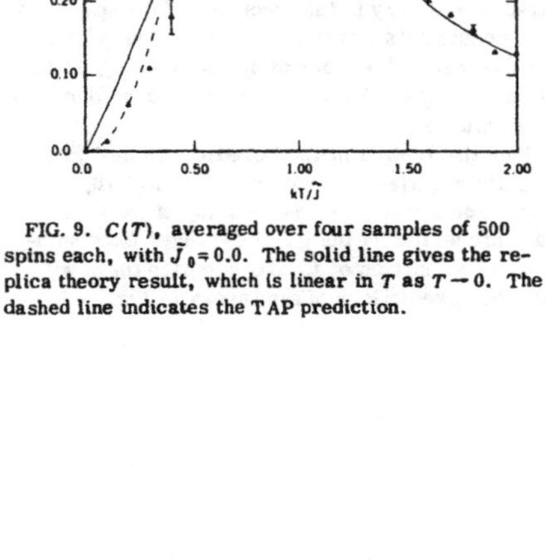

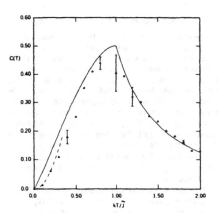

FIG. 9. $C(T)$, averaged over four samples of 500 spins each, with $\tilde{J}_0 = 0.0$. The solid line gives the replica theory result, which is linear in T as $T \to 0$. The dashed line indicates the TAP prediction.

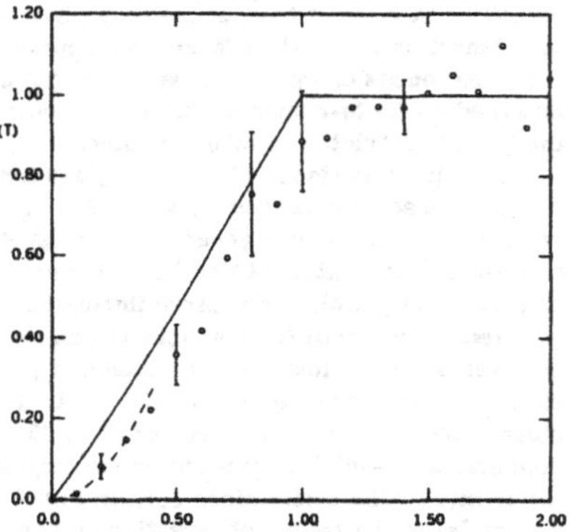

FIG. 10. $T\chi(T)$, averaged over four samples of 500 spins each, with $\tilde{J}_0 = 0.0$. The solid and dashed lines give the replica theory and TAP predictions, respectively.

Once α is determined, predictions are readily obtained for $\chi(T) \sim \alpha(kT/\tilde{J})^2$, $S(T) \sim (\frac{1}{4}\alpha^2)kT^2/\tilde{J}^2$, and $C(T) \sim (\frac{1}{2}\alpha^2)(kT/\tilde{J})^2$.

Some of the glassy features found in actual spin-glasses,[1] e.g., the existence of many metastable energy minima and unusual slow relaxation phenomena, also occur in the computer simulations and can be subjected to "microscopic" examination. We report preliminary results of such a study below.

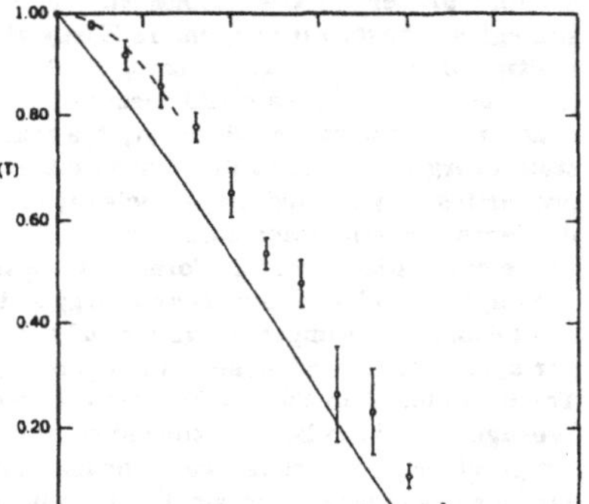

FIG. 11. Edwards-Anderson order parameter $q(T)$, obtained from the data of Fig. 10. Averages were taken over 400 time steps per spin at each temperature. Solid and dashed lines are replica theory and TAP results.

The simplest property to test is the internal energy. Specializing (2.25) to the spin-glass phase gives

$$U(0)/\tilde{J}N = -(2/\pi)^{1/2} \simeq -0.79. \tag{5.1}$$

We wish to check two features of (5.1)—first, that $U(0)N$ scales as $\tilde{J}$ (i.e., as $JN^{-1/2}$), and second the value of the coefficient. One can construct upper bounds to $U(0)$ which show that $U(0)/\tilde{J}N$ must be extensive. Consider the following "dynamic programming"[17] construction of an approximate ground state: The spins are arbitrarily numbered 1–N. The alignment of spin 1 is arbitrary. The orientation of spin 2 is chosen to make $J_{12}S_1S_2 > 0$ so that the contribution $U_2 \equiv -J_{12}S_1S_2$ to the internal energy is negative. The orientation of spin 3 is then chosen to make $U_3 \equiv -(J_{13}S_1 + J_{23}S_2)S_3 < 0$, and so forth. The internal energy of the approximate ground state constructed in this way is obtained by summing all the U_n, and gives an upper bound to the actual $U(0)$. The average value of U_2 is $-J(2/\pi)^{1/2}$, and that of U_n is $-J(2n/\pi)^{1/2}$. The expectation value of the sum of all contributions gives the bound

$$\frac{U(0)}{\tilde{J}N} \leq -\left(\frac{2}{\pi}\right)^{1/2} N^{-3/2} \sum_{n=1}^{N-1} n^{1/2} \sim -\frac{2}{3}\left(\frac{2}{\pi}\right)^{1/2} \simeq -0.53, \tag{5.2}$$

which is $\frac{2}{3}$ of the replica prediction

A fairly elaborate procedure was followed in the calculations to ensure accurate estimates of $U(0)$ for finite samples. The heuristic precautions described below were taken because the computational effort necessary to prove that no better ground state exists increases rapidly with increasing N. Attempts to improve upon the bound (5.2) by constructing the m lowest-energy configurations of $n+1$ spins, and so forth, proved ineffective. For $N = 800$, keeping track of the lowest 100 states as each spin was added gave no significant improvement over (5.2). Only for samples with fewer than 50 spins were apparently convergent estimates obtained, and this required keeping track of 2000 or more intermediate configurations. In the language of computational complexity theory,[17] finding the ground state of a spin-glass has features in common with the "NP-complete" problems, which always require exp(N) effort to solve in the worst case. What is surprising is that all samples prove to be "worst cases."

For each sample, many random starting arrangements of spins were constructed, and for each starting configuration a deterministic procedure, analogous to the method of steepest descents, was followed to reach a minimum energy. At each step a list of partial exchange energies

$\sum_j J_{ij}S_iS_j$ is constructed. The spin with the highest energy is flipped (if that energy is positive), and the list corrected. If no spin can gain energy by flipping, pairs of spins are searched to find a pair which will lower the energy of the system if they flip simultaneously. If such a pair is found they are flipped and the search over single spins continues. If not, selected triples and quartets of spins with small individual exchange fields are also searched. Such searches were carried out for 60–80 starting configurations in each Monte Carlo sample, keeping the best ground state encountered. Small further improvements in the ground-state energy were obtained by slightly disordering the best ground states ("warming" them to a temperature 2 or 3 times T_g for a few time steps per spin) and repeating the descent procedure.

The data obtained, plotted in Fig. 4, confirm that the ground-state energy scales as $\tilde{J}$, the central feature of the Edwards-Anderson picture. However, the coefficient given in (5.1) lies below the values actually found in samples of 40–800 spins, and the discrepancy appears to be outside the range of possible error. Fig. 4 suggests that the calculated $U(0)/\tilde{J}N$, extrapolated to infinite sample size, must lie somewhere between -0.75 and -0.77. This agrees with the -0.755 ± 0.01 which TAP report from their Monte Carlo calculations. The replica theory predicts too low a ground-state energy [since the negative entropy implies $dF(T)/dT > 0$ as $T \to 0$], but the magnitude of the discrepancy is only 3%–5%.

Various portions of the descent algorithm were used on four 500-spin samples with $\tilde{J}_0 = 0$ to determine their relative efficacy. The lowest energies obtained using only single spin flips and 80 random starting configurations of spins ranged from -0.71 to -0.74. Searching also for spin pairs which could flip narrowed this range to -0.72 to -0.74 by lowering the higher-energy states. The further improvement from considering three and four-spin processes was of order 0.001. The warmup process made a more significant contribution. The samples were warmed briefly to $3T_g$, then relaxed into the ground state, then warmed to 1.5 T_g and cooled, and finally warmed to $0.75T_g$ and cooled. After this process, all four were found to have minimum energies between -0.75 and -0.76.

The discrepancy between Monte Carlo results for $U(0)$ and the prediction (2.25) decreases with increasing $\tilde{J}_0$, as is seen in Fig. 5. Although (2.25) predicts that $U(0)$ is independent of $\tilde{J}_0$ in the spin-glass phase, since $q(0)$ and $m(0)$ are, the calculations show $U(0)$ decreasing slightly as the ferromagnetic phase is approached. There is no evidence for a discontinuous change in the derivative of $U(0)$ with respect to $\tilde{J}_0$ at the phase boundary in agreement with (2.25). (There does appear to be a discontinuous change in slope in the $U(0)$ data[18] obtained on finite-dimensional spin-glasses with the bonds restricted to take the values $\pm J_0$.)

More direct evidence of the spin-glass ferromagnet phase boundary is shown in Fig. 6, which compares the observed ground-state magnetization (in a small external field) with the predictions of (2.21) and (2.22). Very large fluctuations in the magnetization, both from sample to sample and between different low-energy states of a given sample, were observed for $1.0 \leq \tilde{J} \leq 1.4$. The predicted phase boundary occurs at $\pi^{1/2} \approx 1.25$, the concentration which in fact shows the largest scatter in $m(0)$. The fluctuations seen in Fig. 6 have a parallel in the recent observation by Vannimenus and Toulouse[19] that (for two-dimensional Ising models with interactions of random sign but uniform magnitude) the energy cost of forming a domain wall around a region of reversed spins becomes very small close to the spin-glass ferromagnet phase boundary. For $\tilde{J} \geq 1.5$, agreement between theory and experiment in Fig. 6 appears excellent.

VI. STATICS FOR $T \neq 0$

To study the properties of an infinite-ranged spin-glass at finite temperatures, we have performed Monte Carlo simulations on four samples of 500 spins, each with $\tilde{J}_0 = 0.0$, taking data at temperatures from zero to $2T_g$. These are described below and compared with the predictions of Sec. II. Within the replica treatment, thermodynamic properties are independent of $\tilde{J}_0$ in the spin-glass phase, since $\tilde{J}_0$ enters Ξ only when multiplied by $m(T)$, which is zero, so study of this one set of samples would seem sufficient to characterize the phase. However, the ground-state energies plotted in Fig. 5 show that some properties may be modified sufficiently close to the ferromagnetic phase boundary.

The calculations were performed by starting each system in its lowest known energy state, and letting each sample evolve for 400 time steps per spin at an ascending series of temperatures. The data plotted in Figs. 6–10 were obtained by averaging over the last 280 time steps at each temperature. The results were compared with averages over larger and smaller time intervals to ensure that equilibrium was reached. The error bars on selected data points in Figs. 6–10 indicate the variation from sample to sample (rms deviation). Some Monte Carlo runs with larger or smaller samples, or much longer averaging times were also performed as checks.

The internal energy $U(T)$ obtained in this way is compared with the prediction of (2.18) and (2.24) in Fig. 7. The discrepancy seen at $T=0$ in Figs. 5 and 6 persists up to roughly 0.5 T_g. The Monte Carlo observations at T_g and above agree with the theory to within numerical uncertainty. The entropy has been obtained by integrating $T^{-1}dU(T)$ up to $2T_g$, using the data in Fig. 7, and matching to the high-temperature limit, $S \sim k\ln 2 - \tilde{J}^2/4kT^2$, of (2.29). Figure 8 compares the computed entropy with the result of substituting the solution to (2.18) into (2.29), and with the TAP low-temperature prediction.

$S(T_g) \approx k(\ln 2 - \frac{1}{4})$ as predicted. The magnetic entropy extracted at temperatures above T_g in the present model is therefore less than is observed experimentally,[1,20] or in calculations on finite dimensional spin-glass models.[2,18] Below ≈ 0.5 T_g, agreement between Monte Carlo and the replica theory becomes poor. The entropy predicted by (2.29) goes negative at $T=0.25$ T_g, while the computed $S(T)$ appears to go to zero with zero slope. The TAP T^2 dependence and coefficient gives a good account of the Monte Carlo entropy for $T \lesssim 0.3T_c$.

This behavior for $S(T)$ is consistent with a specific heat which increases from zero no faster than as T^2. In Fig. 9 are plotted the Monte Carlo results for $C(T)$, obtained by averaging energy fluctuations over the final 280 time steps out of 400 at each temperature. The results agree fairly closely with the theory (2.26) above T_g, but lie consistently below the prediction of (2.27) for $T \lesssim T_g$. Figure 9 suggests that at low temperatures, $C(T)$ tends to zero with zero slope and in qualitative agreement with TAP, but the data are too crude to distinguish $C \propto T^2$ from an activated form, $C \propto \exp(-A/T)$, such as is found in conventional Ising ferromagnets. Similarly, $T\chi(T)$, obtained by averaging

$$T\chi(T) = N^{-1}\left\langle \sum_{ij} S_i S_j \right\rangle, \qquad (5.3)$$

is plotted in Fig. 10. It is found to be ≈ 1 above T_g, and lies below the prediction of (2.28) when $T \lesssim T_g$. The disagreement would be more striking if $\chi(T)$ were plotted. Equation (2.28) gives $\chi(T) \to (2/\pi)^{1/2}$ as $T \to 0$, while the Monte Carlo results are consistent with $\chi(0)=0$.

The differences between simulations and the replica theory in both Figs. 9 and 10 can be summarized by saying that the system appears to be more ordered than the theory would predict. Figure 11, which shows results for $q(T)$ as directly calculated using (2.20) from the simulations, confirms this trend. We see that $q(T)$ lies consistently above the value obtained from (2.18). In both Figs. 10 and 11, the TAP expressions give a good account of the low temperature data.

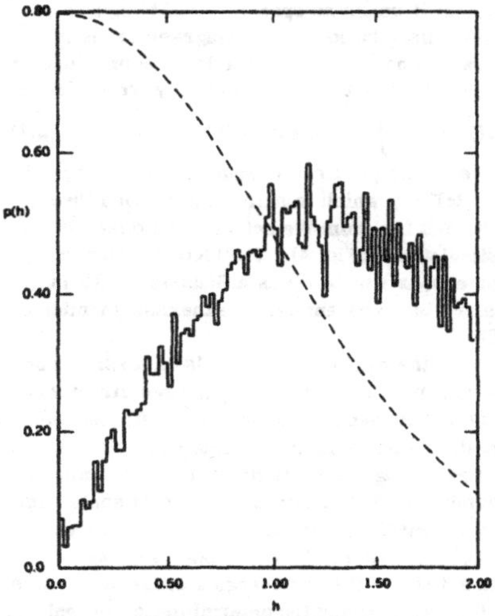

FIG. 12. Distribution of partial exchange energies $p(h)$ as defined in (5.4), averaged over 20 samples of 800 spins each, with $\tilde{J}_0=0.0$. The dashed line indicates the MRF prediction (5.5).

In the course of the ground-state and Monte Carlo calculations, the one-spin excitation energies,

$$h_i \equiv \sum_j J_{ij} S_i S_j, \qquad (5.4)$$

are available and can be used to test TAP's conclusions about $p(h)$ as well as the assumptions of mean-random-field (MRF) treatments of the Marshall-Klein-Brout[21,22] type, which attempt to construct a self-consistent $p(h)$. The distribution $p(h)$ found in the ground states of twenty 800-spin samples, the largest size studied, is plotted in Fig. 12. As TAP predict, $p(h)$ is linear in h for small h. The small positive intercept appears to be an artifact of the finite sample size, and decreases slowly with increasing N.

Mean-random-field theories predict a $p(h)$ which is qualitatively unlike the results in Fig. 12. The major difficulty is that such theories when applied to Ising systems invariably give $p(0) \neq 0$. For the infinite-ranged model, Klein has proposed[23] that the appropriate MRF is

$$p(h) = (2/\pi)^{1/2} \tilde{J} q^{1/2} \exp(-h^2/2\tilde{J}^2 q), \qquad (5.5)$$

arguing that (2.17) and (2.18) should be interpreted as self-consistent equations of the MRF type, e.g.,

$$\langle m \rangle = \int dh p(h) \tanh(\beta h). \qquad (5.6)$$

We plot (5.5) in Fig. 12 as a dashed line, and observe that it bears no special resemblance to the observed distribution. The disagreement is most obvious at small h. Within the MRF approximation one can calculate the internal energy from $p(h)$ by

$$U(T) = -\frac{1}{2}\int dh p(h) h \tanh(\beta h). \tag{5.7}$$

However, this predicts a ground state energy $U(0)/\tilde{J}N = -(2\pi)^{-1/2}$, which is half of the replica theory result, and far from the actual value observed in the simulations (Fig. 4). The identification of the factor $\exp(-x^2)$ in (2.17) as a Gaussian MRF is therefore inconsistent with the thermodynamics of (2.16).

Some of the microscopic details accessible during a computer simulation do not have direct experimental consequences but are nonetheless useful guides to one's intuition. One possibility we explored during the simulations was determining the number of distinct ground states of some samples and searching for the activation pathways which connect them. Two ground states were deemed distinct if their energies differed by more than 0.0001 $\tilde{J}$. Using the deterministic descent procedure described above we first counted the number of distinct local minima found by starting from many randomly generated initial spin configurations in samples with $N = 20$, 30, 40, and 100 spins (20 samples each). The number of ground states reached was: ($N = 20$) 2–6; ($N = 30$) 8.25 $\pm$ 3.5; ($N = 40$) 25.0 $\pm$ 7.5; and ($N = 100$) 415 $\pm$ 19. For the smaller sample sizes, 400 starting configurations were used. For the $N = 100$ samples, after 2000 trials only one or two new local minima were being encountered every 50 trials. Therefore, while in principle the quoted numbers represent lower bounds to the actual number of energy minima, we do not expect large errors.

One might think of the phase space of a spin-glass as consisting of large valleys separated by high activation energies, each valley containing many local minima. The large activation barriers between valleys represent the reversal of large clusters of spins, which will either require high energy or have low entropy (since many spin flips must be coordinated), and thus is an extremely rare event in thermal equilibrium. The local minima are separated by reversals of one or a few spins, which can involve relatively low activation energies. To test this view we started each sample in its lowest-energy state and allowed it to evolve by the usual Monte Carlo dynamics at a constant temperature T while sampling with the descent algorithm for local minima. After each descent, time evolution was resumed from the configuration at temperature T from which the descent was taken. For $T \lesssim 2T_g$, this process yields many fewer distinct minima than did sampling from randomly generated configurations.

To estimate the number of valleys as a function of N, we divide the average number of minima found by searching at $T = T_g$ into the total number of minima found for that sample. The results are: ($N = 20$) ≈ 1; ($N = 30$) ≈ 2.2; ($N = 40$) ≈ 3.5; and ($N = 100$) ≈ 13.0. Both the total number of ground states and the number of valleys appear to increase as some small power of N, rather than as $\propto \exp(N)$, consistent with our finding in Fig. 8 that $S(0) \sim 0$. Since the passes connecting two valleys are apparently inaccessible at T_g and lower temperatures, this picture of the low-temperature states of a spin-glass provides a possible starting point for thinking about remanence and other "glassy" experimental phenomena.

VII. DYNAMICS FOR $T \neq 0$

Study of the dynamical properties of spin-glass models of with infinite-ranged interactions shows additional differences between the spin-glass and conventional ferromagnets. In particular, the behavior of the systems in their low-temperature phases proves to be rather different. The natural dynamics to study for an Ising model is the relaxation process in which spins flip independently but remain in equilibrium with a heat bath at temperature T. Following Glauber,[24] we take the probability for the spin S_i to flip to be given by

$$w(S_i \to -S_i) = \tfrac{1}{2}(1 - S_i \tanh \beta h_i), \tag{5.8}$$

where the unit of time, or the rate at which spin flips are attempted, is set equal to unity, and the change in energy upon flipping the spin S_i is given as $h_i S_i$, with h_i a molecular field. The form (5.8) satisfies detailed balance, and thus assures that the correct equilibrium distribution of microstates is attained. Suzuki and Kubo,[25] in a classic paper, have developed a mean-field theory of relaxation processes governed by (5.8). Their theory has recently been extended to treat the random molecular fields found in the infinite-ranged Ising spin-glass with $\tilde{J}_0 = 0$ by Kinzel and Fischer.[26] To shorten the derivations, we shall follow these two papers closely, although our treatment differs from that of Kinzel and Fischer in an essential step.

Averaging a master equation based on (5.8), Suzuki and Kubo find that the time dependence of the order parameter in the usual ferromagnetic case is governed by

$$\frac{d}{dt}\langle S_i(t)\rangle = -\langle S_i(t)\rangle + \langle \tanh \beta h_i(t)\rangle. \tag{5.9}$$

A mean-field solution to (5.9) is obtained[25] by assuming that unique values of $\langle S(t)\rangle$ and $h(t)$ exist and are proportional

$$\beta h(t) \equiv \lambda \langle S(t) \rangle . \quad (5.10)$$

Then, expanding the tanh in powers of its argument, one obtains

$$\frac{d}{dt} m(t) = -(1-\lambda) m(t) - (\tfrac{1}{3}\lambda^3) m^3(t) , \quad (5.11)$$

valid for temperatures close to or above T_c. Integration of (5.11) is straightforward. For $T > T_c$ the result is

$$m(t, T > T_c) = \frac{(1-\lambda)^{1/2}}{[(1-\lambda+\frac{1}{3}\lambda^3)e^{2(1-\lambda)t} - \frac{1}{3}\lambda^3]^{1/2}} , \quad (5.12)$$

decaying exponentially to zero as $e^{-(1-\lambda)t}$. From (5.10) and (5.12), we identify $\lambda = T_c/T$. As $\lambda \to 1$, the solution exhibits critical slowing-down, changing over to a power law form,

$$m(t, T_c) = (1 + \tfrac{2}{3}t)^{-1/2} \quad (5.13)$$

with the critical exponent $\frac{1}{2}$. Below T_c, exponential relaxation again occurs, this time to a nonzero equilibrium limit, $m(T)$:

$$M(t, T < T_c) = \frac{m(T)}{\{1 - [1 - m(T)^2]e^{-2(\lambda-1)t}\}^{-1/2}} . \quad (5.14)$$

For the spin correlations of interest in the present case, Kinzel and Fischer[26] have used similar arguments to obtain the kinetic equation:

$$\frac{d}{dt}\langle S_i S_j(t)\rangle = -\langle S_i S_j(t)\rangle + \langle S_i \tanh\beta h_j(t)\rangle , \quad (5.15)$$

where S_i here denotes $S_i(t=0)$, and

$$\frac{d}{dt}\langle S_i(t) S_j(t)\rangle = -2\langle S_i(t) S_j(t)\rangle + \langle S_i(t)\tanh\beta h_j(t)\rangle + \langle S_j(t)\tanh\beta h_i(t)\rangle . \quad (5.16)$$

In the steady state, (5.16) determines the equilibrium correlations,

$$\langle S_i S_j\rangle_{eq} = \langle S_i \tanh\beta h_j\rangle_{eq} . \quad (5.17)$$

In a spin-glass, it will not be valid to treat $\langle S_i S_j(t)\rangle$ or $\langle h_j(t)\rangle$ as spatially uniform, since in fact h_j will fluctuate about zero. Also, the expression for $\langle h_j\rangle$ commonly employed when treating static properties in the mean-field approximation, $\langle h_j\rangle = \langle\sum_k J_{jk} S_k\rangle$, gives incorrect results when one attempts to calculate susceptibilities, as Brout and Thomas[27] have pointed out. The remedy was originally noted by Onsager.[28] One must remove from $\langle h_j\rangle$ the field of the extra moment induced on neighboring sites by the presence of the moment $\langle S_j\rangle$. This is

$$\langle \delta S_k\rangle = \chi_{kj}^{(0)}\langle S_j\rangle = \beta J_{kj}\langle S_j\rangle , \quad (5.18)$$

where we have used a simplified expression for $\chi_{kj}^{(0)}$, valid only above the ordering temperature.

Subtracting the effect of (5.18) from $\langle S_k\rangle$ in $\langle h_j\rangle$ leaves

$$\langle h_j\rangle = \sum_k (J_{jk}\langle S_k\rangle - \beta J_{jk}^2\langle S_j\rangle) . \quad (5.19)$$

This expression for $\langle h_j\rangle$ was rediscovered by TAP, who derive it by several independent arguments. They found that without the second term it is impossible to calculate even static properties of a spin-glass correctly in the infinite-ranged limit.[29]

Substituting (5.19) into the kinetic equation (5.15) and keeping only the linear term in the expansion of $\tanh \beta h_j$ leads to

$$\left(1 + \frac{d}{dt} + \sum_k \beta^2 J_{jk}^2\right)\langle S_i S_j(t)\rangle = -\beta \sum_k J_{jk}\langle S_i S_k(t)\rangle . \quad (5.20)$$

Since

$$\sum_k J_{jk}^2 = \tilde{J}^2 + O\left(\frac{1}{N}\right) \quad (5.21)$$

by (2.8b), the left hand side of (5.20) is independent of j. Equation (5.20) can now be solved[26] by expanding the solution in terms of the eigenvectors of the random matrix whose coefficients are J_{jk}. If

$$\beta \sum_k J_{jk} q_k^{(\lambda)} = \lambda q_j^{(\lambda)} , \quad (5.22)$$

the time decay of correlations may be expressed in terms of the independent relaxation frequencies λ.

Thus, if we expand

$$\langle S_i S_j(t)\rangle = N^{-1} \sum_\lambda a_\lambda(t) q_i^{(\lambda)} q_j^{(\lambda)} , \quad (5.23)$$

then

$$a_\lambda(t) = a_\lambda(0) \exp[-(1 + \beta^2\tilde{J}^2 - \lambda)t] . \quad (5.24)$$

The $a_\lambda(0)$ are readily calculated by expanding the equilibrium correlations as in (5.23) and substituting into (5.17). The result is

$$a_\lambda(0) = (1 + \beta^2\tilde{J}^2 - \lambda)^{-1} . \quad (5.25)$$

Note that if the J_{jk} were not random but constant, as in the infinite ranged ferromagnet, the largest eigenvalue of the matrix would be $\tilde{J}_0$, and all other eigenvalues would be zero. Thus, only one mode contributes to the decay of correlations in a ferromagnet in the mean-field limit. In contrast, there is a continuous spectrum of λ for the spin-glass, and its density takes the simple form[30]

$$N^{-1} \sum_\lambda = (2\pi\beta^2\tilde{J}^2)^{-1} \int_{-2\beta\tilde{J}}^{+2\beta\tilde{J}} d\lambda [(2\beta\tilde{J})^2 - \lambda^2]^{1/2} . \quad (5.26)$$

Introducing $x \equiv \lambda/2\beta\bar{J}$ and combining (5.23)–(5.26), we obtain the linearized result, valid for $T > T_c$:

$$\langle S_i S_i(t)\rangle = \exp[-(1-\beta\bar{J})^2 t] \times \int_{-1}^{1} dx \frac{(2/\pi)(1-x^2)^{1/2} e^{-2\beta\bar{J}(1-x)t}}{(1-\beta\bar{J})^2 + 2\beta\bar{J}(1-x)} . \qquad (5.27)$$

In the high-temperature limit, the integral in (5.26) can be evaluated by neglecting terms of order β in the denominator, with the result

$$\langle S_i S_i(t)\rangle \approx e^{-(1-\beta\bar{J})^2 t} e^{-2\beta\bar{J}t} I_1(2\beta t)/\beta t \sim \exp[-(1-\beta\bar{J})^2 t]/[2\pi^{1/2}(\beta t)^{3/2}] . \qquad (5.28)$$

This result, like the ferromagnetic solution (5.12), exhibits critical slowing down as β^{-1} decreases to $\bar{J} = kT_c$. Unlike the ferromagnetic system, the spin-glass has a correlation decay rate proportional to $[(T-T_c)/T]^2$, so the effects of the critical slowing-down should be observable over a wider range of temperatures in the spin-glass than in the ferromagnet.

In studies of dynamical critical phenomena,[31] it is conventional to interpret the characteristic time τ for order parameter relaxation as the ratio

$$\tau \sim \chi^{(e)}(T-T_c)/\Gamma(T-T_c), \qquad (5.29)$$

of the order parameter susceptibility (here, $\chi^{(e)}$) to a friction coefficient Γ. In a mean-field theory, one expects that fluctuations in the order parameter will cause χ to diverge as $(T-T_c)^{-1}$. For spin-glasses, Fisch and Harris[32] have confirmed that this value of the susceptibility exponent is reached at sufficiently high dimensionality, by analysis of series expansions of $\chi^{(e)}$. This implies that $\Gamma \to 0$ as $T \to T_c$ for spin-glasses, which in most mean-field theories, Γ remains finite at T_c. This unusual behavior is consistent, however, with the general picture of the spin-glass transition temperature T_c as the point at which blocking effects (TAP) or "frustration"[33] suddenly set in.

Monte Carlo calculations of $\langle S_i S_j(t)\rangle$, using the Glauber dynamics (5.8) were carried out at several temperatures, both above and below T_c. The results above T_c are compared with the predictions of the linearized theory (5.27) in Figs. 13(a)–13(c). The integral in (5.27) was performed numerically to obtain the plotted curves. Each sample studied in the Monte Carlo simulations had 800 spins, and two samples were considered in each of Figs. 13(a)–13(c). Before beginning the collection of data on $\langle S_i S_j(t)\rangle$, 100–200 time steps per spin were taken to allow the samples to come to equilibrium at the desired temperature. In each of the three cases, the observed decay of correlations was slower than is predicted by the linearized theory,

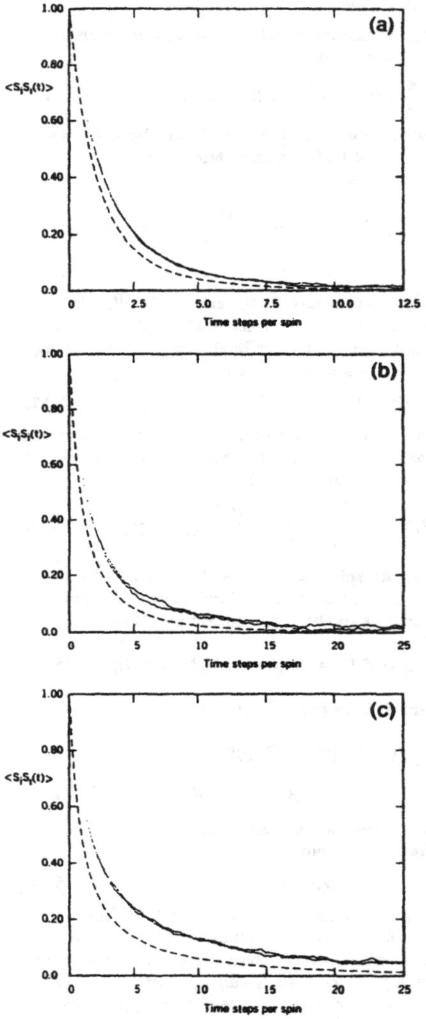

FIG. 13. Dashed lines indicate the decay of spin correlations predicted by the linearized molecular-field theory in (5.27). The closely spaced dots are Monte Carlo data for two samples with 800 spins. Cases shown are (a) $T = 2.0T_c$, (b) $T = 1.5T_c$, and (c) $T = 1.25T_c$.

but the error seen at the highest temperature, $T = 2.0T_c$ [Fig. 13(a)] is fairly small. Agreement is still fair at $T = 1.5T_c$ [Fig. 13(b)], but begins to become poor at $T = 1.25T_c$ [Fig. 13(c)].

Evaluating (5.27) at T_c, where $\beta\bar{J} = 1$, gives[26]

$\langle S_i S_i(t)\rangle \sim t^{-1/2}$, the same asymptotic dependence as seen in the mean-field result (5.13). The similarity is probably coincidental, since (5.27) leaves out the nonlinear restoring term in (5.11) which is the source of the $t^{-1/2}$ limiting behavior in (5.13). It is difficult to add such nonlinear terms to the treatment leading up to (5.27) since these will mix the modes which have been treated as independent. Ma and Rudnick[34] have recently studied spin-glass dynamics in a nonlinear mean-field approximation. Their model, a scalar Landau-Ginzberg-Wilson Hamiltonian with a random molecular field, may be applicable to the present simulations. They find $\langle S_i S_i(t)\rangle \sim t^{-1/2}$ for long times, not only at T_c but also at all lower temperatures. However, their calculation gives $\tau \sim (T - T_c)^{-1}$ above T_c, in contrast to (5.28).

Our Monte Carlo results at and below T_c are consistent with Ma and Rudnick's prediction. At temperatures below T_c, $\langle S_i S_i(t)\rangle$ does not decay exponentially as it would in a ferromagnet [see (5.14)]. Data for $T = T_c$, $0.8T_c$, and $0.6T_c$ are shown in Fig. 14. A plot of $[\langle S_i S_i(t)\rangle^{-2} - 1]$ against t, using the $T = T_c$ data of Fig. 14, gave a straight line for the first 25 time steps per spin. At longer times, the statistical fluctuations due to the finite number of spins swamped any further decrease in $\langle S_i S_i(t)\rangle$. Thus, the data taken at T_c can be described as $\approx (1 + \alpha t)^{-1/2}$ where α is a phenomenological constant which turns out to be not too different from the $\frac{2}{3}$ obtained in (5.13).

Below T_c, an extension of (5.13),

$$\langle S_i S_i(t)\rangle = (1 - q_0)/(1 + \alpha t)^{1/2} + q_0, \qquad (5.30)$$

gives an excellent fit to the data. The heavy circles in Fig. 14 indicate fits of the form (5.30) using $q_0 = 0$, 0.22, and 0.49 for $T = T_c$, $0.8T_c$, and $0.6T_c$, respectively, and values of α between 1.0 and 0.85. These values of q_0 were chosen to agree with the long-time-averaged values of the EA order parameter expected from theory [see (2.18)] and as observed and shown in Fig. 11.

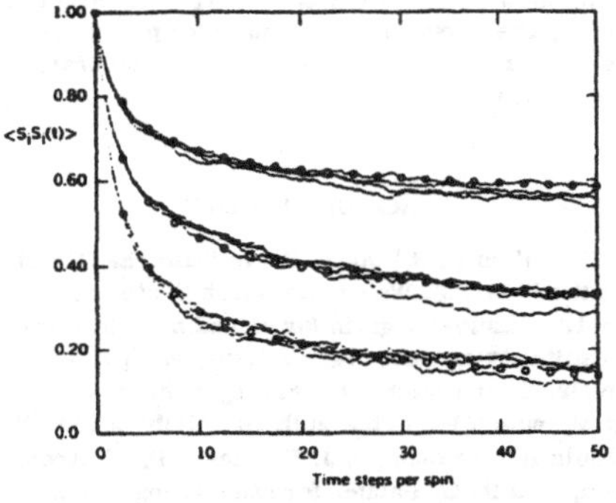

FIG. 14. Monte Carlo data on the time dependence of spin correlations are plotted as in Fig. 13 for three temperatures: (top to bottom) $T = 1.0\,T_c$, $0.8\,T_c$, and $0.6\,T_c$. Four samples of 800 spins each were used at each temperature. The circles indicate the phenomenological expression (5.30), embodying a $t^{-1/2}$ decay.

No choice of parameters in an expression of the form (5.14) gives an adequate fit to the data $T = 0.8$ or $0.6T_c$. If the known long-time limiting values of $q(T)$ are used and the initial slope is taken as a free parameter, (5.14) reaches its limiting value too rapidly. If we force agreement with the data for 25–50 time steps per spin, (5.14) gives too small an initial rate of decay.

Binder[2] has described a similar slow decay of the EA order parameter in Monte Carlo studies in two- and three-dimensional Ising models with random exchange interactions governed by a Gaussian distribution, but he did not assign a functional form to the decay. Since the EA order parameter does not relax rapidly to its equilibrium value at temperatures $\ll T_c$ as the magnetization, governed by (5.14), would for a ferromagnet, Binder has questioned whether some other order parameter might be constructed with more conventional behavior.

We do not think this likely. In the infinite-ranged spin-glasses, we have demonstrated that q is the order parameter which uniquely characterizes the spin-glass phase. The Monte Carlo simulations show that even in this mean-field limit, $q(t)$ exhibits its nonexponential relaxation at all $T \leq T_c$. The linearized analysis of dynamics suggests that the existence of a continuous spectrum of relaxation rates extending to zero is sufficient to introduce power law decays of correlations. Thus, the unusual time dependence of the EA order parameter seems to be a signature of the spin-glass state.

VIII. CONCLUSIONS

In this paper we have investigated the properties of spin-glass models with infinite-ranged random exchange interactions, both analytically and by means of computer simulation experiments. The discussion has been presented mainly, but not exclusively, in terms of Ising spins. Consideration has been given to four theoretical approaches: (i) a replication procedure[3,5] in which the random system is mapped at the outset into a limiting case of a fictitious pure system with extra spin labels and higher-order interactions, (ii) high-temperature series expansion,[8] (iii) a mean-field theory allowing for a different mean field on each site and deferring all averaging to the end of the calculation,[8] and (iv) a MRF approximation.[21–23] Only the first three of these approaches claim any degree of rigor, the fourth being heuristic.

In this paper we give a detailed derivation of the first approach, and briefly review the others. The second and third theories are discussed at length in Ref. 8. Computer experiments presented are of three kinds: (a) investigation of the structure of the low-energy states of the system, (b) Monte Carlo simulation of the equilibrium thermodynamics, and (c) Monte Carlo simulation of dynamics.

The replica procedure has been solved subject only to making an interchange of limits (the thermodynamic limit, $N\rightarrow\infty$, and a limit on the number of replicas, $n\rightarrow 0$). The procedure predicts a phase diagram with two types of magnetically ordered phases, ferromagnet and spin-glass. The spin-glass to paramagnet transition manifests itself by cusps in the zero-field susceptibility (which is rounded in the presence of an external field) and in the specific heat. At all but the lowest temperatures, the predicted thermodynamic functions are physical but at low temperatures the procedure yields a finite negative entropy.

The high-temperature series expansion can be summed exactly in the thermodynamic limit and predicts phase transitions from paramagnet to ordered phase at the same temperatures as found by the replica method. In the paramagnetic phase, analysis of the high-temperature series in zero magnetic field confirms all the corresponding equilibrium thermodynamic predictions of the replica procedure. The third theoretical approach[8] (TAP) has been studied in detail only for a particular case of the general model which can be treated by the replica procedure, namely the case with mean exchange and external field equal to zero. Approximate solutions within this approach, believed reliable in the thermodynamic limit for temperatures above and close to the spin-glass transition temperature T_c are in complete accord with the replica results above T_c and agree to the leading order in $(T_c - T)$ immediately below the transition. At low temperatures, however, the TAP procedure when coupled with an ansatz for the solution[8] (based upon limited computer simulations and physical intuition) leads to results somewhat different from those of the replica method. In particular, it does not exhibit the unphysical negative entropy.

Monte Carlo simulations were performed on samples of up to 800 spins with infinite-ranged interactions to test the predictions of the various theories and to provide quasimicroscopic information about the low-temperature phase of a spin-glass. The ground-state energy predicted by the replica method lies slightly lower than the Monte Carlo results, the difference exceeding the probable error in the simulations. The discrepancy between the predicted internal energy and Monte Carlo observations becomes insignificant at temperatures greater than $0.5T_c$, and at all temperatures when the mean value of the exchange interactions was sufficiently great for the system to remain ferromagnetic at the lowest temperatures.

The entropy was determined by integrating the internal energy found by simulation. To within the accuracy of simulations $S(T=0)=0$. The absolute difference between the replica theory prediction for $S(T)$ and observation becomes small for $T \geq 0.5T_c$. The TAP expression for the limiting behavior of $S(T)$ at low temperatures gives good agreement with the Monte Carlo results. Simulations of the specific heat and the susceptibility are also in good agreement with the TAP ansatz.

An attempt was made to quantify the degeneracy of the spin-glass ground state by counting the number of distinct local energy minima. The minima were found to occur in groups, which may be thought of as large "valleys" in phase space. Both the number of minima and the number of valleys appears to increase with N, the number of spins, as some small power of N. They therefore do not give rise to a finite entropy at $T=0$.

The distribution of exchange fields in the spin-glass ground state was obtained in the course of the simulations and compared with the distribution assumed in the mean-random-field approximation. The two distributions prove to be very different.

Linearized random kinetic equations for the decay of spin correlations above T_c are derived and solved. They give good agreement with Monte Carlo studies of spin relaxation for $T \gtrsim 1.5T_c$. At and below T_c the decay of $\langle S_i S_i(t)\rangle$ to its long time limit, the Edwards-Anderson order parameter q is slower than exponential. It can be accurately described by a $t^{-1/2}$ law.

ACKNOWLEDGMENTS

One of us (D.S.) would like to thank the Directors of the IBM T.J. Watson Research Center and the Institut Laue-Langevin for extending to him the hospitality of their organizations, and particularly to Dr. S. von Molnár for arranging his visit to IBM, which led to the authors' collaboration. We would like to thank D. J. Thouless, P. W. Anderson, and R. G. Palmer for stimulating discussions of their work prior to its publication. Conversations with T. C. Lubensky, E. Pytte, and J. Rudnick, and preprints received from G. Toulouse, A. P. Young, and K. Binder are also gratefully acknowledged.

APPENDIX A

The matrix Λ of Eq. (2.13) is conveniently expressed in the basis $(x^\alpha, y^{\alpha\beta})$, the general member of which we shall denote by z^ν, where ν runs from 1 to $\frac{1}{2}n(n+1)$. The $\mu\nu$ matrix element of Λ_n is $\partial^2 g(\{z\})/\partial z^\mu \partial z^\nu$ where $-N g(\{z\})$ is the exponent of the exponential integrand of (2.10). The derivatives are evaluated with all x^α, y^α set equal to x_n, y_n, and expressed in terms of m_n, q_n using (2.12). Thus,

$$\frac{\partial^2 g}{\partial x^\alpha \partial x^\beta} = \delta_{\alpha\beta}\left(1-\frac{\bar{J}_0}{kT}\right) -(1-\delta_{\alpha\beta})\left(\frac{\bar{J}_0}{kT}\right)q_n + \left(\frac{\bar{J}_0}{kT}\right)m_n^2, \tag{A1}$$

$$\begin{aligned}\frac{\partial^2 g}{\partial y^{(\alpha\beta)}\partial y^{(\gamma\delta)}} &= \delta_{(\alpha\beta)(\gamma\delta)}\left[1-\left(\frac{\tilde{J}}{kT}\right)^2\right]-(1-\delta_{(\alpha\beta)(\gamma\delta)})\\ &\quad\times\left(\frac{\tilde{J}}{kT}\right)^2 q_n(\delta_{\alpha\gamma}+\delta_{\alpha\delta}+\delta_{\beta\gamma}+\delta_{\beta\delta})\\ &\quad-(1-\delta_{\alpha\gamma})(1-\delta_{\alpha\delta})(1-\delta_{\beta\gamma})(1-\delta_{\beta\delta})\\ &\quad\times\left(\frac{\tilde{J}}{kT}\right)^2\langle S^\alpha S^\beta S^\gamma S^\delta\rangle_n + \left(\frac{\tilde{J}}{kT}\right)^2 q_n^2,\end{aligned} \tag{A2}$$

$$\begin{aligned}\frac{\partial^2 g}{\partial y^{(\alpha\beta)}\partial x^\gamma} &= -\left(\frac{\bar{J}_0}{kT}\right)^{1/2}\left(\frac{\tilde{J}}{kT}\right)[(\delta_{\alpha\gamma}+\delta_{\beta\gamma})m_n\\ &\quad-(1-\delta_{\alpha\gamma}-\delta_{\beta\gamma})\langle S^\alpha S^\beta S^\gamma\rangle_n + m_n q_n],\end{aligned} \tag{A3}$$

where

$$\langle S^\alpha S^\beta S^\gamma\rangle_n = \frac{\int dz(2\pi)^{-1/2}\exp(-\frac{1}{2}z^2)\tanh^3\Xi\cosh^n\Xi}{\int dz(2\pi)^{-1/2}\exp(-\frac{1}{2}z^2)\cosh^n\Xi}, \tag{A4}$$

and

$$\langle S^\alpha S^\beta S^\gamma S^\delta\rangle_n = \frac{\int \frac{dz}{(2\pi)^{1/2}}\exp(-\frac{1}{2}z^2)\tanh^4\Xi\cosh^n\Xi}{\int \frac{dz}{(2\pi)^{1/2}}\exp(-\frac{1}{2}z^2)\cosh^n\Xi} \tag{A5}$$

APPENDIX B

In this Appendix, we demonstrate explicitly that

$$m = \langle\langle S_i\rangle\rangle_{\rm av}, \tag{2.19}$$

$$q = \langle\langle S_i\rangle^2\rangle_{\rm av}, \tag{2.20}$$

where m and q are the averages introduced in Sec. II, and defined by

$$m \equiv \lim_{n\to 0}\langle S_i^\alpha\rangle_n \operatorname{Tr}_n \exp(-\beta\mathcal{H}_n^{\rm eff}), \tag{B1}$$

$$q \equiv \lim_{n\to 0}\langle S_i^\alpha S_i^\beta\rangle_n \operatorname{Tr}_n \exp(-\beta\mathcal{H}_n^{\rm eff});\quad \alpha\neq\beta, \tag{B2}$$

where $\langle\ \rangle_n$ denotes an average in the n-replica system characterized by the effective Hamiltonian defined by

$$\exp(-\beta\mathcal{H}_n^{\rm eff}) \equiv \int\prod_{ij} dJ_{ij}\, P(J_{ij})\exp\left(-\beta\sum_\alpha \mathcal{H}^\alpha\right).$$

This result is true for any distribution $P(J_{ij})$, and not restricted to the infinite-ranged models discussed in this paper.

The thermal average of the spin at any site i is given by

$$\langle S_i\rangle = [\operatorname{Tr} S_i \exp(-\beta\mathcal{H})]/[\operatorname{Tr}\exp(-\beta\mathcal{H})]. \tag{B3}$$

Thus,

$$\begin{aligned}\sum_i\langle S_i\rangle &= \beta^{-1}\left[\frac{\partial}{\partial h}\ln\operatorname{Tr}\exp\left(-\beta\mathcal{H}+\beta h\sum_i S_i\right)\right]_{h=0}\\ &= \beta^{-1}\frac{\partial}{\partial h}\lim_{n\to 0} n^{-1}\\ &\quad\times\left[\operatorname{Tr}_n\exp\left(-\beta\sum_\alpha\mathcal{H}^\alpha+\beta h\sum_{i,\alpha}S_i^\alpha\right)-1\right]_{h=0}.\end{aligned} \tag{B4}$$

Averaging over J_{ij} we therefore find

$$\langle\langle S_i\rangle\rangle_{\rm av} = (N\beta)^{-1}\lim_{n\to 0} n^{-1}\frac{\partial}{\partial h}\times\left[\operatorname{Tr}_n\exp\left(-\beta\mathcal{H}_n^{\rm eff}+\beta h\sum_{i,\alpha}S_i^\alpha\right)-1\right]_{h=0}, \tag{B5}$$

where the effective Hamiltonian was given in (2.6):

$$\mathcal{H}_n^{\rm eff} = -\sum_{(ij)}\left(\sum_\alpha S_i^\alpha S_j^\alpha J_0 + \sum_{\alpha\beta} S_i^\alpha S_j^\alpha S_i^\beta S_j^\beta J^2/2kT\right) - H\sum_{i,\alpha}S_i^\alpha\ . \tag{B6}$$

Thus,

$$\begin{aligned}\langle\langle S_i\rangle\rangle_{\rm av} &= \lim_{n\to 0}\operatorname{Tr}_n(nN)^{-1}\sum_{i,\alpha}S_i^\alpha\exp(-\beta\mathcal{H}_n^{\rm eff})\\ &= \lim_{n\to 0}\langle S_i^\alpha\rangle_n \operatorname{Tr}\exp(-\beta\mathcal{H}_n^{\rm eff}) = m\ .\end{aligned} \tag{B7}$$

Similarly,

$$\sum_i\langle S_i\rangle^2 = \beta^{-1}\frac{\partial}{\partial h_{\mu\nu}}\ln\operatorname{Tr}_{\mu\nu}\exp\left(-\beta(\mathcal{H}^\mu+\mathcal{H}^\nu)+\beta h_{\mu\nu}\sum_i S_i^\mu S_i^\nu\right), \tag{B8}$$

where μ, ν label distinct replicas,

$$\sum_i \langle S_i\rangle^2 = \beta^{-1}\frac{\partial}{\partial h_{\mu\nu}} \lim_{n\to 0} n^{-1}\,\mathrm{Tr}_{2n} \times \exp\Big(-\beta \sum_\alpha (\mathcal{H}^{\mu\alpha}+\mathcal{H}^{\nu\alpha}) + \beta h_{\mu\nu}\sum_{i,\alpha} S_i^{\mu\alpha}S_i^{\nu\alpha}\Big). \tag{B9}$$

Averaging over J_{ij} and explicitly performing the $h_{\mu\nu}$ differentiation we obtain

$$\begin{aligned}\langle\langle S_i\rangle^2\rangle_d &= \lim_{n\to 0}\mathrm{Tr}_{2n}(nN)^{-1}\sum_{i,\alpha} S_i^{\mu\alpha}S_i^{\nu\alpha}\exp(-\beta\mathcal{H}_{2n}^{\mathrm{eff}})\\ &= \lim_{n\to 0}\langle S_i^\alpha S_i^\beta\rangle_{2n}\,\mathrm{Tr}_{2n}\exp(-\beta\mathcal{H}_{2n}^{\mathrm{eff}}) \quad (\alpha\neq\beta)\\ &= \lim_{n\to 0} q_{2n}\,\mathrm{Tr}_{2n}\exp(-\beta\mathcal{H}_{2n}^{\mathrm{eff}})\\ &= \lim_{n\to 0} q_n\,\mathrm{Tr}_n\exp(-\beta\mathcal{H}_n^{\mathrm{eff}}) = q\,. \end{aligned} \tag{B12}$$

---

*Permanent address.

†Present address.

[1]For reviews of experimental work, see V. Cannella and J. A. Mydosh, AIP Conf. Proc. 18, 651 (1974); J. A. Mydosh, *Amorphous Magnetism II*, edited by R. A. Levy and R. Hasegawa (Plenum, New York, 1977), p. 73.

[2]For reviews of theoretical work, see K. H. Fisher, Physica (Utr.) 86–88, 813 (1976); K. Binder, *Festkoerperprobleme XVII (Advances in Solid State Physics)*, edited by J. Treusch (Vieweg, Braunschweig, 1977), p. 55; D. Sherrington, AIP Conf. Proc. 29, 224 (1975); and report for Aussois Conference on Glasses and Spin Glasses, March, 1977 (unpublished).

[3]S. F. Edwards and P. W. Anderson, J. Phys. F 5, 965 (1975). We refer to this paper as EA.

[4]K. H. Fisher, Phys. Rev. Lett. 34, 1438 (1975).

[5]D. Sherrington and S. Kirkpatrick, Phys. Rev. Lett. 35, 1792 (1975).

[6]D. Sherrington and B. W. Southern, J. Phys. F 5, L49 (1975).

[7]See, e.g., H. E. Stanley, *Phase Transitions and Critical Phenomena* (Oxford, U. P., New York, 1971), p. 91.

[8]D. J. Thouless, P. W. Anderson, and R. G. Palmer, Philos. Mag. 35, 593 (1977). We refer to this paper as TAP.

[9]M. Kac, Trondheim Theoretical Physics Seminar, Nordita Publ. No. 286, 1968 (unpublished); and T.-F. Lin, J. Math. Phys. 11, 1584 (1970).

[10]S. F. Edwards, *Proceedings of the Third International Conference on Amorphous Materials, 1970*, edited by R. W. Douglass and B. Ellis (Wiley, New York, 1972), p. 279; *Polymer Networks*, edited by A. J. Chompff and S. Newman (Plenum, New York, 1971), p. 83.

[11]G. Grinstein, Ph.D. thesis (Harvard University, 1974) (unpublished); G. Grinstein and A. H. Luther, Phys. Rev. B 13, 1329 (1976).

[12]V. J. Emery, Phys. Rev. B 11, 239 (1975).

[13]A. B. Harris, T. C. Lubensky, and J.-H. Chen, Phys. Rev. Lett. 36, 415 (1976); J.-H. Chen and T. C. Lubensky, Phys. Rev. B 16, 2106 (1977) [see, also R. G. Priest and T. C. Lubensky *ibid*. 13, 4159 (1976)] have used the replica method as the basis for a renormalization group calculation for a spin-glass in $6-\epsilon$ dimensions. Their free energy differs from (2.16), but is also a maximum, not a minimum, at the physical solution.

[14]J. M. Kosterlitz, D. J. Thouless, and R. C. Jones, Phys. Rev. Lett. 36, 1217 (1976).

[15]Diagrams with more complex topology are of order N^{-1} or less. Diagrams with mixed single and double bonds all fall into this category.

[16]S. Kirkpatrick and C. M. Varma, Solid State Commun. (to be published).

[17]E. L. Lawlor, *Combinatorial Optimization* (Holt, Rinehart and Winston, New York, 1976).

[18]S. Kirkpatrick, Phys. Rev. B 16, 4630 (1977).

[19]J. Vannimenus and G. Toulouse, J. Phys. C 10, L537 (1977).

[20]J. M. D. Coey and S. von Molnar, Solid State Commun. 24, 167 (1977).

[21]W. Marshall, Phys. Rev. 118, 1519 (1960).

[22]M. W. Klein and R. Brout, Phys. Rev. 132, 124 (1963).

[23]M. W. Klein, Phys. Rev. B 14, 5008 (1976).

[24]R. J. Glauber, J. Math. Phys. 4, 294 (1963).

[25]M. Suzuki and R. Kubo, J. Phys. Soc. Jpn. 24, 51 (1968).

[26]W. Kinzel and K. H. Fischer, Solid State Commun. 23, 687 (1977).

[27]R. Brout and H. Thomas, Physics 3, 317 (1967).

[28]W. Kinzel and K. H. Fischer (see note added in proof to Ref.

[29]Kinzel and Fischer (see note added in proof to Ref. 26) have observed that the term $(\beta\tilde{J})^2$ on the left-hand side of the kinetic equation (5.20) is the modification of their formalism needed to obtain the correct T_c. They were guided to this observation by the appearance of such a $(\beta\tilde{J})^2$ correction term in the spherical model calculation of Ref. 14. Similarly, B. Brout and H. Thomas (Ref. 27) note that the spherical model differs from the conventional mean-field theory by adding just those terms necessary to satisfy the fluctuation-dissipation theorem. The physical content of this otherwise mysterious correction is evident from the discussion accompanying (5.18) and (5.19).

[30]M. L. Mehta, *Random Matrices and the Statistical Theory of Energy Levels* (Academic, New York, 1967), p. 240; S. F. Edwards and R. C. Jones, J. Phys. A 9, 1595 (1976).

[31]P. Hohenberg and B. I. Halperin, Revs. Mod. Phys. 49, 435 (1977).

[32]R. Fisch and A. B. Harris, Phys. Rev. Lett. 38, 785 (1977).

[33]G. Toulouse, Commun. Phys. 2, 115 (1977).

[34]S.-K. Ma and J. Rudnick, Phys. Rev. Lett. 40, 589 (1978).

J. Phys. A: Math. Gen., Vol. 11, No. 5, 1978. Printed in Great Britain

Stability of the Sherrington–Kirkpatrick solution of a spin glass model

J R L de Almeida†‡ and D J Thouless

Department of Mathematical Physics, University of Birmingham, Birmingham B15 2TT, UK

Received 1 December 1977

Abstract. The stationary point used by Sherrington and Kirkpatrick in their evaluation of the free energy of a spin glass by the method of steepest descent is examined carefully. It is found that, although this point is a maximum of the integrand at high temperatures, it is not a maximum in the spin glass phase nor in the ferromagnetic phase at low temperatures. The instability persists in the presence of a magnetic field. Results are given for the limit of stability both for a partly ferromagnetic interaction in the absence of an external field and for a purely random interaction in the presence of a field.

1. Introduction

Experiments on dilute alloys of magnetic impurities in a non-magnetic metal and on other disordered or amorphous magnetic systems (see Mydosh 1977 for a review) have led to the suggestion that there is a spin glass phase of such systems, in which the spins are aligned in fixed but random directions below some critical temperature T_c. Edwards and Anderson (1975) developed a theory of the spin glass which explained some of the observed features of the spin glass phase, such as the cusp in magnetic susceptibility at T_c, but left some other features, such as the extreme sensitivity of this cusp to field strength and frequency, unexplained. For a system with energy of the form

$$H = -\sum_{(ij)} J_{ij} S_i S_j, \qquad S_i = \pm 1 \tag{1}$$

where the J_{ij} are distributed randomly, the theory is essentially a mean field theory in which the quantity

$$q = \langle \langle S_i \rangle_{\text{th}}^2 \rangle_J \tag{2}$$

is studied. The thermal average of S_i at a given site is squared, and the average of this square over the distribution of the J_{ij} gives q. This work also makes use of the replica trick, in which the logarithm of the partition function Z, whose average must be calculated to find the free energy is evaluated by finding the partition function of n replicas of the system, which is Z^n, and then using the limiting process

$$\ln Z = \lim_{n \to 0} n^{-1}(Z^n - 1). \tag{3}$$

† On leave from Departmento de Fisica da U.F.Pe, Recife, Brasil.
‡ Partially supported by CNPq (Brazilian Government).

0305-4770/78/0005-0983$01.00

Sherrington and Kirkpatrick (1975), which we refer to as SK, introduced a model for which one might expect mean field theory to be exact, since each of the N spins in the system was taken to interact with all the other spins. The J_{ij} were taken to be independent random variables with a common mean J_0/N and a common variance J^2/N. This choice for the N dependence of the mean and variance ensures that the energy per spin remains finite when N becomes infinite. This model appeared to be exactly solvable when the replica trick was used, and the solution reproduced many of the desirable features of the Edwards–Anderson model. However, it gave a negative entropy at low temperatures, which cannot be correct for a model with discrete Ising model spins. The authors suggested that this resulted from an improper interchange of the two limits $n \to 0$ and $N \to \infty$, and that the consequences were confined to low temperatures.

Subsequent work on this model by Thouless *et al* (1977), which avoided the replica trick confirmed that the SK solution was correct above and near T_c, but found very different behaviour at low temperatures. The differences all seem to be rdlated to the fact that the mean field on a particular spin has a normal distribution in the SK solution, but should have a different probability distribution at low temperatures. In that paper no general solution was found for intermediate temperatures, and, although the magnetic susceptibility was found, no study was made of the effect of a non-zero magnetic field.

The SK model is worth further study both because of the information it may give about the hazards of the replica trick and because a sound mean field theory is a useful starting point for more detailed theories. Harris *et al* (1976) have used renormalisation group methods for this type of problem, making use of the fact that the behaviour is supposed to be classical in six-dimensional space, and so it would be useful to understand what is contained in the 'classical' theory. Fisch and Harris (1977) have used power series methods to study the behaviour of q in a similar model in $6-\epsilon$ dimensions, and find an anomalous behaviour for 4 dimensions. This work also calls in question whether q is the right order parameter to study.

In this paper we show that there is an apparent instability in the SK solution. Not only is this instability present in the absence of an external field for all temperatures below T_c, but it exists for the non-zero field, where the SK solution is analytic. We can therefore trace out the spin glass phase boundary as a function of magnetic field. The instability also exists at low temperatures in the ferromagnetic phase, if J is non-zero. The nature of the instability suggests that the symmetry between replicas should be broken in the spin glass phase, but we have not been able to exploit this idea to calculate the properties of the spin glass phase.

2. Instability of the Sherrington–Kirkpatrick solution

By using the replica trick, Sherrington and Kirkpatrick (1975) were able to perform the averaging over the J_{ij} and the sum over sites and so express the free energy in the form

$$F = -kT \lim_{n\to 0} n^{-1} \left[\exp\left(\frac{J^2 Nn}{4(kT)^2}\right) \int \left[\prod_{(\alpha\beta)} \left(\frac{N}{2\pi}\right)^{1/2} \mathrm{d}y^{(\alpha\beta)}\right) \left(\prod_{\alpha} \left(\frac{N}{2\pi}\right)^{1/2} \mathrm{d}x^{\alpha} \right]$$

$$\times \exp\left\{ -N \sum_{(\alpha\beta)} \frac{1}{2} y^{(\alpha\beta)2} - N \sum_{\alpha} \frac{1}{2} (x^{\alpha})^2 \right.$$

$$+N \ln \operatorname{Tr} \exp\left[\frac{J}{kT}\sum_{(\alpha\beta)} y^{(\alpha\beta)} S^\alpha S^\beta + \left(\frac{J_0}{kT}\right)^{1/2} \sum_\alpha x^\alpha s^\alpha\right]\Bigg\} - 1\Bigg] \tag{4}$$

where the indices α, β run from 1 to n and refer to the replica number, $(\alpha\beta)$ denotes distinct pairs with $\alpha \neq \beta$, and the trace is over the 2^n values of the $S^\alpha = \pm 1$. If it is assumed that the thermodynamic limit $N \to \infty$ can be taken before the limit $n \to 0$, then the method of steepest descent can be used, and the value of the integral over the $y^{(\alpha\beta)}$ and x^α is equal to the value of the integrand where the exponent has its maximum value. This leads to the result

$$\frac{F}{N} = -\frac{J^2}{4kT} - kT \lim_{n\to 0} n^{-1} \max\Bigg\{-\sum_{(\alpha\beta)} \frac{1}{2} y^{(\alpha\beta)2} - \sum_\alpha \frac{1}{2} x^{(\alpha)2}$$
$$+ \ln \operatorname{Tr} \exp\left[\frac{J}{kT}\sum_{(\alpha\beta)} y^{(\alpha\beta)} S^\alpha S^\beta + \left(\frac{J_0}{kT}\right)^{1/2} \sum_\alpha x^\alpha S^\alpha\right]\Bigg\}. \tag{5}$$

There is a stationary point of the expression in braces with all x^α and $y^{(\alpha\beta)}$ zero, and this solution gives the paramagnetic phase. There may also be non-trivial stationary points given by

$$y^{(\alpha\beta)} = \frac{J}{kT}\langle S^\alpha S^\beta\rangle = \frac{Jq}{kT},$$
$$x^\alpha = \left(\frac{J_0}{kT}\right)^{1/2} \langle S^\alpha\rangle = \left(\frac{J_0}{kT}\right)^{1/2} m \tag{6}$$

where q has the same meaning as in equation (2) and m is the magnetisation, where

$$q = \frac{1}{(2\pi)^{1/2}} \int \mathrm{d}z\, \mathrm{e}^{-\frac{1}{2}z^2} \tanh^2\left(\frac{Jq^{1/2}}{kT} z + \frac{J_0 m}{kT}\right),$$
$$m = \frac{1}{(2\pi)^{1/2}} \int \mathrm{d}z\, \mathrm{e}^{-\frac{1}{2}z^2} \tanh\left(\frac{Jq^{1/2}}{kT} z + \frac{J_0 m}{kT}\right). \tag{7}$$

In the spin glass phase q is non-zero and m is zero, and this solution exists for $T < J/k$, while in the ferromagnetic phase m is also non-zero.

To examine the question of whether equations (6) and (7) give a maximum of the expression (5) we write

$$x^\alpha = x + \epsilon^\alpha, \qquad y^{(\alpha\beta)} = y + \eta^{(\alpha\beta)} \tag{8}$$

where x and y are the values given in equation (6), and then (5) can be expanded up to second order in the quantities ϵ^α and $\eta^{(\alpha\beta)}$. To this order the deviation of the expression in braces from its stationary value is equal to $-\frac{1}{2}\Delta$, where

$$\Delta = \sum_{\alpha,\beta}\left(\delta_{\alpha,\beta} - \frac{J_0}{kT}(\langle S^\alpha S^\beta\rangle - \langle S^\alpha\rangle\langle S^\beta\rangle)\right)\epsilon^\alpha\epsilon^\beta$$
$$+ \frac{2JJ_0^{1/2}}{(kT)^{3/2}} \sum_{\delta,(\alpha\beta)} (\langle S^\delta\rangle\langle S^\alpha S^\beta\rangle - \langle S^\alpha S^\beta S^\delta\rangle)\epsilon^\delta \eta^{(\alpha\beta)}$$
$$+ \sum_{(\alpha\beta),(\gamma\delta)} \left[\delta_{(\alpha\beta)(\gamma\delta)} - \left(\frac{J}{kT}\right)^2 (\langle S^\alpha S^\beta S^\gamma S^\delta\rangle - \langle S^\alpha S^\beta\rangle\langle S^\gamma S^\delta\rangle)\right]\eta^{(\alpha\beta)}\eta^{(\gamma\delta)}. \tag{9}$$

This quadratic form should be positive definite for a stable solution of the problem.

The matrix G associated with this quadratic form has seven different types of matrix element. The coefficients of the ϵ have the form

$$G_{\alpha\alpha}=1-(J_0/kT)(1-\langle S^\alpha\rangle^2)=A,$$
$$G_{\alpha\beta}=-(J_0/kT)(\langle S^\alpha S^\beta\rangle-\langle S^\alpha\rangle^2)=B. \tag{10}$$

The coefficients of the η have the form

$$G_{(\alpha\beta)(\alpha\beta)}=1-(J/kT)^2(1-\langle S^\alpha S^\beta\rangle^2)=P,$$
$$G_{(\alpha\beta)(\alpha\gamma)}=-(J/kT)^2(\langle S^\beta S^\gamma\rangle-\langle S^\alpha S^\beta\rangle^2)=Q, \tag{11}$$
$$G_{(\alpha\beta)(\gamma\delta)}=-(J/kT)^2(\langle S^\alpha S^\beta S^\gamma S^\delta\rangle-\langle S^\alpha S^\beta\rangle^2)=R.$$

The cross terms have the form

$$G_{\alpha(\alpha\beta)}=G_{(\alpha\beta)\alpha}=JJ_0^{1/2}(kT)^{-\frac{3}{2}}(\langle S^\alpha\rangle\langle S^\alpha S^\beta\rangle-\langle S^\beta\rangle)=C,$$
$$G_{\gamma(\alpha\beta)}=G_{(\alpha\beta)\gamma}=JJ_0^{1/2}(kT)^{-\frac{3}{2}}(\langle S^\gamma\rangle\langle S^\alpha S^\beta\rangle-\langle S^\alpha S^\beta S^\gamma\rangle)=D. \tag{12}$$

The expectation values of spin operators that occur in these expressions are m and q, defined in equation (7), and two closely related quantities

$$t=\langle S^\alpha S^\beta S^\gamma\rangle=\frac{1}{(2\pi)^{1/2}}\int \mathrm{d}z\ \mathrm{e}^{-\frac{1}{2}z^2}\tanh^3\left(\frac{Jq^{1/2}}{kT}z+\frac{J_0m}{kT}\right),$$
$$r=\langle S^\alpha S^\beta S^\gamma S^\delta\rangle=\frac{1}{(2\pi)^{1/2}}\int \mathrm{d}z\ \mathrm{e}^{-\frac{1}{2}z^2}\tanh^4\left(\frac{Jq^{1/2}}{kT}z+\frac{J_0m}{kT}\right). \tag{13}$$

In the paramagnetic phase m, q, t and r are zero, and so the matrix is diagonal. The conditions for stability against ferromagnetism and spin glass formation are $A>0$ and $P>0$, or $kT>J_0$ and $kT>J$, in agreement with SK. For the other phases it is necessary to find the eigenvalues of the matrix. Because of the symmetry of the matrix under permutation of the indices a complete set of eigenvectors can be found for general values of n, and these are given in the appendix, so there is no problem in making the analytic continuation to $n=0$. There are at most five distinct eigenvalues. The eigenvectors that are symmetric under interchange of indices give (see equation (A.4)) for $n=0$

$$\lambda=\tfrac{1}{2}\{(A-B+P-4Q+3R)\pm[(A-B-P+4Q-3R)^2-8(C-D)^2]^{1/2}\}. \tag{14}$$

Eigenvectors that are symmetric under interchange of all but one of the indices give two more eigenvalues for general n but for $n=0$ these reduce to equation (14) (see equation (A.7)). Finally there are eigenvectors symmetric under interchange of all but two of the indices, and these give rise to the eigenvalue (see equation (A.9))

$$\lambda=P-2Q+R. \tag{15}$$

The eigenvalues given in equation (14) can be related to the free energy given by SK. Comparison of our equations (10), (11) and (12) with equations (9) and (10) of SK shows that

$$A-B=(J_0N)^{-1}\partial^2F/\partial m^2,$$
$$P-4Q+3R=-(2kT/NJ^2)\partial^2F/\partial q^2, \tag{16}$$
$$C-D=-(kT/N^2J^2J_0)^{1/2}\partial^2F/\partial m\partial q.$$

The condition that the product of the eigenvalues given in equation (14) is positive is equivalent to the condition that the SK solution gives a saddle point of the free energy, and this seems to be the case. We have not found any region in which the SK solution gives negative eigenvalues in equation (14), and the zeros give the phase boundaries given by SK.

The condition that the eigenvalue given by equation (15) is positive can be written in the form

$$\left(\frac{kT}{J}\right)^2 > \frac{1}{(2\pi)^{1/2}} \int \mathrm{d}z\, \mathrm{e}^{-\frac{1}{2}z^2} \operatorname{sech}^4\left(\frac{Jq^{1/2}z}{kT} + \frac{J_0 m}{kT}\right), \tag{17}$$

by using equations (11), (7) and (13). This inequality is satisfied in the paramagnetic region, where $kT > J$, but appears to be violated everywhere in the spin glass region. Close to $T = J/k$, q is small, and the inequality becomes

$$\left(\frac{kT}{J}\right)^2 > 1 - 2q\left(\frac{J}{kT}\right)^2 + 7q^2\left(\frac{J}{kT}\right)^4 - \ldots \tag{18}$$

while equation (7) gives

$$q = q\left(\frac{J}{kT}\right)^2 - 2q^2\left(\frac{J}{kT}\right)^4 + \frac{17}{3}q^3\left(\frac{J}{kT}\right)^6 - \ldots, \tag{19}$$

and substitution of this in (18) shows that the inequality is violated by terms of order q^2. At very low temperatures q is close to unity and the right-hand-side of the inequality (17) is of order kT/J, so it is certainly not satisfied.

In the ferromagnetic phase this inequality is satisfied for high temperatures, but it is violated at low temperatures. The line of instability obtained by numerical evaluation is shown in figure 1; it should be noticed that even for J_0 much greater than J the inequality (17) is violated at low temperatures, since for m and q close to unity and T small it gives

$$kT > \tfrac{4}{3}(2\pi)^{-1/2} J \exp(-J_0^2/2J^2). \tag{20}$$

The instability of the SK solution in this region becomes less surprising when it is noticed that the SK expression for the entropy of the ferromagnetic phase has a limit $-(Nk/2\pi)\exp(-J_0^2 m^2/J^2)$ at zero temperature. It should also be noticed that the

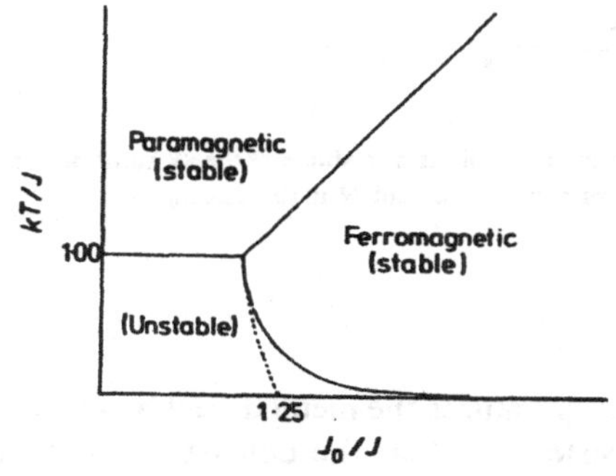

Figure 1. Phase diagram showing the limits of stability of the SK solution in the absence of a magnetic field. The broken curve is the SK phase boundary between the spin glass and ferromagnetic phases, which lies entirely in the unstable region.

instability of the ferromagnetic phase occurs for a non-zero magnetisation, and so presumably the spin glass phase can have non-zero magnetisation when J_0 is non-zero.

Very similar arguments can be applied to this system in the presence of a magnetic field. We have carried out a detailed study for the case in which J_0 is zero. The appropriate equations are obtained by replacing J_0m by H everywhere, so that the stability condition (17) and the equation for q become

$$\left(\frac{kT}{J}\right)^2 > \frac{1}{(2\pi)^{1/2}} \int \mathrm{d}z\, \mathrm{e}^{-\frac{1}{2}z^2} \operatorname{sech}^4\left(\frac{Jq^{1/2}z}{kT} + \frac{H}{kT}\right),$$
$$q = \frac{1}{(2\pi)^{1/2}} \int \mathrm{d}z\, \mathrm{e}^{-\frac{1}{2}z^2} \tanh^2\left(\frac{Jq^{1/2}z}{kT} + \frac{H}{kT}\right). \qquad (21)$$

Although there is no phase transition in the SK solution for non-zero field, these equations give a line of instability for all values of the field H. For small values of H this occurs for T close to $T_c = J/k$ and q small, and a power series expansion can be used to get the condition

$$H^2 > (4J^2/3)(1 - T/T_c)^3, \qquad (22)$$

while for large fields T is small and q close to unity, so that the condition becomes

$$kT > \tfrac{4}{3}(2\pi)^{-1/2} J \exp(-H^2/2J^2), \qquad (23)$$

in close analogy with (20). Again it is not surprising that the SK solution should be unstable in the presence of a field, since the zero temperature limit of the entropy is $-(Nk/2\pi)\exp(-H^2/J^2)$. The result of a numerical evaluation of the line of instability given by (21) is shown in figure 2.

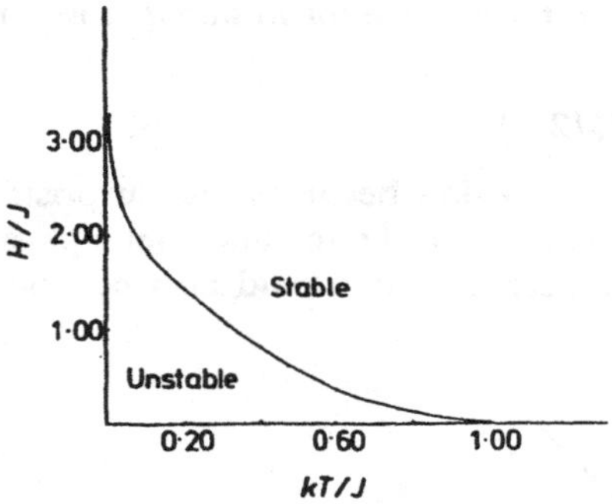

Figure 2. Phase diagram showing the limit of stability of the SK solution for the paramagnetic phase in the presence of a magnetic field H in the case $J_0 = 0$.

3. Discussion

We have shown that while at high temperatures the method used by SK to evaluate the free energy of a spin glass is consistent, in that the dominating extremum of the integrand is indeed a maximum, in the spin glass phase and in the low temperature part of the ferromagnetic phase it is not a maximum. This is consistent with the observatition made by SK that their solution must be wrong at low temperatures because their entropy is negative. This method has enabled us to trace the instability of the paramagnetic phase in a magnetic field and of the ferromagnetic phase, but we have not been able to find an alternative solution for the spin glass phase. The nature of the instability may be significant, in that it breaks the symmetry between the replicas, but it is not obvious how to handle such a broken symmetry in a zero-dimensional space.

Acknowledgments

We are very grateful to Dr J M Kosterlitz for numerous helpful discussions of this work, and the Dr R C Jones for his critical reading of the manuscript.

Appendix

To find the eigenvalues of the matrix G whose elements are given by equations (10), (11) and (12), it is necessary to exploit the symmetry of the matrix elements under permutation of the n indices. The order of the matrix is $\frac{1}{2}n(n+1)$, and since the matrix is real symmetric this is the number of linearly independent eigenvectors to be found. It turns out that by considering three symmetry classes we can find the complete set of eigenvectors and the five distinct eigenvalues.

The eigenvectors μ of G have the form

$$\mu = \begin{pmatrix} \{\epsilon^{(\alpha)}\} \\ \{\eta^{(\alpha\beta)}\} \end{pmatrix} \qquad \alpha, \beta = 1, 2, 3, \ldots, n \tag{A.1}$$

where $\{\epsilon^{(\alpha)}\}$ and $\{\eta^{(\alpha\beta)}\}$ are column-vectors with n and $\frac{1}{2}n(n-1)$ elements respectively. To find the solutions of the eigenvalue equation $G\mu = \lambda\mu$ we first consider the vector μ_1 with elements given by

$$\epsilon^{(\alpha)} = a, \qquad \text{all } \alpha; \qquad \eta^{(\alpha\beta)} = b, \qquad \text{all } (\alpha\beta). \tag{A.2}$$

Substitution of this in the eigenvalue equation gives

$$\begin{aligned} &[A+(n-1)B-\lambda]a+[(n-1)C+\tfrac{1}{2}(n-1)(n-2)D]b=0, \\ &[2C+(n-2)D]a+[P+2(n-2)Q+\tfrac{1}{2}(n-2)(n-3)R-\lambda]b=0, \end{aligned} \tag{A.3}$$

from which we get the two non-degenerate eigenvalues

$$\begin{aligned} \lambda = \tfrac{1}{2}\big[&[A+(n-1)B+P+2(n-2)Q+\tfrac{1}{2}(n-2)(n-3)R] \\ &\pm\{[A+(n-1)B-P-2(n-2)Q-\tfrac{1}{2}(n-2)(n-3)R]^2 \\ &+2(n-1)[2C+(n-2)D]^2\}^{1/2}\big] \end{aligned} \tag{A.4}$$

Next we can consider the vectors μ_2 of the form

$$\begin{aligned} &\epsilon^{(\alpha)} = a, \quad \text{for } \alpha = \theta, \qquad &&\epsilon^{(\alpha)} = b, \quad \text{for } \alpha \neq \theta \\ &\eta^{(\alpha\beta)} = c, \quad \text{for } \alpha \text{ or } \beta = \theta, \qquad &&\eta^{(\alpha\beta)} = d, \quad \text{for } a, \beta \neq \theta. \end{aligned} \tag{A.5}$$

These vectors span a $2n$-dimensional invariant subspace, and therefore yield $2n$ eigenvectors, including those already obtained. To ensure orthogonality to the eigenvectors μ_1, we take $a=(1-n)b$, $c=(1-\frac{1}{2}n)d$, and the eigenvalue equation then become

$$\begin{aligned}&(A-\lambda-B)a+(n-1)(C-D)c=0,\\&\frac{n-2}{n-1}(C-D)a+[P+(n-4)Q-(n-3)R-\lambda]c=0,\end{aligned}\tag{A.6}$$

which give two further eigenvalues

$$\begin{aligned}\lambda=\tfrac{1}{2}\{&[A-B+P+(n-4)Q-(n-3)R]\\&\pm\{[A-B-P-(n-4)Q+(n-3)R]^2+4(n-2)(C-D)^2\}^{1/2}\},\end{aligned}\tag{A.7}$$

each with degeneracy $n-1$.

Finally, the entire space can be spanned with vectors of the form

$$\begin{aligned}&\epsilon^{(\alpha)}=a,\quad \text{for } \alpha=\theta \text{ or } \nu, \qquad \epsilon^{(\alpha)}=b,\quad \text{for } \alpha\neq\theta,\nu\\&\eta^{(\theta\nu)}=c,\quad \eta^{(\theta\alpha)}=\eta^{(\nu\alpha)}=d,\quad \text{for } \alpha\neq\theta,\nu, \qquad \eta^{(\alpha\beta)}=e,\quad \text{for } \alpha,\beta\neq\theta,\nu.\end{aligned}\tag{A.8}$$

Orthogonality to the eigenvectors already found imposes the conditions $a=b=0$, $c=(2-n)d$, $d=\frac{1}{2}(3-n)e$. Substitution of such a vector in the eigenvalue equation gives the equation

$$\lambda=P-2Q+R,\tag{A.9}$$

with degeneracy $\frac{1}{2}n(n-3)$. These five eigenvalues are distinct in general although for $n=0$ equations (A.4) and (A.7) coincide, while for $n=1$ and $n=2$ equation (A.9) coincides with one root of (A.4) and (A.7) respectively.

References

Edwards S F and Anderson P W 1975 *J. Phys. F: Metal Phys.* **5** 965–74
Fisch R and Harris A B 1977 *Phys. Rev. Lett.* **38** 785–7
Harris A B, Lubensky T C and Chen J H 1976 *Phys. Rev. Lett.* **36** 415–8
Mydosh J A 1977 *Proc. 2nd Int. Symp. on Amorphous Magnetism, Troy 1976* eds R A Levy and R Hasegawa (New York: Plenum) pp 73–83
Sherrington D and Kirkpatrick S 1975 *Phys. Rev. Lett.* **35** 1792–6
Thouless D J, Anderson P W and Palmer R G 1977 *Phil. Mag.* **35** 593–601

PHILOSOPHICAL MAGAZINE, 1977, VOL. 35, NO. 3, 593–601

Solution of 'Solvable model of a spin glass'

By D. J. THOULESS
Department of Mathematical Physics, University of Birmingham, Birmingham, England

and P. W. ANDERSON†‡ and R. G. PALMER
Department of Physics, Princeton University, Princeton, New Jersey 08540, U.S.A.‡

[Received 12 October 1976]

ABSTRACT

The Sherrington–Kirkpatrick model of a spin glass is solved by a mean field technique which is probably exact in the limit of infinite range interactions. At and above T_c the solution is identical to that obtained by Sherrington and Kirkpatrick (1975) using the $n \to 0$ replica method, but below T_c the new result exhibits several differences and remains physical down to $T=0$.

§ 1. INTRODUCTION

Sherrington and Kirkpatrick (1975) (SK) have proposed an idealized model of a spin glass which apparently allows an exact formal solution. Unfortunately, the solution is non-physical at low temperatures, leading in particular to a negative zero-point entropy. We present here a new solution of the SK model which behaves sensibly at low temperatures, while agreeing with the SK solution at and above the critical temperature T_c. Our analysis is based on a high temperature expansion, supplemented below T_c by a mean field theory which takes into account not only the average spin on each site, but also the mean square fluctuation from this average.

The Sherrington–Kirkpatrick Hamiltonian

$$\mathscr{H} = -\sum_{(ij)} J_{ij} S_i S_j \tag{1}$$

describes N Ising spins ($S_i = \pm 1$) interacting in pairs (ij) via infinite-range Gaussian-random exchange interactions :

$$\text{Prob}\,(J_{ij}) \propto \exp\left(\frac{-ZJ_{ij}^2}{2\tilde{J}^2}\right) \tag{2}$$

with a variance $\tilde{J}^2/Z$ where Z is the number of neighbours of each spin, presumed effectively infinite ; we work in the limit $N \gtrsim Z \gg 1$. The $Z^{-1/2}$ dependence of the interactions is necessary to ensure a sensible thermodynamic limit. We consider only the case in which J_{ij} has zero mean, setting SK's J_0 parameter to zero. We also set $k_B = 1$ and $\beta = 1/T$ throughout.

---

† Also at Bell Laboratories, Murray Hill, New Jersey 07974, U.S.A.

‡ Work at Princeton University partially supported by National Science Foundation Grant No. DMR76–00886.

The SK solution follows the approach of Edwards and Anderson (1975) using the well-known trick of replacing $\ln Z$ by $\lim_{n\to 0}(Z^n-1)/n$ and regarding Z^n (for integer n) as the partition function of n replicas of the original system. This allows one to perform the average over the J distribution *before* taking the spin trace. It seems necessary, however, to use the thermodynamic limit $N\to\infty$ before taking $n\to 0$, and it is this improper reversal of limits that leads to SK's erroneous solution (R. G. Palmer, to be published). We therefore avoid the replica method and turn to a different approach.

§ 2. The high temperature region

For $T>T_c$ we make a high temperature series expansion for the free energy, using the standard identity

$$\exp(\beta J_{ij}S_iS_j)=\cosh\beta J_{ij}(1+S_iS_j\tanh\beta J_{ij}). \tag{3}$$

Thus

$$\begin{aligned}-\beta F&=\langle\ln \operatorname{Tr}\exp(-\beta\mathscr{H})\rangle_J\\&=\langle\ln\prod_{(ij)}\cosh\beta J_{ij}\rangle_J+\langle\ln\operatorname{Tr}\prod_{(ij)}(1+T_{ij}S_iS_j)\rangle_J\\&=N\beta^2J^2/4+0(N/Z)\\&\quad+\langle\ln\operatorname{Tr}(1+\sum_{(ij)}T_{ij}S_iS_j+\tfrac{1}{2}\sum_{(ij)\neq(kl)}T_{ij}T_{kl}S_iS_jS_kS_l\ldots)\rangle_J,\end{aligned}\tag{4}$$

where $T_{ij}=\tanh\beta J_{ij}$. The expansion may be analysed diagramatically (each line representing a T_{ij}), noting the following conditions for a non-vanishing diagram :

(*a*) There must be an even number of lines at each vertex.
(*b*) No line may be double *before* taking the logarithm.
(*c*) Every line must be double *after* taking the logarithm (because $\langle J\rangle=0$).

We find no terms of order N (except the trivial $N\ln 2$), and a summable series in order N/Z, consisting of simple polygons (fig. 1 (*a*)) which become double (fig. 1 (*b*)) after taking the logarithm. We thus obtain

$$F=Nf_0+(N/Z)f_1+\text{lower order},$$

Fig. 1

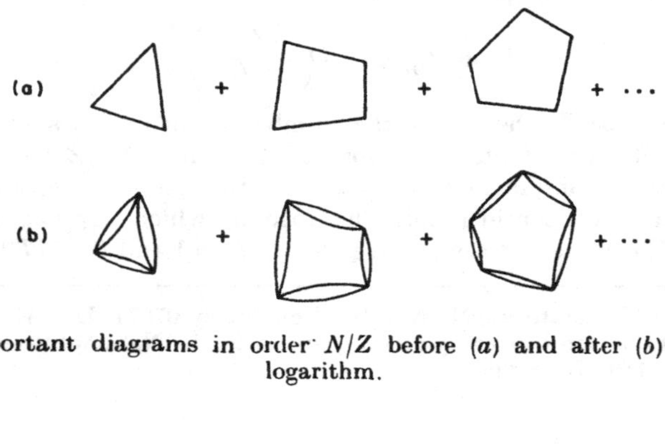

The most important diagrams in order N/Z before (*a*) and after (*b*) taking the logarithm.

with

$$f_0 = -T \ln 2 - \tilde{J}^2/4T, \tag{5}$$

and

$$f_1 = -\tfrac{1}{4}T \ln (1-\beta^2\tilde{J}^2) + \text{non-singular part.}$$

The extensive part of the free energy is identical to the SK result for $T \geqslant T_c$. The contribution of order $N/Z \sim 1$ is negligible in the thermodynamic limit for $T > \tilde{J}$, but diverges as $T = T_c = \tilde{J}$ signaling the transition found by Edwards and Anderson (1975) and SK. We note that the divergent part f_1 is intrinsically positive, in contrast to the corresponding result of conventional mean field theory. This reflects the fact that the spin glass transition is a *blocking* effect of interference among the different interactions ; the free energy below T_c is *greater* than an analytic continuation of the high temperature result.

§ 3. Derivation of the mean field equation

Below T_c we must introduce a mean field in order to reconverge the series for F. We employ the usual identity

$$\operatorname{Tr} \exp(-\beta\mathscr{H}) = \operatorname{Tr} \exp(-\beta\mathscr{H}_0)\langle \exp(\beta\mathscr{H}_0 - \beta\mathscr{H})\rangle_{\mathscr{H}_0}, \tag{6}$$

where $\mathscr{H}_0$ is a soluble mean field Hamiltonian which is to be used in evaluating the diagrams generated by $\exp(\beta\mathscr{H}_0 - \beta\mathscr{H})$. An obvious ansatz is

$$(\mathscr{H}_0 - \mathscr{H})_{ij} = J_{ij}(S_i - m_i)(S_j - m_j) \tag{7}$$

so that

$$(\mathscr{H}_0)_{ij} = J_{ij}(m_i m_j - m_i S_j - m_j S_i)$$

where m_i is the mean spin on the ith site, to be determined self-consistently by the condition

$$\langle S_i \rangle_{\mathscr{H}_0} = m_i. \tag{8}$$

Ignoring the perturbation $\mathscr{H} - \mathscr{H}_0$ leads to the appealing (but incorrect) mean field equation

$$h_i = \sum_j J_{ij} m_j = T \tanh^{-1} m_i \tag{9}$$

which would imply a critical temperature of $2\tilde{J}$, since the largest eigenvalue of a Gaussian-random matrix (Mehta 1967) is $(J_\lambda)_{\max} = 2\sqrt{(NJ^2)} = 2\tilde{J}$. However, the series generated by $\exp(\beta\mathscr{H}_0 - \beta\mathscr{H})$ is still divergent with this choice for $\mathscr{H}_0$, showing that it is essential to consider correlations between spin fluctuations on different sites.

It is possible to proceed diagramatically (Thouless, Anderson, Lieb and Palmer, unpublished report), removing the most divergent diagrams by manipulating the $\mathscr{H}_0/\mathscr{H}$ separation. However, it is simpler, and perhaps more physical, to observe that the set of diagrams contributing in order N are just those which would remain on a ' Bethe lattice ' or Cayley tree ; all diagrams with no loops. Any diagram containing a closed ring is necessarily of order N/Z or less, since the internal connection reduces the number of site summations by one. The Bethe method (1935) is exact for the Ising model on a Cayley tree, and should therefore solve this problem wherever the order N/Z terms are convergent (and hence ignorable).

In the Bethe method, we consider a 'cluster' of a central site 0 and all its neighbours j. On the neighbours j we assume mean fields h_j which, for a Cayley tree, are the only effect *their* neighbours can have on them. Using the smallness of J_{0j} ($\propto Z^{-1/2}$), it is easy to arrive at the following expressions for m_0 and m_j:

$$\left.\begin{aligned} m_0 &= \tanh \beta \sum_j J_{0j} \tanh \beta h_j \\ m_j &= \tanh \beta h_j + m_0 \beta J_{0j}(1-\tanh^2 \beta h_j). \end{aligned}\right\} \tag{10}$$

We may now eliminate the h_js (again using the smallness of J_{0j}), obtaining the fundamental equation

$$\sum_j J_{0j} m_j - m_0 \beta \sum_j J_{0j}^2(1-m_j^2) = T \tanh^{-1} m_0 \tag{11}$$

which supplants the incorrect eqn. (9), and must, of course, be valid for any choice of site 0. The correction term proportional to m_0 is more readily understood upon realizing that $\beta(1-m_j^2)$ is the single-site susceptibility, χ_j, as may easily be proved. Equation (11) may thus be written

$$m_0 = \tanh \beta \sum_j J_{0j}(m_j - m_0 J_{0j} \chi_j) \tag{12}$$

and the second term on the right-hand side is seen as the response of site j to the mean spin on site 0; this must be *removed* from m_j when computing m_0.

The corresponding free energy is not easily obtained from the Bethe method, and we therefore present it as a *fait accompli*, originally derived by diagram expansion:

$$\begin{aligned} F_{\mathrm{MF}} = &- \sum_{(ij)} J_{ij} m_i m_j - \tfrac{1}{2}\beta \sum_{(ij)} J_{ij}^2(1-m_i^2)(1-m_j^2) \\ &+ \tfrac{1}{2}T \sum_i [(1+m_i) \ln \tfrac{1}{2}(1+m_i) + (1-m_i) \ln \tfrac{1}{2}(1-m_i)] \end{aligned} \tag{13}$$

As it must, direct differentiation of eqn. (13) gives eqn. (11). Additionally, eqn. (13) is quite physically transparent: the first term is the internal energy of a frozen lattice; the second term is the correlation energy of the fluctuations, and is just the $N\tilde{J}^2/4T$ term of eqn. (5), modified for the effective 'freedom', $1-m_i^2$, of each spin; and the last term is the entropy of a set of Ising spins constrained to have means m_i.

We emphasize that the Bethe method, the use of the Cayley tree, and the resulting eqns. (11) and (13) are only meaningful if the terms of order N/Z (and lower) are convergent. Evaluation of these terms is rather awkward and will be discussed in detail elsewhere. The only simple region is near T_c, where we find

$$\tilde{J}(1-\overline{m^2}) \leqslant T \quad (T \sim T_c) \tag{14}$$

as the convergence criterion for the N/Z diagrams, and hence as a validity condition for our mean field theory.

Our problem is now reduced to finding solutions to the mean field eqn. (11), subject to the convergence condition (14), and then to using eqn. (13) to obtain the thermodynamics. This programme is not much easier than the original problem, since J_{ij} is a random matrix, and eqns. (11) and (13) hold for an individual realization, *not* for an ensemble average. We have been able to find solutions both near T_c and near $T=0$. In both cases the solutions involve some numerical conjectures checked by machine simulation, so that while we are reasonably certain of the general form of the solution in both regions, they are far from complete analyses. We also encounter some ' coincidences ' which require further investigation. Details of our solutions will be given elsewhere, and we attempt here only a general description of the methods.

§ 4. The critical region

For T near T_c we expect m_i to be small and similar to the eigenvector M_i belonging to the largest eigenvalue $(J_\lambda)_{\max}=2\tilde{J}$ of the matrix J_{ij} :

$$\sum_j J_{ij}M_j=2\tilde{J}M_i \tag{15}$$

We first linearize eqn. (11), approximating $\sum_j J_{ij}{}^2\chi_j$ by $\tilde{J}^2\bar{\chi}$:

$$\sum_j J_{ij}m_j=\beta\tilde{J}^2(1-\overline{m^2})m_i+T(m_i+m_i{}^3/3+m_i{}^5/5+\ldots). \tag{16}$$

We then expand m_i about M_i

$$m_i=M_i+\delta m_i, \tag{17}$$

chosing the R.M.S. amplitude

$$q=\overline{M_i{}^2} \tag{18}$$

of M_i such that M_i is orthogonal to δm_i. The components M_i have a Gaussian distribution, as may be proved from the invariance of a Gaussian-random matrix (with suitable diagonal elements) under orthogonal transformations. Using this fact to take a scalar product of eqn. (16) with M_i, we obtain

$$(2\tilde{J}-\beta\tilde{J}^2-T)q=(T-\beta\tilde{J}^2)q^2+3Tq^3+T\sum_i M_i{}^3\delta m_i+0(q^4). \tag{19}$$

The term in δm_i is essential—there is no solution without it—but is difficult to estimate. Analysing the projection of eqn. (16) orthogonal to M_i by a combination of eigenvector expansions and numerical estimates, we find finally

$$(2\tilde{J}-\tilde{J}^2/T-T)q-(T-\tilde{J}^2/T)q^2+(2T^2/\tilde{J}-3T)q^3=0. \tag{20}$$

Near $T_c=\tilde{J}$ this equation has a double zero at

$$q=\overline{m^2}=1-T/T_c \tag{21}$$

which gives the mean field free energy, eqn. (13), a *saddle point* at T_c.

Near T_c the form of F_{MF} is very complicated and quite unlike that for typical phase transitions (fig. 2), and it is not at all surprising that the Edwards–Anderson and SK continuations come out on a wrong branch of the free energy function. It is important to note that F_{MF} is *not* a genuine

Fig. 2

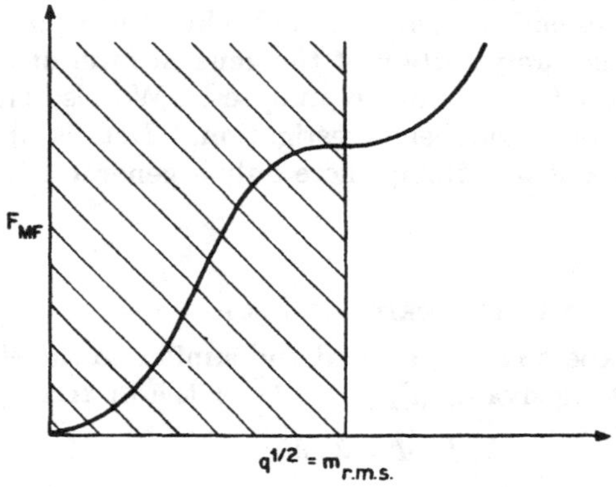

The form of the mean field free energy for T slightly below T_c. The N/Z terms diverge in the shaded region.

free energy functional in the Ginzburg–Landau sense, in that the convergence condition (14) restricts the freedom of q, and in particular eliminates the spurious minimum at $q=0$ as soon as T falls below T_c. This behaviour, the q^3 term, and the saddle point in F_{MF}, are very reminiscent of the heuristic free energy functional of Harris, Lubensky and Chen (1976).

As far as we can see, our solution deviates only in higher order from SK near T_c. The cusp of the specific heat is the same, as is the T-dependence of $\overline{m^2}=q$.

§ 5. The low temperature region

In the low temperature regime, $T \ll T_c$, our analysis is based on the probability distributions of the fields $h_i = \sum_j J_{ij} m_j$ and the mean spins, m_i. At $T=0$ the mean field equation obviously selects a self-consistent lowest energy solution of

$$m_i = \text{sign}\,(h_i), \tag{22}$$

and we have generated a large number of such solutions numerically to investigate the distribution of h_i. We find that Prob $(|h_i|)$—hence written $p(h)$—becomes linear for small h as $N\to\infty$ (there is a finite offset $p(0)$ at finite N). As a by-product of this study, we find a ground state energy of $U_0 = -(0{\cdot}755 \pm 0{\cdot}010)\tilde{J}N$, which is certainly different from SK's value of $U_0 = -(2/\pi)^{1/2}\tilde{J}N = -0{\cdot}80\tilde{J}N$.

To derive the low temperature thermodynamics we assume

$$\lim_{h\to 0} p(h) = h/H^2 \tag{23}$$

and

$$q = \overline{m^2} = 1 - \alpha(T/\tilde{J})^2 \quad (T \ll T_c), \tag{24}$$

where H and α are parameters to be determined later. Equation (24) is easily justified *a posteriori*. Again approximating the $J_{ij}^2\chi_j$ term, m_i and h_i are related by

$$h_i = \alpha T m_i + T \tanh^{-1} m_i \tag{25}$$

and the definition

$$m^2 = \int_0^\infty m^2(h) p(h)\, dh \tag{26}$$

leads after some integration to

$$H^2/\tilde{J}^2 = \tfrac{1}{4}\alpha + (2 \ln 2 + 1)/3 + \ln 2/\alpha, \tag{27}$$

which leaves only one unknown parameter, α. The minimum acceptable value for H is $H = 1{\cdot}28\tilde{J}$ and we believe on the basis of our numerical work and general considerations that H is actually equal to this limiting value, giving

$$\alpha = 2\sqrt{\ln 2} = 1{\cdot}665. \tag{28}$$

Fig. 3

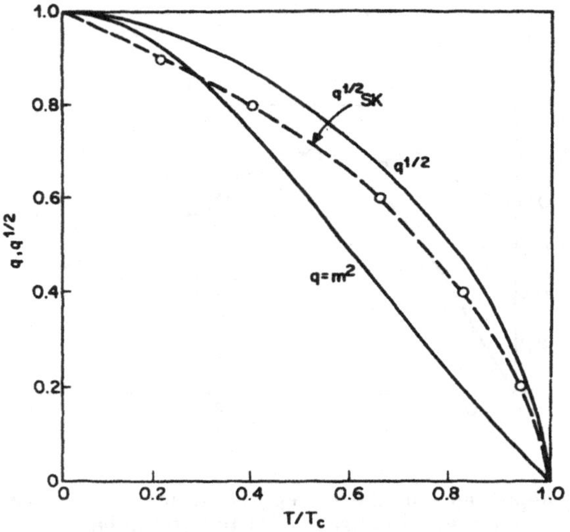

The order parameter $q = m^2$ as a function of temperature. The circles and broken line are the results of SK.

Figure 3 shows the resulting order parameter, q, fitted smoothly to the SK result near T_c. The convergence criterion, eqn. (14), is easily satisfied at low temperatures, but there are corrections to this criterion away from T_c. We suspect, but have not yet proved, that the solution coincides with the *true* convergence criterion, and the F_{MF} has the saddle point form sketched in fig. 2 at *all* temperatures below T_c, thus giving a line of critical points.

We may now calculate the entropy from eqn. (13):

$$S=-(1/2T^2)\sum J_{ij}^2(1-m_i^2)(1-m_j^2)$$
$$-\sum_i\left[\left(\frac{1+m_i}{2}\right)\ln\left(\frac{1+m_i}{2}\right)+\left(\frac{1-m_i}{2}\right)\ln\left(\frac{1-m_i}{2}\right)\right]$$
$$=-\frac{N\alpha^2}{4}(T/\tilde{J})^2-N\int_0^\infty[\ldots]p(h)\,dh$$
$$=0\cdot770N(T/\tilde{J})^2. \tag{29}$$

The low temperature specific heat is thus quadratic:

$$C=1\cdot54N(T/\tilde{J})^2. \tag{30}$$

We have performed some finite temperature Monte Carlo simulations of this spin glass model and find a specific heat consistent with this result.

The low temperature susceptibility

$$\chi=\bar{\chi}_j=1\cdot665T/\tilde{J} \tag{31}$$

is linear in T, in contrast to SK's which has a finite zero-temperature intercept (fig. 4).

Fig. 4

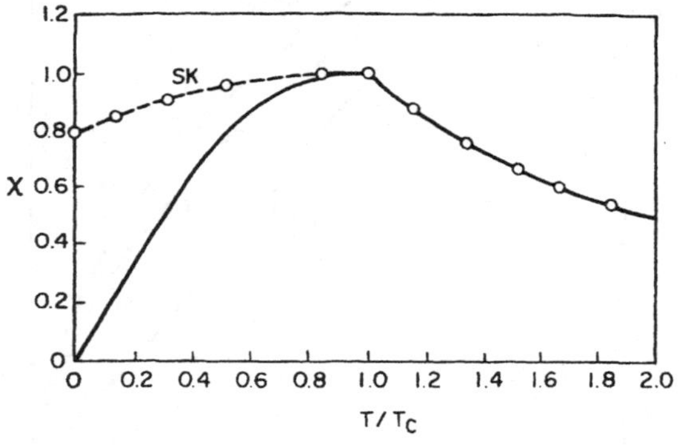

The temperature dependence of the susceptibility, according to the present work (solid line) and SK (circles and broken line).

§ 6. Conclusion

In conclusion, we believe that we have shown that a *consistent* mean field theory of the Sherrington–Kirkpatrick 'solvable model' can be constructed. We believe that this mean field theory represents the actual thermodynamic behaviour of the model accurately to order $1/Z$. The infinite range interactions seem necessary at present to make the model tractable, but also make it somewhat unrealistic. We therefore caution others against any literal comparison of these model results with experiment, but emphasize that our solution strongly supports the essential conclusions of the Edwards–Anderson spin glass theory, that a sharp thermodynamic transition into a totally randomly ordered state can and does occur.

Acknowledgments

We wish to thank Elliot Lieb, David Sherrington, and Scott Kirkpatrick for many illuminating and helpful discussions, with particular gratitude to Kirkpatrick for keeping us abreast of his numerical work, which in several cases confirmed that referred to in the text.

We are also grateful to the Aspen Center for Physics for hospitality and assistance in the final preparation of this manuscript.

References

Bethe, H., 1935, *Proc. R. Soc.* A, **151,** 540; see Fowler, R. H., and Guggenheim, E. A., 1960, *Statistical Thermodynamics* (Cambridge University Press) for a good discussion.

Edwards, S. F., and Anderson, P. W., 1975, *J. Phys.* F, **5,** 965.

Harris, A. B., Lubensky, T. C., and Chen, J.-H., 1976, *Phys. Rev. Lett.*, **36,** 415.

Mehta, M. L., 1967, *Random Matrices and the Statistical Theory of Energy Levels* (London, New York: Academic Press).

Sherrington, D., and Kirkpatrick, S., 1975, *Phys. Rev. Lett.*, **35,** 1792.

J. Phys. C: Solid St. Phys., 13 (1980) L469–76. Printed in Great Britain

LETTER TO THE EDITOR

Metastable states in spin glasses

A J Bray and M A Moore
Department of Theoretical Physics, The University, Manchester M13 9PL, UK

Received 17 April 1980

Abstract. The number of solutions of the equations of Thouless, Anderson and Palmer is obtained as a function of temperature. The density of solutions with a given free energy is calculated for free energies greater than a (temperature-dependent) critical value.

Despite substantial effort during the past five years, a satisfactory mean field theory of spin glasses has not yet appeared. The replica method (Edwards and Anderson 1975, Sherrington and Kirkpatrick 1975 (SK)) requires, for an adequate description of the low-temperature phase, a breaking of the replica symmetry (de Almeida and Thouless 1978, Blandin 1978, Bray and Moore 1978, Parisi 1979). An alternative approach is provided by the mean field equations of Thouless, Anderson and Palmer (1977, referred to as TAP), which are exact for the long-range interactions of the SK model. In this Letter we show that there are a large number of solutions (of order $\exp(\alpha N)$, where N is the number of Ising spins in the system) of the TAP equations below T_c. One is interested in the distribution of solutions over free energy, and averages of observables over solutions or over solutions of given free energy. Since the solutions possess a range of free energies, all but those with the lowest free energy correspond to metastable states. We shall find that solutions with free energies exceeding a (temperature-dependent) critical value are *uncorrelated*, and their distribution over free energy is calculated exactly. Solutions of lower free energy are correlated, with the Edwards–Anderson order parameter (here called $\hat{q}$) as a measure of the correlation. Our approach is generalisable (at least for $T = 0$) to two- and three-dimensional systems with realistic interactions. The numerical work of Morgenstern and Binder (1980) suggests that for such systems the critical energy might coincide with the ground state energy.

We start from the set of N TAP equations for the magnetisation m_i of the ith spin:

$$G_i \equiv \tanh^{-1} m_i + \beta^2 J^2 (1-q) m_i - \beta \sum_j J_{ij} m_j = 0$$

$$\equiv g(m_i) - \beta \sum_j J_{ij} m_j \tag{1}$$

with their associated free energy (divided by Nk_BT)

$$f = -(\beta/N) \sum_{(ij)} J_{ij} m_i m_j - \tfrac{1}{4}\beta^2 J^2 (1-q)^2$$

$$+ (1/2N) \sum_i \{(1+m_i)\ln[\tfrac{1}{2}(1+m_i)] + (1-m_i)\ln[\tfrac{1}{2}(1-m_i)]\} \tag{2}$$

0022-3719/80/190469+08 $01.50

where $\sum_{(ij)}$ means a sum over distinct pairs, $q = N^{-1}\sum_i m_i^2$ and J_{ij} is a random exchange interaction with probability distribution

$$P(J_{ij}) = (N/2\pi J^2)^{1/2} \exp(-NJ_{ij}^2/2J^2).$$

Substituting for $\sum_j J_{ij} m_j$ in equation (2) from equation (1), we express f as a sum of single-site terms:

$$f = N^{-1}\sum_i f(m_i) = N^{-1}\sum_i [-\ln 2 - \tfrac{1}{4}\beta^2 J^2(1-q^2) + \tfrac{1}{2}m_i \tanh^{-1} m_i + \tfrac{1}{2}\ln(1-m_i^2)]. \tag{3}$$

The density of solutions associated with a particular free energy f is therefore

$$N_s(f) = N^2 \int_0^1 \mathrm{d}q \int_{-1}^1 \prod_i (\mathrm{d}m_i)\, \delta\Big(Nq - \sum_i m_i^2\Big)\, \delta\Big(Nf - \sum_i f(m_i)\Big) \times \prod_i \delta(G_i)|\det \mathbf{A}| \tag{4}$$

where $|\det \mathbf{A}|$ is the Jacobian normalising the delta functions:

$$A_{ij} = \partial G_i/\partial m_j = [(1-m_i^2)^{-1} + \beta^2 J^2(1-q)]\,\delta_{ij} - \beta J_{ij} \equiv a_i \delta_{ij} - \beta J_{ij}. \tag{5}$$

(A term $-2\beta^2 J^2 m_i m_j/N$, which comes from differentiating q in equation (1), is negligible as $N \to \infty$ and has been dropped from equation (5)). Since we will find that the determinant is always positive, we will drop the modulus signs henceforth. Introducing integral representations for the delta functions gives

$$N_s(f) = N^2 \int_0^1 \mathrm{d}q \int_{-\mathrm{i}\infty}^{\mathrm{i}\infty} \frac{\mathrm{d}\lambda}{2\pi\mathrm{i}} \int_{-\mathrm{i}\infty}^{\mathrm{i}\infty} \frac{\mathrm{d}u}{2\pi\mathrm{i}} \int_{-\mathrm{i}\infty}^{\mathrm{i}\infty} \prod_i \left(\frac{\mathrm{d}x_i}{2\pi\mathrm{i}}\right) \int_{-1}^1 \prod_i (\mathrm{d}m_i) \exp\Big\{-N(\lambda q + uf) + \lambda\sum_i m_i^2 + u\sum_i f(m_i) + \sum_i x_i g(m_i) - \beta\sum_{(ij)} J_{ij}(x_i m_j + x_j m_i)\Big\} \times \det \mathbf{A}\{J_{ij}\}. \tag{6}$$

We wish now to average over the bond distribution. Since we anticipate that $N_s(f) \sim \exp(\alpha N)$, we should strictly average the extensive quantity $\ln N_s(f)$. This can be done by introducing replicas (see later). However, for the region (of free energies) in which the solutions are uncorrelated, the two types of averaging lead to the same final result. Therefore we construct the 'direct average'

$$\langle N_s(f)\rangle_J = \int \prod_{(ij)} (\mathrm{d}J_{ij} P(J_{ij}))\, N_s(f). \tag{7}$$

Terms involving J_{ij} are of the form

$$\int_{-\infty}^{\infty} \prod_{(ij)} [\mathrm{d}J_{ij}(N/2\pi J^2)^{1/2}] \exp\Big[-N\sum_{(ij)} J_{ij}^2/2J^2 - \beta\sum_{(ij)} J_{ij}(x_i m_j + x_j m_i)\Big] \det \mathbf{A}\{J_{ij}\}$$
$$= \exp\Big[(\beta^2 J^2/2N)\sum_{(ij)}(x_i m_j + x_j m_i)^2\Big] \langle \det \mathbf{A}\{J_{ij} - (\beta J^2/N) \times (x_i m_j + x_j m_i)\}\rangle_J \tag{8}$$

where the second line follows from the first by a simple translation of each integration variable. This translation is irrelevant as far as the final average in equation (8) is concerned, since it introduces a term of order N^{-1} into the matrix element A_{ij} (such terms are negligible and were already dropped from equation (5)). Thus equation (8) may be rewritten

$$\exp\left[\tfrac{1}{2}\beta^2 J^2 q \sum_i x_i^2 + (\beta^2 J^2/2N)\left(\sum_i x_i m_i\right)^2\right]\langle \det \mathbf{A}\{J_{ij}\}\rangle_J. \tag{9}$$

To compute the average of the determinant one can introduce replicas and use the representation

$$\det \mathbf{A} = \int_{-\infty}^{\infty} \prod_{i,\alpha}\left[\frac{d\xi_{i\alpha}}{(2\pi)^{1/2}}\right]\exp\left(-\tfrac{1}{2}\sum_{i,j,\alpha}\xi_{i\alpha}A_{ij}\xi_{j\alpha}\right) \tag{10}$$

where the replica labels α run from 1 to m, and analytic continuation to $m = -2$ is required at the end of the calculation. The J_{ij} integrals are gaussian and give

$$\langle \det \mathbf{A}\rangle_J = \int_{-\infty}^{\infty} \prod_{i,\alpha}\left[\frac{d\xi_{i\alpha}}{(2\pi)^{1/2}}\right]\exp\left[-\tfrac{1}{2}\sum_{i,\alpha} a_i\xi_{i\alpha}^2 + \frac{\beta^2 J^2}{4N}\sum_\alpha\left(\sum_i \xi_{i\alpha}^2\right)^2 + \frac{\beta^2 J^2}{2N}\sum_{\alpha<\beta}\left(\sum_i \xi_{i\alpha}\xi_{i\beta}\right)^2\right]. \tag{11}$$

The squared terms are simplified by the Hubbard–Stratonovich identity

$$\exp(a^2/2) = \int_{-\infty}^{\infty} (dx/\sqrt{2\pi})\exp(-x^2/2 + ax).$$

$$\langle \det \mathbf{A}\rangle_J = \int_{-\infty}^{\infty}\prod_\alpha\left[\left(\frac{N}{\pi}\right)^{1/2} dR_\alpha\right]\int_{-\infty}^{\infty}\prod_{\alpha<\beta}\left[\left(\frac{N}{2\pi}\right)^{1/2} dT_{\alpha\beta}\right]\int_{-\infty}^{\infty}\prod_{i,\alpha}\left(\frac{d\xi_{i\alpha}}{\sqrt{2\pi}}\right)$$
$$\times \exp\left\{-\tfrac{1}{2}\sum_{i,\alpha} a_i\xi_{i\alpha}^2 - N\sum_\alpha R_\alpha^2 - \frac{N}{2}\sum_{\alpha<\beta} T_{\alpha\beta}^2 + \beta J\sum_{i,\alpha} R_\alpha \xi_{i\alpha}^2 + \beta J\sum_{i,\alpha<\beta} T_{\alpha\beta}\xi_{i\alpha}\xi_{i\beta}\right\}. \tag{12}$$

The integrals over $\{R_\alpha\}$ and $\{T_{\alpha\beta}\}$ are eventually performed by steepest descents. We adopt the solution $R_\alpha = R$ (for all α), $T_{\alpha\beta} = 0$ (for all (α, β)). One can show (details will be presented elsewhere) that this is the stable stationary point. With this choice the integrals over the $\xi_{i\alpha}$ are trivial and yield, after setting $m = -2$ and dropping multiplicative prefactors,

$$\langle \det \mathbf{A}\rangle_J = \prod_i (a_i - 2\beta J R)\exp(2NR^2) \tag{13}$$

with R to be determined variationally.

Assembling the various terms, using a further Hubbard–Stratonovich identity to simplify the term in $(\sum_i x_i m_i)^2$ in equation (9), and dropping multiplicative prefactors, yields

$$\langle N_s(f)\rangle_J = \max \int_{-1}^{1}\prod_i (dm_i)\int_{-i\infty}^{i\infty}\prod_i\left(\frac{dx_i}{2\pi i}\right)\exp\left\{N(-\lambda q - uf - \tfrac{1}{2}V^2 + 2R^2) + \tfrac{1}{2}\beta^2 J^2 q\sum_i x_i^2 + \sum_i x_i \tilde{g}(m_i) + \lambda\sum_i m_i^2 + u\sum_i f(m_i)\right\}\prod_i (a_i - 2\beta J R) \tag{14}$$

where $\tilde{g}(m_i) = g(m_i) + \beta J V m_i$ and max indicates the maximum over the variables q, λ, u, V, R. Finally we set $V = -\beta J(1-q) - \Delta/\beta J$ and $2R = \beta J(1-q) - B/\beta J$, and integrate over the x_i to obtain the final result

$$\langle N_s(f)\rangle_J = \max \exp\{N[-\lambda q - uf - (B+\Delta)(1-q) + (B^2 - \Delta^2)/2\beta^2 J^2 + \ln I]\} \tag{15}$$

where

$$I = \int_{-1}^{1} \frac{dm}{(2\pi)^{1/2}\beta J q^{1/2}} \left(\frac{1}{1-m^2} + B\right) \exp\left[-\frac{(\tanh^{-1} m - \Delta m)^2}{2\beta^2 J^2 q} + \lambda m^2 + uf(m)\right] \tag{16}$$

and the maximum is taken over the five variables q, λ, u, Δ, B. The use of steepest descents is justified, in the limit $N \to \infty$, by the factor N inside the exponent in equation (15). The five stationarity equations become, after some manipulation,

$$\begin{aligned} q &= \langle m^2 \rangle, \qquad f = \langle f(m) \rangle \\ 0 &= B\{1 - \beta^2 J^2 \langle (1-m^2)^2/[1 + B(1-m^2)]\rangle\} \\ \Delta &= -\tfrac{1}{2}\beta^2 J^2 (1-q) + \langle m \tanh^{-1} m \rangle / 2q \\ \lambda &= B + \Delta - [1 - \langle (\tanh^{-1} m - \Delta m)^2 \rangle / \beta^2 J^2 q]/2q \end{aligned} \tag{17}$$

where angular brackets mean an average over a probability distribution for m given by the integrand of equation (16) divided by I. The solutions of Sherrington and Kirkpatrick (1975) and of Sommers (1978) correspond to $B = \Delta = 0 = u = \lambda$ and to $B + \Delta = 0 = u = \lambda$ respectively. For both solutions, equation (14) gives $\langle N_s(f)\rangle_J = 1$. Both solutions are known to be unstable (de Almeida and Thouless 1978, Bray and Moore 1980, de Dominicis and Garel 1979).

Before proceeding further we observe that the third of equations (17) admits the

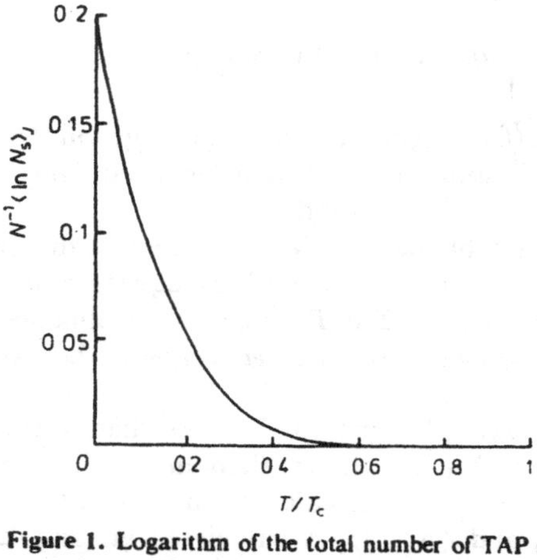

Figure 1. Logarithm of the total number of TAP solutions, divided by N, as a function of temperature. For T close to T_c, $N^{-1}\langle \ln N_s \rangle_J = \frac{8}{81} t^6 + O(t^7)$, where $t = 1 - T/T_c$, and the curve is therefore indistinguishable from the temperature axis for $T \gtrsim 0{\cdot}6\, T_c$.

solution $B = 0$. This is the solution we adopt since it may be shown (details to be presented elsewhere) that the other choice leads to an unstable stationary point. For $B = 0$, our expression for the determinant, equation (13), becomes

$$\langle \det \mathbf{A} \rangle_J = \prod_i (1 - m_i^2)^{-1} \exp(\tfrac{1}{2} N\beta^2 J^2 (1 - q)^2)$$

a positive quantity. This justifies *a posteriori* our earlier assumption that the modulus signs on the Jacobian can be dropped. The positivity of det **A** suggests that nearly all solutions of the TAP equations are *minima* of the TAP free energy. This lends credence to our identification of TAP solutions as metastable states. At $T = 0$ the TAP equations reduce to $m_i = \mathrm{sgn}(\sum_j J_{ij} m_j)$, and the identification is unambiguous in the sense that all solutions are stable against single spin-flips.

For the case $B = 0$, we have solved equations (17) numerically over the entire temperature range $0 \leqslant T \leqslant T_c$. The total number of solutions N_s is obtained by setting $u = 0$ (this removes the constraint imposed by the second delta function in equation (4)). The result is plotted in the form $N^{-1}\langle \ln N_s \rangle_J$ against T/T_c in figure 1 (the justification for placing the average outside the logarithm is given later). The variables q, λ, Δ, f thus obtained from equations (17) are appropriate to averages over all the TAP solutions. Close to T_c, analytic solutions may be obtained:

$$\begin{aligned} N^{-1}\langle \ln N_s \rangle_J &= (8/81)t^6 + O(t^7) \\ q &= t + t^2 - \tfrac{5}{9}t^3 + O(t^4) \\ \Delta &= \tfrac{2}{3}t^2 + \tfrac{10}{9}t^3 + O(t^4) \\ \lambda &= \tfrac{2}{3}t^2 + \tfrac{8}{9}t^3 + O(t^4) \\ f &= -\ln 2 - \tfrac{1}{4}\beta^2 J^2 + \tfrac{1}{6}t^3 + \tfrac{11}{24}t^4 + (79/90)t^5 + O(t^6) \end{aligned} \tag{18}$$

where $t = 1 - T/T_c$. These values are, in fact, characteristic of the overwhelming majority of all TAP solutions, since the integral for the total number

$$N_s = \int \mathrm{d}f\, N_s(f) = \int \mathrm{d}f \exp\{N(N^{-1} \ln N_s(f))\}$$

is dominated by a single value of f. (The equilibrium free energy, on the other hand, the lowest free energy for which TAP solutions exist, is determined from $\langle \ln N_s(f) \rangle_J = 0$ and cannot be calculated with the present methods.)

At $T = 0$, the number of metastable states may be obtained as the zero-temperature limit of the present theory, or by working directly with the equations $m_i = \mathrm{sgn}(\sum_j J_{ij} m_j)$. Either method gives $N^{-1}\langle \ln N_s \rangle_J = 0{\cdot}1992$ at $T = 0$, a result obtained independently by Tanaka and Edwards (1980) and by de Dominicis *et al* (1980), who have also obtained some of equations (18).

Consider now the effect of taking 'logarithmic' rather than 'direct' averages—in principle we should calculate $\langle \ln N_s(f) \rangle_J$ rather than $\ln \langle N_s(f) \rangle_J$, since $\ln N_s(f)$ is proportional to the size N of the system. The calculation utilises the standard replica trick, $\ln N_s = \lim_{n\to 0}(N_s^n - 1)/n$, and $\langle N_s^n(f) \rangle_J$ is calculated via simple generalisation of equation (4). Assuming no replica symmetry-breaking, one needs to introduce three new order parameters $\hat{q}, \eta, \rho$ which couple to quantities which are off-diagonal in the replica space. Details of the calculation will be presented elsewhere. The final result is (setting $B = 0$ as before)

$$N^{-1}\langle \ln N_s(f)\rangle_J = \max\Big\{-\Delta(1-q) - \Delta^2/2\beta^2J^2 + \rho[\beta J(1-q) + \Delta/\beta J] - \lambda q - uf - \tfrac{1}{2}\eta(q-\hat{q}) + \iint_{-\infty}^{\infty} \frac{dx\,dy}{2\pi} \exp[-\tfrac{1}{2}(x^2+y^2)] \ln I\Big\} \tag{19}$$

where

$$I = \int_{-1}^{1} \frac{dm}{(2\pi)^{1/2}\beta J(q-\hat{q})^{1/2}} \frac{1}{1-m^2} \exp\Big\{-\frac{(\tanh^{-1}m - \Delta m + \beta J\hat{q}^{1/2}x)^2}{2\beta^2J^2(q-\hat{q})} + \lambda m^2 + uf(m) + \frac{m}{(\hat{q})^{1/2}}[\rho x + (\eta\hat{q} - \rho^2)^{1/2}y]\Big\}. \tag{20}$$

This reduces to our previous result, equations (15) and (16), with $B = 0$, when $\eta, \rho, \hat{q}$ are set zero (in that order). The order parameters q and $\hat{q}$ have the physical significance

$$q = \langle\langle m_i^2\rangle_s\rangle_J, \qquad \hat{q} = \langle\langle m_i\rangle_s^2\rangle_J \tag{21}$$

where $\langle\ \rangle_s$ means an *average over solutions* with the given free energy f. *Thus it is* $\hat{q}$, *rather than* q, *which should be regarded as the Edwards–Anderson order parameter* $\langle\langle S_i\rangle^2\rangle_J$ *for this problem.*

Setting to zero the derivatives of $N^{-1}\langle \ln N_s(f)\rangle_J$ with respect to $q, \Delta, \lambda, u, \hat{q}, \eta, \rho$ determines these parameters as a function of f. There are trivial solutions with $\hat{q} = q$ corresponding once more to the solutions of Sherrington and Kirkpatrick and of Sommers, although the interpretation is now different (for example, evaluation of $N^{-1}\langle \ln N_s\rangle_J$ for the Sommers solution now gives a *negative* result). Solution of the seven coupled equations for the non-trivial solutions is a formidable task, even numerically. We believe, however, that replica symmetry breaking will be required as soon as the off-diagonal order parameters $\hat{q}, \eta, \rho$ become non-zero.

The free energy below which off-diagonal order develops can be determined by analysing the stability of equations (17) against off-diagonal fluctuations. Choosing u (instead of f) as independent variable for convenience we find (details elsewhere) that, provided u exceeds a critical value $u_c = -\frac{5}{6}t + O(t^2)$, or equivalently that f satisfies

$$f \geqslant f_c = -\ln 2 - \tfrac{1}{4}\beta^2J^2 + \tfrac{1}{6}t^3 + \tfrac{11}{24}t^4 + (133/180)\,t^5 + O(t^6)$$

there is no off-diagonal ordering. This justifies our previous analysis of the case $u = 0$ (and that of Tanaka and Edwards 1980). The vanishing of the off-diagonal order parameters has the implication, via equation (20), that the TAP solutions for $f \geqslant f_c$ are *uncorrelated*. For $f < f_c$, the off-diagonal order parameters become non-zero and correlations develop between solutions. Close to T_c one finds, for $u \geqslant u_c$,

$$N^{-1}\langle \ln N_s(f)\rangle_J = \tfrac{8}{81}t^6 - \tfrac{1}{12}u^2t^4$$
$$q = t + t^2 - \tfrac{5}{9}t^3 + \tfrac{1}{3}ut^2$$
$$f = -\ln 2 - \tfrac{1}{4}\beta^2J^2 + \tfrac{1}{6}t^3 + \tfrac{11}{24}t^4 + (79/90)t^5 + \tfrac{1}{6}ut^4.$$

If one assumes that there is no breaking of the replica symmetry, one can solve equations (19) and (20) for $u < u_c$ and choose u such that $\langle \ln N_s(f)\rangle_J = 0$, corresponding to the lowest free energy consistent with the existence of TAP solutions. One then finds

$q = t + O(t^2)$ and $\hat{q} = 0{\cdot}3471t + O(t^2)$, whereas intuitively one expects $\hat{q} = q$ for this lowest free energy. This is the origin of our belief that replica symmetry breaking is needed in this region, although a detailed stability analysis is needed to verify this.

For $T = 0$, we have calculated $\langle N_s(E)\rangle_J$, where $N(E)$ is the density of metastable states with energy NE, by averaging over solutions of the equations $m_i = \mathrm{sgn}(\sum_j J_{ij} m_j)$. A stability analysis shows that off-diagonal order parameters vanish (i.e. the metastable states are uncorrelated) for $E \geqslant E_c = -0{\cdot}672J$, and $N^{-1}\langle \ln N_s(E_c)\rangle_J = 0{\cdot}1254$. The energy for which the density of metastable states is maximal is $E_m = -0{\cdot}506J$, and $N^{-1}\langle \ln N_s(E_m)\rangle_J = 0{\cdot}1992$. The maximum energy of metastable states is $E_u = -0{\cdot}286J$. The complete function $N^{-1}\langle \ln N_s(E)\rangle_J$ is given in figure 2. If the theory without off-diagonal order parameters is continued into the unstable region (broken curve in figure 2) one finds that $N^{-1} \ln\langle N_s(E)\rangle_J$ vanishes at $E_l = -0{\cdot}791J$.

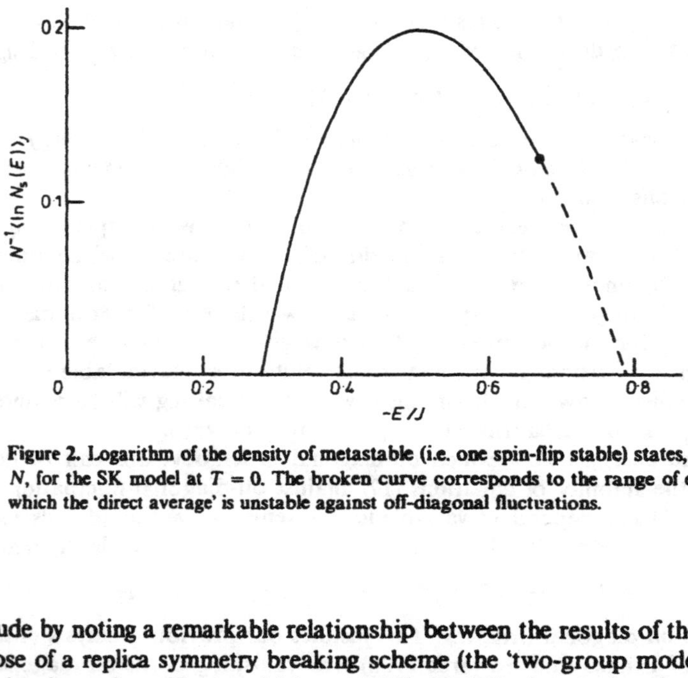

Figure 2. Logarithm of the density of metastable (i.e. one spin-flip stable) states, divided by N, for the SK model at $T = 0$. The broken curve corresponds to the range of energies for which the 'direct average' is unstable against off-diagonal fluctuations.

We conclude by noting a remarkable relationship between the results of the present work and those of a replica symmetry breaking scheme (the 'two-group model') introduced earlier by the authors (Bray and Moore 1978). If $\langle Z^n\rangle_J$ is calculated within the 'two-group model' (Z is the partition function of the SK model) and n is set to zero, one obtains not unity but a function identical to $\langle N_s\rangle_J$, the number of solutions of the TAP equations calculated above. We do not as yet understand the significance, if any, of this result.

During the final stages of this work we learned of the work of de Dominicis *et al* (1980). Their results agree with ours where they overlap, but the detailed results presented in figures 1 and 2, and the conclusion that the 'direct average' gives exact results for a range of free energies, are new here. We thank them for sending us a preprint of this work, and also T Garel for a copy of his thesis. We also wish to thank M G James and R Banach for useful discussions.

References

de Almeida J R L and Thouless D J 1978 *J. Phys. A: Math. Gen.* 11 983
Blandin A 1978 *J. Physique* 39 C6 1499
Bray A J and Moore M A 1978 *Phys. Rev. Lett.* 41 1068
—— 1980 *J. Phys. C: Solid St. Phys.* 13 419
de Dominicis C, Gabay M, Garel T and Orland H 1980 *J. Physique* submitted
de Dominicis C and Garel T 1979 *J. Physique* 40 L575
Edwards S F and Anderson P W 1975 *J. Phys. F: Metal Phys.* 5 965
Morgenstern I and Binder K 1980 preprint
Parisi G 1979 *Phys. Rev. Lett.* 43 1754
Sherrington D and Kirkpatrick S 1975 *Phys. Rev. Lett.* 32 1792
Sommers H-J 1978 *Z. Phys.* B31 301
Tanaka F and Edwards S F 1980 preprints
Thouless D J, Anderson P W and Palmer R G 1977 *Phil. Mag.* 35 593

Volume 73A, number 3 PHYSICS LETTERS 17 September 1979

TOWARD A MEAN FIELD THEORY FOR SPIN GLASSES

G. PARISI
INFN, Laboratori Nazionali di Frascati, Italy

Received 23 April 1979
Revised manuscript received 26 June 1979

We find an approximate solution of the Sherrington–Kirkpatrick model for spin glasses; the internal energy and the specific heat are in very good agreement with the computer simulations, the zero temperature entropy is unfortunately negative, although it is very small.

To find the exact solution of the Sherrington–Kirkpatrick model [1] for spin glasses is a long standing problem; the simple minded application of the replica theory [2] and of the saddle point method solves correctly the model in the high temperature phase, however, in the low temperature phase it disagrees with the computer simulations [3] and the calculated entropy is negative at low temperature. (In this model the true entropy is always nonnegative.)

It has been argued [4] that this discrepancy is due to a wrong choice of the saddle point: the model is invariant under permutations of the replicas, the saddle point used in refs. [1,2] is invariant under this symmetry, however, the saddle point which gives the leading contribution in the thermodynamic limit is not invariant and the replica symmetry is spontaneously broken.

Various patterns of symmetry breaking have been proposed [5,6]; the aim of this note is to present a new one: the novelty is that the pattern is treated as a variational parameter. As a first approximation we consider only very simple patterns of symmetry breaking; doing so, we do not find the exact solution of the model, but we obtain a substantial improvement over the solution with unbroken replica symmetry [1].

For completeness we recall that the SK model consists of N Ising spins interacting via a random infinite range hamiltonian. The distribution of the values of the bonds is a gaussian with variance $1/N$. Using the saddle point method in the thermodynamic limit ($N \to \infty$) it has been argued that the free energy density F is given by

$$\beta F = \tfrac{1}{4}\beta^2 + \max\left[\lim_{n\to 0}\left(\frac{1}{n}\sum_{a,b}\tfrac{1}{4}Y_{a,b}^2 - \frac{1}{n}\ln \mathrm{Tr}\exp\left(\beta\sum_{a,b}Y_{a,b}S_aS_b\right)\right)\right], \qquad (1)$$

where $\beta = 1/kT$, $Y_{a,b}$ is an $n \times n$ matrix, zero on the diagonal, Tr stands for the sum over all the possible values of the n spin variables S_a, the maximum is taken over all the possible choices of the matrix $Y_{a,b}$ and the indices a and b run from 1 to n. The values of the expressions in the limit $n = 0$ are defined as the analytic continuation in n from integer n.

In deriving eq. (1) one has made an unjustified exchange of the two limits: $n \to 0$ and $N \to \infty$ [7], however, we stick to the hypothesis that eq. (1) is correct, provided that the maximum is taken only after the analytic continuation in n and not vice versa.

It is not evident how to parametrize a matrix in a zero-dimensional space. A possible prescription is to parametrize the matrix $Y_{a,b}$ for positive integer n and to continue analytically the final results up to $n = 0$. The number of possible parametrizations is infinite and it is not to clear how to find the one which maximizes eq. (1). Lacking better information we can only proceed by trial and error. In this note we study the following ansatz, which is a generalization of the one proposed in ref. [5]:

$$Y_{a,b} = \beta(p+t), \quad \text{if } I(a/m) = I(b/m),$$

$$Y_{a,b} = \beta p, \quad \text{if } I(a/m) \neq I(b/m), \tag{2}$$

where $I(x)$ is an integer-valued function ($I(x)$ is the smallest integer greater than or equal to x) and m is an integer such that n/m is also integer. For $m = 2$ we recover the proposal of ref. [5]. In the limit $n = 0$ one finds:

$$\beta F(p,t,m)$$
$$= -\tfrac{1}{4}\beta^2[1 + mp^2 + (1-m)(p+t)^2 - 2(p+t)]$$
$$+ \ln 2 - (2\pi)^{-1/2} \int_{-\infty}^{+\infty} dz \left\{ \exp(-\tfrac{1}{2}z^2) \right.$$
$$\times m^{-1} \ln\left[(2\pi)^{-1/2} \int_{-\infty}^{+\infty} dy \exp(-\tfrac{1}{2}y^2) \right.$$
$$\left.\left. \times \text{ch}^m(\beta p^{1/2} z + \beta t^{1/2} y)\right]\right\}. \tag{3}$$

At fixed value of m, we must find the values of p and t which maximize eq. (3); there is no reason for considering only integer values of m: eq. (3) can be easily continued to noninteger values of m. If we demand that at high temperature $p = t = 0$ is a maximum of F, we find that m must belong to the interval 0–1. Having fixed the region of variation of m, we look for the maximum of F as function of p, t and m. At the maximum we must have:

$$\partial F/\partial m = \partial F/\partial p = \partial F/\partial t. \tag{4}$$

The system of eq. (4) has a solution with $t = 0$ (no breaking of replica symmetry) but it corresponds to a saddle point. The solution with $t \neq 0$ has a higher value of the free energy and it is a maximun. Near the critical temperature ($T_c = 1$) one finds:

$$p = \tfrac{1}{3}\tau + O(\tau^2), \quad t = \tfrac{2}{3}\tau + O(\tau^2),$$

$$m = \tau + O(\tau^2), \quad \tau = T_c - T. \tag{5}$$

At lower values of the temperature the analytic solution is very difficult to obtain and the maximum of eq. (3) has been found numerically using a computer: the internal energy $U(T)$, the specific heat $C(T)$ and the entropy $S(T)$ have been calculated. The functions $U(T)$ and $C(T)$ are in excellent agreement with the computer simulations of ref. [3]: at all temperatures the calculated values are inside the error bars of the computer simulations. For example, in the region $0.2 < T < 0.4$ the specific heat is quadratic and we have

$$S(T) \approx \tfrac{1}{2}A^2T^2, \quad C(T) \approx AT^2, \quad U(T) \approx \bar{U} + \tfrac{1}{3}AT^3,$$

$$A \approx 1.4, \quad \bar{U} = -0.764. \tag{6}$$

The Monte Carlo simulations are well represented by similar expressions where $\bar{U} = -0.76 \pm 0.01$ and the value of A is compatible with the value suggested in ref. [8] ($A = 2 \ln 2 \approx 1.39$). In the same region the theory without breaking of the replica symmetry ($t = 0$) does not agree with the Monte Carlo simulations; it predicts:

$$S(T) \approx S(0) + BT, \quad C(T) \approx BT,$$

$$U(T) \approx U(0) + \tfrac{1}{2}BT^2,$$

$$B \approx 0.53, \quad S(0) \approx -0.16, \quad U(0) \approx -0.798. \tag{7}$$

At lower values of T our calculations still agree with the Monte Carlo simulations for the functions $U(T)$ and $C(T)$; unfortunately, our entropy becomes negative for $T \leqslant 0.1$ and we obtain:

$$S(0) \approx -0.01, \quad U(0) \approx -0.765. \tag{8}$$

The obtained value of $U(0)$ is quite good; the negative value of the entropy is not acceptable: it indicates that we have found only an approximate solution of the SK model and that more elaborate patterns of symmetry breaking are needed to produce the exact solution of the model.

The solution with two order parameters (p and t) works much better than the solution with only one order parameter (i.e. $t = 0$). It is quite likely that an infinite number of order parameters is needed in the correct treatment (the generalization of eq. (2) is rather trivial) and that the neglected order parameters have small effects at not too small temperature. This letter should be considered as a tentative step toward the solution of the variational problem eq. (1).

It is a pleasure to acknowledge stimulating discussions with C. De Dominicis, C. Natoli, L. Peliti and N. Sourlas.

Volume 73A, number 3 PHYSICS LETTERS 17 September 1979

References

[1] D. Sherrington and S. Kirkpatrick, Phys. Rev. Lett. 35 (1975) 1972.
[2] S.F. Edwards and P.W. Anderson, J. Phys. F5 (1975) 965.
[3] D. Sherrington and S. Kirkpatrick, Phys. Rev. B17 (1978) 4385.
[4] J.R.L. de Almeida and D.J. Thouless, J. Phys. A11 (1978) 983;
F. Pytte and J. Rudnik, Phys. Rev. B19 (1979) 3603.
[5] A. Blandin, J. Physique C6 (1978) 1568.
[6] A.J. Bray and M.A. Moore, Phys. Rev. Lett. 41 (1978) 1069.
[7] J.L. Van Hemmen and R.G. Palmer, J. Phys. A12 (1979) 567.
[8] D.J. Thouless, P.W. Anderson and R.G. Palmer, Phys. Mag. 35 (1977) 593.

J. Phys. A: Math. Gen. 13 (1980) L115–L121. Printed in Great Britain

LETTER TO THE EDITOR

A sequence of approximated solutions to the S–K model for spin glasses

G Parisi

Istituto Nazionale de Fisica Nucleare, Laboratori Nazionali di Frascati, Casella Postale 13, 0004 Frascati, Roma, Italy

Received 4 January 1980

Abstract. In the framework of the new version of the replica theory, we compute a sequence of approximated solutions to the Sherrington–Kirkpatrick model of spin glasses.

It has recently been shown that, in the replica approach to spin glasses (Edwards and Anderson 1975), if the replica symmetry is broken (de Almeida and Thouless 1978, Pytte and Rudnik 1979), as happens in the spin glass phase at low magnetic field, the local order parameter is a function $q(x)$ defined on the interval 0–1 (Parisi 1980b, c). If the replica symmetry is unbroken, the function $q(x)$ is a constant.

The S–K model for spin glasses (Sherrington and Kirkpatrick 1975) is supposed to be soluble in the mean field approximation (the range of the interaction is infinite) and it is a good testing ground for this approach.

We derive here a convergent sequence of approximations to the free energy of the S–K model; excellent agreement is obtained with the computer simulations of Sherrington and Kirkpatrick (1978). The zero-temperature entropy is consistent with zero, while the zero-temperature internal energy is estimated to be

$$U(0) = -0{\cdot}7633 \pm 10^{-4}.$$

The computer simulations give $U(0) = -0{\cdot}76 \pm 0{\cdot}01$.

As in the conventional approach, we use the replica trick to integrate over the random spin couplings (Sherrington and Kirkpatrick 1975); in the saddle-point approximation (which is supposed to become exact in the thermodynamic limit) one finds that the free energy density F_R, as a function of the magnetic field h, is

$$F_R = TF_T(Q^0), \qquad \partial F_T/\partial Q_{ab}|_{Q=Q^0} = 0, \qquad \beta = 1/T,$$

$$F_T(Q) = -\frac{\beta^2}{4} + \lim_{n\to 0}\frac{1}{n}\left\{\frac{1}{4}\sum_a^n{}_1 \sum_b^n{}_1 \beta^2 Q_{a,b}^2 - \ln\left[\operatorname{Tr}\exp\left(\sum_{a=1}^n \sum_{b=1}^n \beta^2 Q_{a,b} S_a S_b + \beta h \sum_{a=1}^n S_a\right)\right]\right\}, \tag{1}$$

where Tr stands for the sum over all the 2^n possible values of the Ising spin variables S_a, and $Q_{a,b}$ is an $n\times n$ matrix, identically zero on the diagonal ($Q_{a,a} = 0$). In the limit $n\to 0$, Q becomes a 0×0 matrix, which is not a well defined object; the standard solution to this problem consists of writing the matrix Q as a function of some parameters q_i for generic integer n: at fixed q_i the free energy is analytically continued in n up to the point $n=0$ (Blandin 1978, Bray and Moore 1978, Palmer and Van Hemmer 1979).

Equation (2) does not fix the matrix Q^0 uniquely; for positive integer n the saddle-point method gives the correct result only if the function $F_T(Q)$ has a minimum at the point Q^0. This condition implies that the Hessian matrix

$$H_{(a,b):(c,d)} = \partial^2 F_T / \partial Q_{a,b} \partial Q_{c,d} \tag{2}$$

has positive eigenvalues: if we consider variations which leave the matrix $Q_{a,b}$ symmetric and zero on the diagonal, the matrix H will act on a space of dimensions $n(n-1)/2$, whose axes are labelled by a pair of indices (a, b), with $a \neq b$. If $n<1$, the dimensions of the space on which H acts (the number of independent components of Q) becomes negative. Now, it has been remarked that in this unusual situation, in order to apply the saddle-point method correctly, the eigenvalues of H must be non-negative (de Almeida and Thouless 1978, Pytte and Rudnik 1979, Bray and Moore 1978); unfortunately the positivity of H (i.e. the positivity of its eigenvalues) does not imply that F_T is a minimum as a function of the q_i. For example, if we restrict ourselves to studying the problem in the subspace of matrices having the form $Q_{a,b}=q$, the condition of positivity of the eigenvalues of H restricted in this subspace implies that F_T must be a maximum and not a minimum as a function of q; this happens because

$$(1/n)\,\mathrm{Tr}\,Q^2 = (n-1)q^2 \tag{3}$$

becomes negative definite for $n<1$.

In the general case we cannot say if F_T should be maximised or minimised as a function of the parameters q_i; however, if we restrict ourselves to studying the problem in a subspace in which $\mathrm{Tr}(Q^2)/n$ is negative definite, we must maximise and not minimise F_T as a function of q_i.

The condition that the matrix H does not have negative eigenvalues in a subspace does not imply that H has no negative eigenvalue. Indeed it has been found (de Almeida and Thouless 1978, Pytte and Rudnik 1979) that there are negative eigenvalues of H, if we choose the replica symmetric solution ($Q^0_{a,b}=q$). It is necessary to look for other solutions of equation (2) where the matrix Q^0 has a non-trivial dependence on the indices. The space of 0×0 matrices is a very large space (infinite dimensional) and we do not know how to write the generic matrix of this space; at the present moment the only viable approach consists of doing a simple ansatz for the matrix Q^0 and studying the problem in a smaller space; at the end of the computation one should compute the eigenvalues of the Hessian in order to check if the eigenvalues of H are positive.

It has been suggested that the following parametrisation should be considered (Parisi 1980a):

$$Q_{a,b}=q_i \qquad \textit{If}: I\left(\frac{a}{m_i}\right) \neq I\left(\frac{b}{m_i}\right) \text{ and } I\left(\frac{a}{m_{i+1}}\right) = I\left(\frac{b}{m_{i+1}}\right), \qquad i=0, K, \tag{4}$$

where the m_i are integer numbers such that m_{i+1}/m_i is an integer ($i=1,K$) with $m_0=1$ and $m_{K+1}=n$; $I(x)$ is an integer valued function: its value is the smallest integer greater than or equal to x. The matrix Q depends on $K+1$ real parameters (q_i) and K integer parameters (m_i); if we call M_K the space of matrices having the form dictated by equation (4), it is easy to see that $M_{K+1} \supset M_K$. It has been suggested that if n is not integer, there is no reason to restrict ourselves to the case where the m_i are integers and we can treat the variable m_i as a real number: the free energy will be computed as the analytic continuation from the integer m_i.

Let us restrict ourselves to the space M_K^+ where the m_i satisfy the following condition:

$$m_i > m_{i+1}. \tag{5}$$

To work in M_K^+ presents two advantages: $(1/n)\,\mathrm{Tr}\,Q^2$ is negative definite,

$$\lim_{n\to 0}\frac{1}{n}\,\mathrm{Tr}\,Q^2 = -\sum_{0}^{K}{}_i\,(m_i - m_{i+1})q_i^2, \tag{6}$$

and it is possible to associate with the matrix Q a function $q(x)$ on the interval 0–1 defined by

$$q(x) = q_i, \qquad m_{i+1} \leq x \leq m_i. \tag{7}$$

For finite K the function $q(x)$ is piecewise constant and for $K \to \infty$ we can obtain a smooth function. Equation (6) implies that F_T must be maximised as a function of the q_i. If $K = 0$, $q(x)$ is a constant function; we recover the traditional approach with unbroken replica symmetry.

It is easy to show that the internal energy and the susceptibility are given by

$$U = -\beta/2 \int_0^1 (1 - q^2(x))\,\mathrm{d}x, \qquad \chi = \beta \int_0^1 (1 - q(x))\,\mathrm{d}x. \tag{8}$$

The identification of the Edwards–Anderson order parameter $q_{\mathrm{EA}} = \langle\langle\sigma\rangle^2\rangle$ is not easy in this framework; this difficulty is also present in the approach of Blandin (1978) and Blandin *et al* (1979), which corresponds in our language to the case $K = 1$, integer m_1. They have suggested that

$$q_{\mathrm{EA}} = \lim_{\epsilon\to 0} Q_{a,b}(\epsilon), \tag{9}$$

where $Q_{a,b}$ has been computed after we have added to the argument of the exponential in equation (1) a term proportional to $\epsilon S_a S_b$. This term is an infinitesimal breaking of the replica symmetry; it removes the ambiguity that would be present in equation (9) for $\epsilon = 0$. If we apply this suggestion to our case, we find

$$q_{\mathrm{EA}} = \max_x q(x). \tag{10}$$

The derivation of equation (10) is far from being rigorous, and it should be justified by a more careful analysis; equations (8) and (10) together show that the breaking of the replica symmetry is connected with the failure of the relation $\chi = \beta(1 - q_{\mathrm{EA}})$ (Fisher 1975).

Analytic results can be obtained near the critical temperature ($T_c = 1$) (Parisi 1980a); the maximum of F_T is located at $K = \infty$, and for finite K the errors in the free energy and in the magnetic susceptibility decrease respectively like $(2K+1)^{-4}$ and $(2K+1)^{-2}$. The bulk of the corrections for going from $K = 0$ to $K = \infty$ are obtained also for K as small as 1. Numerical results for $K = 1$ (Parisi 1979, 1980b) are indeed in good agreement with the computer simulation. In this Letter we report on the numerical results for $K = 2$ and we present a formalism which allows us to obtain the results also for higher values of K.

Using the Gaussian integral representation to disentangle the sum over different spins, $F_T(Q)$ can be written as a K-fold integral: in the case $K=2$ we find

$$F_T(Q)=-\tfrac{1}{4}\beta^2\Big(1+\int_0^1 q^2(x)\,dx-2q(1)\Big)$$
$$-\int_{-\infty}^{+\infty} dz_0\, G_{q_0}(z_0)\ln\Big[\Big[\int_{-\infty}^{+\infty} dz_1\, G_{q_1-q_0}(z_1-z_0)$$
$$\times\Big(\int_{-\infty}^{+\infty} dz_2\, G_{q_2-q_1}(z_2-z_1)\cosh^{m_2}(\beta z_2+\beta h)\Big)^{m_2/m_1}\Big]^{1/m_1}, \tag{11}$$
$$G_q(z)=(2\pi q)^{-1/2}\exp(-z^2/2q).$$

Using the fact that $G_q(z)$ is the Green function of the heat equation, equation (11) can be formally written as

$$F_T(Q)=-\tfrac{1}{4}\beta^2(1+S_0^1 q^2(x)\,dx-2q(1))$$
$$-C_{q_0}\ln\{C_{q_1-q_0}[C_{q_2-q_1}\exp(m_2 f_S(z+h))]^{m_2/m_1}\}^{1/m_1}|_{z=0}, \tag{12}$$
$$f_S^{(h)}=\ln[2\cosh(\beta h)],\qquad C_q=\exp(\tfrac{1}{2}q\,d^2/dz^2).$$

The numerical evaluation of equation (11) is rather simple; by maximising the free energy as a function of the parameters q_i and m_i, one finds the results shown in table 1.

Table 1. We show the zero temperature entropy, the internal energy and susceptibility for $K=0, 1, 2$.

K	$S(0)$	$U(0)$	$\chi(0)$
0	−0·16	−0·798	0·80
1	−0·01	−0·7652	0·95
2	−0·004	−0·7636	0·98

As expected, the convergence with increasing K is fairly fast. The absolute value of the negative entropy decreases with increasing K and it is quite likely that $S(0)=0$ for $K=\infty$. The problem of negative zero-temperature entropy, which plagues the conventional approach to spin glasses, seems to be absent here; this result strongly suggests that in the limit $K\to\infty$ one obtains the correct solution of the S–K model.

When $T\to 0$, the q_i have a finite limit, while the x_i are proportional to T. As an example we show in figure 1 the function $q(x)$ in the approximations $K=1, 2$ for $T=0{\cdot}3$.

The values we obtain for the magnetic susceptibility do not agree with the results of the Monte Carlo simulations of Sherrington and Kirkpatrick (1978); however, in their computations they have implicitly assumed the validity of the Fisher relation $\chi=\beta(1-q_{EA})$, which is not valid in this approach; on the other hand, the value we obtain for q_{EA}, using equation (10), is in good agreement with their Monte Carlo simulations.

According to Thouless *et al* (1977), in the low-temperature region we have

$$S(T)\simeq\beta T^2,\qquad \dot q_{EA}\simeq 1-\alpha T^2,\qquad \beta=\tfrac{1}{4}\alpha^2=\ln 2. \tag{13}$$

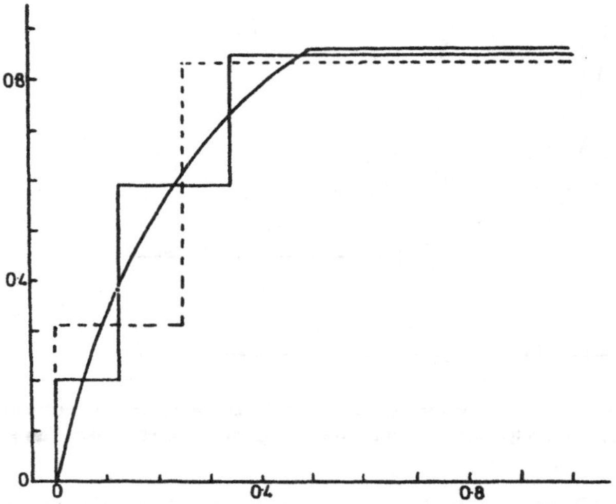

Figure 1. The dashed line and the full line are the functions $q(x)$ in the approximations $K = 1$ and $K = 2$, respectively. The full curve is an educated guess for the true function $q(x)$.

In order to see if equation (13) holds in our approach, we have plotted in figures 2 and 3 the functions

$$s(T) = S(T)/T^2, \qquad r(T) = (1 - q_{EA}(T))/[T^2(2-T)]. \tag{14}$$

The approximation of keeping $q(x)$ piecewise constant worsens with decreasing T, and the division by T^2 enhances the errors in our approximation (the difference between q_{EA} in the two cases $K = 1$ and $K = 2$ is always less than 0·015), so we cannot expect that for finite K the functions s and r have a finite limit for $T \to 0$. In the

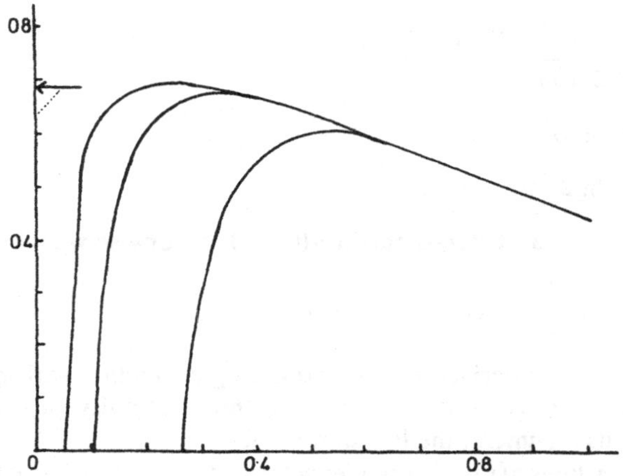

Figure 2. The three curves are from below the functions $s(T)$ in the approximations $K = 0$, 1 and 2, respectively. The arrow is the zero-temperature prediction of Thouless *et al* (1977).

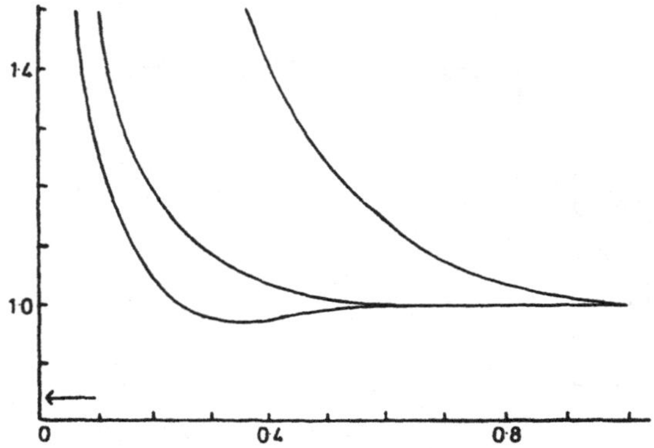

Figure 3. The three curves are from above the functions $r(T)$ in the approximations $K=0, 1$ and 2, respectively. The arrow is the zero-temperature prediction of Thouless *et al* (1977).

intermediate T region our results are in qualitative agreement with equation (10), although we are unable to extract the values of α and β, using such a low value of K.

From explicit computations done near the critical temperature, we know that in the limit $K\to\infty$, $q(x)$ becomes a smooth function (Parisi 1980a, b); it is straightforward to write the generalisation of equations (11) and (12) to an arbitrary K: in the limit $K\to\infty$ one finds that

$$F_R=\max_{q(x)} TF_T[q], \tag{15}$$

$$F_T[q]=-\frac{1}{4}\beta^2\Big(1+\int_0^1 q^2(x)\,\mathrm{d}x+2q(1)\Big)+\tilde{F}_T[q],$$

$$\tilde{F}_T[q]=-f(0,h),$$

where the function $f(x,h)$ satisfies the following nonlinear differential equation:

$$\frac{\partial f}{\partial x}=-\frac{1}{2}\frac{\mathrm{d}q}{\mathrm{d}x}\left[\frac{\partial^2 f}{\partial h^2}+x\left(\frac{\partial f}{\partial h}\right)^2\right] \tag{16}$$

with the boundary condition

$$f(1,h)=\ln[2\cosh(\beta h)]. \tag{17}$$

Equation (15) is correct as it stands only if $q(0)=0$, otherwise we would have

$$\tilde{F}_T[q]=-\int_{-\infty}^{+\infty}\mathrm{d}z\, G_{q(0)}(z)f(0,z+h). \tag{18}$$

Equation (18) can also be derived by approximating a function with $q(0)\neq 0$ by a sequence of functions with $q(0)=0$, i.e. by using the continuity of the functional $\tilde{F}_T[q]$ with respect to its argument, the function $q(x)$.

The shape of the functions $q(x)$ in figure 1, exact results near the critical temperature and preliminary results for high values of K strongly suggest that $q(0)=0$, if the magnetic field h is equal to zero; (an analysis of equation (15) shows that this is possible only if $\chi=1$).

If the function $q(x)$ is monotonically increasing, we can define the function $x(q)$; in this case equations (15)–(17) simplify to

$$\tilde{F}_T(q) = -f(q, h)|_{q=0}, \qquad f(q, h)|_{q=q_{EA}} = \ln[2\cosh(\beta h)],$$
$$\frac{\partial f}{\partial q} = -\frac{1}{2}\left[\frac{\partial^2 f}{\partial h^2} + x(q)\left(\frac{\partial f}{\partial h}\right)^2\right]. \qquad (19)$$

The conventional treatment of the model, i.e. no breaking of the replica symmetry, corresponds to $x(q) = 0$. It would be very interesting to understand if and how equation (19) can be derived, starting from the TAP equations of Thouless *et al* (1977).

The method presented in this Letter enables us to compute thermodynamic properties of the S–K model with arbitrary precision (it would be quite interesting to do analytic computations near $T = 0$). The main unsolved problem consists in the computation of the eigenvalues of the Hessian near our solution, in order to verify that non-negative eigenvalues are present. For finite K we expect the presence of negative eigenvalues; their absolute value should decrease with K, and become zero only for infinite K, leaving to us a massless mode (i.e. a zero eigenvalue), the so-called replicon (Bray and Moore 1978). Knowledge of the Hessian is also needed to compute the corrections to the saddle-point approximation: it would be the key step toward the application of this formalism to more realistic models of spin glasses.

References

Blandin A 1978 *J. Physique* C6 1578
Blandin A, Gabay M and Garel T 1979 *Orsay Preprint* Phys. Sol. 79/39
Bray A J and Moore M A 1978 *Phys. Rev. Lett.* 41 1069
de Almeida J R L and Thouless D J 1978 *J. Phys. A: Math. Gen.* 11 983
Edwards S F and Anderson P W 1975 *J. Phys. F: Metal Phys.* 5 965
Fisher K H 1975 *Phys. Rev. Lett.* 34 1438
Palmer R G and Van Hemmer L 1979 *J. Phys. A: Math. Gen.* 12 567
Parisi G 1979 *Phys. Lett.* A 73 207
—— 1980a *Phys. Rev. Lett.* to be published
—— 1980b *J. Phys. A: Math. Gen.* 13 1101–12
—— 1980c *J. Phys. A: Math. Gen.* 13 in the press
Pytte F and Rudnik J 1979 *Phys. Rev.* B 19 3603
Sherrington D and Kirkpatrick S 1975 *Phys. Rev. Lett.* 35 1972
—— 1978 *Phys. Rev.* B 17 4385
Thouless D J, Anderson P W and Palmer R G 1977 *Phil. Mag.* 35 593

J. Phys. A: Math. Gen. 13 (1980) 1101–1112. Printed in Great Britain

The order parameter for spin glasses: A function on the interval 0–1

G Parisi
INFN–Laboratori Nazionali di Frascati, Cas. Postale 13, 00044, Frascati, Italy

Received 31 July 1979

Abstract. We study the breaking of the replica symmetry in spin glasses. We find that the order parameter is a function on the interval 0–1. This approach is used to study the Sherrington–Kirkpatrick model. Exact results are obtained near the critical temperature. Approximated results at all the temperatures are in excellent agreement with the computer simulations at zero external magnetic field.

1. Introduction

Magnetic transitions in ferromagnetic or antiferromagnetic materials are well understood theoretically; one of the most interesting open problems is the nature of the transitions in spin glasses (i.e. systems which are neither ferromagnetic nor antiferromagnetic, because the sign of the exchange interaction changes randomly from bond to bond). Spin glasses are the simplest amorphous materials we can study; we face the problem of finding the order parameter which is appropriate to describe the onset of ordering in a disordered medium.

The general framework in which we study this problem is the replica theory (Edwards and Anderson 1975). The main idea is rather simple: the free energy (F_R) of a random system can be written as

$$F_{\mathrm{R}} = \int \mathrm{d}[J]\mu[J]F[J] \tag{1}$$

where J stands for the random variables, $\mu[J]$ is their probability measure (normalised to 1) and $F[J]$ is the J-dependent free energy. In spin glasses:

$$\begin{aligned} \beta F[J] &= -\ln \int \prod_{i=1}^{N} \rho(\sigma_i)\, \mathrm{d}\sigma_i \exp(-\beta H[J,\sigma]) = -\ln Z[J] \\ H[J,\sigma] &= \sum_{i,k=1}^{N} J_{ik}\sigma_i\sigma_k \end{aligned} \tag{2}$$

where σ_i are the spin variables, $\rho(\sigma)$ is their distribution (in the Ising model $\rho(\sigma) = \delta(\sigma^2 - 1)$) and N is the number of spins.

Equations (1) and (2) are not easy to study under this form, because F_R is not written as the integral of the exponential of an Hamiltonian, as normally happens. In order to

0305-4470/80/031101+12\$01.00

present the problem under a more familiar form, it is useful to introduce the function

$$Z_n = \frac{1}{n}\int \mathrm{d}[J]\mu[J]Z^n[J]. \tag{3}$$

Obviously one has:

$$\beta F_R = -\lim_{n\to 0}\left(Z_n - \frac{1}{n}\right). \tag{4}$$

Now, for integer n, Z_n can be written as:

$$Z_n = \int \mathrm{d}[J]\mu[J]\int \prod_{1}^{n}{}_{\alpha} \prod_{1}^{N}{}_{i}\, \rho(\sigma_i^\alpha)\,\mathrm{d}\sigma_i^\alpha \exp\left(-\beta \sum_{1}^{n}{}_{\alpha} H[J,\sigma^\alpha]\right) \tag{5}$$

where σ_i^α are $n\times N$ spin variables.

Equation (5) is the partition function of n identical replicas of the same system, interacting with the same J-dependent Hamiltonian.

The strategy consists in finding the partition function Z_n for generic integer n and finally performing the analytic continuation up to the point $n=0$. In this way one is led to introduce, as an order parameter, the $n\times n$ matrix:

$$Q_i^{\alpha,\beta} = \langle \sigma_i^\alpha \sigma_i^\beta \rangle \qquad \alpha \neq \beta \tag{6}$$

and a physical order parameter:

$$\bar{q} = \frac{1}{N}\sum_{1}^{N}{}_{i} \langle\langle \sigma_i \rangle^2 \rangle \tag{7}$$

where the internal bracket indicates the thermodynamic expectation value at fixed J, while the external bracket indicates the mean value over J.

In the high-temperature phase $(1/N)\,\Sigma_i\, Q_i^{\alpha\beta} \equiv Q_{\alpha\beta} = 0$, while in the spin glass phase $Q_{\alpha\beta} \neq 0$. In the standard treatment it is assumed that $Q_{\alpha\beta} = q$ independently from α and β. This possibility is the only one symmetric under permutations of the replicas. In this scheme $\bar{q} = q$.

In order to test the correctness of this approach, it is useful to investigate a model (the S–K model) (Sherrington and Kirkpatrick 1975) in which the mean field approximation should be exact; this model consists of N Ising spins interacting one with all the others with a random Gaussian interaction $(\langle J_{ik}^2\rangle = 1/N)$. Assuming that $Q_{\alpha\beta} = q$, the model can be solved, using the saddle point method when $N\to\infty$. One finds that

$$\beta F_R(T) = \max F_T(q)$$

$$F_T(q) = -\frac{\beta^2}{4}(1+q)^2 + \ln 2$$

$$-(2\pi)^{-1/2}\int \mathrm{d}z\left\{\exp\left(-\frac{z^2}{2}\right)\ln[\cosh(\beta q^{1/2} z)]\right\} \qquad \beta = 1/T. \tag{8}$$

A transition is present at $T=1$ and $q\neq 0$ when $T<1$.

From the knowledge of $F_R(T)$ other thermodynamical quantities, like the specific heat $(U(T))$ and the entropy $(S(T))$, can be calculated. However the results disagree with the computer simulations (Sherrington and Kirkpatrick 1978) for $N=500$, extrapolated up to $N=\infty$ (e.g. the computer simulations give $U(0) = -0{\cdot}76 \pm 0{\cdot}01$ while this analytic method gives $U(0) = -\sqrt{2/\pi} \simeq -0{\cdot}80$). The situation worsens if we consider the entropy: by definition $S(T)$ is non-negative and equation (8) implies a negative value of S at low temperatures ($S(0) \simeq -0{\cdot}17$ while $S(\infty) = \ln 2 \simeq 0{\cdot}69$).

The origin of this failure remained unexplained for some time: it is possible to blame the exchange of limits $n \to 0$ with $N \to \infty$ (Van Hemmer and Palmer 1979) but no constructive approach can be found to avoid this difficulty.

It has been finally remarked (de Almeida and Thouless 1978, Pytte and Rudnik 1979) that the correct expression is

$$F_{\mathrm{R}} = T \max F_T(Q)$$

$$F_T(Q) = -\frac{\beta^2}{4} + \ln 2 + \lim_{n\to 0}\{\tfrac{1}{4}\textstyle\sum_{\alpha,\beta}\beta^2 Q^2_{\alpha,\beta} - \ln[\mathrm{Tr}\exp(\sum_{\alpha,\beta}\beta^2 Q_{\alpha,\beta}S_\alpha S_\beta)]\}/n \tag{9}$$

where Tr stands for the sum over all the 2^n possible values of the n Ising spin variables S_α and the maximum is taken over all possible matrices $Q_{\alpha,\beta}$. Equation (8) is correct only if $F(Q)$ has its maximum at a symmetric point; in reality the symmetric point is only a saddle point. This can be seen by computing the eigenvalues of the matrix $M_{\alpha\beta;\gamma\delta}$ defined by

$$\begin{aligned} &F_T(Q) = F(Q^0) + \Delta_{\alpha\beta} M_{\alpha\beta;\gamma\delta}\Delta_{\gamma\delta} + \mathrm{O}(\Delta^3) \\ &\Delta = Q_{\alpha\beta} - Q^0_{\alpha\beta} \qquad Q^0_{\alpha\beta} = q^0 \end{aligned} \tag{10}$$

where q^0 maximises equation (8).

A straightforward computation shows that the matrix M has negative eigenvalues for $T<1$. The replica symmetry invariant point does not maximise $F(Q)$ and replica symmetry must be broken: we have to look for solutions of equation (9) which are not symmetric in α and β.

We face the rather difficult problem of parametrising an $n \times n$ matrix in the limit $n = 0$. To work directly in zero-dimensional space is rather difficult; to circumvent this problem we will define also the matrix $Q_{\alpha\beta}$ by analytic continuation. We define an $n \times n$ matrix $Q^{(n)}_{\alpha\beta}$ which depends on a set of parameters $\{q_i, m_i\}$ (e.g. the q_i are the elements of the matrix and the m_i describe the form of the matrix). With a suitable choice of the parametrisation $F_T(Q^{(n)})$ can be extended to an analytic function of n (at this end we need that $Q^{(n)}_{\alpha,\beta}$ is defined only for n multiples of a fixed integer) and the maximum of $F_T(Q)$ should be taken over all the possible parametrisations.

It is evident that the number of different parametrisations is unbounded and the space of $O \otimes O$ matrices with these definitions is an infinite dimensional space.

The search for a maximum is not simple in such a big space. We have been guided by the following three requirements:

$$\left[\lim_{n\to 0}\frac{1}{n}\textstyle\sum_{\alpha,\beta} Q^2_{\alpha,\beta}\right] < \infty \tag{11a}$$

$$\sum_{1}^{N}{}_{\beta}\, Q_{\alpha\beta} = \sum_{1}^{N}{}_{\beta}\, Q_{\gamma\beta} \qquad \alpha \neq \gamma \tag{11b}$$

$$-\lim_{n\to 0}\frac{1}{n}\textstyle\sum_{\alpha,\beta} Q^2_{\alpha,\beta} \geq 0. \tag{11c}$$

Requirement (11*a*) comes from the condition that F_{R} must be finite; the eigenvectors with negative eigenvalue of the matrix M satisfy requirement (11*b*): it is natural to look for a maximum of $F_T(Q)$ in the space spanned by these vectors; at high temperatures we want the maximum of $F_T(Q)$ located at $Q_{\alpha,\beta}=0$ and this happens only if condition (11c) is satisfied. We recall that the saddle point method can be applied only if the matrix M has non-negative eigenvalues: this condition implies that the function $F(Q)$ must be maximised, when it is restricted to the subspace where the condition (11c) is identically satisfied.

Although the requirements (11) do not fix the symmetry breaking pattern, they exclude those previously proposed (Blandin 1978, Bray and Moore 1978). In the first case requirement (11c) is not satisfied, in the second case requirement (11a) is violated. In this paper we investigate the simplest parametrisations of the matrix $Q_{\alpha,\beta}$ satisfying requirements (11).

In § 2 we describe the parametrisations we propose and we show that a function $q(x)$ defined on the interval 0–1 is naturally associated to each parametrisation of the class we consider. In this approach the order parameter belongs to $L^2(0, 1)$. If replica symmetry is unbroken $q(x)$ is a constant. In simple approximation schemes $q(x)$ is a piecewise constant function which takes only a finite number of values. A direct interpretation of $q(x)$ is lacking although it may have the meaning of probability distribution. This point deserves more accurate investigations.

In § 3 we apply this approach to the study of the S–K model near T_c. In § 4 we show how a very simple-minded approximation ($q(x)$ takes only two values) is sufficient to obtain a substantial improvement with respect to the situation with unbroken replica symmetry for the S–K model at all the temperatures (e.g. we obtain $U(0)\simeq -0{\cdot}765$ and $S(0)\simeq -0{\cdot}01$).

2. The parametrisation

In this paper we will study the following parametrisation of the matrix $Q_{\alpha,\beta}$ (Parisi 1979b):

$$\begin{aligned} &Q_{\alpha,\alpha}=0 \\ &Q_{\alpha,\beta}=q_i \qquad \text{if } I(\alpha/m_i)\neq I(\beta/m_i) \text{ and } I(\alpha/m_{i+1})=I(\beta/m_{i+1}) \end{aligned} \tag{12}$$

where $q_i (i=0, K)$ are real numbers and m_i ($i=1, K$) are integer numbers such that m_{i-1}/m_i is an integer ($i\geq 1$). (We let $m_0=1$, $m_{K+1}=n$.)

The matrix $Q_{\alpha,\beta}$ depends on $K+1$ real parameters (the q_i) and on K integer parameters (the m_i). For $n=8$, $K=2$, $m_1=2$, $m_2=4$, we have:

$$Q_{\alpha,\beta}=\left|\begin{array}{cccccccc} 0 & q_0 & q_1 & q_1 & & & & \\ q_0 & 0 & q_1 & q_1 & & q_2 & & \\ q_1 & q_1 & 0 & q_0 & & & & \\ q_1 & q_1 & q_0 & 0 & & & & \\ & & & & 0 & q_0 & q_1 & q_1 \\ & & & & q_0 & 0 & q_1 & q_1 \\ & q_2 & & & q_1 & q_1 & 0 & q_0 \\ & & & & q_1 & q_1 & q_0 & 0 \end{array}\right| \tag{13}$$

We do not have any serious argument to justify the ansatz equation (12) (apart from the requirement (11)). Its main virtue is its simplicity. It is not evident *a priori* if the solution of the variational problem, equation (9), has the form dictated by equation

(12). The only possible justification of the ansatz equation (12) is its ability to reproduce the results of the computer simulations, as we shall see in § 4.

We must now continue the matrix $Q_{\alpha\beta}$ up to $n=0$. In doing so it is not evident if the m_i must remain integers. We suppose that for non-integer n, no conditions on the m_i, are present, i.e. they can be arbitrary real numbers (Parisi 1979a). However, we want conditions (11) to be satisfied.

Conditions (11*a*) and (11*b*) are identically satisfied while condition (11*c*) implies

$$1 \geqslant m_1 \geqslant m_2 \ldots m_K \geqslant 0. \tag{14}$$

Equation (14) follows from the relation:

$$\lim_{n\to 0} \frac{1}{n} \sum_{\alpha,\beta} Q^2_{\alpha\beta} = \sum_{0}^{K}{}_{i} (m_i - m_{i+1}) q_i^2. \tag{15}$$

The scheme of Blandin (1978) is $K=1$, $m_1=2$ and obviously does not satisfy condition (11*c*).

It is natural to define the function $q^{(K)}(x)$ as

$$q^{(K)}(x) = q_i \qquad \text{if } m_i < x < m_{i+1}. \tag{16}$$

By definition we have:

$$-\lim_{n\to 0} \frac{1}{n} \sum_{\alpha,\beta} Q^2_{\alpha,\beta} = \int_0^1 \mathrm{d}x (q^{(K)}(x))^2. \tag{17}$$

$q^{(K)}(x)$ is a piecewise function which takes at most $K+1$ different values. In the limit $K\to\infty$, we obtain a generical function of $L^2(0,1)$. In the next section we will argue that the maximum of equation (9) is reached in the limit $K\to\infty$.

At this stage it is unclear if the sequence of functions $q^{(K)}(x)$ converges toward a function $q(x)$ when $K\to\infty$. We shall verify that this happens in an explicit example in the next section.

3. Analytic results near T_c

Near the critical temperature T_c ($T_c=1$) the matrix $Q_{\alpha\beta}$ is small (proportional to $\tau = T_c - T$) so that it is reasonable to expand it in powers of Q.

One finds (Bray and Moore 1978, Pytte and Rudnik 1979):

$$F_T(Q) = \lim_{n\to 0} (-\tau \operatorname{Tr} Q^2 + \tfrac{1}{3} \operatorname{Tr} Q^3 + y \sum_{\alpha,\beta} Q^4_{\alpha,\beta} + \mathrm{O}(Q^4))/n \tag{18}$$

where Tr is the standard trace in the n-dimensional vector space. Among the various terms of fourth order we have written the only one which is responsible for the breaking of the replica symmetry.

Indeed, if $y \geqslant 0$ the symmetric solution would be a maximum and not a saddle point. In the S–K model y is negative and replica symmetry is broken. We will study in detail the case $y=-\frac{1}{4}$ and look for a maximum of $F(Q)$ with $Q_{\alpha,\beta} = \mathrm{O}(\tau)$.

After some algebra one finds that

$$F_T(Q) = \int_0^1 \mathrm{d}x \left(+\tau q^2(x) + \tfrac{1}{4} q^4(x) - \tfrac{1}{3} x q^3(x) - q^2(x) \int_x^1 q(y)\, \mathrm{d}y \right) \tag{19}$$

where the parametrisation (12) has been used and the function $q(x)$ is defined by equation (6) (for simplicity we have written $q^{(K)}(x)$ as $q(x)$).

Equation (19) can also be written using the parameters q_i and m_i as:

$$F_T(q_i, m_i) = \sum_0^N {}_i (m_i - m_{i+1})[+\tau q_i^2 + \tfrac{1}{4}q_i^4 - \tfrac{1}{3}(2m_i - m_{i+1})q_i^3 + q_i \sum_{i+1}^N {}_j (m_j - m_{j+1})q_j^2]. \quad (20)$$

At fixed K we look for a local maximum of $F(Q)$, under the conditions $q_i = O(\tau)$. One finds:

$$\begin{aligned} q_0 &= \tau + C^{(K)}\tau^2 + O(\tau^3) \\ q_i &= B_i^{(K)}\tau + O(\tau^2) \\ m_i &= L_i^{(K)}\tau + O(\tau^2). \end{aligned} \quad (21)$$

After some painful algebra one obtains:

$$\begin{aligned} C^{(K)} &= \frac{3}{2} - \frac{1}{(2K+1)^2} \qquad B_i^K = \frac{2(K-i)+1}{2K+1} \\ L_i^{(K)} &= \frac{6i}{2K+1}. \end{aligned} \quad (22)$$

When $K \to \infty$ the function $q^{(K)}(x)$ converges toward:

$$\begin{aligned} q(x) &= \frac{x}{3} + O(\tau^2) \qquad \text{if } x < 3\tau \\ q(x) &= \tau + O(\tau^2) \qquad \text{if } x > 3\tau. \end{aligned} \quad (23)$$

In figure 1 we have shown the function $q^{(K)}(x)$ taking only the terms of $O(\tau)$, for $K = 1, 4$ and ∞.

It would be tempting to interpret $q(x)$ as the mean value of the parameter q (equation (7)) inside a cluster of size xN, but the rationale for this interpretation is rather mysterious.

If we consider the internal energy $U(\tau) = dF/d\tau$, we find that

$$U(\tau) = \int_0^1 q^2(x)\,dx = \tau^2 + \tau^3 + U_4^{(K)}\tau^4 + O(\tau^5) \quad (24)$$

where:

$$U_4^{(K)} = \frac{q}{4} - \frac{1}{(2K+1)^4}. \quad (25)$$

It is remarkable that $U_4^{(K)}$ for $K = 1$ differs from the exact result by less than 1%; we expect rather good results at all temperatures from the approximation $K = 1$; this expectation is confirmed from the results of the next section.

For completeness we also write the result:

$$\begin{aligned} -\sum_2^n {}_\beta Q_{\alpha\beta} &\equiv \tilde{q} \equiv \int_0^1 q(x)\,dx = \tau + q_2^{(K)}\tau^2 + O(\tau^3) \\ q_2^{(K)} &= \frac{1}{2(2K+1)^2}. \end{aligned} \quad (26)$$

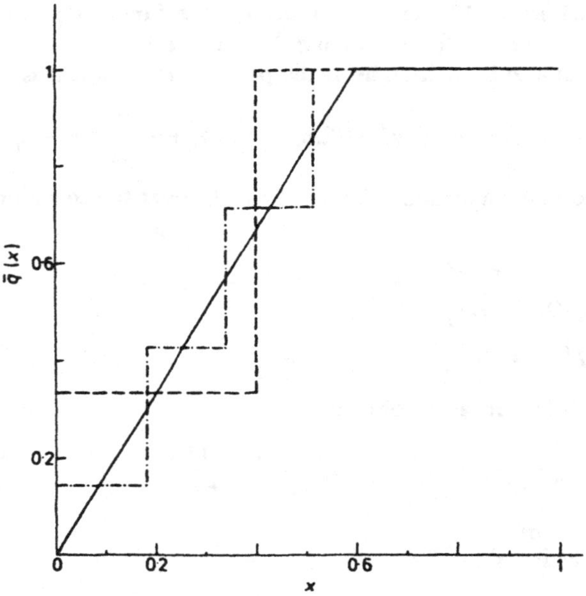

Figure 1. The broken curve, chain curve and full curve are, respectively, the function $q^{(K)}(x)$ for $K = 1, 4$ and ∞ in units of τ.

At this stage it is unclear which of the two equations is correct for q (equation (7)):

$$\begin{aligned} \bar{q} &= q(1) = \tau + \tfrac{3}{2}\tau^2 + O(\tau^3) \\ \bar{q} &= \tilde{q} = \tau + O(\tau^3). \end{aligned} \tag{27}$$

This ambiguity can be clarified by studying the magnetic properties of a spin glass; this task goes beyond the limits of this paper and it will be dealt with in a future publication.

4. All temperatures

In the previous section we have seen that the approximation $K = 1$ gives very good results near $T_c = 1$. We study it now at all temperatures.

One finds (Parisi 1979a):

$$\begin{aligned} \beta F(p, tm) = &-\frac{\beta^2}{4}[1 + mp^2 + (1-m)(p+t)^2 - 2(p+t)] \\ &+ \ln 2 - (2\pi)^{-1/2} \int dz \left\{ \exp\left(-\frac{z^2}{2}\right) m^{-1} \right. \\ &\left. \times \ln\left[(2\pi)^{-1/2} \int dy \exp\left(-\frac{y^2}{2}\right) \cosh^m(\beta p^{1/2} z + \beta t^{1/2} y)\right]\right\} \end{aligned} \tag{28}$$

where $q_1 = p$ and $q_0 = p + t$.

If $m=0$ or $t=0$ we recover the result without breaking of the replica symmetry ($K=0$) (equation (8)), where $q=p+t$.

The internal energy is given by:

$$U(\tau)=-\beta(1-q^2)/2 \qquad q^2\equiv mp^2+(1-m)(p+t)^2. \tag{29}$$

We must now maximise equation (28) as a function of p, t and m. This has been done on a computer using a standard minimisation program.

One finds that for $T>T_c=1$:

$$p=t=0.$$

For $T<1$, p, t and m are all different from zero and the $K=1$ free energy is always greater than the $K=0$ free energy. In figures 2, 3 and 4 we show, respectively, the internal energy, the specific heat and the entropy as functions of τ, both for $K=0$ and $K=1$. As expected, the difference between the two approximations is relevant only for $T<0{\cdot}5$. For comparison we plot also the low-temperature $C(T)$ and $S(T)$ obtained using a different approach (Thouless *et al* 1977).

The entropy is negative for $T<1$ and $S(0)$ is negative although quite small ($S(0)=-0{\cdot}01$); we expect that $S(0)=0$ only for infinite K. A substantial improvement has been obtained with respect to $K=0$. The computation of the entropy for $K=2$ would be rather long, but straightforward.

The values shown in figures 1, 2, 3 and 4 are in excellent agreement with the computer simulations (e.g. $U(0)=-0{\cdot}765$, while the computer simulations suggest $U(0)=-0{\cdot}76\pm0{\cdot}01$) (Sherrington and Kirkpatrick 1978).

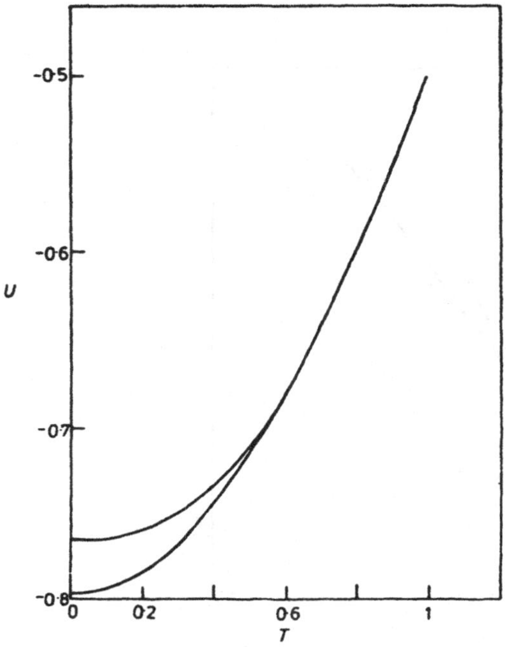

Figure 2. The lower and upper curves are, respectively, the internal energy $U(T)$ for $K=0$ and $K=1$.

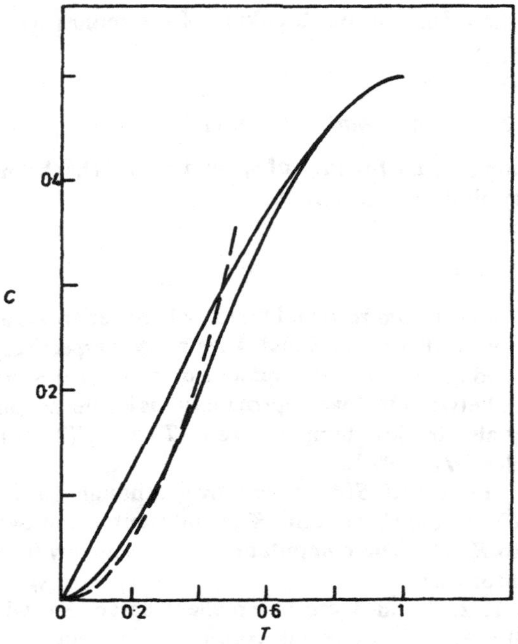

Figure 3. The lower and the upper curves are, respectively, the specific heat $C(T)$ for $K = 1$ and $K = 0$. The broken curve is the prediction $C(T) = 2 \ln 2T^2 + O(T^3)$ (Thouless *et al* 1977).

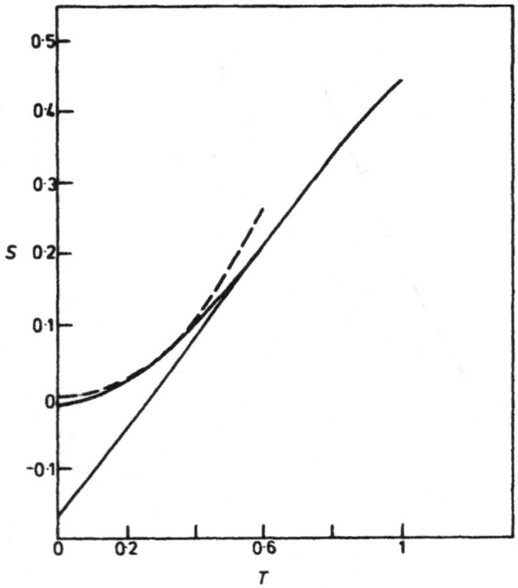

Figure 4. The lower and the upper curves are, respectively, the entropy $S(T)$ for $K = 0$ and $K = 1$. The broken curve is the prediction $S(T) = \ln 2T^2 + O(T^3)$ (Thouless *et al* 1977).

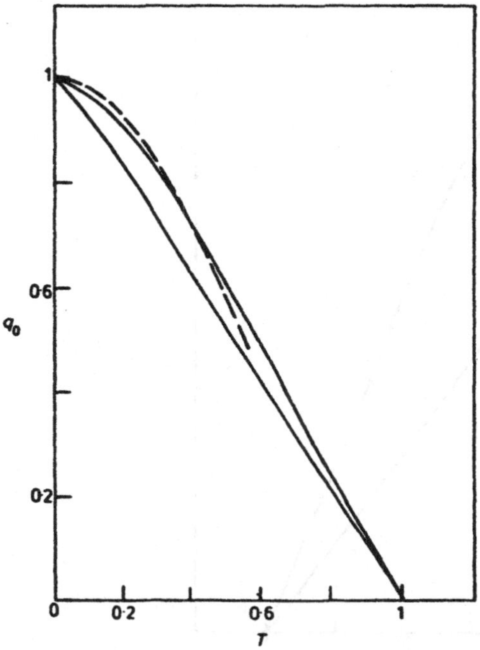

Figure 5. The lower and the upper curves are, respectively, the parameter q_0 as a function of T for $K=0$ and $K=1$. Just for comparison the broken curve is the prediction for the function $\bar{q}(T)$ (Thouless *et al* 1977): $\bar{q}(T)=1-2(\ln 2)^{1/2}T^2+O(T^3)$.

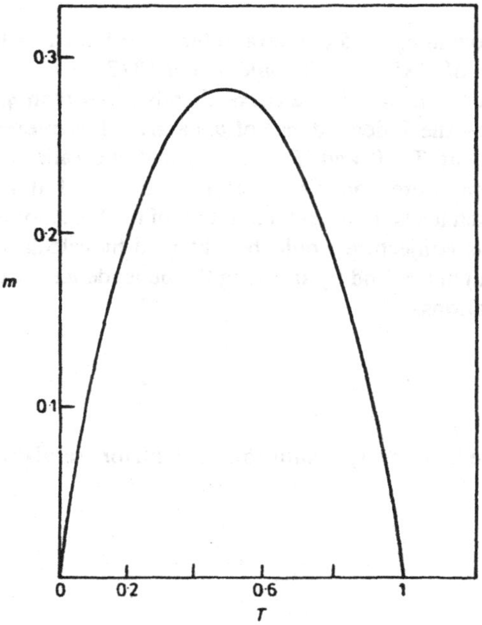

Figure 6. The parameter m_1 for $K=1$ as a function of T.

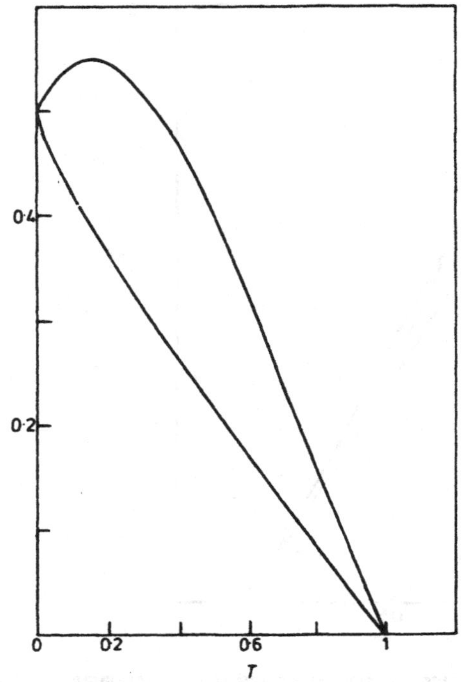

Figure 7. The lower and the upper curves are, respectively, the parameters t and p for $K = 1$ as functions of T.

For completeness we show in figure 5 the parameter q_0 both for $K = 0$ and 1. The broken line is the prediction of TAP for q (Thouless *et al* 1977).

The value of $\tilde{q}$ is not shown: it would be a curve slightly lower than q_0 for $K = 0$.

In figures 6 and 7 we show the T dependence of p, t and m. It is interesting to note that m becomes zero both at $T = 0$ and $T = 1$, and that the ratio t/p decreases monotonously with the temperature from $t/p = 2$ at $T = 1$ to $t/p = 1$ at $T = 0$.

It seems that this approach leads to the exact solution of the S–K model in the limit $K \to \infty$; a crucial test of this conjecture would be obtained by calculating the thermodynamic quantities for higher K and by studying the dependence on the magnetic field of the computer simulations.

Acknowledgments

The author is grateful to C de Dominicis, C Natoli and L Peliti for useful discussions and suggestions.

References

de Almeida J R L and Thouless D J 1978 *J. Physique* **A11** 983
Blandin J 1978 *J. Physique* **C6** 1578

Bray A J and Moore M A 1978 *Phys. Rev. Lett.* 41 1069
Edwards S F and Anderson P W 1975 *J. Physique* F5 965
Parisi G 1979a *Phys. Lett.* A 73 203
—— 1979b *Phys. Rev. Lett.* submitted
Pytte F and Rudnik J 1979 *Phys. Rev.* B 19 3603
Sherrington D and Kirkpatrick S 1975 *Phys. Rev. Lett.* 35 1972
—— 1978 *Phys. Rev.* B 17 4385
Thouless D J, Anderson P W and Palmer R G 1977 *Phil. Mag.* 35 593
Van Hemmen J L and Palmer R G 1979 *J. Physique* A12 567

J. Phys. A: Math. Gen. 13 (1980) 1887–1895. Printed in Great Britain

Magnetic properties of spin glasses in a new mean field theory

G Parisi
INFN–Laboratori Nazionali di Frascati, 00044 Frascati, Italy

Received 7 August 1979

Abstract. We study the magnetic properties of spin glasses in a recently proposed mean field theory; in this approach the replica symmetry is broken and the order parameter is a function ($q(x)$) on the interval 0–1. Exact results at the critical temperature and approximated results at all the temperatures are derived. The comparison with the computer simulations is briefly presented.

1. Introduction

In previous papers (Parisi 1979a, b, 1980) we have proposed a new mean field theory for spin glasses in the framework of the replica theory: the local order parameter is a function ($q(x)$) defined on the interval 0–1. If $q(x)$ is constant, the replica symmetry (i.e. the permutations among different replicas) is exact and we recover the standard mean field theory (Edwards and Anderson 1975); if $q(x)$ is x-dependent, the replica symmetry is broken.

This scheme has been successfully applied to the study of the properties of the S–K model (Sherrington and Kirkpatrick 1975) at zero magnetic field. This model is quite interesting: it is believed that the correct mean field theory should give exact results, so it is a good testing ground for different approaches. In this note we use the same techniques to study the magnetic properties of the S–K model also at $h \neq 0$. In perfect agreement with the results of de Almeida and Thouless (1978), we find that for high values of h, the replica symmetry is exact and at a temperature-dependent critical value (h_c) of the magnetic field h a transition is present from the regime where $q(x)$ is x-dependent, to the regime where $q(x) = \text{constant} \neq 0$.

In § 2 we recall the formalism of Parisi (1980), which has been cast in a more compact form. In § 3 exact results are obtained for the S–K model near the critical temperature. Approximate results are obtained at all temperatures in § 4. In § 5 we compare our results with the computer simulations (Sherrington and Kirkpatrick 1978), and we also discuss the problem of computing the 'physical' order parameter q_{ph} defined by

$$q_{ph} = \langle\langle m\rangle^2\rangle \tag{1}$$

where the inner bracket indicates the thermodynamic expectation value over the spin variables, while the outer bracket indicates the mean over the random spin couplings.

0305-4470/80/051887+09\$01.50

2. Algebraic preliminaries

The order parameter in the replica theory approach to spin glasses is an $n \times n$ matrix ($Q_{\alpha\beta}$) in the limit $n=0$. These matrices are defined as analytic continuation in n from integer n up to $n=0$. It is not a simple job to write down the generical matrix of this infinite-dimensional space. We will consider only a very restricted class of matrices, those which can be written in the form:

$$
\begin{aligned}
&Q_{\alpha\alpha} = \tilde{q} \\
&Q_{\alpha\beta} = q_i \qquad \text{if } I(\alpha/m_i) \neq I(\beta/m_i), I(\alpha/m_{i+1}) = I(\beta/m_{i+1})
\end{aligned} \tag{2}
$$

where q_i are $(k+1)$ real parameters ($i=0, k$), m_i are $(k+2)$ integer parameters ($i=0, k+1$), and the ratios m_{i+1}/m_i are also integer numbers; by definition we have $m_0=1$ and $m_{k+1}=n$. The integer-valued function $I(x)$ is equal to the smallest integer greater than or equal to x (e.g. $I(0{\cdot}5)=1$).

This parametrisation of the matrix $Q_{\alpha\beta}$ is a generalisation of the one introduced by Blandin (1978) and Blandin *et al* (1979). The motivations for this particular choice of parametrisation are discussed in Parisi (1980).

In the spin glasses the order parameter $Q_{\alpha\beta}$ is zero on the diagonal, so that $\tilde{q}=0$. We prefer to consider the slightly more general case ($\tilde{q} \neq 0$); indeed the matrices defined in equation (2) form an algebra closed under addition and multiplication; in the rest of this section we will study the properties of this algebra.

It is crucial to remark that if n is not a positive integer, there is no reason to have integer m_i; in the most interesting case, they satisfy (for $n=0$) the inequalities

$$1 \geq m_1 \geq m_2 \geq \ldots \geq m_k \geq m_{k+1} \equiv 0. \tag{3}$$

If the inequalities (3) are satisfied we can represent the matrix $Q_{\alpha\beta}$ with a pair $[\tilde{q}, q(x)]$, $q(x)$ being a piecewise constant function on the interval 0–1, defined by:

$$q(x) = q_i \qquad \text{for } m_i > x > m_{i+1} (i=1, k). \tag{4}$$

In this representation the addition and the multiplication take a rather simple form. Let us define:

$$
\begin{aligned}
&A_{\alpha,\beta} \leftrightarrow [\tilde{a}, a(x)] \\
&B_{\alpha,\beta} \leftrightarrow [\tilde{b}, b(x)] \\
&C_{\alpha,\beta} \leftrightarrow [\tilde{c}, c(x)]
\end{aligned} \tag{5}
$$

where we denote by the double arrow the canonical representation (4). We have

$$
\begin{aligned}
&A+B=C \leftrightarrow \begin{cases} \tilde{c} = \tilde{a} + \tilde{b} \\ c(x) = a(x) + b(x) \end{cases} \\
&AB = C \leftrightarrow \begin{cases} \tilde{c} = \tilde{a}\tilde{b} - \langle ab \rangle \\ c(x) = (\tilde{b} - \langle b \rangle) a(x) + (\tilde{a} - \langle a \rangle) b(x) + \int_0^x [a(x) - a(y)][b(x) - b(y)]\, dy \end{cases}
\end{aligned} \tag{6}
$$

where

$$\langle a\rangle = \int_0^1 a(x)\,dx \qquad \langle b\rangle = \int_0^1 b(x)\,dx$$
$$\langle ab\rangle = \int_0^1 a(x)b(x)\,dx. \tag{7}$$

The addition rule is trivial while some algebra is needed to verify the multiplication rule. One also finds that

$$\lim_{n\to 0}\frac{1}{n}\,\mathrm{Tr}(A) = \tilde{a}$$
$$\lim_{n\to 0}\frac{1}{n}\sum_{\alpha,\beta}(A_{\alpha,\beta})^l = \tilde{a}^l - \langle a^l\rangle. \tag{8}$$

By continuity equations (5)–(8) can be extended to the case where $q(x)$ is an arbitrary (not piecewise constant) continuous function, this last case being the most interesting for spin glasses.

3. The Sherrington–Kirkpatrick model near T_c

In the S–K model the free energy ($F(T)$) is supposed to be given by:

$$F(T) = \max_{\{Q\}} F[Q]$$
$$\beta F[Q] = \lim_{n\to 0}\frac{1}{n}\left\{-\frac{n}{4}\beta^2 + \tfrac{1}{4}\beta^2\,\mathrm{Tr}(Q^2) - \ln\left[\sum_{S_\alpha=\pm 1}\exp(-\beta^2 S_\alpha Q_{\alpha\beta} S_\beta)\right]\right\} \tag{9}$$

where the sum runs over the 2^n configurations of the n spin variables S_α, and the maximum is taken over all the zero-dimensional matrices, zero on the diagonal. We suppose that the matrix Q, which maximises $F(Q)$, has the form described in equation (2); this hypothesis can be verified by checking that the maximum of $F[Q]$ restricted on the matrices of the form (2) is a real maximum and not a saddle point. This can be done by computing the eigenvalue of the second derivative of F (de Almeida and Thouless 1978, Pytte and Rudnik 1979), but this task goes beyond the aim of this paper and it is postponed to further investigations.

It is not simple to write $F[Q]$ as a functional of $q(x)$; in this section we restrict ourselves to the case where Q is small, i.e. T is near to the critical temperature ($T_c = 1$).

In this situation $F(Q)$ may be approximated by

$$F(Q) = -\lim_{n\to 0}\frac{1}{n}\frac{1}{2}\left(\tau\,\mathrm{Tr}(Q^2) - \tfrac{1}{3}\mathrm{Tr}(Q^3) + \frac{y}{4}\sum_{\alpha,\beta} Q^4 + \ldots + h^2\sum_{\alpha,\beta} Q_{\alpha\beta}\right) + F(Q)\Big|_{Q=0} \tag{10}$$

where we have retained the only term of order Q^4 which is responsible for the breaking of the replica symmetry (Bray and Moore 1978, Pytte and Rudnik 1979); we have also neglected higher-order terms in the magnetic field.

The equations for a stationary point of F are:

$$0 = \frac{\partial F}{\partial Q_{\alpha\beta}} = -2\tau Q_{\alpha\beta} - y(Q_{\alpha\beta})^3 + (Q^2)_{\alpha\beta} - h^2 = 0. \tag{11}$$

Using the ansatz equation (2) for the matrix Q, one finds

$$2q(x)[\tau-\bar{q}]+yq^3=\int_0^x [q(x)-q(y)]^2-h^2$$
$$\bar{q}=\int_0^1 q(x)\,\mathrm{d}x. \tag{12}$$

Differentiating with respect to x twice one obtains

$$q'(x)(3yq(x)-x)=0. \tag{13}$$

Let us consider firstly $h=0$. For simplicity we restrict ourselves to the case $q(x)\geq 0$. Obviously, if $y<0$, the only solution is the replica symmetric one, where

$$q(x)=q_s \qquad q_s=\tau+\frac{y}{2}q_s^2. \tag{14}$$

If $y>0$ there is also another solution:

$$q(x)=\frac{x}{3y} \qquad x\leq x_1$$
$$q(x)=q(1)=\frac{x_1}{3y} \qquad x\geq x_1 \tag{15}$$
$$\bar{q}=\tau.$$

The value of $q(1)$ can be found by computing q as a function of $q(1)$ and imposing the last condition:

$$\bar{q}=(1-\tfrac{1}{2}x_1)q(1)=\tau. \tag{16}$$

One finds

$$q(1)=\tau+\tfrac{3}{2}yq^2(1) \qquad x_1=3y\tau+\tfrac{1}{2}x_1^2. \tag{17}$$

Notice that

$$\bar{q}<q_s<q(1). \tag{18}$$

The second solution has a higher value of the free energy. For Ising spin y is positive ($y=\frac{2}{3}$); the correct solution is (15) and replica symmetry is broken.

The inclusion of higher orders in Q is long but straightforward and a systematic expansion near T_c is possible.

It may be interesting to note that at this order

$$q(0)=0. \tag{19}$$

A preliminary analysis shows that equation (19) is an exact statement which remains valid at all orders in τ.

Let us consider now the case $h^2\geq 0$ and let us put $y=\frac{2}{3}$. The magnetic susceptibility is given for small h by

$$\chi=\beta(1-\bar{q}). \tag{20}$$

The symmetric solution is always possible:

$$q(x)=q_s \qquad 2q_s(\tau-q_s)+\tfrac{2}{3}q_s^3=h^2. \tag{21}$$

The non-trivial solution of equation (12) is

$$\begin{aligned} q(x) &= q(0) && 0 \leq x \leq x_0 \\ q(x) &= 2x && x_0 \leq x \leq x_1 \\ q(x) &= q(1) && x_1 \leq x \leq 1 \end{aligned} \tag{22}$$

where

$$\begin{aligned} &q(0) = (\tfrac{3}{4}h^2)^{1/3} \qquad x_0 = \tfrac{1}{2}q(0) \qquad x_1 = \tfrac{1}{2}q(1) \\ &\bar{q} = \tau + q^2(0) = \tau + (\tfrac{3}{4})^{2/3}h^{4/3} \end{aligned} \tag{23}$$

where $q(1)$ is fixed by the last condition on $\bar{q}$.

The solution (23) make sense only if $x_1 > x_0$. We find that $x_1 = x_0$ when $h = h_c$ where

$$q^c(0) = \tau + \tfrac{3}{2}(q^c(0))^2 \qquad h^c = [\tfrac{4}{3}(q_0^c)^3]^{1/2}. \tag{24}$$

Equation (23) implies that h^c is of order $\tau^{3/2}$ in agreement with previous computations (de Almeida and Thouless 1978).

For $h > h_c$ only the symmetric solution is possible; for $h < h_c$ the asymmetric solution is favoured. At $h = h_c$ we have a second-order transition characterised by the breaking of the replica symmetry.

In the low-field region we find that the magnetic susceptibility is given by

$$\chi(h) = 1 - (\tfrac{3}{4})^{2/3}h^{4/3}. \tag{25}$$

The second derivative of the susceptibility is divergent for $h \to 0$:

$$\mathrm{d}^2\chi/\mathrm{d}h^2 \sim h^{-2/3}. \tag{26}$$

The singular behaviour of the susceptibility for small fields is connected to the fact that $q(0) = 0$ for $h = 0$, and seems to be stable against the addition of higher-order terms in the free energy. Equation (26) is a prediction peculiar to this approach and it would be very interesting to check it directly in the computer simulations. $\mathrm{d}^2\chi/\mathrm{d}h^2$ is also the $\bar{q}$ susceptibility, which behaves like $(\bar{q} - \tau)^{-1/2}$. More precisely we can define an effective free energy $F_{\mathrm{ef}}[\bar{q}]$ if we write

$$q(x) = \bar{q} + p(x) \qquad \int_0^1 p(x) = 0 \qquad F_{\mathrm{ef}}[\bar{q}] = \max_{p(x)} F[q] \tag{27}$$

where the maximum is taken over all the functions p at fixed $\bar{q}$. We find

$$F_{\mathrm{ef}}(\bar{q}) = (\bar{q} - \bar{q}_0)^{5/2} \tag{28}$$

where $\bar{q}_0$ is the value at zero magnetic field.

We have obtained some of the results of Thouless *et al* (1977), in particular the existence of a forbidden region for $\bar{q} < q_0$ and the infinite value of the $\bar{q}$ susceptibility.

The method presented in this section can be used only near to the critical temperature; at lower values of T a different approach is needed. This is the subject of the next section. We note, however, that if we write $F(T) = F_s(T) + \frac{2}{45}(T_c - T)^5 + O(T_c - T)^6$, and we neglect higher orders in $T - T_c$, we find $U(0) \sim -0{\cdot}753$ and $S(0) \sim 0{\cdot}06$, which is an improvement with respect to the standard treatment.

4. At all temperatures

A rather simple-minded approximation which works rather well at all temperatures consists in approximating the function $q(x)$ with a function taking only two values:

$$q(x)=q_0 \qquad x<m$$
$$q(x)=q_1 \qquad x>m. \tag{29}$$

Excellent results are obtained at zero magnetic field especially in the region $T\geqslant 0{\cdot}2$ (Parisi 1979a). In this case the functional $F[Q]$ can be simply written as:

$$\beta F(q_0,q_1,m)=-\frac{\beta^2}{4}[1+mq_0^2+(1-m)q_1^2-2q_1]$$
$$+\ln 2-(2\pi)^{-1/2}\int dz\left\{\exp(-z^2/2)m^{-1}\ln\left[(2\pi)^{-1/2}\right.\right.$$
$$\left.\left.\times\int dy\,\exp(-y^2/2)(ch\tilde{H})^m\right]\right\} \tag{30}$$

$$\tilde{H}=\beta(q_0^{1/2}z+t^{1/2}y+h) \qquad t=q_1-q_0.$$

The maximum of (29) can easily be found numerically. The replica symmetry is broken if

$$(q_1-q_0)m(1-m)\neq 0. \tag{31}$$

For all fields $h<h_c(T)$ and $T<T_c=1$ equation (31) holds at the maximum. For $T>T_c$, $q_1=q_0=0$.

Equation (20) implies that for small h

$$\chi(h)=\beta[1-q_0m-q_1(1-m)]. \tag{32}$$

In figure 1, we show χ against the temperature for $h=0$. For $T>1$, the antiferromagnetic result $\chi=1/T$ holds. For T less than T_c the upper curve is the calculated

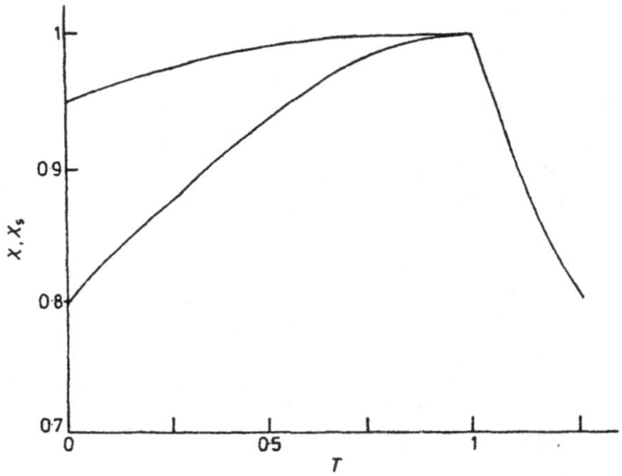

Figure 1. The higher curve is the zero-field susceptibility (χ) plotted against temperature. The lower curve is the prediction of the replica symmetric approach (χ_s).

susceptibility while the lower curve is χ_s, the susceptibility in the conventional replica symmetric treatment of the model ($\chi_s = \beta(1 - q_s)$).

As a typical example of the behaviour of the system in an external magnetic field h, we show in figures 2–4 various quantities as functions of h at $T = 0{\cdot}3$. At this temperature $h_c \simeq 1$. No unexpected phenomenon is present; in this approximation the singular behaviour of $\chi(h)$ at $h = 0$ is absent.

We notice that both $\bar{q}$ and $q(1) = q_1$ are increasing functions of h; $\bar{q}$ behaves quite smoothly at the transition point, while dq_1/dh is discontinuous.

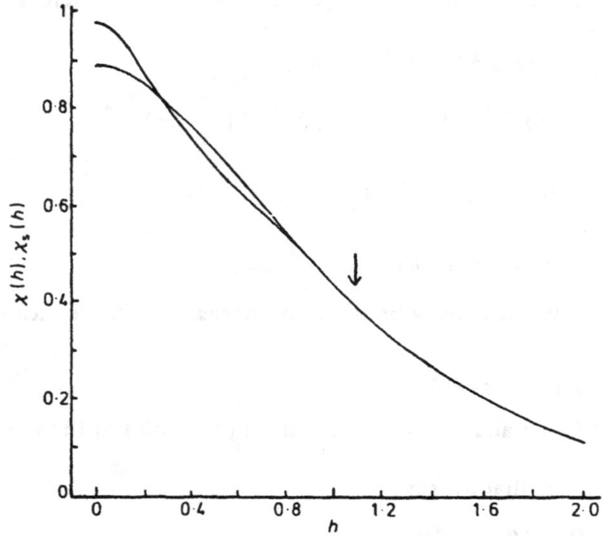

Figure 2. We plot the magnetic susceptibility $\chi(h)$ against h for $T = 0{\cdot}3$. The curve, which is higher at $h = 0$, is the result of our calculation; the other curve is $\chi_s(h)$. The two curves coincide for $h \geq h_c \simeq 1$, which is indicated by an arrow.

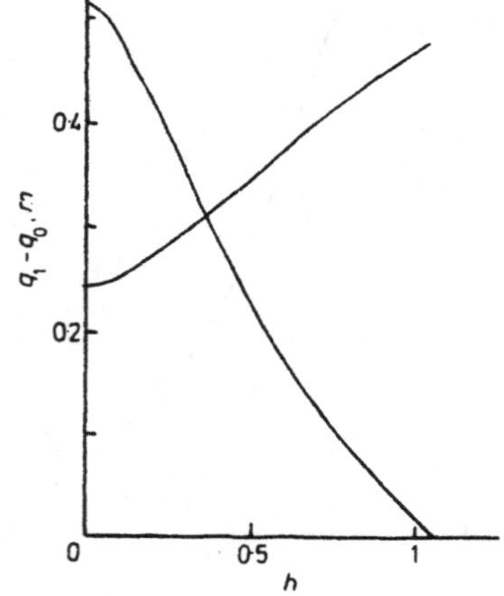

Figure 3. The decreasing and the increasing curve are, respectively, $q_1 - q_0$ and m as functions of h. At $h = h_c$, $q_1 - q_0 = 0$ and $m \neq 0$.

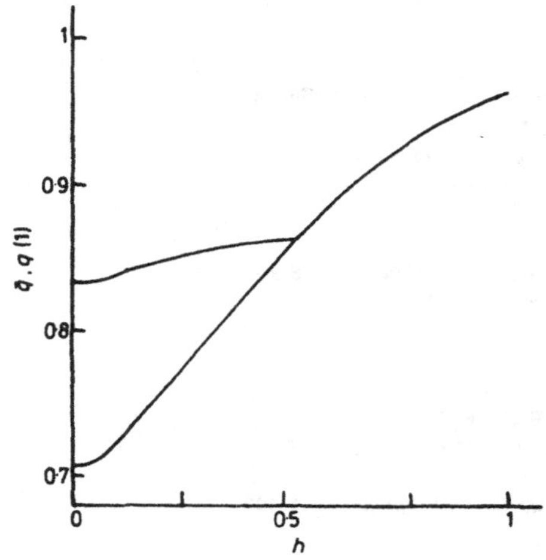

Figure 4. The lower and upper curves are, respectively, $\bar{q}$ and $q(1)$ as functions of h. They coincide for $h \geqslant h_c$.

5. Discussion

The expert reader has probably realised that our results for the susceptibility do not agree with the computer simulations of Sherrington and Kirkpatrick (1978). These authors find that at $h = 0$, $\chi < \chi_s$ ($\chi = 0$ at $T = 0$), i.e. the opposite of our results.

The origin of this discrepancy is not clear: we notice that the inequality $F > F_s$ (Chalupa 1978) implies that $\chi > \chi_s$, at least in the mean. $\chi \neq 0$ at $T = 0$ is not in variance with a quadratic specific heat only if very large size clusters are relevant: the onset of equilibrium in a large-scale cluster is a slow phenomenon and it may be invisible in a not long enough Monte Carlo approach (Fernandez and Medina 1979).

It is not known if hysteresis or remanence is present in the S–K model; if that happens one should be very careful in extracting the susceptibility from Monte Carlo data. Unfortunately no accurate simulations exist at non-zero magnetic field; the zero-field susceptibility has been extracted from the spin–spin correlations. A direct computation of the susceptibility would be welcomed.

In this approach it is unclear how to compute the physical order parameter q_{ph} (defined in equation (1)). A simple-minded argument gives

$$q_{ph} = \max_x q(x). \tag{33}$$

Equation (33) does agree with the computer simulations. If equation (33) is correct, when the replica symmetry is broken:

$$q_{ph} \neq \bar{q} \qquad \chi \neq \beta(1 - q_{ph}). \tag{34}$$

It is suggested that one should consider the validity of equation (34) as a signal for the breaking of the replica symmetry. Unfortunately the arguments leading to equation (33) are not very strong. The soundness of equation (33) may be investigated by studying the time-dependent correlations (De Dominicis 1979). However, we insist that good-quality computer simulations at non-zero h would be very useful to clarify the situation.

References

de Almeida J R L and Thouless D J 1978 *J. Phys. A: Math. Gen.* **11** 983
Blandin A 1978 *J. Physique* **39** C6 1499
Blandin A, Gabay M and Garel T 1979 *Orsay Preprint*
Bray A J and Moore M A 1978 *Phys. Rev. Lett.* **41** 1068
Chalupa J 1978 *Phys. Rev.* B **17** 4335
De Dominicis C 1979 *Saclay Preprint*
Edwards S F and Anderson P W 1975 *J. Phys. F: Metal Phys.* **5** 965
Fernandez J F and Medina R 1979 *Phys. Rev.* B **19** 3561
Parisi G 1979a *Phys. Lett.* A **73** 203
—— 1979b *Phys. Rev. Lett.* **43** 1574
—— 1980 *J. Phys. A: Math. Gen.* **13**
Pytte E and Rudnick 1979 *Phys. Rev.* B **19** 3603
Sherrington D and Kirkpatrick S 1975 *Phys. Rev. Lett.* **35** 1972
—— 1978 *Phys. Rev.* B **17** 4385
Thouless D J, Anderson P W and Palmer R 1977 *Phil. Mag.* **35** 593

PHYSICAL REVIEW LETTERS

VOLUME 49 2 AUGUST 1982 NUMBER 5

Lack of Ergodicity in the Infinite-Range Ising Spin-Glass

N. D. Mackenzie and A. P. Young
Department of Mathematics, Imperial College, London SW7 2BZ, United Kingdom
(Received 26 May 1982)

The size dependence of slow relaxation processes in the infinite-range Ising spin-glass is investigated by computer simulation. Below the transition temperature, relaxation of variables which do not change sign under inversion of the spins is complete by a time τ, where $\ln\tau \propto N^{1/4}$ and so diverges as N, the number of spins, tends to infinity. The "ergodic time" τ_{eg} satisfies $\ln\tau_{eg} \propto N^{1/2}$. These results are consistent with a physical picture where barriers between free-energy minima in phase space have a height proportional to the square root of the number of spins to be flipped.

PACS numbers: 05.50.+q, 64.60.My, 75.10.Hk

Spin-glass systems are characterized by very slow relaxation[1] below the freezing temperature. Controversy[1,2] has centered on whether this is due to a sharp phase transition or a more gradual increase in relaxation times. A simplified model which does have a transition because the range of interactions is infinite has been proposed by Sherrington and Kirkpatrick[3] (SK). Using different lines of argument various authors[4-6] have suggested that relaxation times in the SK model diverge when the number of spins, N, tends to infinity and $T < T_c$, the transition temperature. This could arise from free-energy barriers, separating minima in phase space, whose height diverges in the thermodynamic limit. It is tempting[7] to associate these minima with solutions of the mean-field equations of Thouless, Anderson, and Palmer[8] (TAP), which are known to have an enormous number of solutions.[9] An infinite system would stay close to one minimum at all times because it can never get over the infinite barrier surrounding it. By contrast statistical mechanics (ensemble average) sums over all minima with an appropriate weight. An infinite system would therefore be nonergodic because time and ensemble averages would give different results.[7]

This picture, while intuitively reasonable, rests on the assumption that relaxation times diverge for $N \to \infty$. Here we present results of Monte Carlo simulations which show directly this increase of relaxation times with system size. Furthermore we are able to quantify the relaxation processes in some detail. Our main conclusions are as follows.

(i) For excitations which do *not* involve turning over the whole system from the vicinity of one ground state to the "time-reversed" ground state, the spectrum of relaxation times extends to a rather well defined maximum value τ such that (at $T = 0.4T_c$)

$$\ln\tau = 2.58N^{1/4} - 0.66 \tag{1}$$

(see Fig. 1). Lack of ergodicity then follows because $\ln\tau \to \infty$ as $N \to \infty$.

(ii) The slow relaxation of a correlation function $q^{(2)}(t)$, defined in Eq. (6) below, has been studied in detail for different sizes, mainly at $T = 0.4T_c$. When plotted versus $\ln t/\ln\tau$ all our results for $q^{(2)}(t) - q^{(2)}(\infty)$ lie on a single "uni-

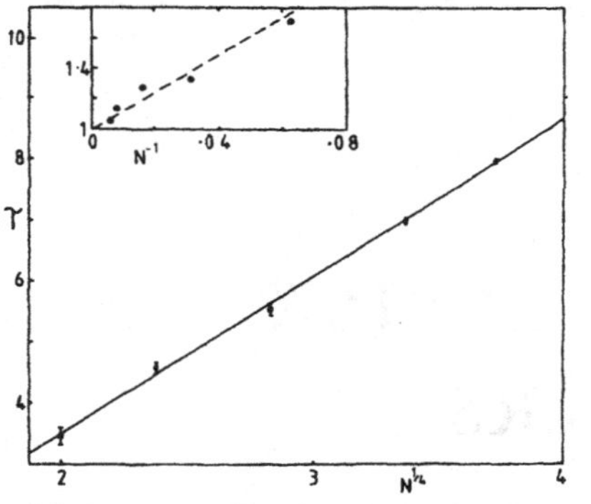

FIG. 1. A plot of $\ln\tau$ against $N^{1/4}$ for several values of N between 16 and 192 at $T=0.4$. Apart from $N=192$, for which only one run was performed, the error bars, which represent one standard deviation, are obtained from the variance of four separate runs. The straight line is a least-squares fit and is given by Eq. (1). Inset: the scale factors for the vertical axis in Fig. 2. The dashed line is a guide to the eye.

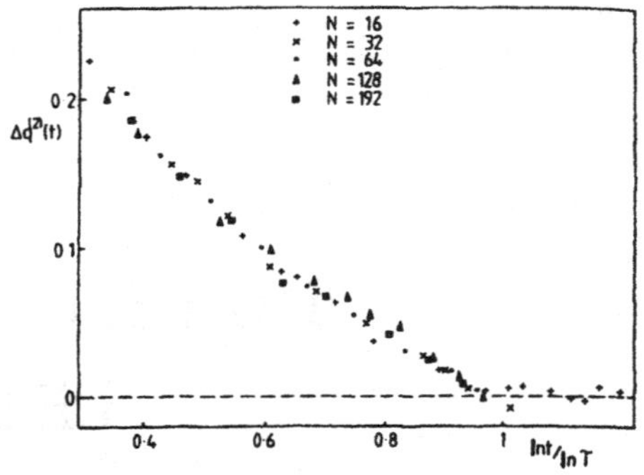

FIG. 2. A plot of $\Delta q^{(2)}(t)=q^{(2)}(t)-q^{(2)}(\infty)$ against $\ln t/\ln\tau$ where $\ln\tau$ is shown in Fig. 1 and the vertical axis has been multiplied by an amount shown in the inset in Fig. 1. The temperature is $T=0.4$. All the data appear to lie on a single universal curve with a change in slope at $\ln t=\ln\tau$.

versal curve" shown in Fig. 2, provided a small rescaling of the vertical axis (inset in Fig. 1) is made for the smaller sizes.

(iii) It was previously argued[5] that excitations from one minimum to another involve turning over a large number of spins, ΔN, where $\Delta N \propto N^{1/2}$. If we assume that $\ln\tau$ is proportional to a free-energy barrier height, Δf, then we have from Eq. (1)

$$\Delta f \propto \Delta N^{1/2}. \quad (2)$$

In other words the barrier between two minima is proportional to the square root of the number of spins which have to be turned over to go between the minima.

(iv) There are also excitations which turn over all the spins. These control the long-time behavior of the standard correlation function $q(t)$ defined in Eq. (5) below. The time for these processes to occur, which we call the ergodic time τ_{eg}, diverges even for a ferromagnet and gives rise to spontaneous symmetry breaking. In mean-field theory $\ln\tau_{eg}\propto N$ for a ferromagnet[10] but for the SK model our calculations rule out the first power of N and are consistent with

$$\ln\tau_{eg}\propto N^{1/2}, \quad (3)$$

as shown in Fig. 3. This result is equivalent to Eq. (2) since here $\Delta N \sim N$. Notice that $\ln\tau_{eg} \gg \ln\tau$ for large N.

We now discuss our calculations and results in more detail.

The SK model is described by the Hamiltonian

$$H=-\sum_{i<j} J_{ij}S_iS_j, \quad (4)$$

where $S_i=\pm 1$ is an Ising spin, $i=1,\ldots,N$, and

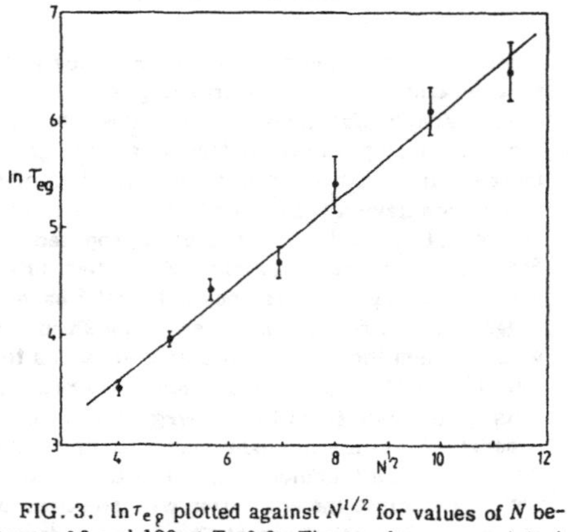

FIG. 3. $\ln\tau_{eg}$ plotted against $N^{1/2}$ for values of N between 16 and 128 at $T=0.6$. The results are consistent with $\ln\tau_{eg}\propto N^{1/2}$. The errors bars correspond to one standard deviation and are obtained from several different runs.

the J_{ij} are independent random variables with a Gaussian probability distribution of width[3,5] $(N-1)^{-1}$, the same for all pairs of spins. In the thermodynamic limit there is a transition at $T_c=1$, in these units. We shall not include a magnetic field. The spins are flipped by the "heat bath" Monte Carlo procedure,[11] and time is given in units of Monte Carlo steps per spin.

Starting from an initial spin arrangement the simulation proceeds for a time t_0 in order to equilibrate the system. One needs $t_0 \gtrsim \tau$ for the energy to relax and we check *a posteriori* that this condition is satisfied. The simulation then continues and we calculate the correlation functions

$$q(t)=N^{-1}\sum_i \langle S_i(t_0)S_i(t+t_0)\rangle_J \qquad (5)$$

and

$$q^{(2)}(t)=\frac{2}{N(N-1)}\sum_{i<j}\langle S_i(t_0)S_j(t_0)\times S_i(t+t_0)S_j(t+t_0)\rangle_J. \qquad (6)$$

Here $\langle\cdots\rangle_J$ denotes an average over samples. For the calculations of $q^{(2)}(t)$ presented here the number of samples, N_s, satisfies $NN_s \simeq 100\,000$, which is necessary in order to get good statistics.

The Hamiltonian in Eq. (1) is invariant under the "time-reversal" operation $S_i \rightarrow -S_i$ for all i. $q(t)$ changes sign if all the spins flip over between t_0 and t_0+t. It will only reach its equilibrium value, which is zero because no symmetry-breaking field is applied,[12] for times longer than the ergodic time τ_{eg}. On the other hand $q^{(2)}(t)$ follows the fluctuations of a pair of spins S_iS_j ($i \neq j$), which does not change sign under inversion of all the spins. Hence it is insensitive to fluctuations which turn over the whole system and will reach its (nonzero) equilibrium value $q^{(2)}$ for times greater than some value, τ, which will be much less than τ_{eg}. Hence

$$q^{(2)}(t) \rightarrow q^{(2)}=\langle\langle S_iS_j\rangle_T{}^2\rangle_J \qquad (7)$$

for $\ln t > \ln\tau$, where $\langle\cdots\rangle_T$ denotes a statistical mechanics average. In fact $q^{(2)}$ is related to the energy per spin, $U(T)$, by[5,13] $q^{(2)}=1-2T|U(T)|$. This is useful because $U(T)$ is obtained at time t_0, the end of the equilibration process, and so we know to what value $q^{(2)}(t)$ is relaxing as $t\rightarrow\infty$.

In order to study the size dependence of $q^{(2)}(t)$ carefully we have mainly considered the single temperature $T=0.4$. Plotting $\Delta q^{(2)}(t)=q^{(2)}(t)-q^{(2)}$ against $\ln t$ we obtain zero for $\ln t$ greater than a certain value $\ln\tau$, which is rather well defined since a change in slope is observed at this point. Furthermore $\ln\tau$ is found to increase with system size, as shown in Fig. 1, and is accurately given by Eq. (1) for $16 \leq N \leq 192$. If one assumes the form $\ln\tau = aN^x+b$ and allows the exponent x to vary we find $x=0.27\pm0.10$. We strongly suspect that the exact value is $x=\frac{1}{4}$.

If $\Delta q^{(2)}(t)$ is plotted against $\ln t/\ln\tau$ all the data for different sizes at $T=0.4$ appear to lie on a single universal curve, as shown in Fig. 2. A small rescaling of the vertical axis is also necessary but this scale factor tends to unity for large N as shown in the inset in Fig. 1. The universal curve is almost a straight line, with negative slope, up to $\ln t/\ln\tau=1$. At this point there is apparently an abrupt change of slope and $\Delta q^{(2)}(t)=0$ for $\ln t/\ln\tau>1$. In terms of barriers these results can be interpreted as an almost constant density of barrier heights up to some critical value Δf_c where $\beta\Delta f_c \simeq \ln\tau$, beyond which there are no more barriers. A $\ln t$ dependence is also seen in many remanance magnetization experiments[14] and in simulations of $q(t)$ on short-range models.[15] Earlier simulations[16] on the SK model report a $t^{-1/2}$ variation of $q(t)$ but this is for shorter times and larger samples, and so represents fluctuations in the vicinity of one minimum. Our results provide evidence for Sompolinsky's[4] idea that there is a spectrum of relaxation times which all diverge in the thermodynamic limit.

We have also studied the times at which the system turns over from the vicinity of a ground state to the "time-reversed" one. If the time between the successive flips of the whole system is Δt, then we define $\ln\tau_{eg}=\langle\langle\ln\Delta t\rangle_S\rangle_J$, where $\langle\cdots\rangle_S$ denotes an average over all flips for a given sample. Results for $\ln\tau_{eg}$ at $T=0.6$ are shown in Fig. 3 and are consistent with an $N^{1/2}$ behavior. Since these large flips involve of order N spins the data in Fig. 3 agree with the free-energy barrier formula, Eq. (2). One of us (A.P.Y.) and, independently, Morgenstein[17] have also obtained Eq. (2) for a nearest-neighbor spin-glass model in two dimensions but with ΔN, and hence Δf, finite in the thermodynamic limit, implying the absence of a transition in that model.

Since $\ln\tau \ll \ln\tau_{eg}$ for large N the earlier stages of relaxation of $q(t)$ are not affected by reversals of the whole system and for $\ln t \ll \ln\tau_{eg}$ we find $[q(t)]^2=q^{(2)}(t)$ apart from expected differences of order $N^{-1/2}$. Consequently a single "order parameter" $q(t)$ is adequate up to times where reversal of the whole system starts to occur.

To conclude, we feel that the SK model is qualitatively understood and that the central feature is the many minima in phase space, corresponding to TAP solutions, which become infinitely long lived in the thermodynamic limit. They are therefore strictly speaking stable rather than metastable states. If a uniform field, H, is included the SK solution is correct above the Almeida-Thouless[18] instability line in the H-T plane and there is only one TAP solution[9] in this region. We therefore expect that $\ln\tau \propto N^{1/4}$ everywhere below this line, but with a coefficient which vanishes as the line is approached, and we anticipate that above the instability line $\ln\tau$ will saturate at a finite value as $N \to \infty$. This conjecture will be verified in future calculations.

One of us (N.D.M.) would like to thank the Science and Engineering Research Council of Great Britain (SERC) for a research studentship. The computations were mainly performed on the Cray-1S at Daresbury Laboratory. We would like to thank the SERC for use of this facility. We also thank I. Mclenaghan and D. Sherrington for useful discussions.

[1]For example, see J. L. Tholence, Solid State Commun. 35, 113 (1980); A. Malozemoff and Y. Imry, Phys. Rev. B 24, 489 (1981).

[2]I. Morgenstern and K. Binder, Phys. Rev. Lett. 43, 1615 (1979), and Phys. Rev. B 22, 288 (1980), and Z. Phys. B 39, 227 (1980); R. Fisch and A. B. Harris, Phys. Rev. Lett. 38, 785 (1977); A. J. Bray and M. A. Moore, J. Phys. C 12, 79 (1979).

[3]D. Sherrington and S. Kirkpatrick, Phys. Rev. Lett. 35, 1792 (1975).

[4]H. Sompolinsky, Phys. Rev. Lett. 47, 935 (1981).

[5]A. P. Young and S. Kirkpatrick, Phys. Rev. B 25, 440 (1982).

[6]G. Toulouse, in "Anderson Localization," Proceedings of the 1981 Taniguchi Symposium, Springer Series in Solid-State Sciences Vol. 39 (Springer, Berlin, to be published); J. Hertz, A. Khurana, and M. Puoskari, to be published; J. Hertz, to be published.

[7]A. P. Young, J. Phys. C 14, L1085 (1981).

[8]D. J. Thouless, P. W. Anderson, and R. Palmer, Philos. Mag. 35, 593 (1977).

[9]A. J. Bray and M. A. Moore, J. Phys. C 13, L469 (1980); C. de Dominicis, M. Gabay, T. Garel, and H. Orland, J. Phys. (Paris) 41, 923 (1980); F. Tanaka and S. F. Edwards, J. Phys. F 10, 2471 (1980).

[10]R. B. Griffiths, C. Y. Wang, and J. S. Langer, Phys. Rev. 149, 301 (1966).

[11]K. Binder, in *Monte-Carlo Methods in Statistical Physics*, edited by K. Binder (Springer-Verlag, New York, 1979); M. Creutz, L. Jacobs, and C. Rebbi, Phys. Rev. B 20, 1915 (1979).

[12]We leave aside the interesting question of whether q would still be zero in the limit of a very small uniform field being applied.

[13]A. J. Bray and M. A. Moore, J. Phys. C 13, 419 (1980).

[14]C. N. Guy, J. Phys. F 8, 1309 (1978).

[15]K. Binder and K. Schröder, Phys. Rev. B 14, 2142 (1976).

[16]S. Kirkpatrick and D. Sherrington, Phys. Rev. B 17, 4384 (1978).

[17]I. Morgenstern, private communication.

[18]J. R. L. de Almeida and D. J. Thouless, J. Phys. A 11, 983 (1978).

Reduction of the Fokker-Planck Equation with an Absorbing or Reflecting Boundary to the Diffusion Equation and the Radiation Boundary Condition

K. Razi Naqvi, K. J. Mork, and S. Waldenstrøm
Department of Physics, University of Trondheim, N-7055 Dragvoll, Norway
(Received 22 March 1982)

It is shown, from microscopic considerations based on the Fokker-Planck equation, that the boundary condition (used in conjunction with the diffusion equation) in which the particle density is set to zero on a perfectly absorbing surface is untenable, and, for the first time, the boundary condition for any plane (partially or perfectly) absorbing surface is derived.

PACS numbers: 05.20.Dd, 05.40.+j, 28.20.-v, 82.40.-g

By treating Brownian motion as a simply Markoff process in phase space, Klein,[1] Kramers,[2] and Chandrasekhar[3] demonstrated (independently) that $f(x,v,t)$, the distribution function of a free Brownian particle of mass m, satisfies the so-called field-free Fokker-Planck equation

$$\left(\frac{\partial}{\partial t}+v\frac{\partial}{\partial x}\right)f=\beta\frac{\partial}{\partial v}\left(v+\frac{q}{\beta}\frac{\partial}{\partial v}\right)f, \tag{1}$$

in which β is the friction coefficient, $q/\beta = kT/m$,

PHYSICAL REVIEW B VOLUME 27, NUMBER 1 1 JANUARY 1983

Eigenvalues of the stability matrix for Parisi solution of the long-range spin-glass

C. De Dominicis and I. Kondor*
Service de Physique Théorique, Centre d'Etudes Nucléaires, Saclay, 91191 Gif-sur-Yvette, France
(Received 21 June 1982)

We study, near T_c, the stability of Parisi's solution for the long-range spin-glass. In addition to the discrete, "longitudinal" spectrum found by Thouless, de Almeida, and Kosterlitz, we find "transverse" bands depending on one or two continuous parameters, and a host of zero modes occupying most of the parameter space. All eigenvalues are non-negative, proving that Parisi's solution is marginally stable.

In the spin-glass riddle the primary question to be settled is the stability of various "mean-field" solutions. The replica-independent Sherrington-Kirkpatrick[1] (SK) ansatz $q_{\alpha\beta}=q$, for the order-parameter matrix, was shown to be unstable by Almeida and Thouless[2] (AT), which led to a search of solutions breaking the replica symmetry. After several attempts[3-5] Parisi[6] proposed a most promising scheme where the order parameter q was replaced by a function $q(x)$ on the unit interval.[7] Its stability was investigated, near T_c, by Thouless, Almeida, and Kosterlitz[8] (TAK) who found it at best marginally stable. Their analysis was however confined to "longitudinal" fluctuations, whereas the space of fluctuations is vastly larger and besides, the most dangerous fluctuations lie in the "transverse" directions (i.e., into configurations with additional symmetry breaking). Indeed, the AT work has shown the existence of three families of eigenvalues: A unique $\lambda^{(1)}$ of order $\tau\equiv(T-T_c)/T_c$, a $(n-1)$ degenerate $\lambda^{(2)}$ (reducing to $\lambda^{(1)}$ as the replica number n vanishes), and a small *negative* $\lambda^{(3)}$, of order τ^2 and $n(n-3)/2$ degenerate. The corresponding eigenvectors $f_{\alpha\beta}$ [column vectors with $n(n-1)/2$ components, which can be conveniently thought of as real, symmetric matrices with zero diagonal elements] are $f^{(1)}_{\alpha\beta}$, a constant, i.e., a purely longitudinal vector; $f^{(2)}_{\alpha\beta}$, a two-valued matrix taking a constant value except on one arbitrarily distinguished row (and column) θ where it takes a different constant value; $f^{(3)}_{\alpha\beta}$, a three-valued matrix with two arbitrarily distinguished rows (columns) θ_1,θ_2. It turns out that TAK analysis amounts to a generalization of $f^{(1)}$, whereas AT work hints that the dangerous mode lies in the transverse direction ($f^{(3)}$ in particular).

In this work we follow the strategy used by Parisi who constructed the space of $q(x)$ by considering the limit of a discrete ansatz sequence $q^R_{\alpha\beta}$, $R=0,1,2,\ldots$. We construct equations for eigenvalues and eigenfunctions of the Hessian around the above sequence. We find that the generalization of $f^{(2)}$ ($f^{(3)}$) is a set of functions of two (three) variables. For simplicity we have kept to zero magnetic field and used the Parisi[9] approximation for the $q_{\alpha\beta}$ Lagrangian. The result exhibits *all eigenvalues as non-negative, which proves that the Parisi solution is marginally stable against all fluctuations*, longitudinal and transverse. The second family $\lambda^{(2)}(\kappa)$, $0\leq\kappa\leq1$ is made of bands pinned on the TAK spectrum $\lambda^{(1)}\equiv\lambda^{(2)}(0)$. The third family is made of a zero mode on most of the parameter space, and of a continuum spanning the range $(0,2\tau^2)$.

(1) A very good representation, near T_c, of the SK free-energy functional, is given by the Parisi model

$$\frac{-f}{T}=\lim_{n\to0}(2n)^{-1}\left[\tau\,\mathrm{tr}q^2+\frac{1}{3}\mathrm{tr}q^3+\left[\frac{y_p}{4}\right]\sum_{\alpha\neq\beta}q^4_{\alpha\beta}\right] , \quad (1)$$

where $\alpha,\beta=1,2\ldots n$. The eigenvalues of the Hessian of (1) are determined by

$$0=(\lambda+2\tau+3y_pq^2_{\alpha\beta})f_{\alpha\beta}+\sum_{\gamma\neq\alpha,\beta}(q_{\alpha\gamma}f_{\gamma\beta}+q_{\beta\gamma}f_{\gamma\alpha}) , \quad (2)$$

where in the following we keep the standard value $y_p=\frac{2}{3}$. The AT solution of (2) around the SK stationarity point now plays for us, in some loose sense, the role of an unperturbed problem in the search for a solution around the Parisi stationarity point. Switching in the "perturbation" (i.e., the replica symmetry breaking) lifts the high degeneracy, and rotates the AT eigenvectors, but these can still be constructed in close analogy with AT. The simplest matrices $f^{(1)}_{\alpha\beta}$ solving (2) have the same hierarchical structure as Parisi's $q^R_{\alpha\beta}$. Substituting this ansatz into (2) yields a system of linear equations which, in the continuous limit $R\to\infty$, goes over into precisely the TAK integral equation.[10] Its solutions have the same breakpoint $x_1=2\tau+O(\tau^2)$ as $q(x)$ [beyond which $q(x)=x_1/2$], i.e., $f(x)=-\sin\omega x$ for $x<x_1$ and $f(x)=$constant for $x\geq x_1$, corresponding to the discrete set of eigenvalues $\lambda^{(1)}=\omega^{-2}$ obtained from [11]

$$\cot\omega x_1=\omega(1-x_1) . \quad (3)$$

This spectrum contains one "large mass"

$\lambda^{(1)} = 2\tau + \cdots$ and a sequence of small (but stable) masses

$$\lambda_m^{(1)} = \frac{4\tau^2}{m^2\pi^2} + \cdots, \quad m = 1, 2, \ldots \quad (4)$$

(2) We need now enter deeper into Parisi's hierarchical block procedure.[6,10] Consider, for example, the α row index ($\alpha = 1, 2 \ldots n$). It may be replaced by the sequence of hierarchical block numbers $(j_0, j_1, j_2 \ldots j_R; a)$, where $j_0 = 1$, $j_1 = 1, 2, \ldots n/m_1$, $j_2 = 1, 2 \ldots m_1/m_2, \ldots, j_R = 1, 2 \ldots m_{R-1}/m_R$ and where $a = 1, 2 \ldots m_R$ labels replicas in the smallest block R (m_l is the number of replicas in each of the m_{l-1}/m_l blocks l). Coordinates of a matrix element $q_{\alpha\beta}$ are now written with the two sequences $(j_0, j_1 \ldots j_R; a)$ and $(l_0, l_1, \ldots l_R; b)$. Let us call *overlap* the *uninterrupted* sequence $j_0 = l_0$, $j_1 = l_1, \ldots j_i = l_i$ but $j_{i+1} \neq l_{i+1}$, or rather the number i, $\alpha \cap_0 \beta = i$. We then have $q_{\alpha\beta} = q_i$. In the continuous limit where $m_i = i/(R+1)$, $R \to \infty$, and $m_i < x < m_i + 1$, then $q_i \to q(x)$. Consider now an eigenvector $f_{\alpha\beta}^{(\theta)}$ of the 2nd family with one distinguished replica θ. We need now know, besides i (or x in the continuous limit), the overlaps $\theta \cap_0 \alpha = k^\alpha$, $\theta \cap_0 \beta = k^\beta$. It is easily observed that if those two numbers are equal $k^\alpha = k^\beta = k$, $k \leq i$, but if $k^\alpha \neq k^\beta$, $i = \min(k^\alpha, k^\beta)$, in which case the other number is $\max(k^\alpha, k^\beta) \equiv k$. The eigenvector is thus now dependent upon two variables i (or x, a measure of the distance to the diagonal) and k (or z, a measure of the distance to θ); $f(x;z)$. This type of ansatz leads to

$$0 = -\lambda^{(2)} f(x;z) + 2q(x) \int_x^1 dt\, f(t;z) + 2\int_0^x dt\, q(t) f(t;z) - 2q(z) \int_z^1 t\, dt \frac{\partial}{\partial t} f(z;t) \quad (5)$$

for $z < x$, and for $x < z$

$$0 = -\left[\lambda^{(2)} + xq(x) + \int_x^z dt\, q(t)\right] f(x;z) + q(x) \int_x^1 dt [f(t;x) + f(t;z)]$$
$$+ \int_0^x dt\, q(t)[f(t;z) + f(t;x)] - \int_z^1 t\, dt \frac{\partial}{\partial t}[q(z) f(x;t) + q(x) f(z,t)] + xq(x) f(x;x) + \int_x^z dt\, q(t) f(x;t) \; . \quad (6)$$

The TAK family is obviously included as a particular solution independent of z. A study of the discrete equations behind (5) and (6) shows that solutions $f(x;z)$ can be adequately parametrized by a breakpoint value κ on z, beyond which $f_\kappa(x; z \geq \kappa)$ is z independent, and that the eigenvalues can be obtained from the one-dimensional integral equation for $(\partial/\partial z) f_\kappa(x;z)|_{z-\kappa-0} \equiv F_\kappa(x)$.

For example, in the region $\kappa < x_1$, $F_\kappa(x)$ is a Gegenbauer function[12] for $0 \leq x < \kappa$, a sine function with a phase shift for $\kappa \leq x < x_1$, and a constant for $x \geq x_1$ (note that F_κ has a jump as x crosses κ). Matching boundary conditions gives the eigenvalues spectrum $\lambda^{(2)} \equiv \lambda_\kappa$.

If $\kappa < x_1$, again, with $\lambda^{(2)} \equiv \omega^{-2}$, one has

$$L(\omega\kappa) = \frac{\tan\omega(x_1 - \kappa) + \omega(1 - x_1)}{1 - \omega(1 - x_1)\tan\omega(x_1 - \kappa)} , \quad (7)$$

where

$$L(y) \equiv |4 - y^2|^{-1/2} C'(\xi)/C(\xi)$$

and

$$\xi \equiv y|4 - y^2|^{-1/2} ,$$

and $C(\xi)$ satisfies

$$(\xi^2 - \epsilon)C'' + 4(\xi C' + C) = 0 \; . \quad (8)$$

Here $\epsilon = \mathrm{sgn}(4 - y^2)$, and the boundary conditions are $C(0) = 0$, $C'(0) = 1$. For $\epsilon = +1$ (8) is a Gegenbauer equation.[12] Solutions of (7) for eigenvalues $\lambda_m^{(2)}(\kappa)$ can be exhibited, for $\kappa << x_1/m$, as

$$\lambda_m^{(2)}(\kappa) = \lambda_m^{(1)} \left[1 + (\tfrac{2}{3})\pi^2 m^2 \left(\frac{\kappa}{x_1}\right)^3 + \cdots \right] , \quad (9)$$

where $\lambda_m^{(1)}$ is given by (4), thus displaying bands pinned at $\lambda_m^{(1)}$. In the other extreme $\kappa >> (x_1/m)$ (but always keeping $\kappa < x_1$) one gets

$$\lambda_m^{(2)}(\kappa) = \kappa^2/(3m\pi)^{2/3} + \cdots \; . \quad (10)$$

As κ increases and reaches the value x_1, (7) becomes

$$L(\omega x_1) = \omega(1 - x_1) \; . \quad (11)$$

It keeps to that form for $\kappa > x_1$, and (11) only adds the large mass $\lambda^{(2)} \sim 2\tau + \cdots$ to the 2nd family spectrum.

Note that *no negative eigenvalue* $\lambda^{(2)} = -\omega^{-2}$ is compatible with (7) or (11). Indeed, in that case, the right- and left-hand sides of (7) (11) remain of *opposite* signs. Since the original matrix is real symmetric no complex solution exists either.

Having solved for $\lambda^{(2)}(\kappa)$ and $F_\kappa(x)$, one can return to (5) and (6) and solve for $(\partial/\partial z) f_\kappa(x;z)$ [and $f_\kappa(x;z)$] thus obtaining the eigenvectors.

(3) Turning to the third family, we need now a complete information on the eigenvector $f_{\alpha\beta}^{(\theta_1\theta_2)}$ with two distinguished replicas θ_1, θ_2. This includes the following: (i) As above the $\alpha \cap_0 \beta$ overlap i; (ii) the overlaps with θ_1, $\theta_1 \cap_0 \alpha = k_1^\alpha$, $\theta_1 \cap_0 \beta = k_1^\beta$ and with θ_2, $\theta_2 \cap_0 \alpha = k_2^\alpha$, $\theta_2 \cap_0 \beta = k_2^\beta$; and (iii)

the overlap $\theta_1 \cap_0 \theta_2 = r$, an external parameter. Again $\max(k_j^\alpha;k_j^\beta) = k_j$, $j = 1, 2$ and the domain of i is determined by $\min(k_j^\alpha,k_j^\beta)$ when $k_j^\alpha \neq k_j^\beta$, and/or by $i \geq k_j^\alpha = k_j^\beta$ when the k_j's are equal. Altogether, taking into account constraints of the hierarchical blocks geometry, and writing x, z_j, ρ for the continuous version of i,k_j,r one is left with only the following sectors:

(i) $f(x;z,z)$, for $z_1 = z_2 \equiv z < \rho$,

(ii) $f(x,z_1,\rho)$ and $f(x;\rho,z_2)$, for $\rho < z_1,z_2$,

where x varies between 0 and 1, and

(iii) $f(\rho;z_1,z_2)$, for $\rho < z_1,z_2$.

For these eigenvectors $f^{(3)}$ one can write a discrete set of coupled linear equations that, in the continuum limit, goes over into six coupled integral equations on sections of the unit cube. The system is fully symmetric in (z_1,z_2) and, as a whole, depends upon the external parameter $0 < \rho < 1$. It shrinks back onto Eqs. (5) and (6) whenever $f^{(3)}$ becomes independent of z_1 (or z_2). The gist of this "dangerous" family lies in sector (iii).

As above, solutions for the eigenvectors can be parametrized by a breakpoint κ (beyond which f is z independent). If $\kappa < \rho$, the system reduces to the second family (it is blind to any distinction between θ_1 and θ_2). If the breakpoint happens above ρ we could in principle introduce κ_1 and κ_2 (for z_1 and z_2, respectively). Assuming for the moment that the solutions are symmetric in (z_1,z_2) we are left with a single breakpoint. Letting $F_\kappa(z_1,z_2) \equiv (\partial^2/\partial z_1 \partial z_2) \times f_\kappa^{(3)}(\rho;z_1,z_2)$ one obtains an equation decoupled from the other five,

$$0 = [-\lambda^{(3)}(\kappa;\rho) + q^2(z_1) + q^2(z_2) - 2q^2(\rho)]F_\kappa(z_1,z_2) + \dot q(z_1)\int_{z_1}^1 t\,dt\,F_\kappa(t;z_2) + \dot q(z_2)\int_{z_2}^1 t\,dt\,F_\kappa(z_1;t)\ . \quad (12)$$

If we let $z_1 = z_2 = \kappa - 0$, then (12) yields either (i) $F_\kappa(\kappa-0;\kappa-0) = 0$ or (ii) $\lambda^{(3)}(\kappa;\rho) = 2[q^2(\kappa) - q^2(\rho)]$. In this last case one obtains, depending upon the relative position of x_1 with respect to ρ and κ,

$$\lambda^{(3)} = \begin{cases} 0, & x_1 < \rho < \kappa \quad (13) \\ \frac{1}{2}(x_1^2 - \rho^2), & \rho < x_1 < \kappa\ , \quad (14) \\ \frac{1}{2}(\kappa^2 - \rho^2), & \rho < \kappa < x_1\ . \quad (15) \end{cases}$$

A study of the system of six coupled integral equations shows that, if $\lambda \neq \lambda^{(3)}$, the system folds back onto the second family [Eqs. (5) and (6)]. It also shows that (15) is not an admissible solution (it corresponds to an identically vanishing eigenvector). This is not a loss of any eigenvalues since (14) and (15) span the same interval $(0, 2\tau^2)$. The same remark applies to possible solutions with asymmetric breakpoints κ_1, κ_2 in sectors (ii) and (iii). Note that the soft mode (13) occupies most of the parameter space (except for a piece of order τ^2) and corresponds to fluctuations disturbing the flat core of $q(x)$. Although we did not mention it, such soft modes also exist in the two previous families.

(4) Equations (3), (7), (11), (13), and (14) give the complete spectrum of eigenvalues. This spectrum in *wholly non-negative*, proving that the Parisi long-range spin-glass solution (at least for the approximate Parisi model used here for simplicity) is *marginally stable*.

In addition to extracting the eigenvalue spectrum from (2) there are good reasons to solve it in full for the eigenvectors. Indeed, that would allow one to construct the one-loop correction to the tree approximation, that is the first correction in the reciprocal of the coordination number. That would reveal both the effect of short-range corrections, and presumably the lower critical dimension that has been the subject of much speculation.

---

*On leave of absence from Institute for Theoretical Physics, Eötvös University, Budapest, Hungary.

[1]D. Sherrington and S. Kirkpatrick, Phys. Rev. Lett. 35, 1792 (1975).
[2]J. R. L. de Almeida and D. J. Thouless, J. Phys. A 11, 983 (1978).
[3]A. J. Bray and M. A. Moore, Phys. Rev. Lett. 41, 1068 (1978); A. Blandin, J. Phys. (Paris) 39, C6-1499 (1978).
[4]A. Blandin, M. Gabay, and T. Garel, J. Phys. C 13, 403 (1980).
[5]C. De Dominicis and T. Garel, J. Phys. Lett. 40, L575 (1979); A. J. Bray and M. A. Moore, J. Phys. 13, 419 (1980).
[6]G. Parisi, J. Phys. A 13, L115 (1980); 13, 1101 (1980); 13, 1887 (1980).
[7]More recent work of H. Sompolinsky [Phys. Rev. Lett. 47, 935 (1981)] parametrizes the solution to SK equations with, besides $q(x)$, an "anomaly" $\Delta(x)$. This formulation is, however, at least at the level of the tree approximation, equivalent to Parisi's [C. De Dominicis, M. Gabay, and B. Duplantier, J. Phys. A 15, L47 (1982); H. J. Sommers (unpublished)].
[8]D. J. Thouless, J. R. L. de Almeida, and J. M. Kosterlitz, J. Phys. C 13, 3272 (1980), Eq. (19).
[9]G. Parisi, Phys. Rev. Lett. 43, 1754 (1979).
[10]Equation (19) of Ref. 8.
[11]J. R. L. de Almeida, Ph.D. thesis (Birmingham, 1980) (unpublished).
[12]See, e.g., W. Magnus, F. Oberhettinger, and R. P. Soni, *Special Functions of Mathematical Physics* (Springer, New York, 1966).

VOLUME 50, NUMBER 24 PHYSICAL REVIEW LETTERS 13 JUNE 1983

Order Parameter for Spin-Glasses

Giorgio Parisi

Università di Roma II, Tor Vergata, Rome, Italy, and Laboratori Nazionali, Istituto Nazionale di Física Nucleare, Frascati, Italy

(Received 1 February 1983)

An order parameter for spin-glasses is defined in a clear physical way: It is a function on the interval 0–1 and it is related to the probability distribution of the overlap of the magnetization in different states of the system. It is shown to coincide with the order parameter introduced by use of the broken replica-symmetry approach.

PACS numbers: 75.50.Kj

The mean-field theory for spin-glasses has been obtained in the framework of the replica approach: In the phase where the replica symmetry is spontaneously broken the order parameter is a function $Q(x)$ defined on the interval 0–1.[1] Although a similar order parameter has been obtained in a dynamical approach,[2] the physical meaning of this order parameter is unclear. In this note I define a physically motivated order parameter which I show to be equal to the order parameter of the conventional replica approach.

We consider an Ising spin system, the total number of spins (N) being sufficiently large so that we stay near the thermodynamic limit. As usual, the statistical expectation values are given by

$$\langle O(\sigma)\rangle = \frac{\sum_{\{\sigma\}} O(\sigma)\exp[-\beta H(\sigma)]}{\sum_{\{\sigma\}}\exp[-\beta H(\sigma)]}, \tag{1}$$

$O(\sigma)$ and $H(\sigma)$ being an observable and the Hamiltonian, respectively.

It is well known that in the presence of spontaneous magnetization Eq. (1) must be modified: It predicts zero magnetization at zero magnetic field for all temperatures. Equation (1) does not describe a pure state (thermodynamic phase) of the system but a mixture of different states. We can decompose it as the sum of pure equilibrium states[3]:

$$\langle\cdots\rangle = \sum_{\alpha=1}^{M} P_\alpha\langle\cdots\rangle_\alpha, \quad \sum_{\alpha=1}^{M} P_\alpha = 1. \tag{2}$$

For each of the pure states (labeled by α) the spontaneous magnetization may be different from zero, while the connected correlation function should go to zero at large distances (clustering).

In the normal ferromagnetic Ising model M is always 1, unless the magnetic field is zero and the temperature sufficiently low. In this case we have two pure states of positive and negative magnetization. For spin-glasses the situation is quite different: There is good evidence that (especially for the infinite-range model) even for a nonzero magnetic field there are many equilibrium states; the space of configurations consists of many valleys separated by high mountains (free-energy barriers) whose height goes to infinity in the infinite-volume limit.[4] Explicit Monte Carlo simulations have shown that in the infinite-range model the system will not change valleys for a time proportional to $\exp(N^{1/4})$.[5] The total number of valleys seems to increase like a power of N, and so they do not contribute to the zero-temperature entropy. If there is not a one to one correspondence between the pure states of the system at two different values of the magnetic field, hysteresis effects are expected: By changing the magnetic field the system will go to an excited metastable state and decay to the true equilibrium state only after a very large time.[6] In the same way the linear-response susceptibility, which according to Fischer[7] is given by

$$\chi_{LR} = \beta\left(1 - \sum_{i=1}^{N} \langle\sigma_i\rangle_\alpha^2/N\right), \tag{3}$$

is different from the total susceptibility

$$\chi = \beta\left(1 - \sum_{i=1}^{N} \langle\sigma_i\rangle^2/N\right), \tag{4}$$

for which the contribution of jumping from one state to another is included.

The existence of many states according to the previous analysis is the main characteristic of the glassy phase and we would like to retain this information in the order parameter.

We first notice that we can characterize the state by the value of the magnetization in each site:

$$m_i^{\alpha} = \langle\sigma_i\rangle_\alpha. \tag{5}$$

For each state we can construct the equivalent

of the Edwards-Anderson order parameter[8]:

$$q_{EA}{}^{\alpha}=\sum_{i=1}^{N}\langle\sigma_i\rangle_\alpha{}^2/N. \qquad (6)$$

It is quite reasonable that in the infinite-volume limit all states will have the same value of $q_{EA}{}^\alpha$: Equation (6) defines an order parameter which does not make reference to a particular pure state. The serious disadvantage of q_{EA} is that it is different from zero also for a normal unfrustrated ferromagnetic or antiferromagnetic material, not only for a spin-glass.

Something more interesting can be obtained by studying the overlap of the magnetizations between two different states:

$$q_{\alpha\beta}=\sum_{i=1}^{N}m_i{}^\alpha m_i{}^\beta/N,$$
$$P(q)=\sum_{\alpha,\beta}P_\alpha P_\beta\delta(q-q_{\alpha\beta}), \qquad (7)$$

where $P(q)$ is the probability distribution of the $q_{\alpha\beta}$. We can introduce the function $x(q)$, where

$$x(q)=\int_{-\infty}^{q}dq\,P(q), \qquad (8)$$

which is monotonic and an inverse function, which is obviously defined in the interval 0–1.

If the function $q(x)$ is constant, we have only pure states which do not differ macroscopically. If the function $q(x)$ is not a constant, macroscopically different pure states must exist; $q(1)$ is identified with the Edwards-Anderson order parameter, while $q(0)$ is the minimum overlap between two states. The linear-response susceptibility (the reversible one) and the true susceptibility (which should be presumably identified with the field-cooled one) are given by

$$\chi_{LR}=\beta[1-q(1)],\quad \chi=\beta[1-\int_0^1 q(x)dx]. \qquad (9)$$

We can thus characterize the glassy phase by a nontrivial dependence of $q(x)$ on x.[9]

Having defined an order parameter having a clear physical meaning, we can now show that this order parameter coincides with the one introduced in the replica approach.[10] The explicit decomposition of a state into its pure components is not a simple operation; a fast way to obtain the function $q(x)$ consists in considering two real replicas of the same system[11] ($\sigma_i{}^a$; $i=1,N$; $a=1,2$) with the Hamiltonian

$$H_2=\sum_{a=1}^{2}H(\sigma^a). \qquad (10)$$

In the infinite-volume limit we have that

$$\langle\exp(y\sum_{i=1}^{N}\sigma_i{}^1\sigma_i{}^2/N)\rangle_2\equiv g(y)\cong\sum_{\alpha=1}^{M}\sum_{\beta=1}^{M}P_\alpha P_\beta\exp(yq_{\alpha\beta})=\int_0^1 dx\,\exp[yq(x)], \qquad (11)$$

where by $\langle\cdots\rangle_2$ we denote the statistical expectation value with the Hamiltonian H_2.[12] Equation (11) is rather interesting. It gives us the possibility of computing $q(x)$ using conventional techniques, like the Monte Carlo method.

In the replica approach one introduces an n times replicated Hamiltonian and the order parameter is an $n\times n$ matrix, which is zero on the diagonal, defined as follows[1]:

$$Q_{ab}=\frac{1}{N}\sum_{i=1}^{N}\langle\sigma_i{}^a\sigma_i{}^b\rangle,\quad a\neq b,\quad Q_{aa}=0. \qquad (12)$$

In the presence of symmetry breaking in replica space the matrix Q_{ab} has a nontrivial dependence on the indices. The naive expression for $g(y)$,

$$g(y)=\exp(yQ_{12}), \qquad (13)$$

must be modified[13] by summing over all the possible ways in which the symmetry may be broken in replica space. We finally obtain

$$g(y)=\frac{1}{n(n-1)}\sum_{\substack{a=1\\ a\neq b}}^{n}\sum_{b=1}^{n}\exp(yQ_{ab}). \qquad (14)$$

In the approach of Ref. 1 the matrix Q_{ab} was characterized by a function $Q(x)$ defined on the interval 0–1; if we go back to the original definitions, we easily find that for $n\to0$

$$g(y)=\int_0^1 dx\,\exp[yQ(x)]. \qquad (15)$$

The functions $Q(x)$ and $q(x)$ are therefore identified, the physical interpretation of $Q(x)$ now being clear.

We notice, *en passant*, that a rather simple expression for $q(0)$ can be obtained if we consider the function $q(h_1,h_2)$ defined as

$$q(h_1,h_2)=\frac{1}{N}\sum_{i=1}^{N}m_i(h_1)m_i(h_2). \qquad (16)$$

Similar arguments, which will be reported in detail elsewhere,[14] tell us that[9]

$$\lim_{h_2\to h_1}q(h_1,h_2)=q(0)\neq q(h_1,h_1)=\int_0^1 dx\,q(x). \qquad (17)$$

In other words $q(0)$ is the average overlap between two states[15] at different but similar magnetic fields.

VOLUME 50, NUMBER 24 PHYSICAL REVIEW LETTERS 13 JUNE 1983

It is now clear that the mysterious breaking of the replica symmetry is just the mathematical transcription of the existence of infinitely many pure thermodynamic states.

The definition of the order parameter $q(x)$ does not make reference to the fact that for spin-glasses the Hamiltonian H is random. It is unclear to me if other amorphous materials (like real glasses or hard spheres at high density) have many different pure states when the volume goes to infinity so that the definition of the order parameter presented here can be successfully extended.

It is a pleasure to thank C. De Dominicis, I. Kondor, and P. Young for a discussion which stimulated this work and many of the theoretical physicists of the Institut des Hautes Etudes Scientifique (IHES) for clarifying discussions. This work has been done while staying at the IHES, whose warm hospitality is gratefully acknowledged. The author is also grateful to R. Ellis for a careful reading of the manuscript.

---

[1]G. Parisi, Phys. Rev. Lett. 43, 1754 (1979), and J. Phys. A 13, L115, 1101, 1887 (1980).

[2]H. Sompolinsky, Phys. Rev. Lett. 47, 935 (1981).

[3]A classical reference for the definition of a pure state is D. Ruelle, *Statistical Mechanics* (Benjamin, New York, 1969); the relevance of pure states for spin-glasses has been recently discussed by A. C. D. van Enter and J. L. van Hemmen, University of Heidelberg Report No. 175, 1982 (unpublished).

[4]The existence of many valleys separated by high mountains has been established in the original paper of S. Kirkpatrick and S. Sherrington, Phys. Rev. B 17, 4384 (1981); further theoretical work on this and related subjects can be found in C. De Dominicis, M. Gabay, H. Orland, and T. Garel, J. Phys. (Paris) 41, 923 (1980); A. J. Bray and M. A. Moore, J. Phys. C 13, L469, L907 (1980); P. Young, J. Phys. C 14, L1085 (1981); R. G. Palmer, Adv. Phys. 31, 669 (1982). The existence of many quasiequilibrium states which become stable in the infinite-volume limit was at the basis of the approach of Ref. 2.

[5]N. D. Mackenzie and A. P. Young, Phys. Rev. Lett. 49, 301 (1982).

[6]G. Parisi, in *Disordered Systems and Localization*, edited by C. Castellani, C. Di Castro, and L. Peliti (Springer-Verlag, Berlin, 1981), and Phys. Rep. 67C, 25 (1980).

[7]K. H. Fischer, Phys. Rev. Lett. 34, 1438 (1975).

[8]S. F. Edwards and P. W. Anderson, J. Phys. F 5, 965 (1975).

[9]By comparing Eqs. (2), (4), and (7), it follows that

$$\frac{1}{N}\sum_{1}^{N}\langle\sigma_i\rangle^2 = \frac{1}{N}\sum_{\alpha,\beta}\sum_{1}^{N} P_\alpha P_\beta m_i^\alpha m_i^\beta = \int dq\, q P(q)$$

$$= \int_0^1 dx\, q(x).$$

[10]It has already been suggested that the order parameter of Ref. 1 should be connected in some way to phase-space distance. See, for example, J. A. Hertz, to be published; H. Sompolinsky, Phys. Rev. B 25, 6860 (1982); C. Dasgupta and H. Sompolinsky, to be published.

[11]A. Blandin, M. Gabay, and T. Garel, J. Phys. C 13, 403 (1980).

[12]The key ingredient for deriving Eq. (11) is the vanishing at large distance of the connected correlation functions (clustering) in the pure states α and β.

[13]A similar technique was used by C. De Dominicis and P. Young, to be published; H. J. Sommers, J. Phys. (Paris), Lett. 43, L719 (1982).

[14]G. Parisi, to be published.

[15]R. G. Palmer and C. M. Pond, J. Phys. F 9, 1451 (1979).

VOLUME 51, NUMBER 13 PHYSICAL REVIEW LETTERS 26 SEPTEMBER 1983

Direct Determination of the Probability Distribution for the Spin-Glass Order Parameter

A. P. Young
Department of Mathematics, Imperial College, London SW72BZ, United Kingdom
(Received 25 July 1983)

The recent interpretation of Parisi's order-parameter function $q(x)$ in terms of a probability distribution for the overlap between magnetizations in different phases is investigated by Monte Carlo computer simulation for the infinite-range Ising spin-glass model. The main features of the solution for $q(x)$ are reproduced, in particular $q(x) \propto x$ as $x \to 0$ and $q'(x) = 0$ at $q = q_{\max}$, the largest value. Finite-size effects prevent one from establishing with certainty whether there is a "plateau," i.e., $q'(x) = 0$ for a range of x.

PACS numbers: 75.30.Kz, 05.50.+q, 64.60.Cn, 75.10.Hk

The replica technique used to be a great mystery when applied in situations[1] where it was necessary to "break replica symmetry." It has been applied particularly to the infinite-range Ising spin-glass model of Sherrington and Kirkpatrick[2] (SK and the most interesting results have been obtained by Parisi.[3] In his theory replica-symmetry breaking shows up as an order-parameter function $q(x)$ for $0 \le x \le 1$. Recently, however, the replica technique has become much better understood following the observation[4] that statistical-mechanics expectation values are obtained by integrals over x; e.g.,

$$q = \langle\langle S_i\rangle_T^2\rangle_J = \int_0^1 q(x)dx$$
$$q^{(2)} = \langle\langle S_iS_j\rangle_T^2\rangle_J = \int_0^1 q^2(x)dx \quad (i \neq j), \tag{1}$$

where $S_i = \pm 1$ is an Ising spin, $i = 1, \ldots, N$, $\langle \ldots \rangle_T$ denotes a statistical-mechanics average for a given set of interactions, and $\langle \ldots \rangle_J$ is an average over interactions. Subsequently it was shown[5] that one can interpret not just integrals over x but $q(x)$ for any particular value of x. The main ingredient in the argument is that the SK model can exist in one of many phases, which are stable for $N \to \infty$. If one defines the magnetization of site i when the system is in phase s by m_i^s and $q^{ss'}$, the overlap between magnetizations, by

$$q^{ss'} = \frac{1}{N}\sum_{i=1}^{N} m_i^s m_i^{s'}, \tag{2}$$

then the derivative of the inverse function, i.e., dx/dg, turns out to be a probability distribution for overlap between magnetizations of solutions, i.e.,

$$\frac{dx}{dq} = W(q) = \langle \sum_{s,s'} P(s)P(s')\delta(q^{ss'} - q)\rangle_J, \tag{3}$$

where $P(s)$ is the Boltzmann weight[4] associated with solution s. In particular, changing the integration variable in Eq. (1) from x to q, one finds that q and $q^{(2)}$ are just the first two moments of the distribution $W(q)$, namely,

$$q = \int q'W(q')dq', \quad q^{(2)} = \int q'^2W(q')dq'. \tag{4}$$

Recently,[6] both q and $q^{(2)}$ have been calculated by computer simulations at a fixed temperature below the transition temperature T_c for several sizes and the results appear to extrapolate to Parisi's values for $N \to \infty$. However, a much more dramatic test of the theory would be to reproduce the entire distribution $W(q)$. This Letter describes numerical simulations of $W(q)$ for several finite sizes which do indeed reproduce many of the features of Parisi's function $q(x)$. Before giving the numerical data it is necessary to describe what Parisi's equations give for $W(q)$.

Rather than plot $q(x)$ against x, as is conventional, I sketch in Fig. 1 the inverse function $x(q)$, since $W(q)$ is just the derivative of this. Above the de Almeida-Thouless[7] (AT) line there is only one phase, the SK solution is correct, and $x(q)$ is a unit step function at $q = q_{SK}$, the SK value, so that

$$W(q) = \delta(q - q_{SK}). \tag{5}$$

For $h \to 0$, on the other hand, Parisi's theory predicts[3,8] that $x(q)$ has a smooth part, starting at the origin and ending at $x = \bar{x}$, $q = q_{\max}$, at which point there is step to $x = 1$, so that

$$W(q) = \overline{W}(q) + (1 - \bar{x})\delta(q - q_{\max}), \tag{6}$$

where $\overline{W}(q)$ is the smooth part of the distribution and has weight $\bar{x}$. Since $x \propto q$ as $q \to 0$, then $\overline{W}(q)$ is finite as $q \to 0$. Obviously $\overline{W}(q)$ is zero for $q > q_{\max}$.

Next I discuss the numerical simulations. The Hamiltonian is given by

$$H = -\sum_{\langle i,j\rangle} J_{ij}S_iS_j - h\sum_i S_i, \tag{7}$$

where the J_{ij} are Gaussian random variables with zero mean and variance[2,9] $J/(N-1)$, and h is a uniform field. For $N \to \infty$ there is a transition in

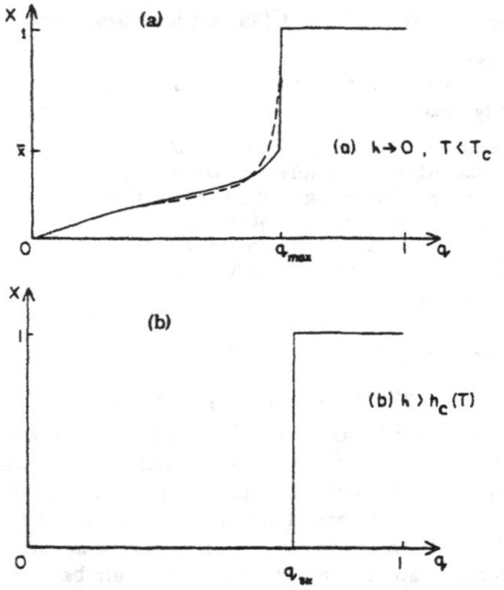

FIG. 1. The solid lines are sketches of the function $x(q)$ from Parisi's theory. In (a), the dashed curve is similar to Parisi's results but with the vertical piece rounded out. The numerical simulations on moderate sizes cannot distinguish between these possibilities. In (b), $h_c(T)$ is the critical de Almeida–Thouless field and q_{SK} is the SK value for the spin-glass order parameter.

zero field at $T_c = J$. It is useful to simulate at the same time two independent samples with the *same* interactions and field. These two samples are simulated up to a time t_0, to equilibrate, before any averaging is done. It is necessary that t_0 is longer than the longest relaxation time,[10] τ, which diverges when $N\to\infty$. However, previous work[11] has obtained the relaxation times at $T=0.4T_c$, $h=0$, and so we shall concentrate on this point in the h-T plane and make sure that t_0 is large enough for each size.

For $t>t_0$ one then calculates

$$Q(t)=\frac{1}{N}\sum_{i=1}^{N} S_i^{1}(t_0+t)S_i^{2}(t_0+t), \tag{8}$$

where the superscripts 1 and 2 on S_i refer to the two identical samples. The distribution of $Q(t)$ is independent of t and is obtained from

$$W(q)=\left\langle \frac{1}{T}\sum_{t=1}^{T}\delta(q-Q(t))\right\rangle_J. \tag{9}$$

With use of an integral representation of the delta function it is straightforward to show that for $N\to\infty$ the distribution is the same as that in Eq. (3). Furthermore, standard statistical arguments predict that if the system exists in just a single phase then, for a large finite system, $W(q)$ is a Gaussian distribution of width of order $N^{-1/2}$, which of course goes over to the delta function of Eq. (5) in the thermodynamic limit.

Above the AT line I find precisely this Gaussian behavior, showing that there is only one phase available to the system. By contrast the results for $W(|q|)$ at $h=0$, $T=0.4T_c$, shown in Fig. 2, have a peak at large $|q|$ but also a long tail extending to $|q|=0$ with a finite weight there. Since the model has time-reversal symmetry $W(q)$ is symmetric[12] and so I plot the distribution against $|q|$. Consider first of all the small-q region in more detail. All sizes show a finite value for $W(0)$ with no sign of this vanishing for $N\to\infty$. In fact, there is some suggestion of an upturn in the curve as $|q|\to 0$ for larger sizes though this is not really outside the statistical errors. Interestingly, it appears that the solution of Parisi's theory does have such an upturn.[8] Since a direct solution of Parisi's equations is very complicated, in order to compare the numerical data with the theory I have used the scaling *Ansatz* of Ref. 8 that $q(x,T)=F(x/T)$ for $x\le\bar{x}$ to determine $q(x)$. The only extra information needed is $q_{max}(T)$. I use the approximate formula

$$q_{max}(T)=1-2\left(\frac{T}{T_c}\right)^2+\left(\frac{T}{T_c}\right)^3, \tag{10}$$

which correctly gives the first three terms in the expansion[13] about $T=T_c$ and correctly gives a T^2 dependence at low temperatures. The resulting prediction for $W(q)$ is shown by the dotted line in Fig. 2. The upturn at low q is more pronounced than in the numerical results.

Next let us look at the region of the peak in $W(q)$. The peak position shifts to smaller q values as N increases, consistent with the value 0.744, from Eq. (10), for $N\to\infty$, although reliable extrapolation is not possible because of uncertainties in the data and because the form of the leading size correction is not known. The peak also becomes narrower as N increases, particularly on the high-q side, indicating that $W(q)$ is strictly zero for q bigger than some q_{max} when $N\to\infty$. However, on the low-q side of the peak the results appear to differ more and more from the approximate analytic solution of Parisi's equations (shown by the dotted line) as N increases. This could indicate that $W(q)$ diverges as $q\to q_{max}$, or in other words, $dq/dx\to 0$ as $q\to q_{max}$. Such a possibility would occur, for instance, if

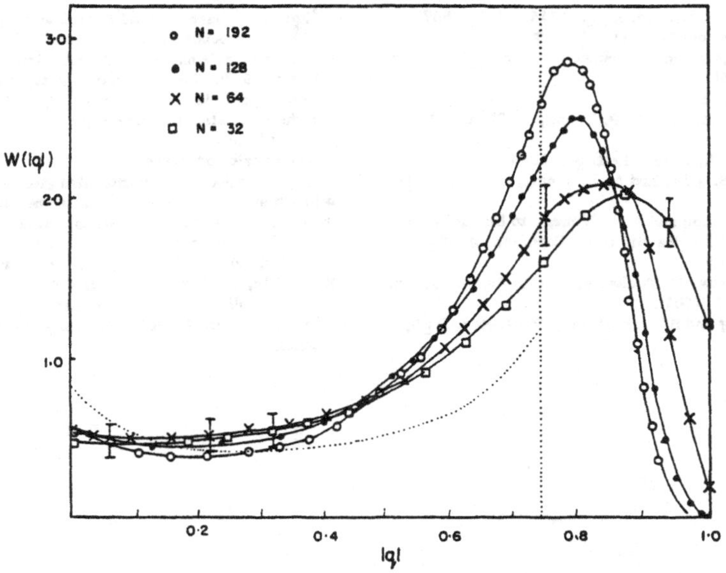

FIG. 2. $W(|q|)$ for $T = 0.4T_c$, $h = 0$. Some typical error bars are shown. The data clearly indicate a tail in $W(|q|)$ extending down to $q = 0$, which corresponds to the existence of many phases some of which have zero overlap between their magnetizations. There is also clear evidence that for $N \to \infty$ a divergence in $W(|q|)$ occurs at a maximum value, q_{max}, beyond which $W(q)$ is zero. The dotted line is obtained from a scaling Ansatz for Parisi's equations, as described in the text. There is a delta function of weight $\frac{1}{7}$ at $q = q_{max} = 0.744$ and a continuous part which has a pronounced upturn as $q \to 0$.

dq/dx is only zero at $x = 1$, i.e., the "plateau" in $q(x)$ is rounded out, as shown by the dashed line in Fig. 1(a). Another possibility is that a plateau region in $q(x)$ does occur but that the rest of the curve joins the plateau region with zero slope. It should be pointed out, though, that the *position* of the peak is also shifting with increasing N and this may account for the data for larger sizes deviating more from the dotted line in Fig. 2. It is possible that for much larger sizes, where the shift in the peak becomes negligible, one would see the numerical results approaching the analytic theory. One cannot really distinguish between these various possibilities from the available data.

To conclude, I have shown that many features of Parisi's order-parameter function are reproduced by the simulations, in particular, $q(x) \propto x$ as $x \to 0$ and $q'(x) = 0$ at $q = q_{max}$. Parisi's theory is therefore at least an excellent approximation to the solution of the SK model and may well be the exact solution. It would help comparison with numerical data if accurate solutions for the functions $q(x)$ were available. The computations also provide strong support for the arguments of Ref. 4 that statistical-mechanics averages are obtained from integrals over x (which differs from Sompolinsky's[14] dynamical interpretation) and for the subsequent interpretation[5] of dx/dq as a probability distribution for overlap between magnetizations of phases. This insight into the physical significance of replica-symmetry breaking may be useful in other situations, for instance a ferromagnet with finite-range interactions in a random field[15] which is of considerable interest at the moment.

The motivation for this work came from a stimulating discussion with H. Sompolinsky. I should also like to thank M. Kosterlitz and J. Vannimenus for helpful discussions.

[1]See, e.g., A. J. Bray and M. A. Moore, Phys. Rev. Lett. 41, 1068 (1978); G. Parisi, Phys. Rev. Lett. 43,

1574 (1979), and J. Phys. A 13, L115, 1101, 1887 (1098), and Philos. Mag. 41, 677 (1980).
[2]D. Sherrington and S. Kirkpatrick, Phys. Rev. Lett. 35, 1792 (1975).
[3]Parisi, Ref. 1.
[4]C. de Dominicis and A. P. Young, J. Phys. A 16, 2063 (1983).
[5]G. Parisi, Phys. Rev. Lett. 50, 1946 (1983); A. Houghton, S. Jain, and A. P. Young, J. Phys. C 16, L375 (1983).
[6]N. D. Mackenzie and A. P. Young, to be published.
[7]J. R. L. de Almeida and D. J. Thouless, J. Phys. A 11, 983 (1978).
[8]J. Vannimenus, G. Toulouse, and G. Parisi, J. Phys. (Paris) 42, 565 (1981).
[9]A. P. Young and S. Kirkpatrick, Phys. Rev. B 25, 440 (1982).
[10]For $h = 0$ there is an additional longer time, τ_{eg}, which corresponds to "turning over" all the spins; see N. D. Mackenzie and A. P. Young, Phys. Rev. Lett. 49, 301 (1982). However, we know that all these fluctuations do is symmetrize $W(q)$, so that I simply run the simulation for time τ and then symmetrize $W(q)$ by hand.
[11]Mackenzie and Young, Ref. 10.
[12]Note that Parisi's solution also gives a symmetric $W(q)$ in strictly zero field because one should average over all equivalent saddle points, including those generated by time-reversal symmetry; see Ref. 4.
[13]D. J. Thouless, J. R. L. de Almeida, and J. M. Kosterlitz, J. Phys. C 13, 3271 (1980).
[14]H. Sompolinsky, Phys. Rev. Lett. 47, 935 (1981).
[15]A. J. Bray and M. A. Moore, unpublished; G. Parisi, unpublished.

J. Physique **45** (1984) 843-854 MAI **1984**, PAGE 843

Classification
Physics Abstracts
75.50K

Replica symmetry breaking and the nature of the spin glass phase

M. Mézard, G. Parisi (+), N. Sourlas, G. Toulouse (*) and M. Virasoro (++)

Laboratoire de Physique Théorique de l'Ecole Normale Supérieure (**)

(+) Universita di Roma II, Tor Vergata, Italy
(*) Laboratoire de Physique de l'Ecole Normale Supérieure, Paris, France

(*Reçu le 15 décembre 1983, accepté le 25 janvier 1984*)

Résumé. — Récemment, l'un d'entre nous a proposé, comme paramètre d'ordre pour les verres de spin, une distribution de próbabilité. Nous montrons que cette probabilité dépend de la réalisation particulière des couplages, même à la limite thermodynamique, et nous étudions sa distribution. Nous montrons aussi que l'espace des états est muni d'une topologie ultramétrique.

Abstract. — A probability distribution has been proposed recently by one of us as an order parameter for spin glasses. We show that this probability depends on the particular realization of the couplings even in the thermodynamic limit, and we study its distribution. We also show that the space of states has an ultrametric topology.

1. Introduction.

The usual approach to investigate the different phases of a physical system starts with the mean field approximation. In the case of spin glasses, even this first step has required a lot of effort [1-5].

The mean field approximation has been formulated by using the infinite range model [2]. The partition function is given by :

$$Z(\beta, J, h) = \sum_{\{\sigma\}} e^{-\beta H(J,h,\sigma)} \tag{1}$$

$$H(J, h, \sigma) = -\sum_{i,j} J_{ij}\, \sigma_i\, \sigma_j - h \sum_i \sigma_i \tag{2}$$

where the σ_i $i = 1, ..., N$ are Ising spins, the sum $\sum_{i,j}$ runs over all pairs of spins, and J_{ij} are random couplings obeying a given probability distribution, which we suppose to be symmetric, with variance $1/\sqrt{N}$.

In order to study the Hamiltonian (2), four main approaches have been used. The replica approach [1-2, 4-5] in which one considers n copies of the same system, averages over the coupling distribution, and at the end takes the limit $n \to 0$. In this way one computes the averages $\overline{O_i(J)}$ of the physical observables $O_i(J)$ (free energy, magnetic susceptibility, correlation functions, etc.) over the coupling distribution. (In this paper we shall always denote the thermal averages by $\langle\ \rangle$ and the averages over the coupling distribution by $\overline{}$.) The other three approaches are the self-consistent field approach [3, 6], the dynamical approach [7] and the numerical simulations [2, 8].

The picture which has emerged from the four above approaches is that the spin glass phase is characterized by the existence of a large number (infinite when $N \to \infty$) of equilibrium states $\alpha = 1, 2, ...$ almost degenerate (free energy valleys separated by free energy barriers becoming infinitely high in the thermodynamic limit).

In a recent paper [10] (to be referred to later as [I]), one of us has proposed an order parameter for the spin glass phase and has shown its connection and interpretation in terms of the many valley picture. In the present paper we further continue this investigation. A short version of the present work has been published elsewhere [11]. The Boltzmann-Gibbs measure is :

$$\langle 0(\sigma) \rangle = \frac{\sum_{\{\sigma\}} 0(\sigma) \exp[-\beta H(\sigma)]}{\sum_{\{\sigma\}} \exp[-\beta H(\sigma)]} \tag{3}$$

(**) Laboratoire Propre du Centre National de la Recherche Scientifique, associé à l'Ecole Normale Supérieure et à l'Université de Paris Sud. Postal Address : 24, rue Lhomond, 75231 Paris Cedex 05, France.

(++) Permanent address : Departimento di Fisica, Universita di Roma I, La Sapienza, Italy.

which can be decomposed as a sum over the pure equilibrium (clustering) states :

$$\langle \cdots \rangle = \sum_\alpha P_\alpha \langle \cdots \rangle_\alpha , \qquad \sum_\alpha P_\alpha = 1 . \tag{4}$$

We may characterize a pure state α of a spin glass by the magnetization $m_i^\alpha = \langle \sigma_i \rangle_\alpha$ at each point i of the system. Following [I] we define the overlap $q^{\alpha\beta}$ of the two pure states α and β :

$$q^{\alpha\beta} = \frac{1}{N} \sum_{i=1}^{N} m_i^\alpha m_i^\beta \tag{5}$$

($q^{\alpha\alpha}$ is the familiar Edwards-Anderson order parameter $q_{\text{E.A.}}^\alpha = \frac{1}{N} \sum_i \langle \sigma_i \rangle_\alpha^2$), and the probability $P_J(q)$ for a pair of states (α, β) to have an overlap q :

$$P_J(q) = \sum_{\alpha,\beta} P_\alpha P_\beta \, \delta(q - q^{\alpha\beta}) . \tag{6}$$

We shall denote by $P(q)$ the average of $P_J(q)$ over the coupling distribution :

$$P(q) = \overline{P_J(q)} . \tag{7}$$

Let us call q_{Max} and q_{Min} the maximum and the minimum possible overlaps between two states at a given temperature T and magnetic field H. Obviously $-1 \leqslant q_{\text{Min}} \leqslant q_{\text{Max}} \leqslant 1$. As in [I] we define

$$x(q) = \int_{-1}^{q} \mathrm{d}q' \, P(q') \tag{8}$$

and $q(x)$ the inverse function of $x(q)$. Because of (6) and (8) : $x(q_{\text{Max}}) = 1$ and $x(q_{\text{Min}}) = 0$.

Let us also define :

$$y(q) = 1 - x(q) = \int_{q}^{1} \mathrm{d}q' \, P(q') \tag{9}$$

which is the probability of two pure states to have an overlap larger than q.

In the replica approach the order parameter is a $n \times n$ matrix Q_{ab}. In the limit $n \to 0$ (if we follow the Parisi pattern of replica symmetry breaking) the matrix Q is characterized by a function $Q(x)$ where $0 \leqslant x \leqslant 1$. It was shown in [I] that $q(x)$ (the inverse function of $x(q)$ defined in (8)), is identical to $Q(x)$, thus giving a physical interpretation to the replica symmetry breaking.

In the familiar case of an homogeneous ferromagnetic system (i.e. in the absence of any disorder), if we start from the high temperature phase and cool down below the Curie temperature, the probabilities P_+ and P_- of arriving to a state of magnetization $+ m(T)$ or $- m(T)$, are well known to depend on the boundary conditions we have imposed on the system. So, even in this simple case, P_α and therefore also $P_J(q)$ are not « good, extensive quantities ». This trivial observation makes one strongly suspect that the same is true in the more complicated spin glass case. This suspicion will become a certainty in chapter 2, since we shall show that $P_J(q)$ depends on the realization of the couplings, even in the thermodynamic limit : it is not a « self averaging » quantity, in the sense that, as far as $P_J(q)$ is concerned, an increasing size of the sample does not imply an average over all disorder configurations.

The same properties are true for

$$Y_J(q) = \int_{q}^{q_{\text{Max}}} \mathrm{d}q' \, P_J(q') , \qquad \overline{Y_J(q)} = y(q) . \tag{10}$$

We show that the probability distribution of $Y_J(q)$ is calculable in the framework of the replica scheme. We find that this distribution is such that the most probable value of Y_J is 1, and therefore differs from its mean value y.

The possibility of such a behaviour for disordered systems has been suspected before [12], but it is the first time, to our knowledge, that this is demonstrated in the context of the S.K. model. The order parameter, far from being a parameter, was shown to be a function, interpreted as a probability law. And now, on top of that, there appears a probability law for this function, i.e. a probability law for a probability law.

In chapter 2 we show another remarkable property of the spin glasses which is the ultrametric topology of the space of states. Taking any three pure spin glass states α, β, γ and computing the three overlaps $q^{\alpha\beta}$, $q^{\beta\gamma}$, $q^{\gamma\alpha}$, we find that at least two of them are equal. Spaces with such a property are called ultrametric spaces [13]. From this property we show that for any value of q, by grouping together all the states with an overlap bigger than q, we separate the space of states into disjoint clusters. Each such cluster is again divided into smaller clusters, by grouping together the states with an overlap bigger than $q' > q$. This procedure can be repeated indefinitely. So we prove that the space of pure states has a hierarchical structure. This hierarchical structure is a characteristic property of ultrametric spaces.

In chapter 4 we compute the cluster distribution. In particular we show that for any value of y in the spin-glass phase (this means for $T < T_g$), the number of clusters is infinite.

Another remarkable property of all the probability distributions we consider is their universality : they depend on the different parameters of the problem (temperature, magnetic field, the particular value of q we are considering) only through the mean value y of $Y_J(q)$.

We should emphasize that all our results are obtained in the framework of the replica symmetry breaking scheme of reference [5], that is in purely static terms.

Our paper is organized as follows. In chapter 2 we recall the replica symmetry breaking mechanism from which we then derive the ultrametric structure of the space of the pure spin glass states. We also show that $P_J(q)$ is not self-averaging in the thermodynamic limit.

In chapter 3 we compute the probability distribution of $Y_J(q)$. In chapter 4 we study the distribution properties of the clusters in the space of the spin glass states.

2. Hierarchical organization of the spin glass states.

We first recall the replica symmetry breaking (R.S.B.) mechanism because in the following we will make an explicit use of it.

The $n \times n$ matrix Q_{ab} is constructed through the following recursive algorithm of successive R.S.B.

a) No symmetry breaking : $Q_{ab}^{(0)} = Q_0$ for $a \neq b$, $Q_{aa} = 0$.

b) First R.S.B. : The $n \times n$ matrix $Q^{(0)}$ is broken into $\frac{n}{m_1} \times \frac{n}{m_1}$ blocks (submatrices) Q^{A_1, B_1}, $A_1, B_1 = 1, \dots, \frac{n}{m_1}$ where the Q^{AB} are $m_1 \times m_1$ matrices. For the non-diagonal submatrices $A_1 \neq B_1$, $Q_{\alpha\beta}^{A_1 B_1} = Q_0$ and $Q_{aa}^{A_1 A_1} = 0$, $Q_{ab}^{A_1 A_1} = Q_1$, $a \neq b$ for the diagonal submatrices. This completes the construction of the matrix $Q^{(1)}$.

c) Second R.S.B. The same procedure of R.S.B. is repeated with the diagonal submatrices $Q^{A_1 A_1}$. They are broken into $\frac{m_1}{m_2} \times \frac{m_1}{m_2}$ submatrices $Q^{A_2 B_2}$, $A_2, B_2 = 1, \dots, \frac{m_1}{m_2}$ and $Q_{ab}^{A_2 B_2} = Q_1$ for $A_2 \neq B_2$, $Q_{ab}^{A_2 A_2} = Q_2$ for $a \neq b$, $Q_{aa}^{A_2 A_2} = 0$.

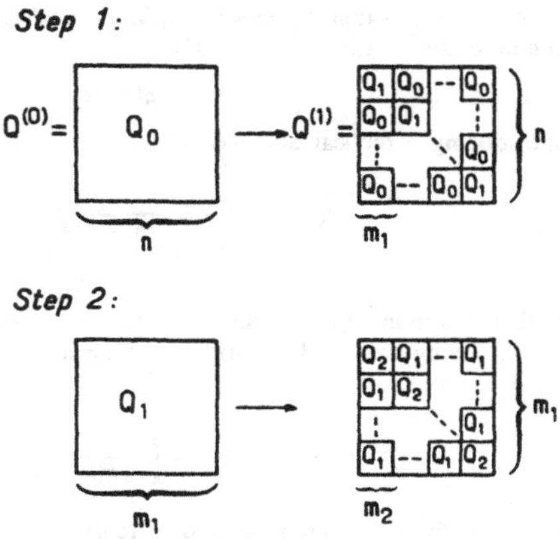

Fig. 1. — Iterative procedure for the construction of the matrix Q_{ab}.

d) The same procedure is repeated indefinitely. The whole process is illustrated in figure 1.

By construction $n \geqslant m_1 \geqslant \cdots \geqslant m_M \geqslant 1$. In the $n \to 0$ limit this becomes $1 \geqslant m_M \geqslant \cdots \geqslant m_1 \geqslant 0$, and in the limit $M \to \infty$, the m_i's become a continuous variable $m_i \to x$, $0 \leqslant x \leqslant 1$, $m_{i+1} \to x + dx$. The Q_i's become the well-known function $Q(x)$.

It was shown in [I] that the characteristic function $g(y)$

$$g(y) = \int dq\, \overline{P_J(q)}\, e^{yq} = \int dq\, P(q)\, e^{yq} \tag{11}$$

can be computed in the replica framework and is given by

$$g(y) = \frac{1}{n(n-1)} \sum_{\substack{a=1 \\ a \neq b}}^{n} \sum_{b=1}^{n} e^{yQ_{ab}} \xrightarrow[n \to 0]{} \int_0^1 dx\, e^{yQ(x)} \tag{12}$$

where the sum is over all pairs of distinct replica indices [14].

Let us now consider any three pure states α_1, α_2, α_3 and $P_J(q_1, q_2, q_3)$ the probability for them to have overlaps $q_1 = q^{\alpha_2 \alpha_3}$, $q_2 = q^{\alpha_3 \alpha_1}$, $q_3 = q^{\alpha_1 \alpha_2}$, respectively. $P_J(q_1, q_2, q_3)$ is obviously symmetric under permutations of its arguments. In order to compute $P_J(q_1, q_2, q_3)$ in the R.S.B. scheme, following [I], we consider the generalized Laplace transform

$$g_J(y_1, y_2, y_3) = \int dq_1\, dq_2\, dq_3 \exp\left(\sum_{i=1}^{3} q_i y_i \right) P_J(q_1, q_2, q_3) \tag{13}$$

and we take three identical copies of the systems, with spins σ_1, σ_2, σ_3 and Hamiltonian $H_3(\sigma_1, \sigma_2, \sigma_3) = H(\sigma_1) + H(\sigma_2) + H(\sigma_3)$. Then

$$g_J(y_1, y_2, y_3) = \left\langle \exp\left[\frac{1}{N} \sum_i \left(y_1\, \sigma_2(i)\, \sigma_3(i) + y_2\, \sigma_3(i)\, \sigma_1(i) + y_3\, \sigma_1(i)\, \sigma_2(i) \right) \right] \right\rangle_3 \tag{14}$$

where $\langle \ \rangle_3$ means that the expectation value is taken with respect to the Hamiltonian H_3. It is then possible to compute the average over the J's

$$g(y_1, y_2, y_3) = \overline{g_J(y_1, y_2, y_3)} . \tag{15}$$

By introducing n replicas and letting $n \to 0$

$$g(y_1, y_2, y_3) = \frac{1}{n(n-1)(n-2)} \sum_{\substack{a=1 \\ a \neq b}}^{n} \sum_{\substack{b=1 \\ b \neq c \\ a \neq c}}^{n} \sum_{c=1}^{n} e^{y_1 Q_{ab}} e^{y_2 Q_{bc}} e^{y_3 Q_{ca}} . \tag{16}$$

As in [I] the sum in (16) must run over all replica indices because of the presence of R.S.B. (In this context see also [14].) Using the R.S.B. algorithm one constructs the matrices $A_{ab}(y) \equiv e^{yQ_{ab}}$ for $a \neq b$ and $A_{aa}(y) = 0$. Then

$$g(y_1, y_2, y_3) = \lim_{n \to 0} \frac{1}{n(n-1)(n-2)} \operatorname{Tr} [A(y_1) A(y_2) A(y_3)] . \tag{17}$$

As usual, in the $n \to 0$ limit one has to replace $m_i \to x$ and $Q_i \to Q(x)$. Thus, after some algebra, we get : $(P(q_1, q_2, q_3) \equiv \overline{P_J(q_1, q_2, q_3)})$

$$P(q_1, q_2, q_3) = \tfrac{1}{2} P(q_1)\, x(q_1)\, \delta(q_1 - q_2)\, \delta(q_1 - q_3) + \\ + \tfrac{1}{2} \{ P(q_1)\, P(q_2)\, \theta(q_1 - q_2)\, \delta(q_2 - q_3) + \text{permutations} \} . \tag{18}$$

This formula is quite interesting. It means that if we take any three states, at least two pairs of them will have the same overlap with probability one. This property is reminiscent of ultrametric spaces [13] ([1]).

Let's now consider two states γ and γ' such that $q^{\alpha\gamma} \geqslant q$, $q^{\alpha\gamma'} \geqslant q$. It follows from (18) that if $q^{\alpha\gamma} = q^{\alpha\gamma'}$ then $q^{\gamma\gamma'} \geqslant q$ and if $q^{\alpha\gamma} \neq q^{\alpha\gamma'}$ then $q^{\gamma\gamma'} = q^{\alpha\gamma}$ or $q^{\gamma\gamma'} = q^{\alpha\gamma'}$.

It follows from this property that the states are organized in non overlapping clusters.

Let α and β be two pure states of our system, $I_\alpha(q)$ the set of states γ which have an overlap with α bigger or equal to q ($q^{\alpha\gamma} \geqslant q$ for every γ) and similarly for $I_\beta(q)$ ($\delta \in I_\beta(q) \Rightarrow q^{\beta\delta} \geqslant q$). It follows from the previous remark that for any pair of states γ, γ' belonging to $I_\alpha(q)$ (or to $I_\beta(q)$), $q^{\gamma\gamma'} \geqslant q$. It is a consequence of (18) that two sets $I_\alpha(q)$ and $I_\beta(q)$ are always either identical or disjoint because if there existed a pure state γ, belonging to both $I_\alpha(q)$ and $I_\beta(q)$, i.e. $q^{\alpha\gamma} \geqslant q$ and $q^{\beta\gamma} \geqslant q$, then by equation (18) one would get $q^{\alpha\beta} \geqslant q$.

The following hierarchical structure of the states of the spin glass phase emerges from the previous remarks. For any q, $q_{\text{Min}} \leqslant q \leqslant q_{\text{Max}}$, the states are organized into disjoint clusters, such that any pair (α, β) of states inside the same cluster has an overlap $q^{\alpha\beta} \geqslant q$. Now one can again divide each of the previous clusters into disjoint smaller clusters by choosing a q', $q < q' \leqslant q_{\text{Max}}$ and grouping together the states with overlap bigger than q'. This procedure can be repeated indefinitely. We have proved that the space of pure states has a hierarchical structure (characteristic of ultrametric spaces).

In order to see whether the probabilities $\overline{P_J(q)}$ approach a definite limit or fluctuate when the number N of spins becomes infinite, we will now compute $\overline{P_J(q_1) P_J(q_2)} - P(q_1) P(q_2)$. One way of computing $\overline{P_J(q_1) P_J(q_2)}$ is to consider four pure states $\alpha_1, \alpha_2, \alpha_3, \alpha_4$, compute the averaged over J probability $P(q_1, q_2)$ to have $q^{\alpha_1\alpha_2} = q_1$, $q^{\alpha_3\alpha_4} = q_2$. The computation of $P(q_1, q_2)$ is similar to the computation of $P(q_1, q_2, q_3)$: we consider the generalized Laplace transform $g(y_1, y_2)$ of $P(q_1, q_2)$ and use again the replica trick.

We finally get

$$g(y_1, y_2) = \int dq_1\, dq_2\, e^{y_1 q_1 + y_2 q_2} P(q_1, q_2)$$
$$= \int dq_1\, dq_2\, e^{y_1 q_1 + y_2 q_2} \overline{P_J(q_1)\, P_J(q_2)}$$

([1]) The definition of an ultrametric space requires the definition of a distance. A natural choice is

$$d(\alpha, \beta) = \frac{1}{N} \sum_i (m_i^\alpha - m_i^\beta)^2 .$$

As $q^{\alpha\alpha} = q_{\text{E.A.}}$ for almost all states α (see (47)), one has simply $d(\alpha, \beta) = 2(q_{\text{E.A.}} - q^{\alpha\beta})$. The characteristic property of ultrametric spaces is that if one takes any three points of the space, they form an isosceles triangle, with the two equal angles larger than or equal to the third one.

$$= \frac{1}{n(n-1)(n-2)(n-3)} \sum_{a=1}^{n} \sum_{b=1}^{n} \sum_{c=1}^{n} \sum_{d=1}^{n} e^{y_1 Q_{ab} + y_2 Q_{cd}} \tag{19}$$

where the sum is restricted to ensembles of replicas a, b, c, d which are all different.

Taking the $n \to 0$ limit we get :

$$\overline{P_J(q_1)\, P_J(q_2)} = \frac{1}{3} P(q_1)\, \delta(q_1 - q_2) + \frac{2}{3} P(q_1)\, P(q_2) . \tag{20}$$

Formula (20) is a direct and manifest proof that $P_J(q)$ (and therefore $x_J(q)$) fluctuates with J even after the thermodynamic limit is taken.

In the next section we will compute the probability distribution of $Y_J(q)$.

We should emphasize that all our results have been obtained within the framework of the replica symmetry breaking scheme of reference [5].

3. Reconstruction of the probability distribution.

As the structure of the ensemble of pure states depends on the realization of the couplings, the function $P_J(q)$ fluctuates and one would like to know the probability distribution of this function. This could, in principle, be studied from the moments $\overline{P_J(q_1)\, P_J(q_2) \ldots P_J(q_k)}$, but it turns out that the direct computation of these moments with the method we have sketched before is a difficult task, as soon as k gets larger than 3 or 4.

We shall see in the following that it is much easier to calculate the moments $\overline{Y_J(q)^k}$ (all the $Y_J(q)$ taken at the same value of q). The second moment is already contained in equation (20)

$$\overline{Y_J(q)^2} = \int_q^{q_{\text{Max}}} \mathrm{d}q_1 \int_q^{q_{\text{Max}}} \mathrm{d}q_2\, \overline{P_J(q_1)\, P_J(q_2)} = \frac{1}{3} y(q) + \frac{2}{3} y(q)^2 . \tag{21}$$

We have computed similarly

$$\overline{Y_J(q)^3} = \frac{1}{5\,!!} (3\, y(q) + 7\, y(q)^2 + 5\, y(q)^3) . \tag{22}$$

The remarkable property of these two equations is that the n'th moment of $Y_J(q)$ ($n = 1, 2, 3$) is expressed in terms only of its mean value $y(q)$ (it is in fact a n'th degree polynomial in y). In particular there is no coupling between different values of q. Those properties are true for any n in the R.S.B. scheme of reference [5] and are due to the ultrametric topology of the replica space itself, whose signature has been seen in other properties [18].

In fact we can interpret Q_{ab} as the overlap between replicas a and b :

— With the first symmetry breaking, the n replicas are organized in n/m_1 clusters of size m_1. The overlap between two replicas within the same cluster is Q_1 while the overlap between two replicas belonging to different clusters is Q_0. As $Q(x)$ is a monotonous function, Q_1 is larger than Q_0 and hence the overlap of the replicas within a cluster is larger than the overlap between replicas in different clusters, as it should be.

— Performing the second breaking, one sees that the replicas inside one cluster of size m_1 are themselves grouped in subclusters of size m_2, etc.

The structure of replica space can be pictorially described as in figure 2. It is an ultrametric space where the sizes of all the clusters at a given scale Q_k are the same and are equal to m_k (²).

Now let us suppose that we want to compute a quantity such as $\overline{Y_J(q)^p}$ which involves only one scale of distances q between pure states. As was shown in the previous section, this quantity can be obtained by using the replica formalism : in the precise case of $\overline{Y_J(q)^p}$, one must choose p pairs of replicas such that

— all the replicas be distinct

— the overlap between the two replicas of a pair be larger than q.

From the ultrametric topology of replica space, the choice of a scale of distances q naturally induces a partition of the n replicas into n/m disjoint clusters of size m, such that the overlap of replicas inside the same cluster be larger than q, while replicas in different clusters have an overlap smaller than q. So if we are interested in only one scale q, we can forget about all the other structures of the replica space. This means that we can compute $\overline{Y_J(q)^p}$ by applying only one replica symmetry breaking (i.e. from the matrix $Q_{ab}^{(1)}$ constructed in the

(²) As Q_{aa} is taken equal to zero, the ultrametric structure is true, strictly speaking, only if one doesn't consider the self overlap of a replica.

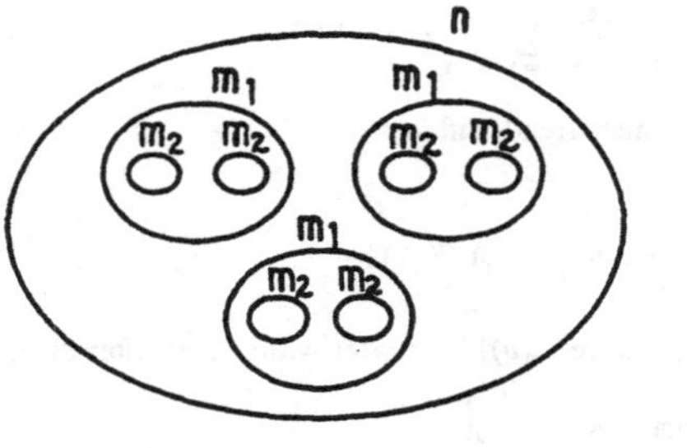

Fig. 2. — Representation of the ultrametric structure of replica space.

previous chapter) provided $Q_0 < q < Q_1$. With full R.S.B., it is known that the numbers m_k, in the limit $n \to 0$, are equal to the value of x for which $q(x) = q_k$. So the right value of m for this single R.S.B. is $m = x(q) = 1 - y(q)$.

With a single breaking of the replica symmetry, the moments can be computed much more easily. Let us for instance compute in detail the second one, defined as :

$$\overline{Y_J(q)^2} = \overline{\sum_{\alpha,\beta,\gamma,\delta} P_\alpha P_\beta \theta(q^{\alpha\beta} - q) P_\gamma P_\delta \theta(q^{\gamma\delta} - q)} . \tag{23}$$

α, β, γ, δ label the pure states, and θ is the usual step function. As explained in section 2, one obtains, with the replica formalism :

$$\overline{Y_J(q)^2} = \frac{1}{n(n-1)(n-2)(n-3)} \sum_{a,b,c,d}{}' \theta(Q_{ab} - q) \theta(Q_{cd} - q)$$

where the symbol $\sum'$ means that the four replica indices a, b, c, d, must be different.

At the scale q, the replicas are grouped into clusters of size m, and one has two possibilities :

— Either the two pairs of replicas are in the same cluster, which we describe as :

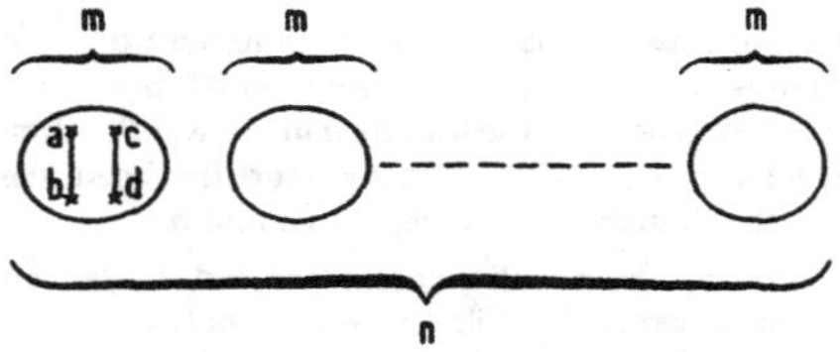

This contribution to the sum is $\frac{n(m-1)(m-2)(m-3)}{n(n-1)(n-2)(n-3)}$ which gives, as $n \to 0$, $m \to 1 - y(q)$: $\frac{1}{3!} y(1+y)(2+y)$.

— Or the two pairs are in different clusters :

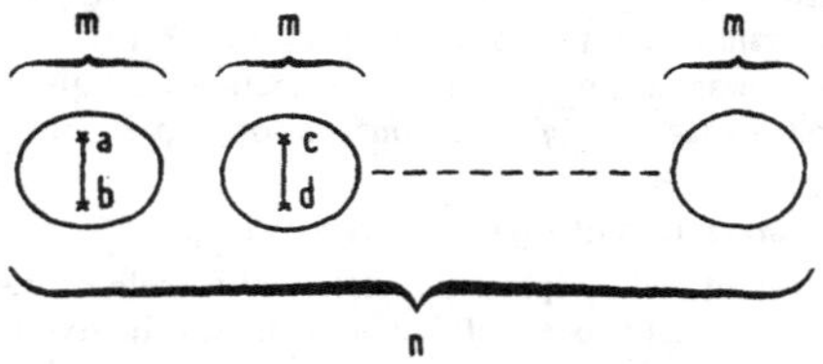

which gives a contribution : $\frac{n(m-1)(n-m)(m-1)}{n(n-1)(n-2)(n-3)} = \frac{1}{3!} y^2(1-y)$. So the final result for $\overline{Y_J(q)^2}$ is (we now omit the explicit reference to the scale q) :

$$\overline{Y_J^2} = \frac{1}{3}(y + 2y^2) .$$

One can write down the rules which give the general moment $\overline{Y_J^p}$:

— One must take p pairs of replicas and distribute them in all the possible distinct ways between the clusters. The diagram obtained :

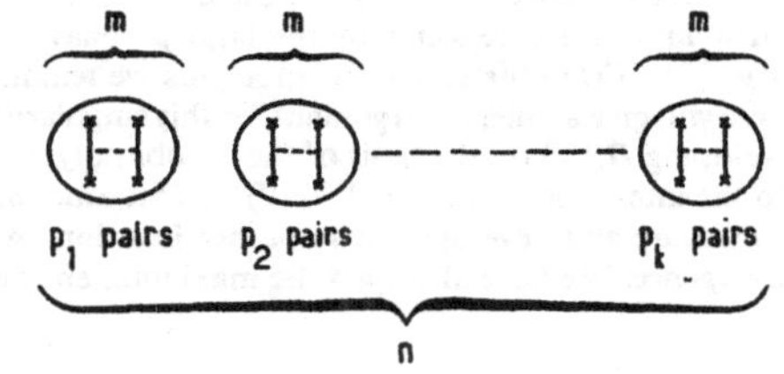

gives a contribution

$$S \frac{p\,!\,(k-1)\,!}{(2\,p-1)\,!\,p_1\,!\dots p_k\,!}(1-y)^{k-1}\,y^k[(1+y)(2+y)]^{N_2}\,[(3+y)(4+y)]^{N_3}\dots \tag{24}$$

where

— $k \equiv N_1$ is the number of clusters containing at least one pair
— N_r is the number of clusters containing at least r pairs
— S is the symmetry factor of the diagram, given by :

$$S = \frac{1}{\prod_{r=1}^{p}(N_r - N_{r+1})\,!}\,. \tag{25}$$

With these rules we have computed the first seven moments :

$$\overline{Y_J^3} = \frac{1}{5\,!!}[3\,y + 7\,y^2 + 5\,y^3]$$

$$\overline{Y_J^4} = \frac{1}{7\,!!}[15\,y + 39\,y^2 + 37\,y^3 + 14\,y^4]$$

$$\overline{Y_J^5} = \frac{1}{9\,!!}[105\,y + 296\,y^2 + 326\,y^3 + 176\,y^4 + 42\,y^5]$$

$$\overline{Y_J^6} = \frac{1}{11\,!!}[945\,y + 2\,838\,y^2 + 3\,458\,y^3 + 2\,228\,y^4 + 794\,y^5 + 132\,y^6]$$

$$\overline{Y_J^7} = \frac{1}{13\,!!}[10\,395\,y + 32\,859\,y^2 + 43\,191\,y^3 + 31\,235\,y^4 + 13\,553\,y^5 + 3\,473\,y^6 + 429\,y^7]\,(^3)\,. \tag{26}$$

---

(3) We have noticed that, up to the 7th moment, the following formulas are true :

$$\overline{Y_J^K} \sim \frac{1}{2\,K-1}\,y \qquad y \to 0$$

$$\overline{Y_J^K} \sim \frac{2^K}{(K+1)\,!}\,y^K \qquad y \to \infty$$

$$\overline{Y_J^K} = \frac{(-1)^K}{(2\,K-1)\,!!} \qquad y = -1$$

$$\overline{Y_J^K} = \frac{2.3^{K-1}(-1)^K}{(2\,K-1)\,!!} \qquad y = -2\,.$$

We have also shown that, up to $K = 3$, the polynomials $R_K(y)$ which give $\overline{Y_J^K}$:

$$\overline{Y_J^K} \equiv \frac{1}{(2\,K-1)\,!!}\,R_K(y)$$

are simply related to the number of diagrams of different types (planar or non planar) of a free field matrix theory : considering a free field theory for $N \times N$ real symmetric matrices M_{ab}, with the constraint that the diagonal elements $M_{aa} = 0$, one finds that, up to $K = 3$:

$$\frac{1}{N(N-1)}\langle\,\mathrm{Tr}\,(M^{2K})\,\rangle = R_K(y)\,, \quad \text{where} \quad y = N - 1\,.$$

We have not been able to obtain a general formula for the coefficients of the polynomial giving $\overline{Y^p_y}$. Instead we have numerically reconstructed the probability distribution of the variable Y at a given q : $\Pi_y(Y)$ (the index y is here to remind us that this distribution depends on $y(q)$ only).

In order to compute $\Pi_y(Y)$ from its moments, it would be useful to control eventual singularities. In particular, the behaviour of $\Pi_y(Y)$ around $Y \simeq 1$ is reflected by the large p behaviour of the moments $\mu_p \equiv \overline{Y^p_y}$ (if $\Pi_y(Y) \sim (1 - Y)^{-\alpha}$, then $\mu_p \sim p^{-1+\alpha}$). From the seven first moments we tentatively conclude to a singularity of the type $\Pi_y(Y) \sim (1 - Y)^{-y}$. (We will give a different argument for this singularity in the next section.) We have inverted the moments μ_p by developing $\Pi_y(Y)$ on the basis of the Jacobi polynomials which are orthogonal on the interval [0, 1] with respect to the integration measure $(1 - Y)^{-y}$. This amounts to factorizing the $(1 - Y)^{-y}$ singularity into the integration measure and developing a smoother function on the polynomials. A good test of the method was its rapid convergence. We have also tried the maximum entropy method with similar results (see Fig. 3) [17].

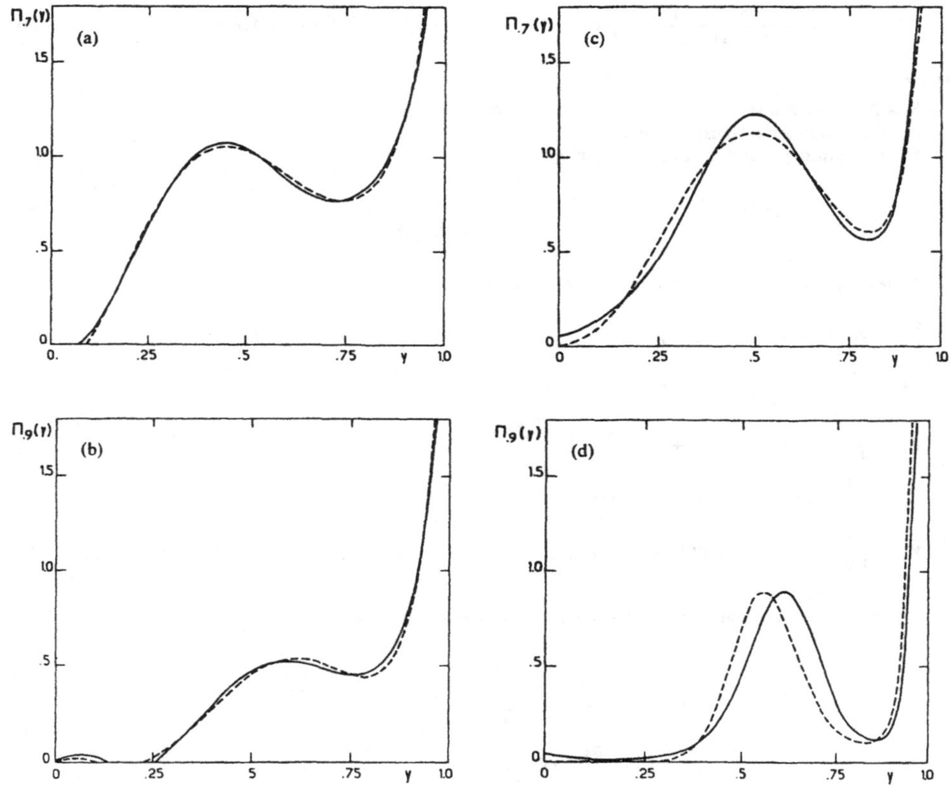

Fig. 3. — (a) Probability distribution of the variable Y, for the average value $y = 0.7$, reconstructed by projecting $\Pi(Y)$ on the Jacobi polynomials. The dashed curve is the probability obtained by inverting the first 6 moments while the full line is obtained from the first 7 moments.
(b) Probability distribution of the variable Y, for the average value $y = 0.9$, reconstructed by projecting $\Pi(Y)$ on the Jacobi polynomials. The dashed curve is the probability obtained by inverting the first 6 moments while the full line is obtained from the first 7 moments.
(c) Probability distribution of the variable Y, for the average value $y = 0.7$, reconstructed from the maximum entropy method. The dashed curve is the probability obtained by inverting the first 6 moments while the full line is obtained from the first 7 moments.
(d) Probability distribution of the variable Y, for the average value $y = 0.9$, reconstructed from the maximum entropy method. The dashed curve is the probability obtained by inverting the first 6 moments while the full line is obtained from the first 7 moments.

The Π_y we have obtained are plotted in figure 3 for various values of y. It is interesting to notice that the average value y of Y is very different from its most probable value $y_{\text{M.P.}}$, which is in general equal to 1. In this case if one keeps to the most probable value, one gets a curve $y_{\text{M.P.}}(q) = 1$ for every q, which is a replica symmetric behaviour, while the average value $y(q)$ obtained with the replica symmetry breaking is very different from $y_{\text{M.P.}}$ It has been emphasized many times that this kind of behaviour often appears in disordered systems [12].

On the other hand the most important property of this probability distribution of Y at a given q is the fact that it does not depend on anything else than $y(q)$. This explains why the integrated probability y (or equivalently x) is so interesting : in terms of this variable, there exists a universal behaviour of the glassy phase in the following sense : for a given value of the parameters (temperature and magnetic field) T_1, H_1, and a given scale of overlap q_1, the probability distribution of Y is entirely determined by y. If one changes the parameters to values T_2, H_2, there exists a change of scale of overlaps $q_1 \to q_2$ such that the probability distribution of $Y(q_2)$ in this second system be exactly identical to the probability distribution of $Y(q_1)$ in the first system; the rescaling in overlaps is given by : $y_{T_1,H_1}(q_1) = y_{T_2,H_2}(q_2)$. The universality comes from the fact that the whole dependence on H and T is through the function $y_{T,H}(q)$. But in order to compute the dependence of y on q, T, H, we must minimize the free energy and for this the entire sequence of R.S.B.'s is required.

Finally let us emphasize the crucial role played by the ultrametric topology of the replica space : because of this topology, a quantity which involves p scales of distances $q_1, \ldots, q_p$ can be computed with p explicit breakings of the replica symmetry. This allowed us to compute the moments $\overline{Y_J(q)^r}$ to an arbitrarily large order, with a single replica symmetry breaking.

4. Distribution properties of the clusters of spin glass states.

We have shown that the space of pure states has a hierarchical structure that consists of clusters contained in clusters. In this section we will try to give more information on this structure, especially on the numbers and sizes of the clusters.

As in the previous section, given a certain scale q, we group into clusters all the states that have an overlap larger than q, defining in this way a partition of the ensemble of pure states into K clusters. Let us call W_I the weight of the Ith cluster :

$$W_I = \sum_{\alpha \in I} P_\alpha \tag{27}$$

obviously

$$\sum_I W_I = 1 \tag{28}$$

We are interested in the distribution of the weights W_I. (It is clear that the knowledge of all the W_I's for every scale q completely characterizes the space of states.) In order to take advantage of the universality demonstrated in the previous section we shall define the scale through the variable $y \equiv y(q)$.

Let $f_J(W, y)\,\mathrm{d}W$ be the number of clusters that have weights W_I between W and $W + \mathrm{d}W$, for a given distribution of couplings J :

$$f_J(W, y) = \sum_I \delta(W - W_I). \tag{29}$$

The average function $\overline{f_J(W, y)}$ can be computed in the following way : the moments $M_k = \int_0^1 W^k\, \overline{f_J(W, y)}$ are given by :

$$M_k = \overline{\sum_I W_I^k}$$
$$= \overline{\sum_I \left(\sum_{\alpha_1 \in I} \cdots \sum_{\alpha_k \in I} P_{\alpha_1} \ldots P_{\alpha_k} \right)}. \tag{30}$$

This is the total probability that k states are in the same cluster :

$$M_k = \overline{\sum_{\alpha_1} \cdots \sum_{\alpha_k} P_{\alpha_1} \ldots P_{\alpha_k}\, \theta(q^{\alpha_1 \alpha_2} - q)\, \theta(q^{\alpha_1 \alpha_3} - q) \ldots \theta(q^{\alpha_1 \alpha_k} - q)}. \tag{31}$$

In the same way as in the previous section, this quantity can be calculated by going to replica space, with one R.S.B. at the scale q : M_k is equal to the probability of choosing k *different replicas*, all belonging to the same cluster (in replica space) :

$$M_k = \frac{1}{n(n-1)\ldots(n-k+1)}\, n(m-1)\ldots(m-k+1). \tag{32}$$

In the limit $n \to 0$, $m \to 1 - y$, we get :

$$M_k = \frac{\Gamma(k+y-1)}{\Gamma(y) \times \Gamma(k)}. \tag{33}$$

It turns out that these moments can be inverted, yielding :

$$\overline{f_J(W, y)} = \frac{W^{y-2}(1-W)^{-y}}{\Gamma(y)\,\Gamma(1-y)}. \tag{34}$$

A number of physical conclusions can be obtained from this expression :

1) The average multiplicity of clusters is infinite for any value of y :

$$\int_0^1 \overline{f_J(W, y)}\, \mathrm{d}W = \infty .$$

2) The infinite number of clusters is concentrated around $W = 0$, and their overall weight is infinitesimal : indeed, if one introduces a cut-off ε in the region of small weights, one gets an average number of clusters :

$$K_\varepsilon \equiv \int_\varepsilon^1 \overline{f_J(W, y)}\, \mathrm{d}W \sim \frac{\varepsilon^{y-1}}{(1-y)\,\Gamma(y)\,\Gamma(1-y)} \quad \begin{pmatrix} \varepsilon \ll 1 \\ y \neq 1 \end{pmatrix} \tag{35}$$

which diverges as $\varepsilon \to 0$, but their total weight

$$\int_\varepsilon^1 W\, \overline{f_J(W, y)}\, \mathrm{d}W \sim 1 - \frac{\varepsilon^y}{\Gamma(1+y)\,\Gamma(1-y)} \quad \begin{pmatrix} \varepsilon \ll 1 \\ y \neq 1 \end{pmatrix} \tag{36}$$

goes to one. For each J, the total number of clusters having a weight larger than ε is an integer between 0 and $1/\varepsilon$. But the average number K_ε is much smaller since it behaves like ε^{y-1}.

3) y is precisely the average size of clusters : this is in fact nothing but the definition of y, and this result is valid for each configuration of the couplings, since :

$$\begin{aligned} y_J &= \sum_\alpha P_\alpha \sum_\beta P_\beta\, \theta(q^{\alpha\beta} - q) \\ &= \sum_\alpha P_\alpha W_{I \ni \alpha} = \sum_I W_I^2 . \end{aligned} \tag{37}$$

4) The average probability, when choosing a state at random, that it be in a cluster of weight W is $W\overline{f_J(W, y)}$. When $y \to 1$, this probability is strongly peaked around $W = 1$. (The integrated probability in the interval [0, 1/2], and in any finite interval that does not contain 1, goes to zero when $y \to 1$.) But for each J, there can be at most one cluster in the interval]1/2, 1] (since $\sum_I W_I = 1$), and therefore in the limit $y \to 1$ there is, for each J, one large isolated cluster that dominates.

5) The same kind of argument explains the nature of the singularity at $Y = 1$ of the function $\Pi_y(Y)$ computed in chapter 3 : $\Pi_y(Y)$ is defined as :

$$\Pi_{y(q)}(Y) = \overline{\delta(y_J(q) - Y)} = \overline{\delta\left(\sum_I W_I^2 - Y\right)} . \tag{38}$$

For Y near to 1, the configurations which contribute are those for which one cluster dominates :

$$\Pi_{y(q)}(Y) = \overline{\delta(W_M^2 - Y)}, \qquad (Y \to 1) \tag{39}$$

where W_M is the weight of the largest cluster. Thus from (34) :

$$\Pi_y(Y) \sim \frac{2^{y-1}}{\Gamma(y)\,\Gamma(1-y)} (1-Y)^{-y}, \qquad (Y \to 1). \tag{40}$$

One can go further in this analysis by computing the fluctuations $\overline{f_J(W_1, y)\, f_J(W_2, y)}$:

$$\begin{aligned} \overline{f_J(W_1, y)\, f_J(W_2, y)} &= \overline{\sum_I \delta(W_I - W_1) \sum_{I'} \delta(W_{I'} - W_2)} \\ &= \delta(W_1 - W_2)\, \overline{f_J(W_1, y)} + \overline{\sum_{I \neq I'} \delta(W_I - W_1)\, \delta(W_{I'} - W_2)} . \end{aligned} \tag{41}$$

We have to compute only the term in which $I \neq I'$. This can be done in the same way as before : one computes all the moments (in W_1 and W_2) by going to replica space. The result is :

$$\overline{\sum_{I \neq I'} \delta(W_I - W_1)\,\delta(W_{I'} - W_2)} = \frac{(1-y)\,\theta(1 - W_1 - W_2)\,(W_1 W_2)^{y-2}\,(1 - W_1 - W_2)^{1-2y}}{\Gamma(y)\,\Gamma(y)\,\Gamma(2-2y)}. \quad (42)$$

This result can be generalized :

$$\overline{\sum_{I_1,\ldots,I_k}' \prod_{l=1}^{k} \delta(W_{I_l} - W_l)} = \frac{(1-y)^{k-1}\,\Gamma(k)}{\Gamma(y)^k\,\Gamma(k-ky)}\,\theta\left(1 - \sum_{l=1}^{k} W_l\right)\left(\prod_{l=1}^{k} W_l\right)^{y-2}\left(1 - \sum_{l=1}^{k} W_l\right)^{k-ky-1} \quad (43)$$

(The Σ' means that all the indices $I_1, \ldots, I_k$ are different.)

One can deduce that the probability distribution of the number of clusters having a weight larger than ε is not peaked :

$$\frac{\overline{K_\varepsilon^2} - \overline{K_\varepsilon}^2}{\overline{K_\varepsilon}^2} = \frac{(1-y)\,\Gamma(1-y)^2}{\Gamma(2-2y)} - 1 \sim 1 \qquad \begin{pmatrix} \varepsilon \ll 1 \\ y \neq 1 \end{pmatrix}. \quad (44)$$

We have a rather clear picture of the structure of the ensemble of pure states : there is an infinity of states in small clusters, carrying a total weight also small, and as soon as y goes near to one, there is one isolated cluster that dominates.

5. Conclusions.

In the previous sections we have shown how the information contained in Parisi's R.S.B. scheme can be decoded to derive the structure of the space of states of the system. We showed that the fluctuations with J_{ik} do not disappear even in the $N \to \infty$ limit. Henceforth we gave examples about how the distribution probability for these fluctuations can be estimated by inverting the moments.

Our calculations suggest a new set of numerical simulations whereby this dependence would be explicitly checked. This would be a crucial test of our results. It would be also interesting to see whether our predictions survive in a finite dimensional system with short range forces.

We again stress the « universality » of the results of our calculations. Through the change of variable

$$q \to y\,, \qquad y = \int_q^{q_{\mathrm{Max}}} P(q')\,\mathrm{d}q'\,. \quad (45)$$

We were able to eliminate all references to the particular order parameter $q(x)$, the stationary point of Pari's free energy. As a consequence the two following probability distributions are equal :

(*a*) the probability distribution of the Boltzmann-Gibbs factor of the pure states $\alpha(P_\alpha)$ at $T_0 \leqslant T_g$;

(*b*) the probability distribution of the W_I of the clusters at $T_1 < T_0$ defined through the equation :

$$W_I = \sum_{\alpha \in I} P_\alpha \quad \text{where} \quad \alpha, \alpha' \in I \quad \text{if} \quad q^{\alpha\alpha'} \geqslant q_{\mathrm{E.A.}}(T_0)\,.$$

The additional information that the order parameter $q(x)$ has a plateau, i.e.

$$q(x, T) = q_{\mathrm{E.A.}}(T) \qquad x \geqslant x_0(T)$$

implies that, up to sets of zero probability, all states must have

$$q^{\alpha\alpha}(T) \leqslant q_{\mathrm{E.A.}}(T)\,.$$

On the other hand, the choice of scale $q = q_{\mathrm{E.A.}}(T)$ induces a partition of the states into clusters. Each cluster must contain at most one state α, and the only states which contribute are those with $q^{\alpha\alpha} \geqslant q_{\mathrm{E.A.}}(T)$. But we have proven that, whatever the scale of the partition into clusters $I : \sum_I W_I = 1$. Hence we obtain for all α :

$$q^{\alpha\alpha} = q_{\mathrm{E.A.}} \quad \text{(with probability one)}\,. \quad (46)$$

Then from formula (37), one finds :

$$\sum_\alpha P_\alpha^2 = \lim_{q \to q_{\mathrm{E.A.}}} y(q)\,. \quad (47)$$

If the function $q(x)$ has a plateau, as it is commonly believed, the right hand side of this equation is nothing but the length of this plateau. As $P_\alpha < 1$, one can then conclude that a few states α dominate the sum $\sum_\alpha P_\alpha^2$.

Several questions remain open :

a) are there physical observables such that their infinite volume limits do not fluctuate with J ?

b) what happens when one adds corrections to the mean field approximation ? In this context we remark that it has recently been proved [15] that the free energy is self-averaging in the thermodynamic limit for short range interactions. This has also been shown in the S.K. model, and the finite volume corrections have been computed [16].

The hierarchical structure (ultrametric topology) of states was demonstrated for the J average but it obviously applies for every realization of J. The equality of the probability distributions mentioned above suggests the following picture : when heating a spin glass from zero temperature up to T_g we go through a series of micro phase transitions characterized by the melting of two or more states into one state at the higher temperature. Furthermore we should have

$$\frac{-NF_{\alpha}(T+\varepsilon)}{kT} = \ln \sum_{\alpha'} \exp \frac{-NF_{\alpha'}(T-\varepsilon)}{kT} + K \tag{48}$$

where K is in dependent of α, and F_{α} is the free energy density of the state α.

Finally we point out that the ultrametric structure allows the definition of a non ergodic Brownian motion. Indeed, if in an ultrametric space a point can jump up to a distance δ in a single step, after N steps it can arrive only at distance δ. (This follows from the well known fact that two overlapping spheres coincide in an ultrametric space.) We think that this fact will have interesting consequences for the dynamical approach to equilibrium of the spin glass.

References

[1] Edwards, S. F. and Anderson, P. W., *J. Phys. F* **5** (1975) 965.

[2] Sherrington, D. and Kirkpatrick, S., *Phys. Rev. Lett.* **32** (1975) 1792.

[3] Thouless, D. J., Anderson, P. W. and Palmer, R., *Philos. Mag.* **35** (1977) 593.

[4] De Almeida, J. R. L. and Thouless, D. J., *J. Phys. A* **11** (1978) 983.

[5] Parisi, G., *Phys. Rev. Lett.* **43** (1979) 1754 and *J. Phys. A* **13** (1980), L 117, 1101, 1887.

[6] De Dominicis, C., Gabay, M., Garel, T. and Orland, H., *J. Physique* **41** (1980) 923 ;
Bray, A. J. and Moore, M. A., *J. Phys. C* **13** (1980) L 469, L 907.

[7] Sompolinsky, H. and Zippelius, A., *Phys. Rev. Lett.* **47** (1981) 359 ; *Phys. Rev. B* **25** (1982) 6860.

[8] See for example : Mackenzie, N. D. and Young, A. P., *Phys. Rev. Lett.* **49** (1982) 301 ;
Young, A. P., *Phys. Rev. Lett.* **51** (1983) 1206.

[9] For recent reviews : Parisi, G., lectures given at the 1982 Les Houches Summer School ; Parisi, G., in *Disordered Systems and Localization* (Springer Verlag) 1981 ; Toulouse G., contribution to the Heidelberg Colloquium on spin glasses, Lecture Notes in Physics (Springer Verlag) **192** (1983) 2 ;
De Dominicis, C., contribution to the Heidelberg Colloquium on spin glasses, Lecture Notes in Physics (Springer Verlag) **192** (1983) 103.

[10] Parisi, G., *Phys. Rev. Lett.* **50** (1983) 1946.

[11] Mezard, M., Parisi, G., Sourlas, N., Toulouse, G., Virasoro, M., LPTENS preprint 83/39 (1983).

[12] Anderson, P. W., in Les Houches 1978, Ill condensed matter (North Holland) ;
Derrida, B., *Phys. Rep.* **67** (1980) 29.

[13] Bourbaki, N., *Espaces Vectoriels Topologiques*. Paris 1966. We thank R. Rammal for pointing this to us.

[14] De Dominicis, C. and Young, A. P., *J. Phys. A* **16** (1983) 2063.

[15] Khanin, K. M. and Sinai, Y. G., *J. Stat. Phys.* **20** (1979) 573.

[16] Kondor, I., Saclay preprint (1983).

[17] Mead, L. R. and Papanicolaou, N., Washington University, St Louis preprint 1983.

[18] De Dominicis, C. and Kondor, I., *Phys. Rev. B* **27** (1983) 606.

J. Physique **46** (1985) 1293-1307 AOÛT **1985**, PAGE 1293

Classification
Physics Abstracts
75.50K

The microstructure of ultrametricity

M. Mézard[+] and M. A. Virasoro[*]

Dipartimento di Fisica, Università di Roma « La Sapienza », 00185 Roma, Italy

(*Reçu le 18 février 1985, accepté le 2 avril 1985*)

Résumé. — L'organisation hiérarchique des états purs d'un verre de spin S.K. (l'ultramétricité) est analysée en termes de la distribution des aimantations locales. Nous montrons que chaque état pur α définit une distance ultramétrique $D_\alpha(i, j)$ entre les N *sites*. Etant donnés deux états α, β dont le recouvrement est q, il y a une distance minimale d_m telle que, pour chaque paire de sites i, j vérifiant $D_\alpha(i, j) \geqslant d_m$, les deux distances D_α et D_β coincident. Il en résulte qu'on peut faire une partition des sites en cellules disjointes à l'intérieur desquelles l'aimantation totale est la même pour tous les sites ayant un overlap mutuel q. Pour cette même famille d'états nous définissons un « ancêtre » qui a, à l'intérieur de chaque cellule, une aimantation locale constante et égale à l'aimantation moyenne des descendants. Les ancêtres vérifient les équations de type T.A.P. La dépendance fonctionnelle de l'aimantation locale en termes du champ local est donnée par la solution de l'équation de diffusion dans l'espace des x qui reçoit une interprétation purement statique.

Abstract. — The hierarchical organization of the pure states of a S.K. spin glass (ultrametricity) is analysed in terms of self-averaging distributions of local magnetizations. We show that every pure state α defines an ultrametric distance $D_\alpha(i, j)$ among the N *sites*. Given two states α, β with overlap q there is a minimum distance d_m such that for two sites i, j with $D_\alpha(i, j) \geqslant d_m$ the two distances D_α and D_β coincide. It follows that the sites can be partitioned in disjoint cells inside which the total magnetization is the same for all the states with mutual overlap q. For this same family of states we then define an « ancestor » that has, inside each cell, constant local magnetization equal to the average magnetization of the descendants. The ancestors satisfy mean field like equations. The functional dependence of the local magnetization in terms of the local field is given by the solution of the diffusion equation in x space which is given a purely static interpretation.

1. Introduction.

Recent progress in the study of the mean field theory of spin glasses is related to the physical interpretation of replica symmetry breaking (R.S.B.) as describing the breaking of ergodicity and the existence of many pure equilibrium states in the spin glass phase. It was shown by Parisi [1] that the order parameter function $q(x)$ which he introduced for describing R.S.B. is related to the distribution of the overlaps (which measure the distances in phase space) between these equilibrium states.

The nature of the spin glass phase is best understood by looking at the geometry of the space of equilibrium states. It was found that this space has a special hierarchical topology characterized by ultrametricity [2], and that the detailed structure of the space, for instance the order parameter function, is not self-averaging, which means that it depends on the special realization of the couplings in the sample, even in the thermodynamic limit [2, 3].

The ultrametric topology of the space of equilibrium states of a spin glass deserves special attention. Such an organization might exist in other systems with frustration and disorder, and it should have consequences in such fields as optimization problems [4] or neural networks [5]. In spin glasses it was found by a direct inspection of the triangles in the space of equilibrium states (all triangles turn out to be either equilateral or isoceles with a shorter third side, which is characteristic of ultrametric spaces), but its physical interpretation (how do the local magnetizations at site i in state α, m_i^α, manage to build up such a strange space ?) was not exhibited. This paper is devoted to the understanding of this microstructure.

A simple representation of an ultrametric space is a genealogical tree (Fig. 1). The different states, α, β, ... of the system are the extremities of the branches of the

($^+$) On leave from Laboratoire de Physique Théorique, Ecole Normale Supérieure, Paris, France.

(*) Also INFN, Sezione di Roma.

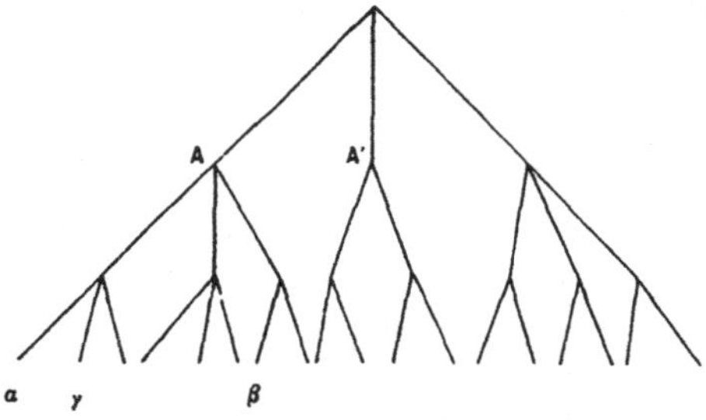

Fig. 1. — The tree of the states. The different states α, β, γ... are the extremities of the branches of the tree. The distance between two states is a monotonic function of the number of steps one has to climb along the tree to find a common ancestor.

tree. The overlap between two states depends only on their closest common ancestor : the higher one must go in the tree to find this ancestor, the lower the overlap q, the larger the distance between the states.

The microstructure deduced in this paper can be schematically summarized as follows :

1) All the pure states $\alpha, \beta, \gamma, \dots$ which have a given common ancestor (point A in Fig. 1) share a common property : there exists a partition of the N sites into disjoint macroscopic cells $\mathcal{C}_1, \dots \mathcal{C}_k$ such that the average magnetization of each of these states $\alpha, \beta, \gamma, \dots$ in every cell $\mathcal{C}_l$ is the same, $\mathcal{M}_l$, within a given resolution $\Delta\mathcal{M}$:

$$\mathcal{M}_l \leqslant \frac{1}{|\mathcal{C}_l|}\sum_{i\in\mathcal{C}_l} m_i^\alpha = \frac{1}{|\mathcal{C}_l|}\sum_{i\in\mathcal{C}_l} m_i^\beta = \dots \leqslant \mathcal{M}_l + \Delta\mathcal{M} \tag{1}$$

(here $|\mathcal{C}_l|$ is the volume, i.e. the number of sites, of the cell $\mathcal{C}_l$, $\sum_l |\mathcal{C}_l| = N$).

2) In every two states $\alpha, \beta, \dots$, of which A is the *closest* common ancestor ($q^{\alpha\beta} = q$), the local magnetizations inside each cell $\mathcal{C}_l$ are completely uncorrelated, and have the same distribution. Hence the overlap $q^{\alpha\beta}$ is :

$$q^{\alpha\beta} = \frac{1}{N}\sum_i m_i^\alpha m_i^\beta = \sum_l \frac{|\mathcal{C}_l|}{N}(\mathcal{M}_l)^2 = q . \tag{2}$$

3) Considering two states α, γ which link below A ($q^{\alpha\gamma} = q' > q$, see Fig. 1), their magnetizations in each cell $\mathcal{C}_l$ are correlated in a simple way : there exists a subpartition of $\mathcal{C}_l$ into subcells $\mathcal{C}_{l,1}, \dots \mathcal{C}_{l,l'}$ such that α and γ have the same average magnetization $\mathcal{M}_{l,l'}$ into each cell $\mathcal{C}_{l,l'}$ (always within a given resolution $\Delta\mathcal{M}$), with the obvious properties :

$$\sum_{l'} |\mathcal{C}_{l,l'}| = |\mathcal{C}_l| ; \quad \sum_{l'} \frac{|\mathcal{C}_{l,l'}|}{|\mathcal{C}_l|}\mathcal{M}_{l,l'} = \mathcal{M}_l \tag{3}$$

and the local magnetizations of α and γ inside each $\mathcal{C}_{l,l'}$ are uncorrelated and have the same distribution, which implies that :

$$q^{\alpha\gamma} = \sum_l \frac{|\mathcal{C}_l|}{N}\sum_{l'}\frac{|\mathcal{C}_{l,l'}|}{|\mathcal{C}_l|}(\mathcal{M}_{l,l'})^2 = q' > q . \tag{4}$$

The details of this structure (volumes of the cells and of the subcells, values of the magnetizations $\mathcal{M}_l$ and $\mathcal{M}_{l,l'}$, distribution of local magnetizations inside each cell, ...) are explained in section 3. Their evolutions as functions of the overlap q of the ancestor A are described by differential equations similar to those studied by Parisi and others to obtain the free energy of the system, which receive here a clear interpretation in purely static terms.

On our way to derive the previous properties we have computed the distribution of local magnetizations in each state α :

$$\mathcal{T}_\alpha(m) = 1/N\left[\sum_i \delta(m_i^\alpha - m)\right] \tag{5}$$

and the correlations of local magnetizations in different states $\alpha_1, \alpha_2, \dots, \alpha_k$, which read for $k = 2$:

$$\mathcal{T}_{\alpha,\beta}(m, m') = 1/N\left[\sum_i \delta(m_i^\alpha - m)\,\delta(m_i^\beta - m')\right]. \tag{6}$$

It may be surprising that such quantities can be computed since there is no known way of computing the thermal average of a given observable $\mathcal{O}$ in one give pure state α : $\langle \mathcal{O} \rangle_\alpha$. In fact in the replica method one can compute only :

$$\overline{\langle \mathcal{O} \rangle} = \overline{\sum_\alpha P_\alpha \langle \mathcal{O} \rangle_\alpha} \tag{7}$$

(where $\langle \mathcal{O} \rangle$ is the Gibbs average, decomposed as a sum over pure states [1], and $\overline{(\ )}$ denotes the average over the random couplings), and many quantities, among which the weights P_α, depend on the state α, and on the sample [2, 6]. However we will look for quantities which are both sample independent (self averaging) and state independent (we shall call them reproducible). It turns out that many observables are reproducible (a general sufficient criterion for reproducibility will be given in (14)), among which :

— The distribution of local magnetizations $\mathcal{T}_\alpha(m)$ is independent of α :

$$\mathcal{T}_\alpha(m) = \mathcal{T}(m) . \tag{8}$$

— The correlation of local magnetizations in two different states α, β depends only on the overlap $q^{\alpha\beta} = (1/N)\sum_i m_i^\alpha m_i^\beta$ between them :

$$\mathcal{T}_{\alpha,\beta}(m, m') = \mathcal{T}_{q^{\alpha\beta}}(m, m') . \tag{9}$$

The reproducible quantities such as $\mathcal{T}(m)$ and $\mathcal{T}_q(m, m')$ can then be computed by the replica method, along lines similar to those introduced by Parisi [1].

The organization of the paper is as follows :

In section 2 we explain the general formalism which enables us to prove the reproducibility and to compute local averages like the distribution of local magnetization. This section is rather lengthy and technical, since it is intended to provide the reader all the basic computational techniques used in the replica solution of the mean field theory of spin glasses. Some of the results have already appeared in other works [7-9]. The reader only interested in the results should skip this section.

In section 3 we analyse the correlation of the distribution of local magnetizations between different pure states, and we deduce from this analysis the structure in cells of the space, together with the properties of the cells at different scales.

Section 4 is devoted to the direct computation of the average couplings between the cells. We deduce from it a set of T.A.P. like equations which are shown to be compatible with the usual T.A.P. equations together with the assumption of the existence of the cells.

Conclusions and perspectives are summarized in section 5.

2. Distribution of local magnetizations-reproducibility.

In this section we explain the basic techniques used to prove the reproducibility of certain observables and to compute their averages. First let us introduce our notations and normalizations : we work in the mean field theory of spin glasses defined by the Sherrington-Kirkpatrick (S.K.) Hamiltonian

$$H_{\text{S.K.}}(J, \sigma) = -\sum_{i<j} J_{ij}\,\sigma_i\,\sigma_j - h\sum_i \sigma_i \qquad (10)$$

$\sigma_i = \pm 1$ are N Ising spin variables, and the infinite range couplings J_{ij} are independent random variables with a Gaussian distribution of mean zero and variance $\overline{J_{ij}^2} = 1/N$.

The replica method enables one to perform the quenched average over the random couplings, by introducing at each site i n replicas of the spin variable : σ_i^a, $a = 1, \dots, n$. The quenched free energy at temperature $1/\beta$ is then :

$$\frac{\overline{\ln Z}}{N} = \lim_{n\to 0}\left(\frac{\beta^2}{4} - \frac{\beta^2}{4}\,\frac{1}{n}\sum_{a\neq b} Q_{ab}^2 + \frac{1}{n}\ln \operatorname{Tr}_{\{\sigma\}} e^{-\beta H(Q,\sigma)}\right) \qquad (11)$$

where Q_{ab} is a $n \times n$ symmetric matrix determined by a saddle point condition. In the following we shall always use the form of Q_{ab} found by Parisi [10], which describes replica symmetry breaking in the sense that the matrix elements Q_{ab} are not all equal. This R.S.B. scheme has been shown to be stable [11], and gives results in agreement with all the numerical computations on the S.K. model today [12, 13]. Our precise notations for Q_{ab} are the standard ones, as described for instance in [2].

In (11), $H(Q, \sigma)$ is a one site Hamiltonian coupling the n replicas given by :

$$H(Q, \sigma) = -\frac{\beta^2}{2}\sum_{a\neq b} Q_{ab}\,\sigma_a\,\sigma_b - \beta h\sum_a \sigma_a\,. \qquad (12)$$

The techniques we use enable one to compute products of averages in different states at fixed distances, of the generic form :

$$A = \overline{\sum_{\alpha_1,\dots,\alpha_k} P_{\alpha_1}\dots P_{\alpha_k}\,\langle \mathcal{O}(\sigma_i^{\alpha_1},\dots,\sigma_i^{\alpha_k})\rangle_{\alpha_1,\dots,\alpha_k} \times \prod_{1\le r<s\le k}\delta(q^{\alpha_r\alpha_s} - q^{rs})} \qquad (13)$$

where $\mathcal{O}$ is any observable which is an extensive function of the local spin variables $\sigma_i^{\alpha_s}$ in any of the k-states. A class of reproducible observables is given by the ones which we call $1/N$ dominant. They are arbitrary functions of the following invariants :

$$\lim_{N\to\infty}\frac{1}{N}\sum_i (\sigma_i^{\alpha_{j_1}})(\sigma_i^{\alpha_{j_2}})\dots(\sigma_i^{\alpha_{j_p}}),\qquad j_1,\dots,j_p \in \{1,\dots,k\}\,. \qquad (14)$$

We are leaving out those functions which in the limit $N \to \infty$ are not scaled as in (14), such as for instance the susceptibility which is a double sum over sites but is divided by only one power of N. We can generalize this class by including invariants which depend on the matrix J_{ij}. In this case and for the purpose of classifying them as $1/N$ dominant or not, J_{ij} may be replaced by $1/N\ \sigma_i^{\alpha_0}\,\sigma_j^{\alpha_0}$ (with α_0 an arbitrary singled out state) thereby reducing it to products of invariants of the previous kind. Notice that with this definition the Boltzmann Gibbs weights (and the P_α) are not $1/N$ dominant.

We will prove that for all $1/N$ dominant observables the average value A factorizes :

$$A = \overline{\sum_{\alpha_1,\dots,\alpha_k} P_{\alpha_1}\dots P_{\alpha_k}\prod_{1\le r<s\le k}\delta(q^{\alpha_r\alpha_s} - q^{rs})}\; X \qquad (15)$$

where X is reproducible and self averaging and measures the observable $\mathcal{O}$ for any k-uple of states that have fixed mutual overlaps q^{rs}.

In this section we will discuss the $1/N$ dominant observables without the J_{ij} (in section 4 we shall use invariants including the J_{ij}'s). Without loss of generality we can consider the observable

$$\mathcal{O}_\eta = \exp\frac{\eta}{N}\sum_i \sigma_i^{\alpha_1}\dots\sigma_i^{\alpha_k}\,. \qquad (16)$$

The first step in the computation of A is to express it as a usual statistical mechanics average using the method introduced in [1] :

$$A = \overline{\left\langle \mathcal{O}_\eta \prod_{1\le r<s\le k}\delta\left(\frac{1}{N}\sum_i \sigma_i^r\,\sigma_i^s - q^{rs}\right)\right\rangle_{(k)}} \qquad (17)$$

where the average $\langle \quad \rangle_{(k)}$ is the standard Gibbs average taken for a system of k identical non interacting copies of the original system, with Hamiltonian :

$$H^{(k)} = \sum_{r=1}^{k} H_{\text{S.K.}}(J, \sigma^r) . \tag{18}$$

The average in (17) can then be computed by any method of statistical mechanics, for instance by Monte Carlo [12]. Here we shall use the replica method to do the average over the disorder in (17). The systems 1, ..., k are any systems among n, and one should average over the different choices [14]. This gives :

$$A = \lim_{n \to 0} \frac{1}{n(n-1)\dots(n-k+1)} \sum_{a_1,\dots,a_k}' \mathrm{Tr}_{\{\sigma^a\}} \times$$

$$\times \exp\left[-\beta \sum_{r=1}^{k} H_{\text{S.K.}}(J, \sigma^r) + \frac{\eta}{N} \sum_i \sigma_i^{a_1} \dots \sigma_i^{a_k} \right]$$

$$\times \prod_{1 \leqslant r < s \leqslant k} \delta\left(\frac{1}{N} \sum_i \sigma_i^{a_r} \sigma_i^{a_s} - q^{rs} \right) \tag{19}$$

where the $\sum'_{a_1,\dots,a_k}$ is a sum over the indices $a_r = 1, \dots, n$, with the constraint that all the indices be different from each other.

Formulae (17) to (19) are the basic steps expressing the physical quantity A defined in (13) as a trace over spin variables in a replicated system. We shall not repeat them for each new quantity we compute, but just summarize these steps by the expression « ... going into replica space... ».

The evaluation of (19) is very cumbersome but relatively straightforward.

We shall show how to do it in one very simple example and from it induce the general result. Let us therefore approximate $q(x)$ by

$$q(x) = \begin{cases} q_0 & 0 \leqslant x \leqslant x_1 \\ q_1 & x_1 \leqslant x \leqslant x_2 \\ q_2 & x_2 \leqslant x \leqslant 1 \end{cases} \tag{20}$$

and let us consider the special case

$$O_\eta = \exp \frac{\eta}{N} \sum_i \sigma_i^1 \sigma_i^2 \sigma_i^3 \tag{21}$$

with fixed overlaps :

$q^{12} = q_2$ and $q^{23} = q_1$ so that $q^{31} = q_1$ is enforced by ultrametricity.

Notice that q_2 is the maximal overlap so that the replicas a_1 and a_2 are in the same state. Therefore we are calculating :

$$\overline{\sum_{\alpha,\beta} P_\alpha^2 P_\beta \, \delta(q^{\alpha\beta} - q_1) \exp \frac{\eta}{N} \sum_i (m_i^\alpha)^2 m_i^\beta} . \tag{22}$$

Going into replica space we obtain :

$$A = \lim_{n \to 0} \frac{1}{n(n-1)(n-2)} \sum_{a_1 a_2 a_3}' \exp N \frac{\beta^2}{4} \left(n - \sum_{a \neq b} Q_{ab}^2 \right) \times$$

$$\times \delta_{Q_{a_1 a_2}, q_2} \delta_{Q_{a_1 a_3}, q_1} \left(\mathrm{Tr}_{\{\sigma^a\}} \left\{ \exp\left[-\beta H(Q, \sigma) + \frac{\eta}{N} \sigma^{a_1} \sigma^{a_2} \sigma^{a_3} \right] \right\} \right)^N \tag{23}$$

where $H(Q, \sigma)$ is the one site Hamiltonian defined in (12).

The contribution to A from each triplet of replica indices a_1, a_2, a_3 satisfying the constraints $Q_{a_1 a_2} = q_2$, $Q_{a_1 a_3} = q_1$ is the same, we call it A_2.

Using

$$\exp \frac{\eta}{N} \sigma^{a_1} \sigma^{a_2} \sigma^{a_3} = \cosh \frac{\eta}{N} + \sigma^{a_1} \sigma^{a_2} \sigma^{a_3} \sinh \frac{\eta}{N} \tag{24}$$

we obtain :

$$A = A_1 \times A_2$$

$$A_1 = \lim_{n \to 0} \frac{1}{n(n-1)(n-2)} \sum_{a_1 a_2 a_3}' \delta_{Q_{a_1 a_2}, q_2} \delta_{Q_{a_1 a_3}, q_1}$$

$$A_2 = \left(\cosh \frac{\eta}{N} \mathrm{Tr}\, \mathrm{e}^{-\beta H(Q,\sigma)} + \sinh \frac{\eta}{N} \mathrm{Tr}\, \sigma^r \sigma^s \sigma^t \mathrm{e}^{-\beta H(Q,\sigma)} \right)^N \tag{25}$$

with $Q_{rs} = q_2$, $Q_{rt} = q_1$.

The first term is simply the probability of finding 3 states with specified overlaps :

$$A_1 = \overline{\sum_{\alpha,\beta} P_\alpha^2 P_\beta \, \delta_{q^{\alpha\beta}, q_1}} . \tag{26}$$

In the second term we take the limits $N \to \infty, n \to 0$. Using :

$$\lim_{n \to 0} \operatorname{Tr} e^{-\beta H} = 1 ;$$

$$\lim_{n \to 0} \operatorname{Tr} \sigma^r \sigma^s \sigma^t e^{-\beta H} = \langle\!\langle \sigma^r \sigma^s \sigma^t \rangle\!\rangle \tag{27}$$

we find :

$$A_2 = \exp \eta \langle\!\langle \sigma^r \sigma^s \sigma^t \rangle\!\rangle . \tag{28}$$

The factorization $A = A_1 A_2$ is an example of the general factorization valid for all $1/N$ dominant observables as announced in (15). As the result (28) for A_2 is valid whatever η, we find finally in this simple example that $1/N \sum_i (m_i^\alpha)^2 m_i^\beta$ is independent of the states α, β which are chosen, provided that $q^{\alpha\beta}$ is fixed equal to q_1. Its value is given by the one site average :

$$\frac{1}{N} \sum_i (m_i^\alpha)^2 m_i^\beta = \langle\!\langle \sigma^r \sigma^s \sigma^t \rangle\!\rangle \quad \text{where} \quad Q_{rs} = q_2 , \quad Q_{rt} = q_1 . \tag{29}$$

The same kind of argument applies to any $1/N$ dominant observable which doesn't contain the J_{ij}'s. It enables one to prove that these observables are reproducible once all the mutual overlaps between the states are fixed, and their values are given in general by products of one site averages.

We must now calculate

$$\langle\!\langle \sigma^r \sigma^s \sigma^t \rangle\!\rangle . \tag{30}$$

The general technique we shall use is inspired by the work of de Almeida and Lage [7] who first computed the distribution of local magnetization. It takes full advantage of the ultrametric structure of the matrix Q_{ab} in Parisi's R.S.B., which allows an easy description of this matrix as a tree (Fig. 2) : replica indices are the extrema of the branches, and the value of the matrix element Q_{ab} depends only on the closest common ancestor to a and b. The number of iterations in Parisi's R.S.B. scheme is the number of branching levels in the tree (3 in our case, M in general). At level k, the matrix element is q_k, and the number of branchings for each branch is n_k/n_{k+1}. Let us first discuss our « tree method » on the example.

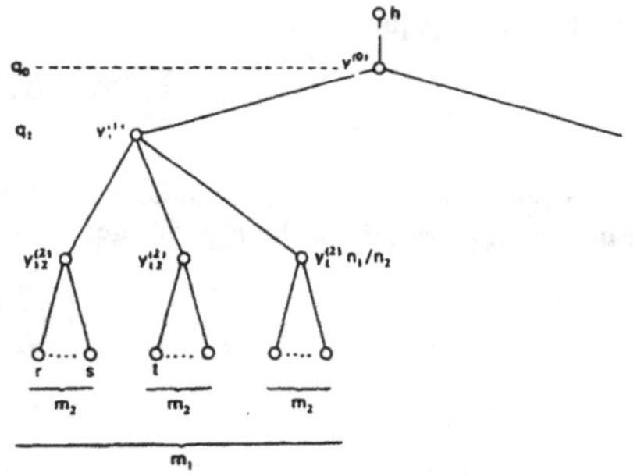

Fig. 2. — The regular tree of replica. The diagram describes the Parisi matrix Q_{ab} with three replica symmetry breakings.

The method [10, 15, 7, 8, 9] can be more easily described if we change the labelling of the replicas using $M = 3$ (number of branching levels) indices instead of just one (see Fig. 2) :

$$a \to u, v, w \qquad 1 \leqslant u \leqslant \frac{n_0}{n_1}$$

$$1 \leqslant a \leqslant n_0 = n \qquad 1 \leqslant v \leqslant \frac{n_1}{n_2} \tag{31}$$

$$1 \leqslant w \leqslant \frac{n_2}{n_3}$$

while $r = (1, 1, 1)$; $s = (1, 1, 2)$; $t = (1, 2, 1)$.

Then :

$$Q_{ab} \sigma^a \sigma^b = q_0 \Big(\sum_{u,v,w} \sigma^{uvw} \Big)^2 + (q_1 - q_0) \sum_u \Big(\sum_{v,w} \sigma^{uvw} \Big)^2 + (q_2 - q_1) \sum_{u,v} \Big(\sum_w \sigma^{uvw} \Big)^2 - q_2 \sum_{uvw} (\sigma^{uvw})^2 . \tag{32}$$

To calculate the trace over the σ we use the Hubbard Stratanovich transformation for each of the terms in (32). We introduce :

$$y^{(0)} ; \quad y_u^{(1)} ; \quad y_{u,v}^{(2)} \tag{33}$$

so that :

$$\langle\!\langle \sigma^r \sigma^s \sigma^t \rangle\!\rangle = \int \frac{dy^{(0)}}{\sqrt{2 \pi q_0}} \prod_u \left(\frac{dy_u^{(1)}}{\sqrt{2 \pi (q_1 - q_0)}} \prod_v \frac{dy_{u,v}^{(2)}}{\sqrt{2 \pi (q_2 - q_1)}} \right) \times$$

$$\times \exp\left(- \frac{(y^{(0)} - h)^2}{2 q_0} - \sum_u \frac{(y_u^{(1)} - y^{(0)})^2}{2(q_1 - q_0)} - \sum_{u,v} \frac{(y_{u,v}^{(2)} - y_u^{(1)})^2}{2(q_2 - q_1)} \right) \times$$

$$\times \operatorname{Tr}_{\{\sigma^{uvw}\}} \left[\sigma^{111} \sigma^{112} \sigma^{121} \exp \beta \sum_{uvw} y_{uv}^{(2)} \sigma^{uvw} \right] . \tag{34}$$

The $\mathrm{Tr}_{\{\sigma^{uvw}\}}$ is equal to :

$$\prod_{u,v} (\cosh \beta y_{u,v}^{(2)})^{n_2} \tanh^2 \beta y_{1,1}^{(2)} \tanh \beta y_{1,2}^{(2)} . \tag{35}$$

The integrations over the y's are done in successive steps, going up the tree in figure 2. At the first step we integrate over n/n_2 variables, the $y_{u,v}^{(2)}$. We get :

$$\left[\prod_{u=2}^{n/n_1} T(y_u^{(1)})\right] \cdot [T'(y_1^{(1)})] \tag{36}$$

with :

$$T(y_u^{(1)}) = \left(\int \frac{dy}{\sqrt{2\pi(q_2 - q_1)}} \exp\left[-\frac{(y - y_u^{(1)})^2}{2(q_2 - q_1)}\right] \cosh^{n_2} \beta y\right)^{\frac{n_1}{n_2}}$$

$$T'(y_1^{(1)}) = \int \frac{dy}{\sqrt{2\pi(q_2 - q_1)}} \exp - \frac{(y - y_1^{(1)})^2}{2(q_2 - q_1)} \cosh^{n_2} \beta y \tanh \beta y \times$$

$$\times \int \frac{dy'}{\sqrt{2\pi(q_2 - q_1)}} \exp - \frac{(y' - y_1^{(1)})^2}{2(q_2 - q_1)} \cosh^{n_2} \beta y' \tanh^2 \beta y'$$

$$\times \left(\int \frac{dy''}{\sqrt{2\pi(q_2 - q_1)}} \exp - \frac{(y'' - y_1^{(1)})^2}{2(q_2 - q_1)} \cosh^{n_2} \beta y''\right)^{\frac{n_1}{n_2} - 2} . \tag{37}$$

At the second step we integrate over the $y_u^{(1)}$ and get :

$$\left(\int \frac{dy^{(1)}}{\sqrt{2\pi(q_1 - q_0)}} \exp - \frac{(y^{(1)} - y^{(0)})^2}{2(q_1 - q_0)} T(y^{(1)})\right)^{\frac{n}{n_1} - 1} \times \left(\int \frac{dy^{(1)}}{\sqrt{2\pi(q_1 - q_0)}} \exp - \frac{(y^{(1)} - y^{(0)})^2}{2(q_1 - q_0)} T'(y^{(1)})\right). \tag{38}$$

The final step is to integrate over $y^{(0)}$. The appearance of the diffusion kernel :

$$\frac{1}{\sqrt{2\pi\,\Delta q}} \exp - \frac{(\Delta y)^2}{2\,\Delta q} \tag{39}$$

shows that the calculation can be seen as the result of a branched diffusion process along the tree. This mathematical interpretation is useful to extract the general rules that allow one to arrive to a formula like (38) reading it directly from the tree :

a) The whole tree is an operator which acts on n functions of y to produce a single function of y;

b) Each oriented line on the tree joining two branching points situated at overlaps q_k, q_{k-1} acts on one function of y at q_k to produce one function at q_{k-1} :

$$(C_{q_k - q_{k-1}}[f])(y) = \frac{1}{\sqrt{2\pi(q_k - q_{k-1})}} \int \exp\left[-\frac{(y - y')^2}{2(q_k - q_{k-1})}\right] f(y')\, dy' ; \tag{40}$$

c) At each branching point the functions arriving to it are simply multiplied at the same value of y;

d) At the lower end we begin either from $\cosh \beta y . \tanh \beta y$ for those replicas whose spin average value are calculated, $\cosh \beta y$ for the rest ;

e) At the upper end we identify $y = h$ (external magnetic field).

These rules are sufficient to do all computations of one site averages. However in practice they can be simplified, and the limit $n \to 0$ (tree with 0 branch !) can be taken more explicitly, through the following reformulation.

In a given computation of $\langle\langle\, \sigma^{a_1} \ldots \sigma^{a_k} \,\rangle\rangle$, there is always a *finite* number k of privileged replica indices, and k remains finite even when $n \to 0$.

We call privileged branches those to which the privileged replica indices are attached (see Fig. 3). Climbing up these privileged branches means going through a new diffusion process (which is the diffusion process we defined before, interrupted at each step by the grafting of normal branches), until two or more privileged branches join.

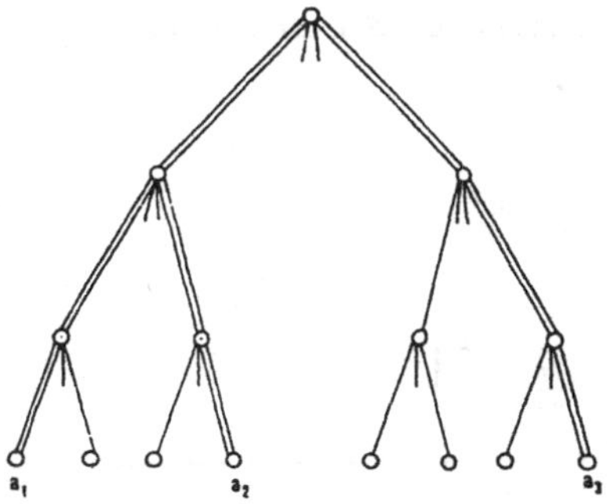

Fig. 3. — The tree of privileged branches. Double lines represent privileged branches. In the figure we chose k = number of privileged states = 3. The single lines correspond to the normal lines that end in the privileged ones. The $n \to 0$ limit can be done once one has summed over all graftings of normal lines.

Once one has identified this new diffusion process, one is able to take the limit $n \to 0$ explicitly, since one can work only on the « backbone » of the tree formed of the (finite number of) privileged branches.

As we saw before, a normal branch is an operator which acts on the initial function

$$z_M(y) = (\cosh \beta y)^{m_M} \tag{41}$$

and gives at each new level when climbing up the tree a new function, with the recursion relation

$$z_{k-1}(y) = (C_{q_k - q_{k-1}}[z_k(y)])^{\frac{n_{k-1}}{n_k}} \tag{42}$$

One can take explicitly the limit $n \to 0$, in which case the inequalities $n = n_0 \geqslant n_1 \dots \geqslant n_M \geqslant 1$ must be reversed : $0 = x_0 \leqslant x_1 \dots \leqslant x_M \leqslant 1$. (We keep the notation x_i for the value of n_i in the $n \to 0$ limit, so that $0 \leqslant x_i \leqslant 1$.)

The evolution along a privileged branch is inferred directly from the general rules a) to e) given above. Going from level k to $k - 1$, there is a grafting of $n_{k-1}/n_k - 1$ normal branches on the privileged one, which leads to a recursion relation :

$$f_{k-1}(y) = [C_{q_k - q_{k-1}} z_k(y)]^{\frac{n_{k-1}}{n_k} - 1} \cdot [C_{q_k - q_{k-1}} f_k(y)] . \tag{43}$$

Introducing $f_k(y) = \dfrac{f_k(y)}{z_k(y)}$, we obtain :

$$f_{k-1} = \frac{C_{q_k - q_{k-1}}(z_k f_k)}{C_{q_k - q_{k-1}}(z_k)} \equiv \mathcal{T}_{q_{k-1}, q_k}[f_k] . \tag{44}$$

This defined the evolution operator along a privileged branch. If such a branch goes without encountering another privileged branch from a level $q_{l'}$ to $q_l < q_{l'}$, it corresponds to a diffusion kernel :

$$\mathcal{T}_{q_l, q_{l'}} = \mathcal{T}_{q_l, q_{l+1}} \mathcal{T}_{q_{l+1}, q_{l+2}} \cdots \mathcal{T}_{q_{l'-1}, q_{l'}} . \tag{45}$$

We obtain finally the general rules for the computation of a one site average :

I) Draw the backbone tree of privileged branches.
II) To each privileged replica associate $f_M = \tanh \beta y$.
III) Go up in the tree, applying the operator $\mathcal{T}$ along each branch.
IV) At a vertex, multiply the incoming functions f.
V) At the upper end (level q_0) apply the operator $C_{q_0} \equiv \mathcal{T}_{0, q_0}$.

The example we worked out before reads :

$$\langle\langle \sigma^r \sigma^s \sigma^t \rangle\rangle \equiv \mathcal{T}_{0,q_1}[(\mathcal{T}_{q_1,q_2}(\tanh \beta y)).(\mathcal{T}_{q_1,q_2}(\tanh^2 \beta y))] . \tag{46}$$

More generally, computing this same $\langle\langle \sigma^r \sigma^s \sigma^t \rangle\rangle$ for $q^{rs} = q$, $q^{rt} = q' > q$, in the full R.S.B. scheme gives $\langle\langle \sigma^r \sigma^s \sigma^t \rangle\rangle = \mathcal{T}_{0,q}[(\mathcal{T}_{q,q_M}(\tanh \beta y)).(\mathcal{T}_{q,q'}[\mathcal{T}_{q',q_M} \tanh \beta y]^2)]$.

It is also useful to take the continuum limit, i.e. an infinite number of R.S.B. In this limit we write $x_k = x$; $x_{k-1} = x - \mathrm{d}x$, and the above recursive relations read :

— for $z : z_k(y) \to \tilde{z}(x, y)$

$$-\frac{\partial \tilde{z}}{\partial x} = \frac{1}{2}\dot{q}(x)\frac{\partial^2 \tilde{z}}{\partial y^2} - \frac{\tilde{z}}{x}\operatorname{Log}\tilde{z}; \tag{47}$$

— for a privileged branch : $f_k(y) \to \tilde{f}(x, y)$

$$-\frac{\partial \tilde{f}}{\partial x} = \frac{1}{2}\dot{q}(x)\left[\frac{\partial^2 \tilde{f}}{\partial y^2} + \frac{\partial \ln \tilde{z}}{\partial y}\frac{\partial \tilde{f}}{\partial y}\right]. \tag{48}$$

3. Local magnetization distributions-organization in clusters.

The distribution of local magnetization in state α :

$$\mathcal{T}_\alpha(m) = \frac{1}{N}\sum_i \delta(m_i^\alpha - m) \tag{49}$$

can be deduced from its moments :

$$M_\alpha^{(k)} = \int \mathcal{T}_\alpha(m)\, m^k\, \mathrm{d}m = \frac{1}{m}\sum_i (m_i^\alpha)^k. \tag{50}$$

Using the methods of the previous section, one finds that these moments are reproducible, with :

$$M_\alpha^{(k)} = \langle\langle \sigma^{a_1} \dots \sigma^{a_k} \rangle\rangle ; \quad q^{a_i a_j} = q_M , \quad 1 \leqslant i < j \leqslant k . \tag{51}$$

This one site average thus corresponds to a simple tree with only one privileged branch, so that :

$$M_\alpha^{(k)} = \mathcal{T}_{0,q_M}((\tanh \beta y)^k) \tag{52}$$

and :

$$\mathcal{T}_\alpha(m) = \mathcal{T}(m) = \mathcal{T}_{0,q_M}[\delta(m - \tanh \beta y)]. \tag{53}$$

Let us now compute the correlations between the magnetizations in states α and β :

$$\mathcal{T}_{\alpha,\beta}(m, m') = \frac{1}{N}\sum_i \delta(m_i^\alpha - m)\,\delta(m_i^\beta - m'). \tag{54}$$

From the double moments :

$$M_{\alpha,\beta}^{(k,l)} = \int \mathcal{T}_{\alpha,\beta}(m, m')\, m^k\, m'^l\, \mathrm{d}m\, \mathrm{d}m' \tag{55}$$

we find that :

— The moments $M^{(k,l)}$ and hence $\mathcal{T}_{\alpha,\beta}$ are reproducible between pairs of states α, β having a fixed overlap $q^{\alpha\beta} = q$:

$$M_{\alpha,\beta}^{(k,l)} = \tilde{M}_q^{(k,l)}; \quad \mathcal{T}_{\alpha,\beta}(m, m') = \mathcal{T}_q(m, m'). \tag{56}$$

— The values of the moments are:

$$M_q^{(k,l)} = \langle\langle \sigma^{a_1} \dots \sigma^{a_k} \sigma^{b_1} \dots \sigma^{b_l} \rangle\rangle \tag{57}$$
$$q^{a_i a_j} = q^{b_i b_j} = q_M \qquad 1 \leqslant i < j \leqslant k$$
$$q^{a_i b_i} = q < q_M .$$

This is again easily computed with the tree method : there are two privileged branches which join at level q, so that

$$M_q^{(k,l)} = \mathcal{T}_{0,q}\{[\mathcal{T}_{q,q_M}(\tanh \beta y)^k][\mathcal{T}_{q,q_M}(\tanh \beta y)^l]\} \tag{58}$$

and

$$\mathcal{T}_q(m, m') = \mathcal{T}_{0,q}\{[\mathcal{T}_{q,q_M}\,\delta(m - \tanh \beta y)][\mathcal{T}_{q,q_M}\,\delta(m' - \tanh \beta y)]\}. \tag{59}$$

It is now clear how these results can be generalized to more states. The correlations between the local magnetizations in different states are always reproducible when all the mutual overlaps of the states are fixed. From the structure of the corresponding tree one can derive the analog of (58).

Let us now quote two more examples which will be useful in the following physical discussion. If $\alpha_1 \dots \alpha_k$ are k states which have all mutual overlaps $q^{\alpha_i \alpha_j} = q$, one finds :

$$\mathcal{T}_{\alpha_1 \dots \alpha_k}(m_1, \dots, m_k) \equiv \frac{1}{N}\sum_i \delta(m_i^{\alpha_1} - m_1) \dots \delta(m_i^{\alpha_k} - m_k) =$$
$$= \mathcal{T}_{0,q}\{[\mathcal{T}_{q,q_M}\,\delta(m_1 - \tanh \beta y)] \dots [\mathcal{T}_{q,q_M}\,\delta(m_k - \tanh \beta y)]\}. \tag{60}$$

If α, β, γ are three states with $q^{\alpha\beta} = q'$, $q^{\alpha\gamma} = q^{\beta\gamma} = q < q'$, one finds :

$$\mathcal{T}_{\alpha\beta\gamma}(m, m', m'') \equiv \frac{1}{N} \sum_i \delta(m_i^\alpha - m)\, \delta(m_i^\beta - m')\, \delta(m_i^\gamma - m'') = \mathcal{T}_{0,q'} \{ [\mathcal{T}_{q',q_M}\, \delta(m'' - \tanh \beta y)] \times \times [\mathcal{T}_{q',q} \{ [\mathcal{T}_{q,q_M}\, \delta(m - \tanh \beta y)] . [\mathcal{T}_{q,q_M}\, \delta(m' - \tanh \beta y)] \}] \} . \tag{61}$$

Our notations here are in terms of the evolution operators $\mathcal{T}_{q,q'}$, which act on functions of y to give other functions of y, as explained in section 2. It will be useful to use notations a bit more explicit, taking advantage of the fact that these operators are linear operators. We write :

$$\begin{aligned} \tilde{\mathcal{T}}_q(m, y) &= \mathcal{T}_{q,q_M}\, \delta(m - \tanh \beta y) \\ \mathcal{T}_{0,q}\, f(y) &= \int \mathrm{d}y\, N_q(y)\, f(y) . \end{aligned} \tag{62}$$

The correlation function between k equidistant states at distance Q can then be written as :

$$\mathcal{T}_Q(m_1, ..., m_k) = \int \mathrm{d}y\, N_Q(y)\, \tilde{\mathcal{T}}_Q(m_1, y) \dots \tilde{\mathcal{T}}_Q(m_k, y) \tag{63}$$

where $\tilde{\mathcal{T}}_q(m, y)$ appears in the distribution of local magnetizations in one state :

$$\mathcal{T}(m) = \int \mathrm{d}y\, N_q(y)\, \tilde{\mathcal{T}}_q(m, y) \tag{64}$$

for all q, $q_0 \leqslant q \leqslant q_M$, for instance $q = Q$.

Furthermore one can deduce from the definitions of N and $\tilde{\mathcal{T}}$ the following properties [9] :

$$\begin{aligned} &\forall q : N_q(y) \geqslant 0 ; \qquad \int \mathrm{d}y\, N_q(y) = 1 \\ &\forall q : \tilde{\mathcal{T}}_q(m, y) \geqslant 0 ; \qquad \int \mathrm{d}m\, \tilde{\mathcal{T}}_q(m, y) = 1 \end{aligned} \tag{65}$$

which indicate that N_q and $\tilde{\mathcal{T}}_q$ are probability distributions of the variables y and m respectively.

The formulae (63) to (65) are the starting point of our physical discussion. They show that at any scale Q, there exists a random variable y with a distribution $N_Q(y)$ such that, when y is fixed, the distribution of local magnetizations in any k equidistant states at distance Q are independent of each other, with distribution $\tilde{\mathcal{T}}_Q(m, y)$. We claim that to each of the N sites of the lattice can be given a value of the parameter y. This is not obvious at this point and will be proven below through the analysis of the « ancestor state ». It will appear even more evident in section 4. Let us now examine the consequences of this assertion :

— at any scale Q, the sites where the variable y takes values between y and $y + \Delta y$ form a cell, $\mathcal{C}_y$;

— the volume of this cell is $| \mathcal{C}_y | = (N_Q(y)\, \Delta y) . N$;

— inside the cell, the local magnetizations in k equidistant states at distance Q are independent random variables with the same distribution $\tilde{\mathcal{T}}_Q(m, y)$:

$$\begin{aligned} \frac{1}{|\mathcal{C}_y|} \sum_{i \in \mathcal{C}_y} \delta(m_i^{\alpha_1} - m_1) \dots \delta(m_i^{\alpha_k} - m_k) &= \tilde{\mathcal{T}}_Q(m_1, y) \dots \tilde{\mathcal{T}}_Q(m_k, y) \\ \tilde{\mathcal{T}}_Q(m, y) &= \frac{1}{|\mathcal{C}_y|} \sum_{i \in \mathcal{C}_y} \delta(m_i^\alpha - m) . \end{aligned} \tag{66}$$

The different cells $\mathcal{C}_k$ are shared by all the states which are equidistant at distance Q. In the ultrametric tree of the states, a partition of cells is associated with each branching at the scale Q (points A, A' ... of Fig. 1). The cells associated with the point A may not be the same as those associated with A' (they are in fact in general different), but they have the same volumes, and the same distribution of local magnetizations inside them. A very useful concept is that of the ancestor state : to each branching point in the tree of the states, such as A in figure 1, we associate a configuration of local magnetizations $m_i^{(A)}$ which is uniform in the corresponding cells $\mathcal{C}_y^{(A)}$. This is not a pure state of the system, we call it the ancestor state. It satisfies :

$$m_i^{(A)} = m_Q(y) , \qquad i \in \mathcal{C}_y^{(A)} , \tag{67}$$

where $m_Q(y) = \int \tilde{\mathfrak{T}}_Q(m, y)\, m\, dm$ is the average magnetization in the cell $\mathcal{C}_y^{(A)}$ for the pure states which are descendants of A.

The ancestor state has a global magnetization equal to the global magnetization in any state :

$$\sum_y \frac{|\mathcal{C}_y^{(A)}|}{N} m_Q(y) = \int N_Q(y)\, dy \int \tilde{\mathfrak{T}}_Q(m, y)\, m\, dm = \int \mathfrak{T}(m)\, m\, dm \equiv M \tag{68}$$

and a self overlap equal to Q, the level of the tree where the point A lies :

$$\sum_y \frac{|\mathcal{C}_y^{(A)}|}{N} [m_Q(y)]^2 = \int N_Q(y)\, dy [m_Q(y)]^2 = Q. \tag{69}$$

There is in fact a way to obtain the ancestor state from its descendants : it can be shown using the methods of [2] that the number of descendants of a particular ancestor A at a finite Δq is always infinite in the thermodynamic limit. Hence it is always possible to choose a family F of $\mathcal{N}$ descendents of A, $\alpha_1, \ldots, \alpha_{\mathcal{N}}$, which are all equidistant at distance Q, with $\mathcal{N} \gg 1$. Defining :

$$m_i^F = \frac{1}{\mathcal{N}} \sum_{k=1}^{\mathcal{N}} m_i^{\alpha_k} \tag{70}$$

one obtains from (63) :

$$\forall r : \frac{1}{N} \sum_{i=1}^{N} (m_i^F)^r = \int dy\, N_Q(y)\, [m_Q(y)]^r \tag{71}$$

which proves that the variable y labels the sites where $m_i = m_Q(y)$; it is easily proven from (63) again that these sites are the same whatever the choice of the family F of $\mathcal{N}$' equidistant descendants of A. They define a unique cell which is exactly the domain where the magnetization of the ancestor is uniform with value $m_Q(y)$.

Finally let us explain what happens when we change from a scale Q to $Q' > Q$. Let us consider three states $\alpha_1, \alpha_2, \alpha_3$ such that $q^{\alpha_1\alpha_2} = q^{\alpha_1\alpha_3} = Q$, $q^{\alpha_2\alpha_3} = Q'$. The correlation of local magnetizations in these three states depends only on Q and Q', it can be written as :

$$\mathfrak{T}_{Q,Q'}(m_1, m_2, m_3) = \int dy\, N_Q(y)\, \tilde{\mathfrak{T}}_Q(m_1, y) \cdot \int dy'\, G(Q, Q', y, y')\, \tilde{\mathfrak{T}}_Q(m_2, y')\, \tilde{\mathfrak{T}}_Q(m_3, y') \tag{72}$$

where $G(Q, Q', y, y')$ is the integral kernel associated with the evolution operator $\mathfrak{T}_{Q,Q'}$ defined in (45).

This formula, which can be obviously generalized to the correlations of any k states at distances Q, l states at distance Q' shows that :

— At the scale $Q' > Q$, any cell $\mathcal{C}_y$ of the scale Q can be cut into subcells $\mathcal{C}_{y,y'}$.

— The probability that a site in the cell $\mathcal{C}_y$ be in a definite subcell $\mathcal{C}_{y,y'}$ is :

$$\frac{|\mathcal{C}_{y,y'}|}{|\mathcal{C}_y|} = G(Q, Q', y, y')\, \Delta y'. \tag{73}$$

— States at distance Q' have uncorrelated local magnetizations inside each subcell $\mathcal{C}_{y,y'}$, with the same distribution $\tilde{\mathfrak{T}}_{Q'}(m, y')$, while states at distance Q have uncorrelated local magnetizations inside the whole cell $\mathcal{C}_y$.

We notice that the coherence of this picture is enforced by the following identities :

— The subcells form a partition of the cell :

$$\int dy'\, G(Q, Q', y, y') = 1.$$

— The distribution of magnetizations in the cell is induced by the distribution in the subcells :

$$\int dy'\, G(Q, Q', y, y')\, \tilde{\mathfrak{T}}_{Q'}(m, y') = \tilde{\mathfrak{T}}_Q(m, y).$$

Of course, the direct analysis of the structure at the scale Q' shows cells $\mathcal{C}'_{y'}$ of volume $N_{Q'}(y').(\Delta y').N$ where the distribution of magnetizations is $\tilde{\mathfrak{T}}_{Q'}(m, y')$. These cells $\mathcal{C}'_{y'}$ are reunions of the $\mathcal{C}_{y,y'}$ corresponding to different y :

$$\mathcal{C}_{y'} = \bigcup_y \mathcal{C}_{y,y'} ;$$

$$|\mathcal{C}'_{y'}| = N \cdot \int dy\, N_Q(y)\, G(Q, Q', y, y') \cdot \Delta y'. \tag{74}$$

Thus a direct study at the scale Q' makes one loose the information about the cells at the scale Q. The maximal information is obtained from the sequence of domains embedded into other domains at all the scales starting from the lowest one ($Q = Q_{\min}$).

4. Couplings between the cells and T.A.P. equations.

In this section we study in details the structure of the cells and the way in which they are coupled one to another. As we saw before, it is crucial in order not to loose any information to keep trace of the partitioning into subcells at *every scale*. Therefore we introduce here a resolution on the function $q(x)$, assuming that it can be approximated by a stepwise constant function with M replica symmetry breakings :

$$q(x) = q_l , \quad x_l \leqslant x \leqslant x_{l+1} \tag{75}$$

where $l = 0, ..., M$; $x_0 = 0$, $x_{M+1} = 1$.

The resolution in the parameters y necessary for the definition of the cells is also self understood in the following.

The discussion in the previous section tells us that :

— Any pure state α is characterized by a number of cells $\mathcal{C}^{\alpha}_{y_0 y_1 \ldots y_M}$ inside which the magnetization m_i^{α} is constant and takes the value $\tanh \beta y_M$.

— Any ancestor state A living at the scale (age) q_l is characterized by cells $\mathcal{C}^{(A)}_{y_0 y_1 \ldots y_l}$ inside which the magnetization $m_i^{(A)}$ is uniform and takes the value $m_{q_l}(y_l)$.

— The cells $\mathcal{C}^{\alpha}$ of all the descendants α of a given ancestor A are disjoint subcells of the $\mathcal{C}^{A}$ and :

$$\bigcup_{y_{l+1}, \ldots, y_M} \mathcal{C}^{\alpha}_{y_0 \ldots y_l y_{l+1} \ldots y_M} = \mathcal{C}^{(A)}_{y_0 \ldots y_l} . \tag{76}$$

The set of all spin configurations with such an arrangement can be constructed in the following way :

a) At every site of the lattice imagine simultaneous independent stochastic processes according to the Langevin's equation [18] :

$$\frac{\mathrm{d}y_i}{\mathrm{d}x}(x) = \beta x \frac{\mathrm{d}q}{\mathrm{d}x} m_{q(x)}(y_i) + z_i(x) \sqrt{\dot{q}(x)} \tag{77}$$

where $z(x)$ is a Gaussian noise with

$$\overline{z_i(x)} = 0$$
$$\overline{z_i(x)\, z_j(x')} = \delta(x - x')\, \delta_{ij} \tag{78}$$

and the initial condition is $y_i(0) = h_{\text{ext}}$, $q(0) = 0$, $q(0^+) = q_0$.

b) Assign the magnetization $m_i = \tanh [\beta y_i(1)]$ to each site.

The discretization we were referring to above is simply the discretization of the Langevin's equation. The ramification of the tree is a trivial consequence of the possible random choices that one can make at every step of x. The ancestor results from integrating the Langevin equation up to x_0 and assigning the magnetization $m_{q(x_0)}(y_i(x_0))$ at each site i.

However it is important to notice that only a small part of the configurations so generated are pure states of the system.

A case which is particularly interesting is the ancestor at the smallest possible scale $q_{\min} = q_0$. It is a common ancestor to all the pure states of the system (we call it grand ancestor G-A) and thus its cells $\mathcal{C}^{G-A}_{y_0}$ are common to all the states, they form a universal partition of the system. (This partition is also the only one which exists above the critical temperature in a field, where $q(x) = q_0$, $0 \leqslant x \leqslant 1$ and there is only one state.)

In order to understand the tree of states better, we have computed the average couplings between the cells associated with a given ancestor. First let us consider the case of the largest cells, the ones of the grand ancestor. We define :

$$J_{y_0, y_0'} = \frac{N}{|\mathcal{C}_{y_0}|\,|\mathcal{C}_{y_0'}|} \sum_{\substack{i \in \mathcal{C}^{G-A}_{y_0} \\ j \in \mathcal{C}^{G-A}_{y_0'}}} J_{ij} \tag{79}$$

These couplings between the $\mathcal{C}^{G-A}_{y_0}$ can be computed from the reproducible and self averaging functions :

$$M_{s,t} = \frac{1}{N} \sum_{i,j} m_i^{\alpha_1} \ldots m_i^{\alpha_s} J_{ij}\, m_j^{\beta_1} \ldots m_j^{\beta_t} \tag{80}$$

where the α's and the β's are $s + t$ pure states which are all equidistant at distance q_0. Proceeding as in section 3 by taking a large family $\mathcal{N}$ of such states and averaging over the possible choices of α_i, we get :

$$\frac{1}{\mathcal{N}} \sum_{\alpha=1}^{\mathcal{N}} m_i^{\alpha} = m_i^{G-A}$$

$$M_{s,t} \underset{N \to \infty}{\sim} \frac{1}{N} \sum_{i,j} [m_i^{G-A}]^s J_{ij} [m_j^{G-A}]^t \tag{81}$$

which reads

$$M_{s,t} = \sum_{y_0, y_0'} \frac{|\mathcal{C}^{G-A}_{y_0}|\,|\mathcal{C}^{G-A}_{y_0'}|}{N^2} [m_{q_0}(y_0)]^s [m_{q_0}(y_0')]^t J_{y_0 y_0'} . \tag{82}$$

We see that the $J_{y_0, y_0'}$ are self-averaging quantities which can be inferred from the $M_{s,t}$.

The quantities $M_{s,t}$ are computed in the appendix. One obtains :

$$J_{y_0, y_0'} = m_{q_0}(y_0') \frac{y_0 - h}{q_0} + m_{q_0}(y_0) \frac{y_0' - h}{q_0} . \tag{83}$$

Before analysing this formula let us derive the corresponding formula for the domains of an ancestor at a scale $q > q_0$. For instance at the scale q_1, one must compute the functions :

$$M_{\substack{s,t \\ u,v}} = \frac{1}{N} \sum_{i,j} m_i^{\alpha_1} \ldots m_i^{\alpha_s} m_i^{\gamma_1} \ldots m_i^{\gamma_u} J_{ij}\, m_j^{\beta_1} \ldots m_j^{\beta_t} m_j^{\delta_1} \ldots m_j^{\delta_v} \tag{84}$$

where $\alpha_1, ..., \alpha_s, \beta_1, ..., \beta_t, \gamma_1, \delta_1$ are $s + t + 2$ states all equidistant at distance q_0, $\gamma_1, ..., \gamma_u$ are all equidistant at distance q_1, and $\delta_1 ... \delta_p$ are all equidistant at distance q_1. One has then :

$$M_{s,t \atop u,v} = \frac{1}{N^2} \sum_{y_0, y_0'} |\mathcal{C}_{y_0}^{G-A}| |\mathcal{C}_{y_0'}^{G-A}| m_{q_0}(y_0)^s m_{q_0}(y_0')^t \times \sum_{y_1, y_1'} \frac{|\mathcal{C}_{y_0,y_1}^{(A)}| |\mathcal{C}_{y_0',y_1'}^{(A)}|}{|\mathcal{C}_{y_0}^{G-A}| |\mathcal{C}_{y_0'}^{G-A}|} J_{y_0 y_1, y_0' y_1'}^{A} \cdot m_{q_1}(y_1)^u m_{q_1}(y_1')^v \tag{85}$$

where :

$$J_{y_0 y_1, y_0' y_1'}^{A} \equiv \frac{N}{|\mathcal{C}_{y_0,y_1}^{(A)}| |\mathcal{C}_{y_0',y_1'}^{(A)}|} \sum_{\substack{i \in \mathcal{C}_{y_0,y_1}^{(A)} \\ j \in \mathcal{C}_{y_0',y_1'}^{(A)}}} J_{ij} . \tag{86}$$

A first striking result is obtained from the reproducibility of the moments $M_{s,t \atop u,v}$. Although the subcells $\mathcal{C}_{y_0,y_1}^{(A)}$ associated with different ancestors at the scale q_1 depend on the ancestor, the average coupling between them, $J_{y_0 y_1, y_0' y_1'}^{A}$ is independent of A. Using the result of the computation of $M_{s,t \atop u,v}$ done in the appendix, we find :

$$J_{y_0 y_1, y_0' y_1'} = J_{y_0, y_0'} + \frac{y_1' - y_0'}{q_1 - q_0} (m_{q_1}(y_1) - m_{q_0}(y_0)) + \frac{y_1 - y_0}{q_1 - q_0} (m_{q_1}(y_1') - m_{q_0}(y_0')) - \\ - \beta x_1 (m_{q_1}(y_1) m_{q_1}(y_1') - m_{q_0}(y_0) m_{q_0}(y_0')) . \tag{87}$$

This formula can be generalized to the couplings between domains at an arbitrary scale. Once again, these couplings are independent of the ancestor state which is considered ; the result is best expressed by the recursive relation :

$$J_{y_0 \dots y_l, y_0' \dots y_l'} = J_{y_0 \dots y_{l-1}, y_0' \dots y_{l-1}'} + \frac{y_l - y_{l-1}}{q_l - q_{l-1}} (m_{q_l}(y_l') - m_{q_{l-1}}(y_{l-1}')) + \frac{y_l' - y_{l-1}'}{q_l - q_{l-1}} (m_{q_l}(y_l) - m_{q_{l-1}}(y_{l-1})) - \\ - \beta x_l [m_{q_l}(y_l) m_{q_l}(y_l') - m_{q_{l-1}}(y_{l-1}) m_{q_{l-1}}(y_{l-1}')] . \tag{88}$$

Several properties can be deduced from equations (87)-(88).

a) There exist in the ancestor states T.A.P.-like equations [16] :

$$\sum_{y_0', \dots, y_l'} \frac{|\mathcal{C}_{y_0' \dots y_l'}|}{N} J_{y_0 \dots y_l, y_0' \dots y_l'} m_{q_l}(y_l') = y_l - h + m_{q_l}(y_l) \sum_{y_0', \dots, y_l'} \frac{|\mathcal{C}_{y_0' \dots y_l'}|}{N} \frac{\partial}{\partial y_l'} m_{q_l}(y_l') \tag{89}$$

in which y_l is the local field, and creates a local magnetization $m_{q_l}(y_l)$. This suggests new ways of approaching the solution of the S.K. model by solving these equations first for the ancestor and then for the descendants.

b) The matrix of the couplings between the cells possesses the remarkable property :

$$\sum_{y_{l+1}' \dots y_M'} J_{y_0 \dots y_M, y_0' \dots y_M'} |\mathcal{C}_{y_0' \dots y_M'}| = \\ = J_{y_0 \dots y_l, y_0' \dots y_l'} |\mathcal{C}_{y_0' \dots y_l'}| \tag{90}$$

whatever the value of $y_{l+1}, ..., y_M$. Such a structure should be found in a large random matrix, remembering that from (86) the $J_{y_0 \dots y_M, y_0' \dots y_M'}$ are themselves averages of J_{ij}'s over *macroscopic* cells.

c) In the limit where the magnetic field is large, one can identify easily the universal cells of the grand-ancestor state. They are regions where the quantity :

$$G_i = \sum_j J_{ij} \tag{91}$$

is uniform.

In general G_i is a random variable whose site distribution follows from the central limit theorem and is a Gaussian with width unity :

$$\frac{1}{N} \sum_i \delta(G_i - G) = \frac{1}{\sqrt{2\pi}} \exp(-G^2/2) . \tag{92}$$

We have computed the distribution of G_i inside a given cell $\mathcal{C}_{y_0}^{G-A}$. Its average value follows from (83) :

$$G_{y_0} = \frac{1}{|\mathcal{C}_{y_0}^{G-A}|} \sum_{i \in \mathcal{C}_{y_0}^{G-A}} G_i = M \cdot \frac{y_0 - h}{q_0} . \tag{93}$$

A more elaborate computation based on techniques introduced in [17] enables one to prove that the distribution of G_i inside the cell $\mathcal{C}_{y_0}^{G-A}$ is in fact a Gaussian, of average G_{y_0}, and of width

$$\frac{1}{|\mathcal{C}_{y_0}^{G-A}|} \sum_{i \in \mathcal{C}_{y_0}^{G-A}} (G_i - G_{y_0})^2 = 1 - \frac{M^2}{q_0} . \tag{94}$$

For general values of the temperature and the magnetic

field $M^2 \leqslant q_0$, which implies that the G_i fluctuate inside each cell of the grand ancestor. On the other hand, for $h \to \infty$ the fluctuations vanish, and sites having the same value of G_i are in the same cell $\mathcal{C}^{G-A}$

d) There is no frustration in the interactions between the cells of the grand-ancestor :

$$J_{y_0, y_0'} \overset{h \to 0}{\sim} 2 \left[\frac{\partial}{\partial y} m_{q_0}(y) \right] \Bigg|_{y=0} \frac{y_0}{\sqrt{q_0}} \frac{y_0'}{\sqrt{q_0}} \tag{95}$$

which is of the Mattis type.

5. Conclusions.

We have shown that the surprisingly high level of organization of the space of equilibrium states of a spin glass, reflected by ultrametricity, can be translated directly into a hierarchical organization among the sites. In fact the sequence of cells imbedded into one another associated with each equilibrium state can be used to define a distance between sites, and for such a distance the real space of sites is ultrametric.

It is very interesting to observe that all the states within a given distance share a common sequence of cells : this allows for a coarsed grained description of a spin glass which can be then refined as much as one wants. Certainly this should have interesting consequences in the use of spin glass-like models to build up associative memories [19]. For instance at the crudest level of description, one is given only the cells (of the « grand ancestor state ») which are common to all the states. These cells are the only ones which exist in a field above T_c. An attractive conjecture would be that there exist some way of crossing the de Almeida-Thouless line such that these cells are left invariant, the system developing new cells inside them as the temperature is lowered.

The Fokker-Planck like differential equation which appears naturally in Parisi's solution receives a purely static interpretation : it describes the sizes of the subcells within each cell at every scale.

The average couplings within the cells are precisely the ones which are required for the existence of generalized TAP equations for the ancestor states which have constant magnetization inside each cell. Furthermore, they possess a hierarchical structure which is reminiscent of ultrametricity : the couplings $J_{y_0 y_1 \dots y_l, y_0' y_1' \dots y_l'}$ between the cells $\mathcal{C}_{y_0 y_1 \dots y_l}$ and $\mathcal{C}_{y_0' y_1' \dots y_l'}$ are such that $\sum_{y_k' \dots y_l'} J_{y_0 \dots y_l, y_0' \dots y_l'}$ is independent of $y_k, \dots, y_l$, $(0 < k \leqslant l)$.

This means that for instance the average coupling between a subcell $\mathcal{C}_{y_0 y_1}$ of the cell $\mathcal{C}_{y_0}$, and another cell $\mathcal{C}_{y_0'}$ is independent of y_1, i.e. independent of the subcell ! These are the results we obtain by working out the detailed consequences of Parisi's solution. They open the way to a direct test of the validity of this solution : one must find out whether such a structure can indeed be found in a large random matrix.

Finally we would like to point out that, in the process of concentrating on self averaging and reproducible quantities, one must leave aside the weights P_α of the equilibrium states. This paper is in that sense complementary to reference [6] where this aspect is addressed.

Acknowledgments.

We gratefully acknowledge useful conversations with C. de Dominicis, G. Parisi, N. Sourlas and G. Toulouse.

Part of this work was done at the Laboratoire de Physique Théorique de l'Ecole Normale Supérieure. MAV thanks all the members of the Laboratoire for the warm hospitality.

Appendix.

In this appendix we compute the reproducible correlation functions defined in section 4, from which one can deduce the coupling between the domains.

Let us start with the function $M_{s,t}$. The average :

$$\sum_{\substack{\alpha_1, \dots, \alpha_s \\ \beta_1, \dots, \beta_t}} \overline{P_{\alpha_1} \dots P_{\alpha_s} P_{\beta_1} \dots P_{\beta_t} M_{s,t}^{\alpha_1 \dots \alpha_s, \beta_1 \dots \beta_t} \prod_{1 \leqslant i < j \leqslant s} \delta(q^{\alpha_i \alpha_j} - q_0)} \times$$
$$\times \overline{\prod_{1 \leqslant i < j \leqslant t} \delta(q^{\beta_i \beta_j} - q_0) \prod_{1 \leqslant i \leqslant s;\, 1 \leqslant j \leqslant t} \delta(q^{\alpha_i \beta_j} - q_0)} \tag{A.1}$$

is expressed in replica space as :

$$\lim_{n \to 0} \frac{1}{n(n-1) \dots (n-(s+t)+1)} \sum_{a_1 \dots a_s, b_1 \dots b_t}{}' \frac{1}{N} \sum_{i,j} \overline{\langle \sigma_i^{a_1} \dots \sigma_i^{a_s} J_{ij} \sigma_j^{a_{s+1}} \dots \sigma_j^{a_{s+t}} \rangle} \tag{A.2}$$

where the replica indices a_i, b_j are all distinct, with $Q_{a_i a_j} = Q_{b_i b_j} = Q_{a_i b_j} = q_0$. As was explained in [17], the J_{ij} can be integrated by parts which amounts to replacing it by $\frac{\beta}{N} \sum_{c=1}^{n} \sigma_i^c \sigma_j^c$. Finally the reproducible part can be factored out of (A.2) as in section 2, it gives :

$$M_{s,t} = \beta \sum_{c=1}^{n} \langle\langle \sigma^{a_1} \dots \sigma^{a_s} \sigma^c \rangle\rangle \langle\langle \sigma^{b_1} \dots \sigma^{b_t} \sigma^c \rangle\rangle \tag{A.3}$$

where $a_1 \dots a_s$, $b_1 \dots b_t$ are any $s + t$ replica indices all different from each other, such that : $Q_{a_i a_j} = Q_{b_i b_j} = Q_{a_i b_j} = q_0$.

We have explained in section 2 how one can compute such quantities as $\langle\langle \sigma^{a_1} \dots \sigma^{a_s} \sigma^c \rangle\rangle$ using the « tree method » when all the overlaps $Q_{a_i c}$ are fixed. For $Q_{a_i c} = q_l$, one finds :

$$\langle\langle \sigma^{a_1} \dots \sigma^{a_s} \sigma^c \rangle\rangle = \int \mathrm{d}y_0 \, N_{q_0}(y_0) \, m_{q_0}(y_0)^{s-1} \int \mathrm{d}y_l \, G(q_0, q_l, y_0, y_l) \, m_{q_l}(y_l)^2 . \tag{A.4}$$

The only problem which is left is the one of counting the number of ways in which c can be chosen with given overlaps $Q_{a_i c}$ and $Q_{b_j c}$ with the other fixed indices, which is simple since for instance $Q_{b_j c} = q_0$ if $Q_{a_i c} = q_l > q_0$. The result is :

$$\frac{1}{\beta} M_{s,t} = s \left\{ \int \mathrm{d}y_0 \, N_{q_0}(y_0) \, m_{q_0}(y_0)^{t+1} \right\} \times$$
$$\times \left\{ \int \mathrm{d}y'_0 \, N_{q_0}(y'_0) \, m_{q_0}(y'_0)^{s-1} \left[1 - \sum_{l=0}^{M} (x_{l+1} - x_l) \, q(0, l, y) \right] \right\} + (s \rightleftarrows t) \tag{A.5}$$

where $(s \rightleftarrows t)$ is the term obtained from the first one by the symmetry operation $\begin{bmatrix} s \to t \\ t \to s \end{bmatrix}$, and

$$q(l', l, y) = \int \mathrm{d}y_l \, G(q_{l'}, q_l, y, y_l) \, [m_{q_l}(y_l)]^2 . \tag{A.6}$$

This can be simplified by use of the formula :

$$\frac{1}{\beta} \frac{\partial}{\partial y} m_{q_l}(y) = 1 - x_l [m_{q_l}(y)]^2 + \sum_{l'=l}^{K} (x_{l'} - x_{l'+1}) \, q(l, l', y) \tag{A.7}$$

which can be proven by induction starting from the case $y = M$.

We have finally

$$M_{s,t} = \int \mathrm{d}y_0 \, \mathrm{d}y'_0 \, N_{q_0}(y_0) \, N_{q_0}(y'_0) \left\{ m_{q_0}(y'_0) \frac{\partial}{\partial y_0} + m_{q_0}(y_0) \frac{\partial}{\partial y'_0} \right\} \{ m_{q_0}(y_0)^s \, m_{q_0}(y'_0)^t \} . \tag{A.8}$$

The computation of the moments $M_{s,t \atop u,v}$ defined in (84) follows the same lines. One finds

$$M_{s,t \atop u,v} = \beta \sum_{e} \langle\langle \sigma^{a_1} \dots \sigma^{a_s} \sigma^{c_1} \dots \sigma^{c_u} \sigma^e \rangle\rangle \langle\langle \sigma^{b_1} \dots \sigma^{b_t} \sigma^{d_1} \dots \sigma^{d_v} \sigma^e \rangle\rangle \tag{A.9}$$

where $a_1 \dots a_s$, $b_1 \dots b_t$, c_1, d_1 all have the same mutual overlap q_0, while $c_1, \dots, c_u$ all have the mutual overlap q_1, and $d_1 \dots d_v$ all have the mutual overlap q_1. The problem of counting the number of ways in which replica e can be chosen with fixed overlaps from the a, b, c, d is a bit more tedious than in the case of $M_{s,t}$, but not difficult. One finds :

$$M_{s,t \atop u,v} = \int \mathrm{d}y_0 \, \mathrm{d}y'_0 \, N_{q_0}(y_0) \, N_{q_0}(y'_0) \int \mathrm{d}y_1 \, \mathrm{d}y'_1 \, G(q_0, q_1, y_0, y_1) \, G(q_0, q_1, y'_0, y'_1) \times$$
$$\times \, m_{q_0}(y_0)^s \, m_{q_0}(y'_0)^t \, m_{q_1}(y_1)^u \, m_{q_1}(y'_1)^v \times \left[\frac{y_0 - h}{q_0} m_{q_0}(y'_0) + \frac{y'_0 - h}{q_0} m_{q_0}(y_0) \right.$$
$$+ \frac{y_0 - y_1}{q_0 - q_1} (m_{q_1}(y'_1) - m_{q_0}(y'_0)) + \frac{y'_0 - y'_1}{q_0 - q_1} (m_{q_1}(y_1) - m_{q_0}(y_0))$$
$$\left. - \beta x_1 (m_{q_1}(y_1) \, m_{q_1}(y'_1) - m_{q_0}(y_0) \, m_{q_0}(y'_0)) \right] . \tag{A.10}$$

References

[1] Parisi, G., *Phys. Rev. Lett.* **50** (1983) 1946.

[2] Mézard, M., Parisi, G., Sourlas, N., Toulouse, G. and Virasoro, M., *Phys. Rev. Lett.* **52** (1984) 1156 and *J. Physique* **45** (1984) 843.

[3] Young, A. P., Bray, A. and Moore, M., *J. Phys.* **C17** (1984) L149.

[4] Kirkpatrick, S. and Toulouse, G., in preparation.
Kirkpatrick, S., Gelatt Jr, C. D., Vecchi, M. P., *Science* **220** (1983) 671.

[5] Hopfield, J. J., *Proc. Nat. Acad. Sci. U.S.A.* **79** (1982) 2554.

[6] Mézard, M., Parisi, G., Virasoro, M. A., *J. Physique Lett.* **46** (1985) L-217.
Derrida, B., Toulouse, G., *J. Physique Lett.* **46** (1985) L-223.

[7] De Almeida, J. R. L., Lage, E. J., *J. Phys.* **C16** (1982) 939.

[8] Goltsev, A. V., *J. Phys.* **A17** (1984) 237.

[9] Sommers, H. J., Dupont, W., *J. Phys.* **C32** (1984) 5785.

[10] Parisi, G., *Phys. Rev. Lett.* **43** (1979) 1754 ; *J. Phys.* **A13** (1980) L115, 1101, 1887.

[11] De Dominicis, C., Kondor, I., *Phys. Rev.* **B27** (1983) 606.

[12] Young, A. P., *Phys. Rev. Lett.* **51** (1983) 1206.

[13] Parga, N., Parisi, G., Virasoro, M. A., *J. Physique Lett.* **45** (1984) L-1063.

[14] De Dominicis, C., Young, A. P., *J. Phys.* **A16** (1983) 2063.

[15] Duplantier, B., *J. Phys.* **A14** (1981) 283.

[16] Sommers, H. J., De Dominicis, C. and Gabay, M., have derived a similar (though not identical) set of mean field equations in the context of successive levels of time averaging, *J. Phys.* **A16** (1983) L-679.

[17] Mézard, M., Parisi, G., *J. Physique Lett.* **45** (1984) L-707.

[18] These equations appear already in [9]. The physical interpretation is however new.

[19] Parga, N., Virasoro, M. A., in preparation.

Tome 46 Nº 6 15 MARS 1985

LE JOURNAL DE PHYSIQUE - LETTRES

J. Physique Lett. **46** (1985) L-217 - L-222 15 MARS 1985, PAGE L-217

Classification
Physics Abstracts
75.50K

Random free energies in spin glasses

M. Mézard (*), G. Parisi (+) and M. A. Virasoro (++)

Dipartimento di Fisica, Università di Roma I, Piazzale A. Moro 2, I-00185, Roma, Italy

(*Reçu le 4 janvier 1985, accepté le 28 janvier 1985*)

Résumé. — Les énergies libres des états purs sont étudiées dans la théorie de champ moyen. On montre qu'elles sont des variables aléatoires indépendantes avec une distribution exponentielle. Les implications sur les fluctuations avec les échantillons sont analysées. La nature physique de la théorie de champ moyen est complètement caractérisée.

Abstract. — The free energies of the pure states in the spin glass phase are studied in the mean field theory. They are shown to be independent random variables with an exponential distribution. Physical implications concerning the fluctuations from sample to sample are worked out. The physical nature of the mean field theory is fully characterized.

The physical understanding of the nature of the spin glass phase has improved recently with the detailed analysis of the solution proposed by one of us for the mean field theory of spin glasses introduced by Sherrington and Kirkpatrick (S.K.) [1]. It is known that the spin glass transition is associated with the breaking of ergodicity and the appearance of an infinite number of pure equilibrium states for the system, unrelated to each other by a symmetry. The spin glass phase can be understood by a geometrical analysis of the space of pure states. Each state α is characterized by its probability P_α, and the value of the local magnetization on each site i : m_i^α. The order parameter [2] is the distribution of overlaps between two pure states α and β chosen with probabilities P_α and P_β : $P(q) = \sum_{\alpha,\beta} P_\alpha P_\beta \, \delta(q_{\alpha\beta} - q)$, where the overlap $q_{\alpha\beta}$ is equal to $(1/N) \sum_i m_i^\alpha m_i^\beta$.

It was further recognized in references [3, 4] that the probabilities of the states depend on the sample, and this was analysed [3] through the explicit computation of the inclusive distribution

(*) On leave from Laboratoire de Physique Théorique de l'Ecole Normale Supérieure, Paris, France.
(+) Università di Roma II, Tor Vergata, Roma and INFN, Sezione di Frascati, Italy.
(++) Also at INFN, Sezione di Roma, Italy.

functions of the probabilities, $f^{(k)}$, defined as :

$$f^{(k)}(P_1, \ldots, P_k) = \overline{\sum_{\alpha_1 \ldots \alpha_k}' \delta(P_{\alpha_1} - P_1) \ldots \delta(P_{\alpha_k} - P_k)} \tag{1}$$

where the sum is carried over all couples of distinct replicas, and $\overline{(\;)}$ denotes the average over the realizations of the couplings.

In this paper we shall show that all these results on the distributions and fluctuations of the probabilities of the states are consequences of one simple property : *the free energies of the pure states are independent random variables.*

Let us make this statement more precise. All the pure states must have, in the thermodynamic limit, the same free energy per spin ($\lim_{N \to \infty} F_\alpha/N$). The probabilities of the different states are related to the $O(1/N)$ corrections f_α/N to the free energy per spin by :

$$P_\alpha = \frac{e^{-\beta f_\alpha}}{\sum_\gamma e^{-\beta f_\gamma}} \tag{2}$$

(β is the inverse of the temperature). We shall prove that, within the solution of the S.K. model proposed in [5], the f_α are independent random variables with an exponential distribution :

$$\mathcal{T}_\rho(f_\alpha) = \rho . \exp \rho(f_\alpha - f_c) . \theta(f_c - f_\alpha) , \tag{3}$$

ρ is a function of the temperature and the external field H and is equal to $\beta(1 - y)$ where y is the width of the right plateau in $q(x)$, $y = \overline{\sum_\alpha P_\alpha^2}$; f_c is a cut-off free energy needed at an intermediate stage. We consider a finite number M of pure states with the distribution $\mathcal{T}_\rho$. In the end we send the cut-off f_c and M to infinity, while keeping a density of levels at any finite free energy fixed :

$$M \, e^{-\rho f_c} = v , \quad \begin{matrix} M \to \infty \\ f_c \to \infty \end{matrix} . \tag{4}$$

All our results are v independent.

In order to prove equation (3), we have computed within this model the inclusive distributions of the probabilities $f^{(k)}$ defined in (1), and verified that the results agree with those of the replica method given in [3]. We shall hereafter sketch the computation of the average number of states of a given probability $f^{(1)}(P)$.

From the relation (2) between the probabilities and the free energies, the expression of the kth moment M_k of $f^{(1)}(P)$ is :

$$M_k = \overline{\sum_\alpha P_\alpha^k} = M \int \prod_{\alpha=1}^{M} (\mathcal{T}_\rho(f_\alpha) \, df_\alpha) \, e^{-\beta k f_1} \left[\sum_{\gamma=1}^{M} e^{-\beta f_\gamma} \right]^{-k} . \tag{5}$$

Using an integral representation of the last term :

$$Z^{-k} = \frac{1}{\Gamma(k)} \int_0^\infty d\lambda \, \lambda^{k-1} \, e^{-\lambda Z} \tag{6}$$

we obtain :

$$M_k = M \int_0^{\infty} \mathrm{d}\lambda \frac{1}{\Gamma(k)} \lambda^{k-1} g(\lambda, k) [g(\lambda, 0)]^{M-1}$$

$$g(\lambda, k) = \int \mathcal{T}_\rho(f) \, \mathrm{d}f \exp[-\beta k f - \lambda \, \mathrm{e}^{-\beta f}] . \tag{7}$$

For $k > 0$ one gets in the limit $f_c \to \infty$:

$$g(\lambda, k) = \frac{\rho}{\beta} \mathrm{e}^{-\rho L} \lambda^{-k+\rho/\beta} \Gamma(k - \rho/\beta) \tag{8}$$

while for $k = 0$, the limit must be taken more carefully and gives (for $\rho < \beta$) :

$$g(\lambda, 0)^M \simeq \exp(-v \lambda^{\rho/\beta} \Gamma(1 - \rho/\beta)) . \tag{9}$$

The value of M_k obtained from equations (7) to (9) is :

$$M_k = \Gamma(k - \rho/\beta) [\Gamma(k) \Gamma(1 - \rho/\beta)]^{-1} \tag{10}$$

which is exactly the kth moment of $f^{(1)}(P)$ given in [3] with the identification : $\rho/\beta = 1 - y$.

The general inclusive distribution $f^{(k)}$ can be computed in the same way. Alternatively one can use the following formula which we have obtained from equations (2) to (4) for the exclusive distribution f^{E} of all the probabilities :

$$f^{\mathrm{E}}(P_1, \dots, P_M) = \frac{1}{M} (1 - y)^{M-1} \delta\left(1 - \sum_{k=1}^{M} P_k\right) P_{\min}^{M(1-y)} \prod_{k=1}^{M} P_k^{-2+y} \tag{11}$$

where $P_{\min}$ is the smallest P_k.

Thus the distribution of probabilities and the fluctuations from sample to sample reflect a simple random process : in each sample the probabilities are chosen according to equations (2) to (4). The dominance of certain states is associated to the finite probabilities of having finite gaps between the lowest levels. Finally we mention that the exponential distribution and the independence of the free energies can be understood more clearly on the « simplest spin glass model » [6], since this model is thermodynamically equivalent to a random (free-) energy model [7]. This model can be studied directly and leads to (3) and (4) with $v = 1/\rho = 1/(2\sqrt{\ln 2})$.

We shall now turn to the generalization of these results for the distribution of clusters of states, and deduce from it the fluctuations of the order parameter function.

Because of ultrametricity one can group together states which have a mutual overlap larger than a given value q, and this defines a partition of the states into nonoverlapping clusters [3]. The weight W_I of a cluster I being defined as the sum of the probabilities of the states it contains, it was shown in [3] that the inclusive distribution of weights at any scale have the same form as the distribution of probabilities of the states, but for a change of the parameter $y = y(q_M)$ into $y(q)$. (The function $y(q)$ is related to the average order parameter function $\overline{P(q)}$ by $y(q) = \int_q^1 \overline{P(q')} \, \mathrm{d}q'$).

Hence the previous results also apply to the weights of the clusters. Defining for each cluster I at the scale q its generalized free energy f_I according to :

$$W_I = \frac{e^{-\beta f_I}}{\sum_{I'} e^{-\beta f_{I'}}} \tag{12}$$

the f_I are independent random variables with an exponential distribution as in equations (2) to (4), but with a value of ρ :

$$\rho(q) = \beta(1 - y(q)) . \tag{13}$$

Considering one cluster at the scale q_1 and its subclusters at the scale $q_2 > q_1$, we have found that the distribution of the free energies $f_1, ..., f_M$ of the M subclusters (M is finite if one imposes a cut-off f_c in free energy, and diverges as in (4) when $f_c \to \infty$) is :

$$\mathcal{T}_{q_1,q_2}(f_1, ..., f_M) = C\mathcal{T}_{\rho(q_2)}(f_1) \; ... \; \mathcal{T}_{\rho(q_2)}(f_M) \times [e^{-\beta f_1} + \cdots + e^{-\beta f_M}]^{\rho(q_1)/\beta} . \tag{14}$$

It is easy to generalize this equation to n clusters at scales $q_1, ..., q_n$.

These results allow for a clear understanding of some statements which appeared before in the literature :

— The universality property [3] : the whole dependence on the temperature, the magnetic field and the choice of the overlap is through a function $\rho_{T,H}(q)$, which is related to the average order parameter through equation (13).

— The content of the PaT approximation [8] from this perspective is that $\rho_{T,H}$ for $H = 0$ does not depend on T. The validity at low temperatures, verified in [9], reflects the fact that $\rho_{T,0}(q)$ has a limit $\rho_{0,0}(q)$, and does not depart too quickly from it at small T.

— Following reference [10] one can consider the states at a given value of T, H and weight them with a factor different from the Boltzmann factor, of the form :

$$P_\alpha^{(u)} = \frac{e^{-\beta u f_\alpha}}{\sum_\gamma e^{-\beta u f_\gamma}} . \tag{15}$$

As $\rho_{T,H}(q)$ and u appear in the formulae only in the ratio $\rho_{T,H}/(\beta u)$, one clearly finds in this case a new average order parameter $y^{(u)}(q)$ related to the normal one by (see Eq. (13)) :

$$1 - y^{(u)}(q) = \frac{1 - y(q)}{u} , \tag{16}$$

in agreement with the result of reference [10]. This different weighting of the states is possible when $\rho_{T,H}(q) < \beta u$ for all q, so that necessarily $u > u_c = 1 - y(q)$, which explains the existence of the critical value of u states in [10], and gives its explicit value.

We now turn to the computation of the fluctuations of the order parameter function. Following [3], we define for each sample a function $P_J(q)$ which depends on the particular realization of the couplings, and the integrated probability :

$$Y_J(q) = \int_q^1 P_J(q') \, dq' . \tag{17}$$

Calling I the clusters at the scale q and W_I their weights, one has simply $Y_J(q) = \sum_I W_I^2$, whose probability distribution $\Pi(Y)$ can be inferred from the exclusive distribution f^E of the weights (11), with $\rho = \beta(1 - y(q))$.

We introduce the auxiliary function $\hat{f}^E(v, W_1, \dots, W_M)$ which differs from f^E only by the fact that the sum $\sum_{i=1}^{M} W_i$ is constrained to be equal to v rather than 1 and we define the characteristic function $F_v(z)$:

$$F_v(z) = \int dW_1 \dots dW_M \, \hat{f}^E(v, W_1, \dots, W_M) \exp\left(-z \sum_{k=1}^{M} W_k^2\right) \tag{18}$$

so that

$$F_1(z) = \int_0^1 \Pi(Y) \, e^{-zY} \, dY . \tag{19}$$

The scaling relation $F_v(z) = F_1(zv^2)/v$, $v > 0$ allows one to write :

$$\int_0^\infty \frac{dv}{v} e^{\varepsilon v} \frac{\partial}{\partial z} F_1(zv^2) = \int_0^\infty dv \, e^{\varepsilon v} \frac{\partial}{\partial z} F_v(z) \tag{20}$$

where $\varepsilon = \pm 1$ and the $\partial/\partial z$ has been introduced to insure the convergence of the integrals at $v = 0$. This gives implicitly the characteristic function F_1 since the right hand side of equation (20) can be computed, and is equal to $\frac{1}{y-1} \frac{\partial}{\partial z} \log(H(z))$, where :

$$H(z) = \int_0^\infty \frac{dP}{\Gamma(y)} P^{-1+y}(-\varepsilon + 2zP) \, e^{\varepsilon P - zP^2} . \tag{21}$$

We have used equations (20) and (21) in two different ways.

1) Taking $\varepsilon = -1$, integrating (20) for z between 0 and z_0, and expanding in powers of z_0, one finds the explicit formula for the moments of Π [11] :

$$\int_0^1 \Pi(Y) \, Y^p \, dY = \frac{1}{(y-1)} \frac{1}{(2p-1)!} \frac{\partial^p}{\partial t^p} \left[\text{Log} \sum_{k=0}^{\infty} \frac{t^k}{k!} \frac{\Gamma(2k-1+y)}{\Gamma(-1+y)} \right] \Bigg|_{t=0} \quad (p \geqslant 1) \tag{22}$$

which agrees with the diagrammatic expansion we gave in [3], and can be used to generate in a fast way a large number of moments.

2) Taking $\varepsilon = 1$ and changing variables in (20) from v to $v\sqrt{Y}$, one gets :

$$\int_0^\infty e^{-zv} g(v) \, dv = \frac{y}{z} \frac{D_{-1-y}(-1/\sqrt{2z})}{D_{+1-y}(-1/\sqrt{2z})} \tag{23}$$

with

$$g(v) = \int_0^1 \Pi(Y) \, e^{\sqrt{v/Y}} \, dY \tag{24}$$

and where D_x are parabolic cylindric functions [12].

The singularity of $\Pi(Y)$ at $Y = 0$ is given by the behaviour of $g(v)$ for $v \to \infty$ which is itself (from Eq. (23)) dominated by the largest number z_0 such that $-1/\sqrt{2z_0}$ is a zero of D_{1-y}. Thus we find that $\Pi(Y)$ has an essential singularity at the origin :

$$\Pi(Y) \overset{Y \to 0}{\sim} \exp - [1/(4\, z_0\, Y)] . \tag{25}$$

This knowledge, together with the known divergence in $(1 - Y)^{-y}$ at $Y = 1$, and the formula (22) used to generate many moments, should allow a safe reconstruction of $\Pi(Y)$.

To conclude, we have shown that within the replica symmetry breaking ansatz proposed in [5], the free energies of the pure equilibrium states in the spin glass phase are independent random variables with an exponential distribution. Similar results are obtained for the cluster distributions. The physical content of the ansatz has now been found. Ultrametricity plus the properties investigated in this paper are not only consequences of the form of the ansatz but are equivalent in the sense that if they are assumed the form of replica symmetry breaking is determined.

Note added : After this work was completed we learned that our colleagues B. Derrida and G. Toulouse have independently obtained some results similar to ours on the fluctuations of the order parameter function.

References

[1] Sherrington, D. and Kirkpatrick, S., *Phys. Rev. Lett.* **32** (1975) 1792.

[2] Parisi, G., *Phys. Rev. Lett.* **50** (1983) 1946.

[3] Mézard, M., Parisi, G., Sourlas, N., Toulouse, G., Virasoro, M., *Phys. Rev. Lett.* **52** (1984) 1156, and *J. Physique* **45** (1984) 843.

[4] Young, A. P., Bray, A. and Moore, M., *J. Phys. C* **17** (1984) L149.

[5] Parisi, G., *Phys. Rev. Lett.* **43** (1979) 1754, and *J. Phys. A* **13** (1980) L115, 1101, 1887.

[6] Gross, D. and Mézard, M., *Nucl. Phys. B* **240** [FS12] (1984) 431.

[7] Derrida, B., *Phys. Rev. Lett.* **45** (1980) 79, and *Phys. Rev. B* **24** (1981) 2613.

[8] Vannimenus, J., Toulouse, G. and Parisi, G., *J. Physique* **42** (1981) 565.

[9] Parga, N., Parisi, G. and Virasoro, M. A., *J. Physique Lett.* **45** (1984) L-1063.

[10] Bray, A., Moore, M. and Young, P., *J. Phys. C* **17** (1984) L155.

[11] This formula has been previously derived directly from the diagrammatic rules given in [3] by D. Gross (private communication).

[12] Gradshteyn, I. S. and Ryzhik, I. M., *Table of Integrals, Series, and Products* (Academic Press, New-York and London) 1965.

EUROPHYSICS LETTERS 15 January 1986
Europhys. Lett., 1 (2), pp. 77-82 (1986)

SK Model: The Replica Solution without Replicas.

M. MÉZARD (*), G. PARISI (**), and M. A. VIRASORO (***)
Dipartimento di fisica, Università di Roma I
piazzale A. Moro 2, 00185 Roma, Italia

(received 14 October, 1985; accepted 29 October 1985)

PACS. 75.50K – Amorphous magnetic material

Abstract. – We introduce a new method, which does not use replicas, from which we recover all the results of the replica symmetry-breaking solution of the Sherrington-Kirkpatrick model.

Since its introduction in the context of spin glasses by EDWARDS and ANDERSON [1], the replica method has been carried to a high degree of sophistication. A solution of the mean-field theory (the SK model [2]) with replica symmetry breaking (RSB) has been proposed [3], and its physical meaning has been fully elucidated recently [4-7]. Although this replica solution builds up a coherent picture and provides us with a powerful method for analysing the equilibrium properties, it is difficult to put it on precise mathematical grounds.

In this paper we introduce an alternative method which does not rely on the replica trick, but leads to the same solution. It can be viewed as an analytic ansatz to solve the mean-field equation of TAP [8]

The basic idea is to go from a SK model with N spins, Σ_N, to one with $N+1$ spins, Σ_{N+1}. We shall make some physical assumptions on the organization of the configurations of Σ_N, inspired from recent results on the meaning of the RSB Ansatz of ref. [3]: the ultrametric organization of the states [6] and the independent exponential distribution of their free energies [7]. Assuming these properties for Σ_N, we shall show that they hold for Σ_{N+1}, and deduce all the other results of the replica treatment: value of the free energy, distribution of the local magnetizations in each state, and shape of the order parameter function.

For completeness, we first briefly review these physical properties.

The spin glass phase is charcterized by an infinite number of equilibrium states α with corresponding free energies F_α and local magnetizations m_i^α [4, 5]. A natural measure of the

(*) On leave from Laboratoire de Physique Théorique de l'E.N.S., Paris.
(**) Università di Roma II, Tor Vergata, and INFN. Frascati.
(***) Also at INFN, Roma.

distance between two states α and β is their overlap

$$q^{\alpha\beta} = \frac{1}{N} \sum_i m_i^\alpha m_i^\beta ,$$

and the order parameter function is the probability distribution of these overlaps [4].

The space of equilibrium states is ultrametric [6]. This essentially means that the states are grouped into clusters: by choosing any scale of overlap q ($< q_M$), the space can be partitioned into nonoverlapping clusters such that two states in the same cluster have an overlap larger than q, while states in different clusters have an overlap smaller than q. The clusters at a scale $q' > q$ are subclusters of those at the scale q.

All the states have the same free energy per spin $F_\alpha/N = F_0$ to leading order in N, but the $O(1/N)$ corrections f_α vary from state to state and determine their probabilities

$$P_\alpha = \exp[-\beta f_\alpha]\left[\sum_\gamma \exp[-\beta f_\gamma]\right]^{-1}$$

(where β is the inverse temperature). The number of states at fixed free energy f is

$$\mathcal{N}_\rho(f - f_s) \sim \exp[\rho(f - f_s)] , \tag{1}$$

where f_s is a free-energy scale. Taking into account the fact that the states are grouped into clusters at the scale q, one finds that the distribution of the states inside the same cluster I is still given by (1), but the free-energy scale f_s^I depends on the cluster. The distribution of these f_s^I is the same as (1) but with a parameter $\rho' < \rho$ [1]. This structre is reminiscent of the generalized random energy model [9].

We can now proceed to analyse what happens when one adds a new spin σ_0 to the system Σ_N of N spins $\{\sigma_1, \ldots, \sigma_N\}$. For simplicity, we shall keep to the case where there is no external magnetic field. The spin σ_0 interacts with the N other ones through a set of couplings K_i which are independent random variables with $\overline{K_i} = 0$ and $\overline{K_i K_j} = 1/N \delta_{ij}$. As we do not change the coupling of Σ_N, they verify $\overline{J^2} = 1/N$ instead of $\overline{J^2} = 1/(N+1)$. The correct rescaling of $\overline{J^2}$ can be absorbed into a rescaling of the temperature, and the change of free energy ΔF we find is related to the free-energy density f and the energy density e through

$$\Delta F = e/2 + f. \tag{2}$$

Let us consider the first stage of symmetry breaking. We suppose that there exist M equilibrium states $\alpha = 1, \ldots, M$ of Σ_N, with

$$q_0 = \frac{1}{N} \sum_i m_i^\alpha m_i^\beta, \quad \alpha \neq \beta; \quad q_1 = \frac{1}{N} \sum_i (m_i^\alpha)^2 , \tag{3}$$

and their free-energy distribution is the $\mathcal{N}_\rho(f)$ of (1). The local field on site 0 in the state α, $h^\alpha = \sum_i k_i m_i^\alpha$, is a random variable which depends on the sample and on the state. The

---

(1) The alert reader might be worried about the compatibility of these formulae with formula (14) of ref. [7]. He should notice that all fomulae in [7] can be multiplied by an arbitrary function of $v = M \exp[-\rho f_c]$ without affecting their validity: this is a global change of energy scale which does not affect the distribution of probabilities. Furthermore formula (14) was at fixed probability of the cluster, while in the present work we fix its energy scale f_s. At f_s fixed, the probability still fluctuates. Taking into account these fluctuations, a detailed computation shows that both formulations are valid.

probability that, choosing a sample, the h^α take values $h^1, \ldots, h^M$ is

$$P(h_1, \ldots, h_M) = \int \prod_i \left(\mathrm{d}K_i \sqrt{\frac{N}{2\pi}} \exp\left[-\frac{N}{2} K_i^2 \right] \right) \prod_{\alpha=1}^{M} \delta(h^\alpha - \sum_i K_i m_i^\alpha); \tag{4}$$

introducing integral representations of the δ-functions, one gets

$$P(h_1, \ldots, h_M) = \int \frac{\mathrm{d}H}{\sqrt{2\pi q_0}} \exp\left[-\frac{H^2}{2q_0} \right] \prod_{\alpha=1}^{M} \left(\frac{\exp[-(h^\alpha - M)^2/2(q_1 - q_0)]}{\sqrt{2\pi(q_1 - q_0)}} \right). \tag{5}$$

Thus there is a common piece $H = (1/M) \sum_\alpha h^\alpha$ which depends on the sample (its distribution is a Gaussian of width $\overline{H^2} = q_0$), around this the h^α are uncorrelated Gaussian variables of width $\overline{(h^\alpha - H)^2} = q_1 - q_0$.

Each state α of Σ_N generates a state of Σ_{N+1} where the magnetization on the new site is $m_0^\alpha = \mathrm{tgh}\,(\beta h^\alpha)$, and the corresponding change of free energy Δf_α is the sum of three pieces:

— the energy of the spin 0 in its local cavity field:

$$\Delta F_1 = -m_0^\alpha h^\alpha = -h^\alpha \,\mathrm{tgh}\,\beta h^\alpha, \tag{6}$$

— the entropy of the new spin:

$$\Delta F_2 = -\frac{1}{\beta} [\ln(2\cosh\beta h^\alpha) - \beta h^\alpha \,\mathrm{tgh}\,\beta h^\alpha], \tag{7}$$

— the change of free energy ΔF_3 due to the rearrangement of the N spins in the presence of spin 0. We shall prove, later on, that

$$\Delta F_3 = -\frac{\beta}{2}(1 - q_M), \tag{8}$$

where q_M is the self-overlap of a state.

For a fixed sample the new distribution of free energies is still the exponential (1), but its scale has been shifted by a factor $\varphi(H, q_1 - q_0, \rho) - (\beta/2)(1 - q_1)$, where

$$\varphi(H, q, \rho) = -\frac{1}{\rho} \ln \int \frac{\mathrm{d}h}{\sqrt{2\pi q}} \exp\left[-\frac{h^2}{2q} \right] [2\cosh\beta h]^{\rho/\beta}, \tag{9}$$

performing the quenched averaged over H gives the average change of free energy:

$$\overline{\Delta F} = \int \frac{\mathrm{d}H}{\sqrt{2\pi q_0}} \exp\left[-\frac{H^2}{2q_0} \right] \varphi(h, q_1 - q_0, \rho) - \frac{\beta}{2}(1 - q_1). \tag{10}$$

The distribution of local field in the states at a fixed new free energy, normalized and averaged over the samples, is

$$P(h) = \int \frac{\mathrm{d}H}{\sqrt{2\pi q_0}} \exp\left[-\frac{H^2}{2q_0} \right] \frac{\exp[-(h - H)^2/2(q_1 - q_0)][2\cosh\beta h]^{\rho/\beta}}{\varphi(H, q_1 - q_0, \rho)}. \tag{11}$$

Finally, in order for the overlaps not to change at order $1/N$ when one adds σ_0, one must impose the consistency equations

$$\begin{cases} q_1 = \int \mathrm{d}h\, P(h)[\mathrm{tgh}\,\beta h]^2\,, \\ q_1 = \int \frac{\mathrm{d}H}{\sqrt{2\pi q_0}} \exp\left[-\frac{H^2}{2q_0}\right] \cdot \\ \qquad \cdot \left[\frac{(\int \mathrm{d}h/\sqrt{2\pi(q_1-q_0)})\exp[-(h-H)^2/2(q_1-q_0)][2\cosh\beta h]^{\rho/\beta}\,\mathrm{tgh}\,\beta h}{\varphi(H, q_1-q_0, \rho)}\right]^2 \end{cases} \tag{12}$$

Using the expression (2) of ΔF, one can check that (10)-(12) are exactly the results of the replica method with one level of RSB [3, 10, 11].

Let us sketch the essential points of the computation at the second level of symmetry breaking. The states are grouped into clusters. The energy scales of the clusters have the distribution $\mathcal{N}_{\rho_1}(f_s^{\mathrm{I}}-f)$, and the distribution of the free energies of the states inside the same cluster I is $\mathcal{N}_{\rho_2}(f-f_s^{\mathrm{I}})$. The self-overlap of a state is q_2, the overlap of two states in the same cluster is q_1, and the overlap of states in different clusters is q_0. The local fields h^α on the site 0 are found to depend on three pieces. A field H which depends only on the sample and is a Gaussian variable with zero average and variance q_0. A field H^{I} which depends on the cluster. For fixed H, the H^{I} are independent Gaussian variables with average H and variance q_1-q_0. A field h^α which depends on the state. For fixed H and H^{I}, the h^α of the various states which belong to the same cluster I are independent Gaussian variables with average H^{I} and variance q_2-q_1.

Adding the new site, one finds the same structure of independent free-energy scales, and independent free energies of the states inside the clusters. The free-energy scale of each cluster is changed as in (10) by a factor $\varphi(H^{\mathrm{I}}, q_2-q_1, \rho_2)-(\beta/2)(1-q_2)$, and the experimental distribution of the free-energy scales is shifted of a factor (for fixed H)

$$\Delta f_s(H) = -\frac{1}{\rho_1}\ln\int\frac{\mathrm{d}H^{\mathrm{I}}}{\sqrt{2\pi(q_1-q_0)}}\cdot \\ \cdot\exp\left[-\frac{(H^{\mathrm{I}}-H)^2}{2(q_1-q_0)}\right]\exp\left[-\rho_1\left[\varphi(H^{\mathrm{I}}, q_2-q_1, \rho_2)-\frac{\beta}{2}(1-q_2)\right]\right]. \tag{13}$$

Averaging finally over the sample-dependent common drift H gives

$$\overline{\Delta F} = \int\frac{\mathrm{d}H}{\sqrt{2\pi q_0}}\exp\left[-\frac{H^2}{2q_0}\right]\Delta f_s(H)\,. \tag{14}$$

This is exactly the result for $\overline{\Delta F}$ in the replica method with two RBS [3, 10, 11]. One can check again that the distribution of the local field and the coherence equations determining the q's are the same as those obtained with replicas. This discussion can be extended in a straightforward way to an arbitrary number of cluster hierarchies.

We still have to demonstrate formula (8) for the change of free energy due to the rearrangement of the N spins. An instructive proof follows from the generalization of our arguments at the level of the configurations.

In a given state α, at a given temperature, the set of relevant configurations has the following properties: the number of configurations of energy $\bar{E}+E$ ($\bar{E}$ is supposed to be of order N and E of order 1) is

$$\mathcal{N}(\bar{E}+E) \sim \exp[s(\bar{E})+\beta E] \tag{15}$$

and the mutual overlap of these configurations is the self-overlap of the state, $q^{\alpha\alpha} = q_M$.

Let us consider a set of M ($\gg 1$) such configurations. Each configuration $\mathscr{C}$ creates a local field $h^{\mathscr{C}}$ on the site 0. Following the same argument as in (5), one finds that, for fixed $\bar{h} = (1/M) \sum_{\mathscr{C}} h^{\mathscr{C}}$, the $h^{\mathscr{C}}$ are independent Gaussian random variables of width $1 - q_M$ and average $\bar{h}$. One has

$$\bar{h} = \sum_i k_i \left[\frac{1}{M} \sum_{\mathscr{C}} \sigma_i^{\mathscr{C}} \right] \sum_i K_i m_i^\alpha , \tag{16}$$

hence $\bar{h}$ is the average magnetic field h^α on site 0 in state α.

Adding the new spin, each configuration gives rise to two configurations with $\sigma_0 = \pm 1$ and energies $E^{\mathscr{C}} \mp h^{\mathscr{C}}$. We can safely assume that the distribution of energies and magnetic field in Σ_N are uncorrelated (in the expressions of the magnetic field on site 0 and of the energies the values of the spins are multiplied, respectively, by K_i and J_{ij} which are statistically independent). Then the distribution of energies (15) in Σ_{N+1} is multiplied by a factor

$$C = \int \frac{dh}{\sqrt{2\pi(1 - q_M)}} \exp\left[- \frac{(h - h^\alpha)^2}{2(1 - q_M)} \right] [\exp[\beta h] + \exp[-\beta h]] . \tag{17}$$

This change of normalization of the exponential has a double origin: a shift of entropy ΔS (shift of the vertical scale), and a shift of $\bar{E}$ into $\bar{E} + \Delta E$, which shifts the horizontal scale. The net change is $C = \exp[\Delta S - \beta \Delta E]$, which proves that the change of free energy is

$$\Delta F = -\frac{1}{\beta} \ln C = -\frac{1}{\beta} \ln[2 \cosh(\beta h_\alpha)] - \frac{\beta}{2}(1 - q_M) . \tag{18}$$

We have recovered all the results of the replica method on purely physical grounds. In fact we start from the same hypotheses as in the replica method (but instead of being hidden in the form of the RSB ansatz they are explicit), and obtain the same results. These results are thus clarified and can be written very easily. For instance the existence of a dip at the origin in the distribution of local fields at low temperature follows from the following fact: in Σ_N there is an exponentially large number of states at a distance h above the ground state, and hence a nonvanishing probability that one of them will be the ground state of Σ_{N+1}, with a large local field.

The resulting picture possesses remarkable properties. The distribution of relevant states inside a cluster is very similar to the distribution of relevant configurations inside one pure state (exponential increase of the number of states with the free energy, and uncorrelated local fields around the average one [11]).

REFERENCES

[1] S. F. EDWARDS and P. W. ANDERSON: *J. Phys. F*, **5**, 965 (1975).

[2] D. SHERRINGTON and S. KIRKPATRICK: *Phys. Rev. Lett.*, **32**, 1792 (1975); S. KIRKPATRICK and D. SHERRINGTON: *Phys. Rev. B*, **27**, 4384 (1978).

[3] G. PARISI: *Phys. Rev. Lett.*, **43**, 1754 (1979); *J. Phys. A*, **13**, L155, 1101, 1887 (1980).

[4] G. PARISI: *Phys. Rev. Lett.*, **50**, 1946 (1983).

[5] A. HOUGHTON, S. JAIN and A. P. YOUNG: *J. Phys. C*, **16** L375 (1983); C. DE DOMINICIS and A. P. YOUNG: *J. Phys. A*, **16**, 2063 (1983).

[6] M. MÉZARD, G. PARISI, N. SOURLAS, G. TOULOUSE and M. VIRASORO: *Phys. Rev. Lett*, 52, 1156 (1884); *J. Phys. (Paris)*, 45, 843 (1984).
[7] M. MÉZARD, G. PARISI and M. A. VIRASORO: *J. Phys. (Paris) Lett.*, 46, L217 (1985).
[8] D. J. THOULESS, P. W. ANDERSON and R. G. PALMER: *Phil. Mag.*, 35, 593 (1977).
[9] B. DERRIDA: *J. Phys. (Paris) Lett.*, 46, L401 (1985).
[10] J. R. L. DE ALMEIDA and E. J. LAGE: *J. Phys. C*, 16, 939 (1982).
[11] M. MÉZARD and M. A. VIRASORO: to appear in *J. Phys. (Paris)*.

PHYSICAL REVIEW B VOLUME 24, NUMBER 5 1 SEPTEMBER 1981

Random-energy model: An exactly solvable model of disordered systems

Bernard Derrida
Service de Physique Théorique, Centre d'Etudes Nucleaires de Saclay F-91190, Gif-sur-Yvette, France
(Received 2 February 1981)

A simple model of disordered systems—the random-energy model—is introduced and solved. This model is the limit of a family of disordered models, when the correlations between the energy levels become negligible. The model exhibits a phase transition and the low-temperature phase is completely frozen. The corrections to the thermodynamic limit are discussed in detail. The magnetic properties are studied, and a constant susceptibility is found at low temperature. The phase diagram in the presence of ferromagnetic pair interactions is described. Many results are qualitatively the same as those of the Sherrington-Kirkpatrick model. The problem of using the replica method is analyzed. Lastly, this random-energy model provides lower bounds for the ground-state energy of a large class of spin-glass models.

I. INTRODUCTION

Recently, many authors have studied the Sherrington-Kirkpatrick model[1] (the SK model). The first interest of the SK model was to try to understand what kind of mean-field theory[2] should be valid for spin-glass models.[3] The second important interest was to know why the replica method used by SK[1] was incorrect. The SK model is simple enough to allow the calculation of all the integer moments $\langle Z^n \rangle$ of the partition function in the thermodynamic limit. Because of the too rapid growth of these moments $\langle Z^n \rangle$ (see Appendix A), the continuation to noninteger values of n is not unique.[4] In order to justify why the continuation $n \rightarrow 0$ used by SK did not provide $\langle \ln Z \rangle$, a breaking of symmetry in the replica space was proposed.[5] Several works[6,7] followed this idea and tried to investigate whether calculations with a broken symmetry in the replica space could lead to the true expression of $\langle \ln Z \rangle$. At present the situation is not yet clear and no simple analytic solution of the SK model has yet been derived. Therefore, it is interesting to study the random-energy model[8] which is simpler but contains most of the difficulties encountered in the SK model.

The random-energy model (RE) describes a system whose energy levels are independent random variables. The model can be defined without specifying any microscopic Hamiltonian. However one can find a family of spin-glass models which generalize the SK model and give in a certain limit the RE model. Because of its simplicity, the moments $\langle Z^n \rangle$ and the average free energy $\langle \ln Z \rangle$ are given by very simple expressions. The first motivation to study this RE model was that the behavior of the $\langle Z^n \rangle$ is the same as in the SK model and that the impossibility of using replicas is even more evident here. Another motivation was to look at the approach to the thermodynamic limit. Thouless, Anderson, and Palmer[2] found that the transition temperature of the SK model could be seen in the $1/N$ correction of the free energy. Moreover the approach to the thermodynamic limit is always a difficulty of numerical simulations. So it was interesting to compare the finite-size corrections of the RE model with those which are conjectured for the SK model. The main result is that the system exhibits a phase transition and that in the low-temperature phase the system is completely frozen. Another conclusion is that the infinite temperature is an accumulation point of critical temperatures where the corrections to the thermodynamic limit change their behavior. So one has to be very careful in the high-temperature expansion for these random systems. The definition of the RE model can be extended to cases where a magnetic field is present or where there are ferromagnetic pair interactions in addition to the random interactions. The phase diagrams are qualitatively the same as in the SK model. Finally, the RE model leads me to consider an approximation to spin-glass models in any dimension and gives lower bounds for the ground-state energies.

The main purpose of this work is to present in a detailed way the derivation of the results announced in a previous Letter.[8] The paper is organized as follows: in Sec. II, the SK model is generalized and the RE model is obtained as a limit of a family of spin-glass models. In Sec. III the definition and a simple solution of the RE model are given. These two sections are just a recall of the arguments contained in Ref. 8. In Sec. IV, the moments $\langle Z^n \rangle$ are calculated and the problem of finding $\langle \ln Z \rangle$ using these moments is discussed. In Sec. V, the approach to the thermodynamic limit is studied. The replica method used for the RE model is described in Sec. VI and surprisingly an unstable saddle point gives the low-temperature free energy. In Secs. VII and VIII, the behavior of the RE model in a magnetic field and in the presence of ferromagnetic pair interactions is studied. In Sec. IX, a lower bound of the ground-state energy of some spin-glass models in finite dimensions is obtained. Lastly in Sec. X, an attempt of expanding spin-glass models around the RE model is presented.

II. RELATED MODELS

In this section, I consider a family of models which generalize the SK model and I explain how the RE model is related to these models. The SK model is defined by the Hamiltonian

$$\mathcal{H} = -\sum_{\langle ij \rangle} J_{ij}\sigma_i\sigma_j \;, \tag{1}$$

where the spins σ_i are Ising spins and where the J_{ij} are random-quenched variables with probability distribution

$$\rho(J_{ij}) = \left[\frac{N}{2\pi J^2}\right]^{1/2} \exp\left[-\frac{J_{ij}^2 N}{2J^2}\right] . \tag{2}$$

The model is infinite ranged because there is an interaction J_{ij} for any pair of spins in the system.

One can generalize the SK model by replacing the random pair interactions in Eq. (1) by random p-spin interactions. The Hamiltonian is then

$$\mathcal{H}_p = -\sum_{i_1,\ldots,i_p} A_{i_1,\ldots,i_p}\sigma_{i_1}\cdots\sigma_{i_p} \;. \tag{3}$$

Here again, the spins are Ising spins and there is an interaction $A_{i_1,\ldots,i_p}$ for any group of p spins in the model. As in the SK model, the probability distribution of $A_{i_1,\ldots,i_p}$ has to be scaled with N in order to ensure an extensive free energy

$$\rho(A_{i_1,\ldots,i_p}) = \left[\frac{N^{p-1}}{\pi J^2 p!}\right]^{1/2} \exp\left[-\frac{(A_{i_1,\ldots,i_p})^2 N^{p-1}}{J^2 p!}\right] . \tag{4}$$

The relation between all these models can be seen in the one-level probability distribution $P(E)$. For a given configuration of the spins $\{\sigma_i^{(1)}\}$, the energy E depends on the interactions $A_{i_1,\ldots,i_p}$ of the Hamiltonian $\mathcal{H}_p$. By definition $P(E)$ is the probability that this configuration has energy E. Using Eqs. (3) and (4), one finds that

$$P(E) \sim \exp\left[-\frac{E^2}{NJ^2}\right] . \tag{5}$$

This result justifies *a posteriori* the distribution chosen in Eq. (4) because all the Hamiltonians $\mathcal{H}_p$ give the same $P(E)$. One can notice that $P(E)$ does not depend on the configuration of spins $\{\sigma_i^{(1)}\}$. One can also define the probability distribution $P_{1,2}(E_1,E_2)$ as the probability that two given configurations of spins $\{\sigma_i^{(1)}\}$ and $\{\sigma_i^{(2)}\}$ have, respectively, energies E_1 and E_2. It turns out that this probability distribution depends only on the distance between the two configurations, namely, on the number Nx of identical spins in the two configurations. One finds

$$P_{1,2}(E_1,E_2) \sim \exp\left[-\frac{(E_1+E_2)^2}{2NJ^2[1+(2x-1)^p]} - \frac{(E_1-E_2)^2}{2NJ^2[1-(2x-1)^p]}\right] . \tag{6}$$

The parameter x is always between 0 and 1. Therefore, when p becomes large, one gets[8]

$$P_{1,2}(E_1,E_2) \sim P(E_1)P(E_2) \;. \tag{7}$$

One can consider also the probability distributions of three or more levels and again when p is large these probability distributions become factorized. So when p is large, the energy levels become independent random variables. The large-p limit must always be taken after the thermodynamic limit $N \to \infty$ and so p is much smaller than N. The RE model is defined as a system of 2^N independent random energy levels distributed according to $P(E)$ given by Eq. (5). One can notice that the SK model which corresponds to $p=2$ lies between the case $p=1$ which is a model of free spins in a random magnetic field and $p=\infty$ which is the RE model. So some of its properties should be well approximated by an interpolation between these two exactly solvable models.

III. DEFINITION OF THE RANDOM-ENERGY MODEL

The random-energy model is defined by the following three properties.

(i) The system has 2^N energy levels E_i.

(ii) These energy levels are distributed according to the probability distribution

$$P(E) = (N\pi J^2)^{-1/2}\exp(-E^2/NJ^2) \;. \tag{8}$$

(iii) The energy levels E_i are independent random variables.

The first two properties are actual features of some spin-glass models like the models defined by the Hamiltonians $\mathcal{H}_p$ in Eq. (3). The third property is specific to this model. It simplifies the model enough to allow us to solve it exactly. However it can be seen as a crude approximation to more realistic models since the correlations between the energy levels are ignored.

One system is given by the choice of the 2^N energy

levels E_i. For each system the partition function Z is

$$Z(\{E_i\}) = \sum_{i=1}^{2^N} \exp\left[-\frac{E_i}{T}\right] \quad (9)$$

and as usual for disordered systems, one wants to calculate the average free energy.

$$\begin{aligned} F &= -T\langle \ln Z\rangle \\ &= -T\int \prod_i [P(E_i)dE_i] \ln Z(\{E_i\}) \ . \end{aligned} \quad (10)$$

The calculation of this average is done in the Appendix B and the results are discussed in Sec. V. Here I just recall the simple argument of Ref. 8 because it will be useful in Secs. VII and VIII.

For one sample of the 2^N energy levels, one can define $n(E)$ the number of energy levels belonging to the interval $(E, E+dE)$. This number fluctuates from one sample to another. It can be calculated by

$$n(E) = \sum_{i=1}^{2^N} y_i \ , \quad (11)$$

where $y_i = 1$ if $E < E_i < E + dE$ and $y_i = 0$ otherwise. The average $\langle n(E)\rangle$ is easily obtained from the probability distribution of the energies E_i:

$$\langle n(E)\rangle = 2^N\langle y_i\rangle = 2^N \exp\left[-\frac{E^2}{NJ^2}\right] A\,dE \quad (12)$$

with $A = (N\pi J^2)^{-1/2}$. In order to have a well-defined energy in the thermodynamic limit dE has to be small enough ($dE \sim N^\alpha$ with $\alpha < 1$). On the other hand, if one wants $n(E)$ to be rather smooth, dE must be large enough. It turns out that one can choose $dE \sim N^\alpha$ for any $\alpha < 1$. Then from Eq. (12) it is clear that there exists a critical energy E_0 defined by

$$E_0/N = J(\ln 2)^{1/2} \ . \quad (13)$$

If $|E| < E_0$, the average number $\langle n(E)\rangle$ of levels is much larger than 1. Because of the statistical independence of the energy levels, the fluctuations of $n(E)$ around its average are of order $\langle n(E)\rangle^{1/2}$ and are small compared with $\langle n(E)\rangle$. Therefore

$$n(E) \sim \langle n(E)\rangle \ . \quad (14)$$

If $|E| > E_0$, the average number $\langle n(E)\rangle$ is much smaller than 1. So for almost all the systems $n(E) = 0$ and with a very small probability which vanishes exponentially with N, $n(E) \geq 1$.

It is then clear that in the thermodynamic limit, if $|E| < E_0$, the entropy $S(E)$ is given by

$$S(E) = N\left[\ln 2 - \left[\frac{E}{NJ}\right]^2\right] \quad (15)$$

and there is no energy level larger than E_0 in absolute value. From the function $S(E)$, one can calculate the average free energy F using the relation $dS/dE = 1/T$. One finds[8]

$$N^{-1}F = \begin{cases} -T\ln 2 - J^2/4T, & \text{if } T > T_c \quad (16) \\ -E_0/N = -J(\ln 2)^{1/2}, & \text{if } T < T_c \ , \quad (17) \end{cases}$$

where

$$T_c = J/2(\ln 2)^{1/2} \ . \quad (18)$$

Equation (16) follows directly from Eq. (15). When the temperature decreases until T_c, the energy E reaches $-E_0$. For lower temperature than T_c, the energy of the system remains E_0 because there is no level of lower energy.

The fact that the average free energy could be obtained by calculating the average entropy is not surprising. The entropy and the free energy are related by a linear relation and so their averages are related by the same linear relation.

IV. MOMENTS

The usual approach of disordered problems by replicas starts by calculating the moments $\langle Z^n\rangle$ of the partition function. Because of the statistical independence of the energies in the RE model, the expression of the moments $\langle Z^n\rangle$ is particularly simple. For the first three moments, one finds

$$\begin{aligned} \langle Z\rangle &= 2^N\langle e^{-E/T}\rangle = 2^N e^{N\lambda^2/4} \ , \\ \langle Z^2\rangle &= 2^N e^{N\lambda^2} + 2^N(2^N-1)e^{N\lambda^2/2} \ , \\ \langle Z^3\rangle &= 2^N e^{9N\lambda^2/4} + 3[2^N(2^N-1)]e^{5N\lambda^2/4} \\ &\quad + 2^N(2^N-1)(2^N-2)e^{3N\lambda^2/4} \ , \end{aligned} \quad (19)$$

where $\lambda^2 = J^2/T^2$.

To calculate any moment $\langle Z^n\rangle$, one can develop the nth power of Z and then average

$$Z^n = \sum_{\{p_i\}} \alpha(p_1, p_2, \ldots, p_{2^N}) \times \exp\left[-\frac{p_1E_1}{T} - \frac{p_2E_2}{T} - , \ldots, - \frac{p_{2^N}E_{2^N}}{T}\right] ,$$

where $\alpha(p_1, p_2, \ldots)$ is a combinatorial factor

$$\alpha(p_1, p_2, \ldots) = \frac{n!}{\prod_i (p_i!)}$$

and where the sum is performed over all the integer partitions of n:

$$p_i \geq 0, \quad \sum_i p_i = n \ .$$

After the average, one can group the terms which

have the same $\{p_i\}$ through a permutation of index. Each integer partition of n is characterized by the numbers ν_p of p_i which are equal to p. One finds at the end

$$\langle Z^n\rangle = \sum_{\{\nu\}} \frac{(2^N)!}{\prod_{p=0}^{n}(\nu_p)!}\,\frac{n!}{\prod_{p=0}^{n}(p!)^{\nu_p}} \times \exp\left[N\sum_{p=0}^{n} p^2\frac{\lambda^2}{4}\nu_p\right] . \qquad (20)$$

where the sum has to be done over all the different choices of the integers ν_p which verify:

$$\nu_p \geq 0 \; ,$$

$$\sum_{p=0}^{n} \nu_p = 2^N \; , \quad \sum_{p=1}^{n} p\nu_p = n \; . \qquad (21)$$

From the exact expression (20), one can derive the asymptotic behavior ($N \to \infty$) of all these moments. First it is easy to replace each term in the sum (20) by its asymptotic behavior. Because of Eq. (21), all the ν_p for $p \geq 1$ are small and ν_0 is of order 2^N. One finds that

$$\frac{(2^N)!}{\nu_0!} \sim 2^{N\sum_{p=1}^{n}\nu_p}$$

and so

$$\langle Z^n\rangle \sim \sum_{\{\nu\}} \frac{n!}{\prod_{p=1}^{n}(\nu_p)!\,\prod_{p=1}^{n}(p!)^{\nu_p}} \times \exp N\left[\sum_{p=1}^{n} \nu_p\left(\ln 2 + p^2\frac{\lambda^2}{4}\right)\right] . \qquad (22)$$

Now we can see what is the dominant term in the sum (22). Because of the constraint (21), one finds that (i) if $n\lambda^2 > 4\ln 2$, the set of ν_p which gives the dominant contribution is $\nu_n = 1$ and $\nu_p = 0$ for $1 \leq p \leq n-1$; and (ii) if $n\lambda^2 < 4\ln 2$, the best choice is $\nu_1 = n$ and $\nu_p = 0$ for $p \geq 2$. Therefore for each moment $\langle Z^n\rangle$ there is a critical temperature

$$T_n = \sqrt{n}\,\frac{J}{2\sqrt{\ln 2}} = \sqrt{n}\,T_c \qquad (23)$$

and the asymptotic behavior of $\langle Z^n\rangle$ is given by

$$\langle Z^n\rangle \sim \begin{cases} \exp Nn\left[\ln 2 + \dfrac{J^2}{4T^2}\right] = \langle Z\rangle^n, & \text{if } T > T_n \; , \quad (24) \\ \exp N\left[\ln 2 + n^2\dfrac{J^2}{4T^2}\right], & \text{if } T < T_n \; . \quad (25) \end{cases}$$

At this stage, it is interesting to make a few remarks.

First, all the integer moments $\langle Z^n\rangle$ for $n \geq 2$ have a transition temperature T_n (see Fig. 1). For the SK model the $\langle Z^n\rangle$ have also a transition temperature.[9] The behavior of these moments are the same in the RE model and in the SK model. For the SK model, Eq. (24) is true whereas Eq. (23) is only valid for large n and Eq. (25) is valid at very low temperature.

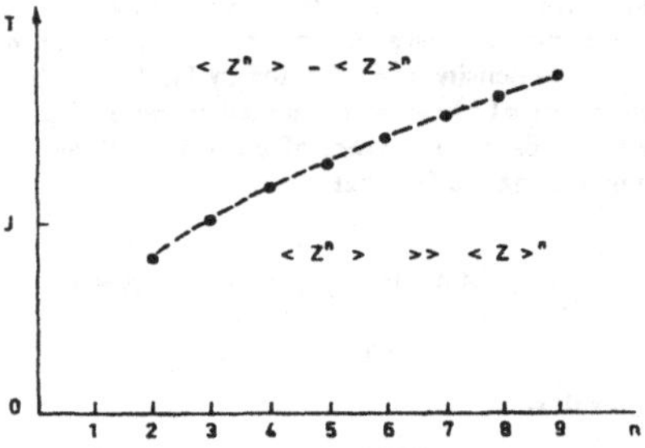

FIG. 1. The critical temperatures $T_n = \sqrt{n}\,T_c$ of the moments $\langle Z^n\rangle$ of the partition function. In the high-temperature region $T > T_n$, $\langle Z^n\rangle \sim \langle Z\rangle^n$. In the low-temperature region $T < T_n$, $\langle Z^n\rangle$ is much larger than $\langle Z\rangle^n$.

The simplest way to compute the average free energy would be to assume that the partition function Z has small fluctuations around its average $\langle Z\rangle$. At infinite temperature, there is no fluctuation at all ($Z = 2^N$) and it seems reasonable that at high temperature, the probability distribution of Z is concentrated around $\langle Z\rangle$. Then $\langle \ln Z\rangle$ could be obtained by

$$\langle \ln Z\rangle = \ln\langle Z\rangle + \left\langle \ln\left[1 + \frac{Z}{\langle Z\rangle} - 1\right]\right\rangle$$

$$= \ln\langle Z\rangle + \sum_{n=2}^{\infty} \frac{(-)^{n+1}}{n}\left\langle\left[\frac{Z}{\langle Z\rangle} - 1\right]^n\right\rangle . \qquad (26)$$

From Eq. (25), it is clear that even at very high temperature, if n is large enough ($n > 4T^2\ln 2/J^2$), $\langle Z^n\rangle$ is given by its low-temperature expression and one has

$$\langle Z^n\rangle \gg \langle Z\rangle^n \; .$$

This proves that Eq. (26) is a divergent series at any temperature except $T = \infty$ and cannot be used to calculate $\langle \ln Z\rangle$.

Expressions (24) and (25) give the behavior of all the moments $\langle Z^n\rangle$. At any finite temperature, these moments for large n behave like the exponential of n^2. It is known that when the growth of the moments is too rapid, there are many distributions which have the same moments.[10] An example where the moments increase in the same way as the moments $\langle Z^n\rangle$ is given in the Appendix A. This example shows that the exact knowledge of the moments is not sufficient in this case to calculate $\langle \ln Z \rangle$.

In the microcanonical ensemble, the quantity of interest is the density $n(E)$ defined by Eq. (11). The moments of $n(E)$ can be calculated as those of Z. One develops the qth power of Eq. (11) and one averages using the fact that

$$\langle (y_i)^{p_i} \rangle = \begin{cases} \int A \exp\left(-\frac{E^2}{NJ^2}\right) dE = \langle y \rangle , & p_i \geq 1 , \\ 1 , & p_i = 0 . \end{cases}$$

The result is

$$\langle [n(E)]^q \rangle = \sum_{\{\nu\}} \frac{(2^N)!}{\prod_{p=0}^{q} (\nu_p)!} \frac{q!}{\prod_{p=0}^{q} (p!)^{\nu_p}} (\langle y \rangle)^{\sum_{p=1}^{q} \nu_p} , \tag{27}$$

where the integers ν_p verify Eq. (21). In the thermodynamic limit, Eq. (27) becomes simpler:

$$\langle [n(E)]^q \rangle \sim \sum_{\{\nu\}} \frac{q!}{\prod_{p=1}^{q} (\nu_p)! \prod_{p=1}^{q} (p!)^{\nu_p}} \langle n(E) \rangle^{\sum_{p=1}^{q} \nu_p} . \tag{28}$$

By looking at the dominant term in Eq. (28), one finds that the energy $E_0 = J\sqrt{\ln 2}$ is a critical energy for all these moments.

If $|E| > E_0$, the dominant contribution comes from $\nu_q = 1$ and $\nu_p = 0$ (for $1 \leq p \leq q-1$), therefore

$$\langle [n(E)]^q \rangle \sim \langle n(E) \rangle \tag{29}$$

if $|E| < E_0$, the choice is $\nu_1 = q$ and $\nu_p = 0$ (for $p \geq 2$) and so

$$\langle [n(E)]^q \rangle \sim \langle n(E) \rangle^q . \tag{30}$$

So, for the RE model, the moments of the density $n(E)$ have a simpler behavior than those of Z. The fact that all the moments of $n(E)$ have the same critical energy indicates that this energy is the critical energy of the quenched system. The growth of the moments $\langle [n(E)]^q \rangle$ is at most an exponential of q because the maximum value of $n(E)$ is always less than 2^N. So there is in principle no problem about the unicity of the distribution of $n(E)$, once we know its moments.

V. FINITE-SIZE CORRECTIONS

The direct calculation of the average free energy is given in the Appendix B. The purpose of this section is only to discuss the results.

The first result is that there is a transition at temperature $T_c = J/2\sqrt{\ln 2}$. Of course, one recovers by averaging $\ln Z$, the extensive part of the free energy given in Eqs. (16) and (17).

At low temperature, the average free energy is

$$\frac{1}{N}\langle F \rangle = -J\sqrt{\ln 2} + \frac{T_c}{2}\frac{\ln N}{N} + \frac{T_c}{2N}\ln(4\pi \ln 2) + (T_c - T)\frac{1}{N}\Gamma'(1) - \frac{T_c}{N}\ln\left[\Gamma\left(1-\frac{T}{T_c}\right)\right] + O\left(\left(\frac{\ln N}{N}\right)^2\right) . \tag{31}$$

In the thermodynamic limit, the specific heat C per spin vanishes below T_c (Fig. 2). This means that the system is completely frozen in the whole low-temperature phase. From Eq. (31), one can compute the $1/N$ corrections to C:

$$C = \frac{1}{N}\frac{T}{T_c}\left[\frac{\Gamma''\left(1-\frac{T}{T_c}\right)}{\Gamma\left(1-\frac{T}{T_c}\right)} - \left(\frac{\Gamma'\left(1-\frac{T}{T_c}\right)}{\Gamma\left(1-\frac{T}{T_c}\right)}\right)^2\right] . \tag{32}$$

This shows that numerical studies which are always done on finite systems would encounter difficulties in predicting a zero specific heat because of the slow convergence ($1/N$) to the thermodynamic limit. In the low-temperature phase, one finds also that the $1/N$ correction is singular at T_c because of the presence of $\ln[\Gamma(1-T/T_c)]$ in the expression (31).

The average ground-state energy E_{GS} is

$$-\frac{1}{N}\langle E_{GS} \rangle = +J\sqrt{\ln 2} - \frac{T_c}{2}\frac{\ln N}{N} - \frac{T_c}{2N}\ln(4\pi \ln 2) - \frac{T_c}{N}\Gamma'(1) + O\left(\frac{1}{N}\right) . \tag{33}$$

The corrections are of order $N^{-1}\ln N$ and are comparable to the N^{-1} (Ref. 1) and the $N^{-1/2}$ (Ref. 11) corrections conjectured for the SK model.

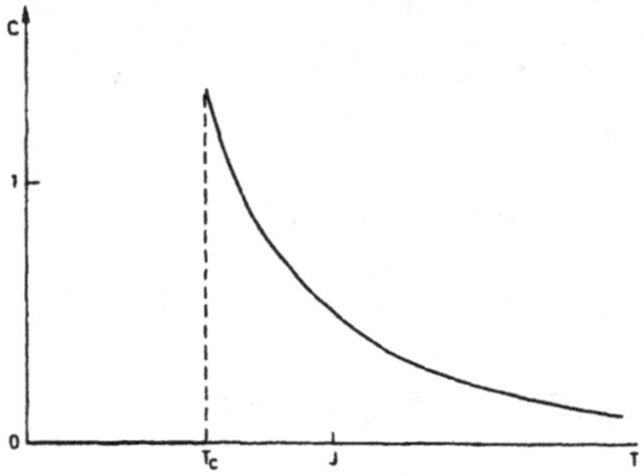

FIG. 2. The specific heat of the random-energy model as a function of temperature.

The ground-state energy depends on the choice of the 2^N energy levels

$$E_{GS} = \min_i (E_1, E_2, \ldots, E_{2^N}) .$$

By calculations similar to those given in the Appendix B, one can find the fluctuations of this ground-state energy

$$\langle E_{GS}^2 \rangle - \langle E_{GS} \rangle^2 = \frac{\pi^2 J^2}{24 \ln 2} + O(1) . \quad (34)$$

These fluctuations of the ground-state energy of the whole system are of order unity and so are rather small. For the random-energy model, the difficulty of finding the ground-state energy by studying finite samples comes more from the slow convergence of the average than from the fluctuations of the ground-state energy.

In the high-temperature phase, the convergence to the thermodynamic limit is more rapid because the corrections decrease exponentially with N. One can notice that at all the temperatures $T_n'(T_n' = \sqrt{2n+1}\, T_c,\ n = 1, 2, \ldots,)$ these corrections are singular. These temperatures are probably the temperatures where the number of terms that one can use in the asymptotic expansion (26) to calculate $\langle \ln Z \rangle$ changes.

VI. REPLICA METHOD

The replica method consists in calculating $\langle \ln Z \rangle$ by the formula

$$\langle \ln Z \rangle = \lim_{n \to 0} \frac{\langle Z^n \rangle - 1}{n} . \quad (35)$$

We have seen in Sec. IV [Eq. (22)] that $\langle Z^n \rangle$ could be written as

$$\langle Z^n \rangle \sim \sum_{\{\nu\}} \frac{n!}{\prod_{p=1}^{n} (\nu_p)! \prod_{p=1}^{n} (p!)^{\nu_p}} \times \exp\left[N \sum_{p=1}^{n} \nu_p \left[\ln 2 + p^2 \frac{\lambda^2}{4} \right] \right] .$$

At high temperature, the dominant term in the sum was $\nu_1 = n$ and $\nu_p = 0$ for $2 \leq p \leq n$. This leads to the expression (24) for $\langle Z^n \rangle$. Then by using Eq. (35) one finds

$$\langle \ln Z \rangle = N \left[\ln 2 + \frac{J^2}{4T^2} \right] ,$$

which is the good expression in the high-temperature phase. At low temperature, the dominant term corresponds to $\nu_n = 1$ and $\nu_p = 0$ for $1 \leq p \leq n-1$. One finds that $\langle Z^n \rangle$ is given by Eq. (25)

$$\langle Z^n \rangle \sim \exp N \left[\ln 2 + n^2 \frac{J^2}{4T^2} \right] .$$

It is then impossible to use Eq. (35) because the $n \to 0$ limit of $\langle Z^n \rangle$ is not 1.

Surprisingly it is possible to recover the low-temperature expression of $\langle \ln Z \rangle$ by looking at another extremum in the sum (22) which is in fact a minimum. The sum (22) has to be done over all the choices of integers ν such that

$$\sum_{p=1}^{n} p \nu_p = n .$$

If one forgets the fact that the p and the ν_p are integer one can find using a Lagrange parameter that any extremum corresponds to all the $\nu_p = 0$ except one.

This nonzero ν_p has to be equal to n/p. We can now find the value of p for which $(n/p)(\ln 2 + p^2 \lambda^2 / 4)$ is extremum. The result is

$$p = \frac{2\sqrt{\ln 2}}{\lambda} = \frac{T}{T_c} , \quad \nu_p = n \frac{T_c}{T} . \quad (36)$$

The corresponding contribution in sum (22) is

$$\exp N n \frac{J}{T} \sqrt{\ln 2} . \quad (37)$$

So by applying the replica formula (35) to the expression (37), one recovers the true expression (17) of $\langle \ln Z \rangle$ below T_c.

This calculation needs two comments: First, the term (37) is not present in the sum (22) below T_c. In Eq. (22) the p and the ν_p are integers verifying $1 \leq p \leq n$ and $0 \leq \nu_p \leq n$ and obviously the choice [Eq. (36)] does not satisfy these conditions. Secondly, even if Eq. (37) were present in Eq. (22), it would be negligible compared with the dominant contribution found in Eq. (25). So for this RE model, a blind calculation where all the constraints are forgotten can lead to the good expression of $\langle \ln Z \rangle$. It would be interesting to justify this calculation and to find an argument which shows that one has to take the $n \to 0$ limit of Eq. (37) below T_c.

The replica method can also be applied in the microcanonical ensemble. It becomes

$$\langle S(E) \rangle = \lim_{q \to 0} \frac{\langle [n(E)]^q \rangle - 1}{q} , \quad (38)$$

where $\langle S(E) \rangle$ is the average entropy at energy E and $n(E)$ is the density defined in Sec. III. In Eqs. (29) and (30) we have found the behavior of all the moments $\langle [n(E)]^q \rangle$.

If $|E| > E_0$, then $\langle n(E) \rangle$ is very small compared with 1 and one has $\langle [n(E)]^q \rangle \sim \langle n(E) \rangle$. By applying Eq. (38), one finds

$$\langle S(E)\rangle = \lim_{q\to 0}\frac{\langle n(E)\rangle - 1}{q} = -\infty$$

so, we find again that there is no level at these energies.

If $|E| < E_0$, then $\langle [n(E)]^q\rangle \sim \langle n(E)\rangle^q$ and Eq. (38) gives

$$\langle S(E)\rangle = \lim_{q\to 0}\frac{\langle n(E)\rangle^q - 1}{q} = \ln[\langle n(E)\rangle] ,$$

which was also established in Sec. III.

We can conclude that although the replica method was rather unsatisfactory for the partition function, it works very well here in the microcanonical ensemble.

VII. EFFECT OF A UNIFORM MAGNETIC FIELD

The definition of the random-energy model given at the beginning of Sec. III is not sufficient to describe its magnetic properties. To define the model in a uniform magnetic field, one has to come back to the Hamiltonians $\mathcal{H}_p$ of Sec. II. In presence of a field H, these Hamiltonians become

$$\mathcal{H}_p' = \mathcal{H}_p - H\sum_{i=1}^{N}\sigma_i . \tag{39}$$

One can calculate again the probability distributions $P(E)$ and $P(E_1,E_2)$. As before, $P(E_1,E_2)$ is factorized and the energy becomes independent when $p\to\infty$. The difference is that now, $P(E)$ depends on the magnetization M of the configuration of spins

$$P(E) = (N\pi J^2)^{-1/2}\exp[-(E+MH)^2/NJ^2] . \tag{40}$$

So the random-energy model is a system of 2^N independent random-energy levels among which $\binom{N}{(N+M)/2}$ have a magnetization M and are distributed according to Eq. (40).

Like in Sec. III, we can calculate the average of the level density $n(E)$:

$$\langle n(E)\rangle = \sum_{M=-N}^{+N}\binom{N}{(N+M)/2}\frac{1}{(\pi NJ^2)^{1/2}} \times\exp\left[-\frac{(E+MH)^2}{NJ^2}\right]dE , \tag{41}$$

where the sum is done over all the possible magnetizations of the system: $M=-N+2p$ with $0\le p\le N$.

In the thermodynamic limit $N\to\infty$, the behavior of $\langle n(E)\rangle$ can be obtained by looking for the dominant term in the sum (41). One finds

$$\frac{1}{N}\ln\langle n(E)\rangle = \max_{-1\le m\le 1}\left[-\left(\frac{1+m}{2}\right)\ln\left(\frac{1+m}{2}\right)-\left(\frac{1-m}{2}\right)\ln\left(\frac{1-m}{2}\right)-(\epsilon+mh)^2\right] , \tag{42}$$

where $\epsilon = E/NJ$, $m=M/N$, and $h=H/J$.

The value of m which gives the maximum is the solution of

$$m=-\tanh[2h(\epsilon+mh)] . \tag{43}$$

Because of the statistical independence of the energy levels, we can use again the argument of Sec. III.

If $(1/N)\ln\langle n(E)\rangle$ is positive, the average number of levels at energy E is very large. The fluctuations of $n(E)$ are small compared with $\langle n(E)\rangle$ and then the average entropy is

$$S(E)=\ln\langle n(E)\rangle . \tag{44}$$

If $(1/N)\ln\langle n(E)\rangle$ is negative, the average $\langle n(E)\rangle$ is much smaller than 1. So with probability 1, there is no level at this energy.

So in a uniform magnetic field H, there is a critical energy $E_0(H)$. If $|E| < E_0(H)$, the entropy is positive and given by Eq. (44) whereas there is no energy level at energies E larger than $E_0(H)$ in absolute value. Using the relation $(1/T)=dS/dE$, one can find the temperature dependence of the physical quantities.

In the high-temperature phase $T > T_c(H)$ the magnetization, the energy and the entropy are given by

$$m=\tanh\frac{H}{T} ,$$
$$\frac{E}{N}=-mH-\frac{J^2}{2T} , \tag{45}$$
$$\frac{S}{N}=\ln 2-\frac{m}{2}\ln\left(\frac{1+m}{1-m}\right)-\tfrac{1}{2}\ln(1-m^2)-\frac{J^2}{4T^2} ,$$

and the critical temperature $T_c(H)$ is the temperature where the entropy vanishes (see Fig. 3).

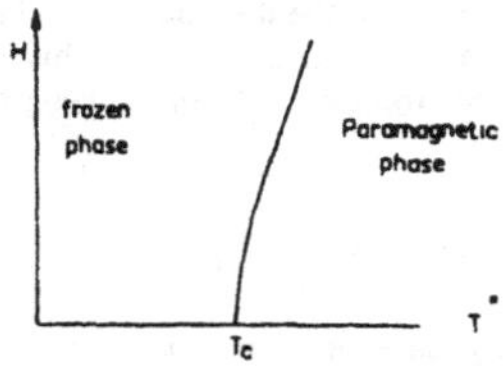

FIG. 3. The phase diagram of the random-energy model in a uniform magnetic field.

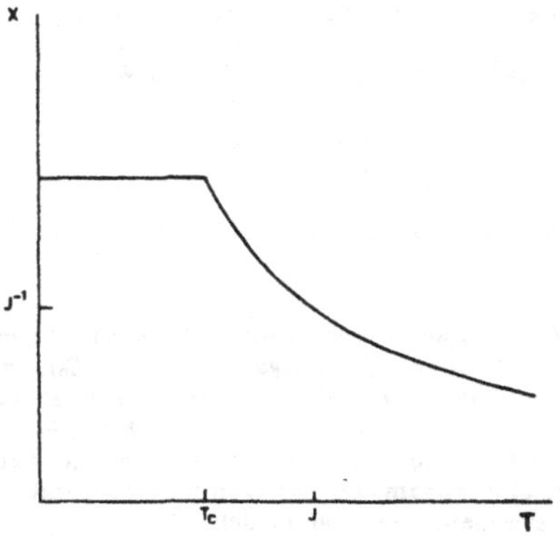

FIG. 4. The magnetic susceptibility in zero magnetic field of the random-energy model.

Below $T_c(H)$, the system is completely frozen in its ground state: for $T < T_c(H)$

$$m = \tanh\left(\frac{H}{T_c(H)}\right) ,$$
$$\frac{E}{N} = -mH - \frac{J^2}{2T_c(H)} , \quad (46)$$
$$\frac{S}{N} = 0 .$$

This gives for the magnetic susceptibility:

$$\chi = \begin{cases} \frac{1}{T}, & \text{for } T > T_c(H) \\ \frac{1}{T_c(H)}, & \text{for } T < T_c(H) . \end{cases} \quad (47)$$

The susceptibility is constant (Fig. 4) in the whole low-temperature phase. A similar result was also predicted by Parisi[12] for the SK model. We can also notice that the magnetization (46) in the low-temperature phase depends only on the magnetic field and not on the temperature. A similar behavior was also proposed as a hypothesis for the SK model[13] and seems to be a good approximation.[14] The important point is that in spite of its simplicity, the RE model gives for the magnetic susceptibility a cusp at the transition and a constant susceptibility at low temperature.

VIII. EFFECT OF FERROMAGNETIC PAIR INTERACTIONS

It is also interesting to study the case where in addition to the random interactions, there are ferromagnetic pair interactions J_0. The Hamiltonians $\mathcal{H}_p$ become

$$\mathcal{H}_p'' = \mathcal{H}_p - \frac{J_0}{N} \sum_{\langle ij \rangle} \sigma_i \sigma_j . \quad (48)$$

Like in the SK model, we have here an interaction J_0/N for any pair of spins in the system. The presence of N^{-1} in the interaction is necessary to ensure an extensive free energy. Here again in the limit $p \to \infty$, the energy levels become independent random variables. The RE model becomes a system of 2^N energy levels among which $\binom{N}{(N+M)/2}$ have a magnetization M and are distributed according to

$$P(E) \sim \exp\left[-\left(E + \frac{M^2 J_0}{2N}\right)^2 \Big/ NJ^2\right] . \quad (49)$$

To obtain the temperature dependence of the free energy F and the magnetization m, the procedure is exactly the same as in Sec. VII. The results are summarized (in Fig 5).

One finds four phases: (i) in the paramagnetic phase

$$m = 0$$
$$F/N = -T\ln 2 - \frac{J^2}{4T} ;$$

(ii) in the ferromagnetic phase

$$m = \tanh\left(\frac{J_0 m}{T}\right)$$
$$F/N = -T\ln 2 - \frac{J^2}{4T} + \frac{J_0 m^2}{2} + \frac{T}{2}\ln(1-m^2) ;$$

(iii) in the frozen phase I

$$m = 0$$
$$F/N = -J\sqrt{\ln 2} ;$$

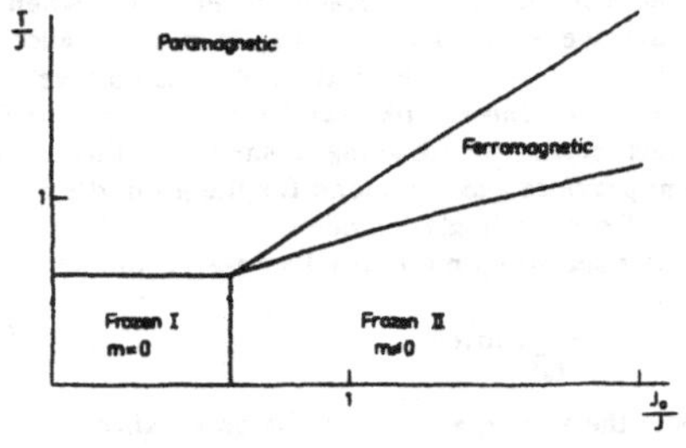

FIG. 5. The phase diagram of the random-energy model in the presence of ferromagnetic pair interactions. At low temperature, the magnetization m in the frozen phase is zero if $J_0 < J/2\sqrt{\ln 2}$ and is a function of J_0 if $J_0 > J/2\sqrt{\ln 2}$.

and (iv) in the frozen phase II the system is completely frozen

$$m = m(T_c(J_0))$$
$$F/N = F(T_c(J_0))/N ,$$

where $T_c(J_0)$ is the transition temperature from the ferromagnetic phase to the frozen phase II.

The lines of transition are between the paramagnetic phase and the frozen phase I $T_c = J/2\sqrt{\ln 2}$; between the paramagnetic phase and the ferromagnetic phase $T_c = J_0$; between the ferromagnetic phase and the frozen phase II. The transition temperature $T_c(J_0)$ is the temperature where the entropy of the ferromagnetic phase vanishes:

$$\frac{1}{N}S(T_c(J_0)) = \ln 2 - \frac{J_0 m^2}{T_c(J_0)} - \frac{1}{2}\ln(1-m^2) - \frac{J^2}{4T_c^2(J_0)} = 0 ,$$

where m is solution of

$$m = \tanh\left[\frac{J_0 m}{T_c(J_0)}\right]$$

between the frozen phase I and the frozen phase II

$$J_0 = J/(2\sqrt{\ln 2}) .$$

The phase diagram found here is qualitatively the same as the one predicted for the SK model.[1] The properties of the RE model in presence of ferromagnetic pair interactions are closely related to its properties in a uniform magnetic field.[15] So all the results of this section could have been obtained directly from those of Sec. VII.

IX. INDEPENDENT ENERGY APPROXIMATION

The main idea of the present paper is that when the average level density $\langle n(E)\rangle$ is much smaller than 1, then, with probability 1, there is no level at this energy. *This remains true for any spin-glass model.* In this section, we are going to show that this argument provides a lower bound for the ground-state energy of some spin-glass models.

Consider now a more usual spin-glass model

$$\mathcal{H} = -\sum_{\langle ij\rangle} J_{ij}\sigma_i\sigma_j , \tag{50}$$

where the spin σ_i are Ising spins on a lattice and the J_{ij} are random nearest-neighbor interactions distributed according to a probability distribution $\rho(J_{ij})$. When this distribution ρ is symmetric

$$\rho(J_{ij}) = \rho(-J_{ij}) \tag{51}$$

it is easy to calculate the average level density

$$\langle n(E)\rangle = 2^N \int \cdots \int \prod_{\langle ij\rangle} \rho(J_{ij})dJ_{ij} \times \delta\left[E - \sum_{\langle ij\rangle} J_{ij}\right] dE . \tag{52}$$

In Eq. (52), we have used the fact that

$$\rho(J_{ij}) = \rho(J_{ij}\sigma_i\sigma_j) .$$

If E_0 is the lowest energy where $N^{-1}\ln\langle n(E)\rangle$ vanishes [$N^{-1}\ln\langle n(E)\rangle$ is negative if $|E| > -E_0$], one knows that with probability 1, there is no level of energy E smaller than E_0. This implies that E_0 is a lower bound for the ground-state energy E_{GS}. Let us look at two examples: the symmetric $\pm J$ model and the symmetric Gaussian model.

For the symmetric $\pm J$ model, the density $\rho(J_{ij})$ is

$$\rho(J_{ij}) = \frac{1}{2}[\delta(J_{ij} - J) + \delta(J_{ij} + J)] . \tag{53}$$

In the thermodynamic limit, one finds

$$N^{-1}\ln\langle n(E)\rangle \sim \left\{\ln 2 + d\ln d - \frac{d}{2}\ln\left[d^2 - \left(\frac{E}{NJ}\right)^2\right] - \frac{E}{2NJ}\ln\left(\frac{dNJ+E}{dNJ-E}\right)\right\} , \tag{54}$$

where the lattice is a d-dimensional cubic lattice. In dimension 2,[16–18] the ground-state energy has been calculated by numerical studies $E_{GS}/NJ = -1.40 \pm 0.01$. The value of E_0 calculated from Eq. (54) is $E_0/NJ = -1.560$. In dimension 3, Kirkpatrick[17] estimated $E_{GS}/NJ = -1.75$ whereas Eq. (54) gives $E_0/NJ = -1.956$. In the high-dimension limit $d \to \infty$, one can also find the analytic expansion of E_0:

$$E_0 = -NJ\sqrt{2d\ln 2}\left[1 - \frac{\ln 2}{6d} + \cdots\right] . \tag{55}$$

For the symmetric Gaussian model, the distribution $\rho(J_{ij})$ is by definition

$$\rho(J_{ij}) = (2\pi J^2)^{1/2}\exp\left[-\left(\frac{J_{ij}^2}{2J^2}\right)\right] .$$

The expression of $N^{-1}\ln\langle n(E)\rangle$ is very simple:

$$N^{-1}\ln\langle n(E)\rangle \sim \left[\ln 2 - \frac{E^2}{2dN^2J^2}\right] .$$

Here again the spins are on a d-dimensional cubic lattice, one finds for E_0:

$$E_0/NJ = -\sqrt{2d\ln 2} . \tag{56}$$

In dimension 2, $E_0/NJ = -1.665$ whereas the result of numerical studies[18] for the ground-state energy is $E_{GS}/NJ = -1.31 \pm 0.01$. In infinite dimension, E_0 is the ground-state energy of the RE model $E_0/NJ = -\sqrt{\ln 2} = -0.8326$ and it is a lower bound for the ground-state energy of the SK model $E_{GS}/NJ \simeq -0.765$. It is also a lower bound for the ground-state energy of the models defined by the Hamiltonians $\mathcal{H}_p$.

So we see that for a large class of models, E_0 is very easy to calculate. E_0 provides a lower bound of the ground-state energy E_{GS} and is a rather good approximation of E_{GS}. For the RE model, it was possible to say that $n(E) \sim \langle n(E)\rangle$ when $\langle n(E)\rangle$ is large. This was justified by the statistical independence of the energy levels. *This cannot be generalized to other spin-glass models.* Because of the correlations between the energy levels, it is always possible that, when $\langle n(E)\rangle$ is large, the probability that $n(E)$ is not zero vanishes exponentially with N. For example, if $n(E)$ could only take two values:

$$n(E) = \begin{cases} 0, & \text{with probability } 1-e^{-bN} \\ e^{aN}, & \text{with probability } e^{-bN} \end{cases}$$

then $\langle n(E)\rangle = e^{(a-b)N}$ which can be very large. However in this example $n(E)=0$ with probability 1 in the thermodynamic limit. This is the reason why E_0 is only a lower bound of E_{GS}.

X. ATTEMPTS OF EXPANSION AROUND THE RE MODEL

At the end of this paper, the open question is to find how to take into account the correlations between the energy levels in realistic spin-glass models. The simplest approach is to study the Hamiltonians $\mathcal{H}_p$ defined by Eqs. (3) and (4) because these correlations are known, and to try a large p expansion. I could not find how to expand for large p the quantities of interest like the average free energy $\langle \ln Z\rangle$ or the average entropy $\langle \ln n(E)\rangle$. I could only expand the moments $\langle Z^n\rangle$ which are of course less interesting quantities. For the Hamiltonians $\mathcal{H}_p$, the calculations of the moments $\langle Z^n\rangle$ can be done using replicas. After the average, one finds

$$\langle Z^n\rangle = \sum_{\{\sigma_i^\alpha\}} \exp\left[\frac{J^2}{4T^2}\frac{p!}{N^{p-1}}\sum_{(i_1,\ldots,i_p)} \left[\sum_{\alpha=1}^{n} \sigma_{i_1}^\alpha \cdots \sigma_{i_p}^\alpha\right]^2\right]. \tag{57}$$

Equation (57) shows that the moments $\langle Z^n\rangle$ are very similar to the partition function of a lattice gauge theory.[19] There is a local symmetry in the effective Hamiltonian H:

$$H = -\frac{J^2}{4T^2}\frac{p!}{N^{p-1}}\sum_{(i_1,\ldots,i_p)}\left[\sum_{\alpha=1}^{n}\sigma_{i_1}^\alpha\cdots\sigma_{i_p}^\alpha\right]^2 .$$

If one changes the sign of all the spins of site i, H remains unchanged: $\{\sigma_i^\alpha\}\to\{\epsilon_i\sigma_i^\alpha\}$ with $\epsilon_i=\pm 1$ for any i. When n is even, there is another gauge symmetry $\{\sigma_i^\alpha\}\to\{\epsilon_\alpha\sigma_i^\alpha\}$ with $\epsilon_\alpha=\pm 1$ for any α. One can make a low-temperature expansion of Eq. (57) for any p:

$$\ln\langle Z^n\rangle \sim N\left[\ln 2 + n^2\frac{J^2}{4T^2} + g(n)\exp\left[-\frac{J^2}{T^2}(n-1)p\right] + \cdots\right], \tag{58}$$

where

$$g(n) = \begin{cases} n, & \text{for } n \geq 3 \\ 1, & \text{for } n = 2 \\ 0, & \text{for } n = 1 . \end{cases}$$

In the expansion (58), it is clear that all the Hamiltonians $\mathcal{H}_p$ give the same behavior of the moments $\langle Z^n\rangle$ at very low temperature. As the continuation of Eq. (25) could not provide $\langle \ln Z\rangle$, it seems difficult to extract the first correction of the free energy from Eq. (58). There is some hope to calculate also the moments $\langle [n(E)]^q\rangle$ for the Hamiltonians $\mathcal{H}_p$ but this requires more complicated calculations.

XI. CONCLUSION

The random-energy model is obviously an extreme simplification of spin-glass problems. Its properties are certainly different from those of real materials and it is hard to expect more than qualitative agreement with experimental data. From a theoretical point of view, it has the advantage to be exactly solvable. It shows the limitations of the replica method. The freezing of the system below its transition can be well understood by looking at the energy-level density. The phase diagrams of the RE model are qualitatively the same as in the SK model. The magnetic susceptibility is constant in the whole low-temperature phase. So the simplification of considering independent random energies has not suppressed the physical interest of the model.

The RE model provides also lower bounds for the ground-state energies of a large class of spin-glass models. It would be interesting to go further than the independent energy approximation by including the correlations between the energy levels. To do so, the simplest models seem to be the Hamiltonians $\mathcal{H}_p$ in the limit $p\to\infty$. It would also be interesting to understand the reason why a blind calculation by the replica method leads to the true free energy. Maybe this calculation has connections with the broken symmetry of replicas introduced by Parisi[7] for the SK model.

ACKNOWLEDGMENTS

I wish to thank the Group de Physique des Solides of the Ecole Normale Supérieure and the Service de Physique Théorique of the Centre d'Etudes Nucléaires Saclay, especially P. Moussa, Y. Pomeau, G. Toulouse, and J. Vannimenus for many stimulating discussions.

APPENDIX A

The question of calculating the average free energy $\langle \ln Z \rangle$ from the expression of the moments $\langle Z^n \rangle$ is not exactly a problem of analytic continuation. The fact that the $\langle Z^n \rangle$ are the moments of a distribution and not only numbers imposes constraints on the continuation to noninteger values of n.

In this appendix, I give an example which illustrates that when the moments $\langle x^n \rangle$ increase too rapidly, the measure is not determined uniquely. Note that the growth of the moments $(\exp \lambda n^2)$ is the same as for the moments $\langle Z^n \rangle$ of the partition function of the RE model. Consider the family of measures $d\mu(x)$ defined by

$$d\mu(x) = \frac{\sum_{n=-\infty}^{+\infty} b^{-n^2/2 - n\nu} \delta(x - b^{n+\nu})}{\sum_{n=-\infty}^{+\infty} b^{-n^2/2 - n\nu}} dx \ . \tag{A1}$$

These measures depend on two parameters (b and ν). Each value of $\nu(-\frac{1}{2} < \nu \leq \frac{1}{2})$ defines a new measure. It is easy to calculate the moments $\langle x^n \rangle$. One finds that for any integer n the moments $\langle x^n \rangle$ do not depend on ν:

$$\langle x^n \rangle = b^{n^2/2} \ . \tag{A2}$$

One can also calculate $\langle \ln x \rangle$

$$\langle \ln x \rangle \sim \nu \ln b, \quad \text{for large } b \ . \tag{A3}$$

So two different measures corresponding to different values of ν give different answers for $\langle \ln x \rangle$ even if all their moments are exactly the same.

APPENDIX B

In this appendix, we obtain the free energy of the random-energy model.

In order to calculate the average of $\ln Z$, one can use an integral representation of the logarithm:

$$\langle \ln Z \rangle = \int_0^\infty \frac{e^{-t} - \langle e^{-tZ} \rangle}{t} dt \ . \tag{B1}$$

After a partial integration, one finds

$$\langle \ln Z \rangle = \int_0^\infty \ln t \, e^{-t} dt - \int_0^\infty \ln t \, e^{-\phi} d\phi \ , \tag{B2}$$

where ϕ is defined by

$$e^{-\phi} = \langle e^{-tZ} \rangle \ . \tag{B3}$$

The only question, now, is to find an explicit expression of t as a function of ϕ, where t and ϕ are related by Eq. (B3).

As the partition function is

$$Z = \sum_{i=1}^{2^N} e^{-\beta E_i} \tag{B4}$$

and the energy levels are random variables with probability distribution

$$p(E_i) = \frac{1}{(\pi N J^2)^{1/2}} \exp\left[-\frac{E_i^2}{N J^2}\right]$$

one has

$$e^{-\phi} = \left[\frac{1}{(\pi N J^2)^{1/2}} \int_{-\infty}^{+\infty} dE \exp\left[-\frac{E^2}{N J^2} - t e^{-\beta E}\right]\right]^{2^N}$$

or

$$\exp\left[-\frac{\phi}{2^N}\right] = \frac{1}{\sqrt{\pi}} \int_{-\infty}^{+\infty} dy \exp(-y^2 - t e^{-\lambda y}) \ , \tag{B5}$$

with

$$\lambda = \sqrt{N} \beta J \ . \tag{B6}$$

Notice that the definition of λ in this appendix is not the same as in the body of the paper.

We are now going to obtain the asymptotic behavior (when λ is large) of the integral in the right-hand side of Eq. (B5).

If we call

$$f(t) = \frac{1}{\sqrt{\pi}} \int_{-\infty}^{+\infty} dy \exp(-y^2 - t e^{-\lambda y}) \tag{B7}$$

we shall prove that: if $-p\lambda^2/2 < \ln t < -(p-1)\lambda^2/2$, where p is an integer then

$$f(t) = 1 - t e^{\lambda^2/4} + \frac{t^2}{2!} e^{\lambda^2} + \cdots + (-)^{p-1} \frac{t^{p-1}}{(p-1)!} e^{(p-1)^2 \lambda^2/4} + \frac{1}{\sqrt{\pi}\lambda} e^{-\ln^2 t/\lambda^2} \left[\Gamma\left(\frac{2\ln t}{\lambda^2}\right) - \frac{1}{\lambda^2} \Gamma''\left(\frac{2\ln t}{\lambda^2}\right) + O\left(\frac{1}{\lambda^4}\right)\right] \ . \tag{B8}$$

Proof:

In Eq. (B7), we can make the change of variable:

$$te^{-\lambda y}=u \ .$$

It follows that:

$$f(t)=\frac{1}{\sqrt{\pi}\lambda}\exp\left[-\frac{\ln^2 t}{\lambda^2}\right]\int_0^\infty \exp\left[\left[\frac{2\ln t}{\lambda^2}-1\right]\ln u-\frac{\ln^2 u}{\lambda^2}-u\right]du \ . \tag{B9}$$

The convergence of the integral is ensured by the presence of the term $-\ln^2 u$ in the exponential. Nevertheless, when $\ln t$ is positive, this term is not necessary and one can expand $\exp(-\ln^2 u/\lambda^2)$ and obtain: If $\ln t>0$:

$$f(t)=\frac{1}{\sqrt{\pi}\lambda}e^{-\ln^2 t/\lambda^2}\left[\Gamma\left(\frac{2\ln t}{\lambda^2}\right)-\frac{1}{\lambda^2}\Gamma''\left(\frac{2\ln t}{\lambda^2}\right)+\cdots+\left(-\frac{1}{\lambda^2}\right)^p\frac{1}{p!}\Gamma^{(2p)}\left(\frac{2\ln t}{\lambda^2}\right)+O\left(\frac{1}{\lambda^{2p+2}}\right)\right] \ . \tag{B10}$$

This proves the formula (B8) when $p=0$.

Let us now consider the case $p\geqslant 1$. By deriving p times Eq. (B7) with respect to t, it is easy to prove that

$$f^{(p)}(t)=(-)^p e^{p^2\lambda^2/4}f(te^{p\lambda^2/2}) \ . \tag{B11}$$

If t is chosen such that: $-p\lambda^2/2<\ln t<-[(p-1)/2]\lambda^2$ then, $\ln(te^{p\lambda^2/2})$ is positive and one can replace $f(te^{p\lambda^2/2})$ in Eq. (B11) by its expansion Eq. (B10). So Eq. (B11) becomes

$$f^{(p)}(t)=(-)^p e^{p^2\lambda^2/4}\exp[-\ln^2(te^{p\lambda^2/2})/\lambda^2]\left[\Gamma\left(\frac{2\ln t}{\lambda^2}+p\right)-\frac{1}{\lambda^2}\Gamma''\left(\frac{2\ln t}{\lambda^2}+p\right)+O\left(\frac{1}{\lambda^4}\right)\right] \ . \tag{B12}$$

One can now find the general solution of the differential equation (B12).

$$f(t)=\alpha_0+\alpha_1 t+\cdots\alpha_{p-1}t^{p-1}+\frac{1}{\sqrt{\pi}\lambda}\exp\left[-\frac{\ln^2 t}{\lambda^2}\right]\left[\Gamma\left(\frac{2\ln t}{\lambda^2}\right)-\frac{1}{\lambda^2}\Gamma''\left(\frac{2\ln t}{\lambda^2}\right)+O\left(\frac{1}{\lambda^4}\right)\right] \ . \tag{B13}$$

where $\alpha_0,\alpha_1,\ldots,\alpha_{p-1}$ are *a priori* the unknown constants of integration of the differential equation.

We can determine these constants using the Taylor expansion of $f(t)$ around $t=0$. From Eqs. (B7) and (B11), one has

$$f^{(p)}(0)=(-)^p e^{p^2\lambda^2/4} \ .$$

So the Taylor expansion of $f(t)$ is

$$f(t)=1-te^{\lambda^2/4}+\cdots+(-)^{p-1}\frac{t^{p-1}}{(p-1)!}e^{(p-1)^2\lambda^2/4}+R_p(t) \ , \tag{B14}$$

where $R_p(t)=(t^p/p!)f^{(p)}(\theta t)$ with $0\leqslant\theta\leqslant 1$. One can easily find a majoration for this rest using the fact that

$$|f^{(p)}(\theta t)|\leqslant|f^{(p)}(0)|=e^{\lambda^2p^2/4} \ .$$

In Eq. (B14), the rest is smaller than any term of the sum when $t<e^{-p\lambda^2/2+\lambda^2/4}$. This means that if $\ln t/\lambda^2<-p/2+\frac{1}{4}$, then

$$f(t)=1-e^{\lambda^2/4}t+\cdots+(-)^{p-1}\frac{e^{\lambda^2(p-1)^2/4}}{(p-1)!}t^{p-1}+O(e^{\lambda^2p^2/4}t^p) \ . \tag{B15}$$

By comparing the two expansions (B13) and (B15) for $-p/2<\ln t/\lambda^2<-p/2+\frac{1}{4}$ one gets:

$$\alpha_i=\frac{f^{(i)}(0)}{i!}$$

and so the expansion (B8) is proved as the α_i are necessarily constant for $-p/2<\ln t/\lambda^2<(-p+1)/2$.

From Eqs. (B5) and (B8), we can now write t as an explicit function of ϕ:

$$\text{If } -\frac{\lambda^2}{2} < \ln t < 0 \tag{B16}$$

$$\exp\left[-\frac{\phi}{2^N}\right] = 1 + \Gamma\left[\frac{2\ln t}{\lambda^2}\right]\frac{1}{\sqrt{\pi}\lambda}\exp\left[\frac{\ln^2 t}{\lambda^2}\right] + \cdots .$$

Therefore one can find $\ln t$ as a function of ϕ for large N

$$\ln t = -\frac{NJ}{T}\sqrt{\ln 2} + \frac{T_c}{T}\left\{\tfrac{1}{2}\ln N + \ln\phi - \ln\left[-\Gamma\left[-\frac{T}{T_c}\right]\right] + \tfrac{1}{2}\ln\pi - \ln\left[\frac{T}{J}\right]\right\} + O(1) , \tag{B17}$$

where

$$T_c = \frac{J}{2\sqrt{\ln 2}} .$$

Using Eq. (B17), one can verify that condition (B16) can be written $T < T_c$ and from Eq. (B2), one gets

$$\langle \ln Z \rangle = \frac{NJ}{T}\sqrt{\ln 2} - \frac{T_c}{2T}\ln N + \left[1 - \frac{T_c}{T}\right]\Gamma'(1) - \frac{1}{2}\frac{T_c}{T}\ln(4\pi\ln 2) + \frac{T_c}{T}\ln\left[\Gamma\left[1 - \frac{T}{T_c}\right]\right] + O(1) . \tag{B18}$$

If

$$-p\lambda^2/2 < \ln t < -(p-1)\lambda^2/2 \text{ with } p \geq 2 \tag{B19}$$

$$\exp\left[-\frac{\phi}{2^N}\right] = 1 - te^{\lambda^2/4} + \frac{t^2}{2!}e^{\lambda^2} + \cdots + (-)^{p-1}\frac{t^{p-1}}{(p-1)!}e^{(p-1)^2\lambda^2/4} + \frac{1}{\sqrt{\pi}\lambda}\exp\left[-\frac{\ln^2 t}{\lambda^2}\right]\Gamma\left[\frac{2\ln t}{\lambda^2}\right] + \cdots .$$

We can again calculate $\ln t$ as a function of ϕ

$$\ln t = -N\ln 2 - \frac{\lambda^2}{4} + \ln\phi + Q_p\left[\frac{\phi}{2^N}, e^{\lambda^2/4}\right]$$

$$+ \frac{1}{\sqrt{\pi}\lambda}\exp\left[-\frac{1}{\lambda^2}\left[N\ln 2 - \frac{\lambda^2}{4}\right]^2 + \frac{2}{\lambda^2}\left[N\ln 2 - \frac{\lambda^2}{4}\right]\ln\phi\right]\Gamma\left[-\frac{1}{2} - \frac{2N\ln 2}{\lambda^2}\right] + \cdots . \tag{B20}$$

where $Q_p(\phi/2^N, e^{\lambda^2/4})$ is a polynomial of its two variables $\phi/2^N, e^{\lambda^2/4}$. For example:

$$Q_2\left[\frac{\phi}{2^N}, e^{\lambda^2/4}\right] = 0 ,$$

$$Q_3\left[\frac{\phi}{2^N}, e^{\lambda^2/4}\right] = \frac{1}{2}\frac{\phi}{2^N}(e^{\lambda^2/2} - 1) ,$$

$$Q_4\left[\frac{\phi}{2^N}, e^{\lambda^2/4}\right] = \frac{1}{2}\frac{\phi}{2^N}(e^{\lambda^2/2} - 1) - \frac{1}{24}\frac{\phi^2}{2^{2N}}(4e^{3\lambda^2/2} - 9e^{\lambda^2} + 6e^{\lambda^2/2} - 1) .$$

Using Eq. (B20), one sees that condition (B19) becomes $\sqrt{2p-3}\,T_c < T < \sqrt{2p-1}\,T_c$. We can now calculate $\langle \ln Z \rangle$ using Eqs. (B2) and (B6)

$$\langle \ln Z \rangle = N\ln 2 + \frac{NJ^2}{4T^2} - \int_0^\infty d\phi e^{-\phi} Q_p\left[\frac{\phi}{2^N}, e^{NJ^2/4T^2}\right]$$

$$+ \frac{2\sqrt{\pi}T}{J\left[\frac{T^2}{T_c^2} + 1\right]\sqrt{N}}\frac{1}{\sin\left[\frac{\pi}{2}\left[\frac{T^2}{T_c^2} + 1\right]\right]}\exp\left[-\frac{NT^2J^2}{16}\left[\frac{1}{T_c^2} - \frac{1}{T^2}\right]^2\right] + \cdots . \tag{B21}$$

The term $\sin\{\pi/2[(T^2/T_c^2)+1]\}$ comes from the product of two Γ functions

$$[\Gamma(z)\Gamma(-z) = -\pi/z \sin\pi z] \ .$$

The integral of the polynomial Q gives a polynomial of two variables (2^{-N} and $e^{NJ^2/4T^2}$). The important fact is that this term is exponentially small in N and is regular at all the temperatures $T_n' = T_c\sqrt{2n+1}$. On the other hand the last term given in the asymptotic expansion (B21) of $\langle \ln Z\rangle$ is singular at these temperatures even if it decreases also exponentially with N. For $n=1,2$ and 3, I could express Eq. (B21) as: for $\sqrt{2n-1}\,T_c < T < \sqrt{2n+1}\,T_c$

$$\langle \ln Z\rangle = N\left[\ln 2 + \frac{J^2}{4T^2}\right] - \frac{1}{2}\left\langle\left[\frac{Z}{\langle Z\rangle}-1\right]^2\right\rangle + \frac{1}{3}\left\langle\left[\frac{Z}{\langle Z\rangle}-1\right]^3\right\rangle + \cdots + \frac{(-)^{n+1}}{n}\left\langle\left[\frac{Z}{\langle Z\rangle}-1\right]^n\right\rangle$$
$$+\frac{2\sqrt{\pi}T}{J\left[\dfrac{T^2}{T_c^2}+1\right]\sqrt{N}}\,\frac{1}{\sin\dfrac{\pi}{2}\left[\dfrac{T^2}{T_c^2}+1\right]}\exp\left[-\frac{NT^2J^2}{16}\left[\frac{1}{T_c^2}-\frac{1}{T^2}\right]^2\right]+\cdots ,$$

by adding to Eq. (B21) terms which are smaller than the terms neglected. This formula is probably true for any $n \geq 1$ but I did not succeed in proving it.

When one approaches T_c from above, the integral of Q in Eq. (B21) is zero and one sees that the first correction to the thermodynamic limit is singular at T_c as it is in the case of the SK model. It is rather striking that all the temperatures $T_n' = \sqrt{2n+1}\,T_c$ appear as singular temperatures for the corrections to the thermodynamic limit. Therefore high-temperature expansions for this random-energy model are expected to be very singular.

[1]D. Sherrington and S. Kirkpatrick, Phys. Rev. Lett. 35, 1972 (1975); Phys. Rev. B 17, 4384 (1978).

[2]D. J. Thouless, P. W. Anderson, and R. G. Palmer, Philos. Mag. 35, 593 (1977).

[3]S. F. Edwards and P. W. Anderson, J. Phys. F 5, 965 (1975).

[4]J. L. van Hemmen and R. G. Palmer, J. Phys. A 12, 563 (1979).

[5]J. R. L. de Almeida and D. J. Thouless, J. Phys. A 11, 983 (1978).

[6]A. Blandin, M. Gabay, and T. Garel, J. Phys. C 13, 403 (1980).

[7]G. Parisi, J. Phys. A 13, 1101 (1980).

[8]B. Derrida, Phys. Rev. Lett. 45, 79 (1980).

[9]D. Sherrington, J. Phys. A 13, 637 (1980).

[10]N. I. Akhiezer, in *The Classical Moment Problem*, edited by D. E. Rutherford (Oliver and Boyd, Edinburgh, 1965), pp. 29–89.

[11]R. G. Palmer and C. M. Pond, J. Phys. F 9, 1451 (1979).

[12]G. Parisi, J. Phys. A 13, 1887 (1980); Philos. Mag. B 41, 677 (1980).

[13]G. Parisi and G. Toulouse, J. Phys. (Paris) Lett. 41, L361 (1980).

[14]J. Vannimenus, G. Toulouse, and G. Parisi, J. Phys. (Paris) 42, 565 (1981).

[15]G. Toulouse, J. Phys. (Paris) Lett. 41, L447 (1980).

[16]J. Vannimenus and G. Toulouse, J. Phys. C 10 L537 (1977).

[17]S. Kirkpatrick, Phys. Rev. B 16, 4630 (1977).

[18]I. Morgenstern and K. Binder, Phys. Rev. B 22, 288 (1980).

[19]J. M. Drouffe and C. Itzykson, Phys. Rep. 38, 133 (1978).

Nuclear Physics B240 [FS12] (1984) 431-452

THE SIMPLEST SPIN GLASS

D.J. GROSS[1] and M. MEZARD

Laboratoire de Physique Théorique de l'Ecole Normale Supérieure, 24 rue Lhomond, 75231 Paris Cedex 05, France*

Received 7 May 1984

We study a system of Ising spins with quenched random infinite ranged p-spin interactions. For $p \to \infty$, we can solve this model exactly either by a direct microcanonical argument, or through the introduction of replicas and Parisi's ultrametric ansatz for replica symmetry breaking, or by means of TAP mean field equations. Although the model is extremely simple it retains the characteristic features of a spin glass. We use it to confirm the methods that have been applied in more complicated situations and to explicitly exhibit the structure of the spin glass phase.

1. Introduction

In recent years much effort has been devoted to the study of the low-temperature behaviour of systems of spins interacting via quenched random couplings – spin glasses [1]. The characteristic feature of such a disordered system is the existence of many states of minimum free energy, separated by very high free energy barriers and unrelated by a symmetry one to another. As a consequence it is believed that in such systems ergodicity can break down, so that the equilibrium state will depend on the initial conditions.

Normally the first step towards understanding the phases of a given system is by means of mean field theory. In the case of spin glasses even mean field theory has proven to be very subtle. An appropriate infinite range spin glass model was proposed by Sherrington and Kirkpatrick (the SK model [2]) many years ago, but its solution has only been recently obtained. By now there is general agreement that the SK model can be solved by means of the "replica method". This method is based initially on a mathematical trick which allows one, by introducing n replicas of the system and taking the $n \to 0$ limit, to replace quenched averages (which are hard) by annealed averages (which are easy). The basic observation, due to Parisi [3], is that the breaking of replica symmetry is physically related to the breakdown of ergodicity in the spin glass phase. Parisi proposed a specific form for this replica symmetry breaking [6], which produces a stable mean field solution and which has

[1] On leave from Princeton University.

* Laboratoire Propre du Centre National de la Recherche Scientifique, associé à l'Ecole Normale Superieure et à l'Université de Paris-Sud.

a natural interpretation in terms of the structure of the space of free energy valleys [3–5].

These results have yielded a consistent picture of the mean field theory of a spin glass. However they rely heavily on a particular replica symmetry breaking scheme. It is not a priori clear what physical principle is responsible for this very specific pattern, which possesses the very special ultra-metric property [5]. The best evidence to date for the validity of Parisi's scheme is its stability [7] and the fact that it agrees with numerical experiments.

A few years ago it was pointed out by Derrida [8], that the SK model could be generalized to models involving p-spin interactions, and that these simplify in the limit of large p. Derrida showed that the $p \to \infty$ SK model is equivalent to a random energy model, which consists of a collection of independent random energy levels. He was then able to solve this model exactly, without recourse to replicas or other potentially dangerous tricks.

In this paper we shall study the generalized p-spin SK model directly, with the aim of testing the methods that have been applied to the usual model and displaying in an explicit fashion the spin glass phase.

Thus we shall apply the replica method to the p-spin model and analyse it within Parisi's hierarchical scheme. When $p \to \infty$ it turns out that the first stage of replica symmetry breaking is exact. Therefore we will obtain the analytic form of the order parameter function $q(x)$, and recover the values of the thermodynamic quantities (free energy, internal energy, magnetization) in agreement with the random energy model. Furthermore we can analyse the structure of the space of free energy valleys in the spin glass phase, following [5], in terms of their statistical weights and the mutual overlap of their spins. The physical interpretation of the order parameter function $q(x)$ in terms of the distribution of weights of the pure states of the system can be subjected to a critical test by evaluating the $1/N$ corrections to the entropy and comparing with Derrida's calculation within the random energy model.

Another standard approach to the SK model is via the mean field equations of Thouless, Anderson and Palmer (the TAP equations) [9]. Again for $p \to \infty$ these simplify enormously. Since the system is totally frozen in the spin glass phase the cumbersome Onsager reaction terms can be neglected. We then can solve the model explicitly by calculating the density of TAP solutions and performing a canonical average over them, without the need to introduce replicas. This approach reinforces the physical picture of the nature of the spin glass phase.

We have attempted to write this paper so that it would be comprehensible to readers that are not spin glass experts, in the hope that the elucidation of the properties of this simplest of all spin glasses can serve as an introduction to the fascinating subject of the spin glass phase.

The structure of the paper is as follows. In sect. 2, we review Derrida's demonstration of the equivalence of the $p \to \infty$ SK model with the random energy model and outline its solution. Sect. 3 is devoted to the replica method and its application to the $p \to \infty$ model. In sect. 4 we study the TAP equations for $p \to \infty$, and use the analysis of their solutions to gain further insight into the structure of the spin glass phase.

2. The random energy model

For the sake of completeness we shall review the argument of Derrida [8] on the equivalence of the p-spin SK model with the random energy model in the limit $p \to \infty$, as well as Derrida's solution of the latter.

The generalized p-spin SK model describes a system of N Ising spins ($\sigma_i = \pm 1$) with infinite range p-spin quenched random interactions. It is defined by the hamiltonian

$$\mathcal{H} = - \sum_{1 \le i_1 < i_2 \cdots < i_p \le N} J_{i_1 i_2 \cdots i_p} \sigma_{i_1} \sigma_{i_2} \ldots \sigma_{i_p} . \tag{1}$$

The interaction strengths are independent random variables which can be taken, for simplicity, to be gaussian. In order for the free energy to be extensive (i.e. proportional to N) the probability distribution of the J's must be scaled as follows:

$$P(J_{i_1 \cdots i_p}) = \sqrt{\frac{N^{p-1}}{\pi p!}} \exp\left[- \frac{(J_{i_1 \cdots i_p})^2 N^{p-1}}{J^2 p!} \right] . \tag{2}$$

For $p = 2$ this reduces to the standard SK model. We shall be interested in particular in the $p \to \infty$ limit of these models, where much simplification occurs. Note that one must be careful to take the $p \to \infty$ limit *after* taking the thermodynamic limit, $N \to \infty$.

Let $\{\sigma_i^{(1)}\}$ denote a given configuration of the spins with energy $\mathcal{H}(\sigma^{(1)})$. This energy depends, of course, on the particular choices of the couplings J. The probability, $P(E)$, that it equals E is given by $P(E) = \overline{\delta(E - \mathcal{H}(\sigma^{(1)}))}$, where $\bar{O}(\langle O \rangle)$ stands for the average over the couplings (the thermodynamic average):

$$\overline{O(J, \sigma)} = \int \prod \mathrm{d}J \, P(J) O(J, \sigma) ,$$
$$\langle O(J, \sigma) \rangle = \frac{1}{Z} \sum_{\sigma_i = \pm 1} \mathrm{e}^{-\beta \mathcal{H}(J, \sigma)} O(J, \sigma) . \tag{3}$$

Since the J have gaussian distribution, $P(E)$ is easily evaluated in the $N \to \infty$ limit to be

$$P(E) = \frac{1}{\sqrt{N \pi J^2}} \exp\left[- \frac{E^2}{J^2 N} \right] . \tag{4}$$

Note that $P(E)$ is independent of p (which justifies the scaling of eq. (2)) and of the spin configuration. This is a consequence of "gauge invariance", namely the fact that $\mathcal{H}(\sigma, J) = \mathcal{H}(\sigma', J')$ and $P(J) = P(J')$, where $J'_{i_1 \cdots i_p} = J_{i_1 \cdots i_p} (\sigma_{i_1} \sigma'_{i_1}) \ldots (\sigma_{i_p} \sigma'_{i_p})$.

Now consider two different spin configurations, $\{\sigma_i^{(1)}\}$ and $\{\sigma_i^{(2)}\}$ and calculate the probability, $P(E_1, E_2)$, that they have energies E_1 and E_2 respectively. Due to the gauge invariance this can only depend on the *overlap*, q, between the two configurations:

$$q^{(1,2)} \equiv \frac{1}{N} \sum_{i=1}^{N} \sigma_i^{(1)} \sigma_i^{(2)} . \tag{5}$$

One finds (as $N \to \infty$)

$$\begin{aligned} P(E_1, E_2, q) &= \overline{\delta(E_1 - \mathcal{H}(\sigma^1))\delta(E_2 - \mathcal{H}(\sigma^2))} \\ &= [N\pi J^2(1+q^p) N\pi J^2(1-q^p)]^{-1/2} \\ &\quad \times \exp\left[-\frac{(E_1+E_2)^2}{2N(1+q^p)J^2} - \frac{(E_1-E_2)^2}{2N(1-q^p)J^2}\right] . \end{aligned} \tag{6}$$

The important point, discovered by Derrida [8], is that *if* $\sigma^{(1)}$ *and* $\sigma^{(2)}$ *are macroscopically distinguishable* ($|q^{(1,2)}| < 1$) *the energies are uncorrelated*, namely

$$P(E_1, E_2, q) \xrightarrow{p\to\infty} P(E_1)P(E_2) \quad (|q| < 1) . \tag{7}$$

Of course when $q = 1$, $P(E_1, E_2, q) = P(E_1)\ \delta(E_1 - E_2)$.

Similarly one can easily show that the probability distribution of n levels $\sigma^{(1)} \ldots \sigma^{(n)}$ with energies $E_1 \ldots E_n$, which can only depend on the overlaps $q^{(i,j)}$, factorizes when all $q^{(i,j)} < 1$:

$$P(E_1, E_2 \ldots E_n; q^{(i,j)}) \xrightarrow{p\to\infty} \prod_{i=1}^{n} P(E_i) \quad (|q^{(i,j)}| < 1) . \tag{8}$$

Therefore in the large $-p$ limit the energy levels become independent random variables. The physics is identical to that of Derrida's random energy model, defined as a system of 2^N independent random energy levels distributed according to eq. (4).

Derrida has solved the random energy model, including the effect of an external magnetic field, as well as the leading $1/N$ corrections to the free energy. For details see ref. [8]. Here we shall only briefly outline the microcanonical derivation of the free energy in zero field.

Since the energy levels are independent random variables the average number of levels, $\langle n(E)\rangle$, of energy E is simply the total number of levels, 2^N, times the probability of finding E:

$$\langle n(E)\rangle = \frac{1}{\sqrt{\pi N J^2}} e^{N[\ln 2 - (E/NJ)^2]} . \tag{9}$$

If $|E| < E_0 = N\sqrt{\ln 2}$ the average number of levels is very large. Since the levels are statistically independent, the fluctuations are of order $1/\sqrt{\langle n(E)\rangle}$ and therefore negligible. Thus $n(E) \sim \langle n(E)\rangle$ for $|E| < E_0$. On the other hand if $|E| > E_0$ there are simply no levels (with probability one). Therefore the entropy is

$$S(E)=N\left[\ln 2-\left(\frac{E}{NJ}\right)^2\right], \quad |E|<E_0. \tag{10}$$

Using $dS/dE=1/T$ one finds that the free energy is

$$\frac{F}{N}=\begin{cases}-T\ln 2-J^2/4T, & T>T_c\\ -\sqrt{\ln 2}, & T<T_c.\end{cases} \tag{11}$$

The critical temperature, T_c, is

$$T_c=1/(2\sqrt{\ln 2}). \tag{12}$$

Below T_c the system gets stuck in the lowest available energy level, $E=-E_0$ and the entropy vanishes. Having completely disposed with the spin configurations, it is not easily seen that this model describes a spin glass. Some evidence is provided by the behaviour of the magnetic susceptibility below T_c, which can be derived by similar arguments [8]. In the following we shall solve the $p\to\infty$ SK model directly and the spin glass nature of the low-temperature phase will be more apparent.

3. Replica symmetry breaking

In this section we shall treat the p-spin generalized SK model defined by eq. (1) (including a magnetic field) directly, and obtain the solution for $p\to\infty$ by the replica method. This model is a nice generalization of the standard SK ($p=2$) model, which shares with it all the essential features which are believed to be responsible for the unusual properties of spin glasses – quenched disorder and frustration*. One expects that the low-temperature phase is a spin glass. The characteristic feature of a spin glass phase is the existence of very many (infinite in the thermodynamic limit) states of minimum free energy (free energy valleys), which are unrelated one to another by any symmetry of the system, and which are separated by very high free energy barriers. In the infinite range model, these barriers are infinitely high and are responsible for the breakdown of ergodicity. Thus the particular valley into which the system will dynamically relax depends sensitively on the initial conditions.

Recently it has been realized that the best way of characterizing the spin glass phase is in terms of the space of equilibrium states (free energy valleys) of the system [3, 4]. Each valley α can be assigned a statistical weight, P_α, determined by its free energy, F_α:

$$P_\alpha=\frac{e^{-F_\alpha/T}}{\sum_\gamma e^{-F_\gamma/T}}. \tag{13}$$

* Note that for odd p the model loses the "time-reversal invariance" which holds for even p, i.e. symmetry under reversal of all the spins. This is of some interest since it yields a situation where the spin glass transition is purer, namely it does not mix with a ferromagnetic transition for zero field.

Because of the breakdown of ergodicity the mean value of any observable O is given by

$$\langle O\rangle = \sum_{\alpha} P_\alpha \langle O\rangle_\alpha , \tag{14}$$

where $\langle O\rangle_\alpha$ is the mean value of O in the valley α. The valleys thus correspond to *pure states* of the system. In contrast to more conventional systems, one expects that there exists an infinite number of such states, unrelated one to another by any symmetry. Furthermore there does not exist any macroscopic way to turn an external field (as one does, say, in a ferromagnetic by applying a magnetic field) in order to pick out a particular pure state. The system is necessarily described by the above *mixture* of pure states.

A measure of the distance in the space of valleys is introduced naturally in the following way: let $m_i^\alpha = \langle \sigma_i \rangle_\alpha$ be the magnetization of the spin i in the valley α. The overlap, $q^{\alpha\beta}$, between two valleys is defined to be

$$q^{\alpha\beta} = \frac{1}{N} \sum_{i=1}^{N} m_i^\alpha m_i^\beta . \tag{15}$$

To describe the structure of this space it is natural to define the probability, $P(q)$, that two valleys, picked at random, have overlap q:

$$P(q) \equiv \sum_{\alpha,\beta} P_\alpha P_\beta \delta(q - q^{\alpha\beta}) , \tag{16}$$

and to characterize the structure of the spin glass by the average of $P(q)$ over the random couplings, $\overline{P(q)}$. In an ordinary Ising model there exists one pure state at high temperature (with $\langle m_i \rangle = 0$) and two pure states at low temperature (with $\langle m_i \rangle^{\pm} \gtrless 0$) with equal probability. Thus for high T we would have $P(q) = \delta(q)$ and for low T, $P(q) = \frac{1}{2}[\delta(q + m^2) + \delta(q - m^2)]$. In the case of a spin glass, however, there are an *infinite number* of pure states – and therefore one expects that q will take many values.

The standard method for performing averages over the quenched couplings is to introduce n replicas of the system, calculate annealed averages and take the $n \to 0$ limit [10]. Thus the average free energy can be obtained as

$$\overline{\ln Z} = \lim_{n\to 0} \frac{1}{n} (\overline{Z^n} - 1) , \tag{17}$$

and $\overline{Z^n}$ can be calculated by introducing n replicas of the system, σ_i^a, $a = 1 \ldots n$. In an Ising-like model (with a symmetric distribution of couplings), once the average over the couplings is performed the effective hamiltonian can only depend on the overlap function of the replicas, Q_{ab}:

$$Q_{ab}(\sigma) \equiv \frac{1}{N} \sum_{i=1}^{N} \sigma_i^a \sigma_i^b . \tag{18}$$

One can relate Q_{ab} to the order parameters, q, that describe the structure of the space of valleys by evaluating the average of $P(q)$ using replicas. One obtains [3]

$$\int \overline{P(q)}\, e^{uq}\, dq = \lim_{n\to 0} \frac{1}{n(n-1)} \sum_{a\neq b} e^{u\langle Q_{ab}\rangle}, \tag{19}$$

where $\langle Q_{ab}\rangle$ is the mean value of the replica overlap matrix.

The effective hamiltonian of Q_{ab} is, of course, symmetric under a permutation of the replica indices. Thus one might have expected that $\langle Q_{ab}\rangle$ would be replica-symmetric, i.e. $\langle Q_{ab}\rangle = Q$ $(a\neq b)$. However, this means that $\bar{P}(q) = \delta(q-Q)$ and therefore there is only a single pure state, with self-overlap (the Edwards–Anderson order parameter) equal to Q. In a true spin glass phase, as for example in the $p=2$ SK model, q ranges over a continuous spectrum. Therefore $\langle Q_{ab}\rangle$ must be characterized by an infinite number of parameters. Consequently the replica symmetry must be drastically broken.

In the $p\to\infty$ model we shall be able to calculate explicitly the function $P(q)$ (this cannot be done in the finite-p case) using the replica method to calculate $\overline{Z^n}$ (we hereafter set $J=1$).

$$\overline{Z^n} = \int \prod dJ_{i_1\cdots i_p} P(J_{i_1\cdots i_p}) \times \mathrm{Tr}_{(\sigma_i^a)} \left[\exp \beta \sum_{a=1}^{n} \left[\sum_{i_1<\cdots<i_p} J_{i_1\cdots i_p} \sigma_{i_1}^a \cdots \sigma_{i_p}^a + h\sum_i \sigma_i^a\right]\right]. \tag{20}$$

One easily obtains

$$\overline{Z^n} = \mathrm{Tr}_{(\sigma_i^a)} \exp\left[\tfrac{1}{4}\beta^2 N\left(n + \sum_{a\neq b} Q_{ab}^p(\sigma)\right) + \beta h \sum_{i,a} \sigma_i^a\right]. \tag{21}$$

The spin trace can be performed by constraining $Q_{ab}(\sigma)$ to equal Q_{ab}, with the aid of a Lagrange multiplier matrix λ_{ab}. One then gets

$$\overline{Z^n} = e^{nN\beta^2/4} \int_{-\infty}^{+\infty} \prod_{a<b} d\,Q_{ab} \int_{-i\infty}^{+i\infty} \prod_{a<b} \frac{d\lambda_{ab}}{2\pi} e^{-NG(Q_{ab},\lambda_{ab})}, \tag{22}$$

$$G(Q_{ab}, \lambda_{ab}) = -\frac{1}{4}\beta^2 \sum_{a\neq b} Q_{ab}^p + \frac{1}{2}\sum_{a\neq b} \lambda_{ab} Q_{ab} - \ln \mathrm{Tr}_{(\sigma_a)} \exp\left[\frac{1}{2}\sum_{a\neq b} \lambda_{ab}\sigma_a\sigma_b + \beta h \sum_a \sigma_a\right]. \tag{23}$$

Unlike the case $p=2$, the effective hamiltonian is not quadratic in Q_{ab}, which therefore cannot be eliminated. In the limit $N\to\infty$, $\overline{Z^n}$ is given by the dominant saddle-point of G, namely mean field theory is exact, and the average free energy is $+\beta\bar{F}/N = \lim_{n\to 0}[G/n - \frac{1}{4}\beta^2]$. Actually one must find the absolute *maximum* of G, not the minimum. This reversal is one of the strange features of the $n\to 0$ limit.

Since the matrix of fluctuations (of Q_{ab} or λ_{ab}) has $\frac{1}{2}n(n-1)$ parameters, it acts, for $n<1$, on a space of negative dimensions. In this situation the role of negative and positive eigenvalues is switched [6] and stability requires that G be maximized!

In order to evaluate G explicitly one must impose some ansatz on the structure of Q_{ab}, and a corresponding structure on λ_{ab}. For example in the high-temperature phase, the replica-symmetric ansatz is reasonable since we expect only one pure state:

$$Q_{ab} = Q\,, \qquad \lambda_{ab} = \lambda, \quad a \neq b. \tag{24}$$

In that case one gets

$$\frac{1}{n} G(Q,\lambda) \stackrel{n\to 0}{=} \tfrac{1}{4}\beta^2 Q^p - \tfrac{1}{2}\lambda Q - \int_{-\infty}^{+\infty} \mathrm{D}z \ln[2\,\mathrm{ch}\,(z\sqrt{\lambda}+\beta h)]\,, \tag{25}$$

where

$$\mathrm{D}z \equiv \frac{\mathrm{d}z}{\sqrt{2\pi}}\,\mathrm{e}^{-z^2/2}\,.$$

The saddle-point equations are

$$\tfrac{1}{2}\beta^2 p Q^{p-1} = \lambda\,, \qquad Q = \int \mathrm{D}z\,\mathrm{th}^2\,(z\sqrt{\lambda}+\beta h)\,. \tag{26}$$

When $p=\infty$ there exists a unique saddle-point for *all* β, h:

$$Q = \mathrm{th}^2\,(\beta h), \quad \lambda = 0\,. \tag{27}$$

The resulting free energy is then calculated from (22) and (25), to be

$$\frac{\bar{F}}{N} = -\frac{1}{4T} - T\ln 2 - T \ln \mathrm{ch}\,\frac{h}{T}\,. \tag{28a}$$

This replica-symmetric solution is indeed stable for large T (we shall derive the precise phase diagram below) and reproduces correctly the value of the thermodynamic quantities in the high-temperature phase of the random energy model [8]. This phase contains a single pure state $\bar{P}(q) = \delta(q - \mathrm{th}^2\,(\beta h))$, whose self-overlap is the square of the magnetization.

The entropy in this phase

$$S = \ln 2 - \frac{1}{4T^2} + \ln \mathrm{ch}\,\frac{h}{T} - \frac{h}{T}\,\mathrm{th}\,\frac{h}{T}\,, \tag{28b}$$

clearly becomes negative for $T \leq T_1(h)$, and therefore there must be a phase transition at some $T_c \geq T_1(h)$. In fact, as is evident from the random energy model, we shall see that $T_c = T_1(h)$.

Unlike the case in the $p=2$ model, the $q = \mathrm{th}^2(\beta h)$ solution is the only replica-symmetric one at all temperature (for $p\to\infty$). In fact, an analysis of the stability of this solution within the complete replica space (à la De Almeida–Thouless [11]) shows that it is always locally stable in zero field, as soon as $p>2$. In this respect, the $p=2$ SK model is somewhat special. The spin glass transition must be (for $p>2$) a first-order one, at least as far as the order parameter function $q(x)$ is concerned. In fact we shall show, in the $p\to\infty$ case, that the Edwards–Anderson order parameter, $q(1)$, jumps from 0 to 1 at T_c. However, since the order parameter is a function, and the discontinuity appears only on a set of zero measures, the transition turns out to be of second order in the thermodynamic sense.

In order to obtain the low-temperature spin glass phase, we must break replica symmetry, allowing Q_{ab} to depend, in general, on an infinite number of parameters. The most general form of such a Q_{ab} is not known. Parisi has given a particular ansatz, which describes a hierarchical breaking of replica symmetry [6]. For $p=2$ this does yield a stable maximum of F and agrees with numerical results. For $p=\infty$ we shall show that it leads to the correct solution. (Note that the equations for a saddle-point of $G(Q,\lambda)$ will force λ_{ab} to have the same structure as Q_{ab}.)

Parisi's ansatz for Q_{ab} can be described by means of the following recursive algorithm:

(i) First breaking: the n replicas are grouped in n/m_1 clusters of m_1 replicas. Any two replicas, $a\neq b$, within the same cluster have overlap $Q_{ab}=q_1$, whereas replicas in different clusters have overlap $Q_{ab}=q_0\leqslant q_1$.

(ii) Second breaking: each cluster of size m_1 is broken up into m_1/m_2 sub-clusters of m_2 spins. Any two replicas, $a\neq b$, in a sub-cluster have overlap $q_2\geqslant q_1$, the other overlaps remain unchanged.

One continues to iterate this procedure, thus obtaining the general k-breaking situation, defined by

$$n\geqslant m_1\geqslant m_2\cdots\geqslant m_k\geqslant 1\,,$$
$$q_k\geqslant q_{k-1}\geqslant\cdots\geqslant q_1\geqslant q_0\,. \tag{29}$$

(Note that to achieve the continuation to $n=0$, one must let the m_i be continuous and reverse the inequalities in (29), i.e. for $n=0$: $0\leqslant m_1\leqslant\cdots\leqslant m_k\leqslant 1$.)

The matrix obtained in the kth step by this procedure is best described by a genealogical tree with k generations, as shown in fig. 1. It can be parametrized by the function $x(q)$ – which equals the fraction of pairs of replicas with overlap $Q_{ab}\leqslant q$. The defining characteristic of Parisi's scheme of replica symmetry breaking is its ultrametric structure. It is clear from the tree that if we consider three distinct replicas a, b, c, then the smallest two of the overlaps Q_{ab}, Q_{bc} and Q_{ac} must be equal.

In the limit of infinite K, q will be continuous, and we can define $q(x)$ to be the inverse of $x(q)$. The physical meaning of $q(x)$ is evident from (19):

$$\int \overline{P(q)}\,e^{uq}\,dq=\int_0^1 dx\,e^{uq(x)}\,,$$
$$\overline{P(q)}=\frac{dx}{dq}\,. \tag{30}$$

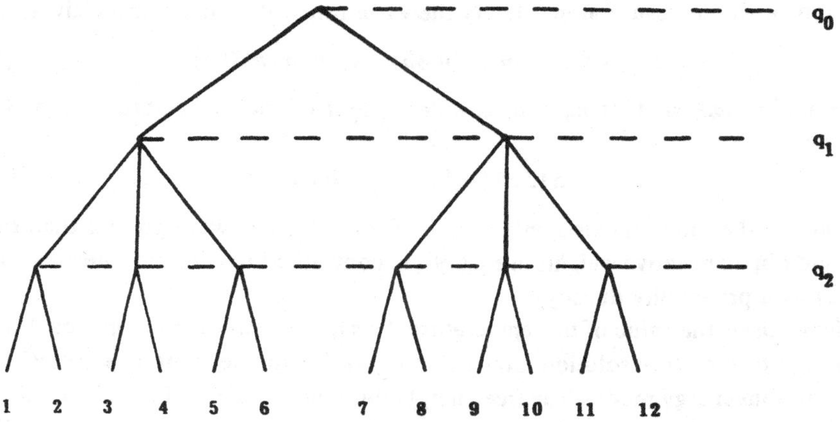

Fig. 1. Parisi's ultrametric ansatz for replica symmetry breaking, described here for $n=12$, $m_1=6$, $m_2=2$. Replica indices $a, b=1, \ldots, 12$, are the extremities of the branches. The value of the matrix element Q_{ab} is q_0, q_1 or q_2, depending on the closest common ancestor of a and b. (For instance $Q_{34}=q_2$, $Q_{57}=q_0 \ldots$.)

Thus $x(q)$, the fraction of pairs of replicas with overlap $\leqslant q$, equals $\int_0^q \bar{P}(q')\,dq'$, the average fraction of pairs of valleys with overlaps $\leqslant q$.

Let us now return to the p-spin SK model and consider the first step ($k=1$) in Parisi's scheme. G is then a function of q_0, q_1, λ_0, λ_1 and $m=m_1$, and is given by

$$\frac{1}{n} G = \ln 2 - \tfrac{1}{4}\beta^2 (m q_0^p + (1-m) q_1^p) + \tfrac{1}{2}(m\lambda_0 q_0 + (1-m)\lambda_1 q_1)$$

$$-\tfrac{1}{2}\lambda_1 + \frac{1}{m}\int \mathrm{D}z_0 \ln \int \mathrm{D}z_1 \,\mathrm{ch}^m (z_0\sqrt{\lambda_0} + z_1\sqrt{\lambda_1 - \lambda_0} + \beta h)\,. \tag{31}$$

For $p \to \infty$ the saddle-point equations are easy to solve. First $\partial G/\partial q_i = 0$ implies

$$\lambda_i = \tfrac{1}{2}\beta^2 p q_i^{p-1}\,. \tag{32}$$

For non-trivial symmetry breaking we must have $q_0 < q_1 \leqslant 1$, thus $\lambda_0 = 0$. If q_1 is also <1 then $\lambda_1 = 0$, in which case we will recover the symmetric solution $q_0 = q_1 = \mathrm{th}^2(\beta h)$. Hence $q_1 = 1$ and $\lambda_1 \sim \infty$.

In this circumstance the double integral in G is easily calculated, and we obtain ($\lambda_1 \sim \infty$, $\lambda_0 \sim 0$)

$$\frac{1}{n} G = -\tfrac{1}{4}\beta^2 (m q_0^p + (1-m) q_1^p) + \tfrac{1}{2}(m\lambda_0 q_0 + (1-m)\lambda_1 q_1) - \tfrac{1}{2}\lambda_1 + \tfrac{1}{2} m \lambda_1$$

$$+ \frac{1}{m} \ln (2\,\mathrm{ch}\,(m\beta h)) - \tfrac{1}{2} m \lambda_0 \,\mathrm{th}^2\,(m\beta h) + \mathrm{O}\,(\lambda_0^2, 1/\lambda_1)\,. \tag{33}$$

Differentiating with respect to λ_i then yields

$$q_0 = \mathrm{th}^2\,(\beta m h)\,, \qquad q_1 = 1\,, \tag{34}$$

consistent with our assumption. Finally the variation with respect to m gives

$$m^2\beta^2 = 4\,[\ln 2 + \ln \mathrm{ch}\,(m\beta h) - m\beta h\,\mathrm{th}\,(m\beta h)]\,. \tag{35}$$

This equation tells us that $m\beta = \beta_c$ is independent of the temperature, and β_c is given by

$$\beta_c^2 = 4[\ln(2\,\mathrm{ch}\,(\beta_c h)) - \beta_c h\,\mathrm{th}\,(\beta_c h)]\,. \tag{36}$$

Since $m \leq 1$, the solution exists only for $T < T_c = 1/\beta_c$ (if m were greater than one, we would obtain negative weights in eq. (19), in contradiction with the interpretation of $P(q)$ as a probability density).

T_c is precisely the value of the temperature $T_1(h)$, at which the entropy, eq. (28b), of the high-temperature solution turns negative, and coincides with the critical line of the random energy model. The free energy obtained for $T < T_c(H)$ can easily be calculated, using the above solution:

$$\frac{F}{N} = -\frac{1}{2T_c} - h\,\mathrm{th}\,\frac{h}{T_c}\,, \tag{37}$$

precisely the result found by Derrida [8], for the low-temperature phase of the random energy model. The magnetization is given by $m = \mathrm{th}\,(h/T_c)$ and the magnetic susceptibility is temperature-independent ($\chi = 1/T_c\,\mathrm{ch}^2\,(h/T_c)$), as is also true in the SK model [6].

In the $p = 2$ SK model one must go to the $k = \infty$ level of replica symmetry breaking. Here *the first breaking of replica symmetry gives the exact answer.* Indeed we prove, in appendix A, that for the general kth-order breaking the only saddle-point is the one derived above. This phase is thus characterized by only 2 values of q: $q(x) = \mathrm{th}^2\,(\beta_c h)\theta(T/T_c - x) + \theta(x - T/T_c)$, and $\overline{P(q)} = (T/T_c)\,\delta(q - \mathrm{th}^2\,(\beta_c h)) + (1 - T/T_c)\delta(q-1)$.

The peak at $q = 1$ means that the self-overlap in a given valley, i.e. the local magnetization, is maximal ($m_i = \pm 1$). Thus in the low-temperature phase the system is completely frozen, within each pure state there are no fluctuations of the magnetization. The peak at $q = \mathrm{th}^2\,(\beta_c h)$ means that two different valleys have an overlap equal to the square of the magnetization, i.e. the valleys are as far apart from one another as they could possibly be.

It might seem that there are only two valleys, however this is not the case. Following [5] we can calculate the distribution of the weights, P_α, of different clusters. Choose an overlap scale, q, and group together all valleys with overlap larger than q, into clusters labelled by I, with weights $P_I = \sum_{\alpha \in I} P_\alpha$. It then follows that the average number of clusters of a given weight P_I is given by

$$f(P) = \overline{\sum_I \delta(P_I - P)} = \frac{P^{y-2}(1-P)^{-y}}{\Gamma(y)\Gamma(1-y)}\,, \tag{38}$$

which is a function of $y(q) = \int_q^1 \overline{P(q')}\,dq'$, the probability that the overlap is greater than q.

If we choose $\mathrm{th}^2(\beta_c h) < q < 1$, then $y(q) = 1 - T/T_c$ and each cluster contains precisely one pure state. Therefore the average number of pure states with weight $P_\alpha = P$ equals

$$f(P) = \overline{\sum_\alpha \delta(P_\alpha - P)} = \frac{P^{-1-T/T_c}(1-P)^{-1+T/T_c}}{\Gamma(T/T_c)\Gamma(1-T/T_c)} . \tag{39}$$

This allows us to calculate the total number of pure states, which equals $N_v = \sum_\alpha 1 = \int_0^1 \mathrm{d}P f(P) = \infty$. The divergence occurs because of the existence of many valleys with small weight ($P \sim 0$). If we introduce a cutoff, $P \geqslant \varepsilon$, for the valley weights, $N_v(\varepsilon)$ blows up as

$$N_v(\varepsilon) \sim (1/\varepsilon)^{T/T_c} . \tag{40}$$

The fact that N_v increases with increasing temperature is somewhat surprising. In general one might expect that as the temperature is lowered, the mean free energy is also lowered and more free energy valleys are explored. In our case, however, the system is frozen for $T \leqslant T_c$, F remains constant for $T \leqslant T_c$, and the only things that change are weights of the different valleys. Valleys with large energies become less significant as T is lowered, and this leads to the decrease of N_v with temperature. This result will be confirmed and explained further in sect. 4.

Even though there are an infinite number of pure states, most have very small weight and are insignificant. In fact the mean weight is given by

$$\overline{\sum_\alpha P_\alpha^2} = 1 - T/T_c . \tag{41}$$

To test the correctness of the above interpretation of the structure of the pure states in the spin glass phase, we can calculate the $1/N$ corrections to the theory and compare with Derrida's calculations within the random energy model [8]. We have seen that since the free energy is frozen for $T \leqslant T_c$, the $O(N)$ contribution to the entropy vanishes: $S/N \xrightarrow{N\to\infty} 0$. Now, if the system is in a mixture of pure states, α, the entropy is given by

$$S = \sum_\alpha P_\alpha S_\alpha - \sum_\alpha P_\alpha \ln P_\alpha , \tag{42}$$

where $S_\alpha = -\partial F_\alpha / \partial T$ is the entropy within the valley α, and $I = -\sum_\alpha P_\alpha \ln P_\alpha$ is the mutual entropy of the valleys, sometimes called the complexity [12]. Since the system is completely frozen within each valley ($m_i = \pm 1$), it is reasonable to assume that each $S_\alpha = 0$ (we cannot rigorously prove this to order $1/N$). Thus the entropy equals the complexity

$$\bar{S} = \bar{I} = -\overline{\sum_\alpha P_\alpha \ln P_\alpha} = -\int_0^1 P \ln P f(P)\, \mathrm{d}P$$

$$= \Gamma'(1) - \frac{\Gamma'(1-T/T_c)}{\Gamma(1-T/T_c)} . \tag{43}$$

The specific heat per spin is then equal to

$$C=\frac{1}{N}\frac{T}{T_c}\left[\frac{\Gamma''(1-T/T_c)}{\Gamma'(1-T/T_c)}-\left(\frac{\Gamma'(1-T/T_c)}{\Gamma(1-T/T_c)}\right)^2\right]. \tag{44}$$

These $1/N$ corrections to F are in complete agreement with the results of [8], and confirm the physical interpretation of replica symmetry breaking developed in [3, 5].

4. The TAP equations

In this section we shall probe the structure of the $p=\infty$ SK model from the point of view of another time-honoured approach to spin glasses – that of the mean field equations of Thouless, Anderson and Palmer (TAP) [9]. Our purpose is not to solve the model for a third time, but rather to gain further insight into the structure of the spin glass phase.

The TAP equations are mean field equations for a particular realization of the hamiltonian (1), which determine the local magnetization $m_i=\langle\sigma_i\rangle_J$. These differ from the naive mean field equations:

$$\text{th}^{-1}m_i=\frac{1}{T}\sum_{i_2<\cdots<i_p} J_{ii_2\cdots i_p}m_{i_2}\ldots m_{i_p}+\frac{h}{T}, \tag{45}$$

by the addition of Onsager-like reaction terms [9]. The modification amounts to subtracting for each m_{i_l} in (45), the part of magnetization due to m_i. However, this is proportional to the susceptibility $\chi_{i_l i_l}=(1-m_{i_l}^2)/T$, and we already know that in the $p=\infty$ model the system is frozen for $T\leq T_c$ and all $m_i=\pm 1$. Therefore we shall simply ignore these corrections. One could, presumably, prove directly from the full TAP equations that m_i had to equal ± 1 for $p=\infty$; we shall simply take this result from the previous solution of the model*.

The TAP equations will then be satisfied in the following fashion: the sum on the right-hand side of (45) will diverge (as $p\to\infty$), as we shall see below, as $\sqrt{p}$, and therefore we will get a solution as long as

$$m_i=\text{sgn}\left(\sum_{i_2<\cdots<i_p} J_{ii_2\cdots i_p}\, m_{i_2}\ldots m_{i_p}\right). \tag{46}$$

This equation is actually valid for any p in the $T\to 0$ limit (for $h=0$), where the m_i are frozen to be ± 1. In our case it holds for all h and $T\leq T_c$. The free energy of a given solution (again for $p=\infty$, where $m_i=\pm 1$) is

$$F\{m_i\}=-\sum_{i_1<i_2\cdots<i_p} J_{i_1 i_2\cdots i_p}m_{i_1}\ldots m_{i_p}-\sum_i hm_i. \tag{47}$$

Due to the extreme simplicity of these equations we shall be able to compute the

* We thank C. de Dominicis for interesting discussions on the generalization of TAP equations to the p-spin SK model.

number of solutions and the free energy exactly, using arguments similar to those used in the microcanonical solution of the random energy model given in sect. 2.

Let us first compute the number of solutions of a given energy $N(E)$, for zero field. We choose a particular configuration $\{m_i\}$ among the 2^N possibilities ($m_i = \pm 1$, $i = 1, \ldots, N$). Then we calculate the probability $P(E, \{m_i\})$ that, when averaged over all choices of couplings $J_{i_1 \cdots i_p}$, this configuration solves (46) and has free energy $F(\{m_i\}) = -E$. Then

$$N(E) = \sum_{\{m_i\}} P(E, \{m_i\}) . \tag{48}$$

Since the distribution of the J's, eq. (2), is invariant under the "gauge transformation"

$$J_{i_1 \cdots i_p} \to \hat{J}_{i_1 \cdots i_p} = J_{i_1 \cdots i_p} m_{i_1} \ldots m_{i_p}, \tag{49}$$

the probability $P(E, \{m_i\})$ is independent of $\{m_i\}$, and equals the probability, $P(E)$, that the J's (with distribution (2)) satisfy

$$\begin{aligned} x_i &= \frac{1}{p} \sum_{i_2 < \cdots < i_p} \hat{J}_{i i_2 \cdots i_p} \geqslant 0 , \quad i = 1, \ldots N, \\ E &= - \sum_{i_1 < \cdots < i_p} \hat{J}_{i_1 \cdots i_p} = - \sum x_i . \end{aligned} \tag{50}$$

The number of solutions will then be given by $2^N P(E)$.

We first calculate the probability that the sums in (50) have values $x_1, \ldots, x_N$:

$$\begin{aligned} P(x_1 \ldots x_N) = \text{const} \int \prod_{i_1 < \cdots < i_p} \mathrm{d}\hat{J}_{i_1 \cdots i_p} \sqrt{\frac{N^{p-1}}{\pi p!}} \exp\left[-\frac{N^{p-1}}{p!} (\hat{J}_{i_1 \cdots i_p})^2 \right] \\ \times \int \mathrm{d}\lambda_1 \ldots \mathrm{d}\lambda_p \exp\left[i \sum_k \lambda_k \left(x_k - \frac{1}{p} \sum_{i_2 < \cdots < i_p} \hat{J}_{k i_2 \cdots i_p} \right) \right] . \end{aligned} \tag{51}$$

The integrations are easily performed with result (for $N \to \infty$)

$$P(x_1 \ldots x_N) = \frac{p^{(N-1)/2}}{\pi^{N/2}} \exp\left[-p \left[\sum_k x_k^2 - \frac{p-1}{Np} \left(\sum_k x_k \right)^2 \right] \right] . \tag{52}$$

From this distribution it follows that the mean value of x_i is of order $1/\sqrt{p}$. The right-hand side of (46) is of order $\sqrt{p}$, and thus these are solutions of the TAP equations as $p \to \infty$.

First let us calculate the total number of solutions, independent of E. This is given by $2^N \int_0^\infty \mathrm{d}x_1 \ldots \mathrm{d}x_N P(x_1 \ldots x_N)$, which, using a gaussian transformation to disentangle the $(\sum_k x_k)^2$ term in (52), can be expressed as

$$\begin{aligned} \mathcal{N} &= 2^N \sqrt{N} \int_{-\infty}^{+\infty} \frac{\mathrm{d}z}{\sqrt{2\pi}} \exp\left[N[-\tfrac{1}{2} z^2 + \ln f(z)] \right] \\ f(z) &= \int_{-z\sqrt{p-1}}^{\infty} \frac{\mathrm{d}x}{\sqrt{2\pi}} \mathrm{e}^{-x^2/2} . \end{aligned} \tag{53}$$

This can be evaluated by saddle-point methods in the $N \to \infty$ limit. We thereby derive that the average number of TAP solutions grows exponentially with N: $\bar{\mathcal{N}} \sim e^{NA}$, where

$$A = \ln 2 - \frac{\mu^2}{2(p-1)} + \ln \int_{-\mu}^{\infty} \frac{dx}{\sqrt{2\pi}} e^{-x^2/2}, \tag{54}$$

and μ is the value of $z\sqrt{p-1}$ at the saddle-point, i.e.

$$\mu = \frac{p-1}{\sqrt{2\pi}} \frac{e^{-\mu^2/2}}{\displaystyle\int_{-\infty}^{\mu} dx\, e^{-x^2/2}/2\pi}. \tag{55}$$

For $p=2$, we recover the well-known result that the average number of TAP solutions, at $T=0$, grows like $e^{0.2N}$ [13, 14]. As p increases so does A and in the $p \to \infty$ limit, it is easily seen that $A \to \ln 2$. Therefore in the $p \to \infty$ SK model the total number of TAP solutions grows like 2^N (more precisely $A \sim 2^N/\sqrt{2 \ln p}$) which means that almost every configuration is a solution of the TAP equations.

What is the physical meaning of these solutions? They, of course, are saddle-points of the TAP free energy. However, in our case, one can say more. Since the system is frozen, $m_i = \pm 1$, we can interpret a solution as a spin configuration which is a local minimum of the energy. This is because the energy can be written as $E = -\sum_{i_1 < \cdots < i_p} J_{i_1 \cdots i_p} m_{i_1} \ldots m_{i_p} = -\sum_{i=1}^{N} x_i$, and all the terms x_i are positive. If we flip any particular spin, $m_k \to -m_k$, the change in energy is $\delta E = +2x_k > 0$. Now of course not all of these local minima will contribute significantly, in fact we expect that the important configurations will be those with the smallest possible energy [4, 15].

To proceed further we must calculate the average number of solutions, $N(E)$, of a given energy $E = -\sum x_i \equiv -N\varepsilon$. Using the same techniques, this can be cast into the form

$$\overline{\mathcal{N}(E)} = \exp[N(\ln 2 + (p-1)\varepsilon^2)] \int_{-\infty}^{+\infty} \frac{d\lambda}{2\pi\sqrt{2}}$$
$$\times \exp\left[N\left[-\tfrac{1}{2}\lambda^2 - i\lambda\varepsilon\sqrt{2p} + \ln \int_{-i\lambda}^{\infty} Dx\right]\right]. \tag{56}$$

Once again the λ integral can be expanded about the saddle-point $\lambda_s = -i\mu$ as $N \to \infty$;

$$\overline{\mathcal{N}(E)} \sim e^{NA(E)},$$
$$A(E) = (p-1)\varepsilon^2 + \ln 2 + \tfrac{1}{2}\mu^2 - \varepsilon\mu\sqrt{2p} + \ln \int_{-\infty}^{\mu} Dx, \tag{57}$$

where μ is determined by the saddle-point equation

$$\varepsilon\sqrt{2p} - \mu = \frac{1}{\sqrt{2\pi}} \frac{e^{-\mu^2/2}}{\displaystyle\int_{-\infty}^{\mu} Dx}. \tag{58}$$

For $p=2$, we recover the results of [13, 14]. As $p\to\infty$, we find

$$A(E)=\begin{cases}\ln 2-\varepsilon^2, & \varepsilon\geqslant 0\\ 0, & \varepsilon<0.\end{cases} \tag{59}$$

This density of TAP solutions, $\mathcal{N}(E)\sim e^{N[\ln 2-\varepsilon^2]}$, is exactly equal (up to terms which are not exponentially large in N and which do vanish as $p\to\infty$) to the density of energy levels in the random energy model. This is not surprising since it equals the product of the total number of configurations, 2^N, times the probability that a configuration has energy $-N\varepsilon$, $e^{-N\varepsilon^2}$, times the probability that a configuration is a solution of the TAP equations, $e^{0.N}/\sqrt{2\ln p}$.

To calculate the free energy one would naively sum over all solutions with a Boltzmann-like weight:

$$Z\sim\sum_{\text{sol.}} e^{-E/T}=\int_{E_{\min}} dE\,\bar{\mathcal{N}}(E)\,e^{-E/T}, \tag{60}$$

where $E_{\min}$ is the minimum value of the energy for which the number of solutions is $\geqslant 1$. In general (for say $p=2$) this procedure is incorrect [13, 14]. It is equivalent to an annealed average over TAP solutions which does not necessarily agree with the correct quenched average of $\ln Z$. One is then led again to the replica method, now applied to TAP solutions [4, 13, 14]. In our case, these complications are unnecessary, since, as we shall see, the different solutions are totally uncorrelated.

We shall calculate the probability $P(E,E',q)$, that two different configurations, $\{m_i\}$ and $\{m_i'\}$, be solutions of the TAP equations with energies E and E' and mutual overlap q. Using the same strategy as above, we gauge-transform and introduce variables

$$x_i=\frac{1}{p}\sum_{i_2<\cdots<i_p} J_{ii_2\cdots i_p},$$

$$y_i=\frac{1}{p}\sum_{i_2<\cdots<i_p} J_{ii_2\cdots i_p}q_{i_2}\cdots q_{i_p}, \tag{61}$$

where $q_i=m_im_i'$, $q=(1/N)\sum_i q_i$. The TAP equations are equivalent (for $p=\infty$ and taking m_i, m_i' to be ± 1) to

$$\forall i=1,\dots N,\quad x_i>0,\quad y_i>0. \tag{62}$$

Thus we are led to compute the probability that x_i and y_i, defined above, take specific values, averaging over the J's. This yields

$$\begin{aligned}P(x_1\dots x_N,y_1\dots y_N,q_i)=\text{const}\times&\int\prod_k d\lambda_k\,d\mu_k\exp\Big[i\sum_k(\lambda_kx_k+\mu_ky_k)\Big]\\&\times\exp\Big[-\frac{1}{4p}\Big\{\sum_k(\lambda_k^2+\mu_k^2+2\lambda_k\mu_kq^{p-1})\\&+\frac{p-1}{N}\sum_{k\neq l}(\lambda_k\lambda_l+\mu_k\mu_lq_kq_l+\lambda_kq_k\mu_lq^{p-1})\Big\}\Big].\end{aligned} \tag{63}$$

This expression is rather unwieldy, however as $p \to \infty$, it drastically simplifies. If $|q|<1$ the q^{p-1} terms vanish, and the distribution factorizes:

$$P(x_1 \dots x_N, y_1 \dots y_N, q) \sim P(x_1 \dots x_N) P(y_1 \dots y_N) \quad (|q|<1) . \tag{64}$$

If $q=1$, then all but a vanishing fraction of the q_i's are equal and

$$P(x_1 \dots x_N, y_1 \dots y_N, q=1) \sim \prod_k \delta(x_k - y_k) P(x_1 \dots x_N) , \tag{65}$$

which simply means that we are considering the same configuration $\{m_i\}=\{m_i'\}$.

Therefore if $\{m_i\}$ and $\{m_i'\}$ are macroscopically different solutions, they are totally uncorrelated. We can then argue, as in sect. 2, that the fluctuation of $\mathcal{N}(E)$ about its mean $\bar{\mathcal{N}}(E)$ is negligible (of order $1/\sqrt{\mathcal{N}(E)}$) as long as $-\sqrt{\ln 2}<E/N<0$ where $\bar{\mathcal{N}}(E)$ is exponentially large, and one can take $\mathcal{N}(E)$ to equal $\bar{\mathcal{N}}(E)$. Thus we apply (60):

$$e^{-\beta F} \sim \int_{-\sqrt{\ln 2}}^{0} d\varepsilon \exp[N(\ln 2 - \varepsilon^2 - \beta\varepsilon)] . \tag{66}$$

Since $\beta > 2\sqrt{\ln 2}$ (recall that we are in the low-temperature phase since we have assumed the system to be frozen with $m_i = \pm 1$), the integral is dominated by the minimal value of ε, yielding $F/N = -\sqrt{\ln 2}$, which is the correct result.

The calculation can easily be generalized for nonvanishing magnetic field h. The distribution $P(x_1, \dots, x_N)$ is independent of h. However we must now calculate the total number of solutions of a given magnetization m, and energy $E = -\sum_i x_i - hm$. This equals:

$$N(E, m) = \binom{N}{\frac{1}{2}(N+m)} \times P(E+hm).$$

By arguments identical to those presented above, one can then compute the free energy in the presence of a field, which will coincide with (37).

We can proceed further and ask which of the many TAP solutions actually contribute to the canonical average (66). These must have a free energy

$$F = -N\sqrt{\ln 2} + \hat{F}, \tag{67}$$

where $\hat{F}$ is finite (relative to N). The number of solutions having this free energy F behaves as

$$\mathcal{N}(\hat{F}) \sim e^{2\sqrt{\ln 2}\hat{F}} = e^{\hat{F}/T_c} . \tag{68}$$

Of course this formula is only valid in the region where $\hat{F} \gg 1$, so that the number of solutions between $\hat{F}$ and $\hat{F}+\delta\hat{F}$ be large, and that the fluctuations of $\mathcal{N}$ be negligible. These solutions can be identified as pure states of weights

$$P_s = \frac{e^{-\beta F_s}}{\sum_{s'} e^{-\beta F_{s'}}} = C\, e^{-\beta F_s} , \tag{69}$$

where C is a temperature-dependent constant. From (68) we can obtain the number of states with a given $\hat{F} \gg 1$, which corresponds to a given $P \ll 1$. This equals

$$f(P) \sim \sum_s \delta(P_s - P)$$
$$\sim \text{const} \times \sum_s e^{F_s/T} \delta\left(F_s + T \ln \frac{P}{C}\right)$$
$$\sim \text{const} \times P^{-1-T/T_c} . \tag{70}$$

This confirms the asymptotic behaviour for small P of the number of pure states with a given weight P given in (39), and thus the temperature dependence of the (infinite) total number of pure states (40), which is dominated by the states with vanishing weights.

We therefore have a clear picture of the $(p = \infty)$ spin glass phase, in which the pure states, free energy valleys, can be identified with the minimal energy solutions of the TAP equations. We can even say something about the overlaps between these states. Since the probability distribution for 2 different solutions factorizes and is independent of the overlap q of the solutions (if $q < 1$), it follows that the mean number of pairs of solutions with overlap q is $\mathcal{N}(q) = 2^N 2^N P(q)$, where $P(q)$ is the probability that, chosen at random, two configurations have overlap q. $P(q)$ clearly has a binomial distribution,

$$P(q) = 2^{-N} \binom{N}{\frac{1}{2}N(1+q)} ,$$

and therefore

$$\mathcal{N}(q) \sim 2^N e^{-Nq^2} \sqrt{\frac{N}{2\pi}} , \tag{71}$$

which is highly peaked about $q = 0$. Therefore the overlap between any two solutions can only be: 1 if the solutions are macroscopically indistinguishable; $O(1/\sqrt{N})$ if they are distinguishable. In that case the distribution function $P(q)$ will have just two δ-function peaks: at $q = 1$ and at $q = 0$. A more careful calculation could probably yield the (temperature-dependent) weights of these δ-functions.

5. Conclusions

The infinite range Ising model with p-spin quenched random interactions is a natural generalization of the SK model, which exhibits, at low temperature, spin glass behaviour. For $p \to \infty$, it can be solved exactly because of its equivalence to the random energy model. We have shown that the exact solution can also be obtained using the techniques that have been applied to the $p = 2$ SK model (and which have produced much of our theoretical understanding of the spin glass phase).

Using the replica method, we have seen that Parisi's ansatz for replica symmetry breaking is valid in the $p \to \infty$ model. The average order parameter function $q(x)$

that we obtain is simply a step function, which means that the first breaking of replica symmetry is exact in this case. The physical interpretation of replica symmetry breaking as a description of the breakdown of ergodicity, and the existence of an infinite number of pure states α with ultrametric topology is confirmed. Indeed, starting from this interpretation, we have computed the mutual entropy of the pure states $-\sum_\alpha P_\alpha \ln P_\alpha$, which gives exactly the leading finite-N corrections to the entropy in the low-temperature phase, in accordance with the direct computation within the framework of the random energy model.

The TAP equations are particularly simple in this model since macroscopically distinct TAP solutions are uncorrelated. Hence they can be solved directly without introducing replicas. The number of solutions of the TAP equations with free energy f is zero for $f < f_{\min} = -\sqrt{\ln 2}$, and exponentially large for $f > f_{\min}$. We find also confirmation, in this case, of the fact that the canonical average over solutions of TAP equations is dominated by the ones which have free energy $f = f_{\min}$, which can then be identified with the pure states of the system [4].

Since the $p \to \infty$ SK model is so much simpler than the $p = 2$ model while it still retains most of the basic properties of a spin glass (especially the existence of an infinite number of pure states unrelated by a symmetry), we think that it constitutes a good starting point for further investigation.

Our first suggestion would be to study the large-p expansion of the infinite range model. One could calculate the free energy and the function $q(x)$ in an expansion about $p = \infty$. It appears that the corrections are of order e^{-p}, and thus the expansion might be rapidly convergent. This might provide analytic insight into the precise structure of the order parameter $q(x)$ for finite p.

Another interesting problem for which the $p \to \infty$ model could provide a simple starting point is that of fluctuations about mean field theory, namely the treatment of the model in finite dimensions. In this regard we would like to point out that the masses that will appear in the propagators of the perturbation theory about mean field have significant p dependence. Since the energy is proportional to Q_{ab}^p, the second derivative with respect to Q is either of order p^2 (if $Q = 1$) or zero (if $Q < 1$). Therefore it is probably possible to take into account consistently the fluctuations (which will be of order $1/p^2$), while neglecting the finite-p corrections to the mean field theory (which are of order e^{-p}). This might yield an extremely simple perturbation theory, which could be analysed to determine the critical behaviour of the theory.

We would like to thank E. Brézin, C. De Dominicis, B. Derrida, I. Kondor, N. Sourlas and G. Toulouse for useful conversations.

Appendix A

Here we show that the general kth-order replica symmetry breaking produces the solution derived above. At kth-order the matrices λ_{ab} and Q_{ab} are given in terms of the parameters $\lambda_0, \lambda_1, \ldots \lambda_k$; $q_0, q_1 \ldots q_k$ and $m_0 = n, m_1, \ldots, m_k, m_{k+1} = 1$, which determine the value of λ_{ab} and Q_{ab} at each level of the tree (fig. 1) and the number of its branches at each generation.

The saddle-point equations will always yield

$$\lambda_i = \tfrac{1}{2}\beta^2 p q_i^{p-1}, \tag{A.1}$$

which implies that $\lambda_i \to 0$ (if $q_i < 1$) or $\lambda_i \to \infty$ (if $q_i = 1$). If all the q's are <1 then all the λ's vanish and we will recover the symmetric, high-temperature solution. Assume therefore that $q_0 \leq q_1 \leq \cdots \leq q_{k-1} < q_k = 1$. We then search for the maximum of $G(q_i, \lambda_i, m_i)$, defined by (23), which in this case equals

$$\frac{1}{n} G(q_i, \lambda_i, m_i) = \sum_{l=0}^{k} (m_l - m_{l+1})[\tfrac{1}{4}\beta^2 q_l^p - \tfrac{1}{2} q_l \lambda_l] + \frac{1}{n} S(\lambda_i, m_i), \tag{A.2}$$

where $S(\lambda_i, m_i)$ is given by

$$\exp S(\lambda_i, m_i) = \mathrm{Tr}_{(\sigma^a)} \exp\left[\frac{1}{2}\sum_{a \neq b} \lambda_{ab}\sigma_a\sigma_b + \beta h \sum_a \sigma_a\right]. \tag{A.3}$$

The evaluation of S is a classic exercise within Parisi's replica symmetry breaking scheme. Using the methods of [6, 16] we find

$$\exp S(\lambda_i, m_i) = \mathrm{e}^{-n\lambda_k/2} \int \mathrm{D}z_0 \left\{ \int \mathrm{D}z_1 \cdots \left[\int \mathrm{D}z_{k-1} I_k^{m_{k-1}/m_k} \right]^{m_{k-2}/m_{k-1}} \cdots \right\}^{m_0/m_1},$$

$$I_k \equiv \int \mathrm{D}z_k \, \mathrm{ch}^{m_k}\left(\sum_{l=0}^{k} z_l \sqrt{\tilde{\lambda}_l} + \beta h \right), \tag{A.4}$$

where $\tilde{\lambda}_0 \equiv \lambda_0$, $\tilde{\lambda}_l \equiv \lambda_l - \lambda_{l-1} (l \geq 1)$ and $\mathrm{D}z \equiv \mathrm{d}z\, \mathrm{e}^{-z^2/2}/\sqrt{2\pi}$.

Although S is rather complicated it simplifies considerably in our case where $\lambda_k \to \infty$ and $\lambda_{i<k} \to 0$. We need only expand S to first order in $\lambda_{i<k}$. We first consider the innermost integral in (A.3):

$$\begin{aligned} I_k = & \int \mathrm{D}z_k \, \mathrm{ch}^{m_k}(z_k\sqrt{\tilde{\lambda}_k} + \beta h) \\ & + m_k \left(\sum_{l=0}^{k-1} z_l \sqrt{\tilde{\lambda}_l} \right) \int \mathrm{D}z_k \, \mathrm{ch}^{m_k}(z_k\sqrt{\tilde{\lambda}_k} + \beta h)\, \mathrm{th}\,(z_k\sqrt{\tilde{\lambda}_k} + \beta h) \\ & + \tfrac{1}{2} m_k \left(\sum_{l=0}^{k-1} z_l \sqrt{\tilde{\lambda}_l} \right)^2 \int \mathrm{D}z_k \, \mathrm{ch}^{m_k}(z_k\sqrt{\tilde{\lambda}_k} + \beta h) \\ & \times [1 + (m_k - 1)\, \mathrm{th}^2\,(z_k\sqrt{\tilde{\lambda}_k} + \beta h)] + \mathrm{O}(\tilde{\lambda}_l^{3/2}). \end{aligned} \tag{A.5}$$

For large λ_k this yields

$$I_k = C\left[1 + m_k \sum_{l=0}^{k-1} z_l \sqrt{\tilde{\lambda}_l}\, \mathrm{th}\,(m_k \beta h) + \tfrac{1}{2} m_k^2 \left(\sum_{l=0}^{k-1} z_l \sqrt{\tilde{\lambda}_l} \right)^2 \right],$$

$$C = 2^{1-m_k}\, \mathrm{e}^{m_k^2 \lambda_k/2}\, \mathrm{ch}\,(m_k \beta h). \tag{A.6}$$

The next step consists of raising I_k to the power m_{k-1}/m_k and integrating over z_{k-1}. This yields

$$I_{k-1}=\int Dz_{k-1}(I_k)^{m_{k-1}/m_k}=C^{m_{k-1}/m_k}\left[1+m_{k-1}\left(\sum_{l=0}^{k-2} z_l\sqrt{\tilde{\lambda}_l}\right)\text{th}\,(m_k\beta h)\right.$$
$$\left.+\tfrac{1}{2}m_{k-1}\left\{\left(\sum_{l=0}^{k-2} z_l\sqrt{\tilde{\lambda}_l}\right)^2+\tilde{\lambda}_{k-1}\right\}\left\{m_k+(m_{k-1}-m_k)\,\text{th}^2\,(m_k\beta h\right\}\right]. \quad (A.7)$$

This can then be iterated, finally yielding

$$e^{n\lambda_k/2+S(\lambda_i,m_i)}=I_0=C^{m_0/m_k}\left[1+\tfrac{1}{2}m_0\sum_{l=0}^{k-1}\tilde{\lambda}_l\{m_k+(m_0-m_k)\,\text{th}^2\,(m_k\beta h)\}\right]. \quad (A.8)$$

This enables us to determine $G(q_i,\lambda_i,m_i)$ to the desired order:

$$\frac{1}{n}G(q_i,\tilde{\lambda}_i,m_i)\approx\sum_{l=0}^{k}(m_l-m_{l+1})[\tfrac{1}{4}\beta^2q_l^p-\tfrac{1}{2}q_l\sum_{r=0}^{l}\tilde{\lambda}_r]$$
$$-\frac{1}{2}\sum_{l=0}^{k}\tilde{\lambda}_l+\frac{1}{2}\sum_{l=0}^{k-1}\tilde{\lambda}_l[m_k+(m_l-m_k)\,\text{th}^2\,(m_k\beta h)]$$
$$+\frac{1}{m_k}[(1-m_k)\ln 2+\tfrac{1}{2}m_k^2\tilde{\lambda}_k+\ln\text{ch}\,(m_k\beta h)]. \quad (A.9)$$

We can now examine the saddle-point equations. The variation with respect to $\tilde{\lambda}_l$ yield

$$q_k=1 \quad (l=k),$$
$$1+\sum_{r=l}^{k}q_r(m_r-m_{r+1})=m_k+(m_l-m_k)\,\text{th}^2\,(m_k\beta h) \quad (l<k). \quad (A.10)$$

If we solve this equation successively for $l=k-1$, $l=k-2,\ldots,l=1$, we find

$$q_{k-1}=q_{k-2}=\cdots=q_0=\text{th}^2\,(m_k\beta h).$$

Thus we recover the previous saddle-point associated with the first stage of replica symmetry breaking. Hence there are no new saddle-points appearing at the higher stages of replica symmetry breaking.

References

[1] In Proc. of the Heidelberg Colloquium on spin glasses, Lecture Notes in Physics 192 (Springer, 1983)
[2] D. Sherrington and S. Kirkpatrick, Phys. Rev. Lett. 32 (1975) 1792;
S. Kirkpatrick and D. Sherrington, Phys. Rev. B17 (1978) 4384
[3] G. Parisi, Phys. Rev. Lett. 50 (1983) 1946
[4] C. de Dominicis and A.P. Young, J. Phys. A16 (1983) 2063
[5] M. Mézard, G. Parisi, N. Sourlas, G. Toulouse and M. Virasoro, Phys. Rev. Lett. 52 (1984) 1156; J. de Phys. 45 (1984) 843

[6] G. Parisi, Phys. Rev. Lett. 43 (1979) 1754; J. Phys. A13 (1980) L115, 1101, 1887
[7] C. de Dominicis and I. Kondor, Phys. Rev. B27 (1983) 606
[8] B. Derrida, Phys. Rev. Lett. 45 (1980) 79; Phys. Rev. B24 (1981) 2613
[9] D.J. Thouless, P.W. Anderson and R.G. Palmer, Phil. Mag. 35 (1977) 593
[10] S.F. Edwards and P.W. Anderson, J. Phys. F5 (1975) 965
[11] J.R.L. de Almeida and D.J. Thouless, J. Phys. A11 (1978) 983
[12] R.G. Palmer, ref. [1]
[13] C. de Dominicis, M. Gabay, T. Garel and H. Orland, J. de Phys. 41 (1980) 923
[14] A.J. Bray and M.A. Moore, J. Phys. C13 (1980) L469
[15] A.J. Bray, M.A. Moore and A.P. Young, J. Phys. C17 (1984) L155
[16] B. Duplantier, J. Phys. A14 (1981) 283

PHYSICAL REVIEW B VOLUME 25, NUMBER 11 1 JUNE 1982

Relaxational dynamics of the Edwards-Anderson model and the mean-field theory of spin-glasses

H. Sompolinsky
Department of Physics, Harvard University, Cambridge, Massachusetts 02138

Annette Zippelius
Physik Department, Technische Universität München, Garching, Federal Republic of Germany

(Received 6 January 1982)

Langevin equations for the relaxation of spin fluctuations in a soft-spin version of the Edwards-Anderson model are used as a starting point for the study of the dynamic and static properties of spin-glasses. An exact uniform Lagrangian for the average dynamic correlation and response functions is derived for arbitrary range of random exchange, using a functional-integral method proposed by De Dominicis. The properties of the Lagrangian are studied in the *mean-field* limit which is realized by considering an *infinite-ranged* random exchange. In this limit, the dynamics are represented by a stochastic equation of motion of a *single* spin with self-consistent (bare) propagator and Gaussian noise. The low-frequency and the static properties of this equation are studied both above and below T_c. Approaching T_c from above, spin fluctuations slow down with a relaxation time proportional to $|T-T_c|^{-1}$ whereas *at* T_c the damping function vanishes as $\omega^{1/2}$. We derive a criterion for dynamic stability below T_c. It is shown that a stable solution necessarily violates the fluctuation-dissipation theorem below T_c. Consequently, the spin-glass order parameters are the time-persistent terms which appear in both the spin correlations and the local response. This is shown to invalidate the treatment of the spin-glass order parameters as purely static quantities. Instead, one has to specify the manner in which they relax in a finite system, along time scales which diverge in the thermodynamic limit. We show that the *finite*-time correlations decay algebraically with time as $t^{-\nu}$ at all temperatures below T_c, with a temperature-dependent exponent ν. Near T_c, ν is given (in the Ising case) as $\nu(T)\sim\frac{1}{2}-\pi^{-1}(1-T/T_c)+\sigma(1-T/T_c)^2$. A tentative calculation of ν at $T=0$ K is presented. We briefly discuss the physical origin of the violation of the fluctuation-dissipation theorem.

I. INTRODUCTION

The low-temperature properties of magnetic systems with quenched random exchange (spin-glasses) have attracted considerable attention in recent years.[1-3] Much of the theoretical work concentrated on the spin-glass (SG) model introduced by Edwards and Anderson[4] (EA); see Sec. II below. They proposed that the system may undergo a phase transition into a state, in which the local spins S_i are frozen in random directions, which can be described by the "EA order parameter,"

$$q_{EA}=[\langle S_i\rangle_T^2]_J\ . \tag{1.1}$$

The symbol $\langle\ \rangle_T$ refers to thermal average in a particular system and $[\ ]_J$ stands for averaging over the random exchange. At present, it is still unclear whether such a phase transition actually occurs in a three-dimensional (3D) EA model.[5-8] Monte Carlo simulations[2] clearly show the appearance of a frozen order at low temperatures in a fashion which is very similar to the observed properties of SG's. However, recent exact results[8] for small samples indicate that the frozen order in 2D or 3D systems is a nonequilibrium phenomenon.[2]

Despite this uncertainty, it is useful to investigate the properties of the SG state in the mean-field limit of the EA model. This limit is realized in the infinite-ranged random-exchange model introduced by Sherrington and Kirkpatrick[9,10] (SK). The SK model has a sharp transition at a finite temperature, but the properties of its low-temperature state are highly nontrivial. The model has been extensively studied by the replica method, which is commonly employed in various problems of quenched disorder. The method consists of calculating the average free energy via the average partition function of n replicated systems, and taking the limit $n \to 0$,

$$[\ln Z]_J = \lim_{n \to 0} ([Z^n]_J - 1)/n \ . \tag{1.2}$$

In this framework, the EA order parameter is identified as

$$q_{\rm EA} = q^{\alpha\beta} = [\langle S_i^\alpha S_i^\beta \rangle_T]_J, \ \ \alpha \neq \beta \tag{1.3}$$

where S_i^α, S_i^β denote (Ising) spins of two different replicas. A straightforward application of this method yields[9,10] a mean-field solution in which the SG phase is characterized by a single order parameter $q = q^{\alpha\beta}$ which obeys a simple self-consistent equation. This solution is unstable, however, below T_c (Refs. 11 and 12) and also yields unphysical negative entropy near $T = 0$.[10] In order to alleviate this instability one needs to break the replica symmetry by introducing order parameters $q^{\alpha\beta}$ which depend on (α, β), but there are infinitely many ways of breaking this symmetry, and the replica theory provides neither a clear criterion of which way is correct, nor a physical insight to the meaning of the various order parameters and the broken symmetry.[13-15] Similar difficulties appear in the replica theory of the short-range EA model below T_c.[16]

An alternative approach has been proposed by Thouless, Anderson, and Palmer[17] (TAP), who derived mean-field equations for the local random magnetizations $\langle S_i \rangle_T$. Although the TAP theory shed further light onto the nature of the SG phase, it has its own difficulties. A straightforward perturbative average of the TAP equations yields Sommer's[18] solution, which has some odd features at low temperatures.[18-20] A nonperturbative method of averaging over the TAP equations gives rise to additional solutions but the physical meaning of these solutions is not clear, and there is still no simple way of finding the solution with the lowest energy.[21,22] Also, an extension of the method to short-range models requires knowledge of the properties of the eigenstates of short-range random matrices, many of which are still not available.[19]

Dynamics have been proposed by Ma and Rudnick[23] and subsequently by other authors[24-27] as an alternative approach to the SG problem. The motivation for this is twofold. Firstly, dynamics provide means for calculating average thermodynamic quantities without using the unphysical $n \to 0$ replica limit. Secondly, one would like to understand not only static properties but also the time-dependent features of the SG state, especially since many of the unique low-temperature properties of real SG's are dynamic in nature.

Although previous dynamic theories correctly predicted the mean-field dynamic behavior above and at T_c, as Monte Carlo simulations of the SK model have shown,[10] they failed below T_c in several crucial aspects. Most importantly, they described the static properties of the SG phase by a single EA order parameter which is defined, in a dynamic framework, as

$$q_{\rm EA} = \lim_{t \to \infty} [\langle S_i(0) S_i(t) \rangle]_J \ . \tag{1.4}$$

Thus they exhibit the same instability that appears in the SK replica solution.[25] Secondly, they arrive at a mean-field-type solution by performing a low-order perturbation calculation in a *short-range* problem. The inadequacy of these approximations even as a proper mean-field theory of SG's, was clearly demonstrated by the fact that they do not distinguish between the cases of random exchange and a local random temperature.[23,24,28]

In this work we present a theory of SG's based on a dynamic approach to the EA model. A brief summary of our results has been reported elsewhere.[29] We use a soft-spin version of the EA model (on a lattice) and define the dynamics of the system by a Langevin equation describing the relaxation of spin fluctuations similar to the time-dependent Ginzburg-Landau model[30] in uniform-spin systems.

A few years ago, De Dominicis[25] showed that the dynamic functional integral method[31-33] of Martin, Siggia, and Rose[31] can be used to average out quenched disorder without using the replica method. Using this formalism we derive a uniform dynamic Lagrangian which generates all average spin correlation and response functions for a general (e.g., short-range) EA model. Of course the price that is paid is the necessity of extracting the statics via a solution of a full dynamic problem. But this "technical" complexity is, in our opinion, worthwhile especially since it turns out that one cannot separate completely the properties of the SG (in the thermodynamic limit) at thermal equilibrium from the nonequilibrium ones.

A study of the properties of the dynamic theory in a short-range EA model will be discussed elsewhere.[34] In the present paper we study its properties in the infinite-range limit where a mean-field solution is exact. We show that, in this limit, the dynamics can be expressed as a stochastic equation of motion of a *single* spin with self-consistent (bare) propagator and noise. Analyzing the equilibrium solution of this equation, we show that the

SG phase is characterized by the appearance of *time-persistent* terms not only in the spin correlations, Eq. (1.4), but also in the average response function, i.e.,

$$\lim_{\omega\to 0} T\chi(\omega) \sim 1-q_{\rm EA}+\Delta\delta_{\omega,0} \,. \tag{1.5}$$

The breakdown of the Fischer[35] relation $\chi=(1-q_{\rm EA})/T$ in the SG phase has been previously derived by various static studies[13,15,18–22] of the SK model. Here we show that the appearance of Δ gives rise to a nonunique thermodynamic limit of the static solution, unless one specifies the *time dependence* in a *finite* system of the spin correlations (and response functions) along time scales which become infinitely long in the thermodynamic limit. This invalidates the treatment of the "frozen" spin correlations as a single static order parameter, Eq. (1.4), and gives rise to a multitude of order parameters and consequently to infinitely many possible static solutions. Using this approach together with some physically plausible assumptions about the time dependence of q and Δ, a *static* solution has been recently constructed[36] which agrees with Parisi's replica results[14] and seems to be the correct mean-field solution of the SK model.

In this paper we proceed to investigate the dynamic properties of the SG phase for frequencies which are small compared to *microscopic* characteristic frequencies but do not vanish as the size of the system approaches infinity. In this scale, the fluctuation-dissipation theorem holds and we are able to express uniquely the low-frequency properties in terms of the moments of the local frozen magnetization (measured in *finite* time). We derive a criterion for the dynamic stability of the various static solutions. In the case of the SK solution, the criterion for dynamic stability is identical to the Almeida-Thouless[11] replica stability condition. The recently derived mean-field static solution has been shown[36] to satisfy the condition for marginal dynamic stability at all temperatures. This is shown here to lead to a power-law decay in time of spin correlations $\sim t^{-\nu}$ with an exponent ν which acquires a universal value $\frac{1}{2}$ at T_c but *decreases* upon cooling below T_c. In the Ising case ν is given, near T_c, as

$$\nu(T) \sim \tfrac{1}{2}-\pi^{-1}(1-T/T_c) + O(1-T/T_c)^2 \,. \tag{1.6}$$

However, calculation of its value at $T\to 0$ K depends on as yet unknown properties of the static solution at very low T.

The outline of the paper is as follows. In Sec. II we define the relaxational dynamic model and derive the effective uniform dynamic Lagrangian for the EA model. In Sec. III an exact self-consistent local equation of motion is derived for the infinite-ranged case. The dynamic properties of this equation at and above T_c are analyzed in Sec. IV. Section V deals with the equilibrium solutions below T_c. The dynamic properties in the SG phase are studied in Sec. VI. Section VII contains concluding remarks and a brief discussion of the breakdown of the fluctuation-dissipation theorem.

Some of the results of this work, in particular, the identification of the Almeida-Thouless instability as a dynamic instability and the appearance of a zero-frequency singularity in the response function, were also found recently by Hertz *et al.*[37] The averaged dynamic Lagrangian discussed in Sec. III has also been derived very recently by Schuster.[38]

II. THE DYNAMIC MODEL

The EA Hamiltonian is

$$H=-\sum_{\langle ij\rangle} J_{ij}S_iS_j \,, \tag{2.1}$$

where $\langle ij\rangle$ means a sum over nearest-neighbor pairs, and the exchange J_{ij} are random variables with a Gaussian distribution,

$$P(J_{ij})=(2\pi z/\tilde{J}^2)^{-1/2} \times \exp[-z(J_{ij}-J_0/z)^2/2\tilde{J}^2] \,. \tag{2.2}$$

The spin variables S_i take the values ± 1, and z is the number of nearest neighbors. The disorder is assumed to be quenched so that the average over J_{ij} has to be carried out on physical observables such as the free energy or spin-spin correlations. We consider here a soft-spin version of the EA model defined by

$$\beta H=\tfrac{1}{2}\sum_{\langle ij\rangle}(r_0\delta_{ij}-2\beta J_{ij})\sigma_i\sigma_j + u\sum_i \sigma_i^4+\sum_i h_i\sigma_i, \quad \beta=1/T \,. \tag{2.3}$$

The length of the soft spin σ_i is allowed to vary continously from $-\infty$ to $+\infty$. The parameters r_0 and u are independent of temperature, and h_i is an external magnetic field divided by temperature. The model with fixed spin length is recovered from (2.3) in the limit $r_0\to -\infty$ and $u\to +\infty$, such that their ratio remains finite.

To study the relaxational dynamics of spin glasses, we propose a simple phenomenological Langevin equation,

$$\begin{aligned}\Gamma_0^{-1}\partial_t\sigma_i(t) &= -\frac{\delta(\beta H)}{\delta\sigma_i(t)}+\xi_i(t)\\ &= \sum_j (r_0\delta_{ij}-\beta J_{ij})\sigma_j(t)\\ &\quad +4u\sigma_i^3(t)+h_i(t)+\xi_i(t)\ . \end{aligned}\tag{2.4}$$

The noise ξ_i is a Gaussian random variable with zero mean and variance

$$\langle \xi_i(t)\xi_j(t')\rangle = \frac{2}{\Gamma_0}\delta_{ij}\delta(t-t')\ , \tag{2.5}$$

which ensures the proper equilibrium distribution and also that locally, the *fluctuation-dissipation theorem* (FDT) holds. The physical quantities are products of spin variables averaged over the noise ξ. Of particular interest are the two-spin correlation

$$C_{ij}(t-t') = \langle \sigma_i(t)\sigma_j(t')\rangle \tag{2.6}$$

and the linear-response function

$$G_{ij}(t-t') = \frac{\partial\langle\sigma_i(t)\rangle}{\partial h_j(t')}\ ,\quad t>t' \tag{2.7}$$

where $\langle\ \rangle$ means average over ξ.

The Fourier transform

$$G_{ij}(\omega) = \int_0^\infty dt\, e^{i\omega t}G_{ij}(t)\ , \tag{2.8}$$

is an analytic function in the upper-half plane, i.e., $\mathrm{Im}\omega>0$. Its real and imaginary parts obey the Kramers-Kronig relations

$$\begin{aligned}\mathrm{Re}G_{ij}(\omega) &= -\int\frac{d\omega'}{\pi}\frac{\mathrm{Im}G_{ij}(\omega')}{\omega-\omega'}\ ,\\ \mathrm{Im}G_{ij}(\omega) &= \int\frac{d\omega'}{\pi}\frac{\mathrm{Re}G_{ij}(\omega')}{\omega-\omega'}\ ,\end{aligned}\tag{2.9}$$

where the principal value of the integrals has to be taken. The FDT reads in the present context

$$\begin{aligned}C_{ij}(\omega) &= \int_{-\infty}^{+\infty} dt\, e^{i\omega t}C_{ij}(t)\\ &= \frac{2}{\omega}\mathrm{Im}G_{ij}(\omega)\ ,\end{aligned}\tag{2.10}$$

and

$$C_{ij}(t=0) = G_{ij}(\omega=0)\ . \tag{2.11}$$

Using the functional integral formulation of De Dominicis[32] and Janssen *et al.*,[33] we define a generating functional for dynamic correlations and response functions,

$$Z\{J_{ij},l_i,\hat{l}_i\} = \int D\sigma D\hat\sigma \exp\left[\int dt\, l_i(t)\sigma_i(t)+i\hat{l}_i(t)\hat\sigma_i(t)+L\{\sigma,\hat\sigma\}\right]\ , \tag{2.12}$$

where

$$L\{\sigma,\hat\sigma\} = \int dt\sum_i i\hat\sigma_i(t)\left[-\Gamma_0^{-1}\partial_t\sigma_i(t)-r_0\sigma_i(t)+\beta\sum_j J_{ij}\sigma_j(t)-4u\sigma_i^3(t)-h_i(t)+\Gamma_0^{-1}i\hat\sigma_i(t)\right]+V\{\sigma\}\ . \tag{2.13}$$

The term V, which arises from the functional Jacobian, is given by[32,33]

$$\begin{aligned}V &= -\tfrac{1}{2}\int dt\sum_i\frac{\delta^2(\beta H)}{\delta\sigma_i^2}\\ &= -\int dt\sum_i[\tfrac{1}{2}r_0+6u\sigma_i^2(t)]\end{aligned}\tag{2.14}$$

and ensures the proper normalization of Z,

$$Z\{J_{ij},l_i=\hat{l}_i=0\}=1\ . \tag{2.15}$$

As usual, $\int D\sigma D\hat\sigma$ means

$$\int_{-\infty}^{+\infty}\prod_{i,t}[d\sigma_i(t)d\hat\sigma_i(t)]\ .$$

The cummulants are given as the coefficients in the Taylor expansion of $\ln Z$ in l and $\hat{l}$

$$\left.\frac{\delta^n\delta^m \ln Z}{\delta\hat{l}_1(\hat{t}_1)\cdots\delta l_m(t_m)}\right|_{l_i=\hat{l}_i=0} = \langle i\hat\sigma_1(\hat{t}_1)\cdots\sigma_m(t_m)\rangle_c\ . \tag{2.16}$$

The auxiliary field $i\hat\sigma_i(t)$, which was introduced by Martin, Siggia, and Rose,[31] acts as a response field $\partial/\partial h_i(t)$. As can be seen from Eqs. (2.12) and (2.13), l_i plays the role of an applied magnetic field, so that all cummulants in (2.16) with $n\geq 1$ are response functions. In particular,

$$\langle i\hat\sigma_j(t')\sigma_i(t)\rangle = G_{ij}(t-t')\ .$$

As required by causality, the response functions vanish if any of the $\hat{t}_i$ are larger than *all* t_i. For more details on the general formalism, see Refs. 25, 32, and 33.

The correlations generated by Eq. (2.16) still depend on the random variables J_{ij}. We are, however, interested in averaged quantities. As noted by De Dominicis,[25] since the generating functional is normalized [see Eq. (2.15)], the quenched average is done directly on Z. This is particularly convenient in cases such as ours in which the Lagrangian is linear in the Gaussian random variable (in our case J_{ij}). A straightforward integration then yields

$$[Z]_J \equiv \int \prod dJ_{ij} P(J_{ij}) Z\{J_{ij}\} = \int D\sigma\, D\hat{\sigma} \exp\left\{ L_0\{\sigma,\hat{\sigma}\} + \frac{\beta J_0}{z} \sum_{\langle ij\rangle} \int dt\, i\hat{\sigma}_i(t)\sigma_j(t) + 2\frac{\beta^2 \tilde{J}^2}{z} \sum_{\langle ij\rangle} \int dt\, dt' [i\hat{\sigma}_i(t)\sigma_j(t') i\hat{\sigma}_i(t')\sigma_j(t) + i\hat{\sigma}_i(t)\sigma_j(t) i\hat{\sigma}_j(t')\sigma_i(t')] \right\}, \tag{2.17}$$

where L_0 is the purely local part of L, i.e.,

$$L_0\{\sigma,\hat{\sigma}\} = \int dt \sum_i [i\hat{\sigma}_i(-\Gamma_0^{-1}\partial_t\sigma_i - r_0\sigma_i - 4u\sigma_i^3 - h_i + i\Gamma_0^{-1}\hat{\sigma}_i) + V\{\sigma\} + i\hat{l}_i\hat{\sigma}_i + l_i\sigma_i] . \tag{2.18}$$

[Note that in deriving (2.17) we use the property $J_{ij}=J_{ji}$.] Thus, integrating out the random exchange, we have introduced a four-spin coupling which is nonlocal in time. This is analogous to the result of averaging of the free energy by the replica method, which generates a four-spin interaction between different replicated systems. The result can be set in a more convenient form (at least for $J_0=0$) by using a Gaussian transformation to decouple the four-spin interactions in (2.17) and introducing four auxiliary fields $Q_\alpha^i(t,t')$, $\alpha=1,\ldots,4$ which are local in space but not in time. This leads to

$$[Z]_J = \int \prod_\alpha^4 DQ_\alpha^i(t,t') \exp\left\{ -\frac{z}{\beta^2 \tilde{J}^2} \int dt\, dt' \sum_{i,j} (K^{-1})_{ij} [Q_1^i(t,t')Q_2^j(t,t') + Q_3^i(t,t')Q_4^j(t,t')] + \ln \int D\sigma\, D\hat{\sigma} \exp L\{\sigma,\hat{\sigma},Q_\alpha\} \right\}, \tag{2.19}$$

where K is the short-range matrix ($K_{ij}=1$ if i,j are nearest neighbors and zero otherwise), and

$$L\{\sigma,\hat{\sigma},Q_\alpha\} = L_0\{\sigma,\hat{\sigma}\} + \tfrac{1}{2} \int dt\, dt' \sum_i [Q_1^i(t,t') i\hat{\sigma}_i(t) i\hat{\sigma}_i(t') + Q_2^i(t,t')\sigma_i(t)\sigma_i(t') + Q_3^i(t,t') i\hat{\sigma}_i(t)\sigma_i(t') + Q_4^i(t,t') i\hat{\sigma}_i(t')\sigma_i(t)] . \tag{2.20}$$

(We have assumed $J_0=0$.)

The results (2.17) and (2.18) can serve as a useful starting point for studying the critical properties of the dynamic and the static properties of short-range spin glasses. Furthermore, the same procedure can be used to derive an averaged generating functional for more complicated dynamic processes. Here, we proceed to study the properties of (2.17) in the mean-field limit. In the following, we will assume for simplicity that $J_0=0$. Generalizing the results to the $J_0 \neq 0$ case is straightforward.

III. EQUATIONS OF MOTION IN THE MEAN-FIELD LIMIT

The mean-field limit is achieved by considering an infinite-ranged J_{ij}, namely, by taking the number of nearest neighbors, z, equal to the number of spins in the systems, N, similar to the static SK model.[9,10] In the infinite-ranged case, Eq. (2.17) can be reduced (in the $N \to \infty$ limit) to a *local* mean-field equation. To do this we write the quartic interaction in Eq. (2.17) as squares of sums of local quantities,

$$N^{-2} \sum_{i \neq j} i\hat{\sigma}_i(t) i\hat{\sigma}_i(t')\sigma_j(t)\sigma_j(t') = \tfrac{1}{4} N^{-2} \left[\sum_i i\hat{\sigma}_i(t) i\hat{\sigma}_i(t') + \sigma_i(t)\sigma_i(t') \right]^2 - \tfrac{1}{4} N^{-2} \left[\sum_i i\hat{\sigma}_i(t) i\hat{\sigma}_i(t') - \sigma_i(t)\sigma_i(t') \right]^2 + O(1/N) ,$$

and similarly for the second quartic term in Eq. (2.17). Thus, decoupling these squares by a Gaussian transformation yields

$$[Z]_J=\int \prod_{\alpha=1}^{4} DQ_\alpha(t,t')\exp\left\{-\frac{N}{\beta^2\tilde{J}^2}\int dt\,dt'[Q_1(t,t')Q_2(t,t')+Q_3(t,t')Q_4(t,t')]\right.$$
$$\left.+\ln\int D\sigma\,D\hat{\sigma}\exp L\{\sigma,\hat{\sigma},Q_\alpha\}\right\}, \tag{3.1}$$

where

$$L\{\sigma,\hat{\sigma},Q_\alpha\}=L_0\{\sigma,\hat{\sigma}\}+\tfrac{1}{2}\int dt\,dt'\sum_i[Q_1(t,t')i\hat{\sigma}_i(t)i\hat{\sigma}_i(t')+Q_2(t,t')\sigma_i(t)\sigma_i(t')$$
$$+Q_3(t,t')i\hat{\sigma}_i(t)\sigma_i(t')+Q_4(t,t')i\hat{\sigma}_i(t')\sigma_i(t)]+O(1)\,. \tag{3.2}$$

In the limit $N\to\infty$ the integration over Q_α can be performed using the method of steepest descent, which amounts to substituting for Q_α, their stationary point values,

$$Q_1^0(t,t')=\frac{\beta^2\tilde{J}^2}{2N}\sum_i\langle\sigma_i(t)\sigma_i(t')\rangle\,, \tag{3.3a}$$

$$Q_2^0(t,t')=\frac{\beta^2\tilde{J}^2}{2N}\sum_i\langle i\hat{\sigma}_i(t)i\hat{\sigma}_i(t')\rangle\,, \tag{3.3b}$$

$$Q_3^0(t,t')=\frac{\beta^2\tilde{J}^2}{2N}\sum_i\langle i\hat{\sigma}_i(t')\sigma_i(t)\rangle\,, \tag{3.3c}$$

$$Q_4^0(t,t')=\frac{\beta^2\tilde{J}^2}{2N}\sum_i\langle i\hat{\sigma}_i(t)\sigma_i(t')\rangle\,. \tag{3.3d}$$

The averages at the right-hand side (rhs) of Eqs. (3.3) are calculated with $L\{\sigma,\hat{\sigma},Q_\alpha^0\}$ of Eq. (3.2), leading to self-consistent equations for Q_α^0.

First we discuss Eq. (3.3b). The correlation $\langle\hat{\sigma}\hat{\sigma}\rangle$ has no simple physical meaning and is indeed identically zero in dynamics of ordinary spin systems.[32,33] It can be seen from the structure of Eqs. (3.1) and (3.2) that

$$Q_2^0=\langle\hat{\sigma}\hat{\sigma}\rangle=0 \tag{3.4}$$

is a self-consistent solution in our case as well, at all temperatures. This solution has the property that it preserves the normalization $[Z]_J=1$. On the other hand, the appearance of a term $Q_2^0\sigma\sigma$ in Eq. (3.2) will generate closed loops which will result in $[Z]_J\sim\exp(N\alpha)$, $\alpha\neq0$. Formally, this can happen as a spontaneous symmetry breaking which violates the original normalization in the limit $N\to\infty$ below T_c. Such a solution will probably invalidate *a posteriori* our averaging procedure, (2.17), which relied crucially on the normalization (2.15). Instead, Z will act as a partition function, and one will have to average over $\ln Z$. This situation is very similar and probably closely related to the analysis by De Dominicis *et al.*[22] and Bray and Moore[23] of the mean-field equations of Thouless, Anderson, and Palmer[18] (TAP). Denoting the average number of the solutions of the TAP equations by $[N_s]_J$ they find that while at $T>T_c$, $[N_s]_J=1$, below T_c there are self-consistent solutions which have the property that $[N_s]_J=\exp(N\alpha)$ with nonzero α. These authors interpret $[N_s]_J$ as the number of metastable states and proceed to calculate averages over these states by a replica method. It is also noted that the TAP solutions with $\alpha\neq0$ have been shown to be related to a replica symmetry breaking solution which violates the normalization $\lim_{n\to0}[Z^n]_J=1$, Z being here the static partition function of the SK model.

However, from the dynamic point of view it is clear that the only physical solution is indeed Eq. (3.4). The reason for that is that the introduction of a vertex $Q_2^0(t,t')\sigma(t)\sigma(t')$ in Eq. (3.2) will lead to *violation of causality*, namely, will yield nonzero contributions to $\langle i\hat{\sigma}(t)\sigma(t')\rangle$ with $t>t'$.[39] Thus we adopt here the self-consistent solution (3.4) which in turn ensures the normalization (2.15) and causality. Thereby we are left with a dynamic local Lagrangian,

$$L\{\sigma_i\hat{\sigma}_i\}=L_0\{\sigma_i,\hat{\sigma}_i\}+\frac{\beta^2\tilde{J}^2}{2}\int dt\,dt'[C(t-t')i\hat{\sigma}_i(t)i\hat{\sigma}_i(t')+2G(t-t')i\hat{\sigma}_i(t)\sigma_i(t')]\,, \tag{3.5}$$

where C and G are the average local correlation and response functions,

$$C(t-t') \equiv [\langle \sigma_i(t)\sigma_i(t')\rangle]_J ,$$
$$G(t-t') \equiv [\langle i\hat{\sigma}_i(t')\sigma_i(t)\rangle]_J , \qquad (3.6)$$

which have to be calculated self-consistently with L. Examining the structure of L_0, one notices that the effect of the spin-glass interaction in the mean-field limit is to modify the inverse bare propagator (via the $\beta^2\tilde{J}^2 G\hat{\sigma}\sigma$ vertex) as well as the width of the Gaussian noise (via the $\beta^2\tilde{J}^2 C\hat{\sigma}\hat{\sigma}$ vertex). Indeed, carrying out the integration over $\hat{\sigma}$ in

$$\int D\sigma_i D\hat{\sigma}_i \exp(L\{\sigma_i\hat{\sigma}_i\}) ,$$

the generating functional can be again expressed by an equation of motion for σ_i, which is, in Fourier representation,

$$\sigma_i(\omega) = G_0(\omega)[\phi_i(\omega)+h_i(\omega)] - 4uG_0(\omega)\int d\omega_1 d\omega_2 \sigma_i(\omega_1)\sigma_i(\omega_2) \times\sigma_i(\omega-\omega_1-\omega_2) . \qquad (3.7)$$

The new effective bare propagator is

$$G_0^{-1}(\omega) = r_0 - i\omega\Gamma_0^{-1} - \beta^2\tilde{J}^2 G(\omega) , \qquad (3.8)$$

and the effective noise ϕ is a Gaussian random variable with width

$$\langle \phi_i(\omega)\phi_i(\omega')\rangle = [2\Gamma_0^{-1} + \beta^2\tilde{J}^2 C(\omega)]\delta(\omega+\omega') . \qquad (3.9)$$

The functions $C(\omega)$ and $G(\omega)$ are the Fourier transforms of the autocorrelations and response, (3.6), and must be calculated self-consistently through Eq. (3.7). Note that the effective local equation of motion is non-Markovian: The noise ϕ is not instantaneous and the bare propagator is nonlocal in time.

IV. DYNAMICS FOR $T \geq T_c$

In this section we investigate the low-frequency behavior of response and correlation functions in the paramagnetic phase. There are no time-persistent correlations and the FDT [Eqs. (2.10), (2.11)] between the full response and correlations is expected to hold. Low-frequency spin fluctuations are conveniently characterized by a generalized damping function $\Gamma(\omega)$, defined as

$$\Gamma^{-1}(\omega) = i\frac{\partial G^{-1}(\omega)}{\partial\omega} . \qquad (4.1)$$

The dynamic-response function obeys a Dyson equation

$$G^{-1}(\omega) = G_0^{-1}(\omega) + \Sigma(\omega) , \qquad (4.2)$$

with a frequency-dependent self-energy $\Sigma(\omega)$. Above T_c, Eq. (4.2) implies

$$\Gamma^{-1}(\omega) = \frac{\Gamma_0^{-1} + i\dfrac{\partial\Sigma}{\partial\omega}}{1-\beta^2\tilde{J}^2 G^2(\omega)} . \qquad (4.3)$$

The denominator represents the renormalization of the damping function due to the random exchange while $i\partial\Sigma/\partial\omega$ is the further renormalization due to the nonlinear coupling u.

We shall first assume that $\partial\Sigma/\partial\omega$ $(\omega=0)$ is finite and discuss the resulting low-frequency behavior. Subsequently we shall show that our assumption is indeed correct. If the low-frequency expansion of the self-energy

$$\Sigma(\omega) \simeq \mathrm{Re}\,\Sigma(0) + i\omega\frac{\partial}{\partial\omega}\mathrm{Im}\,\Sigma(0) , \qquad (4.4)$$

where

$$\frac{\partial}{\partial\omega}\mathrm{Im}\,\Sigma(0) \equiv \frac{\partial}{\partial\omega}\mathrm{Im}\,\Sigma\bigg|_{\omega=0} ,$$

is inserted into Eq. (4.2), we find

$$\mathrm{Re}\,G(\omega) \simeq \frac{r_0 + \mathrm{Re}\,\Sigma(0)}{1+\beta^2\tilde{J}^2|G(\omega)|^2}|G(\omega)|^2 , \qquad (4.5)$$

$$\mathrm{Im}\,G(\omega) \simeq \frac{\Gamma_0^{-1} - \dfrac{\partial}{\partial\omega}\mathrm{Im}\,\Sigma(0)}{1-\beta^2\tilde{J}^2|G(\omega)|^2}|G(\omega)|^2 . \qquad (4.6)$$

In the limit $\omega\to 0$, Eq. (4.6) implies a singularity in the imaginary part of the response function at the transition temperature

$$T_c = \tilde{J}G(0) . \qquad (4.7)$$

The real part of the response function, i.e., the static susceptibility, remains finite, as is expected at the spin-glass transition. For $T > T_c$ the kinetic coefficient Γ has a finite limiting value as $\omega\to 0$:

$$\Gamma^{-1}(\omega=0) = \frac{\Gamma_0^{-1} + \mathrm{Im}\,\dfrac{\partial\Sigma}{\partial\omega}(0)}{1-\beta^2\tilde{J}^2 G^2(0)} . \qquad (4.8)$$

As T approaches T_c, Γ^{-1} $(\omega=0)$ shows critical slowing down,

$$\Gamma^{-1}(0) \propto |\tau|^{-1}, \quad \tau \equiv 1 - T/T_c \qquad (4.9)$$

and spin fluctuations will therefore decay at a rate

$\sim\Gamma G^{-1}\sim|\tau|$. At T_c, we substitute $G^{-1}(\omega)\simeq G^{-1}(0)-i\omega\Gamma^{-1}(\omega)$ in Eq. (4.3), yielding

$$\Gamma(\omega)\sim\omega^{1/2}\,. \tag{4.10}$$

In order to show that $\partial\Sigma/\partial\omega$ is finite, we construct a dynamic perturbation expansion for $\Sigma(\omega)$ similar to a usual time-dependent Ginzburg-Landau model,[29] with the simplification due to the absence of nonlocal interactions. In renormalized perturbation theory the lowest-order contribution is explicitly given by

$$\frac{\partial\Sigma(0)}{\partial\omega}=2(12u)^2\int\frac{d\omega_1}{2\pi}\frac{d\omega_2}{2\pi}C(\omega_1)C(\omega_1-\omega_2)\times\frac{\partial}{\partial\omega_2}\text{Im}\,G(\omega_2)\,. \tag{4.11}$$

Above T_c the frequency integrals are obviously finite. At T_c, $G(\omega)\sim\omega^{1/2}$ and $C(\omega)\sim\omega^{-1/2}$ so that again there is no divergence due to the two integrations over internal frequencies. Since higher order-diagrams contain *at least* two internal integrations of the type of (4.11), they cannot give rise to a divergent $\partial\Sigma/\partial\omega$.

The results (4.9) and (4.10), were found previously by various mean-field approximations to a short-range SG.[23–25] They have also been derived for the SK model using linearized Glauber dynamics. Here we have shown that the nonlinearities do not alter the critical behavior above or at T_c. They do, however, give rise to a finite, temperature-dependent increase of the relaxation rate Γ^{-1} [see Eq. (4.8)]. This is in accordance with the trend seen in the comparison between the Monte Carlo data and the results of the linearized approximation of Ref. 10.

V. STATICS BELOW T_c

The dynamic definition of the EA order parameter is

$$q=\lim_{t\to\infty}[\langle\sigma_i(0)\sigma_i(t)\rangle]_J=\lim_{t\to\infty}C(t)\,. \tag{5.1}$$

It is convenient to define the *finite-time* part of $C(t)$ as

$$\tilde{C}(t)\equiv C(t)-q\,. \tag{5.2}$$

According to (3.9), the noise has now a static component which acts as a random static field to generate time-persistent autocorrelations. We then write the noise ϕ, (3.9), as a sum of two Gaussian noises $\phi=f+z$ where f is defined by the finite-time part of the correlations (3.9); i.e.,

$$\langle f(\omega)f(\omega')\rangle=[2\Gamma_0^{-1}+\beta^2\tilde{J}^2\tilde{C}(\omega)]\,\delta(\omega+\omega')\,, \tag{5.3}$$

and z by the time-persistent part of (3.9),

$$[z(\omega)z(\omega')]=\beta^2\tilde{J}^2q\delta(\omega)\delta(\omega+\omega')\,. \tag{5.4}$$

Substituting this in the equations of motion (3.7) it is readily seen that the self-consistent equations for q is

$$q\delta(\omega)=[\langle\sigma(\omega)\rangle^2]\,, \tag{5.5}$$

where $\langle\ \rangle$ means (here and in the following) average with respect to f keeping z fixed, and [ ] means averaging over the remaining time-persistent noise z.

In order to solve Eq. (5.5), we must determine the relation between $G(\omega)$ and $C(\omega)$. In ordinary phase transitions the *full* response function is related to the finite-time part of C by the FDT which, instead of (2.10), now reads

$$\tilde{C}(\omega)=\frac{2}{\omega}\text{Im}\,G(\omega)\,. \tag{5.6}$$

The static limit of Eq. (5.6) is the Fischer[35] relation $G(0)=[\langle\sigma^2\rangle]-q$. Equation (5.6) can be self-consistently satisfied in our case also. Assuming that (5.6) holds, using Eqs. (3.8) and (5.4) one obtains

$$\langle f(\omega)f(\omega')\rangle=-2\,\text{Im}\,G_0^{-1}(\omega)/\omega\,,$$

which means that the relation (5.6) is indeed obeyed by the bare ($u=0$) correlation and response functions. One then uses the usual diagrammatic expansion to show that the nonlinearity in Eq. (3.7) does not invalidate this relation. Equation (5.6) ensures that the static limit of the solution of (3.7) for $\langle\sigma\rangle$ is exactly the *magnetization* induced in *thermal equilibrium* by a static Gaussian field z. Thus, Eq. (5.5) reads

$$q=\int_{-\infty}^{\infty}\frac{dz}{(2\pi q\beta^2\tilde{J}^2)^{1/2}}\exp(-\tfrac{1}{2}z^2)m^2(z)\,, \tag{5.7}$$

where

$$m(z)=\frac{\int_{-\infty}^{\infty}d\sigma\,\sigma\exp[-\frac{1}{2}G_0^{-1}(0)\sigma^2-u\sigma^4+(z+h)\sigma]}{\int d\sigma\exp[-\frac{1}{2}G_0^{-1}(0)\sigma^2-u\sigma^4+(z+h)\sigma]}\,, \tag{5.8}$$

and in the Ising limit,

$$m(z)=\tanh(z+h)\ . \tag{5.9}$$

This solution, which is identical to the SK solution,[9,10] is, however, unstable below T_c, as will be shown in the next section.

Thus, a stable solution of the mean-field equations necessarily violates the FDT below T_c. This conclusion has been previously reached by various static approaches.[13,15,18–22] As first suggested by Bray and Moore,[13] the violation of the FDT is presumably the consequence of the high degeneracy of the spin-glass free-energy ground states: The FDT describes the response to an external field due to transitions of the system into new ground states in the vicinity of the original one. However, the actual response consists also of transitions among states which are separated by energy barriers which become infinitely high in the $N\rightarrow\infty$ limit. Thus, the full dynamic response consists of two parts: A finite-frequency part which describes the response which is "local" in phase space and obeys the FDT and a part which appears only at $\omega=0$ namely only on the infinitely long time scales which characterize the crossing of the energy barriers between the ground states. Accordingly, we write

$$G(\omega)=\tilde{G}(\omega)+\Delta(\omega)\ , \tag{5.10}$$

where $\tilde{G}(\omega)$ is the finite-frequency response which is related to $\tilde{C}(\omega)$ by

$$\tilde{C}(\omega)=\frac{2}{\omega}\operatorname{Im}\tilde{G}(\omega)\ , \tag{5.11}$$

and

$$\Delta(\omega)=\Delta\delta_{\omega,0}\ . \tag{5.12}$$

$\delta_{\omega,0}$ is defined here as an analytic (complex) function of ω whose real part becomes a Kronecker δ in the limit $N\rightarrow\infty$. We separate also the bare propagator, Eq. (3.8), into a finite-frequency part

$$\tilde{G}_0^{-1}(\omega)=r_0-i\omega\Gamma_0^{-1}-\beta^2\tilde{J}^2\tilde{G}(\omega)\ , \tag{5.13}$$

and a time-persistent part $G_0^{-1}-\tilde{G}_0^{-1}=-\beta^2\tilde{J}^2\Delta(\omega)$. Substituting this in Eq. (3.7) yields, after straightforward algebra, the following equation of motion:

$$\begin{aligned}\sigma(\omega)&=\tilde{G}_0(\omega)\\&=\tilde{G}_0(\omega)[f+H+h(\omega)]\\&\quad-4u\tilde{G}_0(\omega)\int d\omega_1 d\omega_2\sigma(\omega_1)\\&\qquad\times\sigma(\omega_2)\sigma(\omega-\omega_1-\omega_2)\ ,\end{aligned} \tag{5.14}$$

with

$$H=z(\omega)+\beta^2\tilde{J}^2\Delta\delta_{\omega,0}\sigma(\omega)\ . \tag{5.15}$$

Since, however, the area under the curve $\Delta(\omega)$ is vanishingly small, only the part of $\sigma(\omega)$ which is induced by the static noise z, i.e., $\langle\sigma\rangle$, gives a nonzero contribution to H as can be explicitly checked by inspecting diagrams of $\sigma(\omega)$ generated by (5.14). Thus, H is a static random field which is given, in terms of the Gaussian time-persistent noise z, as

$$H(z)=z+\beta^2\tilde{J}^2\Delta\delta_{\omega,0}\langle\sigma\rangle\ . \tag{5.16a}$$

Finally, we note that, since $\langle ff\rangle=2\operatorname{Im}\tilde{G}_0^{-1}(\omega)/\omega$ [see Eqs. (5.3) and (5.11)], $\langle\sigma\rangle$ is again the magnetization induced in thermal equilibrium by H and h, i.e.,

$$\langle\sigma\rangle=m(z)=\frac{\int d\sigma\,\sigma\exp[-\frac{1}{2}\tilde{G}_0^{-1}(0)\sigma^2-u\sigma^4+(H+h)\sigma]}{\int d\sigma\exp[-\frac{1}{2}\tilde{G}_0^{-1}(0)\sigma^2-u\sigma^4+(H+h)\sigma]}\ , \tag{5.16b}$$

and in the Ising limit,

$$m(z)=\tanh(H+h)\ . \tag{5.16c}$$

With this definition of m, the self-consistent equation for q is still given by (5.7) and that of Δ is

$$1-q+\Delta=\left[\frac{\partial m}{\partial h}\right]\ . \tag{5.17}$$

The presence of Δ modifies via Eq. (5.15) the otherwise Gaussian distribution of the local field H. However, the most important consequence is the nonuniqueness of the solution for H. Since both $\delta_{\omega,0}$ and $\langle\sigma\rangle$ are nonzero only at $\omega=0$, the product $\delta_{\omega,0}\langle\sigma\rangle$ in Eq. (5.16a) is ill defined. Actually, the functions $q\delta(\omega)$ and $\Delta\delta_{\omega,0}$ are limits of functions $q_N(\omega)$ and $\Delta_N(\omega)$ which have a finite width at finite N, and the value of $\delta_{\omega,0}\langle\sigma\rangle$ is determined by the convolution of $\Delta_N(\omega)$ and $q_N(\omega)$. Thus, the static solution in the thermodynamic limit depends on the dynamic properties of the finite system in time scales which approach infinity in the limit $N\rightarrow\infty$. For instance, if the frequency width of $\Delta_N(\omega)$ is much larger than that of $q_N(\omega)$, then clearly $\Delta_N(\omega)q_N(\omega)\rightarrow\Delta q\delta(\omega)$ and Δ is coupled to the full magnetization m. In such a case, Eqs. (5.16c) and (5.17) read

$$m(z)=\tanh[z+\beta^2\tilde{J}^2\Delta m(z)+h] , \quad (5.18)$$

$$1-q+\Delta=\left[\frac{(1-m^2)}{1-\beta^2\tilde{J}^2\Delta(1-m^2)}\right] , \quad (5.19)$$

which together with Eq. (5.7) completely specify a solution which is identical to Sommer's solution[18,19] of the SK model. On the other hand, if the width of $\Delta_N(\omega)$ is much smaller than that of $q_N(\omega)$, the coupling between $\Delta(\omega)$ and $\langle\sigma\rangle$ is negligible, and the only self-consistent solution is $\Delta=0$ leading back to the SK solution. Physically, however, neither assumption seems to be right. Both the appearance of Δ and the complete decay of the autocorrelations are the results of crossing the barriers between the various ground states; hence it is plausible that at least part of the time scales of $\Delta_N(\omega)$ and $q_N(\omega)$ are of the same order of magnitude. In such a case, they cannot be represented in the thermodynamic limit by a single number. Indeed, adopting this point of view, a static solution has been recently constructed[36] which has many desired properties and seems to be the correct mean-field theory of the SG transition. Here we proceed to analyze the *finite*-time properties below T_c.

VI. DYNAMICS BELOW T_c

In this section we study the dynamic properties of the system on time scales which are very long compared to the "microscopic" time scale (set by Γ_0^{-1}) but are still finite even in the limit $N\to\infty$. These properties are described by $\tilde{C}(\omega)$ and $\tilde{G}(\omega)$ which by definition vary in a frequency scale which is much larger than that of Δ_N or q_N and have a well-defined zero-frequency limit. It is convenient to introduce the "unaveraged" propagator and correlation functions $\tilde{G}(\omega,z)$ and $\tilde{C}(\omega,z)$ which are derived from Eq. (5.14) before averaging over z, namely,

$$\tilde{G}(\omega,z)=\frac{\partial\langle\sigma(\omega)\rangle}{\partial h(\omega)} , \quad (6.1)$$

$$\tilde{C}(\omega,z)=\langle\delta\sigma(\omega)\delta\sigma(-\omega)\rangle , \quad (6.2)$$

where $\delta\sigma(\omega)\equiv\sigma(\omega)-m(z)$. (Recall that $\langle\ \rangle$ refers to averaging only over the fast noise f.) The averaged quantities are then given as $\tilde{G}(\omega)=[\tilde{G}(\omega,z)]$ and $\tilde{C}(\omega)=[\tilde{C}(\omega,z)]$. The diagrammatic expansion of $\tilde{G}(\omega,z)$ and $\tilde{C}(\omega,z)$ is straightforward. We define a self-energy $\Sigma(\omega,z)$ by

$$\tilde{G}(\omega,z)=1/[\tilde{G}_0^{-1}(\omega)+\Sigma(\omega,z)] . \quad (6.3)$$

Differentiating this equation with respect to ω and using Eq. (5.13) yields for the averaged response $\tilde{G}(\omega)$,

$$\frac{\partial\tilde{G}(\omega)}{\partial\omega}\{1-\beta^2\tilde{J}^2[\tilde{G}^2(\omega,z)]\}$$

$$=\frac{i}{\Gamma_0}[\tilde{G}^2(\omega,z)]-\left[\tilde{G}^2(\omega,z)\frac{\partial\Sigma(\omega,z)}{\partial\omega}\right] . \quad (6.4)$$

In a dynamically stable system, both $\mathrm{Im}\,\partial\tilde{G}/\partial\omega$ and $-\mathrm{Im}\,\partial\Sigma/\partial\omega$ are non-negative, in the limit of low frequency. The imaginary part of (6.4) then implies that

$$T^2/\tilde{J}^2-[\tilde{G}^2(0,z)]$$

$$=T^2/\tilde{J}^2-[(\langle\sigma^2\rangle-\langle\sigma\rangle^2)^2]\geq 0 , \quad (6.5)$$

where the last equality is a consequence of Eq. (5.11). In the Ising case, Eq. (6.5) reads

$$T^2/\tilde{J}^2\geq 1-2q+[m^4] . \quad (6.6)$$

The inequality (6.6) was first derived by Almeida and Thouless[11] as the stability criterion for the SK solution. They have also shown that Eq. (6.6) *with equality* defines a line in the (h,T) plane which separates the high T region, where the SK solution is stable, from the low T "unstable" region. At $h=0$, the SK solution is unstable for all $T\leq T_c$. In fact, near T_c, $q_{\mathrm{SK}}\sim\tau+\frac{1}{3}\tau^2$, $[m^4]\sim 3\tau^2$ which yields

$$T^2/\tilde{J}^2-1+2q_{\mathrm{SK}}-[m^4]_{\mathrm{SK}}\sim-\tfrac{4}{3}\tau^2 .$$

For small fields, the Almeida-Thouless instability line $T_c(h)$ is given by

$$T_c(h)\sim T_c-(\tfrac{3}{4}h^2)^{1/3},\quad T_c(h)\lesssim T_c . \quad (6.7)$$

On the other hand, near zero temperature the equation of the line is

$$T_c(h)\sim\frac{4\tilde{J}}{3\sqrt{2\pi}}\exp(-h^2/2\tilde{J}^2),\quad T_c(h)\simeq 0 . \quad (6.8)$$

It should be noted that Eq. (6.6) was found here to be a criterion of dynamic stability not only for the SK solution but for all possible mean-field solutions.

The above dynamic-stability condition is not necessarily a *sufficient* condition for stability. In fact, Sommer's solution[18,19] satisfies the above stability condition. Near T_c it yields

$$T^2/\tilde{J}^2-1+2q-[m^4]\sim\tfrac{4}{3}\tau^2 ,$$

hence it has a positive kinetic coefficient $\Gamma^{-1}(0)$

which diverges as $T \to T_c^-$, like $\Gamma^{-1}(0) \sim \tau^{-2}$ [compare with Eq. (4.9)]. Nevertheless, based on the discussion at the end of the last section we suspect that that solution is still unstable, as is indicated also by the replica stability analysis of De Dominicis and Garel.[20] Indeed, a complete stability analysis must include the variations of the full dynamic Lagrangian of (3.1) and (3.2) and has not yet been completed. We note however that the recent work[36] on the static mean-field solution shows that the correct solution obeys the condition of *marginal* stability

$$T^2/\tilde J^2 = [(\langle\sigma^2\rangle - \langle\sigma\rangle^2)^2] , \tag{6.9a}$$

or, in the Ising case,

$$T^2/\tilde J^2 = 1 - 2q + [m^4] , \tag{6.9b}$$

at all $T < T_c$ which implies according to Eq. (6.4), a divergence of $\Gamma^{-1}(0)$ below T_c. This is supported by the Monte Carlo results[10] for the SK model which exhibited algebraic rather than exponential decay of dynamic correlations.

In order to study the finite-time properties of the marginally stable solution, we make the ansatz

$$\tilde G(\omega) = \tilde G(\omega=0) + \alpha\,|\,\omega\,|^{\nu} + i\gamma\,|\,\omega\,|^{\nu}\,\mathrm{sgn}\,(\omega) , \tag{6.10}$$

and solve Eq. (6.4) self-consistently for the exponent ν and the constants α and γ. The ratio of the latter is restricted by the Kramers-Kronig relations for the real and imaginary part of $\tilde G(\omega)$. Substituting the ansatz (6.10) into Eq. (2.9) we find

$$\frac{\alpha}{\gamma} + \frac{\gamma}{\alpha} = -\frac{2}{\pi}\int_0^{\infty} \frac{\omega^{\nu-1}}{1+\omega}\, d\omega = -\frac{2}{\pi} B(\nu, 1-\nu) , \tag{6.11}$$

where $B(x,y)$ is the beta function. This equation allows for two solutions

$$\left(\frac{\alpha}{\gamma}\right)_1 = -\tan\frac{\pi\nu}{2} ,$$

and

$$\left(\frac{\alpha}{\gamma}\right)_2 = -\cot\frac{\pi\nu}{2} , \tag{6.12}$$

but only the second one is the correct solution as is shown explicitly below.

To proceed further, we expand the left-hand side (lhs) of Eq. (6.5) for small frequencies

$$2\beta^2 \tilde J^2 \frac{\partial \tilde G(\omega)}{\partial \omega} \left\{ [\tilde G_0^{-1}(\omega) - \tilde G_0^{-1}(0)][\tilde G^{3}(0,z)] + \omega \left[\frac{\partial \Sigma(\omega,z)}{\partial \omega} \tilde G^{3}(0,z) \right] \right\} = \frac{i}{\Gamma_0}[\tilde G^{2}(0,z)] - \left[\tilde G^{2}(0,z) \frac{\partial \Sigma(\omega,z)}{\partial \omega} \right] . \tag{6.13}$$

and examine the perturbation expansion for $\partial\Sigma/\partial\omega$. We first calculate the leading singularity of $\partial\Sigma/\partial\omega$ to lowest order in the coupling u. Later we will sum the contributions to the leading singularity to all orders in u. Some low-order contributions to $\partial\Sigma/\partial\omega$ are shown in Figs. 1 and 2. To lowest order in u^2, only the two diagrams 1(a) and 1(b) contribute to $\partial\Sigma/\partial\omega$. Using the bare response function the contribution of the first diagram is:

$$\frac{\partial \Sigma^a(z,\omega)}{\partial \omega} = -\frac{2}{\pi}(12u)^2 \langle\sigma\rangle^2 \int_{-\infty}^{\infty} d\omega' \frac{2}{\omega'} \mathrm{Im}\, \tilde G_0(\omega') \times \frac{\partial}{\partial \omega} \tilde G_0(\omega - \omega') . \tag{6.14}$$

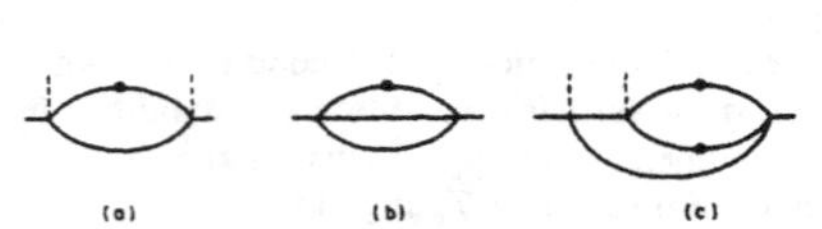

FIG. 1. Some low-order contributions to $\partial\Sigma(\omega,z)/\partial\omega$. A solid line stands for $\tilde G_0(\omega)$, a dot for $\langle ff \rangle$, and a dashed line for $\langle\sigma\rangle$.

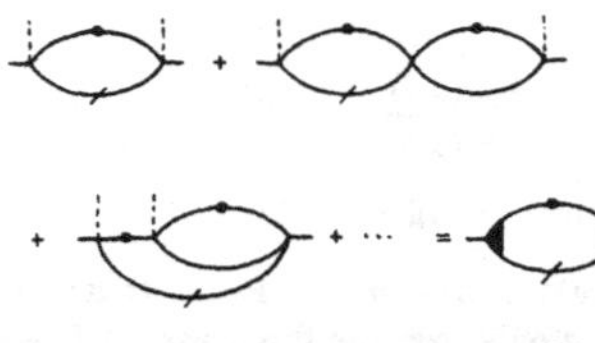

FIG. 2. Most divergent contributions to $\partial\Sigma(\omega,z)/\partial\omega$. A solid line stands for $\tilde G(\omega,z)$, line with dot for $C(\omega,z)$, line with slash for $\partial\tilde G/\partial\omega$, and dashed line for $\langle\sigma\rangle$. The combinatoric weight of the diagrams is *not* displayed.

For $\nu > \frac{1}{2}$, $\partial\Sigma^a/\partial\omega$ is finite, while for $\nu < \frac{1}{2}$ we find $\partial\Sigma/\partial\omega \sim \omega^{2\nu-1}$, and a logarithmic divergence for $\nu=\frac{1}{2}$. Since $\partial\Sigma/\partial\omega$ diverges for all $\nu \leq \frac{1}{2}$, including $\nu=\frac{1}{2}$, whereas the lhs of Eq. (6.11) is finite for $\nu=\frac{1}{2}$, consistency requires that ν be *smaller* than $\frac{1}{2}$ for all temperatures below T_c. Note that at T_c the prefactor of the divergent integral vanishes. The contribution from the diagram in Fig. 1(b) is finite for $\nu > \frac{1}{3}$ and if it is divergent at all, it is less so than $\partial\Sigma^a/\partial\omega$. For the calculation of the leading singularity of the response function this contribution can be neglected. We then obtain from Eq. (6.14) a self-consistent equation for the exponent ν, correct to lowest order in the coupling constant u. To proceed beyond low-order perturbation we notice that the leading divergence $\omega^{2\nu-1}$ is obtained from diagrams which can be separated into two three-point functions by cutting two internal lines: one a correlation function with "internal" frequency ω' and the other one a propagator with frequency $\omega-\omega'$, as illustrated in Fig. 2. The singularity comes from the frequency dependence of these two lines and, hence the frequencies of all other lines can be set to zero. Thus, the sum of all contributions to the leading singularity of $\partial\Sigma/\partial\omega$ has the same frequency integral as (6.14), but with the renormalized propagators and vertices, i.e.,

$$\frac{\partial\Sigma(\omega,z)}{\partial\omega} \sim -\Gamma_3^2(z)\int\frac{d\omega'}{\pi\omega'}\,\mathrm{Im}\,\tilde{G}(\omega',z)\times\frac{\partial}{\partial\omega}\tilde{G}(\omega-\omega',z)\,, \tag{6.15}$$

where

$$\Gamma_3(z)=\lim_{\omega_i\to 0}\langle i\hat{\sigma}(\omega_1)i\hat{\sigma}(\omega_2)\delta\sigma(\omega_3)\rangle\tilde{G}^{-1}(\omega_1,z)\times\tilde{G}^{-1}(\omega_2,z)\tilde{G}^{-1}(\omega_3,z)\,. \tag{6.16}$$

All other diagrams are either finite or less divergent. An example is shown in Fig. 1(c). The inserted bubble diverges (after taking the derivative) as $\omega^{2\nu-1}$. However, the full diagram gives a contribution of $O(\omega^{4\nu-1})$ which is finite if $\nu > \frac{1}{4}$ and in any case, less divergent than $\omega^{2\nu-1}$.

Note that the self-energy itself (not its derivative) is less divergent than $\tilde{G}(\omega)$ [or $\tilde{G}_0(\omega)$]; hence, the low-frequency limit of $\tilde{G}(\omega,z)$ is [see Eqs. (3.8) and (6.3)],

$$\tilde{G}(\omega,z)\sim\tilde{G}(0,z)+\beta^2\tilde{J}^2\tilde{G}^2(0,z)\times[\alpha|\omega|^{\nu}+i\gamma|\omega|^{\nu}\mathrm{sgn}(\omega)]\,. \tag{6.17}$$

Substituting Eq. (6.17) in Eq. (6.15) we obtain for the leading divergence of $\partial\Sigma/\partial\omega$

$$\begin{aligned}\mathrm{Im}\frac{\partial\Sigma(\omega,z)}{\partial\omega}&=-\Gamma_3^2(z)\gamma^2\nu\int\frac{d\omega'}{\pi}|\omega'|^{\nu-1}|\omega-\omega'|^{\nu-1}[\beta\tilde{J}\tilde{G}(0,z)]^4\\&=-\Gamma_3^2(z)\frac{\gamma^2\nu}{\pi}[B(\nu,\nu)+2B(\nu,1-2\nu)]\,|\omega|^{2\nu-1}[\beta\tilde{J}\tilde{G}(0,z)]^4\,,\end{aligned} \tag{6.18a}$$

and similarly,

$$\mathrm{Re}\frac{\partial\Sigma}{\partial\omega}=-\Gamma_3^2(z)\frac{\alpha\gamma\nu}{\pi}B(\nu,\nu)\,|\omega|^{2\nu-1}\times\mathrm{sgn}\,\omega[\beta\tilde{J}\tilde{G}(0,z)]^4\,. \tag{6.18b}$$

To determine the ratio of the two constants α/γ, we consider the *real* part of Eq. (6.13),

$$\frac{\gamma}{\alpha}-\frac{\alpha}{\gamma}=\frac{B(\nu,\nu)}{2\pi}\frac{[\langle\delta\sigma^3\rangle^2]}{[\langle\delta\sigma^2\rangle^2]}\,, \tag{6.19}$$

where again $\omega(\partial\Sigma/\partial\omega)$ has been neglected compared to $\tilde{G}_0^{-1}(\omega)$. Since $B(\nu,\nu)=\Gamma^2(\nu)/\Gamma(2\nu)$, the rhs of Eq. (6.19) is positive definite. Therefore, only the second of the two solutions in Eq. (6.12) is consistent with Eq. (6.1). Inserting this value into (6.19) we obtain an implicit equation for the exponent ν:

$$f(\nu)\equiv 4\pi\cot(\pi\nu)/B(\nu,\nu)=\frac{[\langle\delta\sigma^3\rangle^2]}{[\langle\delta\sigma^2\rangle^3]}\,. \tag{6.20}$$

The function $f(\nu)$ increases monotonically with decreasing ν, such that $f(\frac{1}{2})=0$ and $f(0)=2$.

To calculate explicitly the temperature dependence of the exponent ν we specialize to the Ising case ($\sigma^2=1$). In that case, we have

$$\begin{aligned}\frac{[\langle(\delta\sigma)^3\rangle^2]}{[\langle(\delta\sigma)^2\rangle^3]}&=4\frac{[m^2(1-m^2)^2]}{[(1-m^2)^3]}\\&=4\left(\frac{[\tilde{G}^2]}{[\tilde{G}^3]}-1\right)\\&=4\left(\frac{\tilde{T}^2}{\tilde{J}^2[\tilde{G}^3]}-1\right),\end{aligned} \tag{6.21}$$

where the last equality is a consequence of Eq. (6.9). Close to T_c $q\sim\tau$, $f(\nu)\sim 2\pi(1-2\nu)$; hence

$$\nu(\tau)\simeq\frac{1}{2}-\frac{\tau}{\pi}+O(\tau^2)\ . \tag{6.22}$$

Evaluating $\nu(T)$ at low temperatures requires knowledge of the low-temperature expansion of the spin moments in the marginally stable static solution which is not available at the moment. Instead, we outline here a *tentative* result based on the TAP theory.[17,40] The TAP mean-field equations are

$$\tanh^{-1}m_i=\sum_j \beta J_{ij}m_j-\beta^2\tilde{J}^2(1-q)\ . \tag{6.23}$$

At low temperatures, the deviations of the values of the spin moments from 1 are proportional to T^2 as can be seen from Eq. (6.9). Hence, Eq. (6.23) can be written as

$$\beta\tilde{h}_i=\alpha m_i+\tanh^{-1}m_i,\quad T\sim 0 \tag{6.24}$$

where α is defined by

$$q=1-\alpha(T/\tilde{J})^2\ , \tag{6.25}$$

and $\tilde{h}_i\equiv\sum J_{ij}m_j$ is the local mean field.

Thouless *et al.*[17] and others[10,41] argue that the probability distribution of $\tilde{h}=|\tilde{h}_i|$ behaves in the following manner:

$$P(\tilde{h})\sim\frac{\tilde{h}}{H_0^2},\quad T\sim 0 \tag{6.26}$$

for small $\tilde{h}$. This determines the values of the various spin moments near zero temperature, via the equation,

$$\begin{aligned}1-[m^{2s}]_J&=\int_0^\infty\{1-[m(\tilde{h})]^{2s}\}P(\tilde{h})d\tilde{h}\\&=H_0^{-2}\int_0^1(1-m^{2s})\tilde{h}(m)\frac{d\tilde{h}}{dm}dm\\&=H_0^2T^2\int_0^1(1-m^{2s})(\alpha m+\tanh^{-1}m)\\&\qquad\times\left[\alpha+\frac{1}{1-m^2}\right]dm\ ,\end{aligned} \tag{6.27}$$

which together with Eqs. (6.25) and (6.9) yield $H_0/\tilde{J}\simeq 1.28$ and $\alpha\simeq 1.81$ (see Ref. 40). Equation (6.27) also yields

$$\begin{aligned}[\tilde{G}^3]_J&=[(1-m^2)^3]_J\\&=T^2\left[-\frac{\alpha}{70}+\frac{16}{35}\alpha\ln 2+\frac{8}{15}\ln 2\right.\\&\qquad\left.-\frac{11}{60}+\frac{\alpha^2}{6}\right]\ ,\end{aligned} \tag{6.28}$$

which implies that

$$\nu(T\to 0)=0.25\ . \tag{6.29}$$

We emphasize again that the result (6.29) is only *tentative* since we have not yet proved the validity of Eqs. (6.24) and (6.26) within the dynamic framework.

Finally, we calculate the exponent ν along the Almeida-Thouless line, where the SK solution is marginally stable. Near T_c, $q_{SK}(h)\sim(\frac{3}{4}h^2)^{1/3}$; hence

$$\nu(h)\simeq\frac{1}{2}-\frac{1}{\pi}\left[\frac{3}{4}h^2\right]^{1/3},\quad T\lesssim T_c,\ \beta h\ll 1\ . \tag{6.30}$$

On the other hand, near zero temperature one obtains

$$\begin{aligned}[\tilde{G}^3]&=\int_{-\infty}^{\infty}\frac{dz}{\sqrt{2\pi}}e^{-z^2/2}\operatorname{sech}^6\beta(z+h)\ ,\\ [\tilde{G}^3]&\simeq 16T/(15\sqrt{2\pi}J)\exp(-\tfrac{1}{2}h^2/\tilde{J}^2)\ .\end{aligned} \tag{6.31}$$

Substitution of Eq. (6.31) in Eqs. (6.20) and (6.21) yields $f(\nu)=1$, or

$$\nu(T\to 0)=0.395\ . \tag{6.32}$$

Recently Parisi *et al.*[42] presented a projection hypothesis according to which $q(T,h)=q(T)$ below the Almeida-Thouless line. If this is correct, then according to Eq. (6.9), $[m^4]_J$ also must be independent of h. If we make the further assumption that higher moments and in particular $[m^6]$ are also independent of h, then the result (6.32) should hold also for $T\to 0$, $h=0$, in contradiction with the result (6.29) obtained from the analysis based on the TAP theory.

The result that ν is a function of T below T_c is new. Previous dynamic treatments[23,24] predicted a mean-field value of $\frac{1}{2}$ at all $T\le T_c$. These treatments however were based on a low-order perturbation (in u) and neglected the important singularity which appears in the self-energy below T_c. The result for ν has been derived here from a Langevin equation for a soft-spin version of the SK model. This raises the interesting question whether our results hold also for a Glauber dynamics of an Ising SK model. The Glauber equations of motion have been solved[10] previously in the linearized approximation which is inadequate below T_c. So far we are not aware of a completely satisfactory method of extracting the low-frequency properties from the

Glauber equations below T_c. This issue has been recently addressed by Shastry.[43] He derived from the Glauber equations, mean-field dynamic equations for the time-dependent local magnetizations $m_i(t)$ which reduce in the static limit to the TAP equations. From these equations he concludes that $\nu=\frac{1}{2}$ at all $T \leq T_c$. If this is indeed so, one is inevitably led to the surprising conclusion that the critical behavior of a time-dependent Ginzburg-Landau model in the Ising limit is different in the SG case from that of a Glauber dynamics. However, there are still in our opinion important unanswered questions regarding the correct analytic treatment of the Glauber dynamics of the SK model below T_c.

It should also be mentioned that Monte Carlo simulations of the Glauber dynamics of the SK model yielded[10] $\nu \sim 0.5$ at a temperature range $0.5T_c \leq T \leq T_c$ but the available data is not sufficiently accurate to check the temperature dependence predicted above.

VII. CONCLUDING REMARKS

Our principal result for the relaxational dynamics of SG's in the mean-field limit is the self-consistent local stochastic equation of motion, Eqs. (3.7)–(3.9). As discussed in Sec. III, this solution is a result of the ansatz $Q_2^0=\langle \hat{\sigma}\hat{\sigma}\rangle=0$. Stationary-point solutions with $Q_2^0 \neq 0$ will not in general be reducible to a simple equation of motion. We have argued that solutions with $Q_2^0 \neq 0$ are physically unacceptable, thus justifying our choice. An additional support to this choice stems from the fact that it is consistent with a *static* solution[36] which is probably the correct lowest-energy state. We have also pointed out the analogy between the solution with $Q_2^0 \neq 0$ and the solutions of the TAP equations by Bray and Moore[21] and De Dominicis *et al.*[22] These solutions are described by order parameters other than q and Δ and were associated with a number of metastable states, the logarithm of which is proportional to N. If the relation between the case $Q_2^0 \neq 0$ and these solutions are correct, then it means that the average properties of the TAP equations can be described by q and Δ only with, however, a "careful" treatment of the products of these order parameters. However, in order to prove the relationship between these solutions, one would have to investigate further the properties of Eqs. (3.1)–(3.3) with $Q_2^0 \neq 0$. Also, it should be pointed out that ultimately the neglect of the solution $Q_2^0 \neq 0$ should be justified by a stability analysis of Eq. (3.1), which has not yet been completed by us.

We have shown that the instability of the SK solution below T_c necessarily means that the FDT is violated by the response at $\omega=0$. Thus both correlation and response functions acquire below T_c time-persistent terms denoted by $q\delta(\omega)$ and $\Delta\delta_{\omega,0}$, Eqs. (5.1) and (5.10). As a result, the static local response is

$$G(0)=1-q+\Delta\,, \tag{7.1}$$

with $0<\Delta<q$. Note that the local susceptibility is $G(0)$, since we defined h_i to be an applied field divided by T. The notion of violation of the FDT deserves further comments. Obviously, in a finite system, the FDT is valid, and in particular the equation

$$\frac{1}{N}\sum_i \left[\frac{\partial\langle\sigma_i\rangle_T}{\partial h_i}\right]_J = 1-\frac{1}{N}\sum_i [\langle\sigma_i\rangle_T^2]_J\,, \tag{7.2}$$

must hold for finite N in any configurations of fields. Note that the *apparent* violation of the FDT cannot be attributed simply to nonzero contributions of off-diagonal spin-spin correlations $\langle\sigma_i\sigma_j\rangle_T$, since they do not contribute directly to a true *local* response. One apparent way of interpreting Eq. (7.1) is to conclude that

$$\lim_{N\to\infty}[\langle\sigma_i\rangle_T^2]_J=q-\Delta\,, \tag{7.3}$$

which would then imply that Eq. (7.1) holds only in the presence of some symmetry-breaking fields which are set to zero only after taking the limit $N\to\infty$. This interpretation is in contradiction with the recent static mean-field solution[36] which predicts that, in *thermal equilibrium*,

$$\lim_{N\to\infty}[\langle\sigma_i\rangle_T^2]_J=0 \tag{7.4}$$

at all temperatures while at the same time $G(0)$ is smaller (below T_c) than the Curie value 1. Thus, contrary to the common conception, the anomaly is *not* in the *appearance* of Δ which is simply a consequence of the relaxation of q, but rather in the fact that even in thermal equilibrium $q-\Delta>0$.

A possible way[44] of reconciling Eqs. (7.1), (7.2), and (7.4) is that the average local response that enters the mean-field theory is *not* identical to the lhs of Eq. (7.2) but rather corresponds to

$$G(0)=\lim_{N\to\infty}\frac{1}{N}\sum_i\left[\frac{\delta\langle\sigma_i\rangle_T}{\delta h_i}\right]_J\,, \tag{7.5}$$

where δh_i, in a finite N, are very small but not infinitesimal. As first suggested by Bray and Moore,

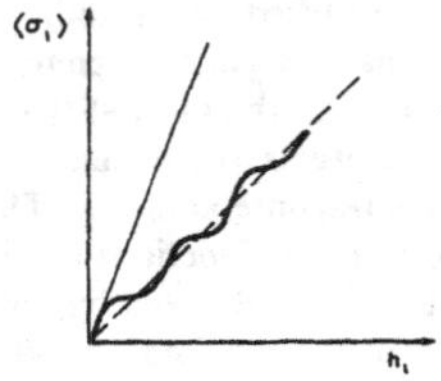

FIG. 3. Schematic plot of the magnetization vs magnetic field demonstrating the possible origin of the apparent violation of the FDT. The slope of the solid straight line is the zero field susceptibility which is equal to 1. The slope of the dashed line is the response $G(0)$, Eq. (7.5), which is equal to $1-q+\Delta$.

it is plausible that the "magnetization versus field" curve contains many steps which correspond to overturning of large clusters of spins. These steps occur as the field changes by an amount which is proportional to some inverse power of N and hence is assumed to be smaller than δh. The actual slope at the origin is 1, but $G(0)$, Eq. (7.5), refers to the slope of the "envelope" of these steps which is smaller than 1; see Fig. 3. This picture has some support from a recent exact solution of the SK model in small samples[45] but, in order to prove it, it is probably necessary to investigate the finite size corrections to the mean-field solution.

ACKNOWLEDGMENTS

We thank B. I. Halperin, P. C. Martin, S.-K. Ma, and S. Kirkpatrick for useful discussions. We have also benefitted from discussions with C. Dasgupta, Y. Lu, A. P. Young, and G. Grinstein. Work at Harvard was supported by the NSF through Material Research Laboratory and Grant No. DMR-77-1020. One of us (H.S.) also acknowledges partial support of a Weizmann Fellowship.

[1]For reviews of theoretical work, see P. W. Anderson, in *Lectures at École de Physics on Ill Condensed Matter, Les Houches, 1978*, edited by R. Balian, R. Maynard, and G. Toulouse (North-Holland, Amsterdam, 1979); A. Blandin, J. Phys. (Paris) 39, C6-1499 (1978).

[2]For review of numerical work, see K. Binder, in *Ordering in Strongly Fluctuating Condensed-Matter Systems*, edited by T. Riste (Plenum, New York, 1979); in *Proceedings of the Enschede Summer School on Fundamental Problems in Statistical Mechanics V*, edited by E. G. D. Cohen (North-Holland, Amsterdam, 1981).

[3]For review of experiments, see J. A. Mydosh, J. Magn. Magn. Mater. 7, 237 (1978); A. P. Murani, J. Phys. (Paris) 39, C6-1517 (1978); J. Appl. Phys. 49, 1604 (1978); H. Maletta, in Proceedings of Nato Advanced Study Institute on Excitation in Disordered System (Plenum, New York, in press).

[4]S. F. Edwards and P. W. Anderson, J. Phys. F 5, 965 (1975).

[5]A. J. Bray and M. A. Moore, J. Phys. C 12, 79 (1979).

[6]P. W. Anderson and C. M. Pond, Phys. Rev. Lett. 40, 903 (1978).

[7]R. Fisch and A. B. Harris, Phys. Rev. Lett. 38, 785 (1977).

[8]I. Morgenstern and K. Binder, Phys. Rev. B 22, 288 (1980).

[9]D. Sherrington and S. Kirkpatrick, Phys. Rev. Lett. 35, 1792 (1975).

[10]S. Kirkpatrick and S. Sherrington, Phys. Rev. B 17, 4384 (1978).

[11]J. R. L. de Almeida and D. J. Thouless, J. Phys. A 11, 983 (1978).

[12]J. R. L. de Almeida, R. C. Jones, J. M. Kosterlitz, and D. J. Thouless, J. Phys. C 11, L871 (1978).

[13]A. J. Bray and M. A. Moore, J. Phys. C 13, 419 (1980).

[14]G. Parisi, Phys. Rev. Lett. 23, 1754 (1979); J. Phys. A 13, L115 (1980); 13, 1887 (1980).

[15]D. J. Thouless, J. R. L. de Almeida, and J. M. Kosterlitz, J. Phys. C 13, 3271 (1980).

[16]E. Pytte and J. S. Rudnick, Phys. Rev. B 19, 3603 (1979); A. J. Bray and M. A. Moore, J. Phys. C 12, 79 (1979).

[17]D. J. Thouless, P. W. Anderson, and R. G. Palmer, Philos. Mag. 35, 593 (1977).

[18]H. J. Sommers, Z. Phys. B 31, 301 (1978); 32, 173 (1979).

[19]H. Sompolinsky, Phys. Rev. B 23, 1371 (1981).

[20]C. De Dominicis and T. Garel, J. Phys. (Paris) 22, L576 (1979).

[21]A. J. Bray and M. A. Moore, J. Phys. C 13, L469 (1980).

[22]C. De Dominicis, M. Gabay, T. Garel, and H. Orland, J. Phys. (Paris) 41, 923 (1980).

[23]S.-K. Ma and J. Rudnick, Phys. Rev. Lett. 40, 589 (1978).

[24]J. A. Hertz and R. A. Klemm, Phys. Rev. Lett. 21, 1397 (1978); 46, 496 (1981).

[25]C. De Dominicis, Phys. Rev. B 18, 4913 (1978).

[26]C. De Dominicis, in *Lecture Notes in Physics*, edited by C. P. Enz (Springer, Berlin, 1979), Vol. 104,

p. 253.

[27]W. Kinzel and K. H. Fischer, Solid State Commun. 23, 687 (1977).

[28]D. Sherrington, Phys. Rev. B 22, 5553 (1980).

[29]H. Sompolinsky and A. Zippelius, Phys. Rev. Lett. 47, 359 (1981).

[30]S.-K. Ma, *Modern Theory of Critical Phenomena* (Benjamin, New York, 1976); P. C. Hohenberg and B. I. Halperin, Rev. Mod. Phys. 49, 435 (1977).

[31]P. C. Martin, E. D. Siggia, and H. A. Rose, Phys. Rev. A 8, 423 (1978).

[32]C. De Dominicis, J. Phys. (Paris) C 1, 247 (1976); C. De Dominicis and L. Peliti, Phys. Rev. B 18, 353 (1978).

[33]H. K. Janssen, Z. Phys. B 23, 377 (1976); R. Bausch, H. K. Janssen, and H. Wagner, *ibid.* 24, 113 (1976).

[34]H. Sompolinsky and A. Zippelius (unpublished).

[35]K. H. Fischer, Phys. Rev. Lett. 34, 1438 (1975).

[36]H. Sompolinsky, Phys. Rev. Lett. 47, 935 (1981).

[37]J. A. Hertz, A. Khurana, and M. Puoskari, Phys. Rev. B 25, 2065 (1982).

[38]H. G. Schuster, Z. Phys. B 45, 99 (1982).

[39]The vertex $Q_2^0\hat{\sigma}\hat{\sigma}$ would not violate *causality* if $Q_2^0(t,t')$ is nonzero only for $t=t'=-\infty$. However, the existence of such a term would necessarily lead to an equilibrium state which is *sensitive* to be *initial* conditions, a situation which is not expected in this system.

[40]A. J. Bray and M. A. Moore, J. Phys. C 12, L441 (1979).

[41]R. G. Palmer and C. M. Pond, J. Phys. F 9, 1451 (1979).

[42]G. Parisi and G. Toulouse, J. Phys. Lett. (Paris) 41, L361 (1980); J. Vannimenus, G. Toulouse, and G. Parisi, J. Phys. (Paris) 42, 565 (1981).

[43]B. S. Shastry (unpublished).

[44]We thank Professor B. I. Halperin and Professor S.-K. Ma for a discussion on this issue.

[45]S. Kirkpatrick and A. P. Young, J. Appl. Phys. 52, 1712 (1981); A. P. Young and S. Kirkpatrick, Phys. Rev. B 25, 440 (1982).

PHYSICAL REVIEW B VOLUME 28, NUMBER 5 1 SEPTEMBER 1983

Role of initial conditions in the mean-field theory of spin-glass dynamics

A. Houghton,* S. Jain, and A. P. Young
Department of Mathematics, Imperial College, London SW7 2BZ, United Kingdom
(Received 4 May 1983)

We discuss the dynamics of the infinite-range Sherrington-Kirkpatrick spin-glass model for which relaxation times diverge when N, the number of spins, tends to infinity. Calculations on a large but finite system are very difficult, so we mimic a large finite system in equilibrium by working with $N=\infty$ and imposing, by hand, a canonical distribution at an initial time. For short times, where no barrier hopping has occurred, we find that the Edwards-Anderson order parameter, $q_{\rm EA}$, is identical to that obtained from an analysis of the mean-field equations of Thouless, Anderson, and Palmer and, with further assumptions, gives $q(x=1)$ in Parisi's theory, in agreement with earlier work. For times longer than the longest relaxation time (of the finite system), true equilibrium is reached and our theory agrees with previous statistical-mechanics calculations using the replica trick. There is no violation of the fluctuation-dissipation theorem.

I. INTRODUCTION

The infinite-range Ising spin-glass model proposed by Sherrington and Kirkpatrick[1] (SK) has proved difficult to solve, contrary to original expectations. It is now clear that below T_c the system can exist in one of many phases, which are separated from each other by barriers whose height diverges in the thermodynamic limit.[2] If we introduce dynamics into the model then the system will stay forever in the phase in which it was originally prepared, assuming external constraints such as temperature and magnetic field are kept fixed. Consequently a complete description would involve characterizing the different phases for each sample which is prohibitively complicated. One would like a statistical description, which can be achieved either from a static (statistical mechanics) or dynamical approach.

The statistical-mechanics formulation generally uses the replica method. This consists of calculating the average free energy via the average partition function of n replicated systems, and taking the limit $n\to 0$, i.e.,

$$\langle \ln Z\rangle_J=\lim_{n\to 0}(\langle Z^n\rangle_J-1)/n\ , \qquad (1.1)$$

where $\langle\ \rangle_J$ indicates an average over the interactions. In this scheme the statistical-mechanics order parameter q, given by

$$q=\langle\langle S_i\rangle_T^2\rangle_J\ , \qquad (1.2)$$

where $\langle\ \rangle_T$ denotes an ensemble (statistical mechanics) average and S_i is the ith spin ($i=1,\ldots,N$), is identified as

$$q=q^{\alpha_0\beta_0}=\lim_{n\to 0}\operatorname{Tr}_{\{S_i\}} S_i^{\alpha_0}S_i^{\beta_0}\left\langle \exp\left[-\beta\sum_{\alpha=1}^{n}H^\alpha\right]\right\rangle_J\ , \qquad (1.3)$$

where $S_i^{\alpha_0}$ and $S_i^{\beta_0}$ denote spins in any distinct pair of replicas and H^α is the Hamiltonian in a particular replica. A straightforward application of this method gives[1] a mean-field solution in which the spin-glass phase is characterized by a single order parameter $q=q^{\alpha\beta}$ which obeys a simple self-consistent equation. However, this solution is unstable below T_c.[3,4] In order to cure this instability one needs to break the replica symmetry by introducing order parameters $q^{\alpha\beta}$ which depend on (α,β).

In the most successful scheme due to Parisi,[5] $q^{\alpha\beta}$ is taken to be a function of a single variable x, $0\le x\le 1$. Furthermore, sums over distinct pairs of replicas become integrals over x, i.e.,

$$\lim_{n\to 0}\frac{1}{n(n-1)}\sum_{\alpha\neq\beta}q^{\alpha\beta}=\int_0^1 q(x)dx\ . \qquad (1.4)$$

The local susceptibility χ_{ii} is given by[6]

$$T\chi_{ii}=1-\int_0^1 q(x)dx\ . \qquad (1.5)$$

However, according to the fluctuation-dissipation theorem (FDT),

$$T\chi_{ii}=1-q\ . \qquad (1.6)$$

De Dominicis and Young[6] (DY) have shown that there is no contradiction between Eqs. (1.5) and (1.6). They pointed out that the effective Hamiltonian in the replica formalism, defined by

$$\exp(-\beta H_{\rm eff})=\left\langle \exp\left[-\beta\sum_{\alpha=1}^{n}H^\alpha\right]\right\rangle_J\ ,$$

is replica symmetric, and so, for every solution $q^{\alpha\beta}$, there are other equivalent solutions obtained by permuting replicas. If these solutions are distinct they must be included as well. Hence, the correct procedure is to average over all distinct solutions, which is clearly equivalent to taking one solution and averaging over all distinct pairs of replicas, i.e.,

$$q=\lim_{n\to 0}\frac{1}{n(n-1)}\sum_{\alpha\neq\beta}q^{\alpha\beta}=\int_0^1 q(x)dx\ , \qquad (1.7)$$

where the $q^{\alpha\beta}$ now refer to one solution. Thus there is no violation of the FDT in Parisi's theory if correctly interpreted. However, this argument does not provide any physical insight into the meaning of the order-parameter function $q(x)$.

The alternative approach, proposed by several authors,[2,7,8] is to study dynamics. Technically this is an attractive choice as it seems to provide a means for calculating average thermodynamic quantities without the use of replicas.[8] The standard time-dependent correlation function is defined to be

$$C(t_1-t_2)=\langle\langle S_i(t_1)S_i(t_2)\rangle_T\rangle_J \ , \tag{1.8}$$

where $S_i(t_1)$ is the value of ith spin at time t_1. This can be written more explicitly as

$$C(t_1-t_2)=\left\langle Z^{-1}\sum_{\{S_j(t_0)\}}\langle e^{-\beta H\{S_j(t_0)\}}S_i(t_1)S_i(t_2)\rangle_\eta\right\rangle_J \ , \tag{1.9}$$

where t_0 is an initial time, $\langle\ \rangle_\eta$ denotes an average over the dynamics for times $t>t_0$ in a sense to be described precisely below, and Z is the partition function. The Edwards-Anderson[9] order parameter is given by the long-time limit of $C(t)$, i.e.,

$$q_{\rm EA}=\lim_{t\to\infty}C(t) \ . \tag{1.10}$$

In all our discussions it is assumed that the thermodynamic limit $(N\to\infty)$ is taken before any other. It is clear then that $q_{\rm EA}$ describes ordering in one valley averaged over valleys, whereas q, Eq. (1.2), includes interference effects between different phases. Hence for a nonergodic system with many phases, such as the SK model, the two order parameters defined above are not the same, i.e., $q\neq q_{\rm EA}$.

The dynamical approach to the problem has been most extensively developed by Sompolinsky[2]; see also Refs. 10 and 11. His theory involves, in addition to the order parameter $q(x)$, an anomaly $\Delta(x)$. These two functions are related by a gauge condition but $q(1)$, $q(0)$, and χ_{ii} are gauge invariant and identical to Parisi's. In addition Sompolinsky provides a plausible interpretation of x in terms of a spectrum of time scales, $x=1$ being the smallest and $x=0$ the largest time scale, all of which diverge in the thermodynamic limit. Consequently $q(1)$ is the order parameter appropriate to one valley averaged over valleys, i.e., $q(1)=q_{\rm EA}$, but, as statistical-mechanics results are obtained on times larger than the longest relaxation time, he finds $q(0)=q$ which, from Eqs. (1.5) and (1.6), is seen to violate the FDT.

In our opinion the standard methods used to discuss dynamics are inadequate for nonergodic systems such as the SK model. It is always tacitly assumed that the Boltzmann distribution in Eq. (1.9) is generated by the dynamics itself starting from an arbitrary state at time t_i earlier than t_0. This clearly requires $t_0-t_i\gg\tau_{\max}$, the largest relaxation time of the system. However, for a nonergodic system $\tau_{\max}$ is infinite so this condition cannot be satisfied. To evaluate the dynamical correlations defined by Eq. (1.9) it is therefore necessary to insert the Boltzmann distribution "by hand." In this paper we shall investigate the consequences of applying this idea to the SK model. A summary of the results of this work has appeared elsewhere.[12]

Unfortunately, as we shall see, one is forced to introduce replicas. We therefore define the order parameters $q^{\alpha_0\beta_0}$ in replica space by

$$q^{\alpha_0\beta_0}=\lim_{t'\to\infty}\langle S_i^{\alpha_0}(t)S_i^{\beta_0}(t+t')\rangle_L \ , \tag{1.11}$$

where the average $\langle\ \rangle_L$ is with respect to a Lagrangian, symmetric in replica space, which will be derived below. Our main conclusions are as follows.

(i) We obtain a self-consistent expression for the diagonal order parameter $q^{\alpha\alpha}$ $(=q_{\rm EA})$ which is identical to that derived by DY from solutions of the Thouless, Anderson, and Palmer (TAP) equations.[13] This expression is rather complicated as it involves the off-diagonal components $q^{\alpha\beta}$, but DY have shown that if one makes the Parisi ansatz for $q^{\alpha\beta}$ then

$$q^{\alpha\alpha}=q_{\rm EA}=q(x=1) \ . \tag{1.12}$$

(ii) If $\alpha\neq\beta$ we find that the correlation function on the right-hand side (rhs) of Eq. (1.11) is time independent and $q^{\alpha\beta}$ is given by the usual statistical-mechanics result, Eq. (1.3). Since L is replica symmetric this implies that the order parameter is given by Eq. (1.7), so there is no violation of the FDT.

(iii) Dynamical effects on finite-time scales do not involve cross coupling between different replicas. Our method gives results identical to those obtained by the "no-replica" formalism of Sompolinsky and Zippelius (SZ).[10,11]

(iv) We find that in general the system keeps its memory of the initial state up to infinite time, but if we make a replica symmetric ansatz for $q^{\alpha\beta}$ the memory of initial conditions is lost. This illustrates the connection between replica symmetry breaking and lack of ergodicity.

(v) The variable x should not be interpreted as a time. Instead, $dx(q)/dq$ is the probability that the overlap between correctly weighted phases is equal to q.[12,14]

The outline of this paper is as follows: In Sec. II we introduce the relaxational dynamic model and derive an effective Lagrangian from which we can compute both static and dynamic equilibrium correlation functions. In Sec. III it is shown that the Lagrangian splits neatly into finite-time and time-persistent parts. Order parameters $q^{\alpha\beta}$ are defined and the Hamiltonian from which they can be determined is derived. In Sec. IV we show that $q^{\alpha\alpha}=q_{\rm EA}=q(x=1)$, where x is the variable introduced in Parisi's replica-symmetry-breaking scheme and we also show that the statistical-mechanics order parameter q is equal to $\int_0 q(x)dx$. It is shown that if initial conditions are not taken into account then the unstable SK solution is the only possible solution. In Sec. V we show that dynamics on a finite-time scale is described by the no-replica formalism of SZ. In Sec. VI we discuss our results.

II. THE DYNAMIC MODEL

The Hamiltonian of the SK model is given by

$$H=-\sum_{i<j} J_{ij}S_iS_j-\sum_i h_iS_i \ , \qquad (2.1)$$

where $S_i=\pm 1$ $(i=1,\ldots,N)$, h_i is an external magnetic field, and the J_{ij} are independent random interactions whose distribution has zero mean and width J/N, the same for all pairs of sites. We set Boltzmann's constant equal to unity and in these units there is a transition in the thermodynamic limit at $T_c=J$. Here we consider a soft-spin version of the SK model defined by

$$\beta H=-\sum_{i<j}\beta J_{ij}S_iS_j+\left[\sum_i (r/2)S_i^2+uS_i^4\right]-\beta\sum_i h_iS_i \ , \qquad (2.2)$$

where $\beta=1/T$. The length of the soft spin S_i is allowed to vary continuously from $-\infty$ to ∞. At the end of the calculation we will let $u\to\infty$ and $r\to-\infty$ in order to recover the length constraint on the spins.

We assume that the dynamics is governed by a Langevin equation

$$\frac{\partial S_i}{\partial t}=-\Gamma\frac{\delta(\beta H)}{\delta S_i}+\eta(t) \ , \qquad (2.3)$$

where Γ is a kinetic coefficient which sets the microscopic time scale and $\eta(t)$ is a Gaussian random noise with zero mean and variance

$$\langle \eta(t_1)\eta(t_2)\rangle=2\Gamma\delta(t_1-t_2) \ . \qquad (2.4)$$

The average denoted $\langle\ \rangle_\eta$ in Eq. (1.9) is an average over this noise. It is convenient to employ the functional integral formalism of Martin, Siggia, and Rose,[15] and see also Bausch *et al.*[16] and De Dominicis[17] to study the dynamics arising from Eq. (2.3). Starting from an arbitrary spin arrangement at time t_0 one formally integrates the equations of motion and averages over the noise. One obtains a path integral $\hat{Z}$, which acts as a generating functional for time-dependent correlation functions. $\hat{Z}$ can also be interpreted as a probability distribution[16] and therefore is normalized to unity if there are no external sources. Consequently $\hat{Z}$ can be averaged directly over the J_{ij} without the need for replicas.[8] The generating functional so obtained is given by SZ,

$$\hat{Z}=\int D[S_i]D[\hat{S}_i]\exp L[S_i,\hat{S}_i] \ , \qquad (2.5)$$

where

$$L[S_i,\hat{S}_i]=\int dt\sum_i i\hat{S}_i(t)\left[-\Gamma^{-1}\dot{S}_i(t)-rS_i(t)+\beta\sum_j J_{ij}S_j(t)-4uS_i^3(t)+\beta h_i(t)+i\Gamma^{-1}\hat{S}_i(t)\right]+V[S_i(t)] \ . \qquad (2.6)$$

Here $\hat{S}_i(t)$ is an auxilliary field and $V[S_i]$, which arises from the functional Jacobean and ensures the correct normalization of $\hat{Z}$, is given by[16,17]

$$V[S_i(t)]=-\int dt\sum_i [(r/2)+6uS_i^2(t)] \ . \qquad (2.7)$$

The functional integral is defined by discretizing the time. One then integrates over $S_i(t)$ for all times later than but not including t_0 because $S_i(t_0)$ is fixed. However, as pointed out in the Introduction, this procedure does not generate the equilibrium correlation functions of the SK model, defined in Eq. (1.9), if $N=\infty$. Sompolinsky[2] has attempted to circumvent this difficulty by working with finite N, but we have reservations about this approach which we will discuss later.

To generate equilibrium correlation functions in the thermodynamic limit $(N=\infty)$ it is necessary to *impose* a Boltzmann distribution at the initial time. Consequently we multiply Eq. (2.5) by $\exp\{-\beta H[S_i(t_0)]\}$ and integrate over the $S_i(t_0)$. Since $\hat{Z}=1$, with no external sources, the result obtained is just the partition function Z. Differentiating $\ln Z$ with respect to external sources now generates both static and dynamic correlation functions. Unfortunately it is also $\ln Z$ which has to be averaged over the J_{ij} and so we are forced to introduce replicas, which was just what one was trying to avoid with the dynamical approach.[7] The generating functional can now be reduced to a convenient form by making use of the methods of SZ. First we integrate out the exchange interactions and then decouple the resultant four spin couplings by introducing seven auxiliary fields $Q_a^{\alpha\beta}(t,t')$ where $a=1,\ldots,7$, and α and β are replica indices. We obtain

$$\langle Z^n\rangle_J=\int\prod_{a=1}^{7} DQ_a^{\alpha\beta}\exp\left[-\frac{2N}{\beta^2J^2}\sum_{\alpha,\beta}\left[\int_{t_0}^{t_f}dt\int_{t_0}^{t_f}dt'[Q_1^{\alpha\beta}(t,t')Q_2^{\alpha\beta}(t,t')+Q_3^{\alpha\beta}(t,t')Q_4^{\alpha\beta}(t,t')]\right.\right.$$
$$\left.\left.+\int_{t_0}^{t_f}dt\frac{Q_5^{\alpha\beta}(t_0,t)Q_6^{\alpha\beta}(t_0,t)}{2}+Q_7^{\alpha\beta}Q_7^{\alpha\beta}\right]\right]$$
$$+\ln\int D[S]D[\hat{S}]D[S(t_0)]\exp L[S,\hat{S},S(t_0),Q_a] \ , \qquad (2.8)$$

where

$$L(S,\hat{S},S(t_0),Q_a)=L_0(S,\hat{S},S(t_0))$$
$$+\sum_i \Big[\int_{t_0}^{t_f} dt \int_{t_0}^{t_f} dt' \{Q_1^{\alpha\beta}(t,t')[i\hat{S}_i^{\alpha}(t)i\hat{S}_i^{\beta}(t')]+Q_2^{\alpha\beta}(t,t')S_i^{\alpha}(t)S_i^{\beta}(t')$$
$$+Q_3^{\alpha\beta}(t,t')i\hat{S}_i^{\alpha}(t)S_i^{\beta}(t')+Q_4^{\alpha\beta}(t,t')S_i^{\alpha}(t)i\hat{S}_i^{\beta}(t')\}$$
$$+\int_{t_0}^{t_f} dt[Q_5^{\alpha\beta}(t_0,t)S_i^{\alpha}(t_0)i\hat{S}_i^{\beta}(t)+Q_6^{\alpha\beta}(t_0,t)S_i^{\alpha}(t_0)S_i^{\beta}(t)]+Q_7^{\alpha\beta}S_i^{\alpha}(t_0)S_i^{\beta}(t_0)\Big], \quad (2.9)$$

and the purely local part of the Lagrangian,

$$L_0(S,\hat{S},S(t_0))=\sum_{\alpha,i}\Big[\int_{t_0}^{t_f} dt(i\hat{S}_i^{\alpha}(t)\{-\Gamma^{-1}S_i^{\alpha}(t)-rS_i^{\alpha}(t)-4u[S_i^{\alpha}(t)]^3+\beta h_i(t)+i\Gamma^{-1}\hat{S}_i^{\alpha}(t)\})$$
$$+\int_{t_0}^{t_F} dt\, V[S_i^{\alpha}(t)]-\{\tfrac{1}{2}r[S_i^{\alpha}(t_0)]^2+u[S_i^{\alpha}(t_0)]^4-\beta h_i S_i^{\alpha}(t_0)\}\Big]. \quad (2.10)$$

We may now use the method of steepest descent, which is exact in the limit $N\to\infty$, to integrate over the Q's, which amounts to replacing $Q_a^{\alpha\beta}(t,t')$ by their stationary point values:

$$Q_1^{\alpha\beta}(t-t')=\frac{\beta^2J^2}{2N}\sum_i\langle S_i^{\alpha}(t)S_i^{\beta}(t')\rangle, \quad (2.11a)$$

$$Q_2^{\alpha\beta}(t-t')=\frac{\beta^2J^2}{2N}\sum_i\langle i\hat{S}_i^{\alpha}(t)i\hat{S}_i^{\beta}(t')\rangle, \quad (2.11b)$$

$$Q_3^{\alpha\beta}(t-t')=\frac{\beta^2J^2}{2N}\sum_i\langle S_i^{\alpha}(t)i\hat{S}_i^{\beta}(t')\rangle, \quad (2.11c)$$

$$Q_4^{\alpha\beta}(t-t')=\frac{\beta^2J^2}{2N}\sum_i\langle iS_i^{\alpha}(t)S_i^{\beta}(t')\rangle, \quad (2.11d)$$

$$Q_5^{\alpha\beta}(t-t_0)=\frac{\beta^2J^2}{N}\sum_i\langle S_i^{\alpha}(t_0)S_i^{\beta}(t)\rangle, \quad (2.11e)$$

$$Q_6^{\alpha\beta}(t-t_0)=\frac{\beta^2J^2}{N}\sum_i\langle S_i^{\alpha}(t_0)i\hat{S}_i^{\beta}(t)\rangle, \quad (2.11f)$$

$$Q_7^{\alpha\beta}=\frac{\beta^2J^2}{2N}\sum_i\langle S_i^{\alpha}(t_0)S_i^{\beta}(t_0)\rangle. \quad (2.11g)$$

The averages on the rhs of Eqs. (2.11) are calculated with the Lagrangian $L[S^\alpha,\hat{S}^\alpha,S^\alpha(t_0),Q_a^{\alpha\beta}]$ leading to self-consistent equations for the Q_a. Because $\exp(-\beta H)$ is a stationary distribution the correlation functions only depend on time differences.

III. REDUCTION OF THE EFFECTIVE LAGRANGIAN

It would appear to be very difficult to self-consistently solve for seven functions which depend on both time and replicas. However, we shall show now that physical arguments simplify the problem considerably.

First, differentiating $n^{-1}\langle Z^n\rangle_J$ with respect to h_j should yield the magnetization at site j, i.e.,

$$m_j=\lim_{n\to 0} n^{-1}\sum_{\alpha=1}^{n}\Big\langle\Big[S_j^{\alpha}(t_0)+\int_{t_0}^{t_f} i\hat{S}_j^{\alpha}(t')dt'\Big]\Big\rangle, \quad (3.1)$$

where the average is with respect to the Lagrangian equations (2.9) and (2.10). Since m_j is given by the first term on the rhs of Eq. (3.1) only, we must have

$$\Big\langle\int_{t_0}^{t_f}\hat{S}_j^{\alpha}(t')dt'\Big\rangle=0$$

for all t_f, which implies

$$\langle\hat{S}_j^{\alpha}(t)\rangle=\langle\hat{S}_j\rangle_T=0. \quad (3.2)$$

Next consider $Q_2^{\alpha\beta}(t-t')$ for $\alpha\neq\beta$,

$$\frac{2N}{\beta^2J^2}Q_2^{\alpha\beta}(t-t')=\langle i\hat{S}_i^{\alpha}(t)i\hat{S}_i^{\beta}(t')\rangle$$
$$=\langle\langle i\hat{S}_i\rangle_T\langle i\hat{S}_i\rangle_T\rangle_J$$
$$=0, \quad (3.3)$$

where the last equality follows from Eq. (3.2). Actually, Eq. (3.2) only implies that $\sum_{\alpha\neq\beta}Q_2^{\alpha\beta}(t-t')=0$ (DY), but we shall make the plausible ansatz that the off-diagonal elements vanish separately. Similar arguments give

$$Q_3^{\alpha\neq\beta}(t-t')=Q_4^{\alpha\neq\beta}(t-t')=Q_6^{\alpha\neq\beta}(t-t')=0. \quad (3.4)$$

Since $\langle i\hat{S}_j^{\alpha}(t')\rangle$ vanishes for all fields one has $\partial/\partial h_j\langle i\hat{S}_j^{\alpha}(t')\rangle=0$, which can be written as

$$\Big\langle i\hat{S}_j^{\alpha}(t')\sum_\beta\Big[S^{\beta}(t_0)+\int_{t_0}^{t} i\hat{S}_j^{\beta}(t'')dt''\Big]\Big\rangle=0. \quad (3.5)$$

Because of Eqs. (3.3) and (3.4) the $\alpha\neq\beta$ terms in Eq. (3.5) vanish so

$$\Big\langle i\hat{S}_j^{\alpha}(t')\Big[S_j^{\alpha}(t_0)+\int_{t_0}^{t} i\hat{S}_j^{\alpha}(t'')dt''\Big]\Big\rangle=0. \quad (3.6)$$

Let us discuss Eq. (3.6). Now $\langle i\hat{S}_j^{\alpha}(t')S_j^{\alpha}(t_0)\rangle$ describes the change in $S_j^{\alpha}(t_0)$ due to a small field applied at time t', and hence vanishes for $t'>t_0$ because of causality. From Eq. (3.6) this implies that $\langle i\hat{S}_j^{\alpha}(t')i\hat{S}_j^{\alpha}(t)\rangle=0$ for $t\geq t_0$ and therefore

$$Q_2^{\alpha\alpha}(t-t')=Q_6^{\alpha\alpha}(t'-t_0)=0. \quad (3.7)$$

If $t'=t_0$ Eq. (3.6) simplifies to

$$\langle i\hat{S}_j^\alpha(t_0)S_j^\alpha(t_0)\rangle+\tfrac{1}{2}\langle i\hat{S}_j^\alpha(t_0)\hat{S}_j^\alpha(t_0)\rangle=0\ , \tag{3.8}$$

and the factor of $\frac{1}{2}$ arises because t_0 is the endpoint of integration (see, for example, the Appendix of Ref. 16). Comparing with Eq. (2.9) and using the definitions given in Eq. (2.11) we see that Q_2 and Q_6 do not contribute to the Lagrangian. The underlying reason is causality. The only SS coupling remaining in the Lagrangian is $Q_7S(t_0)S(t_0)$ which does not violate causality because it only involves spins at the initial time, see Ref. 39 in SZ.

The final simplification is obtained by noting that for $\alpha\neq\beta$,

$$\langle S_i^\alpha(t)S_i^\beta(t')\rangle=\langle\langle S_i(t)\rangle_T\langle S_i(t')\rangle_T\rangle_J\ . \tag{3.9}$$

Since the distribution $\exp(-\beta H)$ is independent of time then so is $\langle S_i(t)\rangle_T$. Hence $Q_1^{\alpha\neq\beta}$ and $Q_5^{\alpha\neq\beta}$ do not depend on time.[18] The description of dynamics on a finite-time scales is, therefore, contained within a single replica. We shall see in Sec. V that our results for the finite-time dynamics are completely equivalent to those given by the no-replica formalism of SZ.

Defining order parameters $q^{\alpha\beta}$ by the long-time limit of the corresponding correlation functions, i.e.,

$$q^{\alpha\beta}=\lim_{t\to\infty}\langle S_i^\alpha(t')S_i^\beta(t'+t)\rangle\ , \tag{3.10}$$

we note that, for $\alpha\neq\beta$ the limit $t\to\infty$ is unnecessary, since the average is independent of t (and of course t'). For $\alpha=\beta$ we write

$$C(t)=\langle S_i^\alpha(t')S_i^\alpha(t'+t)\rangle=\tilde{C}(t)+q^{\alpha\alpha}\ , \tag{3.11}$$

where $\tilde{C}(t)\to 0$ as $t\to\infty$. Note that $q^{\alpha\alpha}$ is just the Edwards-Anderson (EA) order parameter q_{EA} defined by Eq. (1.10).

The Lagrangian, Eq. (2.9), can now be written in a much simpler form, separating out the time-persistent parts for later convenience, as

$$\begin{aligned}L[S_i^\alpha,\hat{S}_i^\alpha,S_i^\alpha(t_0)]={}&L_0[S_i^\alpha,\hat{S}_i^\alpha,S_i(t_0)]\\&+\frac{\beta^2J^2}{2}\Bigg[\sum_\alpha\Bigg\{\int_{t_0}^{t_f}dt\int_{t_0}^{t_f}dt'[\tilde{C}(t-t')i\hat{S}_i^\alpha(t)i\hat{S}_i^\alpha(t')+2G(t-t')i\hat{S}_i^\alpha(t)S_i^\alpha(t')]\\&\qquad+2\int_{t_0}^{t_f}dt\,\tilde{C}(t-t_0)S_i^\alpha(t_0)i\hat{S}_i^\alpha(t)+[C(t=0)-q^{\alpha\alpha}][S_i^\alpha(t_0)]^2\Bigg\}\\&\qquad+\sum_{\alpha,\beta}\left[S_i^\alpha(t_0)+\int_{t_0}^{t_f}i\hat{S}_i^\alpha(t)dt\right]q^{\alpha\beta}\left[S_i^\beta(t_0)+\int_{t_0}^{t_f}i\hat{S}_i^\beta(t)dt\right]\Bigg]\ ,\end{aligned} \tag{3.12}$$

where the response function $G(t)$ is defined by

$$G(t)=\langle i\hat{S}_i^\alpha(t')S_i^\alpha(t'+t)\rangle\ . \tag{3.13}$$

It is now convenient to let $t_0\to-\infty$ and $t_f\to\infty$ so that we can define conventional Fourier transforms. The term in Eq. (3.12) involving $\tilde{C}(t-t_0)$ is then of no importance, because it vanishes except for t close to t_0. Memory of the initial conditions is contained in the terms

$$S_i^\alpha(t_0)q^{\alpha\beta}\left[\int i\hat{S}_i^\beta(t)dt\right]\ .$$

An analogous term appears in Sompolinsky's theory[2] (see also SZ), but there the coefficient of $\int\hat{S}\,dt$ is an "anomaly" related to violation of the FDT. Here we have no such violation and this term arises naturally from memory of the initial conditions.

As pointed out by SZ the term in Eq. (3.12) proportional to $G(t)$ modifies the inverse bare propagator, which becomes

$$[G_0(\omega)]^{-1}=r-i\omega/\Gamma-\beta^2J^2G(\omega) \tag{3.14}$$

and the terms proportional to $\tilde{C}(t)$ give a corresponding memory to the noise, i.e.,

$$\langle\eta(t_1)\eta(t_2)\rangle=2\Gamma(t_1-t_2)\ , \tag{3.15}$$

where

$$\Gamma^{-1}(\omega)=\Gamma^{-1}+\frac{\beta^2J^2}{2}\tilde{C}(\omega) \tag{3.16}$$

such that the FDT still holds for any ω. The term in Eq. (3.12) proportional to $[C(t=0)-q^{\alpha\alpha}][S_i^\alpha(t_0)]^2$ modifies the strength of the quadratic term in the initial Hamiltonian,

$$r\to r-\beta^2J^2G(\omega=0) \tag{3.17}$$

which is just the $\omega=0$ limit of Eq. (3.14).

We absorb these modifications into L_0 and decouple the time-persistent parts by a Gaussian transformation to obtain the dynamic local Lagrangian

$$\begin{aligned}L[S_i^\alpha,\hat{S}_i^\alpha,S_i^\alpha(t_0),z^\alpha]={}&L_0[S_i^\alpha,\hat{S}_i^\alpha,S_i^\alpha(t_0)]\\&-\tfrac{1}{2}\sum_{\alpha,\beta}z^\alpha(q^{-1})^{\alpha\beta}z^\beta\\&+\beta J\sum_\alpha z^\alpha[2\pi i\hat{S}_i^\alpha(\omega=0)+S_i^\alpha(t_0)]\ .\end{aligned} \tag{3.18}$$

Comparing with Eq. (2.10) we see that the time-persistent terms add Jz^α to the field on spin $S_i(t_0)$ and the same field to the spins at later times.

Because the FDT holds, equilibrium (i.e., time-independent) quantities can equally well be obtained from a Hamiltonian. Let us discuss the Hamiltonian from which we can determine averages which involve both $S_i^\alpha(t_0)$ and $S_i^\alpha(t+t_0)$ where t is a fixed large time. We shall now let $u\rightarrow\infty$ and $r\rightarrow-\infty$ so as to recover the length constraint on the spins. If z^α and $S_i^\alpha(t_0)$ are considered fixed then the probability distribution for the $S_i^\alpha(t_0+t)$ labeled σ^α $(=\pm1)$ from now on is

$$\frac{\exp\left[-\beta\sum_\alpha (Jz^\alpha+h)\sigma^\alpha\right]}{\sum_{\{\sigma^\alpha=\pm1\}}\exp\left[-\beta\sum_\alpha (Jz^\alpha+h)\sigma^\alpha\right]} \tag{3.19}$$

Consequently the joint probability distribution for the σ^α and S^α $[\equiv S_i^\alpha(t_0)]$ is

$$\lim_{n\rightarrow 0}(2\pi)^{-n/2}(\det q)^{-1/2}\frac{\int dz^\alpha\exp\left[-\frac{1}{2}\sum_{\alpha,\beta}z^\alpha(q^{-1})^{\alpha\beta}z^\beta+\beta\sum_\alpha (Jz^\alpha+h)(\sigma^\alpha+S^\alpha)\right]}{\prod_\alpha 2\cosh[\beta(Jz^\alpha+h)]}, \tag{3.20}$$

where the $q^{\alpha\beta}$, defined by Eq. (3.10), are to be calculated self-consistently from

$$q^{\alpha\beta}=\langle S^\alpha\sigma^\beta\rangle\ . \tag{3.21}$$

IV. ORDER PARAMETERS

The order parameters defined by Eqs. (3.20) and (3.21) are given by

$$q^{\alpha_0\beta_0}=\lim_{n\rightarrow 0}(2\pi)^{-n/2}(\det q)^{-1/2}\int\left[\prod_\alpha dz^\alpha\right]\exp\left[-\frac{1}{2}\sum_{\alpha,\beta}z^\alpha(q^{-1})^{\alpha\beta}z^\beta\right]$$
$$\times\tanh[\beta(Jz^{\alpha_0}+h)]\tanh[\beta(Jz^{\beta_0}+h)]\prod_\gamma\cosh[\beta(Jz^\gamma+h)] \tag{4.1}$$

For $\alpha_0=\beta_0$ one has $q^{\alpha_0\alpha_0}=q_{\rm EA}$ and Eq. (4.1) is exactly the result obtained by DY from solutions of the TAP equations. This is easily recognized if the integral over $\hat\mu$ in DY Eq. (47) is evaluated. As it stands Eq. (4.1) with $\alpha_0=\beta_0$ is a complicated self-consistent equation for $q^{\alpha_0\alpha_0}$ which involves all the off-diagonal elements $q^{\alpha\beta}$. However, DY have shown that, making Parisi's ansatz for $q^{\alpha\beta}(\alpha\neq\beta)$ one obtains

$$q^{\alpha\alpha}=q_{\rm EA}=q(x=1)\ , \tag{4.2}$$

where $q(x)$ is Parisi's order-parameter function. Sompolinsky's[2] treatment gives the same result.

For $\alpha_0\neq\beta_0$ we can reexpress Eq. (4.1) as

$$q^{\alpha_0\beta_0}=\lim_{n\rightarrow 0}(2\pi)^{-n/2}(\det q)^{-1/2}\int\left[\prod_\alpha dz^\alpha\right]\exp\left[-\frac{1}{2}\sum_{\alpha,\beta}z^\alpha(q^{-1})^{\alpha\beta}z^\beta\right]\sum_{\{S^\alpha=\pm1\}}S^{\alpha_0}S^{\beta_0}\exp\left[\sum_\gamma\beta(Jz^\gamma+h)S^\gamma\right]\ . \tag{4.3}$$

The z integrals are now carried out and one finds the usual statistical-mechanics result

$$q^{\alpha_0\beta_0}=\lim_{n\rightarrow 0}\sum_{\{S^\alpha=\pm1\}}S^{\alpha_0}S^{\beta_0}\exp\left[\frac{\beta^2J^2}{2}\sum_{\alpha\neq\beta}S^\alpha q^{\alpha\beta}S^\beta+\beta h\sum_\alpha S^\alpha\right]\ . \tag{4.4}$$

The terms proportional to $q^{\alpha\alpha}$ vanish because of the length constraint and the limit $n\rightarrow 0$. If replica symmetry is spontaneously broken we may use the arguments of DY to show that the statistical-mechanics order parameter, Eq. (1.2), is given by the average of $q^{\alpha_0\beta_0}$ over all distinct replica pairs, i.e.,

$$q=\lim_{n\rightarrow 0}\frac{1}{n(n-1)}\sum_{\langle\alpha\beta\rangle}q^{\alpha\beta}=\int_0^1 q(x)dx\ . \tag{4.5}$$

The last equality, which is obtained if Parisi's ansatz is made for $q^{\alpha\beta}$, is consistent with the FDT. To complete the picture a direct physical interpretation of the Parisi function $x(q)$ is needed. In Ref. 12 we argued that $x(q)$ should be interpreted as the probability that the overlap between the site magnetizations of two phases of the SK model be less than q. An elegant proof of this result has been given by Parisi.[14]

It is also straightforward to show that

$$\langle S^\alpha S^\beta\rangle=\langle\sigma^\alpha\sigma^\beta\rangle=q^{\alpha\beta} \tag{4.6}$$

for $\alpha\neq\beta$. Following Ref. 19 we call the solution described by Eqs. (3.21), (4.2), (4.4), and (4.6) the "time symmetric" solution. Here this solution arises naturally from the physical argument given above that $\langle S^\alpha(t')S^\beta(t'+t)\rangle$ is independent of t. A different justification for the time symmetric solution is given in Ref. 19.

We emphasize that we have obtained the usual statistical-mechanics results, with the possibility of replica-symmetry-breaking schemes such as Parisi's, only because the initial conditions have been included explicitly. To see this note that omitting the Boltzmann factor at the initial time simply removes the cosh factors from Eq. (4.1). Incorporating this change, and introducing another Gaussian variable one has

$$q^{\alpha_0\beta_0}=\lim_{n\to 0}(2\pi)^{-n}\int\prod_{\alpha}(dz^{\alpha}d\mu^{\alpha})\tanh[\beta(Jz^{\alpha_0}+h)]\tanh[\beta(Jz^{\beta_0}+h)]\exp\left[\frac{\beta^2J^2}{2}\sum_{\alpha,\beta}(i\mu^{\alpha})q^{\alpha\beta}(i\mu^{\beta})+\sum_{\alpha}i\mu^{\alpha}z^{\alpha}\right]. \tag{4.7}$$

Consider $\alpha_0=\beta_0$, and note that the z_α integrals can be carried out to give $2\pi\delta(\mu_\alpha)$ for all $\alpha\neq\alpha_0$. Hence we find

$$q_{\rm EA}=q^{\alpha_0\alpha_0}=\frac{1}{2\pi}\lim_{n\to 0}\int dz\,d\mu\tanh^2[\beta(Jz+h)]\times e^{-q_{\rm EA}\mu^2/2+i\mu z} \tag{4.8}$$

Evaluating the μ integral gives

$$q_{\rm EA}=\int\frac{dz}{(2\pi q_{\rm EA})^{1/2}}e^{-z^2/2q_{\rm EA}}\tanh^2[\beta(Jz+h)], \tag{4.9}$$

which is just the equation for the order parameter $q_{\rm SK}$ in the SK theory. Hence

$$q_{\rm EA}=q^{aa}=q_{\rm SK} \tag{4.10}$$

if we neglect initial conditions. If $\alpha_0\neq\beta_0$ in Eq. (4.7) we obtain an equation which gives $q^{\alpha_0\beta_0}$ in terms of $q_{\rm SK}$. This equation is satisfied if $q^{\alpha_0\beta_0}=q_{\rm SK}$ and we believe this is the only solution. Thus we are forced back to the SK solution if initial conditions are not taken into account. We could have anticipated this result because replicas are unnecessary if initial conditions are neglected[7] and the no-replica formation leads directly to the SK solution.[11]

Conversely if we make a replica symmetric ansatz for $q^{\alpha\beta}$ (i.e., $q^{\alpha\beta}=q$ for all α,β) then the system loses memory of its initial condition. The easiest way to see this is to note that

$$\langle\langle\sigma_i\rangle_T\langle S_i\rangle_T\rangle_J=\lim_{n\to 0}\frac{1}{n(n-1)}\sum_{\alpha\neq\beta}\langle\sigma^{\alpha}S^{\beta}\rangle \tag{4.11}$$

and

$$\langle\langle\sigma_iS_i\rangle_T\rangle_J=\lim_{n\to 0}\frac{1}{n}\sum_{\alpha=1}^{n}\langle\sigma^{\alpha}S^{\alpha}\rangle. \tag{4.12}$$

With a single-order parameter description the rhs of both Eqs. (4.11) and (4.12) are equal to q. Hence $\langle\sigma_iS_i\rangle_T=\langle\sigma_i\rangle_T\langle S_i\rangle_T$, showing that there is no correlation between the σ and S spins, i.e., memory of the initial conditions is lost.

V. DYNAMICS ON FINITE-TIME SCALES

The dynamical part of the Lagrangian Eq. (3.12) is a sum over n-independent replicas of the Lagrangian of the no-replica formalism of SZ. Coupling between replicas occurs only in the time-persistent part of the Lagrangian. In this section we show that our formalism for dynamics on finite-time scales is exactly equivalent to that of SZ if we make the Parisi ansatz for $q^{\alpha\beta}$.

The procedure is to carry out perturbation theory in u. If there were no ordering this would give trivially the same results as SZ because, as noted above the only coupling between replicas is in the time-persistent part. When ordering occurs new vertices which involve spin expectation values appear in the perturbation expansion. It is convenient to discuss the renormalized expansion in which the vertices depend on fully renormalized spin expectation values. With a dynamical approach these expectation values are given by

$$\langle m_i^k\rangle_J=\langle\langle S_i(t_1)S_i(t_2)\cdots S_i(t_k)\rangle_T\rangle_J, \tag{5.1}$$

where $t_{i+1}-t_i\to\infty$ (after the limit $N\to\infty$). Following the argument of Sec. III it is straightforward to show that, in the Ising limit,

$$\langle m_i^k\rangle_J=\lim_{n\to 0}(2\pi)^{-n/2}(\det q)^{-1/2}\int\left[\prod_{\alpha}dz^{\alpha}\right]\exp\left[-\tfrac{1}{2}\sum_{\alpha,\beta}z^{\alpha}(q^{-1})^{\alpha\beta}z^{\beta}\right]\tanh^k[\beta(Jz^{\alpha}+h)]\prod_{\gamma=1}^{n}\cosh[\beta(Jz^{\gamma}+h)]. \tag{5.2}$$

Equation (4.1), with $\alpha_0=\beta_0$, is a special case, $k=2$, of this result. To prove the equivalence of finite-time dynamics in the two theories it is sufficient to prove that $\langle m_i^k\rangle_J$ is the same in the two approaches for all k. The steps necessary to show this are given in the Appendix to DY, where the Parisi (or Sompolinsky) ansatz for $q^{\alpha\beta}$ $(\alpha\neq\beta)$ has been made. In fact it is only necessary to note that in the approach of SZ $\langle m_i^k\rangle_J$ is given by Eq. (A9) of DY, but with the factor $\tanh^2U$ replaced by $\tanh^kU$. Hence the equivalence of our formalism and the no-replica formalism for dynamics on finite-time scales is established.

VI. DISCUSSION

Our results are summarized in Sec. I. The most important is probably Eq. (4.1) with $\alpha_0=\beta_0$ which is exactly the equation obtained for q_{EA} by DY within the TAP formalism. The dynamical derivation has some advantages over the TAP approach. For the latter one assumes that all the TAP solutions are minima (rather than saddle points) and that each solution represents a distinct phase separated from other solutions by an infinite barrier. The dynamical approach discussed here avoids an explicit description of the individual phases and so does not need this assumption.

We have emphasized that the SK model is nonergodic for $N\rightarrow\infty$. A complete description would therefore involve describing each phase, for a given sample, together with information on how these phases merge together or bifurcate as temperature and field are altered. One would then be able to describe completely the irreversibility of the model. Unfortunately this approach appears to be quite impractical.

We have therefore given a statistical description where correlation functions are defined in the standard way, e.g., Eq. (1.9). By inserting the Boltzmann distribution at the initial time all phases are included in principle and averaged over with a weight (DY),

$$P(s)=\frac{1}{Z}\exp(-\beta F_s) \quad (6.1)$$

where s labels the phase, F_s is the free energy of this phase, and Z is the partition function given by $Z=\sum_s \exp(-\beta F_s)$. In fact the exponential factor in Eq. (6.1) has the effect of projecting out those phases with minimum free energy (DY).

Because the SK model is nonergodic one might possibly question the relevance of such an approach. We feel that the statistical description is useful for the following reason. In practice one always has a finite system and one could imagine starting such a system in an arbitrary state at time t_i and measuring the dynamics from time t_0 where $t_0-t_i\gg\tau_{max}$, which is finite for a finite system. The system would then have a Boltzmann distribution at time t_0 and our discussion of subsequent dynamics would be valid up to a time t such that $t-t_0<\tau_{min}$, the time for barrier hopping to start, which also diverges for $N\rightarrow\infty$. In the very long-time limit, $t-t_0\gg\tau_{max}$, our description would again be relevant because the statistical-mechanics results presented here would apply. The intermediate-time range $\tau_{min}<t-t_0<\tau_{max}$, where barrier hopping has started but $q(t-t_0)$ has not yet reached its long-time limit, is very difficult and cannot be treated by our theory.

We have pointed out that our results for statistical mechanics disagree with Sompolinsky's predictions from the no-replica formalism. In his approach saddle-point equations, valid for $N=\infty$, are used to describe a large but finite system including fluctuations over barriers. Since these fluctuations only occur for finite N it is not clear that the saddle-point equations can be used to describe them. We suspect that this is the source of the discrepancy between his results and ours. While it is of great interest to consider a finite system it appears to us to be very difficult, in practice, to describe these large, nonperturbative fluctuations between phases. We should emphasize, however, that we do agree with Sompolinsky's important result that relaxation times diverge in the thermodynamic limit.

ACKNOWLEDGMENTS

We should like to thank C. De Dominicis for helpful discussions. One of us (A.H.) would like to thank the Science and Engineering Research Council (SERC) for financial support and Imperial College for their hospitality during his sabbatical leave from Brown University. He also received financial support from the Materials Research Laboratory at Brown University funded by the National Science Foundation under Grant No. NSF DMR-79-2031. Another one of us (S.J.) would like to thank the SERC for a research studentship.

---

*Permanent address: Department of Physics, Brown University, Providence, Rhode Island 02912.

[1]D. Sherrington and S. Kirkpatrick, Phys. Rev. Lett. 35, 1792 (1975); referred to as SK.

[2]H. Sompolinsky, Phys. Rev. Lett. 47, 935 (1981).

[3]J. R. L. de Almeida and D. J. Thouless, J. Phys. A 11, 983 (1978).

[4]J. R. L. de Almeida, R. C. Jones, J. M. Kosterlitz, and D. J. Thouless, J. Phys. C 11, L871 (1978).

[5]G. Parisi, Phys. Rev. Lett. 23, 1754 (1979); J. Phys. A 13, L115 (1980); J. Phys. A 13, 1887 (1980).

[6]C. De Dominicis and A. P. Young, J. Phys. A 16, 2063 (1983); referred to as DY.

[7]S.-k. Ma and J. Rudnick, Phys. Rev. Lett. 40, 589 (1978).

[8]C. De Dominicis, Phys. Rev. B 18, 493 (1978).

[9]S. F. Edwards and P. W. Anderson, J. Phys. F 5, 965 (1975).

[10]H. Sompolinsky and A. Zippelius, Phys. Rev. Lett. 47, 359 (1981).

[11]H. Sompolinsky and A. Zippelius, Phys. Rev. B 25, 6860 (1982); referred to as SZ.

[12]A. Houghton, S. Jain, and A. P. Young, J. Phys. C 16, L375 (1983).

[13]D. J. Thouless, P. W. Anderson, and R. J. Palmer, Philos. Mag. 35, 593 (1977); referred to as TAP.

[14]G. Parisi, Phys. Rev. Lett. 50, 1946 (1983).

[15]P. C. Martin, E. D. Siggia, and H. A. Rose, Phys. Rev. A 8, 423 (1973).

[16]R. Bausch, H. K. Janssen, and H. Wagner, Z. Phys. 24, 113 (1976).

[17]C. De Dominicis, J. Phys (Paris) Colloq. 37, C1-247 (1976).

[18]Again, strictly speaking, we have only shown that $\sum_{\alpha\neq\beta}Q_1^{\alpha\beta}$ and $\sum_{\alpha\neq\beta}Q_5^{\alpha\beta}$ are time independent.

[19]C. De Dominicis and A. P. Young, J. Phys. C 18, L641 (1983).

J. Phys. F: Metal Phys., 11(1981)261–6. Printed in Great Britain

Magnetic properties of a model spin glass and the failure of linear response theory

F T Bantilan Jr and R G Palmer†
Department of Physics, Duke University, Durham, North Carolina 27706, USA

Received 21 May 1980, in final form 14 July 1980

Abstract. Zero-temperature computer simulations are reported for the Sherrington–Kirkpatrick random Ising model of a spin glass, including an external field. Results are presented for the internal field distribution $P(H)$, and for the ground state energy and magnetisation as functions of field. $P(H)$ has a linear rise from $H = 0$ for all external fields. The zero-temperature susceptibility $\chi(0)$ is close to unity when equilibrium states are examined, in agreement with Parisi's replica symmetry breaking theory and in conflict with linear response theory. The linear response result $\chi(0) = 0$ can be obtained by searching for metastable local energy minima close to the zero-field ground state in configuration space.

Palmer and Pond (1979) have performed zero-temperature computer simulations for the Sherrington–Kirkpatrick (SK) (1975) random Ising model spin glass. We here extend that work to include a non-zero external field, and present results for the internal field distribution, energy and magnetisation as functions of field. Our results for the magnetisation shed light on the failure of linear response theory recently discussed by Bray and Moore (1980) and Parisi (1980b).

The SK model with an external field H_{ext} is described by a Hamiltonian

$$\mathscr{H} = -\sum_{(ij)} J_{ij} S_i S_j - H_{ext} \sum_i S_i \tag{1}$$

involving N Ising spins ($S_i = \pm 1$) interacting in all possible pairs (ij) with a random set of exchange couplings $\{J_{ij}\}$. A particular *sample* has some fixed set $\{J_{ij}\}$, each of the $\frac{1}{2}N(N-1)$ J_{ij} values being chosen independently from a Gaussian probability distribution of mean zero and variance J^2/N. We take $J = 1$.

Our computer simulation is similar to that of Palmer and Pond. We choose a random sample $\{J_{ij}\}$ and a random spin configuration $\{S_i\}$. We then apply a descent algorithm which flips one spin at a time, such that the largest available energy decrement is obtained at each stage, until all spins are aligned with their local field. About 25% of the spins are then flipped and the descent algorithm is applied again. We thus generate a sequence of (at least N) local energy minima, the lowest of which we take as the ground state. Finally, we check for stability against all possible two- or three-spin flips, and perform further descents if necessary; this occurs in about 10% of the samples, but the energy improvements so obtained is always very small.

† Alfred P Sloan fellow.

0305-4608/81/010261 + 06$01.50

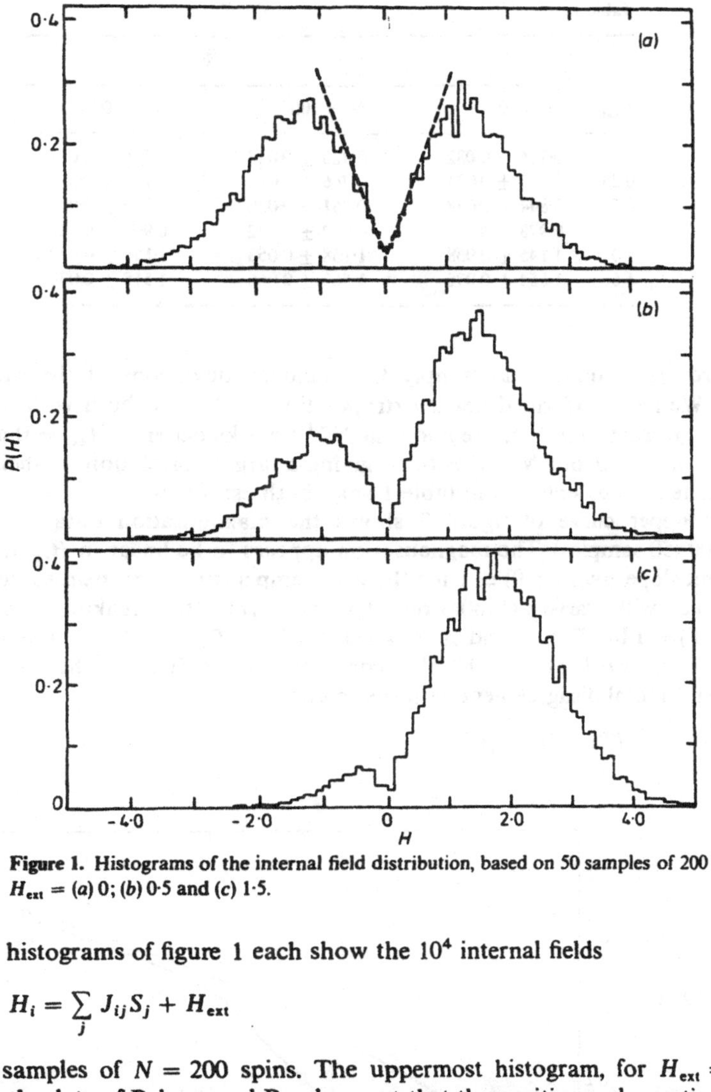

Figure 1. Histograms of the internal field distribution, based on 50 samples of 200 spins at H_{ext} = (*a*) 0; (*b*) 0·5 and (*c*) 1·5.

The $P(H)$ histograms of figure 1 each show the 10^4 internal fields

$$H_i = \sum_j J_{ij} S_j + H_{ext} \tag{2}$$

found in 50 samples of $N = 200$ spins. The uppermost histogram, for $H_{ext} = 0$, is equivalent to the data of Palmer and Pond except that the positive and negative fields have been separated. It shows the characteristic linear rise of $P(H)$ at small $|H|$, from a non-zero value $P(0)$ which vanishes like $N^{-1/2}$ as $N \to \infty$. The other histograms show the effect of turning on an external field, $H_{ext} = 0{\cdot}5$ and $1{\cdot}5$, in the same samples. The linear rise and offset $P(0)$ are retained, but weight is shifted towards the side with fields parallel to H_{ext}. The linear behaviour is consistent with the theory presented by Palmer and Pond (1979), which is easily extended to include a finite external field without change of conclusion. Thermodynamically the linear rise implies a quadratic specific heat ($C_H \propto T^2$ at small T) even with an external field. There is no evidence for the gap suggested by Sommers (1978, 1979).

In table 1, we present the average ground state energy per particle $\overline{E}/\overline{N}$ at each H_{ext} for $N = 50$ (100 samples), $N = 100$ (50 samples), and $N = 200$ (50 samples). The

Table 1.

	$\overline{-E/N}$			
H_{ext}	$N = 50$	$N = 100$	$N = 200$	$N = \infty$
0	0·706 ± 0·032	0·723 ± 0·018	0·738 ± 0·011	0·74–0·78
0·25	0·753 ± 0·031	0·766 ± 0·023	0·776 ± 0·012	0·775–0·81
0·5	0·844 ± 0·038	0·861 ± 0·030	0·864 ± 0·016	0·865–0·895
0·75	0·978 ± 0·045	0·992 ± 0·042	0·993 ± 0·022	0·99–1·02
1·0	1·145 ± 0·058	1·158 ± 0·053	1·154 ± 0·029	1·15–1·18
1·5	1·554 ± 0·076	1·562 ± 0·066	1·555 ± 0·038	1·55–1·575

standard errors quoted are simply the standard deviations of the sample energies found. We have performed linear extrapolation to $N = \infty$ both with respect to N^{-1} and with respect to $N^{-1/2}$. We find that $N^{-1/2}$ works better at $H_{ext} = 0$ and 0·25, as in Palmer and Pond, but N^{-1} is better—giving a larger correlation coefficient—in larger fields. The $N = \infty$ energies in table 1 span both estimates.

The upper curve of figure 2 shows the magnetisation data ($M = \Sigma S_i/N$) for $N = 200$ (50 samples). The magnetisation appears to be linear in H_{ext} for small fields, with unit slope, giving $\chi(0) \simeq 1$ for the zero-temperature, zero-field susceptibility. This agrees well with Parisi's (1980a) recent replica symmetry breaking theory which predicts $\chi(T) = 1$ for $T < T_c$ and $\chi(T) = 1/T$ for $T > T_c$, where $T_c = J = 1$. Parisi's own Monte Carlo work at $T = 0{\cdot}3$ also confirms this prediction. The result $\chi(0) = 1$ is, however, in total disagreement with the relation

$$\chi(T) = (1 - q)/kT \tag{3}$$

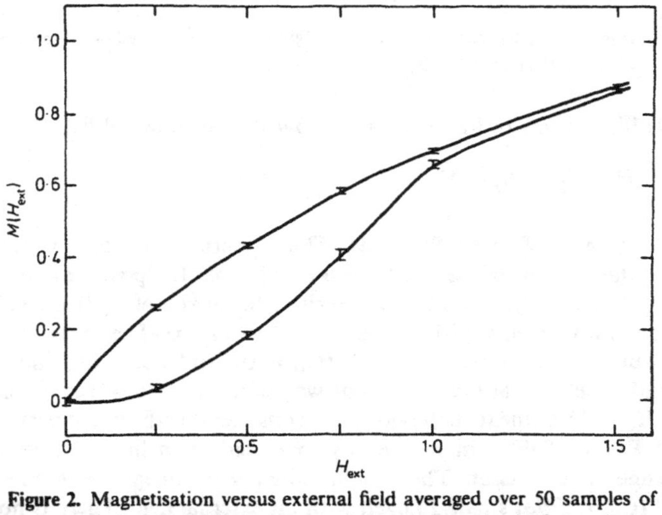

Figure 2. Magnetisation versus external field averaged over 50 samples of 200 spins. The upper curve is for equilibrium states, found independently for each external field. The lower curve shows the effect of applying a field to the zero-field ground state and then descending to a nearby local energy minimum. The error bars denote the standard deviation of the sample values divided by $50^{1/2}$.

when combined with the low-temperature behaviour

$$q \simeq 1 - \alpha T^2 \qquad (\alpha \simeq 1{\cdot}67) \tag{4}$$

found theoretically by Thouless *et al* (1977) and numerically by Kirkpatrick and Sherrington (1978); equations (3) and (4) give $\chi(0) = 0$. Equation (3) is easily derived from linear response theory—simply by differentiating the partition function twice—but Bray and Moore (1980) and Parisi (1980b) have suggested that linear response theory is invalid in this system. One may view the spin glass as having many quasi-degenerate states, unrelated by symmetry and not nearby in configuration space. An infinitesimal change of external field (or perhaps temperature) could then lead to a jump of the true lowest state to a totally different region of configuration space. This could make the partition function non-differentiable in the thermodynamic limit and thus lead to the breakdown of linear response theory.

To examine this hypothesis we have computed the correlation

$$C = \sum_{i=1}^{N} S_i S_i'/N - M M' \tag{5}$$

between the lowest states $\{S_i\}$ and $\{S_i'\}$ in each pair of fields H_{ext} and H'_{ext}, for every $N = 200$ sample. The maximum possible correlation, for given M and M', is

$$C_{max} = 1 - |M - M'| - M M' \tag{6}$$

and in table 2 we present the ratio C/C_{max} (averaged over samples) for each pair of fields. The results are generally quite far from unity, showing a considerable reorganisation of the spin state with a change of field. They are *not* consistent with a picture in which a small increase of H_{ext} simply flips a few more spins up, leaving the majority unchanged.

We have also approached the question from the other side, by finding behaviour that *is* consistent with linear response theory when large jumps in configuration space are effectively forbidden. For each sample, we first find the ground state in zero external field. Each spin is then flipped with probability f, or left unchanged with probability $1 - f$, so that approximately a fraction f of the spins are flipped in all. Finally, we turn on an external field, and apply our usual descent algorithm to find a local energy minimum. The case $f = 1/2$ totally randomises the state and thus corresponds to the previously described random starting point. When f is less than 1/2, the final state tends to be more correlated with the zero-field state, the correlation C increasing as f is reduced. The states found for $f < 1/2$ are generally metastable—higher in energy than the previously found best states for the same sample and field.

Table 2.

	C/C_{max}	H_{ext} 0·25	0·5	0·75	1	1·5
	0	−0·01	−0·04	−0·07	0·06	−0·04
	0·25		0·65	0·52	0·52	0·44
H'_{ext}	0·5			0·65	0·67	0·59
	0·75				0·78	0·74
	1·0					0·76

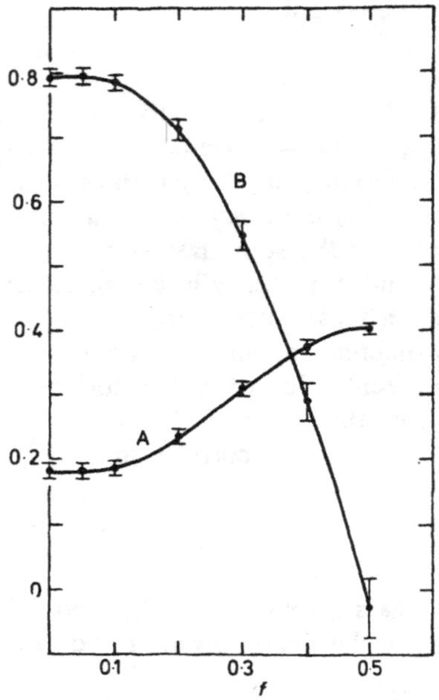

Figure 3. The magnetisation (A), and the correlation (B) (equation (6)) with the zero-field ground state, for states intermediate between the extremes of figure 2. Averaged over 50 samples of 200 spins at $H_{ext} = 0{\cdot}5$. The spins in the zero-field state are flipped with probability f before descent to a local energy minimum.

The lower curve in figure 2 shows the results for $f = 0$, averaged over 50 samples of 200 spins. We note immediately that these data, for metastable states, are consistent with $\chi(0) = 0$, the linear response result. This is quite satisfactory, since the system now never wanders far in configuration space from the zero-field ground state, allowing us the notion of continuity with respect to external field. In experimental terms, we view the lower curve of figure 2 as describing the instantaneous magnetisation resulting from applying a magnetic field *after* cooling to zero temperature, while the upper curve corresponds to cooling with a field already present, or making an infinite-time 'DC' measurement. It is important to realise that it is the upper curve (equilibrium field-cooled states) and not the lower (metastable states) that violates linear response theory.

The effect of varying f is seen in figure 3 at $H_{ext} = 0{\cdot}5$. For $0 < f < 1/2$, the points represent averages from 20 different randomisations of the zero-field ground state for each sample. We see that the magnetisation varies smoothly between the extreme values represented by the curves of figure 2. The correlation C between the final state and the zero-field ground state also behaves smoothly, and falls continuously to zero as f approaches 1/2.

We conceive of the spin glass—and perhaps other disordered systems—as displaying *broken ergodicity* in which a sequence of bifurcations leads to many mutually inaccessible but similar regions of configuration space as the temperature is lowered. A change of external field or temperature can lead to local motion in the bifurcation tree, obeying linear response theory, but not to a discontinuous jump to another region, even if more favourable energetically. Any broken symmetry system with a disconnected order parameter space behaves in the same way, but here there is no known symmetry between the different states. Theories for the spin glass which ignore the broken ergodicity by averaging over these states, and allowing transitions between them, will naturally violate linear response theory and miss much of the essential physics. On the other hand, there is no clear path ahead for circumventing this problem.

We note that the sort of behaviour just discussed is similar to the behaviour of an ordered system at a critical point, where the partition function is also non-differentiable. We tentatively suggest that a spin glass is critical *everywhere* in the ordered region of the phase diagram.

Acknowledgment

We thank David Sherrington for a helpful suggestion.

References

Bray A J and Moore M A 1980 *J. Phys. C: Solid St. Phys.* **13** 419–34
Kirkpatrick S and Sherrington D 1978 *Phys. Rev.* B **17** 4384–403
Palmer R G and Pond C M 1979 *J. Phys. F: Metal Phys.* **9** 1451–9
Parisi G 1980a *J. Phys. A: Math. Gen.* **13** 1101–12
——1980b *Phil. Mag.* to be published
Sherrington D and Kirkpatrick S 1975 *Phys. Rev. Lett.* **35** 1792–6
Sommers H-J 1978 *Z. Phys.* B **31** 301–7
——1979 *Z. Phys.* B **33** 173–80
Thouless D J, Anderson P W and Palmer R G 1977 *Phil. Mag.* **35** 593–601

Part 2
OPTIMIZATION

Chapter VII
COMBINATORIAL OPTIMIZATION PROBLEMS

VII.1 Introduction

Many of the complex problems which one encounters currently can be formalized as combinatorial optimization problems in which one must find the minimum of a certain function depending on many parameters.

As an illustration of how this can be important in *everyday* life, imagine that you want to write the chapter of a book on optimization and statistical physics. You should decide what the ingredients are: classification of optimization problems, Monte Carlo method, analogy cost-energy, etc..., and in order to find an optimal order of presentation you cook up a function which should be minimized. This function includes typically (huge) penalties when a concept is used while it has not been defined before or a bonus for a smooth transition between two topics... Running a simulated annealing algorithm should provide a good presentation. Obviously we have not done this and the only property that the following chapter will share with the result of the above approach is that, due to some "thermal" noise, there might well remain some defects and appearance of a few undefined quantities.

A first example which suggests that optimization and statistical physics might have something in common was briefly described in Chap. 0, where a sociological problem (how to divide a population into two subgroups, trying to manage that friends be in the same group and enemies in different groups) was cast into the search of the ground state for an Ising spin glass. Finding this ground state is a typical hard combinatorial optimization problem.

But complex problems are found in many other places and are often of great practical importance. Examples run from the physical design of computers[1] (partitioning of a large number of circuits between two chips, placement of the circuits, routing of the wires which make the connections between them, ...), the optimization of compilers, etc... to problems in pattern recognition[2], learning in artificial intelligence, prebiotic evolution or the prediction of macromolecule conformations[3]... The transcription of these problems into the form of a "cost function" which must be minimized requires much expertise and a great insight into the precise important features of the problem under study. We shall not say anything about this part of the work which depends very much on the specific problem under consideration.

However after this first step of formalization has been carried out, one must still optimize a cost function which depends generally on many parameters and this can be quite difficult. It has been recognized in the last five years that combinatorial optimization is fundamentally connected to statistical physics in a simple way: Calling the cost function an energy, one must find the ground state of a system which has many degrees of freedom.

This is certainly an interesting question in statistical physics. The reason why this simple analogy has not been exploited much earlier is that physicists have concentrated for years on the behaviour of homogeneous systems which have a large symmetry. In such cases the nature of the ground state can generally be found by a simple analysis of the various possible schemes of spontaneous symmetry breaking. Instead hard optimization problems are generally strongly inhomogeneous, and it is the recent development of new methods and concepts in the statistical physics of disordered systems which has made this connection clear and fruitful[1,4].

In the attempts to understand the low temperature properties of spin glasses, numerical simulations have been performed using the Monte Carlo method. A difficulty is that the thermalization times in the spin glass phase are very large. In order to reach the equilibrium at low temperatures one must be careful to start from high temperatures and cool the system slowly. It has been recognized subsequently that this procedure, called simulated annealing, can be applied efficiently to many optimization problems[4,5]. Algorithms based on this method have developed rapidly and are already being used by engineers for instance in computer design. Finally more recently some theoretical problems on optimization have been attacked using the analytical methods developed in spin glass theory (replica method, *TAP* like equations and cavity method...). Conversely some of the efficient algorithms developed for standard optimization problems can be used in order to find the ground state of physical systems. Much work has been devoted to the study of ground states in spin glasses and random field Ising models. For instance the ground state of a two dimensional spin glass is given by the matching of minimal length of the frustrated plaquettes, which can be found exactly by a polynomial algorithm due to Edmonds[6]. The ground states of the random field Ising model can also be found by a polynomial algorithm in any dimension[7]. Both fields will certainly benefit greatly from these exchanges.

VII.2 Combinatorial Optimization Problems

A simple example of a combinatorial optimization problem, perhaps the most famous one, is the traveling salesman problem (*TSP*)[8]: given a certain set of cities and the distances between them, a traveling salesman must find a tour, as short as possible, in which he visits all the cities and goes back to his starting point. An instance of the problem is a set of cities and distances. In its most ambitious version, the problem is to find an algorithm which is able to find the shortest tour for *any instance* of the problem in a reasonable amount of (computer) time. One algorithm which works consists in looking at all the possible tours and computing their lengths. However one soon realizes that this procedure is too slow to be of any use on a *TSP* of a decent size. The number of possible tours is $N!/2N$ (the factor $2N$ is the degeneracy associated with the starting point of the tour and its direction) and this is prohibitively large even in the reasonable case where $N = 50$ for instance. Unfortunately one doesn't know an algorithm which does very much better than the above exhaustive search yet: it is believed that any algorithm which can give the *exact* shortest tour for *any* instance of the *TSP* must require a computer time which grows exponentially with the number of cities: this problem is an *NP*-complete problem.

Let us make this notion of *NP*-completeness a bit more precise, using the qualitative language of physicists (a rigorous presentation can be found for instance in Ref. 9). Generally an optimization problem is given by:

1. The "domain" of the problem, which is the family of possible instances (e.g. for the *TSP*, any ensemble of points and matrix of distances between them).
2. The rules which define a configuration (for the *TSP*, a configuration is a tour).
3. A "cost function" which allows to compute the cost of any configuration (for the *TSP*, the length of the tour).

An algorithm solves the problem if it is able to find the configuration of lowest cost for any instance. A natural classification among the algorithms is according to the time they take, and the variation of this time with the size of the instance one is solving. A very coarse grained (but useful) distinction is between polynomial algorithms and exponential ones, depending on whether the computer time grows as a power of the size or exponentially. (The size of an instance can be defined as the number of bits which are needed to encode it in a "reasonable"—not redundant—way; for a TSP with N cities, it can be taken as a constant time N (or N^2) depending on how many variables are needed to compute the matrix of $N(N-1)/2$ distances[10]. However this makes no difference for the classification in exponential or polynomial. For this type of reason we do not bother to give more precise definitions to "size" or "time").

The classification of optimization problems is as follows: first there are the "simple" ones which are solved by a polynomial algorithm. They form the class "*P*" of polynomial problems. A much wider class is the "*NP*" class of non deterministic polynomial problems, which can be solved in a polynomial time by a non deterministic algorithm. Essentially this means that if someone gives the optimal configuration, its cost can be computed in a polynomial time[11]. Obviously *NP* contains *P*. Among the *NP* problems one can introduce an order relation. One says that problem *P*1 is at least as hard as *P*2 when the following statement is true: "If *P*1 can be solved in a polynomial time, so can *P*2". *NP* complete problems are *NP* problems which are at least as hard as any other *NP* problem. It has been shown that there exist such problems. To prove that a new problem is *NP* complete, it is then enough to show that it is at least as hard as one of the already known *NP* complete problems[12]. Finally there are problems, even harder than *NP*, which we do not even dare to mention.

It is a trivial consequence of their definition that if one of the *NP* complete problems can be solved in a polynomial time, then all the problems of the huge *NP* class are also polynomial. In this case one would have $P = NP$. So far no polynomial algorithm has been found for an *NP* complete problem, and the question of whether *P* is equal to *NP* is still open, although the general belief is that *NP* complete problems are not polynomial. One must remember however that the difficulty and the *NP* completeness of a problem depends on the space of instances which are allowed: obviously a *TSP* in which all the cities are restricted to lie on the same straight line can be solved by a fast polynomial algorithm.

Many problems of practical interest fall into the *NP* complete class. A typical example is the graph partitioning problem: given a graph with $2N$ vertices and a certain set of edges between them, one must divide the vertices into two groups of equal size and minimize the total number of edges between the groups. This is useful if one wants to partition the circuits of a computer between two chips for instance. It is also closely related to the spin glass problem, as we shall see in the following.

Practically *NP* complete problems require a prohibitive computer time, growing exponentially with the size of the problem (a polynomial algorithm can also have a poor efficiency, but an exponential one is more probably useless). Hence one must change the rules of the game and look for algorithms (named heuristics) which provide an approximate solution of the problem: they find configurations which have a cost nearly equal to the optimal one. This is obviously enough in many applications since the cost function itself is not defined very precisely, but merely represents a rough parametrization of the desired properties of the solution.

VII.3 Optimization and Statistical Mechanics at Zero Temperature

In general good heuristics depend much on the problem under consideration. As one of the general strategies we shall mention the method of iterative improvement. Starting from an initial configuration, one tries to make a "small" modification. If the new configuration so generated has a lower cost, this modification is accepted and the process is iterated. It ends when no further improvement can be found. In the difficult cases such as the *TSP* or the spin glass such a search leads to a local minimum of the cost function, much higher than the true optimal solution. One strategy is to perform this search again and again, starting each time from a different configuration, and to select the best local minimum found.

It is instructive to study this iterative improvement method in the spin glass case in which its limits have been clearly demonstrated. It applies naturally to this problem since the vicinity of a configuration has a simple definition: As "small" modification we shall take a single spin flip. The algorithm consists in picking up a spin (this can be done sequentially or at random) and flipping it if this decreases the energy. It converges very rapidly but one gets stuck in a metastable state. In fact numerical estimates on the *SK* model indicate that in the limit of large samples the energy of the obtained state is $-.70N$ with probability one[13,14] (instead of $-.7633N$ for the ground state). More precisely, the probability of finding an energy density E with this method has a mean equal to $-.70$ and a width which tends to zero as $1/N$ (Ref. 13). Therefore it is clear that one has little chance of finding the true ground state with this method for large N. The large number of metastable states in spin glasses is clearly related to the basic properties of disorder and frustration, and in fact these two properties can be found also as generic elements in hard optimization problems. The travelling salesman is frustrated since it would like to go immediately to the city nearest to the place where he stays, but this strategy clearly fails at the end because of the constraint that he must make a tour of *all* the cities.

A natural way to escape metastable states is to work at finite temperature. The cost function in an optimization problem can be seen as an energy and the partition function of the problem at the "temperature" $T = 1/\beta$ is

$$Z = \sum_{\text{configurations}} \exp(-\beta \times \text{cost(configuration)}). \qquad \text{(VII.1)}$$

This introduction of the canonical ensemble in the new context of optimization is fruitful both from the numerical point of view and from the analytical one, as we shall see in the next sections. Let us draw a table of the equivalence between the two languages:

OPTIMIZATION	STATISTICAL PHYSICS
instance	sample
cost function	energy
optimal configuration	ground state
minimal cost	ground state energy

It will also be useful in the following to introduce the usual thermodynamic quantities such as the internal energy (= averaged cost of the configurations weighted with their Boltzmann Gibbs probabilities), the entropy (= Log of the number of configurations which contribute at a fixed energy), as well as the analogs of some of the basic notions of spin glass theory like the distributions of overlaps or of local fields. Let us now see how the new point of view of statistical physics can lead to effective developments in optimization, as well numerically (Chap. VIII) as analytically (Chap. IX).

References

1. Kirkpatrick *et al.* (1983) Reprint 25.
2. See, e.g.: Geman and Geman (1985); Marr (1982).
3. See the third part of this book and references therein.
4. Kirkpatrick (1981).
5. Cerny (1985); Siarry and Dreyfus (1984).
6. Bieche *et al.* (1980).
7. Angles d'Auriac *et al.* (1985).
8. An introduction to combinatorial optimization can be found for instance in Papadimitriou and Steiglitz (1982); there also exists a specific book on the TSP edited by Lawler *et al.* (1985).
9. Garey and Johnson (1979).
10. These "distances" are not necessarily Euclidean distances: in practical problems, they can represent costs, time, etc..., or various combinations of these parameters.
11. More precise definitions involve the introduction of decision problems which receive a yes/no answer. It is this answer which one must be able to check in polynomial time; see[9].
12. A number of them are described in Ref. 9.
13. Kinzel (1986).
14. Parga and Parisi (1985).

Chapter VIII
SIMULATED ANNEALING

The Monte Carlo method[1] uses the power of computers to simulate the behaviour of a physical system at thermal equilibrium. It generates configurations of the system with a probability given by Gibbs' law

$$P(c) = \exp(-\beta E(c)) \Big/ \sum_{c'} \exp(-\beta E(c')). \qquad \text{(VIII.1)}$$

One algorithm which achieves this was proposed in Ref. 2. Starting from an initial configuration $\mathscr{C}$ one generates through what is called an elementary move a slightly different configuration $\mathscr{C}'$ (e.g. one flips one spin in an Ising magnet, or makes a small displacement of one atom in a gas), and computes the change in energy $\Delta E = E_{\mathscr{C}'} - E_{\mathscr{C}}$. If this difference is negative the change is accepted i.e. $\mathscr{C}'$ is taken as the new configuration. If ΔE is positive the change is accepted with a probability $\exp(-\beta \Delta E)$. This elementary Monte Carlo step is repeated many times. After a large number of steps the system equilibrates and the configurations are generated according to the distribution (VIII.1).

However much care is needed in order to simulate in this way low temperature behaviours, which is precisely the regime in which one is primarily interested for optimization. For instance if one picks up at random an arbitrary initial configuration and runs the above algorithm with a large value of β the system cannot thermalize in any reasonable amount of computer time: it gets blocked in a metastable configuration. (In fact for $\beta = \infty$ the Metropolis algorithm reduces to the iterative improvement method discussed in the previous paragraph). This is well known in real physical systems: In metallurgy this procedure would correspond to taking a melt and quenching it rapidly, in which case it freezes into an amorphous state, very different from the crystalline order which is the stable phase at low temperatures. In order to transit into the crystalline phase the only solution is to heat the system above the freezing temperature and cool it smoothly, being particularly careful of annealing slowly around the freezing temperature.

This same idea applies to Monte Carlo simulations: especially in disordered frustrated systems which possess many metastable states, one must start the Monte Carlo at high temperatures and increase $\beta = 1/T$ slowly, in order to be able to thermalize properly at low temperatures.

This is the essence of the simulated annealing algorithm[3] which can be summarized schematically as follows:

1. Pick an initial configuration.
2. Pick an initial temperature $T = T_i$.
3. Perform $M(T)$ Monte Carlo steps at temperature T.
4. Decrease the temperature and goto 3, until T reaches a final value T_f.

The parameters which must be chosen in order to implement this algorithm are: The "annealing scheme" which is the set of successive temperatures $T_1 = T_i, T_2, \ldots, T_K = T_f$ and the number of Monte Carlo iterations $\mathcal{M}(T_i)$ at each temperature. Another set of parameters is required in the definition of a Monte Carlo step, in order to specify how one chooses the new configuration $\mathcal{C}'$ which one tests (at what distance from $\mathcal{C}$, randomly or through a deterministic process? ...)

Let us mention immediately that in general this will be used as a heuristic. It is possible to implement this algorithm with an extremely slow cooling in order to be sure to find the true ground state, but this is then a very slow process[4]. Practically the algorithm is used with a faster cooling rate and it generates a low lying state, the drawback being that one does not know whether the result is the optimal solution, or how far it is from this solution (although one can get pieces of information by making several runs).

The reason why this algorithm can be efficient in complex optimization problems is that these problems are often characterized by complex energy (i.e. costs) structures, with many metastable states (local minima) and various scales of barriers heights. Working at finite temperature introduces some noise into the problem which enables the system to jump over some energy barriers and to escape local minima. The higher the temperature, the higher the barriers which the system can cross (but at too high temperatures it just wanders around in phase space, paying little attention to the energy...). Clearly the algorithm is efficient if the size of the lowest lying valley is not much smaller than the size of the metastable state, but it runs into trouble in the case of "golf course" like potential: essentially flat everywhere, with a very narrow and deep valley. (In such a case one possible approach could be to modify the cost function and modify the potential—or the elementary moves—in order to get an idea of where the deep valley lies.) Generally speaking, the better the qualitative understanding of the energy landscape of a problem, the better one will be able to adapt the parameters of the algorithm.

Before discussing the choice of these parameters it is worthwhile to emphasize at this point one of the main qualities of this algorithm, namely its generality and versatility. Given a complex problem and assuming that it has been cast into the form of a cost function to optimize, it is generally rather easy to write the above algorithm, at least in its fundamental version. Some parameters can be added to the cost function without any essential change in the structure of the program, and the possibilities of interactive work and easy changes of the program are very useful when one tackles a new problem.

Simulated annealing is maybe not a panacea, and it is probably possible, on any well defined problem, to write a more efficient heuristic. But the flexibility of this algorithm makes it a very powerful tool for practical purposes in which the cost function is not always strictly defined and should be modifiable at will. Another advantage is that it is easy and useful to play interactively with this algorithm: for instance one can simply go from a previous solution to a revised one by heating and cooling. This can be used also to remove some local defects of a solution by locally heating it and then cooling again smoothly. This can be very important in practice.

Various elements should be taken into account in order to choose the parameters efficiently. We are not going to describe them in details but just give a few fundamental ideas[5].

— The choice of elementary moves is important, and it can be quite difficult. A first obvious constraint is that the whole phase space (space of allowed configurations) should be accessible from any configuration through the successive application of elementary moves. This move should also lead typically to a new configuration which has an energy near to one of the starting configuration (this is the practical definition of "nearby configurations" in this context), such that $\beta\Delta E$ be of order one. In problems where there exist algorithms based on the idea of iterative improvement one can take from them the choice of elementary moves. For instance in the *TSP* one often chooses the two bond rearrangements found by Lin[6]. Clearly a larger set of moves (e.g. 3 bond moves in the *TSP*) increases the performance of the algorithm as far as the final result is concerned, but at the price of an increase of the computer time. It is not evident *a priori* whether one should adopt simple moves and do many Monte Carlo iterations or larger moves with less iterations in order to get the best result in a given amount of computer time.

It can be useful to adapt the size of the moves to the temperature, especially if one deals with functions of continuous parameters. For instance in order to find a quadratic minimum, when one reaches the low temperature regime (we suppose that the system has been able to cross barriers at higher temperatures and that it is confined inside one valley near to a quadratic minimum), one should scale the changes as $\sqrt{T}$, which is the width of the potential at an excitation energy T.

— The annealing scheme. A first analysis should be performed both analytically and by a first run with a rapid annealing in order to understand the typical scale of temperatures relevant to the problem (some examples are given in the next section). Then the scheme must be determined interactively by trial and error. It is very useful to follow the evolution of basic thermodynamic functions such as the internal energy which is the average of the energy of the configurations found by the algorithm at a fixed temperature. T_i must be taken high enough such that most of the changes are accepted. Very often the freezing region is well defined and one can scan temperatures rapidly above or below this region, spending most of the time around the freezing temperature. This temperature is signaled by an increase of the specific heat (the fluctuations of the energy). More sophisticated criteria have been proposed, such as the study of the residual entropy[7], but they use quantities which are not easy to determine accurately and are therefore rather time consuming in general, although many refinements to the method can certainly be found on any given problem.

References

1. See the book edited by Binder (1979).
2. Metropolis *et al.* (1953).
3. Kirkpatrick *et al.* (1983) Reprint 25, Kirkpatrick (1981); Cerny (1985); Siarry and Dreyfus (1984).
4. Gidas (1986) and references therein.
5. There is a rapidly increasing amount of literature about simulated annealing which we cannot mention here. A bibliography is being compiled by L. T. Wille (to appear in Journal of Computational and Applied Mathematics).
6. Lin (1965).
7. Kirkpatrick *et al.* (1983) Reprint 25; Ettelaie and Moore (1985).

Chapter IX
ANALYTICAL RESULTS

Interesting questions about optimization problems are not limited to the search for algorithms. A whole set of problems in mathematics consists in predicting properties of optimal or nearly optimal configurations of an optimization problem in which the instances belong to a certain class.

A nice and simple example can be found once again in the *TSP*: Given N points distributed uniformly at random in a square of side unity, what can one say about the length L of the shortest tour? In this case it has been shown[1] that with probability one in the large N limit $L/\sqrt{N}$ tends towards a limit γ, where γ is sample independent (In the spin glass language the length of the ground state is self averaging). γ is known to be bounded by $.62 < \gamma < .93$ (Ref. 1) and is found numerically as $\gamma = .75$ (Ref. 2).

This is typically the kind of problem to which we can apply the methods of spin glass theory. Let us emphasize some essential points. In general statistical physics deals with very large systems in which it is impossible and useless to take care of all the degrees of freedom exactly. Instead the value of macroscopic quantities can be predicted by a statistical analysis of the most probable behaviour of the system, based on ergodicity hypotheses. In large size optimisation problems one can try to use similar ideas. Instead of requiring all the detailed microscopic properties of the ground state configuration, one can first try to perform a macroscopic analysis. The analytic predictions are essentially probabilistic, given a certain probability measure in the space of instances.

Besides its intrinsic interest, this kind of approach can also give some insight into the qualitative understanding of the problem and suggest some ameliorations of heuristics or formulations of new algorithms. The configuration landscape which has been found in the *SK* spin glass with its ultrametic topology is an important piece of information in this respect and it is interesting to study whether a similar picture also appears in other optimization problems[3]. Finally the knowledge of the ground state energy (as well as other quantities), or some bounds on its value, is very useful in order to test the performances of heuristics and compare them.

From the algorithmic point of view, it is naturally the simulated annealing method which most easily take advantage of the statistical physics information one possesses on a problem since it simulates precisely the system at finite temperature. For instance if one knows that the problem has a phase transition at a given temperature T_c this will show up in the Monte Carlo simulation of a (finite size) sample as a strong increase of the thermalization times around T_c. This is the temperature where the annealing should be made most carefully in order to obtain good low temperature configurations.

IX.1 General Thermodynamical Properties and Scalings

In this chapter we shall describe in details the use of statistical physics methods in several well-known optimization problems: matching, *TSP*, assignment, and graph bipartitioning. The first three problems are very different from the spin glass problem. We shall treat them in parallel since they share some common ingredients: One instance is always a set of N points and the matrix of the distances l_{ij} between them. The allowed configurations are obtained by placing bonds between the points with the following constraints (see Fig. IX.1):

— In the *TSP* one must make a tour (i.e. a connected set of bonds) through all the points.

— In the matching problem one must match the points by pairs: two points are matched when they are linked by a bond. On each point there must arrive one and only one bond. (A perfect matching exists only in the case where N is even).

— In the assignment one must visit all the points as in the *TSP*, but without the constraint of connectivity.

In all three cases the problem is to find the set of bonds which has the smallest total length. Note that if the matrix l is not symmetric the bonds are oriented. Hereafter we shall always assume that l is symmetric unless otherwise stated. Formally one searches for a permutation P of the N points: a bond is placed between each point i and $P(i)$, trying to minimize the "energy" (= length)

$$E = \sum_i l_{i,P(i)} \tag{IX.1}$$

and the permutation P must satisfy:

— for the *TSP* it is cyclic,

— for the assignment it is arbitrary,

— for the matching it consists only of cycles of length two, i.e. one has $P(P(i)) = i$ for all $i = 1, \ldots, N$. (The energy in this case is twice the length of the matching, since one has doubled all the bonds). So it is clear that the *TSP* and matching configurations are special cases of assignment.

Although these problems look rather similar at first sight, the matching and assignment are polynomial problems while the *TSP* is *NP* complete. (Obviously the constraint that the travelling salesman must make a *connected* tour is highly non local and complicates his problem).

In order to specify the problem one must describe how the l_{ij} are chosen. For reasons of simplicity and aesthetics two canonical families of problems have been considered so far. In the first, the points are distributed uniformly in a unit hypercube of a d-dimensional Euclidean space (problem $P1$). In the second, the distances l_{ij} are independent random variables identically distributed (problem $P2$). Numerically, as

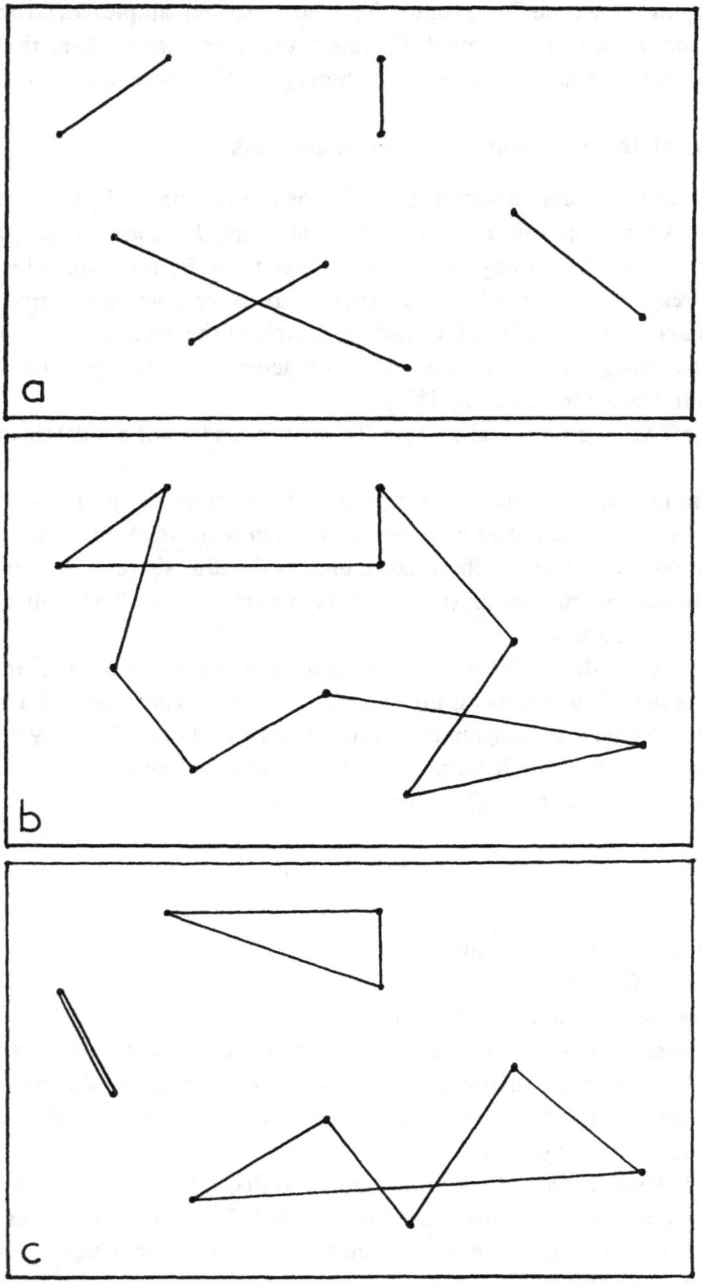

Fig. IX.1. Examples of allowed configurations for one given instance (with Euclidean distances), of the three problems: a) Matching b) Traveling salesman c) Assignment.

far as heuristics are concerned, the Euclidean problem is much simpler because the neighborhood of a point is well defined. One can use the strategy of "divide and conquer", cutting the cube into smaller pieces in which the solutions are found independently, and then pasting together the solutions of the various subpieces[2]. If the sample is large enough and the division does not involve too many subpieces the net error is a term of relative order surface/volume which goes to zero when $N \to \infty$. Analytically the second type of problem is easier to study with the spin glass methods, as we shall see below.

The first puzzle which is encountered currently in optimization problems is the "entropy catastrophe" due to a large number of possible configurations which grows typically as $N!$. Its resolution is not difficult but provides some interesting insights into the physics of the problem. We shall illustrate it on the case of the *TSP*, (the same arguments apply as well to matching or assignment.)

There are $(N)!/2N$ possible tours through N points, and this grows more rapidly than any exponential A^N. Hence the entropy at infinite temperature, equal to the log of the number of configurations, is not extensive

$$S/N \approx c \operatorname{Log} N, \qquad N \to \infty, \tag{IX.2}$$

although the energy at large temperature is of order N. This seems rather unwieldy for a thermodynamical approach to the problem.

However at a finite (properly scaled, see below) temperature the number of relevant tours should be strongly reduced. A point should be connected to two of its "neighbours", i.e. other points at a distance typically of the order of (a few times) the minimal distance between two points. Then the number of possible configurations increases exponentially with N and the entropy is extensive. In the Euclidean case $P1$ this occurs naturally for finite β if the thermodynamic limit is taken with a fixed density of points ($N \to \infty, L \to \infty, N/L^d = \rho = cte$). However in the much studied case where the points are distributed inside a cube of size unity ($N \to \infty$, $L = 1$) the nearest neighbour distance scales as $N^{-1/d}$, hence the length of the interesting matchings scales as $N^{1-1/d}$. In order to select these configurations in the partition function one must scale β as $\beta = \bar{\beta} N^{1/d}$ with $\bar{\beta}$ fixed when $N \to \infty$. An analog rescaling is also necessary in the independent link problem $P2$ (Ref. 4). If $\rho(l)$ is the probability that a link be of length l (in the following we assume that ρ is continuous and that the minimal possible length is zero), the typical distance d_n between neighbours depends only on the behaviour of ρ at the origin. When

$$\rho(l) \approx l^r/r!, \qquad (l \to 0), \tag{IX.3}$$

one has $d_n \approx N^{-1/(r+1)}$ and β should be $\beta = \bar{\beta} N^{1/(r+1)}$.

With these rescalings one can show that the free energy

$$F = -(1/\bar{\beta}) \operatorname{Log}(Z). \tag{IX.4}$$

The energy

$$E = -\partial \, \mathrm{Log}(Z)/\bar{\beta} \tag{IX.5}$$

and the entropy

$$S = \bar{\beta}(E - F) \tag{IX.6}$$

all grow like N for $N \to \infty$, $\bar{\beta}$ fixed. Hence the original average length $E = -\partial \, \mathrm{Log}(Z)/\beta$ scales as $N^{1-1/(r+1)}$ (case $P2$)[4] or $N^{1-1/d}$ (case $P1$)[1]. This explains for instance the $\sqrt{N}$ growth of a TSP tour in a unit square which we described before.

A first insight into the physics of any optimization problem can be obtained simply using the so-called "annealed approximation": instead of computing $\overline{\mathrm{Log}(Z)}$ which is the relevant quantity because of self averageness[4], one computes the annealed average $\mathrm{Log}(\bar{Z})$ which is much easier. This approximation allows one to establish the correct scaling of β with N described above, a piece of information which is also useful in the implementation of simulated annealing. It yields a lower bound on the free energy and on the ground state energy. An upper bound can be found from the study of the performance of a simple algorithm (e.g. the "greedy" algorithm in which the travelling salesman goes at each step to the nearest city not yet visited). The most difficult case is problem $P2$ with $r = 0$, in which the lower bound on the minimal length is a constant while the upper bound grows like $\mathrm{Log}(N)$.

The annealed approximation is correct at high (rescaled) temperatures ($\bar{\beta} \ll 1$), but it breaks down when the temperature $1/\bar{\beta}$ decreases for the same reasons as in spin glasses, and one must resort to more sophisticated methods.

In the next sections we shall study successively the matching, TSP and assignment, using the approach of spin glass theory. The first basic step in this approach is always to find a field theoretic representation of the problem. From this point on, one can apply the replica method and the cavity method. This last method is valid for a given sample (no average over disorder is done) and allows to establish the mean field (TAP like) equations for one sample. Thus it opens the way to the study of new algorithms.

In order to keep the presentation as simple as possible we shall discuss mainly the case of problem $P2$ with $r = 0$. The generalization to other values of r is easy[4,6,15]. The possibility of generalization to Euclidean problems $P1$ will be discussed at the end of this chapter.

IX.2 The Matching Problem

It is natural in all these problems to introduce the (Ising like) link occupation variables $n_{ij} = n_{ji}$, equal to one if the bond between i and j is present and 0 if it is absent. The partition function is in general

$$Z = \sum_{\{n\}} \prod_{i<j} (T_{ij})^{n_{ij}}, \tag{IX.7}$$

where

$$T_{ij} = \exp(-\beta N l_{ij}) \qquad \text{(IX.8)}$$

is the Boltzmann weight associated with the bond $i - j$ when it is occupied, with the correct scaling of β. (To keep notations simple we use hereafter β in place of the variable $\bar{\beta}$ defined above.) The sum $\sum_{\{n\}}$ is over all the n_{ij} which satisfy the suitable constraints. In the case of the matching, these are

$$n_{ij} = n_{ji} = 0, 1$$

$$\sum_{j=1}^{N} n_{ij} = 1, \qquad 1 \le i \le N. \qquad \text{(IX.9)}$$

There exist many representations in terms of local field variables. Here we shall follow the approach of Ref. 6. Other representations can be found for instance in Ref. 7, or hereafter in formula (IX.17). One introduces on each site j an integral representation of the constraints

$$\delta\left(1 - \sum_k n_{jk}\right) = \int_0^{2\pi} d\lambda_j/2\pi \exp\left(i\lambda_j\left(1 - \sum_k n_{jk}\right)\right) \qquad \text{(IX.10)}$$

and then performs the sum over the n_{jk}'s, so that

$$Z_M = \prod_j \left(\int_0^{2\pi} d\lambda_j/2\pi \exp(i\lambda_j)\right) \prod_{j<k} [1 + T_{jk} \exp(-i(\lambda_j + \lambda_k))]. \qquad \text{(IX.11)}$$

The replica evaluation of the quenched average $\overline{Z_M{}^n}$ proceeds as usual. One introduces, on each site j, n replicas $\lambda_j{}^a$ of the variable λ_j, and then averages over the disorder (distribution of l_{ij}). This is explained in Mézard and Parisi, Reprint 26, and we shall skip the details. (Note that in Ref. 6 the number of site was explicitly written as $2N$ while here we call it N—but still assume that it is even.) The order parameters are the thermal averages of products of $\exp(-i\lambda_j{}^a)$ on the same site but in different replicas

$$Q_{a_1 \ldots a_p} = 2/p\beta \langle \exp(-i(\lambda_j^{a_1} + \cdots + \lambda_j^{a_p})) \rangle. \qquad \text{(IX.12)}$$

The simplest approximation is that of replica symmetry in which

$$Q_{a_1 \ldots a_p} = Q_p \qquad p = 1, \ldots, N. \qquad \text{(IX.13)}$$

These Q_p can be seen as the moments as a certain function $f(\varphi)$. After performing the analytic continuation to $n = 0$, one obtains an integral equation for f. The solution (actually derived in terms of some kind of Laplace transform of f) has been found in Ref. 6. The averaged energy $E_M(\beta) = \lim_{n\to 0}(-1/Nn)\, d\overline{Z^n}/d\beta$ can be computed at any temperature. The zero temperature limit gives the length of the optimal matching

$$E_M(\beta = \infty) = \pi^2/12 = .822\ldots. \quad \text{(IX.14)}$$

Rather precise numerical results can be obtained on this problem at zero temperature. Good algorithms have been developed using the fact that it can be written as a problem of linear programming: If one relaxes the constraints $n_{ij} = 0, 1$ to $0 < n_{ij} < 1$, the cost function $\sum_{i<j} n_{ij} l_{ij}$, the constraints $\sum_j n_{ij} = 1$, $0 < n_{ij} < 1$ are all linear functions of the variables n_{ij}. It may happen that the optimal solution of the new problem does not satisfy $n_{ij} = 0, 1$, but this can be taken care of (Ref. 8). In fact the situation is even simpler if one considers the bipartite matching problem. This is the problem in which the $2N$ points are divided into two groups A and B and one must match each point of A to one point of B. The corresponding analytic study parallels the one of the matching and the length of the optimal bipartite matching is predicted to be $\pi^2/6$.

Recent extensive computations[9] on samples of sizes ($2N=$) 100 to 800 have found numerically

$$E_{BM}(\beta = \infty) \simeq 1.642 \pm .004 \quad \text{(IX.15)}$$

in good agreement with the prediction of the replica symmetric solution.

Other detailed properties of the ground state can be determined explicitly in the replica symmetric solution[6]. For instance the distribution of lengths of the occupied bonds has been found analytically and agrees with what is found in the simulations. In a sense it is the equivalent in this problem of the distribution of local fields in the spin glass. So this agreement gives an indication that the replica symmetric solution is correct in this problem. (We recall that, in contrast with this result, the replica symmetric solution of the SK model gives a very bad approximation to the distribution of local fields at low temperature).

The problem of whether the above predictions are exact or not can also be given an analytic answer. The only assumption we made was to look for a replica symmetric saddle point. The coherence of this assumption has been checked[13]. We have found that it is a consistent assumption at any non zero temperature. In other words the replica symmetric saddle point is stable, both in the matching and in the bipartite matching cases. In fact one can fully compute[13] the determinant of fluctuations around the saddle point which gives the leading corrections of order $1/N$ to the ground state energy. The result for the bipartite matching

$$E_{BM}(\beta = \infty) = \pi^2/6 - \frac{1}{N}(\pi^2/12 + 2\zeta(3)) \quad \text{(IX.16)}$$

(ζ is Riemann's zeta function) agrees well with the numerical simulations[9]. To understand what is going on in this problem even better, one would like to establish mean field equations which are valid for one given sample. We shall achieve this with the cavity method of Chap. V (Ref. 10). We start from another field theoretic description of the problem in terms of some spin variables: We introduce on each site i an m-component vector spin $\mathbf{s}_i$ which satisfies $\mathbf{s}_i^2 = m$. The generalized partition function

$$Z(\beta, m) = \int \prod_i (d\mu(\mathbf{s}_i)s_i^1) \prod_\alpha \exp\left(\sum_{i<j} T_{ij} s_i^\alpha s_j^\alpha\right) \qquad \text{(IX.17)}$$

reduces to the partition function of the matching $Z_M(\beta)$ in the limit where $m \to 0$, as can be seen expanding in (IX.17) the integral in series and using[11]

$$\lim_{m \to 0} \int d\mu(\mathbf{s}) \exp(\mathbf{h} \cdot \mathbf{s}) = 1 + \mathbf{h}^2/2. \qquad \text{(IX.18)}$$

Let us consider what happens when one adds a new site, 0, to N existing ones. (Obviously one should keep to even number of sites, i.e. add two sites to the system of N sites. We leave it to the reader to show that the mean field equations below are nevertheless the correct ones.) Assuming that we are working within one pure state, we can neglect the correlations of the N existing spins when we compute their action on the new one as we explained in Chap. V. Hence we can compute the properties of spin 0 from the partition function

$$z \int \prod_{i=0}^{N} (d\mu(\mathbf{s}_i) s_i^1 \exp(\mathbf{h}_i \cdot \mathbf{s}_i)) \exp\left(\sum_j T_{0j} \mathbf{s}_0 \cdot \mathbf{s}_j\right), \qquad \text{(IX.19)}$$

where $h_i^\alpha = h_i \delta^{\alpha,1}$ is an effective field which describes the action on the spin i of the $N - 1$ other ones $j = 1, N, j \neq i$, and $\mathbf{h}_0$ is an auxiliary field parameter. z is equal to

$$z = h_0^1 h_1 \ldots h_N (1 + T_{01}/h_0^1 h_1 + \cdots + T_{0N}/h_0^1 h_N) \qquad \text{(IX.20)}$$

the logarithmic derivative of z with respect to h_0^α at $\mathbf{h} = 0$ is

$$\langle s_0^\alpha \rangle = \delta_{\alpha,1}(T_{01}/h_1 + \cdots + T_{0N}/h_N). \qquad \text{(IX.21)}$$

Obviously the only important component of the spins is the one in direction 1. We call $m_0 = \langle s_0^1 \rangle$. The magnetization $\langle s_i^1 \rangle$ in the *absence* of spin 0 will be named the *cavity magnetization* m_i^c. One has $m_i^c = 1/h_i$. So the magnetization on the new site is a function of the N cavity magnetizations

$$m_0 = \left(\sum_i T_{0i} m_i^c\right)^{-1}. \qquad \text{(IX.22)}$$

(Note that, as $T_{0i} = \exp(-\beta N l_{0i})$, only a finite number of sites contributes effectively to the above sum over i.)

In order to compare formula (IX.22) with the results of the replica method we must average out the samples. Let us call $P(m)$ the distribution of the m's, and $M_p = \int P(m)m^p\,dm$ its pth moment. From (IX.22) we get

$$M_{-1} = \overline{1/m_0} = Ng_1/N\bar{m} = g_1 M_1$$

$$M_{-2} = g_2 M_2 + (g_1 M_1)^2$$

$$M_{-3} = g_3 M_3 + 3g_2 M_2 g_1 M_1 + (g_1 M_1)^3$$

$$\text{etc.} \ldots \tag{IX.23}$$

These equations are precisely the ones obtained from the replica method with the hypothesis of replica symmetry[6], with $Q_p = 2g_p M_p$. Equation (IX.22) has a simple interpretation: the average occupation number of link $i-j$, $\langle n_{ij}\rangle$, is the logarithmic derivative of Z with respect to $\log(T_{ij})$. From (IX.20) one gets

$$\langle n_{0i}\rangle = m_0 T_{0i} m_i^c \tag{IX.24}$$

and (IX.22) expresses the fact that $\sum_i \langle n_{0i}\rangle = 1$.

In order to solve for a given sample, one must relate the true magnetization m_0 (the one we have computed) to the *true* magnetizations m_i (i.e. those computed *in presence* of spin 0), and not to the cavity magnetizations m_i^c. From (IX.20) one finds that m_i is equal to

$$m_i = m_i^c(1 - \langle n_{0i}\rangle) \tag{IX.25}$$

so that the (*TAP* like) equations for a given problem of N sites read

$$\langle n_{ij}\rangle(1 - \langle n_{ij}\rangle) = m_i T_{ij} m_j$$

$$1/m_i = \sum_k T_{ik} m_k (1 - \langle n_{ik}\rangle)^{-1}. \tag{IX.26}$$

These equations are valid at a given temperature $1/\beta$ and for a given sample. An algorithm which solves them can be used to find a near-optimal matching by going to low temperature. (Practically it seems easier to find a solution first at high temperatures and then do some kind of annealing). Preliminary studies seem to indicate that an iterative resolution converges, but more work remains to be done to analyse this new algorithm. The situation might be more favourable in the present case than in the spin glass case[12] since it seems possible to find solutions of Eq. (IX.26) with a small amount of work, which is not the case for the solutions of *TAP* equations (see Chap. II).

IX.3 The Travelling Salesman Problem

A travelling salesman tour can be seen as a self avoiding walk ("*SAW*") on the "lattice" of the bonds between the N points. Hence a field theoretic representation can be found by applying the usual representation of a *SAW* in terms of an m-component vector field in the limit $m \to 0$ (Ref. 11). Using the properties of these fields described in the preceding section, one can show that the partition function

$$Z = \int \prod_i d\mu(\mathbf{s}_i) \exp\left(\gamma \sum_{i<j} T_{ij} \mathbf{s}_i \cdot \mathbf{s}_j\right) \tag{IX.27}$$

is proportional to the *TSP* one in the limits $m \to 0$ (which selects the *SAW*'s), and $\gamma \to \infty$ (which imposes that the *SAW* be of maximal length $= N$)

$$Z_{TSP} = \lim_{m \to 0} 1/m \lim_{\gamma \to \infty} Z/\gamma^N . \tag{IX.28}$$

The replica method has been applied to this problem in Mézard and Parisi, Reprint 28. The order parameters are

$$Q^{\alpha_1 \ldots \alpha}_{a_1 \ldots a_p} = (\gamma/p\beta) \langle s^{\alpha_1}_{i,a_1} \ldots s^{\alpha}_{i,a_p} \rangle \tag{IX.29}$$

where a's are replica indices (going from one to n) and α's are indices of spin components (going from one to m). Although both m and n must go to zero in the end, the natures of the corresponding symmetries are very different. The symmetry in replica space is a symmetry of permutations P_n, while in spin space there is a rotational symmetry O_m. A solution of the saddle-point equations has been found with the following properties:

— it is replica symmetric

$$Q^{\alpha_1 \ldots \alpha}_{a_1 \ldots a_p} = Q^{\alpha_1 \ldots \alpha} . \tag{IX.30}$$

— The rotational symmetry is spontaneously broken, as is natural in the limit $\gamma \to \infty$

$$Q^{\alpha_1 \ldots \alpha} = Q_p \delta^{\alpha_1, 1} \ldots \delta^{\alpha, 1} . \tag{IX.31}$$

An analytic solution for the Q_p's or their generating function has not been found yet. In Ref. 14 a resolution of the saddle-point equations by a high temperature expansion has been performed. The first twenty terms of the expansion of the energy in powers of $\beta = 1/T$ have been found. Standard techniques allow for the estimation of the energy at all temperatures. The result is plotted in Fig. IX.2. In the zero temperature limit the result is

$$E(T = 0) = 2.08 \pm .03 , \tag{IX.32}$$

where the estimated uncertainty is due to the estimation of the zero temperature limit from the (truncated) high temperature series. If this replica symmetric solution is exact, E should be the length of an optimal travelling salesman tour between N points when l_{ij}'s are independent symmetric random variables distributed uniformly on $[0, 1]$, with probability one in the large N limit.

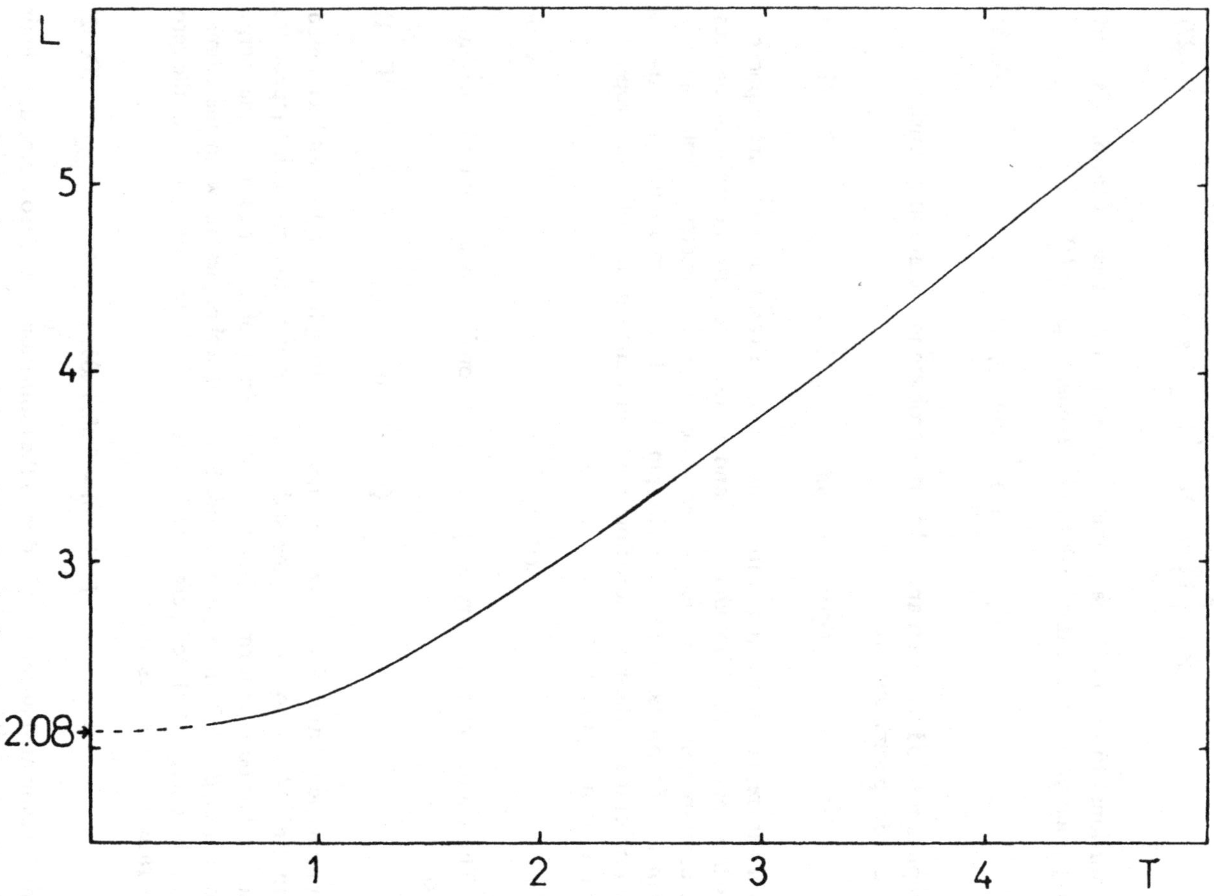

Fig. IX.2. Thermal average of the length versus the temperature for the *TSP* with independent random distances uniformly distributed on [0, 1]: theoretical prediction of the replica symmetric solution.

Table IX.1 Width σ_q of the distribution of overlaps $P(q)$ for different values of N. Numerical results from Ref. 15.

N	60	100	160
σ_q	.054	.040	.032

The distribution of overlaps has also been computed. Again, because of the assumption of replica symmetry, this turns out to be a δ function

$$P(q) = \delta(q - q_0(T)), \tag{IX.33}$$

where q_0 is plotted in Fig. IX.3.

So far numerical estimates of the optimal length[3,15] do not disagree with the prediction (IX.32), but precise numerical data are still lacking. To our knowledge the most precise analysis along these lines so far is the one of Sourlas[15]. With a simulated annealing algorithm he has obtained several near optimal tours for samples of sizes between 60 and 160. The distribution of overlaps $P(q)$ of these configurations (computed without the Boltzmann weights) is very peaked around a certain value $\bar{q}_0$, with a width that decreases with N, as shown in Table 1.

The numerical results for the correlations of $\bar{q}_0$ with the average length of the tours, E, is plotted in Fig. IX.3. The curve is the theoretical prediction obtained through the elimination of T in the analytic results for $E(T)$ and $q_0(T)$. Clearly the replica symmetric theory gives at least a reasonable first estimate of the properties of the system. One should notice however that there are no results corresponding to a very low temperature. (The point of lower energy in Fig. IX.4 corresponds to an effective temperature $T \simeq .85$).

It is also possible, as in the matching, to write down the *TAP*-like mean field equations valid for one given sample. These are a bit more complicated and the reader will find them in Ref. 10. The possibility of using them to find new types of algorithms for the *TSP* is an exciting open question which is presently under study. Of course, as in the matching, there are many fascinating problems which remain to be understood on the analytical side: stability, replica symmetry breaking, etc. One direction is to use the present formulation as a starting point to the study of problems in which the distances are correlated, as the famous Euclidean *TSP*. We shall return to that in the last section of this chapter.

IX.4 The Assignment Problem and its Relations with the Matching and *TSP*

We are not going to describe at length the analytical study of the assignment as we did for the first two problems. Essentially we want to point out the similarities between the assignment and the matching or *TSP*, in order to show how the good algorithms which exist for the assignment can be used in the *TSP*.

To begin with, we keep to the symmetric case $l_{ij} = l_{ji}$ and we suppose that l_{ii} is infinite

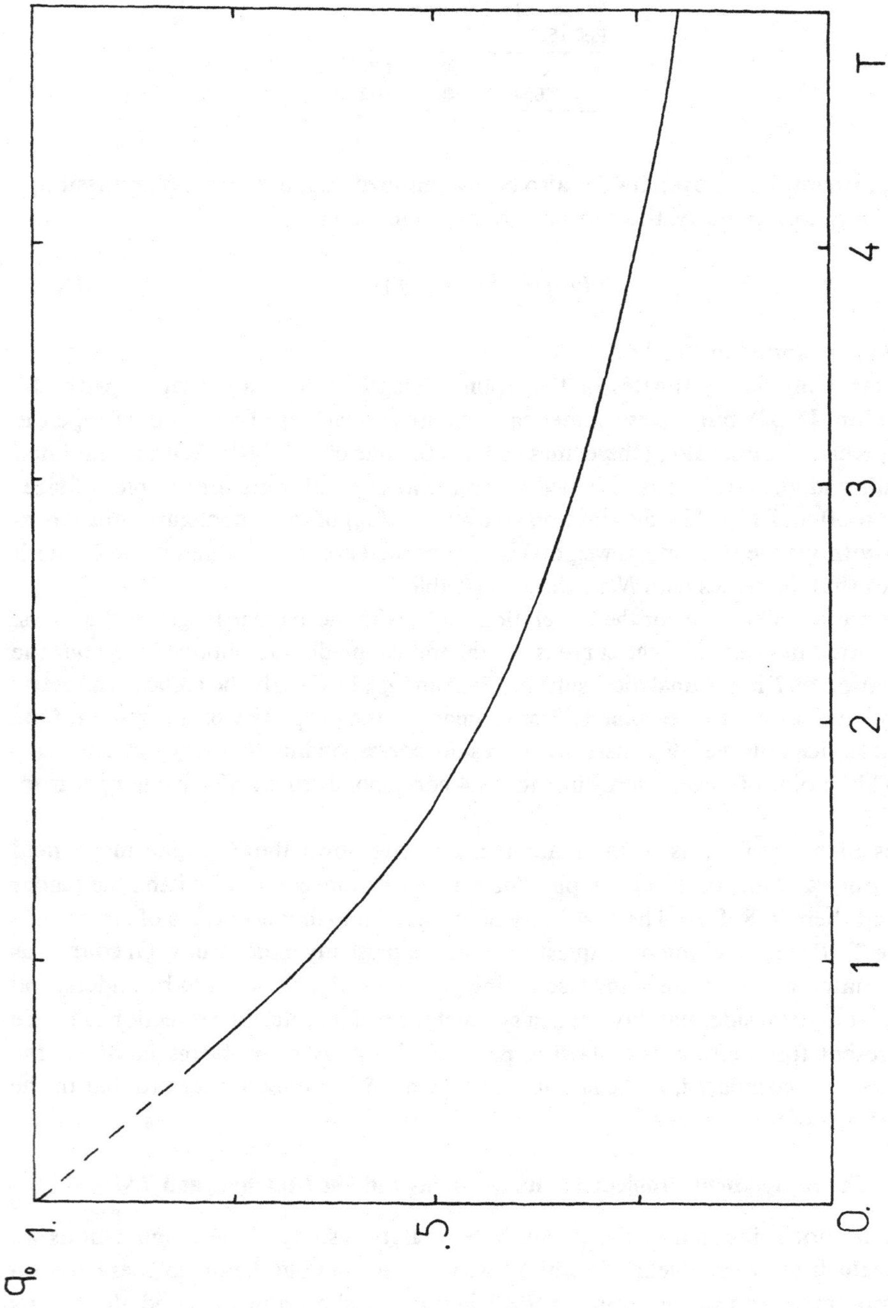

Fig. IX.3. Overlap of the relevant tours versus the temperature for the *TSP* with independent random distances uniformly distributed on [0, 1]: theoretical prediction of the replica symmetric solution.

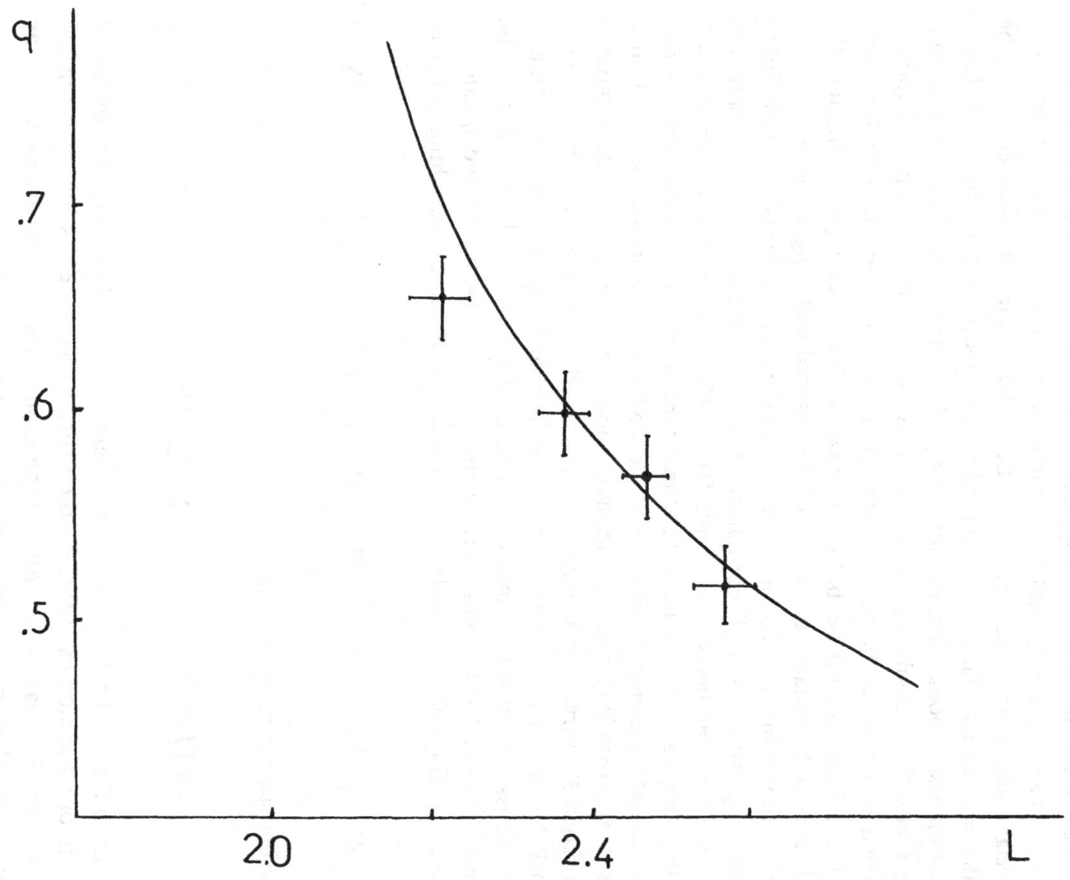

Fig. IX.4. Overlap versus the average length for the *TSP* with independent random distances uniformly distributed on [0, 1]: the points are the results of the numerical simulations, the full curve is the theoretical prediction of the replica symmetric solution obtained by elimination of *T* from the results of Figs. 2, 3. (from Ref. 15)

so that a point cannot be assigned to itself. It is very unlikely that a good assignment will contain a small loop of length larger or equal to three. Let us explain why qualitatively. We have already seen that the only important bonds (those which can be present in a good solution) have a length of order $1/N$. (We shall call them small bonds). Hence a given point typically has a *finite* number of effective neighbours. If one tries to make a small triangular loop starting from this point, the two other points must be chosen among these neighbours, hence there is a finite number of choices of pairs of neighbours, and the probability that the bond between the two neighbours be also small is of order $1/N$. This argument applies to all loops of finite size strictly larger than two. Conversely loops of size two are strongly favoured, hence the optimal assignment must look like a (double) matching. This can be seen numerically: the optimal assignment is often a perfect matching. For certain samples it contains a small number of loops of size larger than three, but in these cases also the asymptotic length of the matching seems to be equal to twice that of the optimal matching, i.e. $\pi^2/6$.

Now imagine that in the assignment one forbids the loops of size two (double links). This is very easy and keeps the linear nature of the problem: instead of imposing that n_{ij} be inside $[0,2]$, one takes them inside $[0,1]$. Then the optimal assignment will contain large loops and eventually (with a small probability) a few small ones. In fact we can show that the asymptotic lengths of the *TSP* and assignments without double links are equal, provided that the hypothesis of spontaneous breaking of the rotational symmetry in (IX.31) is true, and with no hypothesis on the breaking of replica symmetry.

We shall describe this derivation very briefly. The field theoretic representation of this modified assignment is the same as the one used for the matching in Ref. 6. The constraints $\sum_j n_{ij} = 2\,(1 < i < N)$ are written in integral representation and n_{ij} can take the values 0 or 1. The replica method is then applied as in the case of matching and gives

$$\overline{Z^n} = \int \prod dQ_{a_1\ldots a} \exp\left(-N/2 \sum_p (1/g_p) \sum_{a_1<\cdots<a_p} (Q_{a_1\ldots a_p})^2\right) z_{1s}{}^N, \qquad \text{(IX.34)}$$

where the one site partition function is

$$z_{1s} = \int \prod_a d\lambda_a/2\pi \, \exp\left(2i\lambda_a + \sum_p \sum_{a_1\ldots a_p} Q_{a_1\ldots a_p} e^{-i(\lambda_{a_1}+\cdots+\lambda_{a_p})}\right). \qquad \text{(IX.35)}$$

If one assumes that in the *TSP* there is a spontaneous breakdown of the rotational symmetry in one direction as in (IX.31), the expressions of Ref. 14 for the quenched free energy of the *TSP* can be rewritten in the form of the two above formulas. (The combinatorial problem in the evaluation of z_{1s} is the same in both cases: one must expand the exponential in (IX.35) in such a way that each replica index appears exactly twice). Hence our replica symmetric solution of *TSP* applies as well to this modified assignment, which constitutes a choice place to check or infer the prediction $E = 2.08 \pm .03$.

The asymmetric case where $l_{ij} \neq l_{ji}$ is also interesting. It is clear that double bonds can never appear in this case. Indeed it has been shown rigorously that the asymptotic values of the optimal lengths of the assignment and *TSP* coincide, and a heuristic for the *TSP* consists in patching together the (small number) of tours which appear in the optimal assignment[16].

The application of the replica method to this assymmetric case works as before. There are now two sets of constraints $\sum_j n_{ij} = 1$ (links starting from i) and $\sum_j n_{ji} = 1$ (links arriving on i), and $n_{ij} = 0, 1$. A standard computation shows that the free energy in this problem is exactly twice the free energy of the matching. Hence the asymptotic length of the optimal travelling salesman tour in the asymmetric case should be

$$E = \pi^2/6 = 1.64493\ldots, \tag{IX.36}$$

a prediction which seems compatible with numerical results[16].

IX.5 Graph Partitioning

The graph partitioning problem is a well known *NP* complete problem which is much more similar to familiar spin glass models. An instance is a graph: a set of N points (N even) and of some connexions between them. A configuration is a partition of the points into two subsets, say U and D, containing each half of the total number of points. The cost function one tries to minimize is the total number of connexions between U and D.

This is easily mapped onto a model of Ising spins as follows[17,18]: each point i can be in one of the two subsets. We associate to it an Ising spin σ_i which takes value $+1$ if i is in U and value -1 if i is in D. A configuration is a set of spins σ_i with the constraint that U and D have the same size

$$\sum_i \sigma_i = 0. \tag{IX.37}$$

The positions of the connexions can be encoded into a matrix J_{ij} such that $J_{ij} = J$ if i and j are connected and $J_{ij} = 0$, if not. Then the cost of a configuration is

$$C = 1/J \sum_{i<j} J_{ij}(1 - \sigma_i\sigma_j)/2 \tag{IX.38}$$

which can be written as

$$C = C_0 + H(\{J_{ij}\}, \{\sigma_i\})/2J, \tag{IX.39}$$

where $C_0 = (\sum_{i<j} J_{ij})/2J$ is a constant (independent of the configuration and H is a familiar Hamiltonian

$$H(\{J_{ij}\}, \{\sigma_i\}) = -\sum_{i<j} J_{ij}\sigma_i\sigma_j. \tag{IX.40}$$

As J_{ij} takes values 0 or J, H describes a diluted ferromagnet. Frustration comes into this problem from the fact that the allowed configurations must have total magnetization 0, which is a kind of infinite range antiferromagnetic interaction between the spins.

From the physicist point of view this constraint can be taken care of in a "soft" way: all the configurations are allowed, but the ones which have non zero magnetization are strongly disfavoured energetically by adjonction of a new term in the Hamiltonian[18]:

$$H'(\{J_{ij}\},\{\sigma_i\}) = -\sum_{i<j} J_{ij}\sigma_i\sigma_j + (A/N)\cdot\left(\sum_i \sigma_i\right)^2, \qquad A \gg 1. \qquad \text{(IX.41)}$$

Here again in order to make some analytic study one must specify an ensemble of instances. A natural domain is that of random graphs where each connexion is present with a probability p, independently of the other connexions[18]. For N large the average number of neighbours of one point is $\alpha = Np$. We shall consider successively two very different regimes depending on the scaling of p with N.

When N goes to infinity with p kept constant one is dealing with an infinite range problem. With the replica method it has been shown[18] that in the infinite volume limit the order parameter is the usual $n \times n$ matrix $Q_{ab} = \langle \sigma^i{}_a \sigma^i{}_b \rangle$. Furthermore the free energy one must extremize has the same functional dependence on Q_{ab} as in the SK model, apart from trivial proportionality factors. Hence the solution with replica symmetry breaking applies to this problem as well, at any temperature, and the cost of an optimal partitioning in the large N limit is found equal to

$$C = N^2/4p + N^{3/2}[p(1-p)]^{1/2}E_0/2, \qquad \text{(IX.42)}$$

where the leading term of order N^2 is the value of the constant C_0, and the correction due to optimization is the subleading term of order $N^{3/2}$. E_0 is the ground state energy of the SK spin glass, equal to $-.7633$ in the hierarchical solution. This formula is in good agreement with the known numerical results on this problem[18]. Other typical features of the SK spin glass such as ultrametricity or the exponential distribution of free energies of the pure states apply as well to this problem.

This equivalence can be understood simply from the cavity method. Let us give a brief indication on how this works. We shall skip the details and consider only the simple case where there is only one pure state (no replica symmetry breaking). Clearly the whole approach can be made more precise and completed as in Chap. V. The system of N spins σ_i, $i = 1, \ldots, N$ is assumed to be in one given state with zero magnetization and self overlap q. For a given configuration the field on the new site is $h_0 = \sum_i J_{0i}\sigma_i$. (Note that it is also possible to work with the Hamiltonian (IX.41): in this case the penalty term in the Hamiltonian creates some kind of additional global magnetic field in the direction opposite to the global magnetization, in such a way as to insure that this global magnetization vanishes). Its thermal average $\langle h_0 \rangle = \sum_i J_{0i} m_i$ depends on the sample as follows

$$\overline{\langle h_0 \rangle} = pJ \sum_i m_i = 0$$

$$\overline{\langle h_0 \rangle^2} = \sum_{i,j} \overline{J_{0i}J_{0j}} m_i m_j = NJ^2 p(1-p)q.$$

$$\vdots \qquad \text{(IX.43)}$$

It is clear from the start that J is only a bookkeeping variable that can be chosen at will. As in previous cases one must choose the correct scale of energies (or of β) in order to insure the existence of the thermodynamic limit. From (IX.41, 43) we see that J must be of order $N^{-1/2}$. In order to recover the normalizations of the SK model we choose $J^2 = 1/(N(p - p^2))$. Computing higher moments with this scaling one finds that $\langle h_0 \rangle$ is a Gaussian random variable of mean zero and width $\sqrt{q}$. As for the thermal fluctuations, of the type

$$\langle h_0{}^2 \rangle - \langle h_0 \rangle^2 = \sum_{i,j} J_{0i} J_{0j} \langle \sigma_i \sigma_j \rangle_c \qquad \text{(IX.44)}$$

they are Gaussian fluctuations around the (sample dependent) mean $\langle h_0 \rangle$ with a (sample independent) width $1 - q$. Clearly we have exactly recovered the situation of the SK model as we discussed it in Chap. V, and formula (IX.42) follows.

The partitioning of random graphs with a finite connectivity is different[19]. It corresponds to taking the limit $N \to \infty$ with an average number of neighbours $\alpha = Np$ fixed. Going back to (IX.41, 43) we see that the correct scaling of J is N independent. For instance one can take simply $J^2 = 1/\alpha$. Then the higher moments of the distribution of $\langle h_0 \rangle$, $\overline{\langle h_0 \rangle^k}$, are no longer trivial. $\langle h_0 \rangle$ is not a Gaussian variable and its distribution depends on all the higher overlaps of the N spins, $q_p = (1/N) \sum_i m_i{}^p$, and not only on q_2. In the replica method this is reflected in the fact that there is a whole set of order parameters $Q_{ab}, Q_{abc}, \ldots, Q_{12\ldots N}$, as in the cases of the matching or TSP studied before. Even at the replica symmetric level the order parameter is a function: the distribution of local magnetic fields. An integral equation for this distribution can be obtained from the consistency condition of the cavity method as before. If the resulting distribution turns out to be even under $h \to -h$, then the graph partitioning is again equivalent to the spin glass model with $+J$ couplings on the random lattice defined by the graph, as in the problem where the connectivity was infinite. However as the distribution of fields is not Gaussian, it can well be such that the global magnetization vanishes: $\int P(h) \tanh(\beta h) = 0$, but P is not even. In this case the two problems are no longer equivalent. This is precisely what happens for α small enough[19].

With these methods it is not too difficult to obtain the optimal value of the cost at the replica symmetric level. This has been done for instance in Ref. 19 for a variant of graph partitioning in which one considers the ensemble of random graphs with a fixed connectivity. The results are within 5 per cent or so below the precise numerical results of Ref. 20 for this model. The replica symmetry breaking solution has not been exhibited so far, although there is much activity in the study of the stability of the replica symmetric Ansatz[21].

IX.6 Some Remarks

It should be clear at this stage that the fundamental property which allows one to apply the methods of spin glass theory to a given problem is not the fact that this problem have infinite range couplings, but rather the fact that these couplings be weakly correlated variables.

If in addition they are of infinite range, additional simplifications occur: the field on a newly added site is the sum of an infinite number of variables and has a Gaussian distribution. The only parameter is the width of this distribution. In the case of spin models this width is simply related to the overlaps of order two of the N preexisting spins, (of the type $\sum_i \sigma_i \sigma_i'/N$), and does not depend on the overlaps of higher order (of the type $\sum_i \sigma_i \sigma_i' \sigma_i''/N$). In replica language the order parameter is an $n \times n$ matrix Q_{ab}. If there is no breaking of ergodicity it is simply a number.

When the interactions are of finite range but still uncorrelated the field on the new site has a complicated distribution which depends on the overlaps of all orders. As we saw before, even in the absence of ergodicity breaking the order parameter is a function. The replica order parameter is a whole set of tensors $Q_{a_1 \ldots a_k}$, $k = 1, \ldots, n$. The effects of ergodicity breaking and how one must break the replica symmetry in this case have not yet been worked out in any of these problems, as we mentioned at the end of each section in this chapter. Notice that the "random links" TSP and matching problems studied above effectively fall into this second category, although this might not be evident at first sight, since at low temperatures each point has only a finite number of accessible neighbours (see Sec. 1), as soon as the temperature has been scaled properly.

For systems in which the interactions are correlated the kind of solution we have described here should be considered as a zeroth order approximation around which one should be able to expand systematically. A particularly appealing problem is the Euclidean TSP in which the distances are correlated because of triangular inequalities. Suppose we are given N points randomly distributed inside the unit hypercube of a D dimensional space. The Euclidean distance between i and j is

$$l_{ij} = \sqrt{[\vec{x}_i - \vec{x}_j]^2}. \tag{IX.45}$$

One can consider problems in which the cost function is not exactly the sum of the Euclidean distances along the tour of the travelling salesman, but rather

$$E = \sum_{\text{along the tour}} (l_{ij})^k. \tag{IX.46}$$

In this case, given a point, the distribution of distances $d = l^k$ to the other points is of the type

$$P(d) = c d^{(D/k-1)} \tag{IX.47}$$

around $d = 0$. Now if D and k go to infinity with $D/k = r + 1$ fixed, it has been speculated that the triangular inequalities may well be irrelevant[10]. In this case the $D \gg 1$ Euclidean problem would be described by the theory of Sec. IX.3, opening the road to a $1/D$ expansion in order to obtain for instance the value of the optimal length in the $D = 2$, $k = 1$ TSP. But this is still at the level of speculations.

References

1. Beardwood *et al.* (1959).
2. Bonomi and Lutton (1984).

3. Kirkpatrick and Toulouse (1985).
4. Vannimenus and Mézard (1984).
5. In the problem ($P1$), self averageness of the free energy can be derived from the usual Brout argument (see Chap. I), by working with a fixed density of points when $N \to \infty$. This property has also been demonstrated at zero temperature by Beardwood *et al.* (1959). In problem ($P2$) self averageness is clearly seen numerically and can be obtained through replica arguments.
6. Mézard and Parisi (1985) Reprint 26.
7. Orland (1985).
8. See for instance Papadimitriou and Steiglitz (1982).
9. Brunetti, Krauth, Mézard and Parisi (1986, unpublished).
10. Mézard and Parisi (1986a).
11. This is a well-known trick which has been first introduced in the theory of polymers. See e.g. De Gennes (1979) and references therein.
12. A similar approach has been attempted in spin glass problems by Soukoulis *et al.* (1982).
13. Mézard and Parisi, (1987).
14. Mézard and Parisi, (1986) Reprint 28.
15. Sourlas (1986).
16. This is discussed by Karp and Steele in the book "The traveling salesman problem" edited by Lawler *et al.* (1986).
17. Kirkpatrick *et al.* (1983) Reprint 25.
18. Fu and Anderson (1986) Reprint 27.
19. Mézard and Parisi (1986b).
20. Banavar *et al.* (1986).
21. Viana and Bray (1985); De Dominicis and Mottishaw (1986).

Reprints

OPTIMIZATION

13 May 1983, Volume 220, Number 4598

SCIENCE

Optimization by Simulated Annealing

S. Kirkpatrick, C. D. Gelatt, Jr., M. P. Vecchi

Summary. There is a deep and useful connection between statistical mechanics (the behavior of systems with many degrees of freedom in thermal equilibrium at a finite temperature) and multivariate or combinatorial optimization (finding the minimum of a given function depending on many parameters). A detailed analogy with annealing in solids provides a framework for optimization of the properties of very large and complex systems. This connection to statistical mechanics exposes new information and provides an unfamiliar perspective on traditional optimization problems and methods.

In this article we briefly review the central constructs in combinatorial optimization and in statistical mechanics and then develop the similarities between the two fields. We show how the Metropolis algorithm for approximate numerical simulation of the behavior of a many-body system at a finite temperature provides a natural tool for bringing the techniques of statistical mechanics to bear on optimization.

We have applied this point of view to a number of problems arising in optimal design of computers. Applications to partitioning, component placement, and wiring of electronic systems are described in this article. In each context, we introduce the problem and discuss the improvements available from optimization.

Of classic optimization problems, the traveling salesman problem has received the most intensive study. To test the power of simulated annealing, we used the algorithm on traveling salesman problems with as many as several thousand cities. This work is described in a final section, followed by our conclusions.

Combinatorial Optimization

The subject of combinatorial optimization (*1*) consists of a set of problems that are central to the disciplines of computer science and engineering. Research in this area aims at developing efficient techniques for finding minimum or maximum values of a function of very many independent variables (*2*). This function, usually called the cost function or objective function, represents a quantitative measure of the "goodness" of some complex system. The cost function depends on the detailed configuration of the many parts of that system. We are most familiar with optimization problems occurring in the physical design of computers, so examples used below are drawn from that context. The number of variables involved may range up into the tens of thousands.

The classic example, because it is so simply stated, of a combinatorial optimization problem is the traveling salesman problem. Given a list of N cities and a means of calculating the cost of traveling between any two cities, one must plan the salesman's route, which will pass through each city once and return finally to the starting point, minimizing the total cost. Problems with this flavor arise in all areas of scheduling and design. Two subsidiary problems are of general interest: predicting the expected cost of the salesman's optimal route, averaged over some class of typical arrangements of cities, and estimating or obtaining bounds for the computing effort necessary to determine that route.

All exact methods known for determining an optimal route require a computing effort that increases exponentially with N, so that in practice exact solutions can be attempted only on problems involving a few hundred cities or less. The traveling salesman belongs to the large class of NP-complete (nondeterministic polynomial time complete) problems, which has received extensive study in the past 10 years (*3*). No method for exact solution with a computing effort bounded by a power of N has been found for any of these problems, but if such a solution were found, it could be mapped into a procedure for solving all members of the class. It is not known what features of the individual problems in the NP-complete class are the cause of their difficulty.

Since the NP-complete class of problems contains many situations of practical interest, heuristic methods have been developed with computational requirements proportional to small powers of N. Heuristics are rather problem-specific: there is no guarantee that a heuristic procedure for finding near-optimal solutions for one NP-complete problem will be effective for another.

There are two basic strategies for heuristics: "divide-and-conquer" and iterative improvement. In the first, one divides the problem into subproblems of manageable size, then solves the subproblems. The solutions to the subproblems must then be patched back together. For this method to produce very good solutions, the subproblems must be naturally disjoint, and the division made must be an appropriate one, so that errors made in patching do not offset the gains

S. Kirkpatrick and C. D. Gelatt, Jr., are research staff members and M. P. Vecchi was a visiting scientist at IBM Thomas J. Watson Research Center, Yorktown Heights, New York 10598. M. P. Vecchi's present address is Instituto Venezolano de Investigaciones Cientificas, Caracas 1010A, Venezuela.

obtained in applying more powerful methods to the subproblems (4).

In iterative improvement (5, 6), one starts with the system in a known configuration. A standard rearrangement operation is applied to all parts of the system in turn, until a rearranged configuration that improves the cost function is discovered. The rearranged configuration then becomes the new configuration of the system, and the process is continued until no further improvements can be found. Iterative improvement consists of a search in this coordinate space for rearrangement steps which lead downhill. Since this search usually gets stuck in a local but not a global optimum, it is customary to carry out the process several times, starting from different randomly generated configurations, and save the best result.

There is a body of literature analyzing the results to be expected and the computing requirements of common heuristic methods when applied to the most popular problems (1-3). This analysis usually focuses on the worst-case situation—for instance, attempts to bound from above the ratio between the cost obtained by a heuristic method and the exact minimum cost for any member of a family of similarly structured problems. There are relatively few discussions of the average performance of heuristic algorithms, because the analysis is usually more difficult and the nature of the appropriate average to study is not always clear. We will argue that as the size of optimization problems increases, the worst-case analysis of a problem will become increasingly irrelevant, and the average performance of algorithms will dominate the analysis of practical applications. This large number limit is the domain of statistical mechanics.

Statistical Mechanics

Statistical mechanics is the central discipline of condensed matter physics, a body of methods for analyzing aggregate properties of the large numbers of atoms to be found in samples of liquid or solid matter (7). Because the number of atoms is of order 10^{23} per cubic centimeter, only the most probable behavior of the system in thermal equilibrium at a given temperature is observed in experiments. This can be characterized by the average and small fluctuations about the average behavior of the system, when the average is taken over the ensemble of identical systems introduced by Gibbs. In this ensemble, each configuration, defined by the set of atomic positions, $\{r_i\}$, of the system is weighted by its Boltzmann probability factor, $\exp(-E(\{r_i\})/k_BT)$, where $E(\{r_i\})$ is the energy of the configuration, k_B is Boltzmann's constant, and T is temperature.

A fundamental question in statistical mechanics concerns what happens to the system in the limit of low temperature—for example, whether the atoms remain fluid or solidify, and if they solidify, whether they form a crystalline solid or a glass. Ground states and configurations close to them in energy are extremely rare among all the configurations of a macroscopic body, yet they dominate its properties at low temperatures because as T is lowered the Boltzmann distribution collapses into the lowest energy state or states.

As a simplified example, consider the magnetic properties of a chain of atoms whose magnetic moments, μ_i, are allowed to point only "up" or "down," states denoted by $\mu_i = \pm 1$. The interaction energy between two such adjacent spins can be written $J\mu_i\mu_{i+1}$. Interaction between each adjacent pair of spins contributes $\pm J$ to the total energy of the chain. For an N-spin chain, if all configurations are equally likely the interaction energy has a binomial distribution, with the maximum and minimum energies given by $\pm NJ$ and the most probable state having zero energy. In this view, the ground state configurations have statistical weight $\exp(-N/2)$ smaller than the zero-energy configurations. A Boltzmann factor, $\exp(-E/k_BT)$, can offset this if k_BT is smaller than J. If we focus on the problem of finding empirically the system's ground state, this factor is seen to drastically increase the efficiency of such a search.

In practical contexts, low temperature is not a sufficient condition for finding ground states of matter. Experiments that determine the low-temperature state of a material—for example, by growing a single crystal from a melt—are done by careful annealing, first melting the substance, then lowering the temperature slowly, and spending a long time at temperatures in the vicinity of the freezing point. If this is not done, and the substance is allowed to get out of equilibrium, the resulting crystal will have many defects, or the substance may form a glass, with no crystalline order and only metastable, locally optimal structures.

Finding the low-temperature state of a system when a prescription for calculating its energy is given is an optimization problem not unlike those encountered in combinatorial optimization. However, the concept of the temperature of a physical system has no obvious equivalent in the systems being optimized. We will introduce an effective temperature for optimization, and show how one can carry out a simulated annealing process in order to obtain better heuristic solutions to combinatorial optimization problems.

Iterative improvement, commonly applied to such problems, is much like the microscopic rearrangement processes modeled by statistical mechanics, with the cost function playing the role of energy. However, accepting only rearrangements that lower the cost function of the system is like extremely rapid quenching from high temperatures to $T = 0$, so it should not be surprising that resulting solutions are usually metastable. The Metropolis procedure from statistical mechanics provides a generalization of iterative improvement in which controlled uphill steps can also be incorporated in the search for a better solution.

Metropolis *et al.* (8), in the earliest days of scientific computing, introduced a simple algorithm that can be used to provide an efficient simulation of a collection of atoms in equilibrium at a given temperature. In each step of this algorithm, an atom is given a small random displacement and the resulting change, ΔE, in the energy of the system is computed. If $\Delta E \leq 0$, the displacement is accepted, and the configuration with the displaced atom is used as the starting point of the next step. The case $\Delta E > 0$ is treated probabilistically: the probability that the configuration is accepted is $P(\Delta E) = \exp(-\Delta E/k_BT)$. Random numbers uniformly distributed in the interval (0,1) are a convenient means of implementing the random part of the algorithm. One such number is selected and compared with $P(\Delta E)$. If it is less than $P(\Delta E)$, the new configuration is retained; if not, the original configuration is used to start the next step. By repeating the basic step many times, one simulates the thermal motion of atoms in thermal contact with a heat bath at temperature T. This choice of $P(\Delta E)$ has the consequence that the system evolves into a Boltzmann distribution.

Using the cost function in place of the energy and defining configurations by a set of parameters $\{x_i\}$, it is straightforward with the Metropolis procedure to generate a population of configurations of a given optimization problem at some effective temperature. This temperature is simply a control parameter in the same units as the cost function. The simulated annealing process consists of first "melting" the system being optimized at a high effective temperature, then lowering the temperature by slow stages until the system "freezes" and no further changes occur. At each temperature, the simulation must proceed long enough for the system to reach a steady state. The sequence of temperatures and the number of rearrangements of the $\{x_i\}$ attempted to reach equilibrium at each temperature can be considered an annealing schedule.

Annealing, as implemented by the Metropolis procedure, differs from iterative improvement in that the procedure need not get stuck since transitions out of a local optimum are always possible at nonzero temperature. A second and more important feature is that a sort of adaptive divide-and-conquer occurs. Gross features of the eventual state of the system appear at higher temperatures; fine details develop at lower temperatures. This will be discussed with specific examples.

Statistical mechanics contains many useful tricks for extracting properties of a macroscopic system from microscopic averages. Ensemble averages can be obtained from a single generating function, the partition function, Z,

$$Z = \mathrm{Tr}\exp\left(\frac{-E}{k_B T}\right) \qquad (1)$$

in which the trace symbol, Tr, denotes a sum over all possible configurations of the atoms in the sample system. The logarithm of Z, called the free energy, $F(T)$, contains information about the average energy, $<E(T)>$, and also the entropy, $S(T)$, which is the logarithm of the number of configurations contributing to the ensemble at T:

$$-k_B T \ln Z = F(T) = <E(T)> - TS \qquad (2)$$

Boltzmann-weighted ensemble averages are easily expressed in terms of derivatives of F. Thus the average energy is given by

$$<E(T)> = \frac{-d\ln Z}{d(1/k_B T)} \qquad (3)$$

and the rate of change of the energy with respect to the control parameter, T, is related to the size of typical variations in the energy by

$$C(T) = \frac{d<E(T)>}{dT} = \frac{[<E(T)^2> - <E(T)>^2]}{k_B T^2} \qquad (4)$$

In statistical mechanics $C(T)$ is called the specific heat. A large value of C signals a change in the state of order of a system, and can be used in the optimization context to indicate that freezing has begun and hence that very slow cooling is required. It can also be used to determine the entropy by the thermodynamic relation

$$\frac{dS(T)}{dT} = \frac{C(T)}{T} \qquad (5)$$

Integrating Eq. 5 gives

$$S(T) = S(T_1) - \int_{T}^{T_1} \frac{C(T')\,dT'}{T'} \qquad (6)$$

where T_1 is a temperature at which S is known, usually by an approximation valid at high temperatures.

The analogy between cooling a fluid and optimization may fail in one important respect. In ideal fluids all the atoms are alike and the ground state is a regular crystal. A typical optimization problem will contain many distinct, noninterchangeable elements, so a regular solution is unlikely. However, much research in condensed matter physics is directed at systems with quenched-in randomness, in which the atoms are not all alike. An important feature of such systems, termed "frustration," is that interactions favoring different and incompatible kinds of ordering may be simultaneously present (*9*). The magnetic alloys known as "spin glasses," which exhibit competition between ferromagnetic and antiferromagnetic spin ordering, are the best understood example of frustration (*10*). It is now believed that highly frustrated systems like spin glasses have many nearly degenerate random ground states rather than a single ground state with a high degree of symmetry. These systems stand in the same relation to conventional magnets as glasses do to crystals, hence the name.

The physical properties of spin glasses at low temperatures provide a possible guide for understanding the possibilities of optimizing complex systems subject to conflicting (frustrating) constraints.

Physical Design of Computers

The physical design of electronic systems and the methods and simplifications employed to automate this process have been reviewed (*11, 12*). We first provide some background and definitions related to applications of the simulated annealing framework to specific problems that arise in optimal design of computer systems and subsystems. Physical design follows logical design. After the detailed specification of the logic of a system is complete, it is necessary to specify the precise physical realization of the system in a particular technology.

This process is usually divided into several stages. First, the design must be partitioned into groups small enough to fit the available packages, for example, into groups of circuits small enough to fit into a single chip, or into groups of chips and associated discrete components that can fit onto a card or other higher level package. Second, the circuits are assigned specific locations on the chip. This stage is usually called placement. Finally, the circuits are connected by wires formed photolithographically out of a thin metal film, often in several layers. Assigning paths, or routes, to the wires is usually done in two stages. In rough or global wiring, the wires are assigned to regions that represent schematically the capacity of the intended package. In detailed wiring (also called exact embedding), each wire is given a unique complete path. From the detailed wiring results, masks can be generated and chips made.

At each stage of design one wants to optimize the eventual performance of the system without compromising the feasibility of the subsequent design stages. Thus partitioning must be done in such a way that the number of circuits in each partition is small enough to fit easily into the available package, yet the number of signals that must cross partition boundaries (each requiring slow, power-consuming driver circuitry) is minimized. The major focus in placement is on minimizing the length of connections, since this translates into the time required for propagation of signals, and thus into the speed of the finished system. However, the placements with the shortest implied wire lengths may not be wirable, because of the presence of regions in which the wiring is too congested for the packaging technology. Congestion, therefore, should also be anticipated and minimized during the placement process. In wiring, it is desirable to maintain the minimum possible wire lengths while minimizing sources of noise, such as cross talk between adjacent wires. We show in this and the next two sections how these conflicting goals can be combined and made the basis of an automatic optimization procedure.

The tight schedules involved present major obstacles to automation and optimization of large system design, even when computers are employed to speed up the mechanical tasks and reduce the chance of error. Possibilities of feedback, in which early stages of a design are redone to solve problems that became apparent only at later stages, are greatly reduced as the scale of the overall system being designed increases. Op-

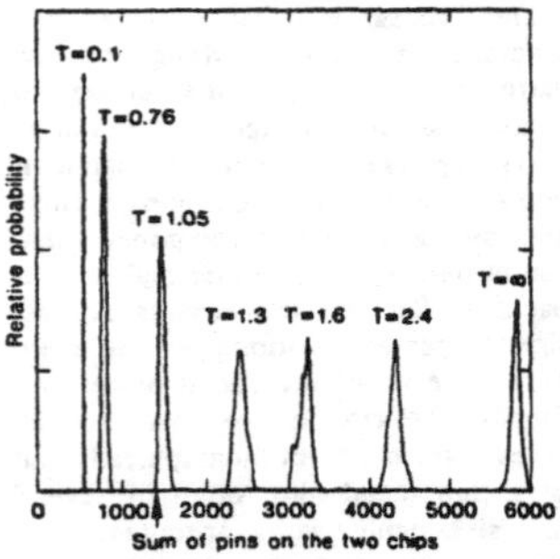

Fig. 1. Distribution of total number of pins required in two-way partition of a microprocessor at various temperatures. Arrow indicates best solution obtained by rapid quenching as opposed to annealing.

timization procedures that can incorporate, even approximately, information about the chance of success of later stages of such complex designs will be increasingly valuable in the limit of very large scale.

System performance is almost always achieved at the expense of design convenience. The partitioning problem provides a clean example of this. Consider N circuits that are to be partitioned between two chips. Propagating a signal across a chip boundary is always slow, so the number of signals required to cross between the two must be minimized. Putting all the circuits on one chip eliminates signal crossings, but usually there is no room. Instead, for later convenience, it is desirable to divide the circuits about equally.

If we have connectivity information in a matrix whose elements $\{a_{ij}\}$ are the number of signals passing between circuits i and j, and we indicate which chip circuit i is placed on by a two-valued variable $\mu_i = \pm 1$, then N_c, the number of signals that must cross a chip boundary is given by $\Sigma_{i>j}(a_{ij}/4)(\mu_i - \mu_j)^2$. Calculating $\Sigma_i \mu_i$ gives the difference between the numbers of circuits on the two chips. Squaring this imbalance and introducing a coefficient, λ, to express the relative costs of imbalance and boundary crossings, we obtain an objective function, f, for the partition problem:

$$f = \sum_{i>j} \left(\lambda - \frac{a_{ij}}{2}\right)\mu_i\mu_j \qquad (7)$$

Reasonable values of λ should satisfy $\lambda \leq z/2$, where z is the average number of circuits connected to a typical circuit (fan-in plus fan-out). Choosing $\lambda = z/2$ implies giving equal weight to changes in the balance and crossing scores.

The objective function f has precisely the form of a Hamiltonian, or energy function, studied in the theory of random magnets, when the common simplifying assumption is made that the spins, μ_i, have only two allowed orientations (up or down), as in the linear chain example of the previous section. It combines local, random, attractive ("ferromagnetic") interactions, resulting from the a_{ij}'s, with a long-range repulsive ("antiferromagnetic") interaction due to λ. No configuration of the $\{\mu_i\}$ can simultaneously satisfy all the interactions, so the system is "frustrated," in the sense formalized by Toulouse (*9*).

If the a_{ij} are completely uncorrelated, it can be shown (*13*) that this Hamiltonian has a spin glass phase at low temperatures. This implies for the associated magnetic problem that there are many degenerate "ground states" of nearly equal energy and no obvious symmetry. The magnetic state of a spin glass is very stable at low temperatures (*14*), so the ground states have energies well below the energies of the random high-temperature states, and transforming one ground state into another will usually require considerable rearrangement. Thus this analogy has several implications for optimization of partition:

1) Even in the presence of frustration, significant improvements over a random starting partition are possible.

2) There will be many good near-optimal solutions, so a stochastic search procedure such as simulated annealing should find some.

3) No one of the ground states is significantly better than the others, so it is not very fruitful to search for the absolute optimum.

In developing Eq. 7 we made several severe simplifications, considering only two-way partitioning and ignoring the fact that most signals connect more than two circuits. Objective functions analogous to f that include both complications are easily constructed. They no longer have the simple quadratic form of Eq. 7, but the qualitative feature, frustration, remains dominant. The form of the Hamiltonian makes no difference in the Metropolis Monte Carlo algorithm. Evaluation of the change in function when a circuit is shifted to a new chip remains rapid as the definition of f becomes more complicated.

It is likely that the a_{ij} are somewhat correlated, since any design has considerable logical structure. Efforts to understand the nature of this structure by analyzing the surface-to-volume ratio of components of electronic systems [as in "Rent's rule" (*15*)] conclude that the

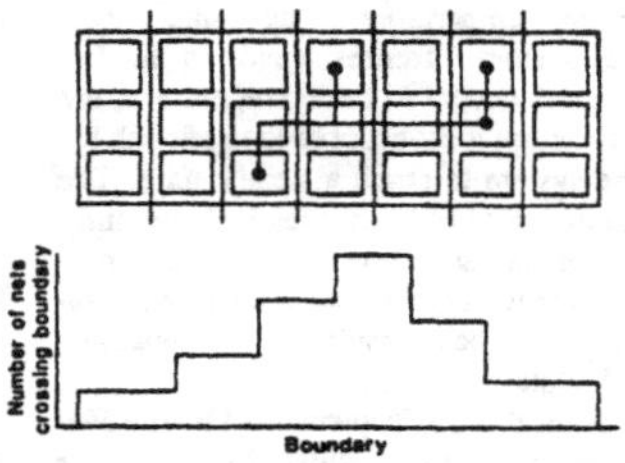

Fig. 2. Construction of a horizontal net-crossing histogram.

circuits in a typical system could be connected with short-range interactions if they were embedded in a space with dimension between two and three. Uncorrelated connections, by contrast, can be thought of as infinite-dimensional, since they are never short-range.

The identification of Eq. 7 as a spin glass Hamiltonian is not affected by the reduction to a two- or three-dimensional problem, as long as $\lambda N = z/2$. The degree of ground state degeneracy increases with decreasing dimensionality. For the uncorrelated model, there are typically of order $N^{1/2}$ nearly degenerate ground states (*14*), while in two and three dimensions, $2^{\alpha N}$, for some small value, α, are expected (*16*). This implies that finding a near-optimum solution should become easier, the lower the effective dimensionality of the problem. The entropy, measurable as shown in Eq. 6, provides a measure of the degeneracy of solutions. $S(T)$ is the logarithm of the number of solutions equal to or better than the average result encountered at temperature T.

As an example of the partitioning problem, we have taken the logic design for a single-chip IBM "370 microprocessor" (*17*) and considered partitioning it into two chips. The original design has approximately 5000 primitive logic gates and 200 external signals (the chip has 200 logic pins). The results of this study are plotted in Fig. 1. If one randomly assigns gates to the two chips, one finds the distribution marked $T = \infty$ for the number of pins required. Each of the two chips (with about 2500 circuits) would need 3000 pins. The other distributions in Fig. 1 show the results of simulated annealing.

Monte Carlo annealing is simple to implement in this case. Each proposed configuration change simply flips a randomly chosen circuit from one chip to the other. The new number of external connections, C, to the two chips is calculated (an external connection is a net with circuits on both chips, or a circuit connected to one of the pins of the original single-chip design), as is the new balance score, B, calculated as in deriving Eq. 7. The objective function analogous to Eq. 7 is

$$f = C + \lambda B \quad (8)$$

where C is the sum of the number of external connections on the two chips and B is the balance score. For this example, $\lambda = 0.01$.

For the annealing schedule we chose to start at a high "temperature," $T_0 = 10$, where essentially all proposed circuit flips are accepted, then cool exponentially, $T_n = (T_1/T_0)^n T_0$, with the ratio $T_1/T_0 = 0.9$. At each temperature enough flips are attempted that either there are ten accepted flips per circuit on the average (for this case, 50,000 accepted flips at each temperature), or the number of attempts exceeds 100 times the number of circuits before ten flips per circuit have been accepted. If the desired number of acceptances is not achieved at three successive temperatures, the system is considered "frozen" and annealing stops.

The finite temperature curves in Fig. 1 show the distribution of pins per chip for the configurations sampled at $T = 2.5$, 1.0, and 0.1. As one would expect from the statistical mechanical analog, the distribution shifts to fewer pins and sharpens as the temperature is decreased. The sharpening is one consequence of the decrease in the number of configurations that contribute to the equilibrium ensemble at the lower temperature. In the language of statistical mechanics, the entropy of the system decreases. For this sample run in the low-temperature limit, the two chips required 353 and 321 pins, respectively. There are 237 nets connecting the two chips (requiring a pin on each chip) in addition to the 200 inputs and outputs of the original chip. The final partition in this example has the circuits exactly evenly distributed between the two partitions. Using a more complicated balance score, which did not penalize imbalance of less than 100 circuits, we found partitions resulting in chips with 271 and 183 pins.

If, instead of slowly cooling, one were to start from a random partition and accept only flips that reduce the objective function (equivalent to setting $T = 0$ in the Metropolis rule), the result is chips with approximately 700 pins (several such runs led to results with 677 to 730 pins). Rapid cooling results in a system frozen into a metastable state far from the optimal configuration. The best result obtained after several rapid quenches is indicated by the arrow in Fig. 1.

Placement

Placement is a further refinement of the logic partitioning process, in which the circuits are given physical positions (*11, 12, 18, 19*). In principle, the two stages could be combined, although this is not often possible in practice. The objectives in placement are to minimize

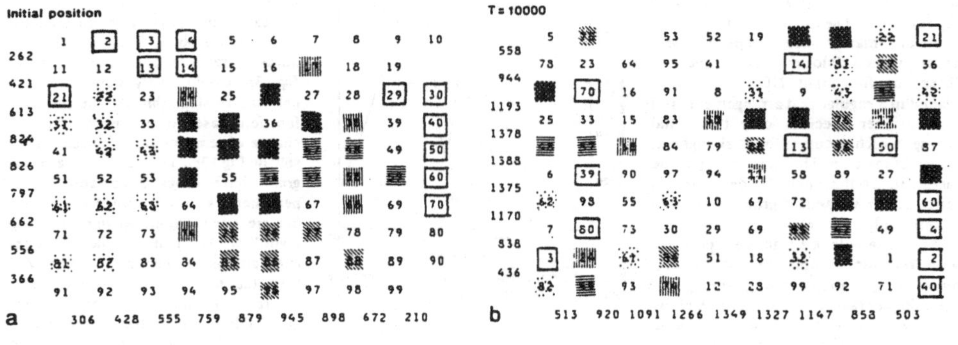

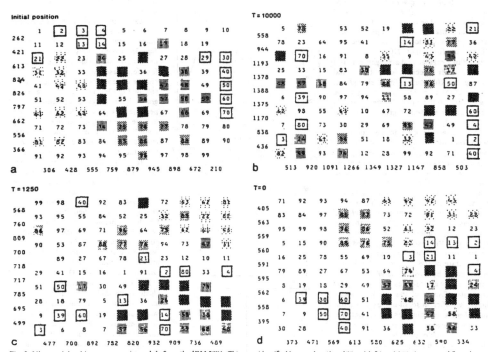

Fig. 3. Ninety-eight chips on a ceramic module from the IBM 3081. Chips are identified by number (1 to 100, with 20 and 100 absent) and function. The dark squares comprise an adder, the three types of squares with ruled lines are chips that control and supply data to the adder, the lightly dotted chips perform logical arithmetic (bitwise AND, OR, and so on), and the open squares denote general-purpose registers, which serve both arithmetic units. The numbers at the left and lower edges of the module image are the vertical and horizontal net-crossing histograms, respectively. (a) Original chip placement; (b) a configuration at $T = 10,000$; (c) $T = 1250$; (d) a zero-temperature result.

signal propagation times or distances while satisfying prescribed electrical constraints, without creating regions so congested that there will not be room later to connect the circuits with actual wire.

Physical design of computers includes several distinct categories of placement problems, depending on the packages involved (*20*). The larger objects to be placed include chips that must reside in a higher level package, such as a printed circuit card or fired ceramic "module" (*21*). These chip carriers must in turn be placed on a backplane or "board," which is simply a very large printed circuit card. The chips seen today contain from tens to tens of thousands of logic circuits, and each chip carrier or board will provide from one to ten thousand interconnections. The partition and placement problems decouple poorly in this situation, since the choice of which chip should carry a given piece of logic will be influenced by the position of that chip.

The simplest placement problems arise in designing chips with structured layout rules. These are called "gate array" or "master slice" chips. In these chips, standard logic circuits, such as three- or four-input NOR's, are preplaced in a regular grid arrangement, and the designer specifies only the signal wiring, which occupies the final, highest, layers of the chip. The circuits may all be identical, or they may be described in terms of a few standard groupings of two or more adjacent cells.

As an example of a placement problem with realistic complexity without too many complications arising from package idiosyncrasies, we consider 98 chips packaged on one multilayer ceramic module of the IBM 3081 processor (*21*). Each chip can be placed on any of 100 sites, in a 10 × 10 grid on the top surface of the module. Information about the connections to be made through the signal-carrying planes of the module is contained in a "netlist," which groups sets of pins that see the same signal.

The state of the system can be briefly represented by a list of the 98 chips with their x and y coordinates, or a list of the contents of each of the 100 legal locations. A sufficient set of moves to use for annealing is interchanges of the contents of two locations. This results in either the interchange of two chips or the interchange of a chip and a vacancy. For more efficient search at low temperatures, it is helpful to allow restrictions on the distance across which an interchange may occur.

To measure congestion at the same time as wire length, we use a convenient intermediate analysis of the layout, a net-crossing histogram. Its construction is summarized in Fig. 2. We divide the package surface by a set of natural boundaries. In this example, we use the boundaries between adjacent rows or columns of chip sites. The histogram then contains the number of nets crossing each boundary. Since at least one wire must be routed across each boundary crossed, the sum of the entries in the histogram of Fig. 2 is the sum of the horizontal extents of the rectangles bounding each net, and is a lower bound to the horizontal wire length required. Constructing a vertical net-crossing histogram and summing its entries gives a similar estimate of the vertical wire length.

The peak of the histogram provides a lower bound to the amount of wire that must be provided in the worst case, since each net requires at least one wiring channel somewhere on the boundary. To combine this information into a single objective function, we introduce a threshold level for each histogram—an amount of wire that will nearly exhaust the available wire capacity—and then sum for all histogram elements that exceed the threshold the square of the excess over threshold. Adding this quantity to the estimated length gives the objective function that was used.

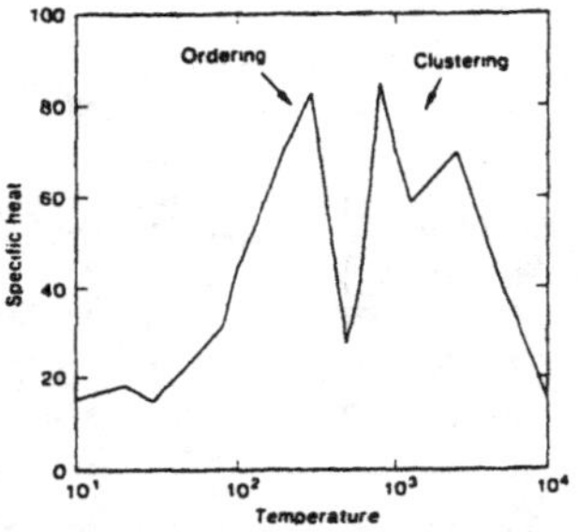

Fig. 4. Specific heat as a function of temperature for the design of Fig. 3, a to d.

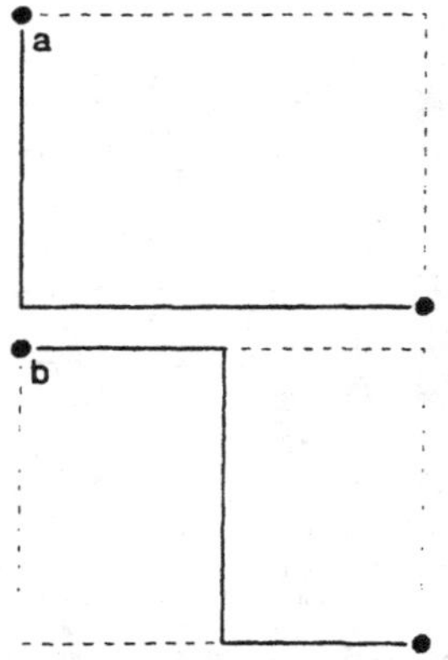

Fig. 5. Examples of (a) L-shaped and (b) Z-shaped wire rearrangements.

Figure 3 shows the stages of a simulated annealing run on the 98-chip module. Figure 3a shows the chip locations from the original design, with vertical and horizontal net-crossing histograms indicated. The different shading patterns distinguish the groups of chips that carry out different functions. Each such group was designed and placed together, usually by a single designer. The net-crossing histograms show that the center of the layout is much more congested than the edges, most likely because the chips known to have the most critical timing constraints were placed in the center of the module to allow the greatest number of other chips to be close to them.

Heating the original design until the chips diffuse about freely quickly produces a random-looking arrangement, Fig. 3b. Cooling very slowly until the chips move sluggishly and the objective function ceases to decrease rapidly with change of temperature produced the result in Fig. 3c. The net-crossing histograms have peaks comparable to the peak heights in the original placement, but are much flatter. At this "freezing point," we find that the functionally related groups of chips have reorganized from the melt, but now are spatially separated in an overall arrangement quite different from the original placement. In the final result, Fig. 3d, the histogram peaks are about 30 percent less than in the original placement. Integrating them, we find that total wire length, estimated in this way, is decreased by about 10 percent. The computing requirements for this example were modest: 250,000 interchanges were attempted, requiring 12 minutes of computation on an IBM 3033.

Between the temperature at which clusters form and freezing starts (Fig. 3c) and the final result (Fig. 3d) there are many further local rearrangements. The functional groups have remained in the same regions, but their shapes and relative alignments continue to change throughout the low-temperature part of the annealing process. This illustrates that the introduction of temperature to the optimization process permits a controlled, adaptive division of the problem through the evolution of natural clusters at the freezing temperature. Early prescription of natural clusters is also a central feature of several sophisticated placement programs used in master slice chip placement (*22, 23*).

A quantity corresponding to the thermodynamic specific heat is defined for this problem by taking the derivative with respect to temperature of the average value of the objective function observed at a given temperature. This is plotted in Fig. 4. Just as a maximum in the specific heat of a fluid indicates the onset of freezing or the formation of clusters, we find specific heat maxima at two temperatures, each indicating a different type of ordering in the problem. The higher temperature peak corresponds to the aggregation of clusters of functionally related objects, driven apart by the congestion term in the scoring. The lower temperature peak indicates the further decrease in wire length obtained by local rearrangements. This sort of measurement can be useful in practice as a means of determining the temperature ranges in which the important rearrangements in the design are occurring, where slower cooling with be helpful.

Wiring

After placement, specific legal routings must be found for the wires needed to connect the circuits. The techniques typically applied to generate such routings are sequential in nature, treating one wire at a time with incomplete information about the positions and effects of the other wires (*11, 24*). Annealing is inherently free of this sequence dependence. In this section we describe a simulated annealing approach to wiring, using the ceramic module of the last section as an example.

Nets with many pins must first be broken into connections—pairs of pins joined by a single continuous wire. This "ordering" of each net is highly dependent on the nature of the circuits being connected and the package technology. Orderings permitting more than two pins to be connected are sometimes allowed, but will not be discussed here.

The usual procedure, given an ordering, is first to construct a coarse-scale routing for each connection from which the ultimate detailed wiring can be completed. Package technologies and structured image chips have prearranged areas of fixed capacity for the wires. For the rough routing to be successful, it must not call for wire densities that exceed this capacity.

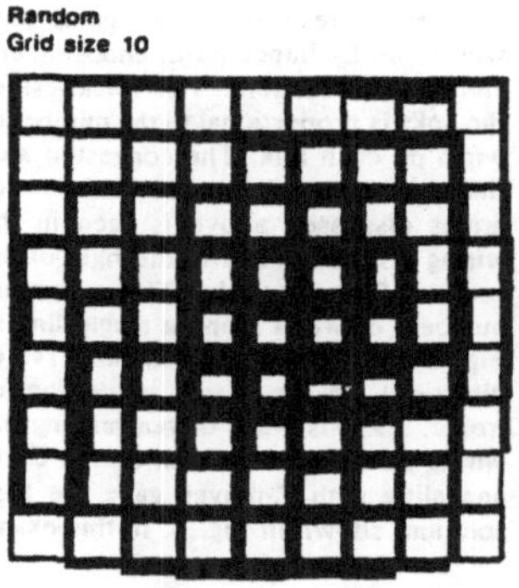

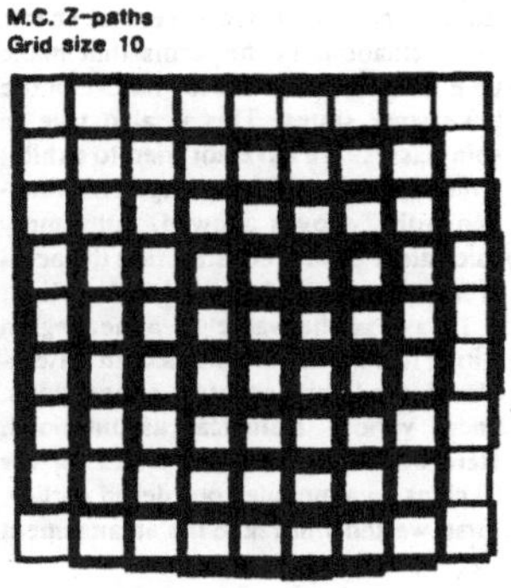

Fig. 6 (left). Wire density in the 98-chip module with the connections randomly assigned to perimeter routes. Chips are in the original placement. Fig. 7 (right). Wire density after simulated annealing of the wire routing, using Z-shaped moves.

We can model the rough routing problem (and even simple cases of detailed embedding) by lumping all actual pin positions into a regular grid of points, which are treated as the sources and sinks of all connections. The wires are then to be routed along the links that connect adjacent grid points.

The objectives in global routing are to minimize wire length and, often, the number of bends in wires, while spreading the wire as evenly as possible to simplify exact embedding and later revision. Wires are to be routed around regions in which wire demand exceeds capacity if possible, so that they will not "overflow," requiring drastic rearrangements of the other wires during exact embedding. Wire bends are costly in packages that confine the north-south and east-west wires to different layers, since each bend requires a connection between two layers. Two classes of moves that maintain the minimum wire length are shown in Fig. 5. In the L-shaped move of Fig. 5a, only the essential bends are permitted, while the Z-shaped move of Fig. 5b introduces one extra bend. We will explore the optimization possible with these two moves.

For a simple objective function that will reward the most balanced arrangement of wire, we calculate the square of the number of wires on each link of the network, sum the squares for all links, and term the result F. If there are N_L links and N_W wires, a global routing program that deals with a high density of wires will attempt to route precisely the average number of wires, N_W/N_L, along each link. In this limit F is bounded below by N_W^2/N_L. One can use the same objective function for a low-density (or high-resolution) limit appropriate for detailed wiring. In that case, all the links have either one or no wires, and links with two or more wires are illegal. For this limit the best possible value of F will be N_W/N_L.

For the L-shaped moves, F has a relatively simple form. Let $\epsilon_{i\nu} = +1$ along the links that connection i has for one orientation, -1 for the other orientation, and 0 otherwise. Let $a_{i\nu}$ be 1 if the ith connection can run through the νth link in either of its two positions, and 0 otherwise. Note that $a_{i\nu}$ is just $\epsilon_{i\nu}^2$. Then if $\mu_i = \pm 1$ indicates which route the ith connection has taken, we obtain for the number of wires along the νth link,

$$n_\nu = \sum_i \frac{a_{i\nu}(\epsilon_{i\nu}\mu_i + 1)}{2} + n_\nu(0) \quad (9)$$

where $n_\nu(0)$ is the contribution from straight wires, which cannot move without increasing their length, or blockages.

Summing the n_ν^2 gives

$$F = \sum_{ij} J_{ij}\mu_i\mu_j + \sum_i h_i\mu_i + \text{constants} \quad (10)$$

which has the form of the Hamiltonian for a random magnetic alloy or spin glass, like that discussed earlier. The "random field," h_i, felt by each movable connection reflects the difference, on the average, between the congestion associated with the two possible paths:

$$h_i = \sum_\nu \epsilon_{i\nu}\,[2n_\nu(0) + \sum_j a_{j\nu}] \quad (11)$$

The interaction between two wires is proportional to the number of links on which the two nets can overlap, its sign depending on their orientation conventions:

$$J_{ij} = \sum_\nu \frac{\epsilon_{i\nu}\epsilon_{j\nu}}{4} \quad (12)$$

Both J_{ij} and h_i vanish, on average, so it is the fluctuations in the terms that make up F which will control the nature of the low-energy states. This is also true in spin glasses. We have not tried to exhibit a functional form for the objective function with Z-moves allowed, but simply calculate it by first constructing the actual amounts of wire found along each link.

To assess the value of annealing in wiring this model, we studied an ensemble of randomly situated connections, under various statistical assumptions. Here we consider routing wires for the 98 chips on a module considered earlier. First, we show in Fig. 6 the arrangement of wire that results from assigning each wire to an L-shaped path, choosing orientations at random. The thickness of the links is proportional to the number of wires on each link. The congested area that gave rise to the peaks in the histograms discussed above is seen in the wiring just below and to the right of the center of the module. The maximum numbers of wires along a single link in Fig. 6 are 173 (x direction) and 143 (y direction), so the design is also anisotropic. Various ways of rearranging the wiring paths were studied. Monte Carlo annealing with Z-moves gave the best solution, shown in Fig. 7. In this example, the largest numbers of wires on a single link are 105 (x) and 96 (y).

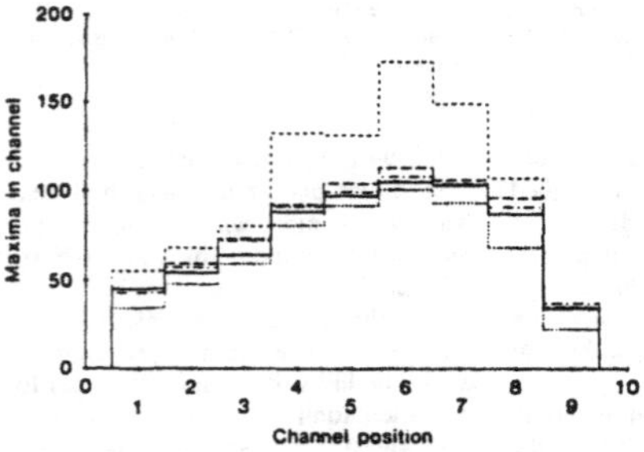

Fig. 8. Histogram of the maximum wire densities within a given column of x-links, for the various methods of routing.

We compare the various methods of improving the wire arrangement by plotting (Fig. 8) the highest wire density found in each column of x-links for each of the methods. The unevenness of the density profiles was already seen when we considered net-crossing histograms as input information to direct placement. The lines shown represent random assignment of wires with L-moves; aligning wires in the direction of least average congestion—that is, along h_i—followed by cooling for one pass at zero T; simulated annealing with L-moves only; and annealing with Z-moves. Finally, the light dashed line shows the optimum result, in which the wires are distributed with all links carrying as close to the average weight as possible. The optimum cannot be attained in this example without stretching wires beyond their minimum length, because the connections are too unevenly arranged. Any method of optimization gives a significant improvement over the estimate obtained by assigning wire routings at random. All reduce the peak wire density on a link by more than 45 percent. Simulated annealing with Z-moves improved the random routing by 57 percent, averaging results for both x and y links.

Fig. 9. Results at four temperatures for a clustered 400-city traveling salesman problem. The points are uniformly distributed in nine regions. (a) $T = 1.2$, $\alpha = 2.0567$; (b) $T = 0.8$, $\alpha = 1.515$; (c) $T = 0.4$, $\alpha = 1.055$; (d) $T = 0.0$, $\alpha = 0.7839$.

Traveling Salesmen

Quantitative analysis of the simulated annealing algorithm or comparison between it and other heuristics requires problems simpler than physical design of computers. There is an extensive literature on algorithms for the traveling salesman problem (*3, 4*), so it provides a natural context for this discussion.

If the cost of travel between two cities is proportional to the distance between them, then each instance of a traveling salesman problem is simply a list of the positions of N cities. For example, an arrangement of N points positioned at random in a square generates one instance. The distance can be calculated in either the Euclidean metric or a "Manhattan" metric, in which the distance between two points is the sum of their separations along the two coordinate axes. The latter is appropriate for physical design applications, and easier to compute, so we will adopt it.

We let the side of the square have length $N^{1/2}$, so that the average distance between each city and its nearest neighbor is independent of N. It can be shown that this choice of length units leaves the optimal tour length per step independent of N, when one averages over many instances, keeping N fixed (*25*). Call this average optimal step length α. To bound α from above, a numerical experiment was performed with the following "greedy" heuristic algorithm. From each city, go to the nearest city not already on the tour. From the Nth city, return directly to the first. In the worst case, the ratio of the length of such a greedy tour to the optimal tour is proportional to $\ln(N)$ (*26*), but on average, we find that its step length is about 1.12. The variance of the greedy step length decreases as $N^{-1/2}$, so the situation envisioned in the worst case analysis is unobservably rare for large N.

To construct a simulated annealing algorithm, we need a means of representing the tour and a means of generating random rearrangements of the tour. Each tour can be described by a permuted list of the numbers 1 to N, which represents the cities. A powerful and general set of moves was introduced by Lin and Kernighan (*27, 28*). Each move consists of reversing the direction in which a section of the tour is traversed. More complicated moves have been used to enhance the searching effectiveness of iterative improvement. We fir l with the adaptive divide-and-conquer effect of annealing at intermediate temperatures that the subsequence reversal moves are sufficient (*29*).

An annealing schedule was determined empirically. The temperature at which segments flow about freely will be of order $N^{1/2}$, since that is the average bond length when the tour is highly random. Temperatures less than 1 should be cold. We were able to anneal into locally optimal solutions with $\alpha \leq 0.95$ for N up to 6000 sites. The largest traveling salesman problem in the plane for which a proved exact solution has been obtained and published (to our knowledge) has 318 points (*30*).

Real cities are not uniformly distributed, but are clumped, with dense and sparse regions. To introduce this feature into an ensemble of traveling salesman problems, albeit in an exaggerated form, we confine the randomly distributed cities to nine distinct regions with empty gaps between them. The temperature gives the simulated annealing method a means of separating out the problem of the coarse structure of the tour from the local details. At temperatures, such as $T = 1.2$ (Fig. 9a), where the small-scale structure of the paths is completely disordered, the longer steps across the gaps are already becoming infrequent and steps joining regions more than one gap are eliminated. The configurations studied below $T = 0.8$ (for instance, Fig. 9b) had the minimal number of long steps, but the detailed arrangement of the long steps continued to change down to $T = 0.4$ (Fig. 9c). Below $T = 0.4$, no further changes in the arrangement of the long steps were seen, but the small-scale structure within each region continued to evolve, with the result shown in Fig. 9d.

Summary and Conclusions

Implementing the appropriate Metropolis algorithm to simulate annealing of a combinatorial optimization problem is straightforward, and easily extended to new problems. Four ingredients are needed: a concise description of a configuration of the system; a random generator of "moves" or rearrangements of the elements in a configuration; a quantitative objective function containing the trade-offs that have to be made; and an annealing schedule of the temperatures and length of times for which the system is to be evolved. The annealing schedule may be developed by trial and error for a given problem, or may consist of just warming the system until it is obviously melted, then cooling in slow stages until diffusion of the components ceases. Inventing the most effective sets of moves and deciding which factors to incorporate into the objective function require insight into the problem being solved and may not be obvious. However, existing methods of iterative improvement can provide natural elements on which to base a simulated annealing algorithm.

The connection with statistical mechanics offers some novel perspectives on familiar optimization problems. Mean field theory for the ordered state at low temperatures may be of use in estimating the average results to be obtained by optimization. The comparison with models of disordered interacting systems gives insight into the ease or difficulty of finding heuristic solutions of the associated optimization problems, and provides a classification more discriminating than the blanket "worst-case" assignment of many optimization problems to the NP-complete category. It appears that for the large optimization problems that arise in current engineering practice a "most probable" or average behavior analysis will be more useful in assessing the value of a heuristic than the traditional worst-case arguments. For such analysis to be useful and accurate, better knowledge of the appropriate ensembles is required.

Freezing, at the temperatures where large clusters form, sets a limit on the energies reachable by a rapidly cooled spin glass. Further energy lowering is possible only by slow annealing. We expect similar freezing effects to limit the effectiveness of the common device of employing iterative improvement repeatedly from different random starting configurations.

Simulated annealing extends two of the most widely used heuristic techniques. The temperature distinguishes classes of rearrangements, so that rearrangements causing large changes in the objective function occur at high temperatures, while the small changes are deferred until low temperatures. This is an adaptive form of the divide-and-conquer approach. Like most iterative improvement schemes, the Metropolis algorithm proceeds in small steps from one configuration to the next, but the temperature keeps the algorithm from getting stuck by permitting uphill moves. Our numerical studies suggest that results of good quality are obtained with annealing schedules in which the amount of computational effort scales as N or as a small power of N. The slow increase of effort with increasing N and the generality of the method give promise that simulated annealing will be a very widely applicable heuristic optimization technique.

Dunham (*5*) has described iterative improvement as the natural framework for heuristic design, calling it "design by natural selection." [See Lin (*6*) for a fuller discussion.] In simulated annealing, we appear to have found a richer framework for the construction of heuristic algorithms, since the extra control provided by introducing a temperature allows us to separate out problems on different scales.

Simulation of the process of arriving at an optimal design by annealing under control of a schedule is an example of an evolutionary process modeled accurately by purely stochastic means. In fact, it may be a better model of selection processes in nature than is iterative improvement. Also, it provides an intriguing instance of "artificial intelligence," in which the computer has arrived almost uninstructed at a solution that might have been thought to require the intervention of human intelligence.

References and Notes

1. E. L. Lawler, *Combinatorial Optimization* (Holt, Rinehart & Winston, New York, 1976).
2. A. V. Aho, J. E. Hopcroft, J. D. Ullman, *The Design and Analysis of Computer Algorithms* (Addison-Wesley, Reading, Mass., 1974).
3. M. R. Garey and D. S. Johnson, *Computers and Intractability: A Guide to the Theory of NP-Completeness* (Freeman, San Francisco, 1979).
4. R. Karp, *Math. Oper. Res.* 2, 209 (1977).
5. B. Dunham, *Synthese* 15, 254 (1963).
6. S. Lin, *Networks* 5, 33 (1975).
7. For a concise and elegant presentation of the basic ideas of statistical mechanics, see E. Shrodinger, *Statistical Thermodynamics* (Cambridge Univ. Press, London, 1946).
8. N. Metropolis, A. Rosenbluth, M. Rosenbluth, A. Teller, E. Teller, *J. Chem. Phys.* 21, 1087 (1953).
9. G. Toulouse, *Commun. Phys.* 2, 115 (1977).

10. For review articles, see C. Castellani, C. DiCastro, L. Peliti, Eds., *Disordered Systems and Localization* (Springer, New York, 1981).
11. J. Soukup, *Proc. IEEE* **69**, 1281 (1981).
12. M. A. Breuer, Ed., *Design Automation of Digital Systems* (Prentice-Hall, Engelwood Cliffs, N.J., 1972).
13. D. Sherrington and S. Kirkpatrick, *Phys. Rev. Lett.* **35**, 1792 (1975); S. Kirkpatrick and D. Sherrington, *Phys. Rev. B* **17**, 4384 (1978).
14. A. P. Young and S. Kirkpatrick, *Phys. Rev. B* **25**, 440 (1982).
15. B. Mandelbrot, *Fractals: Form, Chance, and Dimension* (Freeman, San Francisco, 1979), pp. 237–239.
16. S. Kirkpatrick, *Phys. Rev. B* **16**, 4630 (1977).
17. C. Davis, G. Maley, R. Simmons, H. Stoller, R. Warren, T. Wohr, in *Proceedings of the IEEE International Conference on Circuits and Computers*, N. B. Guy Raboat, Ed. (IEEE, New York, 1980), pp. 669–673.
18. M. A. Hanan, P. K. Wolff, B. J. Agule, *J. Des. Autom. Fault-Tolerant Comput.* **2**, 145 (1978).
19. M. Breuer, *ibid.* **1**, 343 (1977).
20. P. W. Case, M. Correia, W. Gianopulos, W. R. Heller, H. Ofek, T. C. Raymond, R. L. Simek, C. B. Steiglitz, *IBM J. Res. Dev.* **25**, 631 (1981).
21. A. J. Blodgett and D. R. Barbout, *ibid.* **26**, 30 (1982); A. J. Blodgett, in *Proceedings of the Electronics and Computers Conference* (IEEE, New York, 1980), pp. 283–285.
22. K. A. Chen, M. Feuer, K. H. Khokhani, N. Nan, S. Schmidt, in *Proceedings of the 14th IEEE Design Automation Conference* (New Orleans, La., 1977), pp. 298–302.
23. K. W. Lallier, J. B. Hickson, Jr., R. K. Jackson, paper presented at the European Conference on Design Automation, September 1981.
24. D. Hightower, in *Proceedings of the 6th IEEE Design Automation Workshop* (Miami Beach, Fla., June 1969), pp. 1–24.
25. J. Beardwood, J. H. Halton, J. M. Hammersley, *Proc. Cambridge Philos. Soc.* **55**, 299 (1959).
26. D. J. Resenkrantz, R. E. Stearns, P. M. Lewis, *SIAM (Soc. Ind. Appl. Math.) J. Comput.* **6**, 563 (1977).
27. S. Lin, *Bell Syst. Tech. J.* **44**, 2245 (1965).
28. ______ and B. W. Kernighan, *Oper. Res.* **21**, 498 (1973).
29. V. Černy has described an approach to the traveling salesman problem similar to ours in a manuscript received after this article was submitted for publication.
30. H. Crowder and M. W. Padberg, *Manage. Sci.* **26**, 495 (1980).
31. The experience and collaborative efforts of many of our colleagues have been essential to this work. In particular, we thank J. Cooper, W. Donath, B. Dunham, T. Enger, W. Heller, J. Hickson, G. Hsi, D. Jepsen, H. Koch, R. Linsker, C. Mehanian, S. Rothman, and U. Schultz.

J. Physique Lett. **46** (1985) L-771 - L-778 1er SEPTEMBRE **1985**, PAGE L-771

Classification
Physics Abstracts
05.20 — 75.50 — 85.40

Replicas and optimization

M. Mézard (*) and G. Parisi (**)

Dipartimento di Fisica, Università di Roma I, Piazzale A. Moro 2, I-00185 Roma, Italy

(*Reçu le 13 mai 1985, accepté le 3 juillet 1985*)

Résumé. — Nous utilisons la méthode des répliques pour étudier le problème du matching (bipartite) avec des distances aléatoires indépendantes entre les points. Nous proposons une solution, symétrique dans les répliques, qui est en accord avec les valeurs numériques de la longueur minimale et la distribution des longueurs des liens occupés dans la configuration optimale.

Abstract. — We use the replica method to study the (bipartite) weighted matching problem with independent random distances between the points. We propose a replica symmetric solution which fits the numerical values of the minimal length and the distribution of lengths of the occupied links in the optimal configuration.

It has been noticed for some time that combinatorial optimization problems can be formulated as problems in statistical mechanics [1]. For a given instance of an optimization problem, one introduces an artificial temperature $T = 1/\beta$ and a Boltzmann weight on the various possible configurations C given by $e^{-E(C)}$ where $E(C)$ is the cost of the configuration, i.e. the quantity one wants to minimize. $E(C)$ can be called an energy, and in the language of statistical physics one looks for zero temperature properties of the system such as the ground state configuration and its energy.

Physically, the difficulties of optimization problems can be traced back to the absence of translational symmetry and the frustration (in the general sense that in the hard cases it is not possible to get the true ground state by a simple local optimization procedure). These kinds of complications are typically encountered in the statistical physics of disordered systems such as spin glasses, and the recent developments in this field, especially on the infinite range spin-glass [2] (which is an NP complete problem [3]) have risen the hope of a new approach to optimization problems.

Up to now, this approach has been mostly numerical. It has been shown that the Monte Carlo method, combined with a suitable annealing procedure, can be efficiently applied to some optimization problems [1, 4], and that the low lying metastable states (local minima of the cost function of low cost) possess some properties which are reminiscent of those of pure equilibrium states in spin glasses [5, 6].

(*) On leave from Laboratoire de Physique Théorique de l'Ecole Normale Supérieure, Paris.
(**) Università di Roma II, Tor Vergata, Roma and INFN, Sezione di Frascati.

In this paper we will show how the analytical methods developed in the mean field theory of spin glasses, and particularly the replica method, can be used to solve optimization problems. This being a kind of a pioneering paper, we have decided to present the method on a rather simple problem (a polynomial one) the weighted matching [7]. In this problem one is given $2N$ points $i = 1, \ldots, 2N$, with a matrix of distances l_{ij}, and one looks for a matching between the points (a set of N links between two points such that at each point one and only one link arrives) of minimal length.

An instance of this problem is a matrix l_{ij}. Hereafter we shall study the random-link case introduced in [5, 8] where l_{ij} are independent random variables identically distributed with a distribution $\rho(l)$. The configurations will be conveniently characterized by the set of the occupation numbers of the links, $n_{ij} = 0, 1$, which are subject to the constraints :

$$n_{ij} = n_{ji}; \quad \sum_{j=1}^{2N} n_{ij} = 1 \qquad (1 \leqslant i \leqslant 2N). \tag{1}$$

The energy of a configuration is :

$$E(\{ n_{ij} \}) = \sum_{1 \leqslant i < j \leqslant 2N} n_{ij} \, l_{ij} \tag{2}$$

and the partition function is defined as :

$$Z = \sum_{\{n_{ij} = 0,1\}} \left(\prod_{i=1}^{2N} \delta\left(1 - \sum_{j=1}^{2N} n_{ij}\right)\right) e^{-\beta E(\{n_{ij}\})} . \tag{3}$$

Using integral representations of the δ functions and summing over the n_{ij}'s, this can be conveniently written as :

$$Z = \int_0^{2\pi} \frac{d\lambda_1}{2\pi} \cdots \int_0^{2\pi} \frac{d\lambda_{2N}}{2\pi} \exp\left(i \sum_{j=1}^{2N} \lambda_j\right) \prod_{j<k} [1 + e^{-\beta l_{jk}} e^{-i(\lambda_j + \lambda_k)}] . \tag{4}$$

The basic steps which relate this partition function to the ground state properties of the system have been studied in detail in [8] for the travelling salesman problem. They hold similarly in this case and we shall simply quote the most important facts. The interesting regime of temperature is the low temperature one in which β grows with N in such a way as to pick up in (4) configurations involving short links. In this case the only relevant property of the length distribution $\rho(l)$ is its behaviour around $l = 0$. When :

$$\rho(l) \overset{l \to 0}{\sim} l^r \tag{5}$$

the nearest neighbour of a given point i is at a distance of order $N^{-1/(r+1)}$ and one expects the energy to scale as $N^{1-1/(r+1)}$. Indeed, this regime is obtained when β scales as :

$$\beta = \hat{\beta} N^{1/(r+1)} \tag{6}$$

and in that case both the average energy E and the free energy $F = -\dfrac{\partial}{\partial \beta} \log Z$ behave as :

$$E = N^{1-1/(r+1)} \hat{E}; \qquad F = N^{1-1/(r+1)} \hat{F} . \tag{7}$$

One looks for the ground state energy which is the limit of E and F in the limit where $\hat{T} = 1/\hat{\beta} \to 0$.

Clearly, Z in (4), and hence F, still depend on the sample (i.e. of the instance of the l_{ij} one considers). From our experience in the statistical physics of disordered systems we expect however that the free energy in the regime (6) is « self averaging », which means that $\lim_{N \to \infty} F$ exists and is independent of the sample. (In fact if the replica method we use in the following can at all be applied to this problem, it can be used to prove this property so that our method is consistent with respect to the assumption of self averageness). In this case one can deduce F from the value of $\overline{\log Z}$ (in the following $\overline{(\ \ )}$ denotes an average with respect to the realizations of the $\{ l_{ij} \}$).

We shall compute $\overline{\log Z}$ with the replica method which consists in computing first $\overline{Z^n}$, from which $\overline{\log Z}$ is deduced as :

$$\overline{\log Z} = \lim_{n \to 0} \frac{\overline{Z^n} - 1}{n} \tag{8}$$

Z^n is simply the partition function of n uninteracting replicas of the initial system, which is expressed from (4) by introducing on each point i n replicas λ_i^a of the variable λ_i, $a = 1, ..., n$. Then the averages over the distribution of l_{jk} on each link factorize and give (under the conditions (5), (6)) :

$$\int \rho(l_{jk}) \, \mathrm{d}l_{jk} \prod_{a=1}^{n} [1 + \mathrm{e}^{-\beta l_{jk}} \, \mathrm{e}^{-i(\lambda_j^a + \lambda_k^a)}] = 1 + \frac{1}{N} \sum_{p=1}^{n} g_p \sum_{1 \leqslant a_1 < \ldots < a_p \leqslant n} \exp\left(- i \sum_{r=1}^{p} (\lambda_j^{a_r} + \lambda_k^{a_r} \right) \tag{9}$$

where :

$$g_p = (p\hat{\beta})^{-(r+1)} . \tag{10}$$

Using well known properties of the Gaussian integrals, $\overline{Z^n}$ can finally be written as :

$$\overline{Z^n} \underset{N \to \infty}{=} \int \prod_{p=1}^{n} \prod_{a_1 < \ldots < a_p} \frac{\mathrm{d}Q_{a_1 \ldots a_p}}{\sqrt{2 \pi N g_p}} \exp N \left\{ - \frac{1}{2} \sum_{p=1}^{n} \frac{1}{g_p} \sum_{a_1 < \ldots < a_p} (Q_{a_1 \ldots a_p})^2 + 2 \log z \right\} \tag{11}$$

where z is a « one site partition function » :

$$z = \int_0^{2\pi} \frac{\mathrm{d}\lambda'}{2\pi} \cdots \int_0^{2\pi} \frac{\mathrm{d}\lambda^n}{2\pi} \exp\left(i \sum_{a=1}^{n} \lambda^a + \sum_{p=1}^{n} \sum_{a_1 < \ldots < a_p} Q_{a_1 \ldots a_p} \, \mathrm{e}^{-i(\lambda_{a_1} + \ldots + \lambda_{a_p})} \right. . \tag{12}$$

From (11) the free energy F can be obtained in the limit $N \to \infty$ as the result of a saddle point method. Thanks to the introduction of replicas, the sites have decoupled. This is a nice property of these random-link models that they are naturally mean field theories. As in the case of spin glasses, the whole difficulty consists in finding the solution of the saddle point equations :

$$Q_{a_1 \ldots a_p} = 2 \, g_p \langle\!\langle \mathrm{e}^{-i(\lambda_{a_1} + \ldots + \lambda_{a_p})} \rangle\!\rangle \tag{13}$$

where $\langle\!\langle \cdots \rangle\!\rangle$ denotes one site average, with respect to the one site measure defined in (12). This task is in fact harder than in spin glasses since there all the tensors $Q_{a_1 \ldots a_p}$, $1 \leqslant p \leqslant n$ (and not only Q_{ab}) appear as natural order parameters.

We have looked for solutions of (13) in a very restricted space, the replica symmetric one in which :

$$Q_{a_1 \ldots a_p} = Q_p . \tag{14}$$

In order to be able to do the continuation to $n = 0$, one must consider cases in which Q_p depends analytically on p. The quantity one has to extremize in (11) is then :

$$-\frac{1}{2}\sum_{p=1}^{n}\frac{1}{g_p}\frac{n!}{p!(n-p)!}Q_p^2 + 2\log\left[\left(\frac{\partial}{\partial x}\right)^n \exp\left(\sum_{p=1}^{\infty}\frac{x^p Q_p}{p!}\right)\Big|_{x=0}\right] \tag{15}$$

which in the limit $n \to 0$ is equal to :

$$n\left\{-\frac{1}{2}\sum_{p=1}^{\infty}\frac{1}{g_p}(-1)^{p-1}\frac{Q_p^2}{p} + 2\int_{-\infty}^{+\infty} dl\,[e^{-e^l} - e^{-G(l)}]\right\} \tag{16}$$

$$G(l) \equiv \sum_{p=1}^{\infty}(-1)^{p-1}\frac{Q_p\,e^{lp}}{p!}. \tag{17}$$

The saddle point equations can be written as a functional equation for $G(l)$. The solution $G_r(l)$ (r characterizes the distribution of distances (5)) satisfies :

$$G_r(l) = \frac{2}{\beta^{r+1}}\int_{-\infty}^{+\infty} dy\, B_r(l+y)\, e^{-G_r(y)} \tag{18}$$

$$B_r(x) \equiv \sum_{p=1}^{\infty}(-1)^{p-1}\frac{e^{xp}}{p^r(p!)^2}. \tag{19}$$

In terms of this solution the thermodynamic quantities are expressed, from (11) and (16), as :

$$\hat{E}_r = \frac{r+1}{\hat{\beta}}\int_{-\infty}^{+\infty} dl\, G_r(l)\, e^{-G_r(l)} \tag{20}$$

$$\hat{F}_r = \frac{\hat{E}_r}{(r+1)} - \frac{2}{\hat{\beta}}\int_{-\infty}^{+\infty} dl\,[e^{-e^l} - e^{-G_r(l)}]. \tag{21}$$

The limit of zero temperature ($\hat{\beta} \to \infty$) can be directly studied by the introduction of the function $\hat{G}_r(l) = G_r(\hat{\beta} l)$ which verifies :

$$\hat{G}_r(l) = 2\int_{-l}^{\infty} dy\,\frac{(l+y)^r}{r!}\,e^{-\hat{G}_r(y)} \tag{22}$$

$$\hat{E}_r(\hat{T} = 0) = (r+1)\int_{-\infty}^{+\infty} dl\,\hat{G}_r(l)\, e^{-\hat{G}_r(l)}. \tag{23}$$

Equations (18), (22) are easily solved numerically by iteration for any r. The resulting curve of the free energy *versus* temperature, computed for $r = 1$, exhibits the following typical properties : for finite r there is no phase transition but an important freezing phenomenon; the entropy remains nonnegative and goes to zero with T. The ground state energy is :

$$\hat{E}_{r=1}(\hat{T} = 0) = 1.144. \tag{24}$$

The case $r = 0$ is also interesting. It corresponds to a nonvanishing probability of having infinitely short links. The upper bound on the minimal length from the greedy algorithm for instance is of order $\log N$ [8], while the lower bounds are finite when $N \to \infty$. With our Ansatz, equations (18) to (23) can be solved for $r = 0$, although one should notice that the corresponding saddle point values of the intermediate variables Q_p are divergent for p large enough, since $G_0(y)$ grows only linearly with y at $y \to \infty$. Nevertheless $\hat{E}$ and F depend only on G_0 and are well defined. Furthermore the solution of (22) is exactly :

$$\hat{G}_0(l) = \log(1 + e^{2l}) \tag{25}$$

which gives for the ground state energy :

$$\hat{E}_{r=0}(\hat{T} = 0) = \frac{\pi^2}{12}. \tag{26}$$

In order to test the predictions (24) and (26), we have generated the ground states numerically for samples of size ($2N =$) 20, 30 and 60 in the case $r = 1$, and 20, 30, 44 for $r = 0$. The procedure consisted in weakening the constraints from $n_{ij} = 0, 1$ to $n_{ij} \in [0, 1]$ and using a linear programming algorithm. Although this sometimes generates (in less than 30 % of the cases for the values of N we consider) spurious states with odd loops of $n_{ij} = 1/2$, it seems that the energies of these spurious states coincide with those of the true ground states in the large N limit. The results we have obtained for the samples in which our method does not give a spurious ground state are plotted in figure 1. The extrapolation to $N \to \infty$:

$$\hat{E}_{r=1}(\hat{T} = 0) = 1.14 \pm 0.01 ; \qquad \hat{E}_{r=0}(\hat{T} = 0) = 0.825 \pm 0.01 \tag{27}$$

are in good agreement with the predicted values.

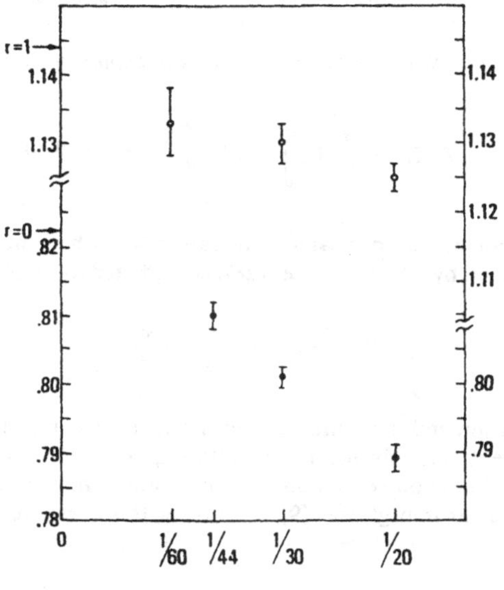

Fig. 1. — The minimal length (times $N^{-r/(r+1)}$) *versus* the inverse of the number of points ($1/2N$), for the cases $r = 0$ (full points, lower scale) and $r = 1$ (open points, upper scale). The predicted values are indicated by the arrows. The number of samples studied is, in the case $r = 0$, 16 000 ($2N = 20, 30$) and 3 800 ($2N = 44$) ; and in the case $r = 1$, 7 000 ($2N = 20$), 1 800 ($2N = 30$) and 330 ($2N = 60$).

As we have not tested the stability of our solution (i.e. the fact that all the eigenvalues of the Hessian in (11) are nonnegative (¹)) we cannot be sure that the Ansatz (14) is correct. Some encouraging indications are the fact that the entropy remains positive, and the agreement of the ground state energies with the results of the simulation. A more stringent test comes from the distribution of the lengths of occupied links in the ground state :

$$P_r(L) = \frac{1}{N} \sum_{i<j} \overline{\langle n_{ij}\, \delta(l_{ij}\, N^{1/(r+1)} - L) \rangle}. \tag{28}$$

This is an interesting information which can be computed with the replica method as follows : we compute the mth moment A_m of $P_r(L)$ by introducing in the partition function an extra weight on each configuration of the type :

$$\exp\left[- \alpha N^{-1+m/(r+1)} \sum_{j<k} n_{jk} (l_{jk})^m \right]. \tag{29}$$

The effect of this weight turns out to change the factors g_p in (10) into :

$$g_p(\alpha) = (p\hat{\beta})^{-(r+1)} \left[1 - \alpha \frac{(r+m)\,!}{r\,!} \frac{p}{N} (p\hat{\beta})^{-m} \right]. \tag{30}$$

Deriving the modified free energy with respect to α, one gets :

$$A_m = \lim_{n \to 0} \frac{1}{2\,n} \sum_{p=1}^{n} (p\hat{\beta})^{r+1-m} \frac{(r+m)\,!}{r\,!} p \sum_{a_1 < \dots < a_p} (Q_{a_1 \dots a_p})^2 \tag{31}$$

within the Ansatz (14), one can invert these moments explicitly to obtain $P_r(L)$. The result at zero temperature is :

$$P_r(L) = \frac{2}{r\,!} L_r \int_{-\infty}^{+\infty} \mathrm{d}l \, \frac{\partial \hat{G}_r}{\partial l} \, \mathrm{e}^{-\hat{G}_r(l) - \hat{G}_r(L-l)}. \tag{32}$$

We have tested this prediction numerically in the case $r = 0$, where the distribution of the lengths of occupied links (rescaled by a factor N, see (28)) is predicted from (25) and (32) to be :

$$P_{r=0}(L) = \frac{L - \mathrm{e}^{-L} \sinh L}{\sinh^2 L}. \tag{33}$$

The predicted histogram and the numerical results are shown in table I and seem to agree.

An intriguing feature of this solution is that, although replica symmetric, the order parameter turns out to be a whole function $G_r(l)$, and one can wonder whether there is a breaking of ergodicity in this problem as in spin glasses [9]. To answer this question, as in (9), we introduce the

---

(¹) *In this case, the number of variables in each tensor $Q_{a_1 \dots a_p}$, $n(n-1) \dots (n-p+1)$, has the sign* $(-1)^{p-1}$ for $n < 1$. Hence the condition of positivity of the eigenvalues of the Hessian corresponds neither to a pure maximum nor to a minimum of (16) in the subspace (17), contrarily to the spin glass case.

Table I. — *Integrated probability distribution of the lengths of the links (multiplied by N) in the optimal configuration in the case r = 0. The first column is the range of integration, the last one is the theoretical prediction obtained from formula* (33). *The average was done on the same number of samples as for the computation of the minimal length (see Fig.* 1). *The numbers between parentheses are the errors on the last digit.*

Interval	N=20	N=30	N=44	Theory
0.0/0.2	.1909 (11)	.1901 (8)	.1895 (15)	.1867
0.2/0.4	.1637 (10)	.1626 (8)	.1593 (14)	.1605
0.4/0.6	.1390 (9)	.1374 (7)	.1364 (13)	.1356
0.6/0.8	.1148 (8)	.1130 (7)	.1141 (12)	.1125
0.8/1.0	.0928 (8)	.0919 (6)	.0921 (11)	.0917
1.0/1.2	.0736 (7)	.0739 (5)	.0739 (9)	.0736
1.2/1.4	.0575 (6)	.0583 (5)	.0581 (9)	.0582
1.4/1.6	.0439 (5)	.0450 (4)	.0459 (7)	.0453
1.6/1.8	.0342 (5)	.0349 (4)	.0346 (6)	.0349
1.8/2.0	.0259 (4)	.0257 (3)	.0264 (6)	.0265
2.0/2.5	.0370 (5)	.0382 (4)	.0386 (7)	.0407
2.5/3.0	.0163 (3)	.0171 (3)	.0182 (5)	.0190
3.0/4.0	.0088 (2)	.0100 (2)	.0110 (4)	.0122
4.0/5.0	.00117 (8)	.00153 (7)	.00167 (5)	.00223
5.0/ ∞	.00010 (2)	.00018 (3)	.00034 (2)	.00045

distribution of overlaps between two real copies of the system which have the same $\{ l_{ij} \}$ but different occupation numbers of the links $\{ n_{ij} \}$ and $\{ n'_{ij} \}$ (²):

$$P(q) = \overline{\left\langle \delta\left(q - \frac{1}{N} \sum_{i<j} n_{ij} n'_{ij} \right) \right\rangle_{(2)}} \tag{34}$$

where $\langle \ \rangle_{(2)}$ is a thermal average with a Boltzmann weight $e^{-\beta[E(\{n_{ij}\}) + E(\{n'_{ij}\})]}$. The characteristic function of $P(q)$ can be computed with the replica method [9]; the final result is :

$$P(q) = \lim_{n\to 0} \frac{1}{n(n-1)} \sum_{a\neq b} \delta\left(q - \frac{1}{2}\left\{ \frac{1}{g_2} Q_{ab}^2 + \sum_{p=1}^{\infty} \frac{1}{g_{p+2}} \sum_{c_1<\ldots<c_p} (Q_{abc_1\ldots c_p})^2 \right. \right). \tag{35}$$

(²) Generalized $P(q)$ functions involving the overlap between three $\left((1/N) \cdot \sum_{i<j} n_{ij} n'_{ij} n''_{ij} \right)$ or more systems can be introduced and computed in a similar way.

A non-trivial $P(q)$ function can be obtained only with replica symmetry breaking. For Ansatz (14), we get :

$$P(q) = \delta(q - q_0); \qquad q_0 = 1 - \int_{-\infty}^{+\infty} \mathrm{d}l \, \frac{\partial^2 G_r}{\partial l^2} \, \mathrm{e}^{-G_r(l)} \tag{36}$$

q_0 is an increasing function of the temperature which goes to one when $\hat{T} \to 0$. Clearly an accurate measurement of $P(q)$ is needed in order to test (36) and to establish or infirm the validity of the replica symmetric Ansatz.

To conclude we would like to make a comment on the generality of the above approach. Although we have described the method in the special case of the weighted matching, we think it can be applied to other problems as well. An interesting variant is the bipartite weighted matching problem [7], which we have solved in the same way. Under the assumption of the symmetry between the two types of points, in this case we recover formulas (18) to (23) exactly, except for the explicit factors 2 in (18) and (22) which should be replaced by 1. It follows that the ground state energy in the bipartite case is $2^{1/(r+1)}$ larger than in the problem we have studied. This bipartite matching is also interesting since the linear programming algorithm does not give any spurious solution. The replica method can also be applied to NP complete problems, and in these cases it will be interesting to see whether one needs replica symmetry breaking.

Acknowledgments.

We gratefully acknowledge many interesting discussions with V. Parisi and M. A. Virasoro.

References

[1] KIRKPATRICK, S., *Lecture Notes in Physics* Vol. **149** (Springer, Berlin) 1981, p. 280 ;
KIRKPATRICK, S., GELATT, C. D. Jr., VECCHI, M. P., *Science* **220** (1983) 671.

[2] For a review, see e.g. : « Heidelberg colloquium on Spin Glasses », *Lecture Notes in Physics* Vol. **192** (Springer, Berlin) 1983 ;
PARISI, G., lectures given at the 1982 Les Houches summer school (North Holland).

[3] BARAHONA, F., *J. Phys. A* **15** (1982) 3241.

[4] SIARRY, P., DREYFUS, G., *J. Physique Lett.* **45** (1984) L-39.

[5] KIRKPATRICK, S., TOULOUSE, G., preprint (1985).

[6] BACHAS, C. P., *Phys. Rev. Lett.* **54** (1985) 53.

[7] PAPADIMITRIOU, C. H., STEIGLITZ, K., *Combinatorial Optimization* (Prentice Hall) 1982.

[8] VANNIMENUS, J., MÉZARD, M., *J. Physique Lett.* **45** (1984) L-1145.

[9] PARISI, G., *Phys. Rev. Lett.* **50** (1983) 1946.

J. Phys. A: Math. Gen. 19 (1986) 1605–1620. Printed in Great Britain

Application of statistical mechanics to NP-complete problems in combinatorial optimisation

Yaotian Fu† and P W Anderson

Department of Physics, Princeton University, Princeton, NJ 08544, USA

Received 4 September 1985

Abstract. Recently developed techniques of the statistical mechanics of random systems are applied to the graph partitioning problem. The averaged cost function is calculated and agrees well with numerical results. The problem bears close resemblance to that of spin glasses. We find a spin glass transition in the system, and the low temperature phase space has an ultrametric structure. This sheds light on the nature of hard computation problems.

1. Introduction

Recent developments in the theory of spin glasses have profound consequences in many branches of science. The application of the replica method [1] enables one to study random systems effectively. The idea of replica symmetry breaking and its interpretation [2] reveals the fascinating phase space structure of spin glasses. This method has far-reaching significance since it enables one to apply statistical mechanics to a system which, technically speaking, does not obey statistical mechanics at all because ergodicity is broken and, worse still, because no *a priori* knowledge about the pattern of this breaking down is available. In order to apply conventional equilibrium statistical mechanics to systems in which ergodicity is absent due to symmetry breaking, one has to know something about the order parameter of the system. A conjugate field is then applied and the partition function calculated. Equilibrium statistical mechanics becomes inadequate without such information. A hidden order parameter is always a headache. The power of the replica symmetry breaking formalism lies in that no such information is needed. The spin glass transition represents symmetry breaking on a higher level, one which has complexity and depth. As one of us anticipated many years ago [3], 'At some point we have to stop talking about decreasing symmetry and start calling it increasing complication'. In the studies of these problems the replica formalism has great promise to become a tool which can be routinely used in the same way that partition functions have been used, and perhaps beyond that.

One area in which this new development of statistical mechanics may have important applications is combinatorial optimisation. Several authors have already discussed the use of the spin glass analogy in this context [4]. In particular, the simulated annealing technique has been successfully applied to solve a number of hard optimisation problems. The performance of this and other techniques as heuristic algorithms has

†Address after September 1985: Department of Physics, 1110 West Green Street, University of Illinois at Urbana-Champaign, Urbana, IL 61801, USA.

been systematically evaluated. Here we wish to discuss the problem from a different angle. Instead of using the spin glass analogy as a practical aid in solving specific optimisation problems, we propose to study the general properties of such problems in the light of recent developments of statistical mechanics. This will include a discussion of the average solution of the problem when it is defined, the structure of solution space, the existence and the nature of phase transitions, and the effective use of local optimisation techniques.

Our work is motivated by the following consideration. Many computational problems, the so-called NP-complete problems [5], have proved difficult to solve. Despite great efforts no effective algorithm has been found for these problems, and there is good reason to believe that such algorithms do not exist. It is therefore highly desirable to seek alternatives. Practical strategies in attacking these problems go under three categories: The first consists of improved exhaustive searching. While it takes exponential time to go through all choices, it is sometimes possible to make decisions early in the process to terminate certain tree searches that are unlikely to be fruitful. This kind of branch and bound technique can benefit from the knowledge of the solution space structure. The second type of algorithm routinely used includes various kinds of heuristics which aim at producing almost optimised solutions at a faster rate. It will be very helpful if one can know something about the expected outcome of the cost function. The third group of techniques is quite unconventional. Anticipating future development of computer designing strategies these techniques may however become very useful. In particular, as observed by Hopfield and co-workers [6], two important features of computing in biological systems are parallelism and analog operation. Phase space structure information will facilitate the effective use of analog techniques and parallel computation becomes easier when local optimisation is possible. If we accept not only the best solution, but also the ones very close to it, a local optimisation may give us good results. In all these cases, analysing the problem using new techniques of statistical mechanics may provide us with valuable information.

Not every NP-complete problem can be analysed in this way. Some problems do not permit a discussion based on the most probable case. A randomly chosen satisfiability problem, for example, is almost always easy to solve, because a random sequence of symbols almost always does not make sense. Even in problems which are not intrinsically decision problems, such as the travelling salesman problem, an answer based on the most probable case, while not useless, is not very interesting, because by its very nature it does not provide a specific answer to a specific problem. Here we encounter one important difference between statistical mechanical problems and optimisation problems. In statistical mechanics we do not have complete information about the system, nor do we demand an answer complete to the minute detail. A prediction in terms of certain macroscopic variables will be quite appropriate. In optimisation problems we do know everything about the specific instance of the problem, and usually we are not content with a 'macroscopic' answer. An answer based on the most probable case, therefore, can only be regarded as a step towards a qualitative understanding of the problem. It may also be a useful aid in designing heuristic algorithms such as simulated annealing. In particular, the possible existence of phase transitions will affect the actual implementation and performance of such algorithms. Such transitions and the accompanying knowledge of the structure of solution space may also play an important role in complexity theory.

In this paper we will apply statistical mechanics to the graph partitioning problem. Apart from its theoretical interests the graph partitioning problem has been studied for a number of practical purposes, ranging from IC chip wiring to memory structure management. This problem is chosen because of its close resemblance to the spin glass problem, and also because many aspects of its solution are known, either theoretically or experimentally. We hope, however, that similar techniques can be applied to other problems as well.

The rest of this paper is organised in the following way. In § 2 we introduce the graph partition problem and define a Hamiltonian formalism for it. We also derive certain aspects of the solution that can be obtained exactly. In § 3 we study the model Hamiltonian by two independent methods, heavily using the results from spin glass theory. An estimation of the cost function is obtained and compared with the numerical results of explicit optimisation. Section 4 contains a study of the phase space structure, again using ideas developed in spin glass theory as a guide. In § 5 we discuss some general problems encountered in applying statistical mechanics to optimisation problems.

2. The model

The graph partitioning problem is specified by a set of vertices $V=\{v_1, v_2, \ldots v_N\}$ and a set of edges $E=\{(v_i, v_j)\}$ with N even. In general some pairs of vertices are connected by edges while others are not. We are now asked to partition the N vertices into two sets V_1 and V_2 of equal size such that the number of edges joining V_1 and V_2 is minimised. This number is defined to be our cost function C.

The graph partition problem is an NP-complete problem [7]. The best algorithm known is due to Kernighan and Lin [8]. Here we study a modified version of the problem. We assume each pair of vertices are connected with probability p independent of whether other pairs are connected (model A in graph theory, see [9]). For large values of N, $\alpha = Np$ is the expectation value of the valence for each vertex. The random graph defined in this way was studied by Erdös and Rényi in their classic work on random networks [10]. One important result is that for large values of N and $\alpha \geq 1$, the largest cluster in the graph has $G(\alpha)N$ vertices where

$$G(\alpha)=1-\frac{1}{\alpha}\sum_{n=1}^{\infty}\frac{n^{n-1}}{n!}(\alpha\, e^{-\alpha})^n. \tag{2.1}$$

One can verify that $G(1)=0$. Hence $\alpha=1$ is the percolation threshold. Also $G(\infty)=1$, showing that the graph becomes completely connected, in which case we expect the cost function to be equal to $N^2/4$, the number of edges joining two sets of size $N/2$ each. Finally if the largest cluster has number of vertices $\leq N/2$ the cost function per vertex number will be zero†. Notice that (this follows from the Lagrange expansion formula, see e.g. [11])

$$\sum_{n=1}^{\infty}\frac{n^{n-1}}{n!}x^n = y \Rightarrow x = y\, e^{-y}$$

we can solve $G(\alpha_c)=\frac{1}{2}$ to get

$$\alpha_c = 2\ln 2 = 1.3863\ldots. \tag{2.2}$$

†Strictly speaking, the cost function is not necessarily zero, since the complement of the largest cluster does not consist of isolated vertices, and the partition may have to go through one of the small clusters in order to maintain the balance of the two subsets. However, the small clusters are expected to have sizes of the order of log N only and in the large N limit C/N will be zero.

We will be interested in calculating the averaged cost function $C(\alpha)$. $C(\alpha)/N=0$ for $\alpha \leq \alpha_c$. For $\alpha > \alpha_c$ we expect

$$C(\alpha)=(pN^2/4)-\Delta(\alpha) \tag{2.3}$$

where the first term is the expected value for a randomly chosen partition scheme which separates all vertices into two sets of $N/2$ each and among the $N^2/4$ edges that might be present only $pN^2/4$ are there. The second term shows improvements due to the optimisation.

A number of authors have tried to estimate C. Bui [12], in particular, has reviewed and improved many of these results. All previous results are in the form of upper and lower bounds. Some of them will be compared with our result in § 3. Here we wish to point out that in this paper an actual optimisation is attempted, that is, we try to look for the cost function of the optimised configuration rather than that of an arbitrary configuration used in 'typical case' or 'worst case' studies which have led to various bounds.

Using the spin glass analogy we can define a Hamiltonian for the system. With each vertex v_i we associate an Ising spin S_i. $S_i=+1$ when v_i belongs to the set V_1 and $S_i=-1$ if v_i is in V_2. Since the two sets have the same size the total spin must be zero

$$\sum_{i=1}^{N} S_i=0. \tag{2.4}$$

Each spin configuration then corresponds to a partition scheme. For each pair of vertices (v_i, v_j) we define a coupling constant J_{ij}. $J_{ij}=J$ if $(v_i, v_j)\in E$ and $J_{ij}=0$ otherwise. Hence $J_{ij}=J$ with independent probability $p=\alpha/N$ and zero otherwise. The Hamiltonian

$$H=-\sum_{i<j} J_{ij}S_iS_j \tag{2.5}$$

is then equal to

$$\begin{aligned} H &= -\tfrac{1}{2}\left(\sum_{i\in V_1, j\in V_1} + \sum_{i\in V_2, j\in V_2} + \sum_{i\in V_1, j\in V_2} + \sum_{i\in V_2, j\in V_1}\right)J_{ij} + \left(\sum_{i\in V_1, j\in V_2} + \sum_{i\in V_2, j\in V_1}\right)J_{ij} \\ &= -\frac{J}{2}[2N(N-1)p/2]+2CJ \end{aligned}$$

or

$$C=\frac{H}{2J}+\frac{N(N-1)p}{4}. \tag{2.6}$$

Therefore to solve the graph partitioning problem is to minimise the Hamiltonian (2.5) under the constraint (2.4). Physically, this is a dilute infinite range ferromagnetic Ising system with a strong antiferromagnetic constraint (2.4). The conflict between these two types of interactions leads to frustration and gives rise to all the interesting properties of this problem.

3. The cost function

In order to calculate the averaged optimised cost function C, we will first calculate the averaged free energy F of the system. The zero temperature free energy should give us C.

We will first study C for very large $\alpha \sim N$ so that $p \sim O(1)$ (model 1). This corresponds to highly connected systems. Systems with small $\alpha \sim O(1)$ (model 2) need special treatment and will be discussed elsewhere. In the following we use $C(p)$ and $C(\alpha)$ to denote the cost functions of these two functions of these two models respectively. Clearly the two models are distinct only in the infinite N limit. To compare with experimental results on finite samples we will take the ratio of the experimental value to the theoretical estimation and extrapolate it to large N. We will evaluate the quality of solutions on the basis of such extrapolations.

3.1. Model 1: p independent of N

We use the replica method to compute F:

$$-\beta F = \lim_{n\to 0} \frac{1}{n}(Z^n - 1) = \lim_{n\to 0}\left[\mathrm{Tr}' \exp\left(-\beta \sum_{\alpha=1}^{n} H_\alpha\right) - 1\right] \tag{3.1}$$

where β is the inverse temperature, α is a replica index which runs from 1 to n, and Tr′ denotes the trace over all spin configurations which satisfy the constraints

$$\sum_{i=1}^{N} S_i^\alpha = 0 \qquad \alpha = 1, 2, \ldots, n. \tag{3.2}$$

F is a self-averaging quantity and in a large N limit it will not depend on the specific choice of J_{ij}. Averaging over the randomness in J_{ij} we obtain

$$[Z^n]_{\mathrm{av}} = (1-p)^{N(N-1)/2}\left[\mathrm{Tr}' \prod_{i<j}\left(1 + p_0 \exp \beta J \sum_{\alpha=1}^{n} S_i^\alpha S_j^\alpha\right)\right] \tag{3.3}$$

where we have introduced $p_0 = p/(1-p)$. The square bracket in (3.3) can be written as

$$\begin{aligned}
&\mathrm{Tr}' \exp \sum_{i<j} \ln\left(1 + p_0 \exp \beta J \sum_{\alpha=1}^{n} S_i^\alpha S_j^\alpha\right) \\
&\quad = \mathrm{Tr}' \exp\left(\frac{N(N-1)}{2}\ln(1+p_0)\right) \exp\left(-N \sum_{l=1}^{\infty} \frac{(J\beta)^l}{2} C_l n^l\right) \\
&\qquad \times \exp\left[\sum_{l=2}^{\infty} \frac{(J\beta)^l}{2} C_l \sum_{[\alpha_l]}\left(\sum_i S_i^{\alpha_1} \ldots S_i^{\alpha_l}\right)^2\right]
\end{aligned} \tag{3.4}$$

where we have expanded the logarithm and the exponential functions in Taylor series, changed the order of summation and rearranged the terms to separate the zeroth-, first- and higher-order terms in the expansion of the exponential. We have also used the constraints (3.2) to set the terms

$$C_1\left(\sum_i S_i^\alpha\right)^2 \tag{3.5}$$

to zero. In (3.4)

$$C_l = \frac{1}{l!}\sum_{m=1}^{\infty} \frac{(-1)^{m-1}}{m} p_0^m m^l \propto p \tag{3.6}$$

and are of order one for $p \sim O(1)$. In particular,

$$C_1 = p_0/(p_0+1) = p \qquad C_2 = \tfrac{1}{2}p_0(p_0+1)^2 = \tfrac{1}{2}p(1-p).$$

Hence

$$C(p) = \tfrac{1}{4}N^2 p + (2J)^{-1} F_1(\beta \to \infty)$$

and

$$-\beta F_1 = \lim_{n\to 0} \frac{1}{n}\left[\mathrm{Tr}' \exp \sum_{l=2}^{\infty} (\beta J)^l \frac{N^2}{2} C_l \sum_{[\alpha_l]} \left(\frac{1}{N}\sum_i S_i^{\alpha_1} \ldots S_i^{\alpha_l}\right)^2 - 1\right]. \tag{3.7}$$

While in this problem J is for bookkeeping purposes only and can take any value, only $J = J_0 N^{-1/2}$ gives us a sensible thermodynamic limit. Keeping the lowest term in $1/N$ (the $l=2$ term), we see this is formally identical to the expression for the free energy of the Sherrington–Kirkpatrick spin glass. The only difference is that here the trace is taken over a subset of spin configurations as determined by the constraints (3.2). These constraints can be replaced by a convenient global soft constraint term

$$\tfrac{1}{2}J_1\left(\sum_i S_i\right)^2 \qquad J_1 > 0 \tag{3.8}$$

in the Hamiltonian. The model can now be solved using standard techniques. Introducing Lagrange multipliers to decouple the quadratic terms we find that the constraint is irrelevant at $T=0$, and the equivalence with the SK spin glass becomes exact in this limit, as shown in appendix 1. Alternatively we can argue that, since the ground states of the SK spin glass do not have finite magnetisation per spin, lifting the constraint at this stage will not affect C/N for large N. The largest contribution to the free energy due to ferromagnetic fluctuations has already been eliminated as in (3.5). Using the known value of the zero temperature energy of the SK spin glass U_0 [2] we have

$$C(p) = \tfrac{1}{4}N^2 p + \tfrac{1}{2}U_0 N^{3/2}[p(1-p)]^{1/2} = \tfrac{1}{4}N^2 p - 0.38 N^{3/2}[p(1-p)]^{1/2}. \tag{3.9}$$

Let us compare this with the known results. The narrowest bounds of C as given in Bui [12] are the following:

$$\tfrac{1}{4}N^2 p - 0.17 N^{3/2}[p(1-p)]^{1/2} > C > \tfrac{1}{4}N^2 p - 0.42 N^{3/2}[p(1-p)]^{1/2}. \tag{3.10}$$

Our result is certainly consistent with this. Bui has also performed optimisation on random graphs generated on a computer. Figure 1 is taken from his work. We see the agreement is very good. Even for $p = 0.01$ and $N = 500$ the agreement is satisfactory [13] (experiments give $C = 207$ while (3.9) gives $C = 203$).

We have repeated the experiment for systems with $N \leq 200$. For each combination of N and p we randomly generate 10 graphs. The Kernighan–Lin algorithm is applied to each one of them. Usually the best result can be found in 20–40 passes. The longest run takes about 20 min CPU time on a VAX 11/750. The results for the 10 graphs are averaged to give $C(p)$. As can be seen from table 1, for $p \geq 0.1$ our estimation is good to within about 10%. There are, however, increasingly large deviations from (3.9) for small p. In particular, (3.9) becomes zero at $\alpha = 2.31\ldots$ which is very far from the threshold (2.2). We also observe large fluctuations from sample to sample for small p.

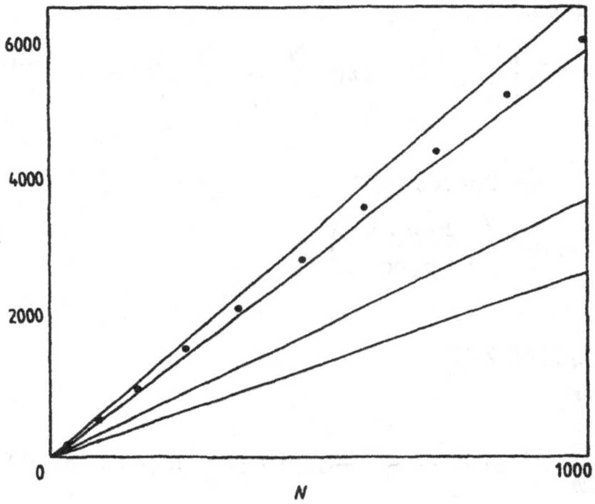

Figure 1. $\Delta = N^2 p/4 - C$ is the improvement due to optimisation, plotted against $N^{3/2}$. $p = \frac{1}{2}$. From bottom: the upper bound, results of the block algorithm, results of Kernighan–Lin algorithm, and the lower bound (from Bui [12]). The upper and lower bounds are given in equation (3.10). Closed circles: predictions of (3.9).

Table 1. C_1/C_2. C_1: average result of 10 different graphs (Kernighan–Lin algorithm). C_2: $N^2p/4 - 0.38N^{3/2}[p(1-p)]^{1/2}$.

	N			
p	50	100	150	200
0.05	5.5	1.5	1.19	1.13
0.1	1.24	1.12	1.11	1.10
0.25	1.03	1.04	1.05	1.02
0.5	1.07	1.03	1.03	1.02

We can estimate C in a different way. Define the density of states

$$\rho(E) = \mathrm{Tr}'\, \delta(E - H) \tag{3.11}$$

and the averaged density of states

$$\rho_{\mathrm{av}}(E) = [\mathrm{Tr}'\, \delta(E - H)]_{\mathrm{av}}. \tag{3.12}$$

The ground state energy can be estimated by solving

$$\int_{-\infty}^{E_0} \rho_{\mathrm{av}}(E)\, \mathrm{d}E = 1. \tag{3.13}$$

Notice that in principle one should calculate E_0 using the density of states (3.11) and then do the average. Here we use the mathematically more tractable (3.13) in the hope that the density of states near E_0 is not too low, in which case (3.13) should not be a bad approximation. In the language of spin glass theory we are taking the annealed average. In general this will produce a lower free energy as we can see from the Peierls inequality:

$$\langle \mathrm{e}^{-\beta F} \rangle \geq \mathrm{e}^{-\beta \langle F \rangle}. \tag{3.14}$$

It is straightforward to calculate

$$\rho_{av}(E) = N_0 \int_{-\infty}^{+\infty} \frac{dx}{2\pi} e^{ixE} \exp\left(\frac{N^2}{2}A - \frac{N}{2}B\right)(1-p)^{N(N-1)/2} \tag{3.15}$$

where

$$A = \ln(1+p_0^2+2p_0\cos x)^{1/2}$$

$$B = A + i\tan^{-1}\left(\frac{p_0 \sin x}{1+p_0\cos x}\right)$$

and

$$N_0 = N!/[(N/2)!]^2.$$

The equation for E_0

$$1 = N_0 \int_{-\infty}^{E_0} dE \int_{-\infty}^{+\infty} \frac{dx}{2\pi} e^{ixE} \exp\left[\frac{(N^2-N)}{4}\ln[1+p(1-p)(-1+\cos x)] - i\tan^{-1}\left(\frac{p_0\sin x}{1+p_0\cos x}\right)\right] \tag{3.16}$$

can be solved using saddle point expansion. Substituting into (2.6) we find the cost function

$$C(p) \geq \tfrac{1}{4}N^2 p - \tfrac{1}{2}(\ln 2)^{1/2} N^{3/2}[p(1-p)]^{1/2} = \tfrac{1}{4}N^2 p - 0.42 N^{3/2}[p(1-p)]^{1/2}. \tag{3.17}$$

Since this result is not restricted to model 1,

$$C(\alpha) \geq (\tfrac{1}{4}\alpha - 0.42\alpha^{1/2})N. \tag{3.18}$$

This is precisely Bui's lower bound.

3.2. Model 2: $p = \alpha/N$, α = constant

Many practical random graphs belong to this category. The treatment of the previous model has to be modified for such graphs, for two independent reasons. A graph with finite valence α is equvalent to a spin system with finite interaction range (the antiferromagnetic interaction responsible for maintaining the balance of the two subsets still has infinite interaction range). Therefore the mean-field theory solution may not be applicable. In addition, one can no longer choose $J = J_0 N^{-1/2}$ to simplify (3.7). Instead one should choose J independent of N, so that the free energy is extensive, and keep all the terms in (3.7). One can see this in a different way. If we use the antiferromagnetic term (3.8) to represent the constraints (3.2), the Hamiltonian is

$$H = \sum_{i<j} (J_1 - J_{ij}) S_i S_j = \sum_{i<j} K_{ij} S_i S_j \tag{3.19}$$

where K_{ij} takes two values J_1 and $J_1 - J$ with probability $1-p$ and p respectively. If p is independent of N and remains finite when $N \to \infty$, one can approximate the distribution of K_{ij} by a Gaussian distribution centred at J_1. This is precisely what we did in keeping the $l = 2$ term in (3.7) while ignoring the rest. As a result, the asymmetry in the distribution of K_{ij} is no longer visible. Indeed the free energy (and the optimised cost function thus obtained) depends on p only through the combination $p(1-p)$. This approximation clearly breaks down for small p. So far, the spin glass problem for finite-range interactions has proved remarkably intractable in all cases, and ours is no exception.

4. Phase space structure and phase transitions

In addition to providing estimations for the most probable outcome of the cost function, the spin glass analogy is useful in analysing the solution space structure. In this section we will discuss this aspect of the problem.

The phase transition in the pure Ising model is associated with the existence of two disconnected minima of the free energy at low temperature. These two minima are related by a global symmetry. To go from one minimum to another one has to flip all N spins in the system. On the other hand, the low temperature phase of a spin glass has many local free energy minima, not related to each other by any symmetry, and the transition between two local minima usually involves flipping many fewer than N spins. It is precisely this property that makes the spin glass analogy relevant in optimisation. If there are too many local minima, sitting very close to each other, the transition between neighbouring minima would involve O(1) spins, there will be no rigidity of the low temperature phases and hence no phase transition (e.g. the infinite-range antiferromagnetic model), and the optimisation will be easy. Computationally non-trivial cases arise when local minima are numerous but not excessively numerous, the distances between them large but not of the order of N. These are features shared by the SK spin glass, and we expect the existence of a spin-glass-like transition in these systems to reflect the difficulty involved in optimisation.

From the viewpoint of heuristics designing the information of phase space structure is also important (for a good discussion of heuristic algorithms, see [14]). The so-called λ-opt solution is an iterative local optimal solution within a distance λ from a given starting point. The starting points are randomly generated and the optimal solutions in their neighbourhood are compared to give the result of optimisation. If, however, one knows something about the distances between local optimal solutions, we can imagine generating the new random starting point by keeping an appropriate distance from the previous one and thus having a good chance to be in the neighbourhood of a different optimal solution. The common features of these solutions, such as the clustering of certain spins, could be used as guidelines in generating new starting points. In such a way we quickly scan the solution space while always trying to adapt the good features of different local optimal solutions. These kinds of intelligent algorithms will be expected to substantially speed up the computation process.

The spin glass theory provides a convenient formalism for this kind of discussion. In this section we will demonstrate the existence of a phase transition in model 1. Throughout the section we work in the neighbourhood of the critical point. The approximation used is not valid for ground state energy calculations.

Keeping the $l=2$ term in (3.7), we have

$$[Z^n]_{\mathrm{av}} = \mathrm{Tr}' \exp\left[\frac{\beta^2 v^2}{2N} \sum_{a<b} \left(\sum_i S_i^a S_i^b\right)^2\right]$$
$$= \int \prod_{a<b} \mathrm{d}Q_{ab} \left(\frac{N}{2\pi\beta^2 v^2}\right)^{1/2} \exp\left(-\frac{N}{2\beta^2 v^2} \sum_{a<b} Q_{ab}^2\right) \mathrm{Tr}' \exp\left(\sum_{a<b} \sum_i Q_{ab} S_i^a S_i^b\right) \tag{4.1}$$

where $v^2 = J_0^2 p(1-p)$, and the restricted trace equals

$$\int_0^{2\pi} \prod_{a=1}^{n} \frac{\mathrm{d}x_a}{2\pi} \left[\mathrm{Tr} \exp\left(\mathrm{i}x_a S^a + \sum_{a<b} Q_{ab} S^a S^b\right)\right]^N \equiv \exp nNf(Q). \tag{4.2}$$

In the high temperature phase the replica symmetry is expected to be unbroken; the integral is dominated by the saddle point

$$Q_{ab} = q \tag{4.3}$$

and (4.2) then equals

$$2^{nN} e^{-nNq/2} \int_0^{2\pi} \frac{dx_a}{2\pi} \left(\int Dz \prod_{a=1}^{n} \cosh(ix_a + \sqrt{q}\, z) \right)^N \tag{4.4}$$

where

$$\int Dz = \int_{-\infty}^{\infty} \frac{dz}{\sqrt{2\pi}} e^{-z^2/2}. \tag{4.5}$$

The large bracket in (4.4) is

$$\int Dz \prod_{a=1}^{n} (\cos x_a \cosh \sqrt{q}\, z + i \sin x_a \sinh \sqrt{q}\, z)$$

$$= \prod_{a=1}^{n} \cos x_a \left(\int Dz \cosh^n \sqrt{q}\, z + \sum_{s=1}^{n} (-1)^s T_s \int Dz \cosh^n \sqrt{q}\, z \tanh^{2s} \sqrt{q}\, z \right) \tag{4.6}$$

$$T_s = \sum_{b_1 \neq b_2 \neq \ldots \neq b_{2s}} \tan x_{b_1} \tan x_{b_2} \ldots \tan x_{b_{2s}}. \tag{4.7}$$

The sth term in (4.6) is of the order of q^s. Near the critical point q is small. We will need $f(q)$ up to the q^3 term only. Writing

$$A = \int Dz \cosh^n \sqrt{q}\, z = 1 + n \int Dz \ln \cosh \sqrt{q}\, z \tag{4.8}$$

and

$$q^s B_s = (-1)^s \int Dz \cosh^n \sqrt{q}\, z \tanh^{2s} \sqrt{q}\, z \tag{4.9}$$

the integral in (4.4) is approximately

$$\left(\int \frac{dx_a}{2\pi} \cos^N x_a \right)^n A^N + \int \frac{dx_a}{2\pi} \prod_{a=1}^{n} \cos^{N-2} x_a \frac{N(N-1)}{2} A^{N-2} (qB_1 T_1 + q^2 B_2 T_2)^2 + O(q^4). \tag{4.10}$$

Since

$$\int T_1 T_2 \prod \cos^{N-2} x_a = 0$$

there is no q^3 contribution from the second term. Because

$$\int T_1^2 \prod \cos^{N-2} x_a = n(n-1) \left(\int_0^{2\pi} \frac{dx}{2\pi} \cos^{N-2} x \right)^{n-2} \left(\int_0^{2\pi} \frac{dx}{2\pi} \cos^{N-4} x \sin^2 x \right)^2 \sim -\frac{n}{e^2 N^2} \tag{4.11}$$

the q^2 contribution is not extensive. We conclude that (4.4) equals

$$\exp nN \left(-\frac{q}{2} + \int Dz \ln \cosh \sqrt{q}\, z + K - \frac{q^2}{2e^2 N} + \ln 2 \right) \tag{4.12}$$

where

$$K=\frac{1}{N}\int_0^{2\pi}\frac{dx}{2\pi}\cos^N x\sim\frac{2}{\sqrt{\pi}\,e}N^{-3/2}. \tag{4.13}$$

Therefore

$$-\frac{\beta F}{N}=\frac{q^2}{4\beta^2 v^2}-\frac{q}{2}+\int Dz\ln\cosh\sqrt{q}\,z. \tag{4.14}$$

The value of the saddle point q is given by the familiar equation

$$\frac{q}{\beta^2 v^2}=\int Dz\tanh^2\sqrt{q}\,z. \tag{4.15}$$

The system undergoes a second-order phase transition at

$$T_c=J_0[p(1-p)]^{1/2}. \tag{4.16}$$

Above T_c, the cost function is equal to N^2p/r. No effect of optimisation will be seen until $T<T_c$. This behaviour was observed in simulated annealing [13].

It is well known [15] that the replica symmetric solution, (4.3) and (4.15), becomes unstable below T_c in the SK spin glass model. The same instability exists in this problem. We can demonstrate this by showing that the Hessian associated with the saddle point has negative eigenvalues. After some tedious but straightforward algebra similar to that of (4.6)-(4.12) (see appendix 2) we obtain results identical to those of de Almeida and Thouless [15] (see also Bray and Moore [16]). It is therefore necessary to break replica symmetry.

In fact, (4.2) can be calculated for any small but arbitrary matrix Q_{ab} and large N. Details of the derivation can be found in appendix 2. It is clear from this calculation that near the transition point the constraints (3.2) are irrelevant; their corrections are all of order $1/N$ compared with the main part. For large N,

$$nf(Q)=\tfrac{1}{4}\operatorname{Tr}Q^2-\tfrac{1}{6}\operatorname{Tr}Q^3+\tfrac{1}{8}\operatorname{Tr}Q^4-\tfrac{1}{4}\sum_{b\neq c}Q_{ab}^2Q_{ac}^2+\tfrac{1}{12}\sum_{a,b}Q_{ab}^4+\ln 2+O(N^{-1}) \tag{4.17}$$

which is the same as that of an SK spin glass. Parisi's solution applies and the ultrametric structure of the solution space follows immediately [2].

We have been unable to establish the irrelevance of the constraints (3.2) at all temperatures. However, in the absence of a second transition the topological structure of solution space should not change drastically. While the quantitative behaviour of the order parameter $q(x)$ in our problem may differ from that of a spin glass at a lower temperature, the qualitative features are the same.

5. Discussion

In this section we discuss some general problems encountered in applying statistical mechanics to optimisation problems.

Statistical mechanics is valid only in the thermodynamic limit. The meaning of such a limit is not clear in an optimisation problem. Intuitively, the system must be sufficiently large, but it is difficult to be more specific. While empirically in every problem there is always one quantity which measures the size of the system most naturally, the choice is by no means unique. Although the thermodynamic functions of physical systems are extensive, the entropy and cost function of an optimisation problem can in principle have quite arbitrary dependence on this equally arbitrarily chosen measure of size.

In general, the free energy of the system can be decomposed into the energy part and the entropy part. At high temperature the entropy is determined by the number of states Ω:

$$S \sim \ln \Omega \propto g(N) \tag{5.1}$$

and hence has a well defined N dependence (although not necessarily a linear dependence). On the other hand, the energy part will depend on certain 'counting variables' (the J in (2.5); the characteristic step length in the travelling salesman problem). These variables are arbitrary, and can be chosen to have the appropritate dependence on N so that the energy part will scale with N in a similar way as that of the entropy part. A generalised thermodynamic limit can then be defined. The free energy is extensive in the sense that the limit

$$\lim_{N\to\infty} F/g(N) \tag{5.2}$$

exists. The ground state energy can then be calculated. Eliminating the counting variables we recover the cost function with the correct N dependence. This procedure also guarantees that the phase transition temperature is N independent and finite.

Another common feature of optimisation problems is the existence of certain constraints. It is customary to replace the constraints by some additive penalty functions in the Hamiltonian (see, for example, the discussion of the travelling salesman problem [6]). This is a convenient device for the practical implementation of optimisations by simulated annealing or other techniques. As a tool for theoretical discussion, however, it has obvious drawbacks. In principle one should tune these penalty functions so that no illegal solution is included. This is achieved by taking the coupling constants in front of the penalty functions to large values, preferably infinity. But this leads to serious problems of exchanging limits. To recover information from a Hamiltonian with too many penalty terms is very difficult. Since the penalty function and the cost function are usually additive in the Hamiltonian, there may be trade-offs between these two terms. For multiple constraints this is a particularly serious problem. Ideally one should use the elements of the appropriate permutation group as the variable. The constraints can be strictly enforced; the permissible solutions form a subgroup. The partition function could then be calculated by a sum over this subgroup. Future development in the relevant mathematical techniques will be welcome.

While our discussion has focused on the theoretical side of the problem, spin glass theory can also help us to deal with specific instances [17]. In particular, the coupling constants of model 2 form a sparse matrix, which can be easily diagonalised. The localised eigenmodes corresponding to high eigenvalues can be first satisfied, producing local clusters. These clusters represent good features of the solution and can be preserved when one proceeds to deal with eigenmodes of lower eigenvalues. Because of frustration, it is impossible to satisfy all modes, and the Ising nature of the spins introduces strong interactions between the modes, making it impossible to carry this program to the end. As the Kernighan–Lin algorithm runs in time of order $N^2 \ln N$, it is also unlikely that one can gain much by this procedure. Nevertheless, since most of the localised modes can be treated locally, it opens new roads to parallel processing. It will be interesting to test these ideas by direct implementation. Some heuristic algorithms actually use ideas of block spin transformation [18] (see also [11]).

Acknowledgments

We thank G Baskaran, D J Gross and A Khurana for many discussions, and T N Bui and D S Johnson for letting us use their unpublished results. A discussion with N Sourlas and D J Sherrington on finite-range problems was useful. One of us (YF) wishes to thank Princeton University for a Charlotte Elisabeth Procter Fellowship. This work was supported in part by the National Science Foundation Grant No DMR-80-20263.

Appendix 1

In this appendix we show that the constraints (3.2) are irrelevant at $T=0$. Introducing the antiferromagnetic term (3.8), we have

$$[Z^n]_{av}=\exp(\beta J_1 n/2)\exp(-n^2\beta^2J_0^2/4)\exp(nN\beta^2J^2/4)$$
$$\times\operatorname{Tr}\exp\left[-\frac{\beta J_1}{2N}\sum_{a=1}^{n}\left(\sum_{i=1}^{N}S_i^a\right)^2+\frac{\beta^2J_0^2}{2N}\sum_{a<b}\left(\sum_i S_i^aS_i^b\right)^2\right]. \quad \text{(A1.1)}$$

The two quadratic terms can be decoupled by two independent Gaussian transformations with auxiliary variables x_a and Q_{ab}. The trace part is

$$\int(\mathrm{D}x_a)(\mathrm{D}Q_{ab})\left[\operatorname{Tr}\exp\left(\mathrm{i}\sum_{a=1}^{n}x_aS^a+\sum_{a<b}Q_{ab}S^aS^b\right)\right]^N \quad \text{(A1.2)}$$

where

$$\mathrm{D}x_a=\prod_a \mathrm{d}x_a\left(\frac{N}{2\pi\beta J_1}\right)^{1/2}\exp\left(-\frac{Nx_a^2}{2\beta J_1}\right)$$
$$\mathrm{D}Q_{ab}=\prod_{a<b}\mathrm{d}Q_{ab}\left(\frac{N}{2\pi\beta^2J_0^2}\right)^{1/2}\exp\left(-\frac{NQ_{ab}^2}{2\beta^2J_0^2}\right). \quad \text{(A1.3)}$$

For simplicity we assume that the replica symmetry is unbroken. (A1.2) can be calculated using saddle point expansion. Standard manipulation leads to

$$\frac{\beta F}{N}\equiv f=-\lim_{n\to 0}\frac{[Z^n]_{av}-1}{n}=\frac{x^2}{2\beta J_1}-\frac{q^2}{4\beta^2J_0^2}-\frac{\beta^2J_0^2}{4}+\frac{q}{2}-\ln 2-\int \mathrm{D}z\ln\cosh(\mathrm{i}x+\sqrt{q}\,z)$$
$$\mathrm{D}z=(\mathrm{d}z/\sqrt{2\pi})\,\mathrm{e}^{-z^2/2}. \quad \text{(A1.4)}$$

The saddle points are given by

$$\partial f/\partial x=0 \qquad \partial f/\partial q=0 \quad \text{(A1.5)}$$

or

$$\frac{x}{\beta J_1}=-\int \mathrm{D}z\frac{\sin 2x}{\cos 2x+\cosh 2\sqrt{q}} \quad \text{(A1.6)}$$

$$\frac{q}{\beta^2J_0^2}=\int \mathrm{D}z\frac{\sinh^2 2\sqrt{q}\,z-\sin^2 2x}{(\cosh 2\sqrt{q}\,z+\cos 2x)^2}. \quad \text{(A1.7)}$$

$x=0$ is the only solution if $q=0$. In general there will be many solutions. We are interested in the local maxima of f. The second derivative

$$\frac{\partial^2 f}{\partial^2 x}=-\frac{1}{\beta J_1}-\int \mathrm{D}z\frac{1+\cosh 2\sqrt{q}\,z\cos 2x}{(\cosh 2\sqrt{q}\,z+\cos 2x)^2} \tag{A1.8}$$

is periodic in x with periodicity π, and diverges at $2x=(2n+1)\pi$, where the right-hand side of (A1.6) jumps from π/q to $-\pi/q$. In the limit β goes to infinity, the local maximal solutions are given by

$$2x_m=2m\pi \qquad |m|\leqslant l \tag{A1.9}$$

where $l\sim\beta J_1/q$ since the right-hand side of (A1.6) is bounded by π/q. We should sum up all their contributions to f. They are

$$\frac{\pi^2}{2\beta J_1}\sum_{m=1}^{l} m^2 \tag{A1.10}$$

and

$$-\int \mathrm{D}z \sum_{m=1}^{l} \ln\cos x_m \tag{A1.11}$$

neither of which contributes to the cost function. The constraints are therefore irrelevant.

Appendix 2

In this appendix we derive (4.17). From (4.2),

$$\begin{aligned}\exp[nNf(Q)]&=\int \mathrm{D}x[\mathrm{Tr}\exp(\tfrac{1}{2}Q_{ab}S^aS^b+\mathrm{i}x_aS^a)]^N\\&=2^{nN}\int \mathrm{D}x\left[\mathrm{Tr}\exp\left(-\frac{1}{2}Q_{ab}\frac{\partial}{\partial x_a}\frac{\partial}{\partial x_b}\right)\prod_{a=1}^{n}\cos x_a\right]^N\end{aligned} \tag{A2.1}$$

where

$$\int \mathrm{D}x\equiv\int_0^{2\pi}\prod_{a=1}^{n}\frac{\mathrm{d}x_a}{2\pi} \tag{A2.2}$$

$$Q_{aa}=0 \qquad a=1,2,\ldots,n \tag{A2.3}$$

and summation over repeated indices is assumed.

We will need $f(Q)$ up to Q^4 order. Defining a set of operators L:

$$L_{ab}=-\frac{1}{2}\frac{\partial}{\partial x_a}\frac{\partial}{\partial x_b} \tag{A2.4}$$

and the function

$$R=\prod_{a=1}^{n}\cos x_a \tag{A2.5}$$

we have

$$[\exp(QL)R]^N=\left(R+(QL)R+\frac{(QL)^2}{2}R+\dots\right)^N$$
$$=R^N+NR^{N-1}\left((QL)R+\frac{(QL)^2}{2}R+\frac{(QL)^3}{3!}R+\frac{(QL)^4}{4!}R+\dots\right)$$
$$+\binom{N}{2}R^{N-2}\left((QL)R+\frac{(QL)^2}{2}R+\frac{(QL)^3}{3!}R\right)^2$$
$$+\binom{N}{3}R^{N-3}\left((QL)R+\frac{(QL)^2}{2}R\right)^3+\binom{N}{4}R^{N-4}[(QL)R]^4+O(Q)^5. \tag{A2.6}$$

Q^0 order:

$$\int \mathrm{D}xR^N=\left(\int_0^{2\pi}\frac{\mathrm{d}x}{2\pi}\cos^N x\right)^n=\left[2\left(\frac{e}{\pi N}\right)^{1/2}\right]^n\equiv W. \tag{A2.7}$$

Q^1 order:

$$-\frac{1}{2}\sum_{c_1\neq c_2}Q_{c_1c_2}\int \mathrm{D}x\left(\prod_{a=1}^{n}\cos^{N-1}x_a\right)\left(\prod_{b\neq c_1,c_2}\cos x_b\right)\sin x_{c_1}\sin x_{c_2}=0. \tag{A2.8}$$

Q^2 order:

$$N\int \mathrm{D}xR^{N-1}\frac{(QL)^2}{2}R+\binom{N}{2}\int \mathrm{D}xR^{N-2}[(QL)R]^2. \tag{A2.9}$$

The integral in the second term:

$$\frac{1}{4}\sum_{c_1\neq c_2,d_1\neq d_2}Q_{c_1c_2}Q_{d_1d_2}\int \mathrm{D}x\prod_{a=1}^{n}\cos^{N-2}x_a\prod_{b\neq c,d}\cos^2 x_b\sin x_{c_1}\dots\sin x_{d_2}$$
$$=\tfrac{1}{2}\operatorname{Tr}Q^2\left(\int_0^{2\pi}\cos^N x\frac{\mathrm{d}x}{2\pi}\right)^{n-2}\left(\int_0^{2\pi}\cos^{N-2}x\sin^2 x\frac{\mathrm{d}x}{2\pi}\right)^2\propto N^{-2}. \tag{A2.10}$$

Here and below we use

$$\int_0^{2\pi}\cos^{N-2l}x\sin^{2l}\mathrm{d}x/2\pi=\sqrt{2/\pi}\,(l-\tfrac{1}{2})^l\,\mathrm{e}^{-l+1/2}(2/N)^{l+1/2}. \tag{A2.11}$$

Therefore the second term in (A2.9) is not extensive and can be neglected. Similarly every term except those in the first square bracket can be neglected. The only extensive contributions are those coming from terms containing

$$\int \mathrm{D}xR^N. \tag{A2.12}$$

The rest of the calculation is straightforward. One notices that the operators $(QL)^l$ must act in such a way as to leave R invariant. This is not possible for the $l=1$ term since the two derivatives must act on different x_a. The rest of the expression can be evaluated using graphical rules similar to those of the linked graph expansion for Ising systems [19], with a common factor

$$W = 1 + n \ln[2(e/\pi N)^{1/2}] \sim 1 \tag{A2.13}$$

from which we obtain (4.17).

References

[1] Edwards S and Anderson P W 1975 *J. Phys. F: Met. Phys.* 5 965
[2] Parisi G 1980 *J. Phys. A: Math. Gen.* 13 L115, 1101, 1887
Mézard M, Parisi G, Sourlas N, Toulouse G and Virasoro M 1984 *Phys. Rev. Lett.* 52 1156
[3] Anderson P W 1972 *Science* 177 393
[4] Kirkpatrick S, Gelatt C D and Vecchi M P 1983 *Science* 220 671
Vannimenus J and Mézard M 1984 *J. Physique Lett.* 45 L1145
Siarry P and Dreyfus M 1984 *J. Physique Lett.* 45 L139
Kirkpatrick S and Toulouse G 1985 *J. Physique* 46 1277
[5] Garey M R and Johnson D S 1979 *Computers and Intractability* (San Francisco: Freeman)
[6] Hopfield J J and Tank D *Bell Laboratory Preprint*
[7] Garey M R, Johnson D S and Graham R L 1977 *SIAM J. Appl. Math.* 32 835
[8] Kernighan B W and Lin S 1970 *Bell Syst. Tech. J.* 49 110
[9] Palmer E M 1985 *Graphical Evolution* (New York: Wiley)
[10] Erdős P and Rényi A 1973 *The Art of Counting* ed J Spencer (Cambridge, MA: MIT)
Bollobás B 1981 *Combinatorics. London Math. Soc. Lecture Notes Ser. 52* ed H N V Temperley (Cambridge: Cambridge University Press)
[11] Whittaker E T and Watson G N 1927 *A Course of Modern Analysis* (Cambridge: Cambridge University Press) p 132
[12] Bui T N 1983 *MIT Laboratory for Computer Science Report* MIT/LCS/TR-287
[13] Aragon C R, Johnson D S, McGeoch L A and Schevon C unpublished
[14] Lin S 1975 *Network* 5 33
[15] de Almeida J R O and Thouless D J 1978 *J. Phys. A: Math. Gen.* 11 983
[16] Bray A J and Moore M A 1978 *Phys. Rev. Lett.* 41 1068
[17] Anderson P W 1979 *Ill Condensed Matter* ed R Balian, R Maynard and G Toulouse (Amsterdam: North-Holland) p 159
Hertz J A, Fleischman L and Anderson P W 1979 *Phys. Rev. Lett.* 43 942
[18] Goldberg M K and Gardner R 1982 *Proc. Silver Jubilee Conf. on Combinatorics, Waterloo*
Goldberg M K and Burstein M 1983 *IEEE Proc. Int. Conf. of Computer Design, Port Chester*
[19] Wortis M 1974 *Phase Transitions and Critical Phenomena* vol 3, ed C Domb and M S Green (New York: Academic) p 113

J. Physique **47** (1986) 1285-1296 AOÛT 1986, PAGE 1285

Classification
Physics Abstracts
05.20 — 75.50 — 85.40

A replica analysis of the travelling salesman problem

M. Mézard (*) and G. Parisi (**)

Dipartimento di Fisica, Universita' di Roma I, Piazzale Aldo Moro 2, 00185 Roma, Italy

(*Reçu le 27 janvier, accepté le 23 avril 1986*)

Résumé. — Nous proposons et analysons une solution symétrique dans les répliques des problèmes de voyageurs de commerce à liens indépendants. Elle fournit des estimations analytiques raisonnables pour les quantités thermodynamiques comme la longueur du chemin le plus court.

Abstract. — We propose and analyse a replica symmetric solution for random link travelling salesman problems. This gives reasonable analytical estimates for thermodynamic quantities such as the length of the shortest path.

1. Introduction.

In trying to understand the theory of spin glasses, a number of methods and concepts have been developed which seem to have possible applications beyond statistical mechanics, particularly in the field of combinatorial optimization.

It was soon realized [1] and demonstrated [2] that the determination of the ground state of an infinite range spin glass is an NP complete problem. As this is probably the NP complete problem which has been most studied by physicists, let us briefly recall the kind of results that can be obtained in this case. The infinite range Ising spin glass is a set of N spins $\sigma_i = \pm 1$ interacting through random couplings J_{ij} which are independent Gaussian random variables with $\overline{J_{ij}} = 0$ and $\overline{J_{ij}^2} = 1/N$. The J's are quenched variables given once for all (a given sample, i.e. a set of J's, is an instance of the problem), and the problem consists in finding the ground state configuration of the spins, that is the one which minimizes the Hamiltonian

$$H = - \sum_{1 \leq i < j \leq N} J_{ij} \sigma_i \sigma_j \qquad (1.1)$$

NP completeness means that there is no known algorithm which can solve every instance of this problem in a time growing less than a power of N, and it is unlikely that such an algorithm can be found [3]. A good heuristic (an algorithm which provides an approximate solution) is the simulated annealing method which consists in sampling the configurations with a Boltzmann-Gibbs probability $e^{-H/T}$ with the Monte Carlo method, and smoothly decreasing the temperature [4]. This is a general and powerful approach for many complex optimization problems [4-7].

On the other hand some kind of analytical information is also available. The ground state energy density E/N (minimum of H/N) converges to a certain value E_0 for large N, independently of the sample (¹).

The replica method allows one to prove that there is a phase transition in this system at the critical temperature $T_c = 1$. Below T_c there is a breaking of ergodicity and one needs replica symmetry breaking. A scheme for such a breaking was proposed in [8], which allows us to compute E_0 precisely

$$E_0 = -0.7633 \pm 0.0001 . \qquad (1.2)$$

This kind of prediction is essentially probabilistic (for large N the value of E/N for a given sample is a Gaussian random variable of mean E_0 and width $\overline{\left(\frac{E}{N} - E_0\right)^2} \sim \frac{1}{N}$), but it can be of direct interest, for instance for testing heuristics.

Although other relevant information can be obtained from the replica solution of this model (e.g. the

(*) On leave from Laboratoire de Physique Théorique de l'ENS, Paris.

(**) Universita' di Roma II, Tor Vergata, and INFN, Sezione di Roma.

(¹) This property of « self averageness » is obtained with the replica method and well established numerically, but has not yet been demonstrated rigorously.

ultrametric topology of the space of equilibrium states below T_c [9] and the correlations of the spin values in these states [10]), no algorithm has been found yet which can use these informations in order to speed the search for the ground state. Naturally the simulated annealing algorithm is the one which takes the greatest advantage of the analytical results of thermodynamics : once one knows the critical temperature T_c (or a freezing temperature in the case where there is no sharp phase transition), one must perform a very slow annealing around T_c, while the regions far above T_c (where the system thermalizes rapidly) or far below T_c (where the system is frozen) can be swept more rapidly.

In this paper we want to show how the replica method can be used generally to get the same kind of information as described above on other NP complete problems. We have chosen to apply it to the travelling salesman problem mainly for aesthetic reasons (it is among the easiest NP complete problems to state, and probably the most studied), and also because some numerical data were previously available. It is also a problem rather far from the spin-glass one. This will exemplify one power of the replica method : the fact that no *a priori* knowledge of the nature of the order and of the order parameter is required.

This method has already been used to predict the ground state energy in the matching problem [11] (a polynomial problem) and in the bipartitioning of infinitely connected random graphs which was shown to be equivalent to the S.K. [12].

In section 2 we describe the problem and its statistical mechanics representation.

In section 3 we propose a replica symmetric solution and analyse it. It is the most technical part of this paper and can be skipped by the reader who does not appreciate the beauty of replica computations.

The results are presented and discussed in section 4.

2. Statistical mechanics of the travelling salesman problem.

The travelling salesman problem (TSP) is the following. Given N points and the distances d_{ij} between any two of them, find the shortest tour, i.e. closed connected path, that goes through all the points.

A popular family of TSP's is the one where d_{ij}'s are Euclidean distances between points distributed randomly in a square. The resulting correlations between the distances allow for very powerful heuristics (for instance a combination of simulated annealing and « divide and conqueer » strategies [4-6]), but they are difficult to take into account in the replica method.

In the more general case the d_{ij} represent complicated costs which do not possess Euclidean correlations. We shall study a family in which good numerical results are far more difficult to achieve than in the Euclidean case, the one where d_{ij}'s are independent random variables with the same distribution $\rho(d)$ [7-13] (we keep to the symmetric case $d_{ij} = d_{ji}$). In this case the low temperature properties, and particularly the length of the shortest tour, depend only on the behaviour of $\rho(d)$ at small d. If :

$$\rho(d) \overset{d \to 0}{\simeq} d^r/r\,! \tag{2.1}$$

the nearest neighbour of any point i is at a distance $d \sim N^{-\frac{1}{r+1}}$ and one can show [13] that the length of the shortest tour is of order $N \times N^{-\frac{1}{r+1}}$. A tricky case is $r = 0$ in which the best known upper bound on the shortest tour is of the order of Log N. This case has also been extensively studied numerically by Kirkpatrick and Toulouse [7], and for these reasons our numerical results in the next sections will be given in this case. We predict a finite length of the shortest path, and not a Log N growth.

In order to extract from the $\frac{(N-1)!}{2}$ tours those which contribute, it was shown in [13] that the partition function which is naturally introduced in the statistical mechanics description of optimization problems must be of the form :

$$Z_{\text{TSP}} = \sum_{\text{tours}} \exp\left(-\beta N^{\frac{1}{r+1}} L_{\text{tour}}\right) \tag{2.2}$$

where β is an inverse « temperature ». The energy density $E = -\frac{1}{N}\frac{\partial \operatorname{Log} Z_{\text{TSP}}}{\partial \beta}$ is $N^{-1+\frac{1}{r+1}}$ times the average length of the tours at the temperature $T = 1/\beta$.

The whole difficulty of the TSP lies in the fact that the sum in (2.2) is over tours, i.e. closed *connected* paths. This constraint of connectivity is highly non-local and difficult to implement with local variables. We use a procedure that has been invented in the theory of polymers [14] and introduce on each site i an m-component spin $\mathbf{S}_i$ of fixed length $\mathbf{S}_i^2 = m$. As was shown by Orland [15], one is then led to study a generalized partition function :

$$Z = \int \prod_i \mathrm{d}\mu(\mathbf{S}_i) \exp\left(\gamma \sum_{i<j} \mathrm{e}^{-\beta N^{\frac{1}{r+1}} d_{ij}} \mathbf{S}_i \cdot \mathbf{S}_j\right) \tag{2.3}$$

$$\mathrm{d}\mu(\mathbf{S}) \equiv C_m \, d^m \mathbf{s}\, \delta(\mathbf{S}^2 - m) \quad \left(\int \mathrm{d}\mu(\mathbf{S}) = 1\right)$$

which reduces to Z_{TSP} in the limit $m \to 0$, $\gamma \to \infty$ [15] :

$$Z_{\text{TSP}} = \lim_{m \to 0} \frac{1}{m} \lim_{\gamma \to \infty} \frac{Z}{\gamma^N} \tag{2.4}$$

as can be shown using the special simplifications of integrals over s in the limit $m \to 0$, summarized by the formula :

$$\int \mathrm{d}\mu(\mathbf{S})\, \mathrm{e}^{\mathbf{h}.\mathbf{S}} \overset{m \to 0}{\simeq} 1 + \frac{\mathbf{h}^2}{2}. \tag{2.5}$$

As we explained in the introduction, extensive thermodynamic quantities such as the free energy $F = -\frac{1}{\beta} \operatorname{Log} Z_{TSP}$ are self averaging, which means that F/N goes for $N \to \infty$ towards a limit which is sample independent. Hence we shall try to compute the average $\overline{\operatorname{Log} Z_{TSP}}$, using the replica method which consists in computing $\overline{Z_{TSP}^n}$ and using :

$$\overline{\operatorname{Log} Z_{TSP}} = \lim_{n \to 0} (\overline{Z_{TSP}^n} - 1)/n . \tag{2.6}$$

From (2.4) one must compute $\overline{Z^n}$ which is equal to :

$$\overline{Z^n} = \int \prod_{i=1}^{N} \prod_{a=1}^{n} \mathrm{d}\mu(\mathbf{S}_i^a) \prod_{i<j} \exp\left[\gamma\, \mathrm{e}^{-\beta N^{\frac{1}{r+1}} d_{ij}} \sum \mathbf{S}_{i,a} \cdot \mathbf{S}_{j,a}\right] \tag{2.7}$$

where we have introduced, on each site i, n replicas of the original spin variable : $\mathbf{S}_i^a$, $a = 1, \ldots, n$.

Because of the statistical independence of the d_{ij}'s and of the introduction of replicas, the averages over each d_{ij} decouple in (2.7) and using :

$$\int_0^\infty \rho(l) \exp\left(a\, \mathrm{e}^{-\beta N^{\frac{1}{r+1}} l}\right) \mathrm{d}l \overset{N \gg 1}{\simeq} 1 + \frac{1}{N} \sum_{p=1}^{\infty} \frac{a^p}{p!} g_p \tag{2.8}$$

where :

$$g_p = (p\beta)^{-(r+1)} \tag{2.9}$$

one finds :

$$\overline{Z^n} = \int \prod_{i=1}^{N} \prod_{a=1}^{n} \mathrm{d}\mu(\mathbf{S}_i^a) \exp\left[\frac{1}{2N} \sum_{p=1}^{\infty} \frac{\gamma^p g_p}{p!} \sum_{\substack{a_1, \ldots, a_p \\ \alpha_1, \ldots, \alpha_p}} \left(\sum_i S_{i,a_1}^{\alpha_1} \ldots S_{i,a_p}^{\alpha_p}\right)^2\right] \tag{2.10}$$

where the a's, going from one to n, are replica indices while the α's, going from one to m, characterize the various spin components. A crucial remark is that in (2.10) the terms where at least two replica indices are equal lead to vanishing contributions in the limit of the TSP (2.4) (which must be taken before the $n \to 0$ limit) as is shown in appendix I. A Gaussian transformation decouples the various sites in (2.10), leading to :

$$\overline{Z_{TSP}^n} = \lim_{m \to 0} (1/m^n) \lim_{\gamma \to \infty} (1/\gamma^{nN}) \int \prod_{p=1}^{n} \prod_{a_1 < \ldots < a_p} \prod_{\alpha_1 \ldots \alpha_p} \mathrm{d}Q_{a_1 \ldots a_p}^{\alpha_1 \ldots \alpha_p} \times$$
$$\times \exp - N\left[\frac{1}{2} \sum_{p=1}^{n} \frac{1}{g_p} \sum_{a_1 < \ldots < a_p} \sum_{\alpha_1 \ldots \alpha_p} (Q_{a_1 \ldots a_p}^{\alpha_1 \ldots \alpha_p})^2 - \operatorname{Log} z\right] \tag{2.11}$$

where z is the one-site partition function :

$$z = \int \prod_{a=1}^{n} \mathrm{d}\mu(\mathbf{S}_a) \exp\left[\sum_{p=1}^{n} \gamma^{p/2} \sum_{a_1 < \ldots < a_p} \sum_{\alpha_1 \ldots \alpha_p} Q_{a_1 \ldots a_p}^{\alpha_1 \ldots \alpha_p} S_{a_1}^{\alpha_1} \ldots S_{a_p}^{\alpha_p}\right]. \tag{2.12}$$

In principle, formulae (2.11) and (2.12) give the free energy of the TSP in the large N limit through a saddle point evaluation of the integral in (2.11). However it is well known from the theory of spin glasses that this saddle point is not easy to find in the limit $n \to 0$. Furthermore there are two other complications here : the appearance of all the order parameters Q_a, Q_{ab}, ..., $Q_{1 \ldots n}$ (similar to the case of the matching [11]), and the presence of a polymeric index α going from one to m, where $m \to 0$. We shall propose and analyse a solution of these saddle point equations in the next section.

Before going to this analysis, let us explain how other quantities, besides the thermodynamic functions, can be deduced from the saddle point values of the Q's.

An interesting information is the average distribution $P(q)$ of overlaps between the relevant tours at a given temperature. We define the overlap $q_{t,t'}$ between two tours t and t' as their number of common bonds, divided by N [7]. The characteristic function $g(y)$ of $P(q)$ is defined as :

$$g(y) = \int P(q)\, \mathrm{e}^{yq}\, \mathrm{d}q = \overline{\sum_{t,t'} \frac{\mathrm{e}^{-\beta N^{\frac{1}{r+1}} (L^t + L^{t'})}}{Z_{TSP}^2} \mathrm{e}^{yq^{t'}_t}} . \tag{2.13}$$

In order to compute $g(y)$ within the present field — theoretical approach, we follow the same method as in spin glasses [16] and introduce two copies of a given instance — two travelling salesmen who must visit the same set of cities. This can be done by using two copies S_i and S_i' of the m component spins, with some suitable link auxiliary variables which count their number of common links. The denominator Z_{TSP}^2 in (2.13) can be obtained with the replica method by introducing $n-2$ independent other replicas of the system : they contribute a factor Z_{TSP}^{n-2} which gives the correct result for $n \to 0$. This whole computation is described in some detail in Appendix II. The final result for $P(q)$ is :

$$P(q) = \lim_{n\to 0} \frac{1}{n(n-1)} \sum_{a\neq b} \delta\left(q - \frac{1}{2}\sum_{p=2}^{\infty} \frac{1}{(p-2)!\, g_p} \sum_{\alpha,\beta,\alpha_3,\dots,\alpha_p} \sum_{\substack{a_3,\dots,a_p \\ \neq a,b}}{}' (Q^{\alpha\beta\alpha_3\dots\alpha_p}_{ab a_3\dots a_p})^2\right) \tag{2.14}$$

where the $Q^{\alpha_1\dots\alpha_p}_{a_1\dots a_p}$ takes its saddle point value This result is very similar to that already found in the matching problem, the only difference being the presence of vector spin indices.

Here also, one can define generalized overlaps between $k > 2$ tours as $1/N$ times the number of bonds common to all of them. Their distribution $P^{(k)}(q)$ is then :

$$P^{(k)}(q) = \lim_{n\to 0} \frac{(n-k)!}{n!} \sum_{a_1\dots a_k}{}' \delta\left(q - \frac{1}{2}\sum_{p=k}^{\infty} \frac{1}{(p-k)!\, g_p} \sum_{\alpha_1\dots\alpha_p} \sum_{\substack{a_{k+1}\dots a_p \\ \neq (a_1\dots a_k)}}{}' (Q^{\alpha_1\dots\alpha_p}_{a_1\dots a_p})^2\right) \tag{2.15}$$

where the $\sum'_{a_{k+1}\dots a_p \neq (a_1\dots a_k)}$ is the sum over all choices of the indices $a_{k+1}, \dots, a_p$, each of them being distinct from all the others and from the $a_1, \dots, a_k$.

Finally let us also describe the computation of the distribution of lengths of the links. This strictly follows the computation of the matching [11], and we include it here for completeness. The important links are those of length $\sim N^{-\frac{1}{r+1}}$, and the distribution $P(L)$ is defined as :

$$P(L) = \frac{1}{Z_{TSP}} \sum_t e^{-\beta L_t} \frac{1}{N} \sum_{(i,j)\in t} \delta\left(L - d_{ij} N^{\frac{1}{r+1}}\right) \tag{2.16}$$

where $\sum_{(i,j)\in t}$ denotes the sum over all the N links (i, j) which belong to the tour t. In order to compute the mth moment M_m of $P(L)$ we introduce a modified TSP partition function defined as :

$$\hat{Z}_{TSP}(\alpha) = \sum_t \exp\left(- \sum_{(i,j)\in t} \left[\beta N^{\frac{1}{r+1}} d_{ij} + \alpha N^{-1+\frac{m}{r+1}} (d_{ij})^m\right]\right) \tag{2.17}$$

so that :

$$M_m = - \left.\frac{\partial \operatorname{Log} \hat{Z}_{TSP}(\alpha)}{\partial \alpha}\right|_{\alpha=0}. \tag{2.18}$$

The $\hat{Z}_{TSP}(\alpha)$ can be computed from a generalized function similar to Z in (2.3), with a simple modification of the weight of each link, from $\gamma \exp\left(-\beta N^{\frac{1}{r+1}} d_{ij}\right)$ to $\gamma \exp\left(-\beta N^{\frac{1}{r+1}} d_{ij} - \alpha N^{-1+\frac{m}{r+1}} d_{ij}^m\right)$.

The corresponding one-link integral in (2.8) is changed into :

$$\int_0^{\infty} \rho(l) \exp\left(a \exp\left[-\beta N^{\frac{1}{r+1}} d_{ij} - \alpha N^{-1+\frac{m}{r+1}} d_{ij}^m\right]\right) \underset{N\to\infty}{\simeq} 1 + \frac{1}{N} \sum_{p=1}^{\infty} \frac{a^p}{p!} (\beta p)^{-(r+1)} \left[1 - p\frac{\alpha}{N}(\beta p)^{-m} \frac{(r+m)!}{r!}\right]. \tag{2.19}$$

So the introduction of the extra weighting proportional to α can be simply absorbed into a change of g_p in (2.11) into

$$g_p \to g_p \left[1 - p\frac{\alpha}{N}(\beta p)^{-m} \frac{(r+m)!}{r!}\right]. \tag{2.20}$$

From (2.18) and (2.20), we get :

$$M_m \equiv \int P(L)\, L^m \, \mathrm{d}L = \lim_{n \to 0} -\frac{1}{n}\frac{\partial}{\partial \alpha}\overline{[\hat{Z}_{\mathrm{TSP}}(\alpha)]^n}\Big|_{\alpha=0} =$$

$$= \lim_{n \to 0} \frac{1}{2n} \sum_{p=1}^{\infty} \frac{(\beta p)^{-m}}{g_p} p \frac{(r+m)!}{r!} \sum_{a_1 < \dots < a_p} \sum_{\alpha_1 \dots \alpha_p} (Q_{a_1 \dots a_p}^{\alpha_1 \dots \alpha_p})^2 \quad (2.21)$$

from which one obtains the distribution of lengths $P(L)$:

$$P(L) = \lim_{n \to 0} \frac{1}{2n} \sum_{p=1}^{\infty} (\beta p)^{2r+2} \frac{pLr}{r!} \mathrm{e}^{-\beta p L} \sum_{a_1 < \dots < a_p} \sum_{\alpha_1 \dots \alpha_p} (Q_{a_1 \dots a_p}^{\alpha_1 \dots \alpha_p})^2 . \quad (2.22)$$

3. The solution within a replica symmetric ansatz

As the full set of saddle point equations is complicated, we have made some hypotheses, *compatible with these equations*, on the type of saddle point we look for.

As for the replica indices we make the assumption of replica symmetry :

$$Q_{a_1 \dots a_p}^{\alpha_1 \dots \alpha_p} = Q^{\alpha_1 \dots \alpha_p} \quad (3.1)$$

independently of the values of the a's.

The spin component indices α must be treated differently since the symmetry in this space is a rotational symmetry instead of the permutation symmetry of the replicas. The problem is more similar to that of vector spin glasses, but in the limit where the dimensionality m of the spin goes to zero. We thus make the simple hypothesis of a spontaneous breaking of the rotational symmetry (for large values of γ) with :

$$Q^{\alpha_1 \dots \alpha_p} = Q_p \, u^{\alpha_1} \dots u^{\alpha_p} \quad (3.2)$$

where $\mathbf{u}$ is a given vector which we normalize to $\mathbf{u}^2 = 1$. (3.1) and (3.2) are two strong hypotheses. We shall postpone their discussion to the next section and first compute the free energy of the TSP within this ansatz.

First of all let us notice that the m^n term necessary to insure the convergence of (2.11) in the TSP limit appears in $\overline{Z^n}$ from the integrals over the Q's. Because of the spontaneous breakdown of the rotational symmetry there is a Goldstone mode associated with arbitrary independent rotations in each replica :

$$Q_{a_1 \dots a_p}^{\alpha_1 \dots \alpha_p} \to (\mathcal{R}_{a_1})^{\alpha_1 \alpha_1'} \dots (\mathcal{R}_{a_p})^{\alpha_p \alpha_p'} Q_{a_1 \dots a_p}^{\alpha_1' \dots \alpha_p'} \quad (3.3)$$

where $\mathcal{R}_1, \dots, \mathcal{R}_n$ are rotation matrices in the m dimensional space of the vectors $\mathbf{S}_a$. As the surface of an m-dimensional unit sphere vanishes linearly with m for $m \to 0$, the integrals over Q's in (2.11) are proportional to m^n for small m.

One must now compute the one-site partition function :

$$z = \int \prod_{a=1}^{n} \mathrm{d}\mu(\mathbf{S}_a) \exp\left[\sum_{p=1}^{n} \gamma^{p/2} \frac{Q_p}{p!} \sum_{a_1 \dots a_p}{}' (\mathbf{S}_{a_1} \cdot \mathbf{u}) \dots (\mathbf{S}_{a_p} \cdot \mathbf{u}) \right] \quad (3.4)$$

where the $\sum'$ means that all the indices must be different from each other. In the TSP limit (2.4) we must pick up the term of order γ^n in z and take its limits $m \to 0$ and then $n \to 0$. In Appendix III we show that the result is (F2) :

$$\lim_{m \to 0} \lim_{\gamma \to \infty} z/\gamma^n \overset{n \to 0}{\simeq} 1 + n \int_0^{\infty} \frac{\mathrm{d}\lambda}{\lambda} \left(\mathrm{e}^{-\lambda} - \int Dz \exp\left[\sum_{p=1}^{\infty} \frac{Q_p}{p!} (i\sqrt{\lambda})^p He_p(z) \right] \right). \quad (3.5)$$

From (2.11) and (3.5) we have the free energy of the TSP :

$$\frac{1}{N} \overline{\mathrm{Log}\, Z_{\mathrm{TSP}}} = -\frac{1}{2} \sum_{p=1}^{\infty} \frac{(-1)^{p-1}}{p g_p} Q_p^2 + \int_0^{\infty} \frac{\mathrm{d}\lambda}{\lambda} \left(\mathrm{e}^{-\lambda} - \int Dz \exp\left[\sum_{p=1}^{\infty} \frac{Q_p}{p!} (i\sqrt{\lambda})^p He_p(z) \right] \right) \quad (3.6)$$

(2) Dz is a Gaussian integration measure and He are Hermite polynomials. These quantities are defined precisely in the appendix.

where the Q's satisfy the saddle point equations :

$$Q_k = (-1)^k k g_k \int_0^{\infty} \frac{d\lambda}{\lambda} \int Dz \int Dt' \frac{[\sqrt{\lambda}(t' + iz)]^k}{k!} \exp\left[\sum_{p=1}^{\infty} \frac{Q_p}{p!}(i\sqrt{\lambda})^p He_p(z)\right]. \quad (3.7)$$

One can introduce a generating function $g(x) = \sum_{k=1}^{\infty} \frac{x^k}{k!} Q_k$, and from (3.7) one gets the following integral equation for g :

$$g(x) = \int_0^{\infty} \frac{d\lambda}{\lambda} \int Dz \int Dt' \sum_{p=1}^{\infty} \frac{(-1)^p p g_p}{(p!)^2} [\sqrt{\lambda}(t' + iz)\ x]^p \exp \int Dt\, g(\sqrt{\lambda}(t + iz)). \quad (3.8)$$

This equation is rather similar to that found in the matching problem [11], apart from the complications due to Hermite polynomials and imaginary terms. Unfortunately the behaviour of the Kernel :

$$\sum_{p=1}^{\infty} \frac{(-1)^p p g_p}{(p!)^2} \int Dt' [\sqrt{\lambda}(t' + iz)\, x]^p \quad (3.9)$$

for large values of the arguments is more complex and we have not been able to find out the correct asymptotic behaviours of g.

In this situation it was impossible to solve the integral equation (3.8) by a simple iterative method : we have used a different method to compute $\overline{Z^n}$, which consisted in a high-temperature expansion of the order parameters.

We shall present the formalism needed to build up this expansion in the general case where $g_p = (p\beta)^{-(r+1)}$, and we shall restrict our numerical computation to the case, $r = 0$. We note :

$$b = \beta^{r+1}. \quad (3.10)$$

In (3.7) we rescale $\lambda \to \frac{\lambda}{Q_1^2}$ and introduce the variables :

$$X_1 = bQ_1^2 ; X_p = \frac{Q_p}{b^{p-1} Q_1^p} \quad (p > 2) \quad (3.11)$$

which satisfy the following equations :

$$X_p = \frac{(-i)^p}{p!\, X_1^p p^r} \int_0^{\infty} \frac{d\lambda}{\lambda} \lambda^{p/2} \int Dz\, He_p(z)\, e^{iz\sqrt{\lambda}} A(z, \lambda) \quad (3.12)$$

$$X_1 = (-i) \int_0^{\infty} \frac{d\lambda}{\lambda} \lambda^{1/2} \int Dz\, He_1(z)\, e^{iz\sqrt{\lambda}} A(z, \lambda) \quad (3.13)$$

with :

$$A(z, \lambda) \equiv \exp \sum_{r=2}^{\infty} b^{r-1} X_r \frac{i^r}{r!} \lambda^{r/2} He_r(z). \quad (3.14)$$

The energy density is expressed in terms of these variables as :

$$E = \frac{1}{2\beta(r+1)}\left[X_1 + \sum_{p=2}^{\infty} (-1)^{p-1} b^{p-1} X_1^p X_p^2 p^r\right] \quad (3.15)$$

Clearly in the high temperature limit $\beta \ll 1$ the function $A(z, \lambda)$ can be approximated by 1, and one gets :

$$X_1 = 2(1 + 0(b)) \quad (3.16)$$

$$X_p = 1/p\,(1 + 0(b)) \quad p > 2. \quad (3.17)$$

In order to compute the high temperature expansion of $2\beta E$ up to the order b^N, one expands each X_i $(1 < i < N)$ up to the order b^{N+1-i}. This can be systematically done from (3.12 to 3.14) by first expanding the function $A(z, \lambda)$ to order b^N, then performing the integrals (3.12, 3.13) to establish the set of $(N + 1)$ equations between the X_i's, and finally solving these equations iteratively.

We have carried out this expansion numerically in the case $r = 0$ up to the order $N = 20$. The coefficients a_k of the series

$$2\beta E = 2 + \sum_{k=0}^{19} a_k \beta^{k+1} \quad (3.18)$$

are given in the first column of the table.

Although the a_k are rational numbers we have represented them on the computer as floating point numbers. In the actual computation we have done, quadruple precision (~ 30 significant digit) has been used; from various checks we know that the error due to rounding increases exponentially with k and only the first 10 digits of a_{19} are correct. This accuracy is far enough for our aims; the whole computation takes a few minutes of Vax 8600. From the expansion of the Q's one can compute other interesting quantities but let us first show in the case of energy how one can use this high temperature expansion to obtain the low temperature properties of the system. It is convenient to perform the following manipulations. We first put aside the first two terms and write :

$$2\,E(\beta) = \frac{2}{\beta} + 1 + 2\,\hat{E}(\beta)\,. \tag{3.19}$$

We then rescale β by a factor $5^{1/3}$ and Borel transform this series, introducing :

$$g(u) = \sum_{k=1}^{19} \frac{a_k}{k\,!}(5^{1/3}\,u)^k\,. \tag{3.20}$$

So that :

$$2\,\hat{E}(\beta) = \int_0^{\infty} \mathrm{d}u\; g[\beta 5^{-1/3}\,u]\,\mathrm{e}^{-u}\,. \tag{3.21}$$

The coefficients $\frac{a_k\,5^{k/3}}{k\,!} = b_k$ are given in the second column of the table. The alternance of three plus signs with three minus signs indicates the presence of complex singularities in u in the three directions $(-1)^{1/3}$. This suggests to change variables from u to x, where

$$u = \frac{x}{(1-x^3)^{\gamma}}\,. \tag{3.22}$$

By trial and error, we have found that $\gamma = \frac{1}{6}$ is a good choice in the sense that the series expansion of the function

$$f(x) = \sum_{k=1}^{19} b_k\left[\frac{x}{(1-x^3)^{\gamma}}\right]^k = \sum_{k=1}^{19} c_k\,x^k \tag{3.23}$$

converges rather well inside the unit circle. The coefficients c_k are given in the third column of the table.

The ground state energy (length of the shortest tour) is :

$$\lim_{\beta\to\infty} E(\beta) = \frac{1}{2} + f(1) \tag{3.24}$$

which can be evaluated directly from (3.23). However the successive approximants to $f(1)$ obtained by adding new terms to its series expansion still have some oscillations. A better method is to compute from (3.19)-(3.21) the function $E(\beta)$ at finite temperature. It turns out the the series is very convergent for $\beta \leqslant 2$ (the relative fluctuations in the values of $E(\beta)$ obtained from successive orders from 10 to 20 are less than 1 %), and the value of temperature $T = 0.5$ is below the freezing region, so that $E(T)$ can be interpolated safely to $T = 0$.

4. Results and discussion.

In this section we present the results obtained within the replica symmetric solution described above. As has been explained in section 3, the high temperature expansion allows for a very precise computation of the

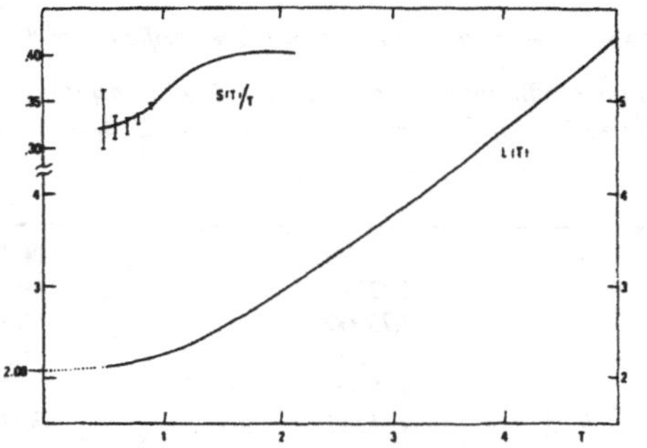

Fig. 1. — Average length L as a function of the temperature T, for $r = 0$. Note that, from (2.2), T is $1/N$ times the « usual » temperature. In the upper left corner (with the upper left scale), the entropy S divided by T, as a function of T. The error bars indicate the fluctuations in the values of S/T obtained from successive orders in the high temperature expansion, between the 10th and the 20th order. These fluctuations are negligible (less than 1 %) in $L(T)$ for $T > 0.5$.

function $E(T)$ (average length of the tours found at a given temperature) for $T \geqslant 0.5$. The result is plotted in figure 1. We have also computed in the same regime the entropy

$$S(T) = \mathrm{Log}\,T + \frac{\hat{E}(T)}{T} - \int_T^{\infty} \frac{\hat{E}(\tau)}{\tau^2}\,\mathrm{d}\tau \tag{4.1}$$

where $\hat{E}$ is the function introduced in (3.19).

It turns out that $\frac{S(T)}{T}$ is nearly constant in the low temperature range, with a value of the order of 0.33 ± 05 (see Fig. 1), the estimate of the error being quite subjective.

This suggests that the specific heat $C = T\frac{\mathrm{d}S}{\mathrm{d}T} = \frac{\mathrm{d}E}{\mathrm{d}T}$ grows linearly at low temperature and so the energy should start quadratically :

$$E(T) = E(0) + \frac{1}{2}(0.33)\,T^2 \tag{4.2}$$

This allows to interpolate the curve $E(T)$ down to $T = 0$. The ground state energy is :

$$E(T = 0) = 2.88 \pm 0.03 \tag{4.3}$$

where the estimation of the error is again subjective.

When the temperature tends to zero the entropy converges to a value near to 0 (e.g. 0 ± 0.03), but it is of course impossible to state whether $S(0)$ is exactly zero or not.

We have also computed the distribution of overlaps between the tours from formula (2.14). Within the

Table I. — *Results for $r = 0$. a_k is the coefficient of β^{k-1} in the high temperature expansion of $2\,\beta E$. $b_k = \frac{a_k 5^{k/3}}{k\,!}$ is the coefficient after rescaling and Borel transforming. C_k are the coefficients of the new series* (3.23) *obtained from* $2\,\hat{E}(\beta)$ *after the change of variable* (3.22). *d_k is the coefficient of β^{k+1} in the high temperature expansion of* $2\,q_o$.

k	a_k	b_k	c_k	d_k
0	1.000000	1.000000	1.000000	2.000000
1	1.222222	2.089971	− 2.089971	0.666667
2	1.750000	2.558516	2.558516	1.166667
3	1.398519	1.165432	1.165432	− 6.044444
4	− 4.881945	− 1.739168	− 1.390840	− 11.60833
5	− 27.53473	− 3.354669	− 2.501830	23.01997
6	− 41.35313	− 1.435872	− 0.8531563	263.3219
7	263.4408	2.234511	1.278257	742.0547
8	1932.493	3.503623	1.276625	− 2201.064
9	3070.568	1.057708	0.0588726	− 29584.90
10	− 34883.26	− 2.054726	− 0.2672521	− 93440.93
11	− 274012.4	− 2.509022	0.0420933	425892.4
12	− 384474.3	− 0.5016609	0.0132258	5825488
13	766676.0	1.315821	− 0.0272207	1.951409×10^{7}
14	6.287688×10^{7}	1.318082	0.116572	$- 1.204130 \times 10^{8}$
15	7.941193×10^{7}	0.1897737	0.0524543	$- 1.723766 \times 10^{9}$
16	$- 2.398438 \times 10^{9}$	− 0.6125602	− 0.0370652	$- 6.411737 \times 10^{9}$
17	$- 2.132721 \times 10^{10}$	− 0.5478927	0.0678134	4.353857×10^{10}
18	$- 3.177768 \times 10^{10}$	− 0.0775535	0.0565673	7.085880×10^{11}
19	9.821437×10^{11}	0.2157200	− 0.0306577	3.200175×10^{12}

replica symmetric ansatz $P(q)$ is a delta function at a value q_0 equal to

$$q_0 = \frac{1}{2} \sum_{p=2}^{\infty} \frac{p-1}{g_p} (-1)^p Q_p^2 . \qquad (4.4)$$

The high temperature expansion of the Q_p's described in section 3 gives a high temperature expansion of q_0, for $r = 0$:

$$2\,q_0 = \sum_{k=0}^{19} d_k\,\beta^{k+1} . \qquad (4.5)$$

The coefficients d_k are given in the fourth column of the table. Assuming that the analytic structure of $q_0(\beta)$ is essentially the same as that of $E(\beta)$. We have performed on the series $q_0(\beta)$ exactly the same manipulations as those we did on $E(\beta)$ in section 3. Again the curve $q_0(T)$ can be obtained precisely for $T = \frac{1}{\beta} \geqslant 0.6$ and is plotted in figure 2. We have interpolated $q_0(T)$ linearly at low temperatures, using the property

$$q_0(T) \overset{T \lesssim 1}{\simeq} 1 - aT \qquad (4.6)$$

which is suggested by the following argument. Keeping $r = 0$, the distribution of lengths of the links in the

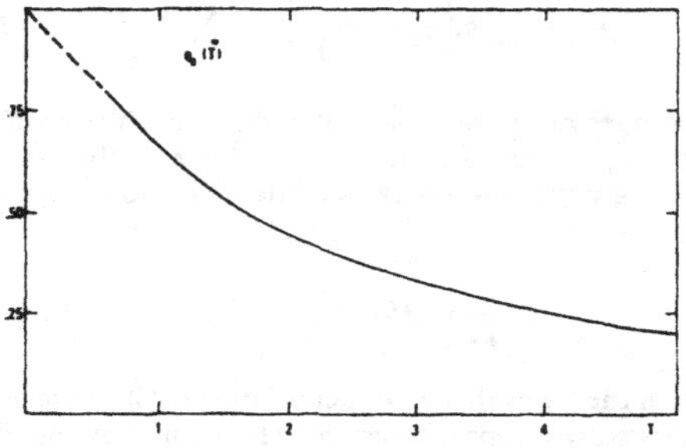

Fig. 2. — The overlap q_0 between relevant tours at temperature T (see (4.5)). The curve for $T > 0.6$ is obtained from the high temperature expansion, and then extrapolated linearly.

relevant tours, $P(L)$, (2.22) reduces in our replica symmetric case to

$$P(L) = \frac{\beta^2}{2} \sum_{p=1}^{\infty} (-1)^{p-1} Q_p^2\, p^2\, e^{-\beta p L} \qquad (4.7)$$

so that

$$q_0 = \int_0^{\infty} P(L)\, dL - \frac{P(L=0)}{\beta} \qquad (4.8)$$

the first term is equal to one (this can be used as a check of the high temperature expansion of the Q_p's), so q_0 starts linearly from one at low temperatures if $P(L = 0) \neq 0$.

Let us now compare these results with existing numerical data. The prediction (4.3) for the length of the shortest path seems to be in rather good agreement with the result of Kirkpatrick and Toulouse [7]. Indeed from their figure 1 one can plot the length *versus* $1/N$. This is done in figure 3. If one discards the data for $N > 48$ in which presumably the ground state has not been found, one finds a nearly linear curve which extrapolates at $N = \infty$ to $E \sim 2.09$. As for the temperature dependence of the length it exhibits typical features, such as the rather narrow freezing region around $T \sim 0.9$, which are also found numerically in [7]. Unfortunately the lack of statistics of [7] in the interesting low temperature range prevents from a really quantitative comparison.

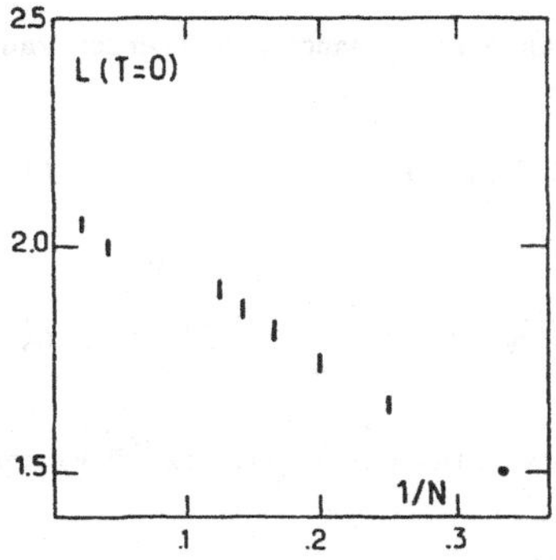

Fig. 3. — Plot of the optimal tour length obtained by Kirkpatrick and Toulouse [7] *versus* $1/N$.

From the theoretical point of view, we want to emphasize that we made two strong hypotheses on the solution $Q_{a_1 \ldots a_p}^{\alpha_1 \ldots \alpha_p}$ of the saddle point equations. The first one is the spontaneous breaking of the rotational symmetry. This seems rather reasonable although we have neglected in this process some potential transverse freezing which is known to occur in vector spin glasses [17].

The second one is the approximation of replica symmetry. This may well be wrong if there is a breaking of ergodicity in this problem. As it stands it is very difficult to study the stability of our solution [18], so the simplest thing one could do would be to analyse precisely the relaxation times and the shape of $P(q)$ in this system with a Monte Carlo at finite temperature.

The length of the ground state is known to be rather well approximated by the replica symmetric ansatz (in the S.K. model it predicts an energy which is too low by 5 %). This might explain the success of [4-3], but clearly more precise numerical data and some theoretical predictions for the case of broken replica symmetry will be welcome.

Acknowledgments.

We are grateful to Scott Kirkpatrick for communicating us some details of his data.

Appendix I.

In this appendix we want to prove that in the TSP limit ($m \to 0$ and $\gamma \to \infty$) for large values of N, the leading contributions arise from the terms in equation (2.10) where all replica indices are different.

Indeed for large N we have

$$\left(\sum_i S_{i,a_1}^{\alpha_1} \ldots S_{i,a_p}^{\alpha_p}\right)^2 \simeq \sum_{i \neq j} S_{i,a_1}^{\alpha_1} S_{j,a_1}^{\alpha_1} \ldots S_{i,a_p}^{\alpha_p} S_{j,a_p}^{\alpha_p} \qquad \text{(A.I.1)}$$

where the terms with $i = j$ give a negligible contribution (proportional to $1/N$).

It is convenient to evaluate equation (2.10) using a diagrammatic high temperature expansion, i.e. by expanding the exponential in powers of its argument. The diagrammatical rules are a simple generalization of the usual rules; we place a « link in replica a_k » between sites i and j whenever the term

$$S_{i,a_1}^{\alpha_k} S_{j,a_k}^{\alpha_k} \qquad \text{(A.I.2)}$$

is present.

The integration over $d\mu(S_{i,a})$ implies that on each site i, for each replica a, there must be exactly an even number of identical spin components $S_{i,a}^{\alpha}$. Hence the diagrams which contribute have the following properties for each replica index a, there are only closed paths of « links in replica a » and if in each replica a there are k_a closed paths, the weight of the diagram is

$$m^{\Sigma k_a}. \qquad \text{(A.I.3)}$$

In the limit $\gamma \to \infty$, we get a factor γ^{nN} only if all the sites in all the replicas will be visited by some path;

k_a must be different from zero for all values of a : the leading contribution (in m) is given when

$$k_a = 1 \qquad \forall a, \tag{A.I.4}$$

(in this case a prefactor m^n is produced).

Having established these results we can prove the main issue of the appendix. Each time that two replica indices are equal in the expression (AI.1) (say $a_1 = a_2$), there is in the replica a_1 a small closed path between i and j; if all the sites in the replica 1 must be visited, we must have $k_1 > 1$ and this term does not contribute in the TSP limit, as stated in the text.

Appendix II.

In this appendix we establish the formula for the distribution of overlaps $P(q)$. For a given sample of the lengths l_{ij}, we introduce two systems of m component spins $\mathbf{S}_i$ and $\mathbf{S}'_i$, and compute :

$$\Xi(x_{ij}) = \int \prod_{i=1}^{N} [d\mu(\mathbf{S}_i)\, d\mu(\mathbf{S}'_i)] \exp\left[\gamma \sum_{i<j} x_{ij}\, e^{-\beta N^{\frac{1}{r+1}} l_{ij}} (\mathbf{S}_i.\mathbf{S}_j + \mathbf{S}'_i.\mathbf{S}'_j)\right] \tag{A.II.1}$$

where x_{ij} are bookkeeping auxiliary link variables which can be chosen for instance as independent random variables with a distribution $P(x)$ such that :

$$\int P(x)\, dx = 1 ; \int P(x)\, x\, dx = 1 ; \int P(x)\, x^2\, dx = 1 + \frac{y}{N}. \tag{A.II.2}$$

In the TSP limit one has

$$\Xi_{\mathrm{TSP}}(x_{ij}) \equiv \lim_{m \to 0} \frac{1}{m^2}\left[\lim_{\gamma \to \infty} \frac{1}{\gamma^{2N}} \Xi(x_{ij})\right] = \frac{1}{Z_{\mathrm{TSP}}^2} \sum_{t,t'} \exp\left[-\beta N^{\frac{1}{r+1}}(L_t + L_{t'})\right] \prod_{(i,j)\in t} x_{ij} \prod_{(i',j')\in t'} x_{i'j'}. \tag{A.II.3}$$

Averaging over the x variables with the distribution (AII.2) clearly gives a factor $\left(1 + \frac{y}{N}\right)$ for each link common to the two tours t and t', so that

$$\int \prod_{i<j} P(x_{ij})\, dx_{ij}\, \Xi_{\mathrm{TSP}}(x_{ij}) = \sum_{t,t'} \frac{\exp\left[-\beta N^{\frac{1}{r+1}}(L_t + L_{t'})\right]}{Z_{\mathrm{TSP}}^2} e^{yq^{tt'}}. \tag{A.II.4}$$

In order to compute the average over the distribution of the lengths in (2.13) with the correct denominator Z_{TSP}^2, one can use as before the replica method. After some work one gets for the characteristic function of $P(q)$ defined in (2.13) :

$$g(y) = \lim_{n \to 0} \frac{1}{n(n-1)} \sum_{a \neq b} \int \prod_{i<j} P(x_{ij})\, dx_{ij} \lim_{m \to 0} \frac{1}{m^n} \lim_{\gamma \to \infty} \frac{1}{\gamma^{nN}} \int \prod_{i=1}^{N} \prod_{c=1}^{n} d\mu(\mathbf{S}_{i,c}) \times$$

$$\times \prod_{i<j} \exp\left[\gamma\, e^{-\beta N^{\frac{1}{r+1}} l_{ij}} \sum_c Z_{ij}^c\, \mathbf{S}_{i,c} \cdot \mathbf{S}_{j,c}\right] \tag{A.II.5}$$

where $Z_{ij}^c = x_{ij}$ if $c = a$ or b, $Z_{ij}^c = 1$ otherwise. The averages over the distributions of l_{ij} and x_{ij} for *one* link, using (2.8), gives :

$$1 + \frac{1}{N} \sum_{p=1}^{\infty} \frac{g_p}{p!} \gamma^p \left[\left(\sum_c \mathbf{S}_{i,c}.\mathbf{S}_{j,c}\right)^p + \frac{y}{N} p(p-1) \left(\sum_c \mathbf{S}_{i,c}.\mathbf{S}_{j,c}\right)^{p-2} \mathbf{S}_{i,a}.\mathbf{S}_{j,a}\, \mathbf{S}_{i,b}.\mathbf{S}_{j,b}\right] \tag{A.II.6}$$

so that :

$$g(y) = \lim_{n \to 0} \frac{1}{n(n-1)} \sum_{a \neq b} \lim_{m \to 0} \frac{1}{m^n} \lim_{\gamma \to \infty} \frac{1}{\gamma^{nN}} \int \prod_{i=1}^{N} \prod_{a=1}^{n} d\mu(\mathbf{S}_{i,a}) \exp \frac{1}{2N} \sum_{p=1}^{\infty} \frac{\gamma^p g_p}{p!} \sum_{\substack{a_1 \ldots a_p \\ \alpha_1 \ldots \alpha_p}} \left(\sum_i S_{i,a_1}^{\alpha_1} \ldots S_{i,a_p}^{\alpha_p}\right)^2 \times$$

$$\exp \frac{y}{2N^2} \sum_{p=1}^{\infty} \frac{\gamma^p g_p}{p!} p(p-1) \sum_{\substack{a_3 \ldots a_p \\ \alpha_3 \ldots \alpha_p}} \left(\sum_i S_{i,a}^{\alpha} S_{i,b}^{\beta} S_{i,a_3}^{\alpha_3} . S_{i,a_p}^{\alpha_p}\right)^2 \tag{A.II.7}$$

Clearly in the first exponential factor the sites can be decoupled by a Gaussian transformation as in (2.10) (2.11), while the second exponential is a term of relative order $1/N$ which doesn't change the saddle point values of the variables. $Q^{\alpha_1 \ldots \alpha_p}_{a_1 \ldots a_p}$.

Using the saddle point equations to express this second exponential in terms of the Q's one gets :

$$g(y) = \lim_{n \to 0} \frac{1}{n(n-1)} \sum_{a \neq b} \exp \frac{y}{2} \sum_{p=2}^{\infty} \frac{1}{p\,!\, g_p} p(p-1) \sum_{\substack{a_3 \ldots a_p \\ \neq a,b}}' \sum_{\alpha_3 \ldots \alpha_p} \sum_{\alpha,\beta} (Q^{\alpha\beta\alpha_3 \ldots \alpha_p}_{aba_3 \ldots a_p})^2 \qquad \text{(A.II.8)}$$

from which the formula (2.14) for $P(q)$ follows immediately.

Appendix III.

We compute the one-site partition function in the replica symmetric ansatz, z, given in (3.4). The quantity $\sum'_{a_1 \ldots a_p} X_{a_1} \ldots X_{a_p}$ (where $X_a = \mathbf{S}^a \cdot \mathbf{u}$) differs from $\left(\sum_a X_a\right)^p$ by terms with equal indices, involving sums of the type $\sum_a X_a^2, \sum_a X_a^3, \ldots$ However from (2.5) the integral over $d\mu(\mathbf{S}_a)$ of a quantity involving at least three spins with the same replica index, a, vanishes in the $m \to 0$ limit. Hence one can forget about the terms $\sum_a X_a^k$, $k \geqslant 3$ in going from $\sum'_{a_1 \ldots a_p} X_{a_1} \ldots X_{a_p}$ to $\sum_{a_1 \ldots a_p} X_{a_1} \ldots X_{a_p}$. We write

$$\sum'_{a_1 \ldots a_p} X_{a_1} \ldots X_{a_p} = \left(\sum_a X_a^2\right)^{p/2} Q_p \left(\frac{\sum_a X_a}{\sqrt{\sum_a X_a^2}} \right) \qquad \text{(A.III.1)}$$

where $Q_p(x)$ is a polynomial of degree p : $Q_p = X^p + q_{p,2} X^{p-2} + \cdots$. From the recurrence relation

$$\sum'_{a_1 \ldots a_p} X_{a_1} \ldots X_{a_p} = \left(\sum'_{a_1 \ldots a_{p-1}} X_{a_1} \ldots X_{a_{p-1}} \right) \left(\sum_b X_b \right) - (p-1) \left(\sum_b X_b^2 \right) \left(\sum'_{a_1 \ldots a_{p-2}} X_{a_1} \ldots X_{a_{p-2}} \right) \qquad \text{(A.III.2)}$$

one finds that $Q_p(X)$ is nothing but the Hermite polynomial $He_p(X)$ [19] :

$$\sum'_{a_1 \ldots a_p} X_{a_1} \ldots X_{a_p} = \left(\sum_a X_a^2\right)^{p/2} He_p \left(\frac{\sum_a X_a}{\sqrt{\sum_a X_a^2}} \right)$$

$$= \int_{-\infty}^{+\infty} Dt \left[\sum_a X_a + it \sqrt{\sum_a X_a^2} \right]^p \qquad \text{(A.III.3)}$$

where

$$Dt = \frac{dt}{\sqrt{2\pi}} e^{-t^2/2} . \qquad \text{(A.III.4)}$$

The presence of Hermite polynomials is not surprising : indeed the subtraction of terms with the same replica index leeds to a combinatorial problem which is the same as that of finding the normal product : φ^p : of a free field φ. In this case the result is well known

$$: \varphi^p : = \langle \varphi^2 \rangle^{p/2} He_p \left(\frac{\varphi}{\sqrt{\langle \varphi^2 \rangle}} \right) . \qquad \text{(A.III.5)}$$

So the partition function z is

$$z = \int \prod_a d\mu(\mathbf{S}_a) \exp \left\{ \sum_{p=1}^{\infty} \frac{y^{p/2}}{p\,!} Q_p \int Dt \left[\sum_a \mathbf{S}_a \cdot \mathbf{u} + it \sqrt{\sum_a (\mathbf{S}_a \cdot \mathbf{u})^2} \right]^p \right\} . \qquad \text{(A.III.6)}$$

We write $X = \sqrt{y} \sum_a \mathbf{S}_a \cdot \mathbf{u}$, $Y = y \sum_a (\mathbf{S}_a \cdot \mathbf{u})^2$, and we impose these constraints through δ functions written in integral representation

$$z = \int_{-\infty}^{+\infty} \frac{\mathrm{d}\hat{X}}{2\pi} \int_{-\infty}^{+\infty} \mathrm{d}X \int_{-\infty}^{+\infty} \frac{\mathrm{d}\hat{Y}}{2\pi} \int_{-\infty}^{+\infty} \mathrm{d}Y \mathrm{e}^{-i\hat{X}X - i\hat{Y}Y} \times$$

$$\times \exp\left\{ \sum_{p=1}^{\infty} \frac{Q_p}{p!} \int Dt[X + it\sqrt{Y}]^p \right\} \prod_a \left[\int \mathrm{d}\mu(\mathbf{S}_a)\, \mathrm{e}^{i\hat{\mathbf{X}}\sqrt{\gamma}\mathbf{S}_a \cdot \mathbf{n} + i\hat{\mathbf{Y}}\gamma(\mathbf{S}_a \cdot \mathbf{n})^2} \right]. \qquad \text{(A.III.7)}$$

The integral over the $\mathbf{S}^a$, in the limit $m \to 0$, is

$$\left\{ 1 + \gamma\left[\frac{(i\hat{\mathbf{X}})^2}{2} + i\hat{\mathbf{Y}} \right] \right\}^n. \qquad \text{(A.III.8)}$$

In the TSP limit we are interested in the term of the order of γ^n of z which is

$$z \sim \gamma^n \left\{ \left[\frac{1}{2} \frac{\partial^2}{\partial X^2} + \frac{\partial}{\partial Y} \right]^n \exp\left[\sum_{p=1}^{\infty} \frac{Q_p}{p!} \int Dt(X + it\sqrt{Y})^p \right] \Bigg|_{X=Y=0} \right\}. \qquad \text{(A.III.9)}$$

Having extracted the γ^n term we can go on and take the limit $n \to 0$. One way to do it is to write

$$\frac{z}{\gamma^n} = \left(\frac{\partial}{\partial \lambda} \right)^n \exp \lambda \left[\frac{1}{2} \frac{\partial^2}{\partial X^2} + \frac{\partial}{\partial Y} \right] \exp\left[\sum_{p=1}^{\infty} \frac{Q_p}{p!} \int Dt[X + it\sqrt{Y}]^p \right] \Bigg|_{\lambda=0}$$

where

$$f(\lambda) = \int Dz \exp\left\{ \sum_{p=1}^{\infty} \frac{Q_p}{p!} \sqrt{\lambda^p}\, He_p(z) \right\} \qquad \text{(A.III.11)}$$

is a function of λ with $f(0) = 1$, so that

$$\gamma^{-n} z = \left(\frac{\partial}{\partial \lambda} \right)^n f(\lambda) \Bigg|_{\lambda=0} \sim 1 + n \int_0^{\infty} \frac{\mathrm{d}\lambda}{\lambda} \left[\mathrm{e}^{-\lambda} - \int Dz \exp\left(\sum_{p=1}^{\infty} \frac{Q_p}{p!} (i\sqrt{\lambda})^p He_p(z) \right) \right]. \qquad \text{(A.III.12)}$$

References

[1] Kirkpatrick, S. and Sherrington, D., *Phys. Rev.* **B 17** (1978) 4384.

[2] Barahona, F., *J. Phys.* **A 15** (1982) 3241.

[3] Garey, M. R. and Johnson, D. S., *Computers and Intractability* (Freeman, San Francisco) 1979.

[4] Kirkpatrick, S., *Lecture Notes in Physics* **149** (Springer, Berlin) 1981, p. 280; Kirkpatrick, S., Gelatt, C. D. Jr., Vecchi, M. P., *Science* **220** (1983) 671.

[5] Siarry, P. and Dreyfus, G., *J. Physique Lett.* **45** (1984) L-39.

[6] Bonomi, E. and Lutton, J. L., *SIAM Rev.* **26** (1984) 551.

[7] Kirkpatrick, S. and Toulouse, G., *J. Physique* **46** (1985) 1277.

[8] Parisi, G., *J. Phys.* **A 13** (1980) L-115, 1101, 1887.

[9] Mézard, M., Parisi, G., Sourlas, N., Toulouse, G. and Virasoro, M. A., *Phys. Rev. Lett.* **52** (1984) 1156, and *J. Physique* **45** (1984) 843.

[10] Mézard, M. and Virasoro, M. A., *J. Physique* **46** (1985) 1293.

[11] Mézard, M. and Parisi, G., *J. Physique Lett.* **46** (1985) L-771.

[12] Fu, Y. and Anderson, P. W., Princeton University preprint (1985).

[13] Vannimenus, J. and Mézard, M., *J. Physique Lett.* **45** (1984) L-1145.

[14] De Gennes, P. G., *Phys. Lett.* **A 38** (1972) 339.

[15] Orland, H., *J. Physique Lett.* **46** (1985) L-763.

[16] Parisi, G., *Phys. Rev. Lett.* **50** (1983) 1946.

[17] Gabay, M. and Toulouse, G., *Phys. Rev. Lett.* **47** (1981) 201.

[18] What we mean is the analog of the computation for spin glasses done in de Almeida, J. R. L. and Thouless, D. J., *J. Phys.* **A 11** (1978) 983.

[19] Abramowitz, M. and Stegun, I. A. ed., *Handbook of mathematical Functions*, National Bureau of Standards. Applied Mathematics Series (1964).

Part 3
BIOLOGICAL APPLICATIONS

Chapter X
INTRODUCTION

It is far from evident that the type of organization present in biosystems should in any way resemble those encountered in physics. Organic matter surely obeys basic physical laws but still, the collective behaviour of large aggregations in typical physical systems could be irrelevant for the understanding of the corresponding behaviour in a biosystem.

For instance, we naturally assumed in this book that parameters in dynamical systems that do not show any regularity can be studied better as a realization of a statistical ensemble. This is a kind of metapostulate of maximal disorder similar to Shannon's maximal entropy principle. It is at the base of all theories of amorphous systems and from there we have seen it invading other fields (i.e. optimization).

But then we have to face a somewhat provocative question. Is it possible that what looks like a goal oriented organization, with subtle interrelationships leading to amazing ways of reaching efficiency can in any way be seen as a maximally disordered system?

Surprisingly, many biologists are giving a tentative affirmative answer to this question. In this respect we personally have found the discourses by Changeux, Purves and Lichtman, Eigen and Kauffman[1] particularly interesting. A partial summary of the arguments is:

i) While it is true that natural evolution leads the system towards better adaptation and more efficient behaviour, a simple estimate shows that during the lifetime of the earth the mutations random walk of a DNA/RNA composed of 10^9 nucleotides could only explore a negligible fraction of the total phase space (10^{60} configurations compared to 4^{10^9} possible arrangements (see Eigen[1]; the rate of nucleic acids replications is estimated as 1/sec).

ii) At a certain high level natural selection works mainly through competition. The appearance of a new particularly advantageous feature in the species may introduce new strategies to survive even disadvantageous mutations, for instance, simply by evading competition moving into a new environment[2]. While part of the biosystem may have been perfectly modelled to perform a function, another part, or even the same part but looked at a different scale, can be analysed assuming randomness. For instance, we can distinguish in the mammalian brain different pieces whose interconnections seem like the result of a careful design. But if we look inside each of these pieces they look much more random.

iii) In fact it is impossible that every single detail of this organization had been optimized because the amount of information needed surpasses a large factor the information content of the DNA (just count, for instance, the information contained in the way 10^9 neurons are connected by 10^{13} synapses).

In other words the biosystem is much larger, has many more variables than the ones that have been put under control by the genetic code. The situation is therefore not very different from the one encountered in statistical mechanics with macroscopic variables under control and microscopic variables being random. It would be rewarding to be able to identify the *relevant* macroscopic variables that control the emergence of a typical behavior as a collective manifestation of the infinite number of microscopic degrees of freedom (the *same* problem is crucial, and unfortunately very difficult, in non-perturbative quantum field theory).

Disorder therefore plays a role. What about frustration? We have stressed several times in this book that a system with frustration may be endowed with a large number of equilibrium states. This looks adequate when trying to model a biosystem (there is nothing more similar to death than a single state system!). Every equilibrium state is stable but only under *small* perturbations. This property is also useful[3].

In this book we would like to discuss just two examples of applications of spin glass theory to biological problems. We have chosen those that make more fundamental use of the properties of a spin glass. This does not address the problem of ultimate relevance. It may turn out that applications to 'protein folding'[4] or to the immunosystem[5] be the most fruitful.

In Chap. XI we will discuss a model for prebiotic evolution originally proposed by Anderson. In Chap. XII we will discuss brain modelling, in particular Hopfield's model for memory and several proposed modifications. In Chap. XIII we will show that Hopfield's model is a particular type of spin glass and can be solved using the Replica method or the cavity method (see Chap. V).

References to Chap. (X–XIII) are listed at the end of Chap. XIII.

Chapter XI
PREBIOTIC EVOLUTION AND SPIN GLASSES

In the study of the origin of life one can distinguish a moment in which, in the primordial ocean, an abundant supply of amino acids and nucleotides already exists while no microorganism subject to natural evolution has yet appeared.

Anderson[6] analyzes this moment to see when, in the transition from inanimate matter to life, long polymers, DNA-like, constituted of simply building blocks and capable of containing information could appear for the first time. The last requirement imposes constraints on the possible structure of the polymers. A periodic ordered polymer contains no information: one should be able to find the polymer in different arrangements. On the other hand, if the linear structure of the polymer is too sensitive to external influences it will change too rapidly in time and no information could be transmitted.

In the large majority of biosystems, as we know them today, the unit responsible for replication is relieved from any other function. Thus there is a specialization of the DNA and the proteins on both sides. This separation, of course, allows a better optimization of possibly two conflicting goals. For instance, proteins take advantage of the three dimensional structure for their enzymatic capability while the simple double stranded linear structure of DNA is ideal for replication. However it has been argued[7,8] that perhaps early polynucleotides had dual roles, performing rudimentary catalytic functions. One advantage of this hypothesis is that it leads to a particular, relatively simple *dynamical model*[7].

One starts with a primordial soup of small polymers which for simplicity are constituted of two building blocks: nucleotides A and B.

An external parameter, for instance, the temperature changes cyclically in such a way that during half of the cycle—the renaturation period—the system tends to test pairs of polymers for complementarity in their corresponding sequences (i.e. for each A there is a B in front) and with a certain probability creates weak bonds across. There are errors in this process, i.e. with smaller probability a weak bond may be created between two A's or two B's. Loose ends are matched again and again until the whole soup is exhausted. During this same period covalent, stronger bonds are created between nucleotides that remained adjacent as a consequence of being weakly bonded to a single polymer (again one assigns a certain *a priori* probability to this process).

During the second half of the cycle (the denaturation part) weak bonds dissolve leaving behind presumably longer polymers. An example could be (== indicate covalent bonds; | indicates a weak bond):

Beginning of the cycle:

Renaturation period

First stage: three small polymers

A==B==A   A==B==B   A==B

Second stage: creation of weak bonds (in this example there are no errors)

```
A==B==A   A==B
   |   |    |
   A==B==B
```

Third stage: creation of covalent bonds

```
A==B==A==A==B
   |   |  |
   A==B==B
```

Denaturation period

First stage: the double structure dissolves into two polymers

A==B==A==A==B   A==B==B

Second stage: stochastic death, to be discussed later.

So defined, the model generates a very rich class of long polymers. The possibility of making errors at the moment of creating a complementary copy amounts to allowing a certain rate of mutations. As a consequence the system drifts continuously: the type of dominating polynucleotide changes from generation to generation. The model lacks differentiation into species and stability.

Interestingly enough, both these features can be implemented by introducing a "death function" similar to a *SK* hamiltonian. This new ingredient simulates selection and ensures stability under small mutations. A "death probability" is introduced which is a function of the polymer linear structure. A and B are translated into values ± 1 for an Ising spin.

$$p(S) = \frac{\exp(\lambda(D[S] + \mu))}{1 + \exp(\lambda(D[S] + \mu))}, \tag{XI.1}$$

where

$$D[S] = \sqrt{2/N(N+1)} \sum_{i>j} J_{ij} s_i s_j \tag{XI.2}$$

with

$$J_{ij} = \pm 1 .$$

For very large λ all polymers with

$$D[S] < -\mu \tag{XI.3}$$

survive, while all others disappear. Eq. (XI.3) defines a microcanonical ensemble. It would correspond to a finite temperature canonical ensemble if μ was of order $\sqrt{N}$. In Ref. 8 it is chosen finite (between $-.7$ and -1 in the numerical simulations) in such a way that the number of relevant configurations that satisfy (XI.3) is ≈ 0.16 of all configurations. In the thermodynamics limit the entropy is $N \ln 2$ minus finite terms. The energy is therefore above any freezing transition. Still the numerical simulations[8] show phenomena reminiscent of a multivalley landscape: which polymers dominate differ from run to run in the model, however all along a single run there are remarkable signs of stability. As there is a continuous increase of the average length, polymers are eventually superceded by longer versions of themselves but if one groups together longer and shorter sequences with similar subsequences into a single 'species', one finds that all along the run (500 generations) the same dominating species are present.

Interestingly enough, this type of "open end" dynamics similar to a flow of lava down a hill is able to distinguish valleys and barriers where equilibrium statistical mechanics would not.

Chapter XII
BRAIN MODELLING

The goal is to build a model of the brain where the pecularities of its performance emerge naturally. A realistic model is unfortunately out of question (and perhaps useless given its potential complexity). Therefore one is led to try to abstract which are the important features responsible for its idiosyncratic behavior.

The development of computers have shown how strikingly their performances differ from those of the animal brain. Where the former performs fast and reliable calculations with very rare—but in general disastrous—errors, the latter is slow and makes a large quantity of nonfatal mistakes. The computer excels at repetitive recursive calculations while our ability for recognizing a friend's face has no match even in the fastest among them[9,10]. And this in spite of the fact that the chips that Nature can afford are much less performant.

In the recent epoch of natural evolution it looks as if the growth of the relative size of the mammalian brain and particularly of the cerebral cortex were driving factors in the evolution[11]. Using (or abusing) metaphors from statistical mechanics, it is as if the size was one of the macroscopic parameters that controls the onset of new emerging behavior, like in a phase transition. The fact of having a large number of independent CPU units certainly gives an advantage.

In addition each one of these units is connected to a large number of other units, receiving and sending information to them.

Therefore a reasonable hypothesis appears. In the process of evolution, Nature has struck upon a simple solution on how to develop an extremely powerful computer with slow and old fashioned chips: just take a large number of them (10^9 in man's brain) and interconnect them to a large extent ($\sim 10^4$: the order of magnitude of synapses between one given neuron and the rest of the brain) in a more or less random way. An additional crucial ingredient is that the system is adaptive, i.e. able to modify the connectivity through a learning process.

This is the working hypothesis (coined "connectionism"[12]) behind brain modelling, Notice that it is part of the hypothesis that not all the anatomical and neurophysiological details are relevant to understand brain's performance. Models are, in a sense, like theoretical laboratories where one can monitor the consequences of a particular hypothesis. Therefore by necessity we want the model to be more amenable to analysis than the brain itself. Different models rather than being exclusive alternatives should be seen as complementing each other.

A first good property shared by such models is that all of them process in parallel. The brain must necessarily process in parallel as witnessed by the time of reaction in front of a danger and the typical time scale of a neuron (1/10 second as compared to a few milliseconds). This means that the brain would have less than 100 computing cycles to work out fully a complicated response. This is possible only if a large number of computing units are working in parallel.

Apart from that it is fair to say that so far the working hypothesis has been neither confirmed nor falsified. We feel that there has been progress, for instance: the idea that recognition is like falling in the basin of attraction of a fixed point or limit cycle will probably remain in any better model; but otherwise the new models could be very different. In addition the models have suggested new foci of attention. For instance, the fact that there must be a limit to the capacity of storing memories. In any case cross fertilization has not occurred yet. Perhaps, from the limited point of view of physics, the most important result is that we have found new solvable models in statistical mechanics where the valley structure can be largely controlled.

The history of this subject has been rather tortuous. Sometimes one feels that one is advancing in circles. Scrutinizing it carefully, a small but steady progress can be detected.

The first two crucial references are quite old: in 1943 McCulloch and Pitts[13] show that with a sufficiently large number of ideal neurons (the type of which will be discussed at length in the next section) interconnected among themselves through excitatory and inhibitory synapsis, one can construct the equivalent of a Universal Turing Machine. Therefore by definition anything that is computable can be computed by such a network. This result is reassuring but frustrating. Nothing else can be inferred about a model that includes a Universal Computer.

The second crucial reference is Hebb's book[14] where he proposes the synaptic junctions as the natural locus for memory. Synapses are to be modified during conditional learning in a simple but specific way (see Sec. XIII.2).

In 1958 the influential proposal of the Perceptron as a generic learning machine is made[15]. The interest in it mysteriously ended with Minsky and Pappert's book[16] where the rather obvious limitations of such an approach were underlined.

In perceptron-like machines the information is forward fed from one layer of neurons to the next layer. There is no feedback circuit, i.e. no information from a later stage projects onto an earlier stage. This is one source of limitations. Recently there has been a revival of interest in these models because several learning algorithm have been found that apply to multilayered machines[17]. Though it might be possible, no one has tried to relate this type of models with the physics of disordered systems.

Instead a new line, closer to statistical mechanics was initiated by Little[18] (1974) and later developed by many authors[19]. In these models one tries to simulate the retrieval of a memory from a partial information about it. In a sense, these models complement the perceptron-like ones because in them feedback is maximized.

Along this line appears Hopfield's model[20] that turns out to be exactly soluble[21] and prolific because it allows for innumerable improvements. This model and its variations will be the subject of the next chapter.

Chapter XIII
THE HOPFIELD MODEL AND ITS VARIATIONS

XIII.1 The Assembly of Neurons as a Spin Glass

The structure and functioning of a neuron is, generally speaking, similar in different parts of the body and in different animals[22]. However, as always, one can find quite different kinds of neurons if one looks for them. As stressed in the previous chapter we are not trying to build *realistic* models of the brain. We allow for a sufficient number of simplifications if they are necessary to build the model.

A very simplified picture of a neuron[23] only partially based on experimental observations is the following: a neuron receives input through its dendrites from upstream neurons, integrates the signals and reacts if the total input during a certain time interval exceeds a threshold value. In that case, it fires a train of equal spikes of electric potential at a rate which is a function of the input current (see Fig. XIII.1 in Hopfield, 1982, Reprint 29). The electric potential impulse (the action potential) propagates along the axon reaching new neurons.

For some neurons, it can be safely assumed that only the frequency of firing encodes useful information, the individual timing of each spike being too sensitive to delays in transmission and noise (see discussion in Ref. 10. For a different point of view see Ref. 24).

The response function: neuron firing frequency vs. total input is similar to a sigmoid function. Hopfield and Tank[10] (1984) have shown that such a smooth profile may play a positive role in complex computations, roughly similar to the effect of finite temperature in simulated annealing[5]. In general, in the past, this function has been approximated by a linear function[19] or a step function[13,18,20]. It is probably correct that nonlinearity is a necessary condition for reliable computational performances. In the following we will modelize the sigmoid curve by a step function. This simplifying assumption has the merit of enabling one to describe a neuron by an Ising spin, so that the analysis of this type of highly simplified model of neural network will eventually enter into the range of applicability of spin glass theory. We mention, *en passant*, that the fixed point equations of Hopfield and Tank are similar to finite temperature TAP equations but *without* the Onsager term and that similar equations appear in Ref. 26 as a way of modelling the noisy (= finite temperature) behavior of Boolean neurons which are supposed to react only through a step function.

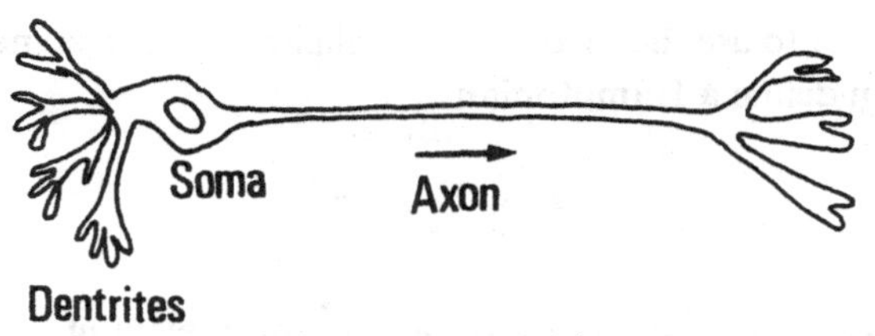

Fig. XIII.1. A schematic picture of a neuron.

Neurons therefore are assumed to be 2-states devices (either quiescent or firing at the maximum rate). The information is transferred from neuron to neuron through the synapses, which can be either excitatory, i.e. a firing upstream neuron lowers the threshold of the downstream one or inhibitory, with the opposite effect. Synapses will then be represented by a positive or negative real number which measures how the firing uphill neuron modifies the threshold of the downstream neuron. The basic ingredients of the modelization are:

s_i = state of the ith neuron ($s_i = -1$, quiescent neuron; $s_i = 1$ firing neuron).

J_{ij} = synapses from the jth neuron to the ith neuron.

h_i = threshold of the ith neuron.

One must add to these a description of the evolution in time of the neural network. Two types of dynamical evolution have been considered: 1) sequential[20] or 2) parallel synchronous[18]. In the first case at any time t the ith neuron is updated according to the equation

$$s_i(t + \delta t) = \text{sign}\left(\sum_j J_{ij} s_j(t) + h_i\right), \qquad \text{(XIII.1)}$$

while when updating the next neuron the new value of s_i at time $t + \delta t$ will have to appear in the right-hand side of the equation. In synchronous dynamics all neurons are updated at the same time.

Presumably the dynamics in the brain lies somehow in the middle between these two extreme cases. Sequential updating usually applies in statistical mechanics and is more natural for disordered systems where there is no clock to control the simultaneity of the updating. On the other hand delays in signal transmissions are relatively large. Therefore sometimes the information about the updating of a spin may not have enough time to arrive to the next one. This effect could be modelled by parallel updating of randomly chosen blocks of spins. When the network approaches asymptotically a stationary state the type of updating becomes progressively irrelevant.

In Ref. 20, Hopfield adds the assumption of symmetric synapses

$$J_{ij} = J_{ji} \quad \text{and} \quad J_{ii} = 0 \qquad \text{(XIII.2)}$$

opening the possibility to use the theory of spin glasses to understand neural networks. In that case one can define a Hamiltonian

$$H = \sum_{i>j} J_{ij} s_i s_j \tag{XIII.3}$$

and Eq. (1) describes a spin system at zero temperature relaxing towards equilibrium. The only possible asymptotic states are fixed points (configurations whose energy is a local minimum). Hopfield has shown through numerical simulations that *near a fixed point* relaxing (2) does not essentially modify the behavior of the system. On the other hand in the more general case, there is no Lyapunov function that decreases monotonically to control the dynamics of the network. Accordingly the asymptotic behavior is much richer and the system may end up in limit cycles or chaotic attractors. The dynamics of such networks is studied in the context of cellular automata, more specifically threshold networks[27]. Very little is known about the general behavior. Let us mention again that with a particular choice of the couplings it can embody a universal calculator.

Real synapses are not symmetric and the richer asymptotic behavior of asymmetric networks are needed to model the myriad of computational capabilities of the brain. Limit cycles have been investigated with parallel synchronous updating[18] and with sequential one[20]. It seems that the existence of long cycles with more than 4 different states necessitates coherent synchronous updating (see Ref. 24 for an alternative way of implementing it). A chaotic attractor seems to be generic for sufficiently random asymmetric couplings[28]. It can be interpreted as a state of confusion that results when the system is unable to recognize a pattern.

The fixed point states are in general interpreted as a successful recognition. Every fixed point has its basin of attraction. When the initial configuration lies sufficiently near to the fixed point the network evolves in the direction of the latter. This models the reconstruction of stored information from partial deteriorated information about it.

XIII.2 The Hebb Rule. Different Proposals

The Hebb rule is a way of incorporating learning into the model. If one is trying to memorize a pattern, encoded as a word of N bits $\{\xi_i = +1, i = 1, N\}$, the synapses are supposed to be modified as follows

$$J_{ij}^{\text{new}} = \lambda J_{ij}^{\text{old}} + \varepsilon \xi_i \xi_j . \tag{XIII.4}$$

In Hopfield's model $\lambda = \varepsilon = 1$. More recently[29,30,31] the rule has been modified so that old memories fade out while new memories can always be learned (choosing $\varepsilon > \lambda$). Thus the model reproduces well-known capabilities of our Short Term Memory (originally the model was rather thought of as applying to the Long Term Memory, see Ref. 32 for the phenomenological definition of both components). As will be discussed later, Hopfield's model goes through a phase transition when the number of memorized patterns surpasses a critical value. As a consequence, there is a true collapse of its ability of retrieving any pattern. This is unnatural. The modification $\varepsilon > \lambda$ leads to a smaller storage capability but with the advantage that it never collapses. Instead when we continuously feed in new patterns it reaches a quasi-stationary regime where only old memories are forgotten.

In a different development the learning rules were modified so as to be more adequate when the system is trying to memorize categorized patterns[33]. Then new information will belong to a category, the characteristics of which might be already known to the system. In this case memories to be stored are by definition ultrametrically ordered. The individual pattern can be denoted ξ_i^{ab} while the class (the ancestor in the spin glass language) will be ξ_i^a with

$$\xi_i^a = \left(\sum_b \xi_i^{ab}\right) \Big/ \left(\sum_b 1\right), \tag{XIII.5}$$

where b denotes the individual pattern belonging to category a.

Then the proposal is to modify the synapses only in that part where the individual differs from the class i.e.

$$J_{ij}^{\text{new}} = J_{ij}^{\text{old}} + \varepsilon(\xi_i^{ab} - \xi_i^a)(\xi_j^{ab} - \xi_j^a) \tag{XIII.6}$$

while if the system has also to learn the category

$$J_{ij}^{\text{new}} = J_{ij}^{\text{old}} + \varepsilon' \xi_i^a \xi_j^a. \tag{XIII.7}$$

In Ref. 33, ε'/ε was chosen as $(1 - q)/q$ where q is the overlap between individuals belonging to the same class. Other choices may however be more interesting as they may lead to situations in which the system is unable to retrieve individuals while it can still recognize the category. Addressing this problem we would like also to mention a related but different proposal[34] and a more general prescription proposed in Ref. 35 where Hebb's rule is essentially abandoned in favour of maximal flexibility in storing words that are correlated.

There is an aspect of the Hebb rule which is too unrealistic. It assumes that *before* any learning the synaptic connections are zero. This is anatomically impossible. Furthermore it reminds one of the tabula rasa philosophy. In simple words, there is no innate structure and everything is a consequence of the external world teaching the biosystem. It is easy to see that an extreme adhesion to this point of view leads to paradoxes. For instance, the newborn receives an infinite amount of information during the first days of its life. Obviously, most of it is irrelevant and the baby should do better not to learn it. But in the tabula rasa point of view he does not yet know what is to be discarded. Toulouse, Dehaene and Changeux[36] have proposed a radical change building a model such that at the beginning the J_{ij} are random uncorrelated variables. Then the innate landscape will be one of the SK model. The biosystem learns essentially only those patterns that are similar to the local minima of the SK hamiltonian. In that case the minima are deepened. Otherwise nothing happens. In this way one expects that the corrected landscape will conserve some of its properties and in particular ultrametricity. Categorization will be imposed upon us as an innate property.

There is another proposal to modify the model. It is well known that in the brain neurons are most of their time quiet. On the other hand, in Hopfield's model as we have defined it, there is symmetry between the states $s_i = +1$ and $s_i = -1$. One can modify it by adding the equivalent of an external magnetic field. In Ref. 37 a different implementation is considered. First of all, instead of the hypercube the configuration space is restricted, via a constraint

$$(1/N) \sum_i s_i = A \qquad \text{(XIII.8)}$$

to a subset that is not a linear space. Therefore all linear combinations of the memorized patterns will fall outside the allowed region. This is important because in the original model such combinations, if they lie near a vertex of the hypercube, could generate a new valley so that during retrieval a *spurious pattern* could be recognized. Second, instead of Eq. (XIII.4) the Hebb rule is changed to

$$J_{ij}^{\text{new}} = J_{ij}^{\text{old}} + \varepsilon(\xi_i^{ab} - A)(\xi_j^{ab} - A). \qquad \text{(XIII.9)}$$

(Compare with Eq. XIII.6). An interesting consequence of this modification is that the information content of each memory is $A \cdot N$ bits.

XIII.3 The Analytical Solution of Hopfield's Model

XIII.3.a The replica method

Amit *et al.*[21] solved the Hopfield model as a Spin Glass system using the replica approach. If we call P the total number of stored memories and N, as usual, the number of spins there are two different regimes:

1) $P/N \to 0$ when $N \to \infty$

In this case there is no need to use replicas to solve the problem. For simultaneous synchronous updating even the dynamics can be solved. One observes that all the original stored patterns can be retrieved if one starts from a pattern sufficiently near to it. On the other hand for arbitrary starting point one retrieves *spurious solutions*, i.e. solutions that do not correspond to any learned pattern. They correspond to linear combinations of binary words that happen to lie near to a vertex of the N-dimensional hypercube (the configuration space). The distance measures the excitation energy. A classification of such solutions was done. The statistical mechanics can also be developed at finite temperature. We refer to Ref. 21a for the details.

2) $P/N = \alpha$ when $N \to \infty$

In this case one needs replicas and only the statistical equilibrium can be analyzed. We again discuss only the results at $T = 0$. There are two phase transitions on α.

For $\alpha \le \alpha_{\text{crit}_1} \approx .05$ there are ground states corresponding to all the stored memories. The former differ from the original input memories in a few percent of the total number of bits. Replica symmetry is broken and as a consequence each state has the fine structure of a full ultrametric tree. In addition there is a so-called spin glass phase defined by the fact that it is orthogonal to all stored patterns. For α larger than the second α_{crit_2} which is of the order of .15 there is a total collapse of the capabilities of the memory. The only state is the spin glass phase. Between α_{crit_1} and α_{crit_2} the spin glass state is the ground state but the retrieval is still efficient.

More recently similar techniques were used to solve the Personnaz *et al.* model[38] as well as other models.

The details of the replica calculation can be studied directly in the original papers[21]. In the next section, we are going to reproduce the results (with the TAP equations of the Hopfield model as a bonus) using the cavity method. The presentation will necessarily be sketchy because it parallels the one done in Chap. V for the *SK* model.

XIII.3.b The cavity method

As we explained before (see Eq. XIII.1 and subsequent paragraphs) Hopfield's model can be described by the spin glass Hamiltonian (XIII.5) with the $J_{ij} = (1/N)\sum_p \xi_i^p \xi_j^p$. During retrieval the system dynamical equations (XIII.1) represent the relaxation towards equilibrium at zero temperature. It will be useful, and it will cost the same effort, to discuss simultaneously the finite temperature case. Some aspects of finite temperature dynamics might be relevant to the brain[26].

In Hopfield's spin glass system the J_{ik} are correlated random variables. Indeed the loop products have expectation values

$$\overline{N^2 J_{ik} J_{ki}} = \overline{N^3 J_{ik} J_{kl} J_{li}} = \cdots = 1. \qquad \text{(XIII.10)}$$

The techniques developed in Chap. V can easily be generalized to this case. Introducing variables μ_p with $p = 1, \ldots, P$ (P = number of stored words)[39] we can rewrite the partition function

$$Z_\xi(\beta) = \int \operatorname{Tr}_{\{s\}} \exp\left(-\frac{\beta}{2}\sum_p \mu_p{}^2 + \frac{\beta}{\sqrt{N}}\sum_{p,i} \mu_p \xi_i{}^p s_i\right) \prod d\mu_p/\sqrt{2\pi/\beta} \qquad \text{(XIII.11)}$$

with $\langle \mu_p \rangle = \langle (1/\sqrt{N}) \sum_i \xi_i{}^p s_i \rangle$.

In this form we have a system involving two types of field variables: the original Ising spins s_i and some continuous variables μ_p with a Gaussian measure. The $\xi_i{}^p$ are the coupling constants describing the interaction of the s_i with μ_p. The latter have a simple physical interpretation, since $\langle \mu_p \rangle$ is the overlap of the spin configuration with the pattern p. We shall find hereafter that in a certain range of values of α and temperature, there can be breaking of ergodicity even in the approximation amounting to no "replica symmetry breaking", and appearance of several type of pure states, in which some of the $\langle \mu_p \rangle$ are non zero in the limit $N \to$ infinity. States in which only one of the μ_p, say $\langle \mu_1 \rangle$, develops a spontaneous expectation value while the other are zero correspond to spin configurations lying near to pattern 1. If such a state exists, it means that pattern 1 can be recognized.

Because of the existence of two types of fields in (XIII.11), one can distinguish in the cavity method two possibilities: adding one spin s_0, or adding one "pattern variable" μ_0. We will have to consider both possibilities in order to get the full self consistency equations of the cavity method. As usual we are going to develop the formulae in the simplest case of one level of replica symmetry breaking so that we assume an infinite number of equilibrium states belonging to a single cluster.

i) Inside each state the number of configurations (labeled by the $N + P$ variables s_i, μ_p) with energies in the interval E, $E + dE$ is

$$d\mathcal{N}^{(N)}(E_N) = \exp\big(\beta(E_N - F_{\alpha(N)})\big) \qquad \text{(XIII.12)}$$

ii) The magnetic field acted by the N spins on the $(N + 1)$th spin is a Gaussian random variable with average value

$$(1/\sqrt{N}) \sum_p \xi_0{}^p \langle \mu_p \rangle_{\alpha(N)} = h_{\alpha(N)} \qquad \text{(XIII.13)}$$

and variance

$$(1/N) \sum_p (\langle \mu_p{}^2 \rangle_{\alpha(N)} - \langle \mu_p \rangle_{\alpha(N)}{}^2) = r_2 - r_1 \,. \qquad \text{(XIII.14)}$$

The parameters r_2 and r_1, defined in terms of the corresponding sums in the LHS of this equation, and r_0 later defined will be eventually determined from selfconsistency as the old q_0 and q_1 in Chap. V (see also Eq. XIII.19).

As in Chap. V we can check that the distribution (XIII.12) is stable when adding the new spin. The normalization changes in such a way that

$$\begin{aligned} F_{\alpha(N+1)} - F_{\alpha(N)} &= -(1/\beta) \ln \int 2 \cosh \beta h \, dp_{r_2 - r_1}(h - h_{\alpha(N)}) \\ &= -(1/\beta) \ln 2 \cosh \beta h_{\alpha(N)} - \beta(r_2 - r_1)/2 \end{aligned} \qquad \text{(XIII.15)}$$

iii) The joint s_0, h distribution at fixed $h_{\alpha(N)}$, E_{N+1} is given by

$$\mathcal{P}(s_0, h)\, dh = \frac{\exp(\beta h s_0) \cdot dp_{r_2 - r_1}(h - h_{\alpha(N)})}{\int 2 \cosh \beta h \cdot dp_{r_2 - r_1}(h - h_{\alpha(N)})} \,. \qquad \text{(XIII.16)}$$

The TAP equation is then

$$\langle s_0 \rangle_{\alpha(N+1)} = \tanh \beta\big(\langle h \rangle_{\alpha(N+1)} - \beta(r_2 - r_1) \langle s_0 \rangle_{\alpha(N+1)}\big), \qquad \text{(XIII.17)}$$

where

$$\langle h\rangle_{\alpha(N+1)} = 1/\sqrt{N}\sum_p \xi_0{}^p\langle\mu_p\rangle_{\alpha(N+1)} = 1/N\sum_{p,i}\xi_0{}^p\xi_i{}^p\langle s_i\rangle_{\alpha(N+1)} \qquad \text{(XIII.18)}$$

TAP equations are complemented by the defining equations

$$r_2 = (1/N)\sum_p\langle\mu_p{}^2\rangle_{\alpha(N+1)} = (P/\beta N)1/N\sum_{i,j,p}\xi_i{}^p\xi_j{}^p\langle s_i s_j\rangle_{\alpha(N+1)}$$

$$r_1 = (1/N)\sum_p(\langle\mu_p\rangle_{\alpha(N+1)})^2 = 1/N\sum_{i,j,p}\xi_i{}^p\xi_j{}^p\langle s_i\rangle_{\alpha(N+1)}\langle s_j\rangle_{\alpha(N+1)}\,. \qquad \text{(XIII.19)}$$

These two order parameters will be computed selfconsistently in the second phase of the cavity method when we will add a new pattern. We will see that the corresponding cavity fields will depend on the q overlaps.

As in Chap. V we now compute the fluctuations of the fields $h_{\alpha(N)}$ from state to state and then from sample to sample.

iv) The states $\alpha(N)$ belong to the unique cluster $\Gamma(N)$ wherein the free energies distribution is given by

$$d\mathcal{N}^{(N)}(F_N) = \exp(\beta x F_N - \beta x F_{\Gamma(N)})\,dF_N\,. \qquad \text{(XIII.20)}$$

Then as in Chap. V

$$F_{\Gamma(N+1)} - F_{\Gamma(N)} = -1/\beta x\cdot\ln\int(2\cosh\beta h_1)^x\,dp_{r_1-r_0}(h_1 - h_{\Gamma(N)}) - \beta(r_2 - r_1)/2\,. \qquad \text{(XIII.21)}$$

In this equation $r_1 - r_0$ measures the variance of the variable $\mu_{p,\alpha(N)}$ inside the cluster $\Gamma(N)$ i.e.

$$r_1 - r_0 = (1/N)\sum_p[\langle\mu_{p,\alpha(N)}{}^2\rangle_{\Gamma(N)} - (\langle\mu_{p,\alpha(N)}\rangle_{\Gamma(N)})^2], \qquad \text{(XIII.22)}$$

where, as in Chap. V, the averages are at fixed $F_{\alpha(N)}$ and therefore the Gibbs-Boltzmann weights should not be included.

v) Finally the average over $h_{\Gamma(N)}$ is the average over the quenched couplings $\xi_0{}^p$

$$h_{\Gamma(N)} = (1/\sqrt{N})\sum_p\xi_0{}^p\mu_{p,\Gamma(N)}\,. \qquad \text{(XIII.23)}$$

The $\xi_0{}^p$ are not Gaussian. The $h_{\Gamma(N)}$ is a realization of a Gaussian random variable if the conditions of validity of the Central Limit Theorem are satisfied. The latter require essentially that the μ_p be "smooth" functions of p. If, on the contrary, a few of them (say $p = 1, \ldots, k$) are large while the rest is smaller and of the same order we have to separate the large terms and apply the Central Limit theorem to the rest

$$h_{\Gamma(N)} = (1/\sqrt{N})\boldsymbol{\xi}_0 \cdot \boldsymbol{\mu}_{\Gamma(N)} + h_2, \tag{XIII.24}$$

where the bold symbols indicate k-dimensional vectors and h_2 is a Gaussian random variable with zero average and variance r_0.

The final expression is

$$\Delta_N F = -\left\langle\left\langle \int dp_{r_0}(h_2 - (1/\sqrt{N})\boldsymbol{\xi}_0 \cdot \boldsymbol{\mu}_{\Gamma(N)})(1/\beta x)\ln\left[\int dp_{r_1-r_0}(h_1 - h_2)(2\cosh\beta h_1)^x\right]\right\rangle\right\rangle$$

$$-\beta(r_2 - r_1)/2, \tag{XIII.25}$$

where the double brackets indicate an average over the ξ distribution. The subindex N in Δ reminds us that we are increasing the value of N. Later we will consider Δ_P.

vi) Calling $Z_N(\beta, \alpha)$ the partition function, the $\Delta_N F$ derived in (XIII.25) is

$$-\beta\Delta_N F = \ln Z_{N+1}(\beta(N+1)/N, \alpha N/(N+1)) - \ln Z_N(\beta, \alpha) \tag{XIII.26}$$

so that using the stability of the thermodynamic limit we obtain

$$\Delta_N F = f(\beta, \alpha) + \partial(\beta f)/\partial\beta - \alpha\partial f/\partial\alpha. \tag{XIII.27}$$

Introducing the new variables

$$u = \beta/\alpha \qquad v = \beta\alpha, \tag{XIII.28}$$

Eq. (26) becomes

$$\Delta_N F = \partial(uf)/\partial u. \tag{XIII.29}$$

vii) The next step is to compute self consistently the values of r_0, r_1 and r_2. This is done by adding a new pattern μ_0 on which the acting field will be

$$h_0 = (1/\sqrt{N})\sum_i \xi_i^0 s_i. \tag{XIII.30}$$

This field inside a state has a Gaussian distribution of width $\sqrt{(1-q_1)}$ so that the joint distribution of h_0 and μ_0 is proportional to

$$dp_{1/\beta}(\mu_0)\,dp_{1-q_1}(h_0 - \langle h_0\rangle_{\alpha(P)})\exp(+\beta\mu_0 h_0). \tag{XIII.31}$$

From this we derive

$$\langle\mu_0\rangle_{\alpha(P+1)} = \langle h_0\rangle_{\alpha(P)}/(1 - \beta(1-q_1)) \tag{XIII.32}$$

and a change of free energy

$$F_{\alpha(P+1)} - F_{\alpha(P)} = -(1/\beta)\ln\left[\int dp_{1-q_1}(h - h_{\alpha(P)})\, dp_{1/\beta}(\mu) e^{\beta\mu h}\right] \tag{XIII.33}$$

$$= 1/(2\beta)\{\ln[1 - \beta(1 - q_1)]\} - \frac{1}{2}\frac{h_{\alpha(P)}{}^2}{1 - \beta(1 - q_1)}, \tag{XIII.34}$$

while in the next level we find

$$(F_{\Gamma(P+1)} - F_{\Gamma(P)}) = \frac{1}{2\beta x}\ln[1 - \beta\{1 - q_1(1 - x) - q_0 x\}]$$
$$+ \frac{1}{2\beta}\left(1 - \frac{1}{x}\right)\ln[1 - \beta\{1 - q_1\}]$$
$$- \frac{1}{2}\frac{h_{\Gamma(P)}{}^2}{1 - \beta\{1 - q_1(1 - x) - q_0 x\}} \tag{XIII.35}$$

The field $h_{\Gamma(P)}$ is equal to

$$h_{\Gamma(P)} = (1/\sqrt{N}) \sum_{i=1}^{N} \xi_i^{P+1} \mu_i^{\Gamma(P)} \tag{XIII.36}$$

and fluctuates solely from sample to sample. The Central Limit Theorem applies with no restrictions with

$$\overline{h_{\Gamma(P)}} = 0 \qquad \overline{h^2{}_{\Gamma(P)}} = q_0 \tag{XIII.37}$$

so that the average of Eq. (35) finally becomes

$$\overline{(F_{\Gamma(P+1)} - F_{\Gamma(P)})}$$
$$= \Delta_P F$$
$$= \frac{1}{2\beta x}\ln([1 - \beta\{1 - q_1(1 - x) - q_0 x\}][1 - \beta\{1 - q_1\}]^{x-1})$$
$$- \frac{1}{2}\frac{q_0}{1 - \beta\{1 - q_1(1 - x) - q_0 x\}} \tag{XIII.38}$$

It is not difficult to guess the general result for a continuous $q(x)$ from this expression.

viii) The $\Delta_P F$ is equal to

$$\Delta_P F = F_N(\beta, (P + 1)\alpha/P) - F_N(\beta, \alpha) = \partial f/\partial\alpha \tag{XIII.39}$$

ix) The self consistency equations for the two types of overlaps r_0, r_1, r_2 and q_0, q_1 are

$$q_1 = \overline{\langle(\langle s_0\rangle_{\alpha(P)})^2\rangle_{\Gamma(P)}}$$

$$q_0 = \overline{(\langle\langle s_0\rangle_{\alpha(P)}\rangle_{\Gamma(P)})^2}$$

$$r_2 = P/N\overline{\langle\langle \mu_0{}^2\rangle_{\alpha(P)}\rangle_{\Gamma(P)}}$$

$$r_1 = P/N\overline{\langle(\langle \mu_0\rangle_{\alpha(P)})^2\rangle_{\Gamma(P)}}$$

$$r_0 = P/N\overline{(\langle\langle \mu_0\rangle_{\alpha(P)}\rangle_{\Gamma(P)})^2}. \tag{XIII.40}$$

The last three equations can be solved explicitly as a consequence of the fact that the integration over the auxiliary μ variables is Gaussian

$$r_2 - r_1 = \frac{\alpha}{\beta[1-\beta\{1-q_1\}]}$$

$$r_1 - r_0 = \frac{\alpha(q_1 - q_0)}{[1-\beta\{1-q_1(1-x)-q_0x\}][1-\beta\{1-q_1\}]}$$

$$r_0 = \frac{\alpha q_0}{[1-\beta\{1-q_1(1-x)-q_0x\}]^2}, \tag{XIII.41}$$

where $\alpha = P/N$.

There are also k selfconsistency equations for the components of the vector $\boldsymbol{\mu}$

$$\langle\boldsymbol{\mu}\rangle_{\Gamma(N)} = \langle\boldsymbol{\xi}_0 s_0\rangle_{\Gamma(N+1)}. \tag{XIII.42}$$

The equations derived here agree with those obtained by Amit *et al.*[21] We refer the reader to those papers for a detailed discussion of their solutions.

XIII.3.c Fluctuations around the mean field

It is again instructive to reproduce the results obtained using the replica method[21].

We then imagine a (N, P) system to which we add either two spins or one spin and one pattern or two patterns. We expect connected correlations to be of order $1/\sqrt{N}$ and accordingly we add terms to the Gaussian distribution of the fields.

a) $N \to N+2$.

We add two spins at points 0 and 0′ where the magnetic fields are

$$h_0 = (1/\sqrt{N})\sum_p \xi_0{}^p\mu_p$$

$$h_{0'} = (1/\sqrt{N})\sum_p \xi_{0'}{}^p\mu_p. \tag{XIII.43}$$

The corresponding random variables h and h' are correlated Gaussian variables (to order $1/\sqrt{N}$) so that the joint probability distribution of h, h', s_0, $s_{0'}$ will be

$$\mathscr{P}(h, h', s_0, s_{0'}) = k \exp(\delta(h - h_{\alpha(N)})(h' - h'_{\alpha(N)}) + \beta h s_0 + \beta h' s_{0'}) \cdot$$
$$\cdot\, dp_{r_2 - r_1}(h - h_{\alpha(N)})\, dp_{r_2 - r_1}(h' - h'_{\alpha(N)}). \qquad \text{(XIII.44)}$$

From this distribution and to first order in δ we derive

$$\langle s_0 s_{0'} \rangle_{c,\alpha(N+2)} = (r_2 - r_1)^2 \delta \beta^2 (1 - \tanh^2 \beta h_{\alpha(N)})(1 - \tanh^2 \beta h'_{\alpha(N)}) \quad \text{(XIII.45)}$$

while by construction

$$\langle (h - h_{\alpha(N)})(h' - h'_{\alpha(N)}) \rangle_{c,\alpha(N)} = (r_2 - r_1)^2 \delta. \qquad \text{(XIII.46)}$$

Using (XIII.43) we can identify

$$\langle (h - h_{\alpha(N)})(h' - h'_{\alpha(N)}) \rangle_{c,\alpha(N)} = (1/N) \sum_{p,p'} \xi_0{}^p \xi_{0'}{}^{p'} \langle \mu_p \mu_{p'} \rangle_c \qquad \text{(XIII.47)}$$

so that finally

$$\langle s_0 s_{0'} \rangle_{c,\alpha(N+2)} = \beta^2 (1 - \tanh^2 \beta h_{\alpha(N)})(1 - \tanh^2 \beta h'_{\alpha(N)}) \cdot$$
$$\cdot (1/N) \sum_{p,p'} \xi_0{}^p \xi_{0'}{}^{p'} \langle \mu_p \mu_{p'} \rangle_c . \qquad \text{(XIII.48)}$$

We observe that the connected correlation function of the s spins is determined as a function of the correlations of the μ 'spins'. Another relation is needed to determine both functions selfconsistently.

b) $P \to P + 2$.

We add two patterns 0 and 0′ on which the two active 'magnetic fields' are

$$\tilde{h}_0 = (1/\sqrt{N}) \sum_i \xi_i{}^0 s_i$$

$$\tilde{h}_{0'} = (1/\sqrt{N}) \sum_i \xi_i{}^{0'} s_i \qquad \text{(XIII.49)}$$

so that instead of Eq. (XIII.44) we will obtain

$$\mathscr{P}(\tilde{h}, \tilde{h}', \mu_0, \mu_{0'}) = k \exp(\tilde{\delta}(\tilde{h} - \tilde{h}_{\alpha(P)})(\tilde{h}' - \tilde{h}'_{\alpha(P)}) + \beta(\tilde{h}\mu_0 + \tilde{h}'\mu_{0'})) \cdot$$
$$\cdot\, dp_{1-q_1}(\tilde{h} - \tilde{h}_{\alpha(P)})\, dp_{1-q_1}(\tilde{h}' - \tilde{h}'_{\alpha(P)})\, dp_{1/\beta}(\mu_0)\, dp_{1/\beta}(\mu_{0'}) \quad \text{(XIII.50)}$$

while instead of (XIII.45) and (XIII.46)

$$\langle \mu_0 \mu_{0'} \rangle_{c,\alpha(P+2)} = \delta(1 - q_1)^2 \frac{\partial \langle \mu_0 \rangle_{\alpha(P)}}{\partial \beta \tilde{h}_{\alpha(P)}} \frac{\partial \langle \mu_{0'} \rangle_{\alpha(P)}}{\partial \beta \tilde{h}'_{\alpha(P)}} \qquad \text{(XIII.51)}$$

$$\langle(\tilde{h}-\tilde{h}_{\alpha(P)})(\tilde{h}'-\tilde{h}'_{\alpha(P)})\rangle_{\alpha(P)}=\delta(1-q_1)^2. \tag{XIII.52}$$

In Eq. (XIII.51) the 'susceptibilities' $\partial\langle\mu\rangle/\partial\tilde{h}$ are easily calculated from

$$\partial\langle\mu\rangle/\partial\tilde{h}=\partial^2\ln\tilde{Z}/(\partial\tilde{h})^2$$

$$\tilde{Z}=\int dp_{1-q_1}(h-\tilde{h})\cdot dp_{1/\beta}(\mu)\cdot\exp(\beta h\mu) \tag{XIII.53}$$

and come out

$$\partial\langle\mu\rangle/\partial\tilde{h}=\beta/(1-\beta(1-q_1)). \tag{XIII.54}$$

We finally derive, instead of Eq. (XIII.48)

$$\langle\mu_0\mu_{0'}\rangle_{c,\alpha(P+2)}=\frac{1}{[1-\beta(1-q_1)]^2}\left(\frac{1}{N}\right)\sum_{i,j}\xi_i{}^0\xi_j{}^{0'}\langle s_is_j\rangle_{c,\alpha(p)}. \tag{XIII.55}$$

This is the second equation needed to derive selfconsistency conditions for the correlation functions. As in Chap. V we square both Eqs. (XIII.55), (XIII.48) and average out the couplings. We then separate the contributions of the diagonal terms and finally derive a system of two linear equations that can be solved to derive

$$\overline{\langle ss'\rangle^2{}_c}=\frac{1}{N}\frac{U}{1-\alpha AB}$$

$$\overline{\langle\mu\mu'\rangle^2{}_c}=\frac{1}{N}\frac{V}{1-\alpha AB}, \tag{XIII.56}$$

where U and V are positive quantities and

$$A=\overline{(\partial\langle s\rangle/\partial h)^2};\qquad B=\overline{(\partial\langle\mu\rangle/\partial\beta h)^2} \tag{XIII.57}$$

or equivalently

$$A=\overline{\beta^2(1-m_0{}^2)^2};\qquad B=[1/(1-\beta(1-q_1))]^2. \tag{XIII.58}$$

The positivity requirement corresponding to Eq. (V.54) is

$$x=1-\alpha\overline{\beta^2(1-m_0{}^2)^2}[1/(1-\beta(1-q_1))]^2\geq 0. \tag{XIII.59}$$

At this point we refer the reader to the discussion following Eq. (V.57). We expect this inequality to be violated for any finite number of replica symmetry breaking levels and we expect it to become an equality in the infinite number of levels limit.

References to Chapters X–XIII

1. Changeux (1983). Chapters 6, 7. Purves and Lichtman (1985). Eigen (1985). Kauffman (1986).
2. Oliverio (1984).
3. Anderson (1985).
4. Frauenfelder (1984).
5. Jerne (1974). There will be a book by Stein dedicated to the connections between biology and spin glasses. (World Scientific to be published).
6. Anderson (1983).
7. see comment and references in: Gilbert (1986).
8. Rokhsar *et al.* (1986).
9. see for instance discussion in Hurlbert and Poggio (1986).
10. Hopfield and Tank (1986).
11. Wilson (1985).
12. Ballard (1986).
13. McCullochs and Pitts (1943).
14. Hebb (1949).
15. Rosenblatt (1961).
16. Minsky and Papert (1969).
17. Hinton *et al.* (1985). Le Cun (1985). Rummelhart *et al.* (1986). Sejnowski and Rosenberg (1986).
18. Little (1974).—and Shaw (1978). See also Caianiello (1961).—and de Luca (1966).
19. In addition to [18] see also Kohonen (1984). Cooper *et al.* (1979). Hinton and Anderson (1981).
20. Hopfield (1982: reprint 29 this book).
21. Amit *et al.* (1985a: reprint 30 this book) and (1985b: reprint 31 this book). Crisanti *et al.* (1986).
22. For a pleasant, readable introduction to the new discoveries in Neurobiology see Churchland (1986).
23. Koch and Poggio (1984).
24. Bienenstock and Von Malsburg (1986).
25. Kirkpatrick *et al.* (1983).
26. Peretto (1984).
27. Goles Chacc (1985).
28. Parisi (1986b: reprint 34 this book).
29. Parisi (1986a).
30. Nadal *et al.* (1986).
31. Mézard *et al.* (1986: reprint 35 this book).
32. Klatzky (1976).
33. Parga and Virasoro (1986: reprint 33 this book).
34. Dotsenko (1985).
35. Personnaz *et al.* (1985).
36. Toulouse *et al.* (1985: reprint 32 this book).
37. Amit *et al.* (1986).
38. Kanter and Sompolinsky (1986).
39. Provost and Vallee (1983).

Reprints

BIOLOGICAL APPLICATIONS

Proc. Natl. Acad. Sci. USA
Vol. 79, pp. 2554–2558, April 1982
Biophysics

Neural networks and physical systems with emergent collective computational abilities

(associative memory/parallel processing/categorization/content-addressable memory/fail-soft devices)

J. J. HOPFIELD

Division of Chemistry and Biology, California Institute of Technology, Pasadena, California 91125; and Bell Laboratories, Murray Hill, New Jersey 07974

Contributed by John J. Hopfield, January 15, 1982

ABSTRACT **Computational properties of use to biological organisms or to the construction of computers can emerge as collective properties of systems having a large number of simple equivalent components (or neurons). The physical meaning of content-addressable memory is described by an appropriate phase space flow of the state of a system. A model of such a system is given, based on aspects of neurobiology but readily adapted to integrated circuits. The collective properties of this model produce a content-addressable memory which correctly yields an entire memory from any subpart of sufficient size. The algorithm for the time evolution of the state of the system is based on asynchronous parallel processing. Additional emergent collective properties include some capacity for generalization, familiarity recognition, categorization, error correction, and time sequence retention. The collective properties are only weakly sensitive to details of the modeling or the failure of individual devices.**

Given the dynamical electrochemical properties of neurons and their interconnections (synapses), we readily understand schemes that use a few neurons to obtain elementary useful biological behavior (1–3). Our understanding of such simple circuits in electronics allows us to plan larger and more complex circuits which are essential to large computers. Because evolution has no such plan, it becomes relevant to ask whether the ability of large collections of neurons to perform "computational" tasks may in part be a spontaneous collective consequence of having a large number of interacting simple neurons.

In physical systems made from a large number of simple elements, interactions among large numbers of elementary components yield collective phenomena such as the stable magnetic orientations and domains in a magnetic system or the vortex patterns in fluid flow. Do analogous collective phenomena in a system of simple interacting neurons have useful "computational" correlates? For example, are the stability of memories, the construction of categories of generalization, or time-sequential memory also emergent properties and collective in origin? This paper examines a new modeling of this old and fundamental question (4–8) and shows that important computational properties spontaneously arise.

All modeling is based on details, and the details of neuroanatomy and neural function are both myriad and incompletely known (9). In many physical systems, the nature of the emergent collective properties is insensitive to the details inserted in the model (e.g., collisions are essential to generate sound waves, but any reasonable interatomic force law will yield appropriate collisions). In the same spirit, I will seek collective properties that are robust against change in the model details.

The model could be readily implemented by integrated circuit hardware. The conclusions suggest the design of a delocalized content-addressable memory or categorizer using extensive asynchronous parallel processing.

The publication costs of this article were defrayed in part by page charge payment. This article must therefore be hereby marked "*advertisement*" in accordance with 18 U. S. C. §1734 solely to indicate this fact.

The general content-addressable memory of a physical system

Suppose that an item stored in memory is "H. A. Kramers & G. H. Wannier *Phys. Rev.* **60**, 252 (1941)." A general content-addressable memory would be capable of retrieving this entire memory item on the basis of sufficient partial information. The input "& Wannier, (1941)" might suffice. An ideal memory could deal with errors and retrieve this reference even from the input "Vannier, (1941)". In computers, only relatively simple forms of content-addressable memory have been made in hardware (10, 11). Sophisticated ideas like error correction in accessing information are usually introduced as software (10).

There are classes of physical systems whose spontaneous behavior can be used as a form of general (and error-correcting) content-addressable memory. Consider the time evolution of a physical system that can be described by a set of general coordinates. A point in state space then represents the instantaneous condition of the system. This state space may be either continuous or discrete (as in the case of N Ising spins).

The equations of motion of the system describe a flow in state space. Various classes of flow patterns are possible, but the systems of use for memory particularly include those that flow toward locally stable points from anywhere within regions around those points. A particle with frictional damping moving in a potential well with two minima exemplifies such a dynamics.

If the flow is not completely deterministic, the description is more complicated. In the two-well problems above, if the frictional force is characterized by a temperature, it must also produce a random driving force. The limit points become small limiting regions, and the stability becomes not absolute. But as long as the stochastic effects are small, the essence of local stable points remains.

Consider a physical system described by many coordinates $X_1 \cdots X_N$, the components of a state vector X. Let the system have locally stable limit points $X_a, X_b, \cdots$. Then, if the system is started sufficiently near any X_a, as at $X = X_a + \Delta$, it will proceed in time until $X \approx X_a$. We can regard the information stored in the system as the vectors $X_a, X_b, \cdots$. The starting point $X = X_a + \Delta$ represents a partial knowledge of the item X_a, and the system then generates the total information X_a.

Any physical system whose dynamics in phase space is dominated by a substantial number of locally stable states to which it is attracted can therefore be regarded as a general content-addressable memory. The physical system will be a potentially useful memory if, in addition, any prescribed set of states can readily be made the stable states of the system.

The model system

The processing devices will be called neurons. Each neuron i has two states like those of McCullough and Pitts (12): $V_i = 0$

("not firing") and $V_i = 1$ ("firing at maximum rate"). When neuron i has a connection made to it from neuron j, the strength of connection is defined as T_{ij}. (Nonconnected neurons have $T_{ij} \equiv 0$.) The instantaneous state of the system is specified by listing the N values of V_i, so it is represented by a binary word of N bits.

The state changes in time according to the following algorithm. For each neuron i there is a fixed threshold U_i. Each neuron i readjusts its state randomly in time but with a mean attempt rate W, setting

$$\begin{matrix} V_i \to 1 \\ V_i \to 0 \end{matrix} \quad \text{if} \quad \sum_{j \neq i} T_{ij}V_j \begin{matrix} > U_i \\ < U_i \end{matrix} \qquad [1]$$

Thus, each neuron randomly and asynchronously evaluates whether it is above or below threshold and readjusts accordingly. (Unless otherwise stated, we choose $U_i = 0$.)

Although this model has superficial similarities to the Perceptron (13, 14) the essential differences are responsible for the new results. First, Perceptrons were modeled chiefly with neural connections in a "forward" direction $A \to B \to C \to D$. The analysis of networks with strong backward coupling $A \rightleftarrows B \rightleftarrows C$ proved intractable. All our interesting results arise as consequences of the strong back-coupling. Second, Perceptron studies usually made a random net of neurons deal directly with a real physical world and did not ask the questions essential to finding the more abstract emergent computational properties. Finally, Perceptron modeling required synchronous neurons like a conventional digital computer. There is no evidence for such global synchrony and, given the delays of nerve signal propagation, there would be no way to use global synchrony effectively. Chiefly computational properties which can exist in spite of asynchrony have interesting implications in biology.

The information storage algorithm

Suppose we wish to store the set of states V^s, $s = 1 \cdots n$. We use the storage prescription (15, 16)

$$T_{ij} = \sum_s (2V_i^s - 1)(2V_j^s - 1) \qquad [2]$$

but with $T_{ii} = 0$. From this definition

$$\sum_j T_{ij}V_j^{s'} = \sum_s (2V_i^s - 1)\left[\sum_j V_j^{s'}(2V_j^s - 1)\right] \equiv H_j^{s'}. \qquad [3]$$

The mean value of the bracketed term in Eq. 3 is 0 unless $s = s'$, for which the mean is $N/2$. This pseudoorthogonality yields

$$\sum_j T_{ij}V_j^{s'} \equiv \langle H_i^{s'} \rangle \approx (2V_i^{s'} - 1)\,N/2 \qquad [4]$$

and is positive if $V_i^{s'} = 1$ and negative if $V_i^{s'} = 0$. Except for the noise coming from the $s \neq s'$ terms, the stored state would always be stable under our processing algorithm.

Such matrices T_{ij} have been used in theories of linear associative nets (15–19) to produce an output pattern from a paired input stimulus, $S_1 \to O_1$. A second association $S_2 \to O_2$ can be simultaneously stored in the same network. But the confusing simulus $0.6\ S_1 + 0.4\ S_2$ will produce a generally meaningless mixed output $0.6\ O_1 + 0.4\ O_2$. Our model, in contrast, will use its strong nonlinearity to make choices, produce categories, and regenerate information and, with high probability, will generate the output O_1 from such a confusing mixed stimulus.

A linear associative net must be connected in a complex way with an external nonlinear logic processor in order to yield true computation (20, 21). Complex circuitry is easy to plan but more difficult to discuss in evolutionary terms. In contrast, our model obtains its emergent computational properties from simple properties of many cells rather than circuitry.

The biological interpretation of the model

Most neurons are capable of generating a train of action potentials—propagating pulses of electrochemical activity—when the average potential across their membrane is held well above its normal resting value. The mean rate at which action potentials are generated is a smooth function of the mean membrane potential, having the general form shown in Fig. 1.

The biological information sent to other neurons often lies in a short-time average of the firing rate (22). When this is so, one can neglect the details of individual action potentials and regard Fig. 1 as a smooth input–output relationship. [Parallel pathways carrying the same information would enhance the ability of the system to extract a short-term average firing rate (23, 24).]

A study of emergent collective effects and spontaneous computation must necessarily focus on the nonlinearity of the input–output relationship. The essence of computation is nonlinear logical operations. The particle interactions that produce true collective effects in particle dynamics come from a nonlinear dependence of forces on positions of the particles. Whereas linear associative networks have emphasized the linear central region (14–19) of Fig. 1, we will replace the input–output relationship by the dot-dash step. Those neurons whose operation is dominantly linear merely provide a pathway of communication between nonlinear neurons. Thus, we consider a network of "on or off" neurons, granting that some of the interconnections may be by way of neurons operating in the linear regime.

Delays in synaptic transmission (of partially stochastic character) and in the transmission of impulses along axons and dendrites produce a delay between the input of a neuron and the generation of an effective output. All such delays have been modeled by a single parameter, the stochastic mean processing time $1/W$.

The input to a particular neuron arises from the current leaks of the synapses to that neuron, which influence the cell mean potential. The synapses are activated by arriving action potentials. The input signal to a cell i can be taken to be

$$\sum_j T_{ij}V_j \qquad [5]$$

where T_{ij} represents the effectiveness of a synapse. Fig. 1 thus

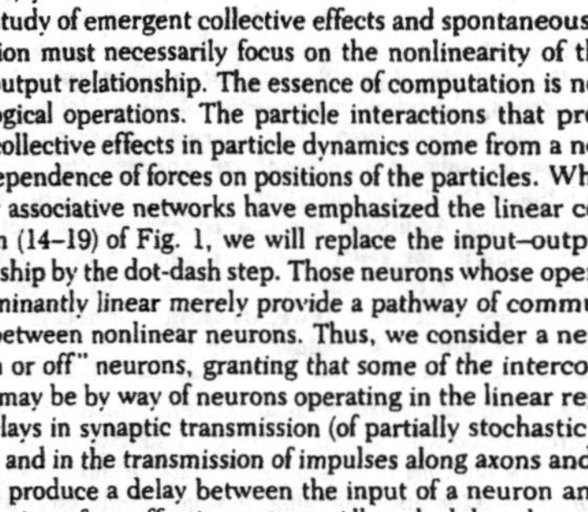

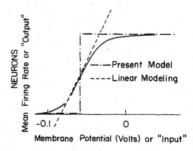

FIG. 1. Firing rate versus membrane voltage for a typical neuron (solid line), dropping to 0 for large negative potentials and saturating for positive potentials. The broken lines show approximations used in modeling.

becomes an input–output relationship for a neuron.

Little, Shaw, and Roney (8, 25, 26) have developed ideas on the collective functioning of neural nets based on "on/off" neurons and synchronous processing. However, in their model the relative timing of action potential spikes was central and resulted in reverberating action potential trains. Our model and theirs have limited formal similarity, although there may be connections at a deeper level.

Most modeling of neural learning networks has been based on synapses of a general type described by Hebb (27) and Eccles (28). The essential ingredient is the modification of T_{ij} by correlations like

$$\Delta T_{ij} = [V_i(t)V_j(t)]_{\text{average}} \qquad [6]$$

where the average is some appropriate calculation over past history. Decay in time and effects of $[V_i(t)]_{\text{avg}}$ or $[V_j(t)]_{\text{avg}}$ are also allowed. Model networks with such synapses (16, 20, 21) can construct the associative T_{ij} of Eq. 2. We will therefore initially assume that such a T_{ij} has been produced by previous experience (or inheritance). The Hebbian property need not reside in single synapses; small groups of cells which produce such a net effect would suffice.

The network of cells we describe performs an abstract calculation and, for applications, the inputs should be appropriately coded. In visual processing, for example, feature extraction should previously have been done. The present modeling might then be related to how an entity or *Gestalt* is remembered or categorized on the basis of inputs representing a collection of its features.

Studies of the collective behaviors of the model

The model has stable limit points. Consider the special case $T_{ij} = T_{ji}$, and define

$$E = -\frac{1}{2}\sum\sum_{i \neq j} T_{ij}V_iV_j \ . \qquad [7]$$

ΔE due to ΔV_i is given by

$$\Delta E = -\Delta V_i \sum_{j \neq i} T_{ij}V_j \ . \qquad [8]$$

Thus, the algorithm for altering V_i causes E to be a monotonically decreasing function. State changes will continue until a least (local) E is reached. This case is isomorphic with an Ising model. T_{ij} provides the role of the exchange coupling, and there is also an external local field at each site. When T_{ij} is symmetric but has a random character (the spin glass) there are known to be many (locally) stable states (29).

Monte Carlo calculations were made on systems of $N = 30$ and $N = 100$, to examine the effect of removing the $T_{ij} = T_{ji}$ restriction. Each element of T_{ij} was chosen as a random number between -1 and 1. The neural architecture of typical cortical regions (30, 31) and also of simple ganglia of invertebrates (32) suggests the importance of 100–10,000 cells with intense mutual interconnections in elementary processing, so our scale of N is slightly small.

The dynamics algorithm was initiated from randomly chosen initial starting configurations. For $N = 30$ the system never displayed an ergodic wandering through state space. Within a time of about $4/W$ it settled into limiting behaviors, the commonest being a stable state. When 50 trials were examined for a particular such random matrix, all would result in one of two or three end states. A few stable states thus collect the flow from most of the initial state space. A simple cycle also occurred occasionally—for example, $\cdots A \to B \to A \to B \cdots$.

The third behavior seen was chaotic wandering in a small region of state space. The Hamming distance between two binary states A and B is defined as the number of places in which the digits are different. The chaotic wandering occurred within a short Hamming distance of one particular state. Statistics were done on the probability p_i of the occurrence of a state in a time of wandering around this minimum, and an entropic measure of the available states M was taken

$$\ln M = -\sum p_i \ln p_i \ . \qquad [9]$$

A value of $M = 25$ was found for $N = 30$. *The flow in phase space produced by this model algorithm has the properties necessary for a physical content-addressable memory* whether or not T_{ij} is symmetric.

Simulations with $N = 100$ were much slower and not quantitatively pursued. They showed qualitative similarity to $N = 30$.

Why should stable limit points or regions persist when $T_{ij} \neq T_{ji}$? If the algorithm at some time changes V_i from 0 to 1 or vice versa, the change of the energy defined in Eq. 7 can be split into two terms, one of which is always negative. The second is identical if T_{ij} is symmetric and is "stochastic" with mean 0 if T_{ij} and T_{ji} are randomly chosen. The algorithm for $T_{ij} \neq T_{ji}$ therefore changes E in a fashion similar to the way E would change in time for a symmetric T_{ij} but with an algorithm corresponding to a finite temperature.

About 0.15 N states can be simultaneously remembered before error in recall is severe. Computer modeling of memory storage according to Eq. 2 was carried out for $N = 30$ and $N = 100$. n random memory states were chosen and the corresponding T_{ij} was generated. If a nervous system preprocessed signals for efficient storage, the preprocessed information would appear random (e.g., the coding sequences of DNA have a random character). The random memory vectors thus simulate efficiently encoded real information, as well as representing our ignorance. The system was started at each assigned nominal memory state, and the state was allowed to evolve until stationary.

Typical results are shown in Fig. 2. The statistics are averages over both the states in a given matrix and different matrices. With $n = 5$, the assigned memory states are almost always stable (and exactly recallable). For $n = 15$, about half of the nominally remembered states evolved to stable states with less than 5 errors, but the rest evolved to states quite different from the starting points.

These results can be understood from an analysis of the effect of the noise terms. In Eq. 3, $H_i^{s'}$ is the "effective field" on neuron i when the state of the system is s', one of the nominal memory states. The expectation value of this sum, Eq. 4, is $\pm N/2$ as appropriate. The $s \neq s'$ summation in Eq. 2 contributes no mean, but has a rms noise of $[(n-1)N/2]^{1/2} \equiv \sigma$. For nN large, this noise is approximately Gaussian and the probability of an error in a single particular bit of a particular memory will be

$$P = \frac{1}{\sqrt{2\pi\sigma^2}} \int_{N/2}^{\infty} e^{-x^2/2\sigma^2}\, dx \qquad [10]$$

For the case $n = 10$, $N = 100$, $P = 0.0091$, the probability that a state had no errors in its 100 bits should be about $e^{-0.91} \approx 0.40$. In the simulation of Fig. 2, the experimental number was 0.6.

The theoretical scaling of n with N at fixed P was demonstrated in the simulations going between $N = 30$ and $N = 100$. The experimental results of half the memories being well retained at $n = 0.15\,N$ and the rest badly retained is expected to

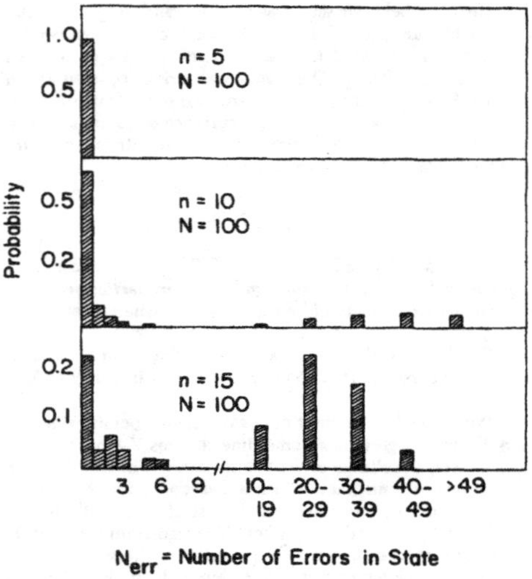

FIG. 2. The probability distribution of the occurrence of errors in the location of the stable states obtained from nominally assigned memories.

be true for all large N. The information storage at a given level of accuracy can be increased by a factor of 2 by a judicious choice of individual neuron thresholds. This choice is equivalent to using variables $\mu_i = \pm 1$, $T_{ij} = \Sigma_s \mu_i^s \mu_j^s$, and a threshold level of 0.

Given some arbitrary starting state, what is the resulting final state (or statistically, states)? To study this, evolutions from randomly chosen initial states were tabulated for $N = 30$ and $n = 5$. From the (inessential) symmetry of the algorithm, if (101110· · ·) is an assigned stable state, (010001· · ·) is also stable. Therefore, the matrices had 10 nominal stable states. Approximately 85% of the trials ended in assigned memories, and 10% ended in stable states of no obvious meaning. An ambiguous 5% landed in stable states very near assigned memories. There was a range of a factor of 20 of the likelihood of finding these 10 states.

The algorithm leads to memories near the starting state. For $N = 30$, $n = 5$, partially random starting states were generated by random modification of known memories. The probability that the final state was that closest to the initial state was studied as a function of the distance between the initial state and the nearest memory state. For distance ≤ 5, the nearest state was reached more than 90% of the time. Beyond that distance, the probability fell off smoothly, dropping to a level of 0.2 (2 times random chance) for a distance of 12.

The phase space flow is apparently dominated by attractors which are the nominally assigned memories, each of which dominates a substantial region around it. The flow is not entirely deterministic, and *the system responds to an ambiguous starting state by a statistical choice* between the memory states it most resembles.

Were it desired to use such a system in an Si-based content-addressable memory, the algorithm should be used and modified to hold the known bits of information while letting the others adjust.

The model was studied by using a "clipped" T_{ij}, replacing T_{ij} in Eq. 3 by ± 1, the algebraic sign of T_{ij}. The purposes were to examine the necessity of a linear synapse supposition (by making a highly nonlinear one) and to examine the efficiency of storage. Only $N(N/2)$ bits of information can possibly be stored in this symmetric matrix. Experimentally, for $N = 100$, $n = 9$, the level of errors was similar to that for the ordinary algorithm at $n = 12$. The signal-to-noise ratio can be evaluated analytically for this clipped algorithm and is reduced by a factor of $(2/\pi)^{1/2}$ compared with the unclipped case. For a fixed error probability, the number of memories must be reduced by $2/\pi$.

With the μ algorithm and the clipped T_{ij}, both analysis and modeling showed that the maximal information stored for $N = 100$ occurred at about $n = 13$. Some errors were present, and the Shannon information stored corresponded to about $N(N/8)$ bits.

New memories can be continually added to T_{ij}. The addition of new memories beyond the capacity overloads the system and makes all memory states irretrievable unless there is a provision for forgetting old memories (16, 27, 28).

The saturation of the possible size of T_{ij} will itself cause forgetting. Let the possible values of T_{ij} be 0, ± 1, ± 2, ± 3, and T_{ij} be freely incremented within this range. If $T_{ij} = 3$, a next increment of $+1$ would be ignored and a next increment of -1 would reduce T_{ij} to 2. When T_{ij} is so constructed, only the recent memory states are retained, with a slightly increased noise level. Memories from the distant past are no longer stable. How far into the past are states remembered depends on the digitizing depth of T_{ij}, and $0, \cdots, \pm 3$ is an appropriate level for $N = 100$. Other schemes can be used to keep too many memories from being simultaneously written, but this particular one is attractive because it requires no delicate balances and is a consequence of natural hardware.

Real neurons need not make synapses both of $i \rightarrow j$ and $j \rightarrow i$. Particular synapses are restricted to one sign of output. We therefore asked whether $T_{ij} = T_{ji}$ is important. Simulations were carried out with only one ij connection: if $T_{ij} \neq 0$, $T_{ji} = 0$. The probability of making errors increased, but the algorithm continued to generate stable minima. A Gaussian noise description of the error rate shows that the signal-to-noise ratio for given n and N should be decreased by the factor $1/\sqrt{2}$, and the simulations were consistent with such a factor. This same analysis shows that the system generally fails in a "soft" fashion, with signal-to-noise ratio and error rate increasing slowly as more synapses fail.

Memories too close to each other are confused and tend to merge. For $N = 100$, a pair of random memories should be separated by 50 ± 5 Hamming units. The case $N = 100$, $n = 8$, was studied with seven random memories and the eighth made up a Hamming distance of only 30, 20, or 10 from one of the other seven memories. At a distance of 30, both similar memories were usually stable. At a distance of 20, the minima were usually distinct but displaced. At a distance of 10, the minima were often fused.

The algorithm categorizes initial states according to the similarity to memory states. With a threshold of 0, the system behaves as a forced categorizer.

The state 00000 · · · is always stable. For a threshold of 0, this stable state is much higher in energy than the stored memory states and very seldom occurs. Adding a uniform threshold in the algorithm is equivalent to raising the effective energy of the stored memories compared to the 0000 state, and 0000 also becomes a likely stable state. The 0000 state is then generated by any initial state that does not resemble adequately closely one of the assigned memories and represents positive recognition that the starting state is not familiar.

Familiarity can be recognized by other means when the memory is drastically overloaded. We examined the case $N = 100$, $n = 500$, in which there is a memory overload of a factor of 25. None of the memory states assigned were stable. The initial rate of processing of a starting state is defined as the number of neuron state readjustments that occur in a time $1/2W$. Familiar and unfamiliar states were distinguishable most of the time at this level of overload on the basis of the initial processing rate, which was faster for unfamiliar states. This kind of familiarity can only be read out of the system by a class of neurons or devices abstracting average properties of the processing group.

For the cases so far considered, the expectation value of T_{ij} was 0 for $i \neq j$. A set of memories can be stored with average correlations, and $\overline{T}_{ij} = C_{ij} \neq 0$ because there is a consistent internal correlation in the memories. If now a partial new state X is stored

$$\Delta T_{ij} = (2X_i - 1)(2X_j - 1) \quad i,j \leq k < N \qquad [11]$$

using only k of the neurons rather than N, an attempt to reconstruct it will generate a stable point for all N neurons. The values of $X_{k+1} \cdots X_N$ that result will be determined primarily from the sign of

$$\sum_{j=1}^{k} c_{ij}\, x_j \qquad [12]$$

and X is completed according to the mean correlations of the other memories. The most effective implementation of this capacity stores a large number of correlated matrices weakly followed by a normal storage of X.

A nonsymmetric T_{ij} can lead to the possibility that a minimum will be only metastable and will be replaced in time by another minimum. Additional nonsymmetric terms which could be easily generated by a minor modification of Hebb synapses

$$\Delta T_{ij} = A \sum_s (2V_i^{s+1} - 1)(2V_j^s - 1) \qquad [13]$$

were added to T_{ij}. When A was judiciously adjusted, the system would spend a while near V_s and then leave and go to a point near V_{s+1}. But sequences longer than four states proved impossible to generate, and even these were not faithfully followed.

Discussion

In the model network each "neuron" has elementary properties, and the network has little structure. Nonetheless, collective computational properties spontaneously arose. Memories are retained as stable entities or *Gestalts* and can be correctly recalled from any reasonably sized subpart. Ambiguities are resolved on a statistical basis. Some capacity for generalization is present, and time ordering of memories can also be encoded. These properties follow from the nature of the flow in phase space produced by the processing algorithm, which does not appear to be strongly dependent on precise details of the modeling. This robustness suggests that similar effects will obtain even when more neurobiological details are added.

Much of the architecture of regions of the brains of higher animals must be made from a proliferation of simple local circuits with well-defined functions. The bridge between simple circuits and the complex computational properties of higher nervous systems may be the spontaneous emergence of new computational capabilities from the collective behavior of large numbers of simple processing elements.

Implementation of a similar model by using integrated circuits would lead to chips which are much less sensitive to element failure and soft-failure than are normal circuits. Such chips would be wasteful of gates but could be made many times larger than standard designs at a given yield. Their asynchronous parallel processing capability would provide rapid solutions to some special classes of computational problems.

The work at California Institute of Technology was supported in part by National Science Foundation Grant DMR-8107494. This is contribution no. 6580 from the Division of Chemistry and Chemical Engineering.

1. Willows, A. O. D., Dorsett, D. A. & Hoyle, G. (1973) *J. Neurobiol.* **4**, 207–237, 255–285.
2. Kristan, W. B. (1980) in *Information Processing in the Nervous System*, eds. Pinsker, H. M. & Willis, W. D. (Raven, New York), 241–261.
3. Knight, B. W. (1975) *Lect. Math. Life Sci.* **5**, 111–144.
4. Smith, D. R. & Davidson, C. H. (1962) *J. Assoc. Comput. Mach.* **9**, 268–279.
5. Harmon, L. D. (1964) in *Neural Theory and Modeling*, ed. Reiss, R. F. (Stanford Univ. Press, Stanford, CA), pp. 23–24.
6. Amari, S.-I. (1977) *Biol. Cybern.* **26**, 175–185.
7. Amari, S.-I. & Akikazu, T. (1978) *Biol. Cybern.* **29**, 127–136.
8. Little, W. A. (1974) *Math. Biosci.* **19**, 101–120.
9. Marr, J. (1969) *J. Physiol.* **202**, 437–470.
10. Kohonen, T. (1980) *Content Addressable Memories* (Springer, New York).
11. Palm, G. (1980) *Biol. Cybern.* **36**, 19–31.
12. McCulloch, W. S. & Pitts, W. (1943) *Bull. Math Biophys.* **5**, 115–133.
13. Minsky, M. & Papert, S. (1969) *Perceptrons: An Introduction to Computational Geometry* (MIT Press, Cambridge, MA).
14. Rosenblatt, F. (1962) *Principles of Perceptrons* (Spartan, Washington, DC).
15. Cooper, L. N. (1973) in *Proceedings of the Nobel Symposium on Collective Properties of Physical Systems*, eds. Lundqvist, B. & Lundqvist, S. (Academic, New York), 252–264.
16. Cooper, L. N., Liberman, F. & Oja, E. (1979) *Biol. Cybern.* **33**, 9–28.
17. Longuet-Higgins, J. C. (1968) *Proc. Roy. Soc. London Ser. B* **171**, 327–334.
18. Longuet-Higgins, J. C. (1968) *Nature (London)* **217**, 104–105.
19. Kohonen, T. (1977) *Associative Memory—A System-Theoretic Approach* (Springer, New York).
20. Willwacher, G. (1976) *Biol. Cybern.* **24**, 181–198.
21. Anderson, J. A. (1977) *Psych. Rev.* **84**, 413–451.
22. Perkel, D. H. & Bullock, T. H. (1969) *Neurosci. Res. Symp. Summ.* **3**, 405–527.
23. John, E. R. (1972) *Science* **177**, 850–864.
24. Roney, K. J., Scheibel, A. B. & Shaw, G. L. (1979) *Brain Res. Rev.* **1**, 225–271.
25. Little, W. A. & Shaw, G. L. (1978) *Math. Biosci.* **39**, 281–289.
26. Shaw, G. L. & Roney, K. J. (1979) *Phys. Rev. Lett.* **74**, 146–150.
27. Hebb, D. O. (1949) *The Organization of Behavior* (Wiley, New York).
28. Eccles, J. G. (1953) *The Neurophysiological Basis of Mind* (Clarendon, Oxford).
29. Kirkpatrick, S. & Sherrington, D. (1978) *Phys. Rev.* **17**, 4384–4403.
30. Mountcastle, V. B. (1978) in *The Mindful Brain*, eds. Edelman, G. M. & Mountcastle, V. B. (MIT Press, Cambridge, MA), pp. 36–41.
31. Goldman, P. S. & Nauta, W. J. H. (1977) *Brain Res.* **122**, 393–413.
32. Kandel, E. R. (1979) *Sci. Am.* **241**, 61–70.

PHYSICAL REVIEW A VOLUME 32, NUMBER 2 AUGUST 1985

Spin-glass models of neural networks

Daniel J. Amit and Hanoch Gutfreund
Racah Institute of Physics, Hebrew University, 91904 Jerusalem, Israel

H. Sompolinsky
Department of Physics, Bar-Ilan University, 52100 Ramat-Gan, Israel
(Received 22 March 1985)

Two dynamical models, proposed by Hopfield and Little to account for the collective behavior of neural networks, are analyzed. The long-time behavior of these models is governed by the statistical mechanics of infinite-range Ising spin-glass Hamiltonians. Certain configurations of the spin system, chosen at random, which serve as memories, are stored in the quenched random couplings. The present analysis is restricted to the case of a finite number p of memorized spin configurations, in the thermodynamic limit. We show that the long-time behavior of the two models is identical, for all temperatures below a transition temperature T_c. The structure of the stable and metastable states is displayed. Below T_c, these systems have $2p$ ground states of the Mattis type: Each one of them is fully correlated with one of the stored patterns. Below $T \sim 0.46T_c$, additional dynamically stable states appear. These metastable states correspond to specific mixings of the embedded patterns. The thermodynamic and dynamic properties of the system in the cases of more general distributions of random memories are discussed.

I. INTRODUCTION

A. Models of neural networks

Recently, a number of models have been proposed which view the human memory as a collective property of large interconnected neural networks. Two such models, one proposed recently by Hopfield[1] and a closely related model, proposed some ten years ago by Little,[2] are the focus of this paper. We briefly review the physiological background of these models. For more details the reader is referred to Refs. 1 and 2 and to a recent study of these models by Peretto.[3] In both models each neuron is viewed as an Ising spin with two possible states: an "up" position or a "down" position depending on whether the neuron has, or has not, fired an electrochemical signal (within an interval of time of the order of a millisecond). The state of the network of N such neurons at time t is defined as the instantaneous configuration of all the spin variables at time t:

$$|\alpha,t\rangle = |S_1^\alpha, S_2^\alpha, \ldots, S_N^\alpha; t\rangle \ . \tag{1.1}$$

The dynamic evolution of these states, in the phase space of 2^N states, is determined by the interactions among the neurons. The neurons are interconnected by synaptic junctions of strength J_{ij}, which determine the contribution of a signal fired by the jth neuron to the postsynaptic potential which acts on the ith neuron. This contribution can be either positive (excitatory synapse) or negative (inhibitory synapse). The potential V_i on each neuron is the sum of all postsynaptic potentials delivered to it in an integrating period of time, of the order of a few milliseconds, i.e.,

$$V_i = \sum_j J_{ij}(S_j + 1) \ . \tag{1.2}$$

In the absence of noise, or external perturbation, each neuron fires a signal if its potential V_i exceeds a threshold value U_i. Thus the stable states of the network will be those configurations in which each of the spin variables S_i is aligned with its molecular field $h_i = V_i - U_i$, i.e.,

$$S_i h_i = S_i(V_i - U_i) > 0 \ . \tag{1.3}$$

It will be assumed throughout the paper that the J_{ij}'s are symmetric, i.e., $J_{ij} = J_{ji}$. In such a case Eq. (1.3) is equivalent to the requirement that the configurations $\{S_i\}$ be local minima (i.e., stable to all single-spin flips) of the Hamiltonian

$$H = -\tfrac{1}{2}\sum_i h_i S_i = -\tfrac{1}{2}\sum_{i,j} J_{ij} S_i S_j \ , \tag{1.4}$$

where it is usually assumed that the threshold potentials satisfy $U_i \simeq \sum_j J_{ij}$. Thus there is no external field term in H. In the presence of noise there is a finite probability of having configurations other than those given by Eq. (1.3). This can be taken into account by introducing an effective temperature $1/\beta$, characterizing the level of noise in the system,[4] as will be described below.

For the network to have a capacity for learning and memory its stable configurations must be correlated with certain configurations, which are determined by the learning process. This is achieved by choosing the interactions J_{ij} to be given by

$$J_{ij} = \frac{1}{N}\sum_{\mu=1}^{p} \xi_i^\mu \xi_j^\mu, \quad i \neq j \ . \tag{1.5}$$

The p sets of $\{\xi_i^\mu\}$ are certain configurations of the network which were fixed by the learning process. The ξ_i^μ are taken to be quenched random variables, assuming the values $+1$ and -1 with equal probabilities. Note that according to Eq. (1.5) every pair of neurons is connected.

The model (1.3)–(1.5) will have the capacity of storage and retrieval of information if indeed the emergent dynamically stable configurations $\{S_i\}$ are correlated with the "learned memories" $\{\xi_i^\mu\}$. This question is at the center of the present study. However, in order to complete the definition of the model one has to prescribe a dynamic mechanism, by which the network evolves from an arbitrary initial condition.

B. The generalized Hopfield model and the Little model

Hopfield's dynamic model is essentially a $T=0$ Monte Carlo[5] (or Glauber) dynamics. Starting from an arbitrary initial configuration the system evolves by a sequence of single-spin flips, involving spins which are misaligned with their instantaneous molecular fields. This process monotonically decreases the value of H, (1.4), and leads eventually to steady states, which are the local minima of (1.6). A natural generalization of this model to a system with noise is to adopt Glauber single-spin dynamics at a finite temperature $1/\beta$. The distribution of configurations (1.1) relaxes, in this case to a Gibbs distribution

$$P\{S\} \propto \exp(-\beta H\{S\}) , \tag{1.6}$$

with H of (1.4). We refer to this finite-temperature model as the generalized Hopfield model. Note that stability of a state to all single-spin flips is not sufficient for dynamic stability at finite temperatures.

In Little's model the probability that the ith spin be in a state S_i' at time $t+\delta t$, given a configuration $\{S_i\}$ of the network at time t, is proportional to

$$P(S_i') = \frac{\exp(-\beta S_i' h_i\{S_i\})}{\exp(-\beta S_i' h_i\{S_i\}) + \exp(+\beta S_i' h_i\{S_i\})} , \tag{1.7}$$

where $h_i = V_i - U_i = \sum_j J_{ij} S_j$, as before. The matrix W, of transition probabilities from the state $|\alpha, t\rangle$ to $|\beta, t+\delta t\rangle$, is just a product of the probabilities (1.7), i.e.,

$$\langle \beta | W | \alpha \rangle = \prod_{i=1}^{N} [\tfrac{1}{2} e^{-\beta h_i^\alpha S_i^\beta} \operatorname{sech}(\beta h_i^\alpha S_i^\beta)] . \tag{1.8}$$

Thus, at each time step all the spins check simultaneously their states against their molecular field. Each step may consist of many, even N, spin flips.

It has been shown by Peretto[3] that as long as J_{ij} is symmetric, the master equation, based on the transition rates given by (1.8), obeys detailed balance and hence leads to a stationary Gibbs distribution of states, $\exp(-\beta \bar{H})$ with the effective Hamiltonian,

$$\bar{H} = -\frac{1}{\beta} \sum_i \ln \left[2\cosh \left| \beta \sum_j J_{ij} S_j \right| \right] . \tag{1.9}$$

The synchronous dynamics of this model seems at first glance to lead to a very different collective behavior from the asynchronous mechanism of Hopfield. The dynamics of real systems is, most probably, in between. Thus it is important to investigate to what extent this difference is relevant.

C. Relationship to models of random magnets

The study of the models described above is interesting not only in the context of models of memory but also in the context of the statistical mechanics of disordered magnetic systems. The Hamiltonian defined in (1.4) and (1.5) is a special case of infinite-range spin glasses[6] where every pair of spins is interacting via a quenched random exchange J_{ij}. In the canonical infinite-ranged spin-glass model, introduced by Sherrington and Kirkpatrick[7] (SK), each J_{ij} is an independent random variable. In this model the disorder leads to the appearance of an infinite number of ground states (in the $N\rightarrow\infty$ limit) with static and dynamic properties, which are very different from the usual ferromagnetic case.

The other extreme is the case of (1.5) with $p=1$, which is an infinite-range Mattis model.[8] Here the disorder can be gauged away, hence it is irrelevant thermodynamically. There are two ground states ($\{S_i\} = \pm\{\xi_i\}$) with no frustration: Each "bond" $S_i S_j J_{ij}$ in the ground state is positive. The model (1.4) and (1.5) with $p>1$ represents an intermediate case. There is always a finite fraction of frustrated bonds. Nevertheless, the correlation between the bonds may be sufficiently strong to yield a structure of broken symmetry phase considerably simpler than that of the SK model. Van Hemmen[9] introduced and solved a related model with $p=2$. His mean-field equation has been generalized to arbitrary p by Provost and Vallee.[10] However, the structure and the properties of the mean-field solutions for general p have not been investigated, nor have the possible existence and the properties of metastable states. These issues are the main focus of this paper.

A similar model was also studied in the context of the mean-field theory of random-axis ferromagnets.[11] In the limit of strong local anisotropy the system is mapped onto a spin-glass Ising model with $J_{ij} \propto \hat{\mathbf{n}}_i \cdot \hat{\mathbf{n}}_j$ where $\hat{\mathbf{n}}_i$ is the direction of the local easy axis. These J_{ij} are similar to Eq. (1.5) but with ξ_i^μ which are the Cartesian components of random unit vectors, rather than independent and discrete random variables. This raises the issue of the sensitivity of the properties of the models to the form of the distribution of ξ_i^μ.

D. Outline and summary of results

In this paper a statistical mechanical study of the Hopfield and the Little models is presented. This study is explicitly restricted to the limit $N\rightarrow\infty$ and finite p. In Sec. II the solutions of the mean-field theory of the Hopfield model are studied. It is shown that at all $T<T_c=1$ the free energy ground states are all Mattis states: Each one of them is correlated with one of the p memories, $\{\xi_i^\mu\}$. At $T=1$, additional mean-field solutions with higher free energy appear. These are *symmetric* states which have equal overlap with several memories.

Section III presents the stability analysis of these solutions. It is shown that as the temperature is decreased below

$$T \simeq 0.461 \tag{1.10}$$

some of these solutions become local minima. At $T=0$ all symmetric saddle points which overlap with an odd number of memories become local minima. These states are truly metastable: They are separated by free energy barriers proportional to N.

In Sec. IV mean-field *asymmetric* solutions which have unequal overlaps on some memories are discussed. They appear only below $T\sim 0.57$ and none of them are stable at temperatures higher than that of (1.10).

The properties of the Little model are studied in Sec. V. We show that the model has the same thermodynamic properties as the Hopfield model, including the properties of the metastable states.

It has been remarked[3] that in the Little model the system may in certain cases get trapped in indefinite oscillations between several configurations, unable to reach the aligned states given by (1.3). We show that with the J_{ij}'s given by (1.5) this does not happen.

In Sec. VI we consider distributions of $\{\xi_i^\mu\}$ other than ± 1. It is shown that the low-temperature properties of the models depend strongly on the details of the distribution of $\{\xi_i^\mu\}$. If the probability density at $\xi_i^\mu=0$ is sufficiently large, the "mixed" states, rather than the Mattis states, become the ground states of the system. On the other hand, certain continuous distributions may eliminate the "mixed" states, leaving the Mattis states as the only dynamically stable states of the system.

II. THE GENERALIZED HOPFIELD MODEL

A. Mean-field theory

We now turn to the investigation of the thermodynamics of the Hamiltonian

$$H=-\tfrac{1}{2}\sum_{\substack{i,j\\ i\neq j}}\left[\left(\frac{1}{N}\sum_{\mu=1}^{p}\xi_i^\mu\xi_j^\mu\right)S_iS_j\right], \tag{2.1}$$

where ξ_i^μ are independent random variables with zero mean. This system will be studied in the limit $N\to\infty$ and finite p. The ensemble averaged free energy density is given by

$$-\beta f(\beta)=\lim_{N\to\infty}\,[N^{-1}\langle\langle \ln \operatorname{Tr}_s \exp(-\beta H)\rangle\rangle]\,, \tag{2.2}$$

where $\beta\equiv 1/T$ (with units in which $k_B=1$). The notation $\langle\langle\cdots\rangle\rangle$ stands for the average over the distribution of $\{\xi_i^\mu\}$. The partition function is rewritten, for a given realization of the ξ's, as

$$Z=\operatorname{Tr}_s\exp(-\beta H)=\exp(-\beta p/2)\operatorname{Tr}_s\exp\left[(\beta/2N)\sum_\mu\left[\sum_i S_i\xi_i^\mu\right]^2\right]$$
$$=(N\beta)^{p/2}e^{-\beta p/2}\int\prod_\mu\frac{dm^\mu}{\sqrt{2\pi}}\exp\left[\frac{-N\beta\mathbf{m}^2}{2}+\sum_i\ln[2\cosh(\beta\mathbf{m}\cdot\boldsymbol{\xi}_i)]\right], \tag{2.3}$$

where a vector notation for the p components of ξ_i^μ and m^μ has been introduced. As long as p remains finite, the integral over m is dominated by its saddle-point value,

$$-\frac{1}{N\beta}\ln Z=\tfrac{1}{2}\mathbf{m}^2-\frac{1}{N\beta}\sum_i\ln[2\cosh(\beta\mathbf{m}\cdot\boldsymbol{\xi}_i)]\,. \tag{2.4}$$

The order parameter $\mathbf{m}$ is determined by the saddle-point equations $\partial\ln Z/\partial m^\mu=0$,

$$\mathbf{m}=\frac{1}{N}\sum_i\boldsymbol{\xi}_i\tanh(\beta\mathbf{m}\cdot\boldsymbol{\xi}_i)\,. \tag{2.5}$$

At any finite N the right-hand sides of Eqs. (2.4) and (2.5) depend on the particular realization of $\{\xi_i^\mu\}$. However, in the limit $N\to\infty$ the random fluctuations are suppressed and both $\ln Z$ and $\mathbf{m}$ are self-averaged, as discussed in Refs. 9 and 10. The sums $(1/N)\sum_i$ are, therefore, replaced by averages over $\{\xi_i\}$, leading to the mean-field equations[10]

$$f(\beta)=\tfrac{1}{2}\mathbf{m}^2-\frac{1}{\beta}\langle\langle\ln[2\cosh(\beta\mathbf{m}\cdot\boldsymbol{\xi})]\rangle\rangle\,, \tag{2.6}$$

$$\mathbf{m}=\langle\langle\boldsymbol{\xi}\tanh(\beta\mathbf{m}\cdot\boldsymbol{\xi})\rangle\rangle\,. \tag{2.7}$$

To interpret the order parameter $\mathbf{m}$ one adds an external source conjugate to $\xi_i^\mu S_i$, to find that $\mathbf{m}$ is just the average overlap between the local magnetization and the ξ's. Explicitly, one has

$$m^\mu=\langle\langle\xi_i^\mu\langle S_i\rangle\rangle\rangle\,, \tag{2.8}$$

where

$$\langle S_i\rangle=\tanh(\beta\mathbf{m}\cdot\boldsymbol{\xi}_i) \tag{2.9}$$

is the *thermal* average of the spin at site i.

The detailed structure of the solutions of Eq. (2.7) is essential for the determination of the correlations of the spin states $\{\langle S_i\rangle\}$ with each of the p quenched "memories" $\{\xi_i^\mu\}$.

B. The Mattis states

We will restrict ourselves in this subsection to the case in which the distribution of ξ_i is given by

$$P\{\xi_i^\mu\}=\prod_{\mu,i}p(\xi_i^\mu)\,,$$
$$p(\xi_i^\mu)=\tfrac{1}{2}\delta(\xi_i^\mu-1)+\tfrac{1}{2}\delta(\xi_i^\mu+1)\,. \tag{2.10}$$

Expanding Eqs. (2.6) and (2.7) in powers of $\mathbf{m}$ one obtains

$$f=-T\ln 2+\tfrac{1}{2}(1-\beta)\mathbf{m}^2+O(m^4)\,, \tag{2.11}$$

$$m^\mu=\beta m^\mu+\tfrac{2}{3}\beta^3(m^\mu)^3-\beta^3 m^\mu\mathbf{m}^2+O(m^4)\,,\quad \mu=1,2,\ldots,p \tag{2.12}$$

from which it is seen that above $T=1$ the only solution is

the paramagnetic state $\mathbf{m}=0$, with $f=-T\ln 2$. This solution becomes unstable below $T_c=1$ where solutions with nonzero m appear. *We will denote by n the dimensionality of* $\mathbf{m}$, *i.e., the number of nonzero components of* $\mathbf{m}$ *in a particular solution below* T_c. It is clear from Eqs. (2.7) and the form (2.10) that permuting the m^μ's or changing the signs of each of the n nonzero components independently generates entirely equivalent solutions. Hence without loss of generality we can restrict ourselves to solutions in which *the first n components* (i.e., m^μ with $\mu=1,2,\ldots,n$) *are positive;* the rest of them are zero.

We first discuss solutions with $n=1$. Assuming $m^\mu=0$ for all $\mu>1$,

$$f=\tfrac{1}{2}(m^1)^2-\frac{1}{\beta}\ln[2\cosh(\beta m^1)]\,, \tag{2.13}$$

$$m^1=\tanh(\beta m^1)\,. \tag{2.14}$$

These are the usual mean-field equations of Ising ferromagnets. Indeed, this solution corresponds to a state in which all the local magnetizations (up to a negligible fraction as $N\to\infty$) are equal to

$$\langle S_i\rangle=\xi_i^1\tanh(\beta m^1)\,. \tag{2.15}$$

This state is thermodynamically equivalent (via a Mattis transformation[8]) to the ferromagnetic state. There are $2p$ equivalent states of this form corresponding to different μ's and different signs of m. We refer to these states as Mattis states.

The Mattis states are the global minima of the free energy both near $T=1$ and $T=0$ and most probably at all $T<1$. Near $T=1$, we show explicitly in the next subsection that the Mattis free energy is the lowest of all other saddle points [see Eq. (2.29)]. At $T=0$, Eqs. (2.13) and (2.14) read

$$\mathbf{m}(T=0)=(1,0,0,\ldots,0)\,, \tag{2.16}$$

$$E(T=0)=-\tfrac{1}{2}\,. \tag{2.17}$$

To show that $-\frac{1}{2}$ is the ground-state energy at $T=0$, note that the general mean-field equations [Eqs. (2.6) and (2.7)] yields as $T\to 0$ the following:

$$E=-\tfrac{1}{2}\mathbf{m}^2\,, \tag{2.18}$$

$$\mathbf{m}=\langle\langle\, \xi\,\mathrm{sgn}(\mathbf{m}\cdot\xi)\rangle\rangle\,. \tag{2.19}$$

We have used here the limits

$$\begin{aligned}&\tanh(\beta\mathbf{m}\cdot\xi)\to\mathrm{sgn}(\mathbf{m}\cdot\xi)\,,\\ &T\ln[2\cosh(\beta\mathbf{m}\cdot\xi)]\to|\mathbf{m}\cdot\xi|\end{aligned} \tag{2.20}$$

which, by defining $\mathrm{sgn}(0)=0$, apply also to the case where $\mathbf{m}\cdot\xi$ may take the value zero, as long as the distribution of ξ is *discrete*. Each component m^μ is, of course, bounded from above by 1. However, $\mathbf{m}$ obeys a stronger bound which is

$$\mathbf{m}^2\le 1 \tag{2.21}$$

with the equality being satisfied only for a one-component $\mathbf{m}$. The bound (2.21) can be derived using Eq. (2.19) and the Schwartz inequality,

$$\begin{aligned}\mathbf{m}^2&=\langle\langle\, |\xi\cdot\mathbf{m}|\,\rangle\rangle\\ &\le[\langle\langle(\xi\cdot\mathbf{m})^2\rangle\rangle]^{1/2}\\ &=\left[\sum_{\mu,\nu=1}^{n} m^\mu m^\nu\langle\langle\xi^\mu\xi^\nu\rangle\rangle\right]^{1/2}=(\mathbf{m}^2)^{1/2}\,,\end{aligned}$$

which implies that $\mathbf{m}^2$ is less than unity for $n>1$ and equals unity for $n=1$.[12]

From the point of view of storage and retrieval of memory these states are ideal, since each of them is fully correlated with one of the "quenched" memories. However, although the Mattis states are the only states which contribute to the thermodynamics of the system, solutions with $n>1$ may also be important for the dynamics, if they are local minima of the free energy.

C. Symmetric solutions

A particularly simple class of solutions of Eqs. (2.7) consists of those in which all n nonzero components of $\mathbf{m}$ are equal in magnitude, i.e.,

$$\mathbf{m}=m_n(1,1,\ldots,1,0,0,\ldots,0)\,, \tag{2.22}$$

where the first n components are unity and the remaining $p-n$ zeros. For a given n there are $\binom{p}{n}2^n$ solutions which are equivalent to that of (2.22). These symmetric solutions are important because they are the only solutions that exist throughout the whole temperature range $T<1$. The transition temperatures for the appearance of asymmetric solutions, in which some of the n components have different magnitudes, are all lower than 1. To see this note that dividing each of the first n equations of (2.12) by m^μ results in $\frac{2}{3}(m^\mu)^2\simeq T-1+\mathbf{m}^2$ *which is independent of* μ. It is straightforward to see that the equality of all nonzero $(m^\mu)^2$ holds order by order in perturbation theory about $T=1$ to all orders. Thus the breaking of the symmetry among $(m^\mu)^2$ occurs below a critical temperature which is less than 1. In fact, we will show in Sec. IV that the critical temperatures for the appearance of the asymmetric solutions are all lower than $T\sim 0.57$.

The mean-field equations for the symmetric states (2.22) are

$$f_n=\frac{n}{2}m_n^2-\frac{1}{\beta}\langle\langle\ln[2\cosh(\beta m_n z_n)]\rangle\rangle\,, \tag{2.23}$$

$$m_n=(1/n)\langle\langle z_n\tanh(\beta m_n z_n)\rangle\rangle\,, \tag{2.24}$$

where

$$z_n^i=\sum_{\mu=1}^{n}\xi_i^\mu\,.$$

The distribution of z_n^i is given, according to Eq. (2.10), by

$$p(z_n)=2^{-n}\begin{vmatrix}n\\k\end{vmatrix}\,, \tag{2.25}$$

where

$$k=(z_n+n)/2$$

is the number of positive ξ_i^μ's contributing to z_n^i. Incidentally, we note that Eq. (2.25) is the distribution of a random walk on a one-dimensional lattice. These solutions correspond to states in which the local magnetization is induced by a molecular field

$$h_i = m_n z_n^i ,$$

i.e.,

$$\langle S_i \rangle = \tanh(\beta m_n z_n^i) . \quad (2.26)$$

In other words, the symmetric solutions with $n>1$ represent states which are equal mixtures of several memories.

To evaluate the solutions near T_c we expand Eqs. (2.23) and (2.24) in powers of m and obtain

$$\bar{f}_n \equiv \beta f_n - \ln 2$$

$$\simeq \frac{n}{2}(T-1)(\beta m_n)^2 + \tfrac{1}{12}(\beta m_n)^4 \langle\langle z_n^4 \rangle\rangle , \quad (2.27)$$

$$m_n \simeq \beta m_n - \frac{\langle\langle z_n^4 \rangle\rangle}{3n}(\beta m_n)^3 . \quad (2.28)$$

Using the equality $\langle\langle z_n^4 \rangle\rangle = n(3n-2)$ one obtains the final results

$$\bar{f}_n \simeq -\frac{3nt^2}{4(3n-2)} , \quad (2.29)$$

$$m_n^2 \simeq \frac{3t}{3n-2} , \quad (2.30)$$

where $t \equiv 1-T$. Thus, $T=1$ is the critical temperature for the appearance of all symmetric solutions. Equations (2.29) imply that near $T=1$ the free energy is *monotonically increasing with n;* the lowest free energy state is $n=1$, namely the Mattis states with

$$m_1^2 \simeq 3t \text{ and } \bar{f}_1 \simeq -\frac{3t^2}{4} .$$

To study the symmetric solutions near $T=0$, we use Eqs. (2.18) and (2.19) to obtain

$$m_n(T=0) = \frac{1}{n}\langle\langle |z_n| \rangle\rangle , \quad (2.31)$$

$$f_n(T=0) = -\tfrac{1}{2} n m_n^2 . \quad (2.32)$$

Using (2.25) one arrives, for even n, at

$$m_{2k} = \frac{1}{2^{2k}}\binom{2k}{k} ,$$

$$f_{2k} = -\frac{2k}{2^{4k+1}}\binom{2k}{k}^2 , \quad k=1,2,\ldots \quad (2.33)$$

and for odd n

$$m_{2k+1} = \frac{1}{2^{2k}}\binom{2k}{k} ,$$

$$f_{2k+1} = -\frac{2k+1}{2^{4k+1}}\binom{2k}{k}^2 , \quad k=0,1,\ldots . \quad (2.34)$$

This sequence of f_n is bounded from below by the ground-state energy $f_1=-0.5$ and from above by $f_2=-0.25$. Moreover, the sequence (2.33) is monotonically decreasing with k, while the sequence (2.34) is monotonically increasing. Both have a common limit ($=-1/\pi$) as $k\to\infty$. This limit coincides with the ground-state energy per spin for a Gaussian distribution of ξ^μ (see Sec. VI). Details of the derivations of these results are left to Appendix A.

In conclusion, we obtain the following ordering of the energies of the symmetric saddle points at $T=0$:

$$f_1 < f_3 < f_5 < \cdots f_\infty \cdots < f_6 < f_4 < f_2 . \quad (2.35)$$

Note that in the "even" solutions there is a finite probability of $z_n=0$. Hence, a finite fraction of the spins remain disordered at all temperatures. This difference between the odd and even solutions manifests itself in the low-temperature value of the Edwards-Anderson order parameter,[13]

$$q_n = \langle\langle \langle S_i \rangle^2 \rangle\rangle = \langle\langle \tanh^2(\beta m_n z_n) \rangle\rangle . \quad (2.36)$$

In the case of odd n the minimum value of $|z|$ is 1. Hence,

$$q_n \simeq 1 - 2p(z_n=1)\exp(-2\beta m_n) \to 1$$
$$\text{as } \beta\to\infty \text{ (or } T\to 0\text{)}, \quad (2.37)$$

whereas for even n one has

$$q_n = 1 - p(z_n=0) \text{ at } T=0 . \quad (2.38)$$

In Sec. III we study the stability properties of the various symmetric saddle points in order to determine their importance to the dynamics. The asymmetric solutions will be studied in Sec. IV.

III. METASTABILITY IN THE GENERALIZED HOPFIELD MODEL

A. The stability matrix of the symmetric solutions

The local stability of the saddle points of f, Eq. (2.6), is determined by the eigenvalues of the matrix $\underline{A}$,

$$A^{\mu\nu} = \frac{\partial^2 f}{\partial m^\mu \partial m^\nu} = \delta^{\mu\nu} - \beta(\delta^{\mu\nu} - Q^{\mu\nu}) , \quad (3.1)$$

with

$$Q^{\mu\nu} \equiv \langle\langle \xi^\mu \xi^\nu \tanh^2(\beta \mathbf{m}\cdot\boldsymbol{\xi}) \rangle\rangle . \quad (3.2)$$

Solutions of Eq. (2.7) are locally stable if all the eigenvalues of A are positive.

The general form of $\underline{A}$ in the case of the symmetric solutions is quite simple. Its diagonal elements are all

$$A^{\mu\mu} = 1 - \beta(1-q) ,$$

where $q = Q^{\mu\mu}$ is given by Eq. (2.36). The off-diagonal elements *with* $\mu,\nu \le n$ are all equal to

$$\beta Q = \beta \langle\langle \xi^1 \xi^2 \tanh^2(\beta m_n z_n) \rangle\rangle \quad (3.3)$$

and all other elements vanish. Recall that we have chosen a solution which has the form (2.22).

The matrix $\underline{A}$ has three groups of eigenvalues: (1) a nondegenerate eigenvalue

$$\lambda_1 = 1-\beta(1-q)+\beta(n-1)Q \tag{3.4}$$

corresponding to "longitudinal" fluctuations in the amplitude m_n; (2) an eigenvalue of degeneracy $p-n$,

$$\lambda_2 = 1-\beta(1-q)\,, \tag{3.5}$$

which corresponds to fluctuations in directions which mix more memories; and (3) an eigenvalue of degeneracy $n-1$,

$$\lambda_3 = 1-\beta(1-q)-\beta Q\,, \tag{3.6}$$

which is associated with fluctuations of anisotropy in the space of the n "occupied" memories. Since Q is positive for all $T<1$ (see Appendix B), the lowest eigenvalue of $\underline{A}$ is λ_3. It is this eigenvalue which determines the stability of the solutions. This is the case for all n except $n=1$, for which Q does not exist. In this case the only eigenvalue is $\lambda_1 = 1-\beta(1-q)$. In the following these eigenvalues are calculated near T_c and near $T=0$.

B. Stability near T_c and near $T=0$

Expanding (2.36) and (3.3) in powers of $t=1-T$ one obtains

$$q \simeq 3nt/(3n-2), \quad Q \simeq 2q/n\,,$$

from which it follows that

$$\lambda_1 \simeq -t+q+(n-1)Q \simeq 2t > 0\,, \tag{3.7}$$

$$\lambda_2 \simeq -t+q \simeq \frac{2t}{3n-2} > 0\,, \tag{3.8}$$

but

$$\lambda_3 \simeq -t+q-Q \simeq \frac{-4t}{3n-2} < 0 \tag{3.9}$$

for all $n>1$. Hence near $T=1$, only the solution with $n=1$ is locally stable; all other solutions are saddle points.

The situation is quite different at lower temperatures. Near $T=0$ the odd-n symmetry solutions order fully, with at most exponentially small deviations [see Eq. (2.37)]. Thus, as $T\to 0$, $q=1$ and $Q=0$ (up to exponentially small corrections) and all eigenvalues equal unity. The solutions are all locally stable. On the other hand, in the even-n solutions the system resists full order, even at $T=0$ [Eq. (2.38)]. Consequently, both λ_2 and λ_3 are proportional to $-\beta$, whereas $\lambda_1 \sim 1$. These results imply that while the even-n symmetric solutions are unstable for all T, the odd-n ones become locally stable below a certain temperature, $0<T_n<1$, given by the vanishing of λ_3, i.e.,

$$-T_n = (q-Q) = \langle\langle (1-\xi^1\xi^2)\tanh^2(\beta m_n z_n)\rangle\rangle\,. \tag{3.10}$$

Numerical solution of this equation yields $T_3=0.461$, $T_5=0.385$, and $T_7=0.345$. It can be seen from Eq. (2.37) that the finite-T corrections, at low T, are exponentially small, as long as $T \ll m_n$. As $n\to\infty$, $m_n \propto 1/\sqrt{n}$ (see, e.g., Appendix A). Thus, for large n, T_n is expected to scale as $1/\sqrt{n}$, which implies that only the odd-n symmetric solutions with $n<T^{-2}$ are stable. This is corroborated by numerical solutions of Eq. (3.10) for large n.

C. Dynamic stability

It has been shown above that the Mattis states are the only stable mean-field solutions near $T=1$. Below $T=0.461$ some of the odd-n symmetric solutions become local minima as well, whereas the even-n ones remain unstable at all T. In addition, at low temperature some of the asymmetric solutions become locally stable as will be discussed in Sec. IV. The significance of the locally stable solutions to the dynamic evolution of the system depends on the basin of attraction in phase space of each of these states and on the energy barriers that separate them from each other or from the ground states. It is quite hard to estimate the basins of attraction, though one generally expects that states higher in free energy have significantly smaller basins of attraction than those of the ground states. The barriers are easier to estimate. Since fluctuations of the free energy per spin about these states are finite, the free energy barriers separating them are all proportional to the size of the system, N. Hence, all local minima of the mean-field free energy are true metastable states. The lifetime of such a metastable state is proportional to $\exp(N\Delta f)$ where Δf is the difference between the free energy per spin of the metastable state and that of the lowest saddle point above it. For instance, the lowest energy path from the $n=3$ state $(\frac{1}{2},\frac{1}{2},\frac{1}{2},0,\ldots,0)$ to the $n=1$ state $(1,0,\ldots,0)$ passes through the $n=2$ saddle point $(\frac{1}{2},\frac{1}{2},0,\ldots,0)$, yielding an energy barrier per spin

$$\Delta f = f_2(T=0)-f_3(T=0)=0.175$$

[see Eqs. (2.33) and (2.34)].

IV. ASYMMETRIC SOLUTIONS OF THE GENERALIZED HOPFIELD MODEL

In Sec. II it was shown that solutions which appear continuously at $T=1$ are symmetric, namely all nonzero components of $\mathbf{m}$ are equal in magnitude. At low temperatures, however, additional saddle points appear[14] which are asymmetric. The appearance of these additional solutions becomes apparent by following the change in stability of the symmetric saddle points as the temperature is reduced. When a particular saddle point changes stability in a certain direction, it does not usually *exchange* stability with another existing symmetric saddle point, which lies in that direction. Instead, a new, asymmetric saddle point between the two existing saddle points appears. The highest temperature where such a change in stability occurs is when the $n=2$ symmetric solution

$$\mathbf{m} = (m,m,0,0,\ldots,0) \tag{4.1}$$

becomes unstable to the mixing of more memories. The eigenvalue that controls this stability is, according to Sec. III,

$$\lambda_2 = 1-\beta(1-q) \tag{4.2}$$

This eigenvalue is positive near $T=1$ [see Eq. (3.8)], is negative for that solution at $T=0$, and changes sign at

$$T_2 \simeq 0.575\,. \tag{4.3}$$

Other symmetric saddle points do not change stabilities in

any direction at that temperature. Instead, a new set of saddle points of the form

$$\mathbf{m}=(m,m,\epsilon,\epsilon,\ldots,\epsilon,0,0,\ldots,0)\,, \tag{4.4}$$

appears, where there are k entries of ϵ and $p-k-2$ entries of zeros. The magnitude of the new components, ϵ, vanishes continuously as T approaches T_2 from below.

At still lower temperatures when other saddle points change stability, additional sets of asymmetric solutions appear. For instance, at each temperature T_{2k} where λ_2 of an $n=2k$ symmetric solution becomes negative, a set of saddle points similar to (4.4) appears:

$$\mathbf{m}=(m,m,\ldots,m,\epsilon,\epsilon,\ldots,\epsilon,0,0,\ldots,0)\,, \tag{4.5}$$

where the first $2k$ components are m, the next l components are ϵ, and the last $p-2k-l$ components are zeros. The temperatures $T_4 \simeq 0.465$ and $T_6 \simeq 0.408$ and T_{2k} decreases as $1/\sqrt{k}$ as $k\to\infty$. Of course, other saddle points with even more complicated asymmetry also develop. As $T\to 0$, some of these asymmetric saddle points merge with the symmetric ones. For instance, the $(m,m,\epsilon,0,0,\ldots,0)$ which appears below T_2 approaches very rapidly the $n=3$ symmetric state $(m,m,m,0,0,\ldots,0)$. On the other hand, many of them do remain distinct even at $T=0$, with energies which are always higher than the $n=3$ symmetric value -0.375. Two examples with $n=5$ and 6 are

$$\mathbf{m}=(\tfrac{1}{2},\tfrac{1}{2},\tfrac{1}{4},\tfrac{1}{4},\tfrac{1}{4},0,0,\ldots,0),\quad E=-0.344 \tag{4.6}$$

$$\mathbf{m}=(\tfrac{3}{8},\tfrac{3}{8},\tfrac{3}{8},\tfrac{3}{8},\tfrac{3}{16},0,0,\ldots,0),\quad E=-0.334 \tag{4.7}$$

which satisfy Eq. (2.19), as can be checked explicitly.

Although most of the asymmetric solutions seem to follow the scenario described above of a continuous appearance below a critical temperature, we have also encountered a few solutions which appear *discontinuously*, for instance, the solution

$$\mathbf{m}=(\tfrac{3}{8},\tfrac{3}{8},\tfrac{1}{4},\tfrac{1}{4},\tfrac{1}{2},0,0,\ldots,0) \tag{4.8}$$

below $T\simeq 0.085$. However, this phenomenon is apparently restricted to very low T. Also, none of the discontinuous solutions that we found were stable.

The stability analysis of the various solutions at finite T is increasingly difficult as the symmetry of the solution decreases. In general, they are unstable as they first appear. At lower temperatures some of them do become metastable, as in the case of the odd-n symmetric solutions. The highest temperature where a metastable asymmetric solution was found was

$$T\sim 0.452$$

just below the temperature $T=0.461$ at which the $n=3$ symmetric state becomes metastable. At $T=0$ the criterion of stability of an arbitrary solution is rather simple. Following the same reasoning as in the case of the symmetric solutions (Sec., III) we note that the solutions are stable if their molecular fields are always finite, i.e., that

$$|\mathbf{m}\cdot\boldsymbol{\xi}|>0 \tag{4.9}$$

holds for all realizations of $\boldsymbol{\xi}$. In such a case all off-diagonal elements [Eq. (3.3)] of the stability matrix are exponentially small as $T\to 0$ and all diagonal elements approach 1. Such states are surrounded by energy barriers proportional to N. Saddle points in which there is a finite probability that $\mathbf{m}\cdot\boldsymbol{\xi}=0$ represent states which do not fully order even at $T=0$. They have some eigenvalues which are proportional to $-\beta$. For instance, the state (4.6) is metastable at $T=0$ because its molecular field is bounded below by $\frac{1}{4}$ and becomes unstable at $T\simeq 0.18$, whereas the solutions (4.7) and (4.8) are unstable even at $T=0$.

A rather central question is whether the system possesses other, "spurious" states, i.e., states which are not separated by barriers of order N, but are nonetheless long lived, at low T. Such states would surely be stable to single-spin flips at $T=0$. The answer is that, in the limit of $N\to\infty$ and p finite, the only states which are stable to all single-spin flips are the true metastable states, namely solutions of the mean-field equations which satisfy (4.9). The argument proceeds as follows. One notes that the condition for the stability of a state $\{S_i\}$ to all single-spin flips is that each spin S_i be aligned with its molecular field, namely that

$$\begin{aligned} S_i &= \operatorname{sgn}\left[(1/N)\sum_{j(\neq i)}\sum_{\mu}\xi_i^{\mu}\xi_j^{\mu}S_j\right] \\ &= \operatorname{sgn}\left[\mathbf{m}\cdot\boldsymbol{\xi}_i-\frac{p}{N}S_i\right], \end{aligned} \tag{4.10}$$

with

$$\begin{aligned} \mathbf{m} &= (1/N)\sum_j \boldsymbol{\xi}_j S_j \\ &= (1/N)\sum_j \boldsymbol{\xi}_j \operatorname{sgn}\left[\mathbf{m}\cdot\boldsymbol{\xi}_j-\frac{p}{N}S_j\right]. \end{aligned} \tag{4.11}$$

As long as $\mathbf{m}\cdot\boldsymbol{\xi}_i$ *has a nonzero lower bound,* the term $(p/N)S_i$ can be neglected and Eqs. (4.10) and (4.11) become precisely the saddle-point equations of the mean-field theory [Eqs. (2.9) and (2.7)] at $T=0$ and have the same set of fluctuations. On the other hand, states which have a finite fraction of sites with $\mathbf{m}\cdot\boldsymbol{\xi}_i=0$ are rendered unstable by the term $-(p/N)S_i$.

V. THERMODYNAMICS OF THE LITTLE MODEL

The synchronous dynamic process introduced by Little leads to a stationary Gibbs distribution of states with the effective Hamiltonian

$$\bar{H}=-\frac{1}{\beta}\sum \ln\left[2\cosh\left[\beta\sum_j J_{ij}S_j\right]\right] \tag{5.1}$$

as was shown by Peretto.[3] In the limit of $T=0$, Eq. (5.1) reads

$$\bar{H} = -\sum_i \left|\sum_j J_{ij} S_j\right| . \quad (5.2)$$

We study the thermodynamic properties of the model (5.1), with the same J_{ij} as in the Hopfield model, i.e., $J_{ij} = N^{-1}\sum_{\mu=1}^{p} \xi_i^\mu \xi_j^\mu$ for $i \neq j$, and $\{\xi_i\}$ is distributed according to Eq. (2.10). The partition function of the Hamiltonian (5.1) can be written as

$$\begin{aligned} Z &= \operatorname*{Tr}_s \exp(-\beta \bar{H}) \\ &= \operatorname*{Tr}_s \int \left[\prod_\mu dm^\mu\right] \exp\left[\sum_i \ln[2\cosh(\beta \xi_i \cdot \mathbf{m})]\right] \delta\left[\mathbf{m} - N^{-1}\sum_i \xi_i S_i\right] \\ &= \left[\frac{\beta N}{2\pi}\right]^p \int \left[\prod_\mu dm^\mu dt^\mu\right] \exp\left[-N\beta \mathbf{t}\cdot\mathbf{m} + \sum_i \ln[2\cosh(\beta \xi_i \cdot \mathbf{m})] + \sum_i \ln[2\cosh(\beta \xi_i \cdot \mathbf{t})]\right] . \end{aligned} \quad (5.3)$$

The contours of integration in the complex planes of m^μ and t^μ are understood to be analytically deformed, so that they pass the saddle point. Using again the self-averaging property of the free energy, the saddle-point equations reduce to

$$\mathbf{m} = \langle\langle \xi \tanh(\beta \mathbf{t}\cdot\xi) \rangle\rangle , \quad (5.4)$$

$$\mathbf{t} = \langle\langle \xi \tanh(\beta \mathbf{m}\cdot\xi) \rangle\rangle , \quad (5.5)$$

and the free energy density at the saddle point is

$$\begin{aligned} f(\beta) = \mathbf{t}\cdot\mathbf{m} &- \frac{1}{\beta}\langle\langle \ln[2\cosh(\beta\xi\cdot\mathbf{m})] \rangle\rangle \\ &- \frac{1}{\beta}\langle\langle \ln[2\cosh(\beta\xi\cdot\mathbf{t})] \rangle\rangle . \end{aligned} \quad (5.6)$$

Although the theory contains two order parameters $\mathbf{t}$ and $\mathbf{m}$, all mean-field solutions obey

$$\mathbf{t} = \mathbf{m} . \quad (5.7)$$

To prove this, we subtract Eqs. (5.5) from (5.4) to obtain

$$\mathbf{m} - \mathbf{t} = \langle\langle \xi[\tanh(\beta\xi\cdot\mathbf{t}) - \tanh(\beta\xi\cdot\mathbf{m})] \rangle\rangle ,$$

from which it follows that

$$\begin{aligned} \sum_\mu (m^\mu - t^\mu)^2 = \langle\langle (\xi\cdot\mathbf{m} - \xi\cdot\mathbf{t})[\tanh(\beta\xi\cdot\mathbf{t}) \\ - \tanh(\beta\xi\cdot\mathbf{m})] \rangle\rangle . \end{aligned}$$

The right-hand side is obviously *nonpositive*, hence the two sides must be zero, implying the equality (5.7). Substituting Eq. (5.7) in Eqs. (5.4)—(5.6) reduces them to

$$f(\beta) = \mathbf{m}^2 - \frac{2}{\beta}\langle\langle \ln[2\cosh(\beta\xi\cdot\mathbf{m})] \rangle\rangle , \quad (5.8)$$

$$\mathbf{m} = \langle\langle \xi \tanh(\beta\xi\cdot\mathbf{m}) \rangle\rangle . \quad (5.9)$$

Comparison of Eqs. (5.8) and (5.9) with Eqs. (2.6) and (2.7) reveals that *the Little free energy is exactly twice the free energy of the Hopfield model, at all T.* Both have the same mean-field equations for $\mathbf{m}$. Thus, below $T=1$ the Little model has the same ground states (i.e., the Mattis states) and saddle points as the generalized Hopfield model. Moreover, as the stability analysis below will show, also the metastability of the saddle points is the same in both models.

In order to perform a stability analysis of the saddle points of (5.3) we have to use the variables appropriate to the rotated contours of integration. These variables are x_μ and y_μ defined via

$$\delta t_\mu = \frac{1}{\sqrt{2}}(x_\mu + iy_\mu), \quad \delta m_\mu = \frac{1}{\sqrt{2}}(x_\mu - iy_\mu) , \quad (5.10)$$

where δt_μ and δm_μ are the deviations of t_μ and m_μ along their contours, from their saddle-point values. In terms of these variables the stability $2p \times 2p$ matrix consists of the following two $p \times p$ blocks:

$$\begin{aligned} (\underline{A}_x)^{\mu\nu} &= \frac{\partial^2 f}{\partial x_\mu \partial x_\nu} \\ &= \delta^{\mu\nu}(1-\beta) + \beta\langle\langle \xi^\mu\xi^\nu \tanh^2(\beta\xi\cdot\mathbf{m}) \rangle\rangle \end{aligned} \quad (5.11)$$

and

$$\begin{aligned} (\underline{A}_y)^{\mu\nu} &= \frac{\partial^2 f}{\partial y_\mu \partial y_\nu} \\ &= \delta^{\mu\nu}(1+\beta) - \beta\langle\langle \xi^\mu\xi^\nu \tanh^2(\beta\xi\cdot\mathbf{m}) \rangle\rangle . \end{aligned} \quad (5.12)$$

The mixed derivatives $\partial^2 f/\partial x_\mu \partial y_\nu$ are zero.

Comparison of Eqs. (5.11) and (5.12) with the stability matrix $\underline{A}$ (3.1) of the generalized Hopfield model reveals that $\underline{A}_x$ is identical to $\underline{A}$ and $\underline{A}_y$ to $2\underline{I} - \underline{A}$. Since the eigenvalues of A are bounded from above by 1, $\underline{A}_y$ is positive definite and the stability of the saddle points is determined by the same stability matrix as in the Hopfield model. Hence, the analysis of Secs. III and IV applies here as well. In particular, the Mattis states are the only locally stable states near $T_c = 1$. Additional local minima appear only below $T = 0.461$.

It has been pointed out[3] that the synchronous dynamics of Little may lead, at $T=0$, to indefinite cycles of transitions between some of the ground states of the effective Hamiltonian, which occur in one or a small number of time steps. This phenomenon is absent in the single-spin dynamics of Hopfield. As an example, consider a d-dimensional hypercubic lattice with nearest-neighbor ferromagnetic interaction, $J > 0$. In a single-spin flip dynamics the system will spend most of its time, at low temperatures, in one of the two ferromagnetic ground states, with a very small probability of hopping from one of these states to the other, in a finite time. On the other hand, the Hamiltonian (5.1) [or (5.2)] has four ground states: the two ferromagnetic states and the two antiferromagnetic states. In each of the antiferromagnetic states every spin is antiparallel to its molecular field $-\sum_j J_{ij}S_j$ and therefore, according to the dynamics implied by Eq. (1.8), wants to flip. Consequently, starting from the antiferromagnetic states, the system will make a transition in a single time step to the time-reversed (antiferromagnetic) state and vice versa.

Could such cycles appear in our case? The answer is no. A necessary condition for these cycles to occur is that some of the ground states of the Hamiltonian $\bar{H}$ contain spins which are antiparallel to their molecular field. However, we have argued here that in the $N\to\infty$ limit the ground states (as well as the metastable states) of the Hamiltonian (5.1) are also local minima of the Hopfield Hamiltonian

$$H=-\tfrac{1}{2}\sum_i\left[S_i\left[\sum_j J_{ij}S_j\right]\right].$$

This necessarily implies that each spin S_i in these states is aligned with $\sum_j J_{ij}S_j=\mathbf{m}\cdot\boldsymbol{\xi}$, as was discussed in Sec. IV. Thus, not only the thermodynamic properties of the two models but also their long-time behavior is the same.

VI. GENERAL DISTRIBUTION OF MEMORIES

A. General distribution of $\{\xi_i^\mu\}$

We consider here a general distribution $P(\xi_i)$ which is invariant under reflections $\xi_i^\mu\to-\xi_i^\mu$ (for each μ separately) and permutations of the components ξ_i^μ. For convenience we will normalize the variance by $\langle\langle(\xi_i^\mu)^2\rangle\rangle=1$. The mean-field equations (2.6) and (2.7) hold, of course, for an arbitrary distribution and imply a phase transition to a broken symmetry state at $T=1$. The Mattis solutions as well as the class of symmetric solutions, of the form (2.22), always exist at all $T<1$. Moreover, expanding Eqs. (2.7) order by order in perturbation about $T=1$ shows that for almost all distributions of the form

$$P(\xi_i)=\prod_\mu p(\xi_i^\mu)\ , \tag{6.1}$$

asymmetric solutions do not exist near $T=1$. The only exception is the Gaussian distribution which will be discussed in Sec. VI B. Nevertheless, the stability of the various solutions as well as the appearance at low T of the asymmetric solutions depend on the form of the probability $p(\xi_i^\mu)$.

We first determine the conditions under which the *Mattis states become unstable* near $T=1$ and 0 for a general distribution of the form (6.1). The mean-field equation for the Mattis state, $m^\mu=0$, for all $\mu>1$ is

$$m^1\equiv m=\langle\langle\xi^1\tanh(\beta m\xi^1)\rangle\rangle\ . \tag{6.2}$$

Generalizing the stability analysis of Sec. III one obtains for the symmetric solutions with a general distribution the following three eigenvalues:

$$\lambda_1=1-\beta(1-\bar{q})+(n-1)\beta Q\ , \tag{6.3}$$

$$\lambda_2=1-\beta(1-q)\ , \tag{6.4}$$

$$\lambda_3=1-\beta(1-\bar{q})-\beta Q\ , \tag{6.5}$$

where $q=\langle\langle\tanh^2(\beta z_n)\rangle\rangle$, $Q=\langle\langle\xi^1\xi^2\tanh^2(\beta z_n)\rangle\rangle$, and $\bar{q}\equiv\langle\langle(\xi^1)^2\tanh^2(\beta z_n)\rangle\rangle$. The variable z_n is $\sum_{\mu=1}^n\xi^\mu$. In the Mattis case ($n=1$) only the eigenvalues $\lambda_1=1-\beta(1-\bar{q})$ and $\lambda_2=1-\beta(1-q)$ exist. A negative λ_2 implies that when $p>1$ the Mattis state is unstable to mixing of more memories. Expanding Eq. (6.2) in powers of $t=1-T_c$ yields $q\cong 3t/\langle\langle(\xi^\mu)^4\rangle\rangle$, $1-\beta(1-q)\cong t[-1+3/\langle\langle(\xi^\mu)^4\rangle\rangle]$, which implies the instability of the Mattis states near T_c if

$$\langle\langle(\xi^\mu)^4\rangle\rangle>3\ . \tag{6.6}$$

The stability at low temperatures depends on the behavior of $p(\xi)$ near the origin. As long as $p(0)=0$, $1-\beta(1-q)\sim 1$ and the Mattis states are stable. This is, however, not necessarily so when $P(0)\neq 0$, in which case we have

$$1-q=\int_{-\infty}^{\infty}d\xi\,p(\xi)\operatorname{sech}^2(\beta m\xi)=\frac{T}{m}\int_{-\infty}^{\infty}d\xi\,p(T\xi/m)\operatorname{sech}^2\xi\simeq\frac{2Tp(0)}{m}\ , \tag{6.7}$$

where

$$m=m(T=0)=\int d\xi\,p(\xi)\,|\xi|\ . \tag{6.8}$$

Thus, the Mattis states are unstable to mixing of more memories at low T, if

$$2p(0)>\langle\langle|\xi|\rangle\rangle\ . \tag{6.9}$$

As an example, consider the following distribution:

$$p(\xi)=\frac{a}{\sqrt{2}}e^{-|\xi|\sqrt{2}}+\frac{1-a}{2}[\delta(\xi-1)+\delta(\xi+1)]\ .$$

Evaluating the inequalities (6.7) and (6.9) one finds that for $a<0.4$, the Mattis states are stable at all $T<1$; for $a>(1+1/\sqrt{2})^{-1}\simeq 0.6$, the Mattis states are unstable at all T; and for $0.4<a<(1+1/\sqrt{2})^{-1}$, they are unstable near T_c and stable at low temperatures.

Using a continuous distribution of $\{\xi_i^\mu\}$ may have a similar effect as increasing temperature. It may smooth the free energy surface and eliminate the rich structure of metastable states and saddle points that exist in the ± 1 case at low temperature, as was described above. We have demonstrated this effect by studying the mean-field equations with a rectangular distribution,

$$p(\xi_i^\mu)=\frac{1}{l},\quad \frac{-l}{2}\le\xi_i^\mu\le\frac{l}{2}\ . \tag{6.10}$$

Investigating the symmetric solutions with $n\le 3$, we have found that the only stable solutions at all T are the $n=1$ states. Furthermore, the $n=2$ and 3 solutions do not change stability in any direction and their free energies do not cross, at all T below 1. These last properties imply that there are no topological constraints which would force the generation at lower temperatures of additional, asymmetric saddle points. Thus, it is quite possible that in this case, or with other continuous distributions, the only dynamically stable states are the Mattis states at all $T<1$ and p.[15]

B. Rotationally invariant distributions

In certain circumstances the appropriate distribution of the random vectors $\boldsymbol{\xi}_i$ is invariant under arbitrary $O(p)$ rotations of $\boldsymbol{\xi}_i$. In the case of random axis ferromagnets the $\boldsymbol{\xi}_i$ is the local direction of the easy axis. In the absence of bulk anisotropy these directions are uniformly distributed on the three-dimensional unit sphere.

In the context of models of memory, rotational invariance emerges in the case of a *Gaussian* distribution,

$$P(\boldsymbol{\xi}_i)=\prod_{\mu} p(\xi_i^{\mu}) ,$$
$$p(\xi)=\frac{1}{\sqrt{2\pi}}\exp(-\xi^2/2) . \tag{6.11}$$

The ξ_i^{μ} will have a Gaussian distribution if, e.g., each one of them is itself a sum of many independent random variables. Rotationally invariant distributions lead to thermodynamic behavior which is qualitatively very different from that of Eq. (2.10). The free energy (2.6) depends in this case only on the amplitude

$$m^2=\frac{1}{p}\sum_{\mu=1}^{p}(m_{\mu})^2$$

which is determined by Eq. (2.7), whereas the direction of $\mathbf{m}$ is left arbitrary. Thus, in the limit $N\to\infty$ the manifold of ground states has a continuous degeneracy, similar to $O(p)$ uniform models.

As an example we work out explicitly the case of a Gaussian distribution, Eq. (6.11). Rotating the μ axes, so that the direction of $\mathbf{m}$ coincides with one of the axes, Eqs. (2.6) and (2.7) read

$$f=(p/2)m^2-\frac{1}{\beta}\int_{-\infty}^{\infty}\frac{d\xi}{\sqrt{2\pi}}e^{-\xi^2/2}\ln[2\cosh(\beta m\sqrt{p}\,\xi)] , \tag{6.12}$$

$$m=\frac{1}{\sqrt{p}}\int_{-\infty}^{\infty}\frac{d\xi}{\sqrt{2\pi}}e^{-\xi^2/2}\xi\tanh(\beta m\sqrt{p}\,\xi) . \tag{6.13}$$

Integrating Eq. (6.13) by parts leads to the relation

$$1=\beta\int_{-\infty}^{\infty}\frac{d\xi}{\sqrt{2\pi}}e^{-\xi^2/2}\mathrm{sech}^2(\beta m\sqrt{p}\,\xi)=\beta(1-q) \tag{6.14}$$

which is an implicit equation for m. It is interesting that in this case the local susceptibility [which is $\beta(1-q)$] is constant below T_c, just as in the infinite-range SK model.

At $T=0$, $m=\sqrt{2/\pi p}$ and $f(T=0)=-1/\pi$. As for fluctuations, the eigenvalue $\lambda_1=1-\beta(1-\bar{q})$ corresponds to amplitude fluctuations and is positive at all $T<1$. On the other hand, there are $p-1$ degenerate modes, with eigenvalue $1-\beta(1-q)$ which, according to Eq. (6.14), is identically zero in the ordered phase. These modes correspond to *transverse* fluctuations (changing the direction of $\mathbf{m}$) and hence are marginal.

The continuous symmetry in these models has been studied in some detail in the context of random axis models.[11] It should be noted that, unlike $O(p)$ uniform models, the continuous degeneracy of the ground state in our case is valid only in the thermodynamic limit. In any finite system there will be N possible directions of $\mathbf{m}$, one of which will be singled out by the fluctuations in $\boldsymbol{\xi}_i$ as the ground state of the system. This state will have, in general, projections on all the original memories.

VII. DISCUSSION

One of the main results of this work is that despite the difference in the dynamic mechanisms of the Little and Hopfield models, the long-time behavior, so essential for the retrieval of memories, is identical in the two models. While the dynamic properties of these models have not been analyzed in this paper, extensive numerical simulations of the two processes have been carried out. They confirm their overall similarity.

Previous numerical studies of these models have noticed the existence of "spurious states," namely stable configurations of the network which deviate significantly from the original embedded patterns. Here we have shown that in the limit of large networks all these spurious states are not random but correspond to well-defined mixtures of several patterns.

We have shown that these metastable states appear only at low temperatures, whereas in the temperature range $0.46<T<1$ only the states which are correlated with single memories are stable. This suggests that thermal noise plays an important role in enhancing the efficiency of these systems. This efficiency also depends rather strongly on the distribution, of the learned information. Using a continuous distribution, rather than a discrete one, may increase or destroy the dynamic stability of the embedded patterns, depending on the details of that distribution. So far only uncorrelated patterns have been discussed. Introducing correlations between them may, of course, change significantly the properties of the system.

Next we address the question of the storage capacity of these model networks. Throughout this work we have assumed the limit of an infinite size (N) network, with a finite number (p) of stored patterns. *In this limit* the structure of the low-lying states, as well as the barriers between them, are not affected by the increase of p. However, this limit represents a rather modest storage capacity. It is important to know whether this capacity can still increase as p becomes of the order of N^x for some positive x. We have studied the case $x=1$,[17] $p=\alpha N$, and found that in this limit the system becomes a spin glass. Yet, for small enough values of α the system preserves its quality as an associative memory.

Finally, we mention a few of the issues that deserve further investigation:

(1) The crossover from the finite p behavior to the spin-glass behavior when p becomes of order N;[17]

(2) the detailed dynamic properties of the Little and Hopfield models, e.g., the rate of relaxation and the sizes of the basins of attraction of the various stable states;

(3) the possible exploitation, in the information theoretic sense, of the long-lived mixed states;

(4) the effect of introducing correlations among the embedded patterns;

(5) the effect of modifying the form of the connections J_{ij}. In particular, adding an asymmetric part to J_{ij} may lead to interesting new dynamic behavior.

ACKNOWLEDGMENTS

We are grateful for stimulating and helpful discussions with J. J. Hopfield, M. Mezard, J. Milnor, and M. Virasoro. The research of D.J.A. and H.S. is supported in part by the Fund of Basic Research administered by the Israel Academy of Science and Humanities.

APPENDIX A

To compute the explicit value of m_n, at $T=0$, Eq. (2.31), we proceed as follows:

$$\langle\langle |z| \rangle\rangle = \left\langle\left\langle z\left[1+\frac{1}{\pi i}\int_{-\infty}^{\infty}\frac{d\theta}{\theta}e^{i\theta z}\right]\right\rangle\right\rangle \tag{A1}$$

$$= \frac{2^{-n}}{\pi}\int\frac{d\theta}{\theta}\left[-\frac{d}{d\theta}\right]\sum_{k=0}^{n}\binom{n}{k}e^{i\theta(2k-n)} = -\frac{2^{-n}}{\pi}\int\frac{d\theta}{\theta}\frac{d}{d\theta}e^{-i\theta n}(1+e^{2i\theta})^n = \frac{n}{\pi}\int\frac{d\theta}{\theta}\sin\theta\cos^{n-1}\theta$$
$$= \frac{n}{2^{2k}}\binom{2k}{k}, \tag{A2}$$

where $k=n/2$ for even n and $k=(n-1)/2$ for odd n. (See Ref. 16, Eq. 3.832.34.)

Next we show that f_{2k}, Eq. (2.33), decreases monotonically with k and f_{2k+1}, Eq. (2.34), increases monotonically with k. To this end we use the identity

$$\binom{2k+2}{k+1} = \frac{2(2k+1)}{k+1}\binom{2k}{k}. \tag{A3}$$

One can write

$$f_{2k+3}-f_{2k+1} = \frac{2k+1}{2^{4k+1}}\binom{2k}{k}\left[1-\frac{(2k+3)(2k+1)}{4(k+1)^2}\right] > 0, \tag{A4}$$

$$f_{2k+2}-f_{2k} = \frac{2k}{2^{4k+1}}\binom{2k}{k}\left[1-\frac{(2k+1)^2}{4k(k+1)}\right] < 0 \tag{A5}$$

which proves our claim.

It is clear that the two sequences, Eqs. (2.33) and (2.34), have a common limit as $k\to\infty$. To calculate this limit we use the Stirling formula

$$n! \sim \sqrt{2\pi}(n+1)^{n+1/2}e^{-(n+1)} \tag{A6}$$

giving, for the asymptotic form of m_n, with even n,

$$m_n = 2^{-n}\binom{n}{n/2} = (2/n\pi)^{1/2}. \tag{A7}$$

The limiting value for f_n is obtained by substituting (A7) in Eq. (2.32). We find

$$\lim_{n\to\infty} f_n = -1/\pi. \tag{A8}$$

APPENDIX B

In the following we show that the off-diagonal elements of the stability matrix $\underline{A}$, Eqs. (3.1) and (3.2), are non-negative, at all T, for all the saddle points given by Eqs. (2.22). Specifically, we will show that

$$B = \langle\langle \xi^1\xi^2\tanh^2(xz_n)\rangle\rangle \tag{B1}$$

is non-negative for all x.

For definiteness we choose $x>0$. Since ξ^μ takes on the values ± 1, z_n is distributed according to

$$p(z_n) = 2^{-n}\binom{n}{k},\quad k=(z_n+n)/2$$

as in (2.25). B can be written as

$$B = 2^{-(n-1)}\sum_{k=0}^{n-2}\binom{n-2}{k}[\tanh^2(2k-n+4)x - \tanh^2(2k-n+2)]. \tag{B2}$$

Note that $\tanh^2 y$ is an increasing function of y for $y>0$ and decreasing for $y<0$. In general, some of the terms on the right-hand side of (B2) are negative, but they are outweighed by the positive terms.

To see this, the sum in (B2) is divided in the following way. (i) Remove the term with $k=n-2$. It is manifestly positive. (ii) For odd n the term with $k=(n-3)/2$ vanishes. (iii) The rest of the sum, comprising an even number of terms, for all n, is split into two sums, each with $[(n-2)/2]$ terms, namely

$$D_1 = \sum_{k=0}^{[(n-2)/2]-1}\binom{n-2}{k}\{\tanh^2[(2k-n+4)x] - \tanh^2[(2k-n+2)x]\}$$

and

$$D_2=\sum_{k=[(n-1)/2]}^{n-3}\binom{n-2}{k}\{\tanh^2[(2k-n+4)x]-\tanh^2[(2k-n+2)x]\}\,.$$

Substituting $n-3-k$ for k in the second sum, it becomes

$$\sum_{k=0}^{[(n-2)/2]-1}\binom{n-2}{k+1}\{\tanh^2[(2k-n+2)x]-\tanh^2[(2k-n+4)x]\}$$

after use has been made of the identities

$$n-3-[(n-1)/2]=[(n-2)/2]-1$$

and

$$\binom{n-2}{n-3-k}=\binom{n-2}{k+1}$$

and of the fact that the square of the hyperbolic function is an even function. B can now be rewritten as

$$B=2^{-(n-1)}\left\{\sum_{k=0}^{[(n-2)/2]-1}\left[\binom{n-2}{k+1}-\binom{n-2}{k}\right](\{\tanh^2[(2k-n+2)x]-\tanh^2[(2k-n+4)x]\}+\{\tanh^2(nx)-\tanh^2[(n-2)x]\})\right\}. \tag{B3}$$

Each term in the sum (B3) is positive, since both small curly brackets are positive, in the range of variation of k. The last term in the large curly brackets is also positive. Hence $B\geq 0$ for all x.

Next we show that all the eigenvalues of $\underline{A}$ are bounded from above by 1. Since we have shown above that Q is always positive, the largest eigenvalue is

$$\lambda_2=1-\beta+\beta[q+(n-1)Q]\,. \tag{B4}$$

[See Eqs. (3.4)–(3.6).] But

$$\langle\langle z_n^2\tanh^2(m_n z_n)\rangle\rangle/n=q+(n-1)Q$$

and hence

$$q+(n-1)Q<\langle\langle z_n^2\rangle\rangle/n=1\,,$$

leading to

$$\beta[q+(n-1)Q]<\beta$$

and consequently $\lambda_i(i=1,2)\leq\lambda_3\leq 1$.

[1] J. J. Hopfield, Proc. Natl. Acad. Sci. USA 79, 2554 (1982); Proc. Natl. Acad. Sci. USA B 1, 3088 (1984).

[2] W. A. Little, Math. Biosci. 19, 101 (1974); W. A. Little and G. L. Shaw, Behav. Biol. 14, 115 (1975); Math. Biosci. 39, 281 (1978).

[3] P. Peretto, Biol. Cybern. 50, 51 (1984).

[4] G. L. Shaw and R. Vasudevan, Math. Biosci. 21, 207 (1974).

[5] R. J. Glauber, J. Math. Phys. 4, 294 (1963); K. Binder, in *Fundamental Problems in Statistical Mechanics V*, edited by E. G. D. Cohen (North-Holland, Amsterdam, 1980), p. 21.

[6] A collection of works on spin glasses can be found in *Heidelberg Colloquium on Spin Glasses*, Vol. 192 of *Lecture Notes in Physics*, edited by J. L. Van Hemmen and I. Morgenstern, (Springer, New York, 1983).

[7] S. Kirkpatrick and D. Sherrington, Phys. Rev. B 17, 4384 (1978).

[8] D. C. Mattis, Phys. Lett. 56A, 421 (1976).

[9] J. L. Van Hemmen, Phys. Rev. Lett. 49, 409 (1982); in Ref. 6, p. 203.

[10] J. P. Provost and G. Vallee, Phys. Rev. Lett. 50, 598 (1983).

[11] C. Jayaprakash and S. Kirkpatrick, Phys. Rev. B 21, 4072 (1980).

[12] In Ref. 3 the same conclusion has been reached, but with the use of the bound $\sum_{\mu=1}^{p}|m^\mu|\leq 1$. This bound, however, is incorrect and, in fact, is not obeyed by most of the saddle points.

[13] S. F. Edwards and P. W. Anderson, J. Phys. F 5, 965 (1975).

[14] We are grateful to M. Virasoro and N. Parga for drawing our attention to the existence of asymmetric solutions.

[15] Details will be published elsewhere.

[16] I. S. Gradshteyn and I. M. Ryzhik, *Tables of Integrals, Series and Products* (Academic, New York, 1965).

[17] D. J. Amit, H. Gutfreund, and H. Sompolinsky (unpublished).

VOLUME 55, NUMBER 14 PHYSICAL REVIEW LETTERS 30 SEPTEMBER 1985

Storing Infinite Numbers of Patterns in a Spin-Glass Model of Neural Networks

Daniel J. Amit and Hanoch Gutfreund
Racah Institute of Physics, Hebrew University, Jerusalem 91904, Israel

and

H. Sompolinsky
Department of Physics, Bar Ilan University, Ramat Gan, Israel
(Received 11 July 1985)

The Hopfield model for a neural network is studied in the limit when the number p of stored patterns increases with the size N of the network, as $p=\alpha N$. It is shown that, despite its spin-glass features, the model exhibits associative memory for $\alpha < \alpha_c$, $\alpha_c \gtrsim 0.14$. This is a result of the existence at low temperature of $2p$ dynamically stable degenerate states, each of which is almost fully correlated with one of the patterns. These states become ground states at $\alpha < 0.05$. The phase diagram of this rich spin-glass is described.

PACS numbers: 87.30.Gy, 64.60.Cn, 75.10.Hk, 89.70.+c

Spin-glass models which exhibit features of learning, memory, and pattern recognition have become the focus of exciting numerical and analytical studies.[1-4] Of particular interest is Hopfield's model of associative memory.[1] In this model of neural network, a given set of patterns is embedded in the "synaptic" interactions between the neurons so as to make these patterns dynamically stable. A crucial issue is the storage capacity of the network. In a previous work,[4] we have established the stability of the embedded patterns, under the severe restriction that the number of stored patterns, p, remains finite as the size of the network, N, approaches infinity. On the other hand, there have been apparently conflicting statements regarding the capacity of the system as p increases to infinity with N.

Hopfield[1] concluded, on the basis of simulations and Gaussian noise arguments, that the system continues to provide associative memory for $p \lesssim \alpha_c N$, at $T=0$, with $\alpha_c \simeq 0.1$–0.2, but degrades rapidly when $p \gtrsim \alpha_c N$. In apparent contrast, Weisbuch[5] and Posner[6] have proved that at $T=0$, the original patterns will be locally stable only if $p < N/(2\ln N)$.

In this Letter we study the model in the limit that p increases to infinity as $p=\alpha N$. We have determined the properties of the stable and metastable states of the system, thereby resolving the issue of its storage capacity. Apart from their relevance to neural networks the results reveal rich statistical mechanical properties which emerge from an unusual intertwining of ferromagnetic (FM) and spin-glass (SG) symmetry breaking. These features disappear for large α and the model approaches the spin-glass model of Kirkpatrick and Sherrington.[7]

The network of N neurons is modeled by N spins, $S_j = \pm 1$. A pattern of neural firings corresponds to a spin configuration. The dynamics is modeled by a serial heat-bath Monte Carlo process,[4] governed by an energy

$$H = -\frac{1}{2}\sum_{i\neq j} J_{ij} S_i S_j \tag{1}$$

at temperature $T\ (=\beta^{-1})$. The couplings (synaptic efficiencies) are constructed of p given spin configurations (patterns), according to

$$J_{ij} = \frac{1}{N}\sum_{\mu=1}^{p} \xi_i^\mu \xi_j^\mu, \tag{2}$$

with $\xi_i^\mu\ (=\pm 1)$ quenched, independent, random variables.[8] They represent the p patterns, which were embedded in the system by a "learning" process. *Retrieval of memory is diagnosed as the dynamical persistence of a pattern, provoked by an external stimulus.* The discussion centers, therefore, around the nature of persistent states of the underlying dynamical process, and hence about the stable states of the free energy associated with H. Of particular pertinence are questions about the correlation of the dynamically stable states with the ξ^μ and the dependence of those correlations on p.

For finite p, as $N \to \infty$, the situation is rather clear[4]: (1) At a critical temperature ($T_c = 1$) the system undergoes a second-order phase transition, from a disordered phase to a phase of $2p$ degenerate free-energy ground states—each one a Mattis[9] state, correlated with one of the embedded patterns $\{\xi_i^\mu\}$. With definition of the overlap of a state with the νth pattern as

$$m^\nu = N^{-1}\sum_i \langle S_i\rangle \xi_i^\nu, \tag{3}$$

where $\langle\ldots\rangle$ denotes a thermal average, the μ Mattis state has $m^\nu = m\delta^{\mu\nu}$. The local magnetization is $\langle S_i\rangle = \xi_i^\mu m$, and $m=\tanh\beta m$. (2) Near T_c, the Mattis states are the only stable states. Below $T=0.46$ additional dynamically stable states appear, corresponding to well-defined mixtures of several patterns. The number of these metastable states increases as T decreases. It also increases (at least exponentially) with increasing p. (3) The $2p$ ground states as well as the metastable (mixture) states are separated by free-energy barriers of $O(N)$.

We now proceed to the case of a finite $\alpha = p/N$. We have evaluated the average free energy per spin, $f = -\langle\langle \ln \mathrm{Tr} \exp(-\beta H)\rangle\rangle / N\beta$, of the Hamiltonian (1),(2) with $p = \alpha N$ by the replica method. Here $\langle\langle \ldots \rangle\rangle$ denotes an average over the quenched disorder $\{\xi_i^\mu\}$. Since the system is fully connected, f can be calculated exactly in the $N \to \infty$ limit by a mean-field theory. Most of our discussion will be within the replica-symmetric theory.[7] The occurrence and implications of replica-symmetry breaking[10] will be discussed at the end.

In the present model, the low-temperature phase will have weak random overlaps with most of the patterns, each of which will be typically of $O(1/\sqrt{N})$. This can be realized from the fact that the zero-temperature energy per spin which is of order unity is just $E = \frac{1}{2}\alpha - \sum_\mu (m^\mu)^2$, where m^μ are the overlaps defined in Eq. (3). However, it is possible that one or a finite number of overlaps condense macroscopically, i.e., retain a finite value as $N \to \infty$. We thus find three sets of order parameters: (i) the *macroscopic* overlaps m^μ with the patterns $\{\xi_i^\nu\}$, $\nu = 1, 2, \ldots, s$. These will be denoted by an s component vector $\mathbf{m}$, where s remains finite as $N \to \infty$. (ii) The total mean square of the random overlaps with the other $p - s$ patterns, denoted by r,

$$r = \alpha^{-1} \sum_{\mu > s}^{N\alpha} \langle\langle (m^\mu)^2 \rangle\rangle. \tag{4}$$

(iii) The Edwards-Anderson[11] order parameter $q = \langle\langle \langle S_i \rangle^2 \rangle\rangle$, which measures the local ordering. In terms of these quantities, f is given by

$$f = \tfrac{1}{2}\mathbf{m}^2 + \tfrac{1}{2}\alpha\left(\beta^{-1}\ln[1-\beta(1-q)] + \frac{(1-\beta)(1-q)}{1-\beta(1-q)} + \beta r(1-q)\right) - \beta^{-1}\langle\langle \ln 2\cosh\beta[(\alpha r)^{1/2} z + \mathbf{m}\cdot\boldsymbol{\xi}]\rangle\rangle. \tag{5}$$

The average $\langle\langle \ldots \rangle\rangle$ in Eq. (5) and in the following stands for averaging over the discrete distribution of $\boldsymbol{\xi}$ ($\xi^\nu = \pm 1$, $\nu = 1, \ldots, s$) and over a Gaussian variable z with zero mean and unit variance. The saddle-point equations for the order parameters are

$$\mathbf{m} = \langle\langle \boldsymbol{\xi} \tanh\beta[(\alpha r)^{1/2} z + \mathbf{m}\cdot\boldsymbol{\xi}]\rangle\rangle, \tag{6}$$

$$q = \langle\langle \tanh^2\beta[(\alpha r)^{1/2} z + \mathbf{m}\cdot\boldsymbol{\xi}]\rangle\rangle, \tag{7}$$

$$r = q[1-\beta(1-q)]^{-2}. \tag{8}$$

Note that the local field consists of two parts: A "ferromagnetic" part $\mathbf{m}$ which results from the s condensed overlaps and a "spin-glass" part $(r\alpha)^{1/2} z$, generated by the sum of the overlaps with the rest of the patterns.

Equations (6)–(8) have two types of solutions which are locally stable to variations in q, r, and $\mathbf{m}$: (1) A solution with $\mathbf{m} = 0$, $q, r \neq 0$. It represents a SG state which does not have a macroscopic overlap with any of the patterns. It does not contribute to associative memory and is truly "spurious." (2) FM solutions with $\mathbf{m} \neq 0$ in addition to q and r. These solutions, which exist for sufficiently small α, make the system useful for associative memory.

The most important FM solutions are characterized by *macroscopic overlaps with a single pattern*, $m^\nu = m\delta^{\nu\rho}$. There are $2N\alpha$ degenerate solutions of this type. As $\alpha \to 0$ they approach the finite-p Mattis states.

Zero temperature.—As $\beta \to \infty$, Eqs. (6)–(8) yield

$$m = \mathrm{erf}[m/(2r\alpha)^{1/2}], \tag{9}$$

$q = 1 - CT$, and $r = (1-C)^{-2}$, with

$$C = (2/\pi r\alpha)^{1/2} \exp(-m^2/2r\alpha). \tag{10}$$

For $\alpha > \alpha_c = 0.138$, there is no solution with $m \neq 0$. As α decreases below α_c, two solutions with $m \neq 0$ appear discontinuously. Out of the two, the one with the larger m is locally stable to variations in $\mathbf{m}$. This solution corresponds to a state which deviates only slightly from the precise stored pattern. The maximum deviation occurs at $\alpha = \alpha_c$, where $m = 0.967$, implying a 1.5% error. The percentage of errors decreases to zero very rapidly with decreasing α, see Fig. 1. The energy spin of these states is (at $T = 0$)

$$E = \tfrac{1}{2}\alpha(1-r) - \tfrac{1}{2}m^2. \tag{11}$$

At $\alpha = \alpha_c$, $E = -0.5014$ and it approaches -0.5 as $\alpha \to 0$. For a fully correlated state ($s_i = \xi_i^\rho$), $m = 1$, $r = 1$, and $E = -0.5$. Thus, at finite α, the system is able to slightly lower its energy by relaxing a small fraction of the spins, to accommodate for fluctuations in the overlaps of the other patterns.

As $\alpha \to 0$, Eq. (9) gives asymptotically, $m \sim 1 - (2\alpha/\pi)^{1/2} \exp(-1/2\alpha)$. The average number of errors is, therefore,

$$N_e \equiv \tfrac{1}{2}N(1-m) = N(\alpha/2\pi)^{1/2} \exp(-1/2\alpha). \tag{12}$$

This result implies that the average fraction of errors

VOLUME 55, NUMBER 14 PHYSICAL REVIEW LETTERS 30 SEPTEMBER 1985

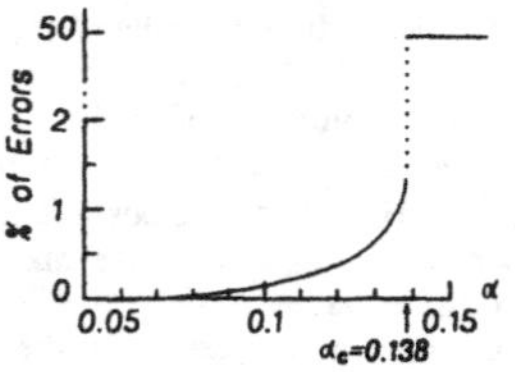

FIG. 1. Average percentage of errors in the FM states, as a function of α at $T=0$.

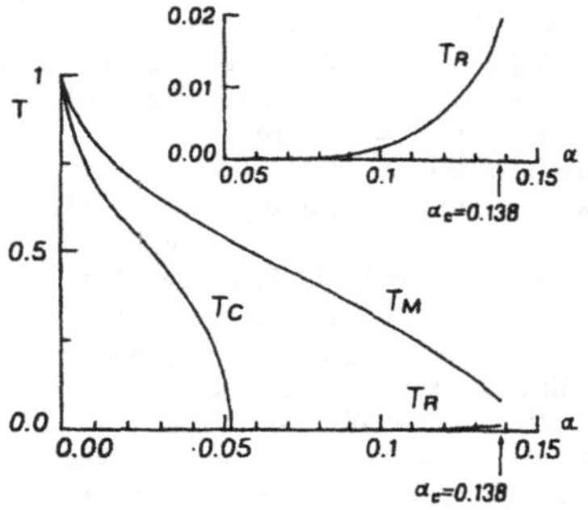

FIG. 2. Plots of critical temperatures of the FM states as a function of α. T_M is the temperature at which FM states first appear. T_c is the first-order transition at which these states become global minima. Replica-symmetry breaking occurs below T_R, which is displayed on an expanded scale in the inset.

N_e/N vanishes as $\alpha \to 0$ but the total number of errors vanishes only for $\alpha < (2\ln N)^{-1}$, in agreement with the results of Refs. 5 and 6. Indeed, substituting $\alpha = (2x\ln N)^{-1}$ in Eq. (12), one finds that $N_e \sim N^{(1-x)}$. Here we have shown that *with allowance for a small fraction of errors, very effectively controlled by small α, the system provides a useful mechanism for associated memory,* at least up to $\alpha_c \simeq 0.14$, in agreement with the estimate of Hopfield.[1]

Note, that even for $\alpha > \alpha_c$, there may still exist spin states, with nonzero m, which are *stable to single spin flips.* These states, however, will decay at finite T much faster than the thermodynamically stable or metastable FM states below α_c, which are surrounded by barriers of order N.

Equations (9) and (10) have a locally stable SG solution, $m=0$, $r=[1+(2/\pi\alpha)^{1/2}]^2$ for all α. Its energy, Eq. (11), equals $E=-1/\pi-(\pi\alpha/2)^{1/2}$. In the $\alpha \to 0$ limit, $E \to -1/\pi$, and $C=\beta(1-q)\sim 1$. This limit coincides with the value of E of the finite-p solutions, which mix n patterns, in the limit of $n \to \infty$.[4] This indicates that, as $p \to \infty$, the numerous states which mix infinitely many patterns merge to form the present SG phase. Comparing the energies of the SG and the FM states we find that the SG energy is lower for $0.051 < \alpha < 0.138$, whereas the FM state becomes the absolute minimum[12] below 0.051.

Finite-temperature.—The Sg phase appears continuously as T decreases via a second-order transition. Expansion of Eqs. (9) and (10) for small q and zero m yields for the SG transition temperature the value of $T_g = 1+\sqrt{\alpha}$. This phase is stable to the development of finite overlaps. The susceptibility of $\mathbf{m}$ with respect to its conjugate field $\mathbf{h}$, $d\mathbf{m}/d\mathbf{h}$, is positive, for all $T < T_g$ in the SG phase. Indeed, above $T=1$ other solutions do not exist for any α. For $\alpha < \alpha_c$, FM states, with a single macroscopic overlap, m, appear discontinuously as T is decreased below $T_M(\alpha)$. The mimimum value of T_M is 0.07, at $\alpha=\alpha_c$. As $\alpha \to 0$, T_M increases to 1 and $m(T_M)$ vanishes, thus approaching the finite-p continuous transition; see Fig. 2.

Near T_M, the FM states are metastable. If $\alpha < 0.051$, they become the global minima of f below a temperature $T_c(\alpha)$. At $T_c(\alpha)$, there is a thermodynamic first-order transition from a SG to a FM state, accompanied by a jump in m, q, and r, and by a latent heat. This transition is similar to that which occurs in a Landau theory of a scalar ϕ^6 model.[13] $T_c(\alpha)$ increases from 0 at $\alpha=0.051$ to unity at $\alpha=0$; see Fig. 2.

Additional FM solutions appear discontinuously for sufficiently small α and T. These solutions are characterized by $\mathbf{m}$ with more than one nonzero component. Some of these "mixture" states are metastable but none is an absolute minimum at any T. This happens first below $\alpha=0.03$, where a locally stable solution with three symmetric overlaps $[\mathbf{m}=m(1,1,1)]$ appears. For example, at $\alpha=0.02$ they appear below $T=0.14$. There are $-(1/3!)(2N\alpha)^3$ degenerate metastable states of this type. In general, the critical values of T and α below which mixture solutions exist decrease with increasing dimensionality s of $\mathbf{m}$, as $\alpha, T_M^2 \propto 1/s$. As $\alpha \to 0$, these solution coincide with the finite-p mixture states, described in Ref. 4. Note the similar role played here by T and α. The presence of either a thermal noise (T) or a static internal one (α) smooths the free-energy surface in $\mathbf{m}$ space, leading to the successive elimination of FM states as either T or α increases.

So far we have assumed that replica symmetry is unbroken. This is clearly incorrect at $T=0$, where the entropy per spin is $S=-\frac{1}{2}\alpha[\ln(1-C)+C/(1-C)]$ with $C \equiv \beta(1-q)$. It is negative for all replica-symmetric solutions, and hence is unphysical. The condition for stability of the replica-symmetric solutions is that the "eigenvalue"

$$\lambda = [1-\beta(1-q)]^2 - \alpha\beta^2\langle\langle \operatorname{sech}^4\beta[(r\alpha)^{1/2}z + \mathbf{m}\cdot\boldsymbol{\xi}]\rangle\rangle \quad (13)$$

be positive. We find that $\lambda < 0$ in the SG solution, for all $T < T_g$. On the other hand, the FM solutions are stable in a finite temperature regime, $T_R < T < T_M$. The instability temperature $T_R(\alpha)$ is shown in Fig. 2, for the states with single overlaps. It decreases very rapidly to zero with α, as

$$T_R(\alpha) \sim (8\alpha/9\pi)^{1/2}\exp(-1/2\alpha).$$

This symmetry breaking is not expected to modify most of our results. The existence, at small α, of SG and FM metastable, or stable, states still holds. Furthermore, the properties of the FM states, e.g., the average number of errors and the energy, are hardly affected, since their replica-symmetry breaking occurs (continuously) only at very low temperatures T_R, where the system is already almost completely ordered ($q \sim 1$). This is also exemplified by the fact that the $T=0$ negative entropy of the replica-symmetric FM states is extremely small. Already at α_c, $S \sim -1.4\times10^{-3}$ in the FM state, compared with -7×10^{-2} in the SG state and -1.6×10^{-1} in the model of Kirkpatrick and Sherrington.[7]

The free energy of the solution with broken replica symmetry is expected to be higher than that of the replica-symmetric one. Hence, the values of $T_c(\alpha)$ and the maximum value of α for which the FM states become the absolute minima will be higher than the above estimates, Fig. 2. Also, with replica-symmetry breaking, the value of α_c, where metastable FM states first appear, may be slightly higher than 0.138.

The most important implication of replica-symmetry breaking at low T concerns the structure of the states of the system. Calculations near $T = T_g$ indicate that the symmetry-breaking scheme is of the same nature as that of the model of Kirkpatrick and Sherrington,[7] with continuous-order functions $q(x)$ and $r(x)$ $(0 \leq x \leq 1)$ replacing q and r.[10] According to the accepted interpretation of this scheme, the following remarkable organization of states, at small α and T, emerges. States are first classified according to their macroscopic overlap $\mathbf{m}$ with the embedded patterns. This already leads to enormous degeneracy, with $2N\alpha$ degenerate states overlapping (almost completely) with a single pattern. States with different macroscopic $\mathbf{m}$ are far apart from each other in phase space, and are separated by barriers of order N. Energy differences between nondegenerate states (e.g., SG and FM) are proportional to N. On a finer level, each of these macroscopically different stable and metastable phases actually represents (at $T=0$) many (an infinity as $N \to \infty$) degenerate states, organized in a hierarchical ultrametric structure.[14] These states have the same macroscopic properties but differ in the organization of their random components (i.e., the location of the "errors"). These macroscopically equivalent states are separated by barriers which are probably not higher than $O(\sqrt{N})$.

We have studied the effect of the application of external fields conjugate to ξ_i. Many new features appear. They will be described in an extended presentation of this work.

We have been stimulated by discussions with A. J. Bray, E. C. Posner, and G. Weisbuch. We thank J. J. Hopfield for helpful comments on the manuscript. The research of two of us (D.J.A. and H.S.) was supported in part by the Fund of Basic Research administered by the Israel Academy of Science and Humanities.

[1]J. J. Hopfield, Proc. Nat. Acad. Sci. USA **79**, 2554 (1982), and **81**, 3088 (1984); J. J. Hopfield, D. I. Feinstein, and R. G. Palmer, Nature **304**, 158 (1983).

[2]W. A. Little, Math. Biosci. **19**, 101 (1974); W. A. Little and G. L. Shaw, Math. Biosci. **39**, 281 (1978).

[3]P. Peretto, Biol. Cybernet. **50**, 51 (1984), and in "Disordered Systems and Biological Organization," Proceedings of the Les Houches Meeting, February 1985, edited by E. Bienenstock *et al.* (to be published); M. Y. Choy and B. A. Huberman, Phys. Rev. A **28**, 1204 (1983).

[4]D. J. Amit, H. Gutfreund, and H. Sompolinsky, Phys. Rev. A **32**, 1007 (1985).

[5]G. Weisbuch, private communication, and in "Disordered Systems and Biological Organization," Proceedings of the Les Houches Meeting, February 1985, edited by E. Bienenstock *et al.* (to be published).

[6]E. C. Posner, private communication.

[7]S. Kirkpatrick and D. Sherrington, Phys. Rev. B **17**, 4384 (1978).

[8]Most of our results depend on the discreteness of the ξ^μ. The model with a Gaussian distribution of ξ^μ was studied by J. P. Provost, to be published, and by A. J. Bray, private communication.

[9]D. C. Mattis, Phys. Lett. **56A**, 421 (1976).

[10]G. Parisi, Phys. Rev. Lett. **50**, 1946 (1983).

[11]S. F. Edwards and P. W. Anderson, J. Phys. F **5**, 965 (1975).

[12]Note that unlike the SG order parameters, $\mathbf{m}$ is determined by the *minimum* of the free energy.

[13]R. Bausch, Z. Phys. **254**, 81 (1972).

[14]M. Mezard, G. Parisi, N. Sourlas, G. Toulouse, and M. Virasoro, J. Phys. (Paris) **45**, 843 (1984).

Proc. Natl. Acad. Sci. USA
Vol. 83, pp. 1695–1698, March 1986
Biophysics

Spin glass model of learning by selection

(Darwinism/categorization/Hebb synapse/ultrametricity/frustration)

GÉRARD TOULOUSE, STANISLAS DEHAENE, AND JEAN-PIERRE CHANGEUX

Unité de Neurobiologie Moléculaire and Laboratoire Associé au Centre National de la Recherche Scientifique, no. 270, Interactions Moléculaires et Cellulaires, Institut Pasteur, 25 rue du Docteur Roux, 75724 Paris Cédex 15, France

Contributed by Jean-Pierre Changeux, October 31, 1985

ABSTRACT **A model of learning by selection is described at the level of neuronal networks. It is formally related to statistical mechanics with the aim to describe memory storage during development and in the adult. Networks with symmetric interactions have been shown to function as content-addressable memories, but the present approach differs from previous instructive models. Four biologically relevant aspects are treated—initial state before learning, synaptic sign changes, hierarchical categorization of stored patterns, and synaptic learning rule. Several of the hypotheses are tested numerically. Starting from the limit case of random connections (spin glass), selection is viewed as pruning of a complex tree of states generated with maximal parsimony of genetic information.**

Aside from the inneist, or preformist, point of view, according to which experience does not cause any significant increase of order in an already highly structured brain organization, two main classes of learning theories have been proposed and discussed (for review see ref. 1). On the empiricist side, the initial state is considered as a *tabula rasa*, and the whole internal organization results from direct instructive prints by the environment. Alternatively, selectionist theories postulate that the increase of internal order associated with experience is indirect (2–8). The organism generates, spontaneously, variable patterns of connections (3) at the sensitive period of development, referred to as "transient redundancy" (6), or variable patterns of activity named prerepresentations (7, 8) in the adult. Interaction with the environment merely selects or selectively stabilizes the preexisting patterns of connections and/or firings that fit with the external input, a step named "resonance" (7, 8). As a correlate of learning, connections between neurons are eliminated and/or the number of accessible firing patterns is reduced.

Several attempts to model learning at the level of large ensembles or "assemblies" of interconnected neurons have been made in quantitative terms mostly with the help of statistical mechanics (9, 10). Their revival is largely due to the introduction by Hopfield (10) of the conceptual simplification that (*i*) if one restricts the interactions between neurons only to symmetric ones, this allows for the introduction of an energy function and, as a consequence, the dynamics of neuronal networks can be viewed as a downhill motion in an energy landscape and (*ii*) then, the reallowance for dissymmetric interactions does not discontinuously upset the picture.

On the other hand, such models still belonged to the empiricist mode of learning with the initial state taken as a flat energy landscape (*tabula rasa*) that becomes progressively structured and complex by direct instructions from the environment.

The publication costs of this article were defrayed in part by page charge payment. This article must therefore be hereby marked "*advertisement*" in accordance with 18 U.S.C. §1734 solely to indicate this fact.

The aim of this communication is to propose a model of learning by selection based on an advance in the statistical mechanics of disordered systems—namely, the theory of spin glasses (11–13). In contrast to the empiricist approach, the initial state is viewed as a complex energy landscape with an abundance of valleys typical of spin glasses with learning consisting of the progressive smoothening and gardening of this landscape. The paper also contains a biological critique of the standard instructive version of the Hopfield model, referred to here in short as the instructive model. The main proposals for a selectionist model of learning are outlined and preliminary numerical results are reported and discussed.

The Activity of Neuronal Networks Described by Statistical Mechanics

The all-or-none firing of a neuron is represented by a spin that can take two values: $S = +1$ (firing), $S = -1$ (rest). A pattern of activity, α, of a network of N neurons is represented by a spin configuration (S_i^α), $i = 1, \ldots, N$, that lies at one of the corners of a hypercube in N-dimensional configuration space. Two patterns of activity, α and β, may then be compared through their overlap, which is an index of proximity or matching in configuration space:

$$q^{\alpha\beta} = \frac{1}{N} \sum_{i=1}^{N} S_i^\alpha S_i^\beta. \quad [1]$$

The neurons interact via binary synapses of synaptic strength, T_{ij}. With the assumption (5) of symmetric interactions $T_{ji} = T_{ij}$, an energy function can be written as follows:

$$E = -\frac{1}{2} \sum_{i \neq j} T_{ij} S_i S_j - \sum_i h_i S_i, \quad [2]$$

where h_i is a local field acting on spin S_i and is often used to represent an external input (yielding an apparent shift of the firing threshold of a neuron). The neuron dynamics is such that, in the absence of probabilistic effects leading to random spontaneous activity, each spin tends to decrease its energy. A stable configuration is, therefore, a local minimum of the energy E. On the other hand, probabilistic effects can be described by introducing a finite temperature (14).

Synaptic modifications have been hitherto often expressed by the learning rule:

$$\Delta T_{ij} \sim \langle S_i S_j \rangle, \quad [3]$$

where the brackets mean some time average. This expression, referred to as the "generalized Hebb rule," differs from the original Hebb rule (15), which may be written as

$$\Delta T_{ij} \sim \left\langle \left(\frac{S_i + 1}{2} \right) \left(\frac{S_j + 1}{2} \right) \right\rangle, \quad [4]$$

and exclusively takes into account reinforcements of excitatory synapses. Rule 3 has attractive features—it is local and formally natural—but it also has undesirable ones—for instance, when neuron i makes inhibitory synapses with neuron

j, rule 3 would predict a modification of synaptic strength T_{ij} and eventually a reversal from inhibitory to excitatory, if none of the neurons is firing and the synapses are silent.

Instructive Models of Learning

Instructive models of learning (10) postulate that, in the initial state, the interactions between neurons are vanishingly small and the energy landscape is flat (*tabula rasa*). Storage into memory of an activity pattern α, where $S_i = \mu_i^\alpha$, results from the following synaptic modification,

$$\Delta T_{ij} = \frac{1}{N} \mu_i^\alpha \mu_j^\alpha, \qquad [5]$$

and the network is said to have learned M patterns, $\alpha = 1, \ldots, M$, when the interactions have been set to

$$T_{ij} = \frac{1}{N} \sum_{\alpha=1}^{M} \mu_i^\alpha \mu_j^\alpha, \qquad [6]$$

as a consequence of the successive prints of the M input patterns. With such interactions **6**, the network functions as a distributed, fault-tolerant, content-addressable memory. Starting from any input data, the network configuration rapidly converges toward a local minimum and recognizes the closest stored memory pattern (provided M is not too large and no confusion takes place) (10).

Assuming that the learned patterns are random and uncorrelated, Hopfield (10) has suggested that the maximal storage capacity is $M_c = \gamma N$ (with $\gamma < 1/2$, since each pattern corresponds to N bits of information, and the information is stored in the interactions, with $N^2/2$ of them) and further has shown that loss of recall occurred around $\gamma = 0.15$, an estimate that has been confirmed by subsequent analytical calculations (16, 17).

Such instructive mode of learning, if legitimate and useful for artificial intelligence, does not hold for the brain for the following reasons.

(*i*) As more and more patterns are stored, according to formula **6**, the synaptic patterns keep changing sign. As was already stressed by Hopfield (10), it is the signs of the interactions T_{ij} together with their absolute values that are responsible for the proper shaping of the energy landscape. Thus, storing a new memory amounts largely to reversing the signs of a particular set of synaptic strengths. Yet, no physiological evidence exists of synaptic sign reversal, such as a shift of postsynaptic response from excitatory to inhibitory, as a cellular correlate of learning (1).

(*ii*) Up to now, the ultimate organization of stored patterns in memory space has been viewed as a configuration-space-filling *jardin à la française*, with a regular distribution of the basins of attraction corresponding to the various stored patterns (18). A more hierarchical distribution less prone to confusions, with categorization properties and correlations between stored patterns, appears more appropriate for higher brain functions, even if it is wasteful of configuration space.

(*iii*) The hypothesis of an initial state with vanishing interactions does not take into account the existence of an already connected and functional neuronal network at the moment learning occurs.

Spin Glasses

Spin glasses, by definition, consist of networks of spins with symmetric random (positive and negative) interactions. The energy is simply given by

$$E = -\frac{1}{2} \sum_{i \neq j} T_{ij} S_i S_j. \qquad [7]$$

The mean field theory of spin glasses (valid for a fully connected network and a number of spins N large) is intricate but yields a simple physical picture for the energy landscape. The total number of local minima in configuration space is exponential in N. However, the dominant valleys (their importance is weighed by the Boltzmann factor, which favors low-energy valleys) have positive mutual overlaps. More precisely, to any spin state, time-reversal symmetry associates another state with all spins flipped and the same energy, so that the previous statement holds for each half of the valleys separately. In geometric terms, the dominant valleys of a spin glass lie within a cone, centered at the origin and of right angle in configuration space (one-half of the valleys within one sector of the cone, the other half within the opposite sector). Such a right-angle cone spans a very small fraction of configuration space, which is another way of stating that the dominant valleys are strongly intercorrelated.

Furthermore, the distribution of these valleys possesses an ultrametric structure (11)—i.e., a hierarchical organization of clusters within clusters—in configuration space. A similar ultrametric distribution occurs in taxonomy when species are classified, for instance, according to protein sequence homologies (19).

The spin glass energy landscape thus exhibits, spontaneously, a categorized organization. The appearance of ultrametricity for large heterogeneous assemblies is a remarkable feature, which may be partly understood by realizing that there are fewer bonds ($N^2/2$) than possible spin configurations (2^N), and thus that the energy states have to exhibit some form of correlation. Indeed, if ever random multiple-spin (ternary, quaternary, etc.) interactions are introduced, since they occur in larger combinatorial number than ordinary binary interactions, the energy landscape tends to become more and more rough, and the notion of hills, passes, and valleys eventually disappears (20).

Spin Glass Model of Learning by Selection

The proposal we make here is that the theory and formalism of spin glasses appear particularly adequate to model learning by selection. As discussed (7), selection has been postulated to operate during development on a variable connective organization (3, 6) or, in the adult, on variable patterns of activity named prerepresentations (7, 8). In both cases, a significant (though limited) randomness characterizes the initial state. This legitimizes the modeling of this "fringe" state by a network of N neurons with randomly connected excitatory and inhibitory synapses that would behave as a spin glass.

In brief, a spin glass has an energy landscape with: (*i*) an abundance of valleys and (*ii*) dominant valleys strongly intercorrelated (positive mutual overlaps) in a tree-like fashion.

Item (*i*) gives to this network the property of a "generator of internal diversity" (7, 21)—that is, each valley corresponds to a particular set of active neurons and plays the role of a prerepresentation. Item (*ii*) further indicates a spontaneous categorization of the prerepresentations.

Learning with very small synaptic changes is both advantageous and possible. It is advantageous because it tends to preserve ultrametricity—i.e., the spontaneous hierarchical categorization of the prerepresentations. It is possible because learning by selection involves the stabilization of preexisting valleys instead of creation of new ones. The foremost constraint is interdiction of synaptic sign reversals. Other proposals for the learning rule fall into two categories. The rule should remain local but avoid the inconsistencies mentioned above. In addition, a weighted factor is introduced and contributes to the selection of the input patterns that match the prerepresentations (resonance). This selective factor enters naturally as a time average, if one assumes that the synaptic changes occur during a relaxation time of the configuration initiated by the input pattern. More coherent synaptic modifications will favor input patterns that match with preexisting valleys.

In summary, learning by selection may occur as follows: An input pattern sets an initial configuration that converges toward an attractor of the dynamics (bottom of a valley, i.e., a prerepresentation). The energy of this selected valley is lowered by synaptic modifications (particularly if the learning time is longer than the relaxation time), and its basin of attraction is shifted and enlarged at the expense of other valleys. Starting from a hierarchical distribution of valleys, the learning process can be viewed as pruning of a tree, analogous to that occurring in the course of phylogenesis. As a consequence the whole energy landscape evolves during the learning process, the already stored information influencing the prerepresentations available for the next learning event. Moreover, the constraints on the synaptic modifications give internal rigidity to the system. Not every external stimulus can equally be stored. Selection by the external stimuli among internal prerepresentations has its counterpart in selection by the internal network among external inputs. In a parallel assembly of such networks, one may further speculate that an input pattern will select its memory location, the place where it fits, if any.

A prerequisite of learning by selection is the existence of a nontrivial valley structure prior to the interaction with the outside world. There are only two ways whereby a neuronal network with symmetric binary connections can exhibit such a structure.

One way is via frustration (22). The frustration function $\Phi(c)$ of a closed loop (c) of interacting spins is the product of the interactions around the loop:

$$\Phi(c) = \prod_{(c)} T_{ij}. \quad [8]$$

If $\Phi(c) > 0$, it is possible to find a spin configuration around the loop such that each bond is satisfied. If $\Phi(c) < 0$, this is not possible, and the spin configurations can be at best partially satisfactory. In this latter case, the loop is said to be frustrated. Frustration is a source of metastability and degeneracy. By definition, an unfrustrated network, where all loops are unfrustrated, has only two minima, related by time-reversal symmetry.

The other way to get multiplicity of valleys is via disconnection. A network, broken into p disconnected unfrustrated clusters, has 2^p minima.

The rich valley structure of a long-range, fully connected spin glass stems from frustration. It differs sharply from the valley structure of a set of disconnected clusters, although intermediate cases are conceivable. Indeed, if all the neurons are decoupled, the storage capacity is in some sense maximal, but all the useful properties of a distributed, associative memory are lost.

Any realistic neuronal network model for biological learning by selection should include both frustration and disconnection. Not only is the initial connectivity in central nervous systems far from maximal (see for instance, the anatomical evidence for columns) but the occurrence of synaptic elimination during development is well documented (6, 23). The constrained learning rules, introduced above, preserve frustration because they forbid sign reversals and allow for synaptic elimination.

Numerical Implementations and Results

Our model contains a set of hypotheses that are precise enough to be tested numerically, and we have begun a systematic investigation of their consequences. Some salient results, for small network sizes ranging from $N = 30$ to $N = 200$, are reported. A more elaborate discussion will be presented elsewhere. For the sake of clarity, we have studied separately the effects of each of our basic hypotheses and compared them with the instructive model.

The *Tabula Rasa* Withdrawn. In the initial state, the synapses are set with random signs and an average strength S. It is known from spin glass theory that partial learning of an arbitrary pattern can be obtained with synaptic increments of order $S/\sqrt{N}$ for a complete graph of N neurons. Keeping unchanged the form of the generalized Hebb rule, for the sake of comparison,

$$\Delta T_{ij} = \frac{\varepsilon S}{\sqrt{N}} \mu_i^a \mu_j^a, \quad [9]$$

we have checked that retrieval quality [more precisely, in notations defined below, a normalized index $R = (q_a - q_b)/(1 - q_b)$] is a function of ε, independent of network size N. Furthermore, for $\varepsilon \gtrsim 2.5$, retrieval quality was found to be practically perfect.

As more and more patterns are stored, the strength of a given synapse undergoes a random walk, with steps of length $\varepsilon S/\sqrt{N}$ starting from the initial values $+S$ or $-S$. Whenever the strength of a synapse hits the value zero (an occurrence possible after learning $p \sim \sqrt{N}/\varepsilon$ patterns) it is prevented from changing sign. Two subsequent rules are conceivable, and both have been examined. Either the synapse is altogether eliminated, which is a strong form of the constraint, or its strength is temporarily blocked at zero until it eventually receives an increment of the correct sign, which obviously constitutes a weaker constraint.

In the case of the strong constraint, ruin theory (24) predicts that the fraction of surviving synapses will decay as $1/\sqrt{p}$, for p large (where p is the number of memorized patterns). With the weaker constraint, the fraction of nonvanishing synapses tends toward a constant.

We have defined a global learning index G and studied its variation as a function of p. This learning index is the difference between retrieval overlaps (a retrieval overlap is the overlap between an input pattern and its attractor) measured after learning and before learning, summed over all p patterns. Note that learning an additional pattern modifies the retrieval of previously stored patterns. Thus, the global index has to be completely recalculated after each learning event.

For p small, $G(p)$ is linear in p; for $\varepsilon \gtrsim 2.5$, the slope is the same for the *tabula rasa* condition or the non*tabula rasa* condition. Both curves are also asymptotically linear for p large (with smaller slope) and superimposed, showing a regime where the influence of the initial state has been lost. In the intermediate regime, the two curves differ. In addition, there is a difference between the cases with sign constraints (under weak or strong form) and the case without, which is clearly observed even on the smaller samples ($N = 30$).

Learning Strength and Selectivity. For comparison with previous studies, the values of ε chosen above were so large as to "burn a hole" in the energy landscape, for any input pattern. Such storage is clearly unselective. We have plotted the statistics of retrieval-overlap-after-learning q_a versus retrieval-overlap-before-learning q_b for various values of ε. Starting from $\varepsilon = 0$, for which the curve is obviously along the diagonal $q_a = q_b$, there is a range of values of ε for which marked fluctuations in retrieval quality are observed, before the hole burning regime sets in, with $q_a = 1$.

These results prove the existence of a diversity and an incipient selectivity. Note that, in these simulations as in earlier studies (10, 14, 16, 17), the learned configurations are the input patterns, because no relaxation effects are taken into account. The selectivity in the learning process, resulting from the existence of an initial structured energy landscape, will be enhanced by averaging over time. A learned configuration will then be intermediate between an input pattern and its attractor, and the total amount of synaptic modification will be larger for a matching pattern than for a nonmatching one.

Alternatives to the "Generalized Hebb Rule"

Consistent with current models of regulation of synapse efficacy inspired from the allosteric properties of the acetylcholine receptor (25), one may express the change in the efficacy of a synapse between neurons i and j as a function of the activities of the other neurons k afferent on j, as

$$\Delta T_{ij} \sim \sum_k C^k_{ji} \langle S_i S_k \rangle, \qquad [10]$$

the coefficient C^k_{ji} being determined by chemical and geometrical factors, such as the relative positions of the synapses (i, j) and (k, j) on the dendrites of neuron j. Such a general expression points to the possibility that the printing process does not stabilize with exact precision a given imposed pattern but rather introduces a shift between an input and its trace. However, at this stage, we limit ourselves to a modification of the generalized Hebb rule **3**, which eliminates its most obvious flaws while keeping symmetric interactions. A simple way consists in replacing rule **5** by

$$\Delta T_{ij} = \frac{1}{4N} [3\, \mu_i^\alpha \mu_j^\alpha + (\mu_i^\alpha + \mu_j^\alpha) - 1]. \qquad [11]$$

Then, no synaptic modification occurs if $\mu_i = \mu_j = -1$, as desired. Consequently, every neuron will not be equally stabilized after the storage of one pattern and any stored pattern will have some labile spots.

As a first step, we have looked at the consequences of learning rule **11** in comparison with the generalized Hebb rule **5**, within the instructive model. The new rule has been found to affect the retrieval quality of the Hopfield model significantly. The reduction of the performances is comparable in magnitude to the effect of withdrawing the *tabula rasa* hypothesis (with generalized Hebb rule and without synaptic sign constraints) as described above.

Conclusions

Learning by selection is a generalization to the development of neuronal networks (3, 6) and to higher brain functions (4, 5, 8) of the selectionist (or Darwinist) mechanisms that have already been successfully applied to the evolution of species and antibody biosynthesis (2, 19). The spin glass model described here creates an additional bridge between statistical mechanics and theoretical biology and may offer original theoretical "tools" to quantitatively treat the neuronal bases of highly integrated brain processes. At this stage the model contains a severe restriction in scope due to its limitation to static memory patterns (time sequences and synchronicity effects are beyond present investigation).

One major neurobiological outcome of our model is the description of a memory with a hierarchical, ultrametric, structure which offers possibilities of "categorization" (11) on a rather simple basis—an initial "fringe" state of random synapses yielding a spin glass-like energy landscape and strong learning constraints at the storage level. This does not preclude, but rather complements, a hierarchical categorization at the encoding level (26) that originates, for instance, from a more innate organization of the sensory analyzers at the cortical level with multiple entries of the inputs into a layered architecture. In this framework, our study considers the less genetically determined layers that would then receive partially precategorized inputs.

In conclusion, this learning process can be epitomized as pruning (by selection) instead of packing (by instruction). It is too early yet to predict what will be the most fruitful implementation of this model, but two ideas appear profound and worth stressing. The first idea for the physicist is that selection, par excellence, is pruning of a tree and that the spin glass supplies the tree with parsimony of genetic information. The second idea for the biologist is that random synapses in a neuronal network cannot be equated with a *tabula rasa*.

We acknowledge valuable discussions with D. Amit, E. Bienenstock, J. J. Hopfield, H. Sompolinsky, and M. Virasoro. G.T. thanks the Aspen Center for Physics where part of this work was carried out. Computations were performed on the IBM 4341 of the Centre de Calcul de l'Ecole Normale Supérieure.

1. Marler, P. & Terrace, H., eds. (1984) *The Biology of Learning* (Springer, Berlin).
2. Jerne, N. (1967) in *The Neurosciences: A Study Program*, eds. Quarton, G., Melnechuk, T. & Schmitt, F. O. (The Rockefeller Univ. Press, New York), pp. 200–208.
3. Changeux, J. P., Courrège, P. & Danchin, A. (1973) *Proc. Natl. Acad. Sci. USA* **70**, 2974–2978.
4. Edelman, G. (1978) *The Mindful Brain* (MIT Press, Cambridge, MA).
5. Finkel, L. & Edelman, G. (1985) *Proc. Natl. Acad. Sci. USA* **82**, 1291–1295.
6. Changeux, J. P. & Danchin, A. (1976) *Nature (London)* **264**, 705–712.
7. Changeux, J. P., Heidmann, T. & Patte, P. (1984) in *The Biology of Learning*, eds. Marler, P. & Terrace, H. (Springer, Berlin), pp. 115–133.
8. Heidmann, A., Heidmann, T. & Changeux, J. P. (1984) *C.R. Acad. Sci. Ser. 2*, **299**, 839–844.
9. Little, W. & Shaw, G. (1978) *Math. Biosci.* **39**, 281–290.
10. Hopfield, J. J. (1982) *Proc. Natl. Acad. Sci. USA* **79**, 2554–2558.
11. Mézard, M., Parisi, G., Sourlas, N., Toulouse, G. & Virasoro, M. (1984) *Phys. Rev. Lett.* **52**, 1156–1159.
12. Toulouse, G. (1984) *Helv. Phys. Acta* **57**, 459–469.
13. Mézard, M. & Virasoro, M. (1985) *J. Phys. (Les Ulis, Fr.)* **46**, 1293–1307.
14. Peretto, P. (1984) *Biol. Cybern.* **50**, 51–62.
15. Hebb, D. (1949) *The Organization of Behavior* (Wiley, New York).
16. Amit, D. J., Gutfreund, H. & Sompolinsky, H. (1985) *Phys. Rev. A* **32**, 1007–1018.
17. Amit, D. J., Gutfreund, H. & Sompolinsky, H. (1985) *Phys. Rev. Lett.* **55**, 1530–1533.
18. Hopfield, J. J., Feinstein, D. I. & Palmer, R. G. (1983) *Nature (London)* **304**, 158–159.
19. Ninio, J. (1983) *Molecular Approaches to Evolution* (Princeton Univ. Press, Princeton, NJ).
20. Derrida, B. (1980) *Phys. Rev. Lett.* **45**, 79–82.
21. Stein, D. L. & Anderson, P. W. (1984) *Proc. Natl. Acad. Sci. USA* **81**, 1751–1753.
22. Toulouse, G. (1977) *Commun. Phys.* **2**, 115–119.
23. Cowan, W., Fawcett, J., O'Leary, D. & Stanfield, B. (1984) *Science* **225**, 1258–1265.
24. Feller, W. (1957) *An Introduction to Probability Theory and Its Applications* (Wiley, New York).
25. Heidmann, T. & Changeux, J. P. (1982) *C.R. Acad. Sci. Ser. 2* **295**, 665–670.
26. Virasoro, M. (1985) in *Disordered Systems and Biological Organization*, eds. Bienenstock, E., Fogelman, F. & Weisbuch, G. (Springer, Berlin), in press.

Tome 47 N° 11 NOVEMBRE 1986

LE JOURNAL DE PHYSIQUE

J. Physique 47 (1986) 1857-1864 NOVEMBRE 1986, PAGE 1857

Classification
Physics Abstracts
87.30G — 64.60C — 75.10H — 89.70

The ultrametric organization of memories in a neural network

N. Parga (*) and M. A. Virasoro

International Centre for Theoretical Physics, Trieste
and, Dipartimento di Fisica, Universita' di Roma I « La Sapienza », Roma, Italy

(Reçu le 7 octobre 1985, accepté sous forme définitive le 21 mai 1986)

Résumé. — Dans le modèle de mémoire humaine proposé par Hofpield, les mots à emmagasiner doivent être orthogonaux. Du point de vue de la catégorisation, cette condition est peu commode à moins que ces mots ne soient des prototypes appartenant à des catégories primordiales différentes. Dans ce cas, le modèle doit être complété de façon à pouvoir emmagasiner tout l'arbre hiérarchique : des sous-catégories appartenant à une même catégorie, des éléments appartenant à une sous-catégorie, et ainsi de suite. Nous utilisons des résultats récents sur la théorie du champ moyen des verres de spin pour démontrer que cette réalisation est possible avec une modification minimale de la règle de Hebb. On trouve que la catégorisation est une conséquence naturelle d'une étape de précodage structurée en couches.

Abstract. — In the original formulation of Hopfield's memory model, the learning rule setting the interaction strengths is best suited for orthogonal words. From the point of view of categorization, this feature is not convenient unless we reinterpret these words as primordial categories. But then one has to complete the model so as to be able to store a full hierarchical tree of categories embodying subcategories and so on. We use recent results on the spin glass mean field theories to show that this completion can be done in a natural way with a minimal modification of Hebb's rule for learning. Categorization emerges naturally from an encoding stage structured in layers.

1. Introduction.

Neural networks have been proposed as associative memories to model the behaviour of human long term memory [1, 2]. A neural network is essentially an amorphous aggregate of neurons that, in this context, are idealized as physical two-state devices [3] coupled through a symmetrical matrix J_{ij} that represents the synapses.

During « learning » the J_{ij} are modified by the environment in a time scale T_J larger than the time scale T_S needed by the neurons to adapt to the coupling (retrieval process). So defined the model has the following properties.

i) It stores a certain amount of information (if N is the number of neurons, one can store $< - 0.15\, N$ [2], [17] words of N bits).

ii) The retrieval of the information is such that from a partial, deteriorated, knowledge of one word the system is able to reconstruct the full word (the number of bits that can be so reconstructed decreases as we try to store more words and goes rapidly to zero when one surpasses the storage capability referred to in i).

iii) After storing P words the system — requested to retrieve a particular pattern — may, generically, answer with a « spurious » word, i.e. one that has not been originally stored. The number of spurious words increases at least exponentially with P [4].

iv) The storage prescription adopted in e.g. reference [2] works best if the words are at least approximately orthogonal.

The type of organization of a neural network is natural for a biological system. The idea that the J_{ij} are modified by the environment eliminates the paradox that would follow if the information about the coupling matrix among 10^{14} neurons had to be contained in the DNA [5]. Other types of architecture would be difficult to reconcile with prevalent ideas about the type of order that can evolve from evolution.

From the neural sciences point of view, such a model has the drawback of assuming symmetric synapses. Modifying it with an asymmetric matrix J_{ij} leads us outside the well-known domain of statisti-

(*) Permanent address : Centro Atomico Bariloche, 8400 Bariloche, Argentina.

cal mechanics into the realms of cellular automata where must less is known. As one is trying to attack a problem hitherto overlooked in statistical mechanics, namely, how to constrain a system to have certain states as equilibrium states it is clearly safer to proceed by steps studying first the behaviour of a symmetric neural network.

The major testable predictions of these models fall in the realm of cognitive psychology : we should confront the behaviour of a neural network with what we know about human memory. Comparisons of this type have already been discussed in the literature. For instance the spurious words have been seen as a proof that these models are not just repetitive but have the ability to imagine new representations [6]. The storage capability [2, 7] has also been discussed in this context.

From this point of view, we notice as a serious flaw in the model that words or patterns to be stored have to be encoded in approximately orthogonal vectors just exactly the opposite of the way human memory works. It is apparent that when we try to memorize new information we look for all the possible relationships with previously stored words [8]. If we can classify it, that is place it in a tree of categories we do it with so much eagerness that sometimes we just censor the data so as to cancel any exceptional anomalous features. However, if the word is really orthogonal to all the previously stored ones we have reluctantly to initiate a new category.

We therefore propose a reinterpretation of the patterns discussed in [1, 2, 4, 7] as primordial categories. Then the problem of storage capability is renormalized : how many totally uncorrelated patterns are we able to memorize ?

Another problem of the model, closely related to the previous one, is that when errors in retrieval occur, either because of spurious states or because of the limits in capacity, there is no way to control the quality of these errors. On the other hand, biosystems cannot afford certain errors more than once in a lifetime. A hierarchy among errors is therefore mandatory. This can be automatically implemented if we have classified the patterns in a hierarchial tree of categories. An unimportant error will be to confuse individuals inside a category while a more serious one will be to confuse categories (incidentally it follows that this type of categorization must occur very early in the evolutionary tree : distinguishing between prey and predators is more vital than distinguishing among varieties of predators).

Both problems above will be solved if we could modify the model in such a way that the patterns to be memorized instead of being orthogonal fall into a hierarchical tree. This type of organization (ultrametricity) appears spontaneously in a Sherrington-Kirkpatrick spin-glass [9]. If we assume that the synaptic connections J_{ij} are chosen independently at random with zero average and standard deviation $\sigma = 1/\sqrt{N}$, and look for the solutions of the equation (labelled by α)

$$S_i^\alpha = \operatorname{sign}\left(\sum_j J_{ij} S_j^\alpha\right) \quad \alpha = 1, 2, \ldots \qquad (1.1)$$

whose energies

$$E^\alpha = \sum_{i>j} J_{ij} S_i^\alpha S_j^\alpha \qquad E^1 \leqslant E^2 \leqslant E^3 \ldots \quad (1.2)$$

do not differ too much from the ground state energy

$$\lim_{N\to\infty} (E^\alpha - E^1) = \text{finite}\,, \qquad (1.3)$$

then one can prove that chosen three states at random the two of them that are nearer to each other lie exactly at the same distance of the third. The distance is the natural one : $(d(\alpha, \beta))^2 = \frac{1}{N}\sum_{i=1}^{N} (S_i^\alpha - S_i^\beta)^2$. This property, called ultrametricity, implies that the states can be located in the higher ends of the branches of a tree. The distance between two states is then measured by how much one has to go down along the tree before the two branches converge.

Due to the similarity between the SK spin glass and a neural network there is an exciting possibility that the type of architecture of the network leads spontaneously to categorization.

In this paper we build a detailed model to check whether something like that could happen.

Human memory is an information processing system that can be conveniently analysed in three subprocesses : encoding, storage and retrieval. In the first stage stimuli are encoded into a form compatible with storage. In section 2 we show how such an encoding processing unit should be architectured to generate words that are ultrametric. We show that a system organized in layers so that part of the stimuli enter at different levels of the encoding process naturally leads to a hierarchical organization of the words to be stored.

In section 3 we study the storage stage. We show that a minimal modification of the Hebb rule is required. The J_{ij} that come out have the same structure as those of the SK spin glass [10].

In section 4 we consider the retrieval process. We restrict our analysis to the thermodynamic limit where, N, the number of neurons, goes to ∞, while P/N goes to zero. We do not discuss the problem of storage capability but compare our results on retrieval of spurious state with the ones derived in the Hopfield model [4]. We find that there are two types of spurious states : whether they mix patterns belonging to a same category or mix categories. A hierarchy of errors emerges. In section 5 we discuss our results.

2. Encoding. Generation of ultrametric words.

Approximately orthogonal words can be generated in the following way : one chooses every bit in a word $+1$ or -1 with equal probability. Then two such words $\{ S_i^\alpha \}$ and $\{ S_i^\beta \}$ satisfy

$$\frac{1}{N}\sum_{i=1}^{N} S_i^\alpha S_i^\beta = O\left(\frac{1}{\sqrt{N}}\right), \qquad \alpha \neq \beta . \quad (2.1)$$

If we want to generate ultrametric words we have to proceed differently. The example of spin glasses [10] suggests the following generic procedure. We consider an inhomogeneous Markov process with K time steps (in the figure, $K = 3$). At each step we choose a value for a certain random variable with a probability distribution with parameters depending on the result of the previous step, i.e. we define

$$P_k(y_k \mid y_{k-1})\, \mathrm{d}y_k = \text{probability distribution of } y_k \text{ conditioned to the value } y_{k-1},\ k = 1, \ldots, K . \quad (2.2)$$

The random variables y_k could in general be real continuous but for simplicity we will choose them discrete.

The draw of a y_k value is repeated M times, where M is the number of branchings of the tree we are trying to generate. At the end we will have R choices of y_K (see Fig. 1). Then the value of the spin at site i will be

$$S_i^\alpha = \operatorname{sign} y_K^\alpha \qquad \alpha = (\alpha_1, \alpha_2, \ldots, \alpha_K) . \quad (2.3)$$

The whole procedure is repeated, independently, at every site of the lattice. In this way we generate works which are ultrametric to order $1/\sqrt{N}$.

Instead of distances it is more convenient to work in terms of overlaps i.e.

$$q^{\alpha\beta} = \frac{1}{N}\sum_i S_i^\alpha S_i^\beta \quad (2.4)$$

in the limit $N \to \infty$ the overlaps among the R states are solely dependent on the level where the branches originating in the different states converge. In the example of figure 1

$$q_3 = 1 = \int P_1(y_1)\, \mathrm{d}y_1 \int P_2(y_2 \mid y_1)\, \mathrm{d}y_2 \times \int P_3(y_3 \mid y_2)\, \mathrm{d}y_3 (\operatorname{sign} y_3)^2$$

$$q_2 = \int P_1(y_1)\, \mathrm{d}y_1 \int P(y_2 \mid y_1)\, \mathrm{d}y_2 \times \left[\int P(y_3 \mid y_2) \operatorname{sign} y_3\, \mathrm{d}y_3\right]^2$$

$$q_1 = \int P_1(y_1) \left[\int P_2(y_2 \mid y_1)\, \mathrm{d}y_2 \times \int P(y_3 \mid y_2) \operatorname{sign} y_3\, \mathrm{d}y_3\right]^2 \mathrm{d}y_1 . \quad (2.5)$$

There are many different Markov processes that

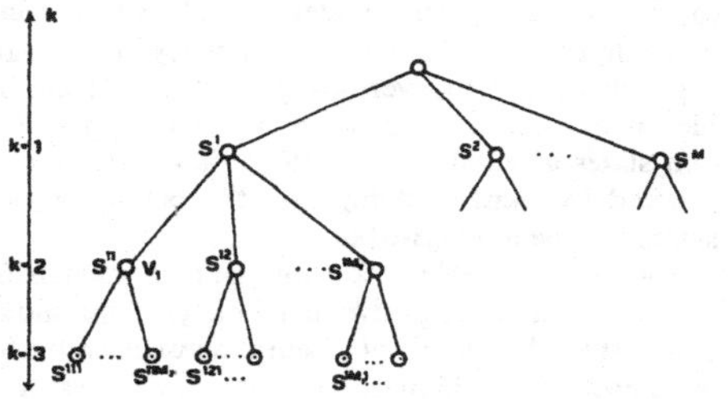

Fig. 1. — An ultrametric tree with three levels. The states lie at the lower ends of the branches. They are parametrized by three superindices that indicate their genealogy. At the other branchings we place the ancestors.

generate the same tree with the same overlaps. The differences manifest themselves in the value of the multiple correlations

$$\frac{1}{N}\sum_i S_i^\alpha S_i^\beta \ldots S_i^\omega . \quad (2.6)$$

We have investigated several different examples, albeit in a non-systematic way. In the following we will after refer to the following two particular cases.

A) At level k, y_k can take only two values $\pm r_k$ with

$$P_k(+) = (\text{probability that } y_k = + r_k) = \frac{1}{2}\left(1 + \frac{y_{k-1}}{r_k}\right)$$

$$P_k(-) = (\text{probability that } y_k = - r_k) = \frac{1}{2}\left(1 - \frac{y_{k-1}}{r_k}\right). \quad (2.7)$$

Furthermore $r_K = 1$ so that $\operatorname{sign} y_K = y_K$. Obviously

$$\overline{y}_k = (\text{average value of } y_k/\text{conditioned to } y_{k-1}) = y_{k-1}$$

$$\overline{y_k^2} = r_k^2 \quad (2.8)$$

Therefore $q_k = r_k^2$.

B) At level k, y_k can take three values $(-1, 0, 1)$ with the following conditional probabilities :

a) if $y_{k-1} = \pm 1$ then $y_k = \pm 1$ with probability one,

b) if $y_{k-1} = 0$ then

$$P_k(y_k = 1) = P_k(y_k = -1) = r_k/2$$

$$P_k(y_k = 0) = 1 - r_k . \quad (2.9)$$

The r_k increase with k and $r_K = 1$. In this case again

$$\bar{y}_k = y_{k-1}\,; \qquad \overline{y_k^2} = r_k \tag{2.10}$$

and therefore $q_k = r_k(1 - q_{k-1}) + q_{k-1}$.

In the context of an ultrametric organization it is useful to define ancestors of a word [10]. If there are M words that have all the same overlap q among them we can define their common ancestors as the configuration which obtains when we average the local value of the spin of all the descendants. The ancestors lie at every branching point and embody the amount of information common to the descendants. It is the prototype of a category and may be different from all the states belonging to the category. In the example of figure 1 the ancestor located at the vertex V_1 has local magnetization

$$m(y_2) = \int P(y_3 \mid y_2)\, \mathrm{d}y_3 \operatorname{sign} y_3 . \tag{2.11}$$

We can define a natural cell structure on the ancestor by grouping all sites with the same value of y_2. Then each descendant knows about this cell structure. In fact, if for one particular word we average the spin value among all sites belonging to a cell then we obtain again $m(y_2)$. The cell structure and the average cell magnetization is all the information that is common to the category.

In the context of neural networks the words or patterns to be stored must be the result of an encoding stage in the biosystem. The question then arises whether the general mechanism that we have discussed to generate ultrametric works is in any sense natural.

For this purpose we remember that any Markov process can be seen as following a discretized Langevin equation so that

$$a(y_k) = b(y_{k-1}) + \mathcal{E}, \tag{2.12}$$

where $\mathcal{E}$ is a random noise. The latter represents the input stimuli. Equation (2.12) suggests an encoding system structured in ordered layers. Each layer receives from the previous one an amount of information (partially elaborated) that determines a set (presumably finite and not too large) of possible roads which the Markov process can take. The new stimuli arriving at that level then chooses among these roads. Interestingly enough the first layer determines the primordial category into which the pattern will be classified. The details are received and coded later.

There is some partial evidence that the perception system has such a layered structure [11]. This suggests the fascinating hypothesis that the most primitive categorization follows directly from a layered perception system without any elaboration by the central nervous system. In more evolved species perhaps the same layered structure has been reproduced for the encoding stage.

3. The storage stage. The Hebb rule.

We discuss in this section how the pattern is stored in the synapses J_{ij}. We will show that our prescription is very similar to the relation that exists in the spin glass between the equilibrium states and the couplings. The prescription is, however, extremely simple and natural so that a reader, not interested in this comparison, may simply jump to equation (3.6).

In reference [10] it was proved that each state (with all its ancestors) defines a clusterization of the sites in cells. Each cell (of a particular state) is labelled by an K-tuple of real parameters where K is the total number of generations. Each cell of an ancestor, K' generations older than the state, is characterized by $K - K'$ real parameters. If we call

$$\mathcal{C}_{y_0 y_1 \dots y_K} = \{\text{ set of sites belonging to the cell } y_0\, y_1 \dots y_K \text{ of the state } \alpha\,\} \tag{3.1}$$

then the cell of its ancestor

$$\mathcal{C}_{y_0 y_1 \dots y_{K-K'}} = \bigcup_{y_{K-K'+1} \dots y_K} \mathcal{C}_{y_0 y_1 \dots y_K} . \tag{3.2}$$

We now define

$$J_{y_0 \dots y_l, y'_0 \dots y'_l} = \sum_{\substack{i \in \mathcal{C}_{y_0 \dots y_l} \\ j \in \mathcal{C}_{y'_0 \dots y'_l}}} J_{ij} . \tag{3.3}$$

Then it was proved that [10]

$$\begin{aligned} J_{y_0 \dots y_l, y'_0 \dots y'_l} &= J_{y_0 \dots y_{l-1}, y'_0 \dots y'_{l-1}} + \\ &+ \left(\frac{y'_l - y'_{l-1}}{q_l - q_{l-1}} - \frac{\rho_l}{2} (m_{q_l}(y'_l) - m_{q_{l-1}}(y'_{l-1})) \right) (m_{q_l}(y_l) - m_{q_{l-1}}(y_{l-1})) \\ &\quad + (m_{q_l}(y'_l) - m_{q_{l-1}}(y'_{l-1})) \left(\frac{y_l - y_{l-1}}{q_l - q_{l-1}} - \frac{\rho_l}{2} (m_{q_l}(y_l) - m_{q_{l-1}}(y_{l-1})) \right) \end{aligned} \tag{3.4}$$

where q_l, q_{l-1} are the overlaps at the considered level, $m_{q_l}(y_l)$ is the average magnetization on the cell and ρ is related to the limit of Parisi's order parameter at $T = 0$.

Based on this information we propose

$$\begin{aligned} NJ_{ij} &= \sum_{\alpha_1 \dots \alpha_l} \frac{(S_i^{\alpha_1 \dots \alpha_l} - S_i^{\alpha_1 \dots \alpha_{l-1}})(S_j^{\alpha_1 \dots \alpha_l} - S_j^{\alpha_1 \dots \alpha_{l-1}})}{q_l - q_{l-1}} + NJ_{ij}^{\text{ancestors}} \\ S_i^{\alpha_1 \dots \alpha_{l-1}} &= \sum_{\alpha_l} S_i^{\alpha_1 \dots \alpha_l} \,\%\, \sum_{\alpha_l} 1 \end{aligned} \tag{3.5}$$

where the labels $\alpha_1 \dots \alpha_{l-1}\, \alpha_l$ indicate the descendants of the ancestor $\alpha_1 \dots \alpha_{l-1}$; q_l is the self-overlap of the descendant while q_{l-1} is the overlap among the descendants of the same ancestor.

In the case of figure 1 we derive

$$NJ_{ij} = \sum_{\alpha\beta\gamma} \frac{(S_i^{\alpha\beta\gamma} - S_i^{\alpha\beta})(S_j^{\alpha\beta\gamma} - S_j^{\alpha\beta})}{1 - q_2} + \sum_{\alpha\beta} \frac{(S_i^{\alpha\beta} - S_i^{\alpha})(S_j^{\alpha\beta} - S_j^{\alpha})}{q_2 - q_1} + \sum_{\alpha} \frac{S_i^{\alpha} S_j^{\alpha}}{q_1}. \tag{3.6}$$

This learning mechanism generalizes the Hebb rule. The original Hebb mechanism is unique in the sense that storing a new word does not require any « thought » or « reflection » on the part of the central nervous system. This is reasonable as long as we are trying to store uncorrelated words. However, it becomes artificial as soon as there is any relational content. The generalization we are proposing, suitable for an ultrametric organization, is natural and fast.

Assume that R words have already been stored and we are trying to store the $R + 1$. If R is sufficiently large, the new word will not, in a first approximation, affect the average configuration corresponding to the ancestors. So at this point there is no need to recall previous information to store the new word. In a second moment, however, the system may have to modify the ancestor taking into account the new pattern. Only then the piece of J_{ij} depending on the ancestors will be modified. The prototype of the category has changed.

4. The retrieval process. The spurious words.

In this section we discuss the ability of the system to recall a stored pattern. For practical reasons we will refer to one particular example of a tree with two generations. There will be P number of ancestors ($\alpha = 1 \dots P$) and P_α number of descendants of ancestor α. Both these numbers can become large but they are negligible with respect to N, the number of sites that tends to infinity. Then

$$NJ_{ij} = \sum_{\alpha=1}^{P} \frac{S_i^\alpha S_j^\alpha}{q} + \sum_{\alpha=1}^{P} \sum_{\beta=1}^{P_\alpha} \frac{(S_i^{\alpha\beta} - S_i^\alpha)(S_j^{\alpha\beta} - S_j^\alpha)}{1 - q}$$
$$J_{ii} = 0 \tag{4.1}$$

The system starts from an initial configuration and relaxes according to the equation [2]

$$S_i(t + \Delta t) = \text{sign}\left(\sum_j J_{ij}\, S_j(t)\right) \tag{4.2}$$

until it reaches an asymptotic state. The updating can be done in an infinite number of different ways from the purely sequential rule, as in usual Monte Carlo algorithms, to a parallel synchronous updating in which all spins are flipped simultaneously [1]. The limit motion in these two cases satisfy the equations

$$\text{sequential} \quad S_i = \text{sign}\left(\sum_j J_{ij}\, S_j\right) \tag{4.3a}$$

$$\text{synchronous} \quad \begin{cases} S_i = \text{sign}\left(\sum_j J_{ij}\, S_j'\right) \\ S_i' = \text{sign}\left(\sum_j J_{ij}\, S_j\right). \end{cases} \tag{4.3b}$$

Solutions of (4.3a) are included among the solutions of (4.3b). A biosystem will act somehow half way between these two extremes. There is no clock to synchronize the updating but delays in signal transmission amount to a certain degree of parallelism. The relevant dynamics therefore corresponds to parallel updating of conveniently chosen subsets of spins. If these subsets are sufficiently random and change in each updating the general solution of (4.3b) cannot remain stable. Therefore a biosystem will eventually evolve into the fixed points of (4.3a).

The presence of random noise is modelled as in statistical mechanics through the introduction of temperature. Equation (4.2) becomes

$$\mu_i(t + \Delta t) = \tanh\left(\beta \sum_j J_{ij}\, \mu_j(t)\right) \tag{4.4}$$

where μ_i is the short time average of S_i.

Equation (4.2) can be conveniently studied by contracting it will all the spin configurations corresponding to words stored in the J_{ij} [12]. Calling

$$m^{\alpha\beta} = \frac{1}{N} \sum_i S_i^{\alpha\beta} S_i, \qquad m^\alpha = \frac{1}{N} \sum_i S_i^\alpha S_i \tag{4.5}$$

we derive

$$m^{\alpha\beta}(t + \Delta t) = \left\langle\!\!\left\langle S^{\alpha\beta}\, \text{sign}\left(\sum_\alpha \frac{S^\alpha\, m^\alpha(t)}{q} + \right.\right.\right.$$
$$\left.\left.\left. + \sum \frac{(S^{\alpha\beta} - S^\alpha)}{1 - q} (m^{\alpha\beta}(t) - m^\alpha(t))\right)\right\rangle\!\!\right\rangle \tag{4.6}$$

where the double brackets indicate the average value with respect to the stochastic variables $S^{\alpha\beta}$ and S^α. There is an energy functional that decreases monotonically during the sequential relaxation

$$E = \sum_{i>j} J_{ij}\, S_i S_j$$
$$= \frac{1}{2} \sum_\alpha \frac{m_\alpha^2}{q} + \frac{1}{2} \sum_{\alpha\beta} \frac{(m_{\alpha\beta} - m_\alpha)^2}{1 - q}. \tag{4.7}$$

The solutions of equation (4.3a) are of the following type :

a) the original patterns or words stored in the system, i.e.

$$m^{\alpha_0\beta_0} = 1 \; ; \; m^{\alpha_0\beta} = q \, , \; \beta \neq \beta_0 \, ; \; m^{\alpha\beta} = 0 \, , \; \alpha \neq \alpha_0 \qquad (4.8)$$

b) spurious solutions which mix descendants of the same ancestor

$$\begin{aligned} m^{\alpha_0\beta_i} &\neq 0 \, , \quad i = 1, ..., p \\ m^{\alpha\beta} &= 0 \, , \quad \alpha \neq \alpha_0 \end{aligned} \qquad (4.9)$$

c) spurious solutions which mix different ancestors

$$m^{\alpha_i\beta_j} \neq 0 \, , \quad j = 1 \dots p_{\alpha_i} \, , \qquad i = 1 \dots p \, . \qquad (4.10)$$

The solutions of the Hopfield model have been analysed in great detail [4]. Unfortunately some of the more interesting properties of the solutions are sensitive to the distribution probability of the S_i^α. If we choose instead

$$NJ_{ij} = \sum_\alpha \xi_i^\alpha \, \xi_j^\alpha \qquad (4.11)$$

where ξ_i^α is a random continuous variable with $\overline{\xi_i^\alpha} = 0$ and $\overline{\xi_i^{\alpha 2}} = 1$, then the number of spurious solutions and their stability properties are modified.

In our case we do not have any « natural » choice for the random process generating the $S_i^{\alpha\beta}$, S_i^α. Therefore we cannot expect to derive detailed results. We must rely much more on semiqualitative arguments. Eventually we should try to optimize storage and retrieval capabilities by choosing a particular ultrametric tree.

Solutions of the type a) and b) are similar to the ones appearing in the Hopfield model. We can argue that a category delimits a cone-like region in the space of configurations with the central axis coinciding with the ancestor. The « words »

$$\{ S_i^{\alpha_0\beta} - S_i^{\alpha_0} \}$$

are vectors which in the $N \to \infty$ limit are orthogonal as in the Hopfield model. Their probability distribution is in general different and this affects the general properties of the spurious solutions. However we can choose it so as to deplete the probability of zero and then we expect the same qualitative behaviour, i.e. :

i) solutions of type a) are stable below a certain critical temperature;

ii) the number of spurious solutions increases at least exponentially with P_α;

iii) the spurious solutions with low energy are essentially linear combinations of the original patterns;

iv) the attraction basin of all spurious solutions increases with P_α, this means that the attraction basin of the spurious solution lies in the frontier region between the original patterns.

For the tree generated by the Markov process B) of section 2 all of these arguments can be rigorously proved because one can derive

$$\frac{m^{\alpha_0\beta} - m^{\alpha_0}}{1 - q} = \left\langle S^{\alpha_0\beta} \, \mathrm{sign} \sum_\beta S^{\alpha_0\beta} \frac{(m^{\alpha_0\beta} - m^{\alpha_0})}{1 - q} \right\rangle \qquad (4.12)$$

an equation identical to the one in the Hopfield model [4].

For the random choice A) there is one particular spurious solution degenerate with the original patterns, namely the ancestor itself

$$S_i = \pm \, S_i^\alpha \, | q |^{-1/2} \, . \qquad (4.13)$$

The solutions of type c) are the more interesting in this model. They can be interpreted as spurious categories. In the introduction we have stressed that one of the motivations to introduce a hierarchy among patterns was the need to introduce a hierarchy among errors during retrieval. In this context we would like to analyse whether the attraction basin of the spurious categories grows exponentially with the number of categories or with the number of patterns. In the latter case something would have to be modified in the model. It is already hard to understand why in Hopfield's model the percentage of stimuli that leads to spurious patterns grows when we store more patterns [1]. In our case spurious categories can be easily interpreted (they classify border case patterns). But it is hard to accept that the number of external stimuli that should end in these spurious categories should grow exponentially with the number of patterns correctly stored in the originally categories.

We have analysed several ultrametric trees, albeit in a non-systematic way. The tree generated by the random process A) of section 2 is more similar to Hopfield's model and therefore facilitates comparisons. We consider the most interesting case :

$$P_\alpha \gg P \qquad (4.14)$$

i.e. number of patterns belonging to a category much larger than the number of categories.

In that case for a large class of input stimuli the system will retrieve a spurious pattern with appreciable overlap on a large set of original patterns. Therefore we can apply the central limit theorem to the distribution of

$$X = \sum_{\alpha\beta} \frac{(m_\alpha^{\alpha\beta} - m^\alpha)}{1 - q} (S^{\alpha\beta} - S^\alpha) \qquad (4.15)$$

---

(1) Unless we see it as a proof that « the more we learn the more we realize we do not know nothing ». However a spurious pattern is not recognized as such so it would be more similar to « the more we learn the more we deceive ourselves ».

with $\overline{X} = 0$ and

$$\overline{X}^2 = \sum_{\alpha\beta} \frac{(m^{\alpha\beta} - m^{\alpha})^2}{1 - q} = r(t) \qquad (4.16)$$

then the equations of motions become

$$m^{\alpha}(t + \Delta t) = \left\langle\!\!\left\langle S^{\alpha} \operatorname{sign}\left(\sum_{\alpha} S^{\alpha} m^{\alpha}(t) + X\right)\right\rangle\!\!\right\rangle \qquad (4.17)$$

$$\sum_{\alpha\beta} v^{\alpha\beta}(t + \Delta t)\, v^{\alpha\beta}(t) = \left(\sum_{\alpha\beta} (v^{\alpha\beta}(t))^2\right)^{1/2} \times$$

$$\times \left\langle \sqrt{\frac{1}{\pi}} \exp - \frac{\left(\sum_{\alpha} S^{\alpha} m^{\alpha}(t)\right)^2}{2\, r(t)\, q^2} \right\rangle$$

with

$$v^{\alpha\beta}(t) = m^{\alpha\beta}(t) - m^{\alpha}(t). \qquad (4.18)$$

In equation (4.17) the average is over the S^{α} and X while in (4.18) there is a simple average over S^{α}.

Equation (4.17) at fixed $X(t)$ is similar to the corresponding equation in Hopfield's model with temperature. The fixed point

$$m^{\alpha} = 0 \qquad (4.19)$$

is stable if $r(t) > \frac{2}{\pi}$. Equation (4.18) implies that if the vector $v^{\alpha\beta}(t)$ stays in the region where the Gaussian approximation holds then its norm will decrease. This reflects in equation (4.17) as a decrease of the effective temperature. Eventually $r \to 0$ or $v^{\alpha\beta}$ is zero except for a few values of α, β. In the first case the fixed points of equation (4.17) will be the ones of the Hopfield model.

This process is reminiscent of a spontaneous simulated annealing [13]. It may actually help in the retrieval of the lowest energy patterns.

5. Discussions and outlook.

In this paper we have shown that it is possible to modify Hopfield's model in such a way that it efficiently stores words organized in categories. However, categorization does not seem to emerge spontaneously as a consequence of the particular type of architecture of a neural network but must be rather superimposed on it. On the other hand, it seems that categorization is connected to the layered character of the encoding stage or even with the perception system itself. In this sense there is interesting evidence coming from the field of cognitive psychology and neurophysiology. Grouping colours into classes is clearly a subjective task. However, there is evidence that the usual classification is universal [14] and that it has to do with the way colours are processed by the visual system [15].

The limits of the storage capability of the model have not been investigated here. We have instead shown that the system is able to distinguish between spurious solutions that belong to well-defined categories and spurious solutions that mix categories. A hierarchy among errors can therefore be introduced.

Another question that remains to be answered is whether there is a best choice for the Markov process that generates the ultrametric tree and its possible connection with the one appearing spontaneously in the SK spin glass model. This interesting problem requires much more work.

Finally we would like to stress that we are aware that the relational context that exists in the human memory is infinitely richer than the one we have been analysing. We believe however that placing a word into a more complicated semantic context requires lots of processing by the central nervous system [16]. Our point is that placing it in an ultrametric tree is almost for free.

Acknowledgments.

We would like to thank M. Mezard for a very fruitful interaction and G. Toulouse for a crucial discussion at the very beginning of this work. Looking backwards we now realize how many of his ideas we have been using freely. We would also like to thank Professor Abdus Salam, the International Atomic Energy Agency and UNESCO for hospitality at the International Centre for Theoretical Physics, Trieste.

References

[1] LITTLE, W. A., *Math. Biosci.* **19** (1974) 101.

[2] HOPFIELD, J. J., *Proc. Natl. Acad. Sci. USA* **79** (1982) 2554.

[3] MCCULLOCH, W. S. and PITTS, W., *Bull. Math. Biophys.* **5** (1943) 115.

[4] AMIT, D. J., GUTFREUND, H. and SOMPOLINSKY, H., *Phys. Rev.* **A 32** (1985) 1007.

[5] See for example : CHANGEUX, J. P., in *Disordered Systems and Biological Organization*. Eds. E. Bienenstock *et al.* (Springer Verlag, Berlin-Heidelberg) 1986.

[6] ANDERSON, J. A., *IEEE Trans. Sys. Man Cybern.* **SMC-13** (1983) 799.

[7] PERETTO, P., *Biol. Cybern.* **50** (1984) 51.

[8] See for example : KLATZKY, R. L., *Human Memory* (W. H. Freeman, San Francisco) 1975.

[9] MEZARD, M., PARISI, G., SOURLAS, N., TOULOUSE, G. and VIRASORO, M. A., *Phys. Rev. Lett.* **52** (1984) 1156 ; *J. Physique* **45** (1984) 843.

[10] MÉZARD, M. and VIRASORO, M. A., *J. Physique* **46** (1985) 1253.

[11] KUFFLER, S. W., NICHOLS, J. G. and MARTIN, A. R.,

From Neuron to Brain, 2nd Ed. (Sinauer Associates Inc., Sunderland, Massachusetts) 1984.

[12] PROVOST, J. P. and VALLÉE, G., *Phys. Rev. Lett.* **50** (1983) 598.

[13] KIRKPATRICK, S., GELATT, Jr. C. D., and VECCHI, M. P., *Science* **220** (1983) 671.

[14] ROSCH, E., On the internal structure of perceptual and semantic categories, in *Cognitive Development and the Acquisition of Language*, Ed. T. E. Moore (Academic Press, New York) 1973.

[15] DE VALOIS, R. L. and JACOBS, G. H., *Science* **162** (1968) 533.

[16] See for example : HINTON, G. E., SEJNOWSKI, T. J. and ACKLEY, D. H., *Cognitive Sci.* **9** (1985) 147.

PERSONNAZ, L., GUYON, I. and DREYFUS, G., Neural network design for efficient information retrieval in *Disordered Systems and Biological Organization*, Eds. E. Bienenstock *et al.* (Springer Verlag, Berlin-Heidelberg) 1986.

[17] AMIT, D. J., GUTFREUND, H. and SOMPOLINSKY, H., *Phys. Rev. Lett.* **55** (1985) 1530.

J. Phys. A: Math. Gen. 19 (1986) L675–L680. Printed in Great Britain

LETTER TO THE EDITOR

Asymmetric neural networks and the process of learning

Giorgio Parisi
Dipartimento di Fisica, II Universita' di Roma 'Tor Vergata', Via Orazio Raimondo, Roma 00173, Italy and INFN, Sezione di Roma, Italy

Received 15 April 1986

Abstract. In this letter we study the influence of a strong asymmetry of the synaptic strengths on the behaviour of a neural network which works as an associative memory. We find that the asymmetry in the synaptic strengths may be crucial for the process of learning.

In order to understand how neural networks are able to learn and to work as associative memories, at the present moment it is convenient to consider very simplified models.

A very interesting and widely studied model is the following: the neurons can be firing or quiescent and are represented by variables S_i which may only take the values ± 1 and the synaptic strengths (the influence of the neuron i on the neuron k) are real numbers $J_{i,k}$. The dynamics is very simple: it can be fully deterministic (the so-called zero temperature limit) or a finite amount of randomness may be present (finite temperature).

In the Hopfield model (Hopfield 1982) the $J_{i,k}$ are symmetric and the process of learning is the following: each time that a pattern μ_i is learnt (the μ_i are also ± 1) the J are changed according to the following formula:

$$J_{i,k}^{\text{new}} = J_{i,k}^{\text{old}} + \mu_i \mu_K \tag{1}$$

which is a generalisation of the original Hebb rule.

The assumption of a symmetric distribution of the J has been criticised for not being realistic (if the neuron i influences the neuron k there is no reason why the neuron k should influence the neuron i). If the Hopfield model were a model for a properly working memory, this criticism would not be very relevant. Indeed, it seems that a not too large amount of asymmetry in the synaptic strengths does not qualitatively change the behaviour of the system (Amit 1985).

The aim of this letter is to show that the performance of the Hopfield model is not that which is required from a good memory; in particular, the learning procedure is quite problematic. In contrast, if the model is strongly modified in such a way that the asymmetry of the synaptic strength plays a crucial role, the memory performs much better and the learning can be done in a simple way. It may be possible for the qualitative differences between the two models to be experimentally observed.

Before presenting the criticism of the Hopfield model we first summarise what is known about this model from both analytic and numerical computations (Amit *et al* 1985, Kinzel 1985, Mezard *et al* 1986a). Let us work under the simplifying hypothesis that N is large (e.g. 100–1000) and that all the inputs are orthogonal and the number of learnt patterns (p) is proportional to N:

$$p = \alpha N. \tag{2}$$

It is normally assumed that an input pattern is presented to the network by forcing the state of the network to coincide with the input pattern. After this moment, the system evolves with its own dynamics and the output of the system is the state of the network after some time. If the network had a perfect memory, each time that the input pattern had some resemblance to one of the memorised patterns, the output should be the memorised pattern.

If α is smaller than a critical value (≈ 0.14) and the input pattern is sufficiently near to one of the memorised patterns, the retrieval procedure is successful with probability 1 when $N \to \infty$ (apart from about a 1-2% discrepancy between the memorised pattern and the output pattern). On the other hand, if the input is not sufficiently near to one of the memorised patterns, the network is confused and it goes into some state which is very far from every memorised pattern. The nearer α is to the critical value 0.14, the nearer the input state must be to one of the learnt patterns in order to avoid confusion. When α becomes greater than the critical value the network is always confused; also, if the input pattern is one of the memorised ones, it becomes confused and goes into a state which is very far from every memorised pattern.

These results hold exactly at zero temperature; if a non-zero amount of noise is allowed and α is not too small (greater than 0.05), the state in which the system stays near one of the learnt patterns is metastable. If the input pattern is near one of the memorised patterns the system goes into a state very similar to the learnt pattern but after a very long time the network becomes confused and it goes into some state which is very far from every memorised pattern.

It has recently been recognised that the state of total confusion may be avoided by a simple modification of the generalised Hebb rule. For example, we can write (Changeaux *et al* 1986, Mezard and Nadal 1986)

$$J_{i,k}^{\text{new}} = (1-\lambda) J_{i,k}^{\text{old}} + \mu_i \mu_K \tag{3}$$

or more generally (Parisi 1986)

$$J_{i,k}^{\text{new}} = g(J_{i,k}^{\text{old}} + \mu_i \mu_K) \tag{4}$$

where g is an appropriate non-linear function such that the J cannot become arbitrarily large (or small).

If the parameter λ is well tuned, only the last αN patterns ($\alpha \approx 0.03$–0.05) which are learnt are correctly memorised, while the others are forgotten. In other words, the memory has a finite capacity and the learning of a new pattern forces the forgetting of an old pattern. In this way the memory never reaches the state of total confusion. This modification of the generalised Hebb rule is crucial if we want to understand how the learning process works in detail.

The main disadvantage of a memory which works according to the Hopfield model is that, in both cases ((a) retrieval of one of the input patterns and (b) confusion), after a short transient time the network goes to a time-independent state and it is not possible to discriminate between the two cases. In other words, the outside world which examines the output is unable to discriminate between a valid output (a) and an unreliable output (b). It is clear that a capability to discriminate between case (a) and case (b) would be very useful: an error can be tolerated, if it is identified as such, but the possibility of having non-recognised errors may affect the reliability of the whole neural system of which our memory is supposed to be one component.

The incapacity to discriminate between case (*a*) and case (*b*) becomes really dramatic if we consider the learning mechanism in detail. The crucial question is the following: does the modification of the synaptic strength (according to a generalised Hebb rule) happen only when the network is requested to do so or can the synaptic strengths always be modified? In other words, does the network learn only when it is instructed to learn or is it the network which decides when to learn?

The first option (learning under request) is clearly possible. For example, some chemical modifications of the environment of the network could trigger the application of the Hebb rule. In this letter we explore the feasibility of the second possibility, i.e. that the Hebb rule is always applied, with the exception of times when the network is confused. The new form of the Hebb rule should be

$$dJ_{i,k}(t)/dt = -\lambda J_{i,k}(t) + S_i(t)S_k(t) \tag{5a}$$

if the network is not in a confused state or

$$dJ_{i,k}(t)/dt = 0 \tag{5b}$$

if the network is in a confused state, where t is the time. The synaptic strengths are modified only when the network is not in a confused state.

In this way the behaviour of our ideal memory should be the following: the network remembers the states in which it has been for a sufficiently long time. In this way the memory learns a pattern if it is forced to stay in the corresponding state a sufficiently long time and, on the other hand, the process of retrieval of a pattern starts when the memory is set in the corresponding state for a short time and after the memory is left free to evolve according to its own laws.

In other words, if a pattern is presented to the memory for a long time it is memorised; if it presented to a memory for a short time, the pattern is searched for in the memory. If the pattern is not found, the memory becomes confused and nothing happens. However, if the retrieval procedure ends with success the state of the neural network coincides with the found pattern for a certain time (during this time the pattern found becomes better memorised); at later times the memory jumps again in the confused states and the content of the memory is not changed until a new pattern is presented.

If we disregard the case of short term memory and we consider only long term memory, this proposal makes sense only if the application of the Hebb rule is inhibited when the network is in a confused state. Indeed, in the absence of input the memory goes into a confused random state and after some time only this state would be memorised if the Hebb rule (equation (5*a*)) were always operating. This problem is solved by the introduction of equation (5*b*).

The Hopfield model satisfies all the necessary requirements. Unfortunately, in this model (as we have already remarked) it is impossible to discriminate between the two cases ((*a*) retrieval of one of the input patterns and (*b*) confusion). Our proposal is that the neural network should work as the Hopfield network with the main difference being that, when the input state is such that it does not lead to the retrieval of one of the learnt states and the network goes into a state of confusion, the state of the network becomes time dependent in a chaotic way.

In this way, after the input has been presented to the system, if the S_i are time independent, the retrieval of one of the input states has been completed; on the other hand, if the S_i are time dependent the network stays in a confused state. In other words, only the outputs that are time independent can be considered as valid outputs and the outputs which depend on time should be disregarded.

If this happens, the Hebb rule can be modified in the following way:

$$dJ_{i,k}(t)/dt = [-\lambda J_{i,k}(t) + \bar{S}_i(t)\bar{S}_k(t)]f(\bar{S}_i(t)\bar{S}_k(t)) \tag{6}$$

$$\bar{S}_i(t) = 1/T \int_0^T d\tau S_i(t-\tau)$$

when $f(x)$ is a function which is practically zero below a threshold (x_t) and is very near to one for x greater than the threshold and $\bar{S}_i(t)$ is the average over a time T of the status of the ith neuron (although $\bar{S}$ depends on T, we have not indicated this dependence in order to simplify the notation).

In other words, the updating of the synaptic strengths (which is at the basis of the learning process) is sensitive only to the average value of the neurons in the most recent past and it happens only if the neurons do not flip from one state to another too quickly.

Having established how our ideal neural network should work, we should discuss how one can realise a network which works in the way we have described. The claim of this letter is that strong asymmetry of the synaptic strengths is needed to reach this goal. We can arrive at this conclusion by studying the Hopfield model in detail.

The simplest form of the dynamic of the neural network is the following; the time is discretised (in a realistic model the time steps are of the order of a millisecond) and in the so-called zero temperature limit the variables S_i are updated by applying the following equation to all the neurons (sequentially or in random order):

$$S_i = \text{sgn}(h_i) \qquad h_i \equiv \sum_{k=1,N} J_{i,k}S_k. \tag{7}$$

At finite temperature some noise is present, equation (6) holds only in an approximate way and we have

$$S_i = \text{sgn}(h_i) \qquad \text{with probability } p_i \equiv 1/[1+\exp(-\beta h_i)] \tag{8}$$

$$S_i = -\text{sgn}(h_i) \qquad \text{with probability } 1-p_i.$$

Obviously in the limit $\beta \to \infty$ we recover equation (7). The great advantage of the Hopfield model is that it can be studied in great detail analytically because its properties coincide with those of a common statistical mechanical system. Due to the symmetry of the synaptic strengths we can define an energy

$$E[S] = \sum_{i,k=1,N} J_{i,k}S_iS_k \tag{9}$$

and the probability distribution of the S at large time is given by the usual Gibbs formula†:

$$P[S] \propto \exp[-\beta E[S]]. \tag{10}$$

† This is not true for parallel updating in which equation (8) is applied to all the neurons simultaneously. It is only true if equation (8) is applied to each neuron at different times as stated in the text (independent updating). Also, parallel updating has a description in statistical mechanics and the differences are not very strong (generally speaking, the result on the large time behaviour is true only if the temperature $(1/\beta)$ is not strictly equal to zero; one should carefully note that the two limits, time going to infinity and temperature going to zero, do not commute).

The zero temperature dynamics corresponds to searching for the minimum of $E[S]$ using the fastest descent algorithm: each neuron flips its state (from firing to quiescent or vice versa) if the energy decreases by doing so. Each time that the neuron is flipped the energy decreases so that we must reach a stable state after a not too large number of steps.

On the other hand, if the neuron strengths are asymmetric ($J_{i,k} \neq J_{k,i}$) the energy function does not exist anymore and the large time behaviour may be much more complicated at zero temperature. For example, we could have that the S_i become a periodic function of the time (limit cycle). It is also possible that the length of this cycle is very large (proportional to $\exp(N)$); in this case we say that the system behaves in a chaotic way.

We can be more quantitative by introducing the order parameter $q(t)$ (Edwards and Anderson 1975, Mezard *et al* 1986b) as

$$q(t) = (1/N) \sum_{i=1,N} [\bar{S}_i(t)]^2. \tag{11}$$

It is well known that at low temperature q is different from zero (also for large average time T) both when the network is near to one of the input states and when the network is confused. However, we would like it that when the network is a confused state q is zero (or small) and consequently the function f in equation (6) is very near to zero and the synaptic strengths are time independent.

We have checked numerically that if the synaptic strengths are random and asymmetric ($J_{i,k}$ is independent from $J_{k,i}$), q remains zero also in the low temperature limit. In other words, in the equivalent of the spin glass phase for an asymmetric network, the system does not order itself in a random direction but has a chaotic behaviour.

There are many ways in which the asymmetry may be introduced in the Hebb rule. For example, a synapse going from neuron i to neuron k may exist with probability p: not all the neurons are connected one with the other (diluted network) and the dilution is done in an asymmetric way. In other words, there are $N \times N$ random variables $C_{i,k}$ ($C_{i,k} = 0$ or 1) such that $\bar{C} = p$: only if $C_{i,k} = 1$ does the synaptic connection exist. If we apply one of the various generalised Hebb rules only to the connections which exist (i.e. $C_{i,k} = 1$), we generate an asymmetric set of J, whose asymmetry is stronger for smaller values of p.

It is clear that the various parameters of the model must be correctly tuned in order to avoid spurious states in which half of the network is near to one of the input states while the other half of the network is near to another input state (in the Hopfield model this is true when $\alpha > 0.03$).

The condition that the network should automatically go into the confused state (also after a successful retrieval operation) is the most delicate point (which we have not investigated numerically). It is very reasonable that (as in the Hopfield model) the ordered state is metastable and after some time it decays into the confused state only if α is sufficiently large. If the memory is blank when we start to use the memory and we learn the first patterns, α is by definition zero and some problems may be present in initialising the memory. It is rather likely that the suggestion (Changeaux *et al* 1985) that the memory is not blank at the initial time will play a crucial role here, although with some imagination one could start to build a theory for the imprinting.

We could also consider more complex models. The synapses are divided in two groups: the Hebb rule applies only to the synapses of the first group while the others are not modified by the learning process. The synapses of the second kind are needed in order to guarantee that, no matter what happens to the synapses of the first kind, the ordered state is always metastable and the network jumps automatically to the chaotic state. For example, it may be possible that only the excitatory synapses participate in the process of learning while the inhibitory synapses are not modified during the learning and are responsible for bringing the system into the chaotic phase. It seems to me that a modification of this kind will strongly increase the robustness of the network.

Summarising, we propose that a neural network learns automatically any input pattern which is presented to it for a long enough time without the need for any explicit chemical order and that a neural network of a long term memory is in a chaotic time-dependent state when it is not active. The correctness of these two proposals (especially of the second) can clearly be experimentally verified. This behaviour of the network is possible only if the synaptic strengths are sufficiently asymmetric. It is difficult at this stage to discriminate between the many possible models for the asymmetry because models (like the one proposed in this letter) in which we assume that the synapses are randomly distributed are probably not realistic because the process of connecting the neurons is not a pure random process but contains a strong deterministic part.

It is a pleasure for me to thank D Amit, M Mezard, J P Nadal and M Virasoro for useful discussions on the subject of this letter.

References

Amit D J 1985 *Jerusalem preprint*
Amit D J, Gutfreund H and Sompolinsky H 1985 *Phys. Rev. Lett.* 55 1930
Changeaux J P, Dehaene S, Nadal J P and Toulouse G 1986 *Europhys. Lett.* to be published
Changeaux J P, Nadal J P and Toulouse G 1985 *ENS, Paris Preprint*
Edwards S F and Anderson P W 1975 *J. Phys. F: Met. Phys.* 5 965
Hopfield J J 1982 *Proc. Natl Acad. Sci. USA* 79 2554
Kinzel W Z 1985 *Z. Phys.* B 60 205
Mezard M and Nadal J P 1986 Private communication
Mezard M, Parisi G and Virasoro M 1986a *Rome preprint*
—— 1986b *The Spin Glass Theory and Beyond* (Singapore: World Scientific) to be published
Parisi G 1984 *Mathematical Physics VII* ed W E Brittin, K E Gustafson and W Wyss (Amsterdam: North-Holland)
—— 1986 *J. Phys. A: Math. Gen.* 19 L617

J. Physique 47 (1986) 1457-1462 SEPTEMBRE 1986, PAGE 1457

Classification
Physics Abstracts
87.30G — 75.10H — 89.70 — 64.60C

Solvable models of working memories

M. Mézard (°), J. P. Nadal (*) and G. Toulouse (+)

Universita di Roma I, Dipartimento di Fisica, Piazzale Aldo Moro 2, 1-00185, Roma, Italy
(*) Groupe de Physique des Solides de l'Ecole Normale Supérieure, 24, rue Lhomond, 75231 Paris Cedex 05, France
(+) E.S.P.C.I., 10, rue Vauquelin, 75231 Paris Cedex 05, France

(*Reçu le 27 mars 1986, accepté le 7 mai 1986*)

Résumé. — Nous considérons une famille de modèles qui généralise le modèle de Hopfield, et qui peut s'étudier de façon analogue. Cette famille englobe des schémas de type palimpseste, dont les propriétés s'apparentent à celles d'une mémoire de travail (mémoire à court terme). En utilisant la méthode des répliques, nous obtenons un formalisme simple qui permet une comparaison détaillée de divers schémas d'apprentissage, et l'étude d'effets variés, tel l'apprentissage par répétition.

Abstract. — We consider a family of models, which generalizes the Hopfield model of neural networks, and can be solved likewise. This family contains palimpsestic schemes, which give memories that behave in a similar way as a working (short-term) memory. The replica method leads to a simple formalism that allows for a detailed comparison between various schemes, and the study of various effects, such as repetitive learning.

Introduction.

Networks of formal neurons provide models for associative memories [1-5] and much numerical and analytical progress has been made recently, especially on the Hopfield model [2, 3]. In particular Amit *et al.* [3] have solved for the thermodynamics of this model, using the replica method with the approximation of replica symmetry. Since then, alternative learning schemes have been proposed [4-6], which avoid the catastrophic deterioration of the memory when the number of stored patterns exceeds a critical value. In these schemes, new patterns may always be learned, at the expense of previously stored patterns which get progressively erased. For this reason, such memories have been called palimpsests, and may provide inspiration for the study of working (short term) memories.

In this paper, we define a family of models, which generalizes the Hopfield model and can be solved likewise. One of the palimpsestic schemes defined in [4] (the marginalist scheme) is a member of the family. Within the replica method, with the approximation of replica symmetry, we obtain analytical results that agree with previous numerical calculations, and allow for a detailed comparison between various schemes. The most interesting case is when one requests a very good retrieval quality of the learned patterns : remarkably, the formulation becomes very simple in this limit.

In section 1, we introduce the family of models. In section 2, we give the solution within the approximation of replica symmetry, and study the zero-temperature limit. In section 3, we consider various effects, such as repetitive learning of a pattern. The main results are summarized in the conclusion.

1. Palimpsestic schemes.

All the models we will consider have the same basic ingredients : the network is made of N interconnected formal neurons S_i. Each neuron can be either in the firing state ($S_i = +1$) or in the quiescent state ($S_i = -1$). The synaptic efficacies T_{ij} contain the information on a set of patterns $S^\mu = (S_i^\mu)_{i=1,N}$ which one wants to memorize. They can be either positive (excitatory) or negative (inhibitory). The network should work as an associative memory : setting the network in a pattern S^μ (or close to S^μ), it relaxes under a suitable dynamics towards a close stationary

(°) Permanent address : Laboratoire de Physique Théorique, Ecole Normale Supérieure, 24, rue Lhomond, 75231 Paris Cedex 05, France.

state. Proximity is measured by the retrieval overlap

$$m = (1/N)\left\langle \sum_i S_i^\mu S_i \right\rangle, \tag{1.1}$$

where the bracket is an average over stochastic noise (thermal averaging). When $m \sim 1$, retrieval is good. Assuming symmetric connections and relaxational dynamics, the (meta)stable states of the network are those of the Hamiltonian

$$H = -(1/2)\sum_{i \neq j} T_{ij} S_i S_j. \tag{1.2}$$

As a first rough criterion of efficiency, a pattern S^μ is said to be memorized (or recognized) if its retrieval quality (1.1) is good.

The models we study differ in their learning schemes. These are the rules which fix the synaptic efficacies for the given set of patterns to be learned. The Hopfield scheme [1] is :

$$i \neq j \quad T_{ij} = (1/N)\sum_{\mu=1}^{p} S_i^\mu S_j^\mu. \tag{1.3}$$

In this model, each pattern is learned with the same acquisition intensity $k = 1/N^2$. In the first palimpsestic scheme proposed in [4], named marginalist scheme, each pattern μ is learned with an intensity $k(\mu)$, which increases exponentially with μ. To avoid this undesirable exponential growth, we consider a slightly different version of this model : once t patterns have been stored, storing a new pattern S^{t+1} is done through the following modification of the synaptic efficacies :

$$T_{ij}(t+1) = \lambda[(\varepsilon/N) S_i^{t+1} S_j^{t+1} + T_{ij}(t)] \tag{1.4}$$

where λ is such that the cumulated intensity (average squared synaptic efficacy) remains fixed :

$$\overline{T_{ij}^2} - \overline{T}_{ij}^2 = 1/N. \tag{1.5}$$

This normalization is convenient for the learning of a large number of patterns. The bar denotes the average over the quenched disorder $\{S_i^\mu\}$. For independent random patterns, this implies

$$\lambda = (1 + \varepsilon^2/N)^{-1/2}. \tag{1.6}$$

This scheme is stationary in the sense that the rule (1.4), (1.6) is time independent. This scheme will also be named marginalist since it is equivalent to the one of [4] through a rescaling of the temperature.

For simplicity of notation, we will use in the following a « time » such that the interval between two consecutive learning events is equal to one « time » unit. Emphasis is given first to the asymptotic regime, after an infinite number of patterns has been learned. Finite time effects will be considered later. Hence, if we denote $\xi^1 = (\xi_i^1)_{i=1,N}$ the last stored pattern, ξ^2 the previously stored one and so on, the model we consider is given by

$$T_{ij} = (\varepsilon/N)\sum_{\mu \geq 1} \lambda^\mu \xi_i^\mu \xi_j^\mu. \tag{1.7}$$

In this scheme, the actual intensities are exponentially decreasing with storage ancestry. In the large N limit, $\lambda \sim \exp - \varepsilon^2/2N$. Thus, for the most recently learned patterns, that is for μ finite (relative to N), the intensity is constant. For $\mu \ll N$, the intensity is vanishingly small : this observation suggests that the network will act like a Hopfield system, with a capacity of order N, the memorized patterns being the most recently stored.

By a straightforward generalization of (1.7), we can actually introduce a whole family of models in which the strength of learning is time dependent :

$$T_{ij} = (1/N)\sum_\mu \Lambda(\mu/N)\, \xi_i^\mu \xi_j^\mu \tag{1.8}$$

where $\Lambda(\mu)$ is any positive function such that

$$\int_0^\infty du\, \Lambda^2(u) = 1. \tag{1.9}$$

The marginalist scheme (1.7) is obtained for $\Lambda = \Lambda_m$:

$$\Lambda_m(\mu) = \varepsilon \exp - \mu\varepsilon^2/2. \tag{1.10}$$

It is particularly instructive to compare a given model with the Hopfield model. This model also belongs to the family (1.8)-(1.9) : it corresponds to learning with constant amplitude between some initial « time » $-N\tau$ up to the present moment. Hence, in our formulation it is given by the function $\Lambda = \Lambda_H$:

$$\begin{array}{ll} 0 \leq u \leq \tau & \Lambda_H(u) = \varepsilon \\ u > \tau & \Lambda_H(u) = 0. \end{array} \tag{1.11}$$

The normalization (1.9) imposes

$$\varepsilon^2 = 1/\tau. \tag{1.12}$$

Within this formulation, the Hopfield scheme is clearly not stationary : as the number $N\tau$ of stored patterns increases, the parameter ε which measures the effective uniform acquisition amplitude decreases.

For the general problem (1.8)-(1.9), the replica method can be used in the very same way as for the Hopfield model, that is following exactly the calculations of reference [3]. In the following, we will discuss the properties of these models within the assumption of replica symmetry : for any function Λ, the entropy at zero temperature is negative, but very small as in the Hopfield case. Thus, replica symmetry breaking effects are expected to be small, at least for recognition properties. Furthermore, we will mainly discuss the zero temperature limit, since it is the low temperature behaviour which is the most interesting.

However, we emphasize that any calculations that have been, or can be, done for the Hopfield model can be simply generalized for a generic function Λ.

2. Replica symmetric solution.

We compute the average free energy per spin

$$f = \overline{(- \text{Log}\,\text{Tr}\exp - \beta H)}/N\beta \tag{2.1}$$

with the replica method. As in [3], we are led to introduce three sets of order parameters : the states we consider are characterized by

i) the macroscopic overlaps m^{μ_i}, $i = 1, s$, s being finite as $N \to \infty$,

ii) the total mean square of the small random overlaps with the other patterns μ, weighted by $\Lambda(\mu/N)$:

$$r = \sum_{\mu \neq \mu_i} \Lambda^2(\mu/N)\overline{\langle (m^\mu)^2 \rangle} \tag{2.2}$$

iii) the Edwards-Anderson order parameter,

$$q = \overline{\langle S_i \rangle^2}. \tag{2.3}$$

One gets the following expression for the free energy

$$f = 1/2 \sum_{j=1}^{s} (m^{\mu_j})^2 \Lambda(\mu_j/N) - (\beta/2)\, r(q-1) + 1/2 \int_0^\infty du\, \Lambda(u) + $$
$$+ (1/2\beta) \int_0^\infty du \{ \text{Log}[1 - \beta\Lambda(u)(1-q)] - q\Lambda(u)/(1 - \beta(1-q)\Lambda(u)) \}$$
$$- (1/\beta) \left\langle\!\left\langle \text{Log}\, 2\cosh \beta\left[z\sqrt{r} + \sum_j \xi^{\mu_j} \Lambda(\mu_j/N)\, m^{\mu_j} \right] \right\rangle\!\right\rangle \tag{2.4}$$

where $\langle\langle . \rangle\rangle$ means averaging over the Gaussian variable z with zero mean and unit variance, and over the discrete distribution of $\{ \xi^{\mu_i} \}$. The values of the order-parameters are determined by the saddle-point equations :

$$m^{\mu_j} = \left\langle\!\left\langle \xi^{\mu_j} \tanh \beta\left[z\sqrt{r} + \sum_{i=1}^{s} \xi^{\mu_i} \Lambda(\mu_i/N)\, m^{\mu_i} \right] \right\rangle\!\right\rangle \tag{2.5}$$

$$q = \left\langle\!\left\langle \tanh^2 \beta\left[z\sqrt{r} + \sum_{i=1}^{s} \xi^{\mu_i} \Lambda(\mu_i/N)\, m^{\mu_i} \right] \right\rangle\!\right\rangle \tag{2.6}$$

$$r = \int_0^\infty du\, q\Lambda^2(u)/[1 - \beta(1-q)\Lambda(u)]^2. \tag{2.7}$$

We shall now concentrate on the zero temperature limit.

An indication of the capability of the network to recognize a given pattern of ancestry $p = \alpha N$, is given by looking at the solutions with a macroscopic overlap with the single pattern p :

$$m^{\nu_i} = m\delta_{\nu_i, p}. \tag{2.8}$$

As $\beta \to \infty$, equations (2.5)-(2.7) yield

$$m = \sqrt{2/\pi} \int_0^x dz\, e^{-z^2/2} \tag{2.9}$$

with

$$x = m\Lambda(\alpha)/\sqrt{r},$$
$$q = 1 - C/\beta, \tag{2.10}$$

with

$$C = (2/\pi r)^{1/2} \exp - x^2/2 \tag{2.11}$$

$$r = \int_0^\infty du\, \Lambda^2(u)/[1 - C\Lambda(u)]^2. \tag{2.12}$$

Finally one obtains two coupled equations for the reduced variables x and C :

$$x\, e^{-x^2/2} = \Lambda(\alpha)\, C \int_0^x dz\, e^{-z^2/2} \tag{2.13}$$

$$e^{-x^2} = \pi/2 \int_0^\infty du\, \Lambda^2(u)/[1 - C\Lambda(u)]^2 \tag{2.14}$$

which, for a given function Λ, can be solved numerically. In particular, for the Hopfield scheme, $\Lambda = \Lambda_H$ (see (1.11), (1.12)), one recovers the equations of reference [3]. Numerical solution of these give the critical values $\alpha_c = 1/\varepsilon_c^2 = 0.138...$, $m_c = 0.97...$

For the marginalist scheme $\Lambda = \Lambda_m$, we find that there is no solution with $m \neq 0$ for

$$\varepsilon < \varepsilon_c = 2.465. \tag{2.15}$$

For $\varepsilon = \varepsilon_c$, there is one (stable) solution with $\alpha = 0$,

and

$$m = 0.933 . \qquad (2.16)$$

For $\varepsilon > \varepsilon_c$, there are two solutions with m non zero, for :

$$0 \leqslant \alpha \leqslant \alpha(\varepsilon)$$

and one finds that the (meta)stable one corresponds to the highest value of m. This state has a retrieval quality $m(\alpha, \varepsilon) > m_c$. The functions $\alpha(\varepsilon)$, $m(\alpha(\varepsilon), \varepsilon)$ and $m(0, \varepsilon)$ are shown in figure 1. $\alpha(\varepsilon)$ is the (stationary) capacity of the memory. It increases with ε until a maximal value at $\varepsilon = \varepsilon_{opt}$

$$\varepsilon_{opt} = 4.108 \qquad (2.17)$$

with a capacity

$$\alpha_{opt} = 0.04895 \qquad (2.18)$$

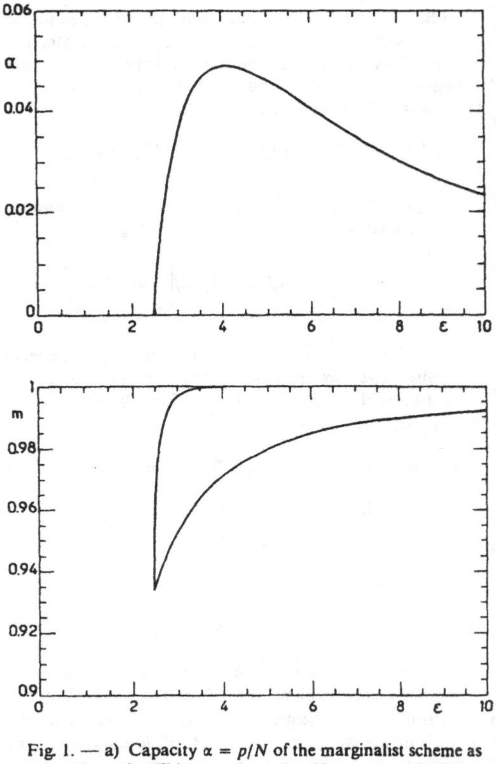

Fig. 1. — a) Capacity $\alpha = p/N$ of the marginalist scheme as a function of ε. This capacity is zero for $\varepsilon < \varepsilon_c = 2.4648...$, and goes through a maximum at $\varepsilon = \varepsilon_{opt} = 4.108$. b) Retrieval quality as a function of ε. The upper curve is the retrieval quality $m(0, \varepsilon)$ of the most recently stored pattern. The lower curve is the retrieval quality $m(\alpha(\varepsilon), \varepsilon)$ of the pattern $p = \alpha(\varepsilon) N$, that is of the most anciently stored, but still memorized, pattern.

and decreases for $\varepsilon > \varepsilon_{opt}$ — the number αN of memorized patterns being 1 in the large ε limit.

It is also of interest to request a retrieval quality at least equal to a given value of m — hence of x, see (2.9). The interesting limit consists in requiring a very good retrieval quality, that is m close to one. In this limit, it is easy to compute the capacity for any function Λ. In particular for models with a tunable parameter, such as the marginalist scheme, one finds explicitly the optimal value of the parameter and the corresponding optimal capacity.

When $m \to 1$, $x \to \infty$ and $C \to 0$. Then one has

$$m = 1 - \sqrt{2/\pi}(1/x)\, e^{-x^2/2} \qquad (2.19)$$

$$\Lambda(\alpha) = x . \qquad (2.20)$$

If Λ is a decreasing function, the capacity α_x is the largest value of α for which (2.20) holds. For the Hopfield scheme Λ_H one recovers

$$\alpha_x = 1/\varepsilon^2 \qquad (2.21)$$

for

$$\varepsilon \geqslant x . \qquad (2.22)$$

The maximal capacity is reached at $\varepsilon = \varepsilon_H$, with the capacity α_H :

$$\varepsilon_H = x , \quad \alpha_H = 1/x^2 . \qquad (2.23)$$

In fact, this capacity α_H is the maximal possible capacity whatever Λ is. This is easily seen in this limit of good retrieval : due to the normalization condition (1.9) on Λ, if Λ is piece-wise continuous, α_x can be equal to $\alpha_H = 1/x^2$ if and only if Λ is constant, equal to x, on an interval of length $1/x^2$, and zero everywhere else. This result confirms the fact, observed in previous numerical simulations [4], that there is a price to pay, viz. a decrease of the capacity, to obtain a working memory, robust to new learning.

For the marginalist scheme, in the limit $m \to 1$, one finds

$$\begin{array}{ll} \varepsilon \geqslant \varepsilon_c = x & \alpha_x(\varepsilon) = (2/\varepsilon^2)\,\mathrm{Log}(\varepsilon/x) \\ \varepsilon < \varepsilon_c & \alpha_x = 0 . \end{array} \qquad (2.24)$$

This gives a maximal value of α_x at a value $\varepsilon_{opt}(x)$:

$$\varepsilon_{opt}(x) = x\sqrt{e} , \qquad (2.25)$$

with a capacity

$$\alpha_{opt}(x) = 1/ex^2 . \qquad (2.26)$$

Note that $\alpha_{opt} = 1/\varepsilon_{opt}^2$, $\varepsilon_{opt} = \sqrt{e\varepsilon_c}$, whatever x is.

In order to compare with the numerical estimations of reference [4], let us suppose that $m > 0.97$. Numerical solution of the equations (2.13), (2.14) gives the exact values

$$\begin{aligned} \varepsilon_c &= 2.529 \\ \varepsilon_{opt} &= 4.088 \\ \alpha_{opt} &= 0.0489 . \end{aligned} \qquad (2.27)$$

Note that these values are close to the values (2.15), (2.17) and (2.18), obtained without restriction on m. The values reported for $\varepsilon_c \sim 2.2$, $\varepsilon_{opt} \sim 4$, were in reasonable agreement with (2.27), but at $N = 100$, a capacity of 0.065 was estimated. The corresponding discrepancy can be attributed mainly to finite size effects : further numerical simulations at $\varepsilon = 4$, for N up to 320 give an estimate 0.059 ± 0.005 for the capacity.

3. Learning by reinforcement, and finite time effects.

Other interesting features can be obtained within our general formulation :

i) the repeated learning of a given pattern with a period L, within a background of random patterns; ii) the learning of a sequence of L patterns, repeated *ad infinitum*; iii) the learning of a number L of patterns starting from tabula rasa at « time » $- L$.

The last problem (iii) will be treated in this section because it appears to be formally very similar to problem (ii). The two first problems will illustrate the well known efficiency of repetitive learning (« to teach is to repeat »). That a learning scheme leads to such effects is not surprising. What is interesting is the possibility, within our formalism, to quantify these effects.

i) We illustrate here the repetitive priming effect. We consider the learning of patterns independently chosen at random, except one given pattern which is learned periodically, with a period $L = gN$.

If $p = \alpha N$ is the last « time » when this pattern has been learned, its actual weight is

$$\Lambda_g(\alpha) = \sum_{l \geq 0} \Lambda(\alpha + lg) . \tag{3.1}$$

Thus instead of the function Λ, we have to consider the fonction Λ_g

$$\begin{aligned} \Lambda_g(\alpha) &= \Lambda(u) \quad u \neq \alpha \bmod g \\ \Lambda_g(\alpha) &= \sum_{l \geq 0} \Lambda(\alpha + lg) . \end{aligned} \tag{3.2}$$

Since Λ_g differs from Λ only at a denumerable set of points, we still have

$$\int_0^{\infty} \Lambda_g^2(u)\, du = 1 . \tag{3.3}$$

Hence we can apply the preceeding formalism for Λ_g. If we look at the states having a macroscopic overlap with the periodically learned pattern, the equations are the same as (2.9)-(2.12), with $\Lambda_g(\alpha)$ instead of $\Lambda(\alpha)$ in (2.10).

Since $\Lambda_g(\alpha) > \Lambda(\alpha)$, we see already that the retrieval quality will be enhanced by this periodic learning. Consider in particular the case $\alpha = g$, which means that we are looking at the retrieval quality just before re-learning. At zero temperature, the maximal value g^* of g, for which the retrieval quality is at least equal to a given value m, is given by

$$\Lambda_{g^*}(g^*) = \Lambda(\alpha_x) . \tag{3.4}$$

We recall that x is related to m *via* (2.19). For the marginalist scheme, this gives the relation

$$g^*(\varepsilon) = (2/\varepsilon^2) \operatorname{Log}\left[1 + \exp(\varepsilon^2 \alpha_x(\varepsilon)/2)\right] . \tag{3.5}$$

This shows that for any ε

$$g^*(\varepsilon) > \alpha_x(\varepsilon) . \tag{3.6}$$

Thus, whereas a given pattern learned only once is retained during a time $N\alpha_x(\varepsilon)$, a pattern learned every N_g patterns is never forgotten provided $g \leq g^*$, where g^* is much greater than $\alpha_x(\varepsilon)$. Even for $\varepsilon < \varepsilon_c(x)$, where $\alpha_x(\varepsilon) = 0$, g^* is finite :

$$\varepsilon < \varepsilon_c(x) \quad g^* = (2/\varepsilon^2) \operatorname{Log} 2 . \tag{3.7}$$

In fact, through the diminution of the noise due to the other stored patterns, the smaller ε, the greater g^*.

ii) Consider now the repeated learning of the same sequence of $L = gN$ patterns. At « time » zero, the weight of the pth pattern is

$$1 \leq p \leq L \qquad \lambda_g(p/N) = \sum_{l \geq 0} \Lambda((p/N) + lg) .$$

To cast this problem into the general formulation, we have to consider the normalized weight Λ_g :

$$\begin{aligned} 0 \leq u \leq g \quad & \Lambda_g(u) = \lambda_g(u) \Big/ \left[\int du\, \lambda_g^2(u)\right]^{1/2} \\ u > g \quad & \Lambda_g(u) = 0 \end{aligned} \tag{3.8}$$

Now the model defined by Λ_g belongs to our general family : the new equations look the same as (2.9)-(2.14), with Λ replaced by Λ_g. For the marginalist scheme, one has

$$0 \leq u \leq g \quad \Lambda_g(u) = \varepsilon\, e^{-\varepsilon^2 u/2} / [1 - e^{-\varepsilon^2 g}]^{1/2} \tag{3.9}$$

In particular, for ε small, i.e.

$$\varepsilon^2 g \ll 1$$

Λ_g is constant for $u \leq g$:

$$0 \leq u \leq g \quad \Lambda_g \simeq 1/\sqrt{g} . \tag{3.10}$$

This means that for ε small enough, the resulting memory is equivalent to the Hopfield limit. Within a palimpsestic scheme, one way to approach the optimal capacity, that is the Hopfield capacity, is to repeat again and again the sequence of patterns to be learned, but with a very low intensity.

iii) Finite « time » effects. The learning of $L = gN$ patterns (starting from tabula rasa) appears to be formally very similar to the learning of a periodic sequence. We want to recover the results of sections 1 and 2 in the infinite g limit : we consider the storing of the pattern of ancestry $p = \alpha N$ with a weight $\Lambda(\alpha)$, for $\alpha \leqslant g$. But to cast this problem into our general formulation, we have to consider the normalized weight $\Lambda_g(\alpha)$, such that (1.9) remains true for any finite g :

$$1 \leqslant p \leqslant L \quad \Lambda_g(p/N) = \Lambda(p/N) \Big/ \left(\int_0^g \mathrm{d}u \, \Lambda^2(u) \right)^{1/2} \tag{3.11}$$

$$u > g \qquad \Lambda_g(u) = 0$$

In the limit $g \to \infty$, Λ_g becomes identical to Λ. For the marginalist scheme, Λ_g is again given by (3.9) : the two problems are thus equivalent.

If one works at a given value of ε, this gives immediately that, for short times, the complete set of patterns is memorized. This is the case until a critical value $g_c(\varepsilon)$. For $g > g_c$, only a fraction of the patterns are memorized, and the capacity decreases from g_c toward its asymptotic value $\alpha(\varepsilon)$. This is illustrated in figure 2 for $\varepsilon = 4$. Solving numerically the equations for Λ_g one finds g_c slightly greater than $1/\varepsilon^2$:

$$g_c(\varepsilon = 4.0) = 0.0672 \tag{3.12}$$

whereas

$$\alpha(\varepsilon = 4.0) = 0.0488 \,. \tag{3.13}$$

While the capacity decreases from $g_c(\varepsilon)$ to $\alpha(\varepsilon)$, the retrieval quality decreases slightly from a little better than 98 % to about 97 %.

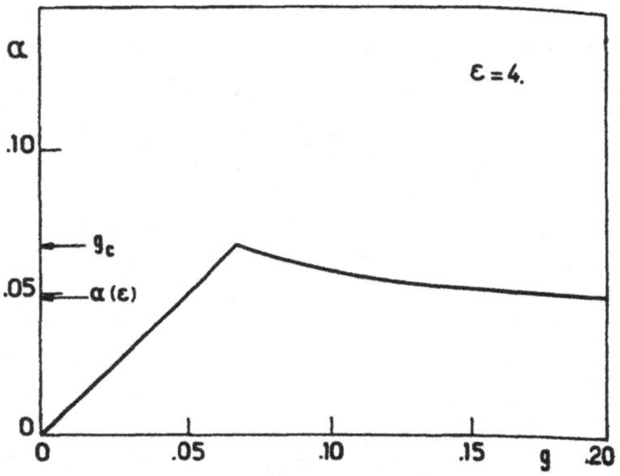

Fig. 2. — Capacity of the marginalist scheme for $\varepsilon = 4.0$, as a function of $g = L/N$, where L is the total number of stored patterns. All the patterns are memorized for g up to $g_c = 0.06719...$ For $g > g_c$, the capacity decreases toward its asymptotic value $\alpha(\varepsilon)$.

3. Conclusion.

In this paper we have shown how one can analyse very simply a whole family of models for working memory. Using the replica symmetry approximation, the resulting formalism is quite simple, and allows one to answer many questions about the behaviour of such memories. In particular we confirm that the capacity of these working memories are lower than the capacity of the Hopfield model. Typical phenomena of human short-term memory, such as the repetitive priming effect, are simply modelized and quantified within our formalism.

The incentive for considering non uniform acquisition intensities came from biology. From the view point of statistical physics, it is clear that everything that has been computed [2, 3] in the case of uniform intensities can be generalized. This paper has only made a first step in this promising direction.

Acknowledgments.

Fruitful discussions with J. P. Changeux and S. Dehaene are gratefully acknowledged.

References

[1] Hopfield, J. J., *Proc. Natl. Acad. Sci.* **79** (1982) 2554 ;
Peretto, P., *Biol. Cybern.* **50** (1984) 51.

[2] Amit, D. J., Gutfreund, H., Sompolinsky, H., *Phys.*
Toulouse, G., Dehaene, S., Changeux, J. P., *Proc. Natl. Acad. Sci.*, **83** (1986) 1695 :
Personnaz, L., Guyon, I., Dreyfus, G., Toulouse, G., *J. Stat. Phys.* **43** (1986) 411 :
Feigelman, M. V., Ioffe, L. B., *Europhys. Lett.* **1** (1986) 197 ;
Sompolinsky, H., preprint.

[3] Amit, D. J., Gutfreund, H., Sompolinsky, H., *Phys. Rev. Lett.* **55** (1985) 1530 ;
Amit, D. J., Gutfreund, H., Sompolinsky, H., preprint (1986).

[4] Nadal, J. P., Toulouse, G., Changeux, J. P., Dehaene, S., *Europhys. Lett.*, **1** (1986) 535.

[5] Parisi, G., preprint.

[6] A suggestion for such learning schemes was already present in a paragraph of the original paper of Hopfield [1] and should have been underlined in [4].

REFERENCES

Amit D. J., H. Gutfreund and H. Sompolinsky, *Phys. Rev.* **A32** (1985a) 1007, reprint 30, this book.

Amit D. J., H. Gutfreund and H. Sompolinsky, *Phys. Rev. Lett.* **55** (1985b) 1530, reprint 31, this book.

Amit D. J., H. Gutfreund and H. Sompolinsky, preprint No. 23, (Racah Institute of Physics, Jerusalem, 1986).

Anderson P. W., *Proc. of Nac. Acad. Sci. USA* **80** (1983) 3386.

Anderson P. W., in "Emerging Synthesis in Science", ed. D. Pines, (Santa Fe Institute, Santa Fe NM, 1984).

Angles d'Auriac J. C., M. Preissman and R. Rammal, *J. Phys. Lett.* **46** (1985) L173.

Athanasiu G. G., C. P. Bachas and W. F. Wolff, *Phys. Rev. Lett.* (1986).

Ballard D. H., *The Behavioral and Brain Sciences* **9** (1986) 67.

Banavar J. R., D. Sherrington and N. Sourlas, to appear in *J. Phys. A* (1986).

Bausch R., H. K. Janssen and H. Wagner, *Z. Phys.* **B24** (1976) 113.

Beardwood J., J. H. Halton and J. M. Hammersley, *Proc. Camb. Philos. Soc.* **55** (1959) 299.

Bhatt R. N. and A. P. Young, *J. Magn. Matter* **54–57** (1986) 191.

Bieche I., R. Maynard, R. Rammal and J. P. Uhry, *J. Phys.* **A13** (1980) 2553.

Bienenstock E. and C. Von Malsburg in "Disordered Systems and Biological Organization", eds. E. Bienenstock, F. Fogelman and G. Weisbuch (Springer, Berlin).

Binder K., ed., "Monte Carlo Methods in Statistical Physics" (Springer, 1979).

Blandin A., *J. Phys. C6* **39** (1978) 1499.

Blandin A., M. Gabay and T. Garel, *J. Phys.* **C13** (1980) 403.

Bonomi E. and J. L. Lutton, *SIAM review* **26** (1984) 551.

Bray A. J. and M. A. Moore, *Phys. Rev. Lett.* **41** (1978) 1068.

Bray A. J. and M. A. Moore, *J. Phys.* **C12** (1979) L441.

Brout R., *Phys. Rev.* **115** (1959) 824.

Caianiello E. R., *J. Theor. Biol.* **2** (1961) 204.

Caianiello E. R. and A. de Luca, *Kybernetic* **3** (1966) 33.

Cerny V., *J. of Optimization Theory and Applications* (1985).

Changeux J. P., "L'Homme Neuronal", (Fayard, Paris, 1983) Chaps. 6, 7; "Neuronal Man" (Pantheon, New York, 1985).

Churchland P. S., "Neurophilosophy", (The MIT Press, Cambridge, Mass., 1986).
Cooper L. N., F. Liberman and E. Oja, *Neuroscience* **33** (1979) 9.
Crisanti A., D. J. Amit and H. Gutfreund, *Europhys. Lett.* **2** (1986) 337.
Dasgupta C. and H. Sompolinsky, *Phys. Rev.* **B27** (1983) 4511.
De Almeida J. R. L. and E. J. S. Lage, *J. Phys.* **C16** (1983) 939.
De Dominicis C., *Phys. Rev.* **B18** (1978) 4913.
De Dominicis C., *Phys. Rep.* **67** (1980) 37.
De Dominicis C., in "Heidelberg colloquium on spin glasses", Morgenstern and Van Hemmen eds., (Springer, 1983).
De Dominicis C. and I. Kondor, in *Lecture Notes in Physics* (Springer Verlag, 1984).
De Dominicis C. and L. Peliti, *Phys. Rev.* **B18** (1978) 353.
De Dominicis C. and A. P. Young, *J. Phys.* **A16** (1983) 2063.
De Dominicis C., M. Gabay, T. Garel and H. Orland, *J.. Phys.* **41** (1980) 923.
De Dominicis C., M. Gabay and H. Orland, *J. Phys. 42* (1981) L523.
De Dominicis C. and P. Mottishaw, Saclay preprints SPhT/86/088,097, (1986).
De Gennes P. G., "Scaling concepts in polymer physics", (Cornell University Press, 1979).
Derrida B and G. Toulouse, *J. Phys. Lett.* **46** (1985) L223.
Derrida B., *J. Phys. Lett.* **46** (1985) L401.
Derrida B. and Flyvjberg, in preparation, (1986).
Derrida B. and E. Gardner, preprint CEA Saclay, (1985).
Doreian P., "Mathematics and the study of the social relation", (Academic Press, New York, 1971) Chap. V.
Dotsenko Vik S., *J. Phys.* **C17** (1985) L1017.
Duplantier B., *J. Phys.* **A14** (1981) 283.
Eigen M., in "Disordered Systems and Biological Organization", eds. E. Bienenstock, F. Fogelman and G. Weisbuch, (Springer, Berlin, 1985) p. 25.
Ettelaie R. and M. A. Moore, *J. Phys. Lett.* **46** (1985) L893.
Frauenfelder H., *Helvetica Physica Acta* **57** (1984) 165.
Gardner E., preprint, (1985).
Garey M. R. and D. S. Johnson, "Computers and Intractability" (Freeman, New York, 1979).
Geman S. and D. Geman, *IEEE Proc. Pattern Analysis and Machine Intelligence 6* (1985).
Gidas B., in "Disordered systems and biological organization", eds. Bienenstock *et al.* (Springer, 1986).
Gilbert W., Nature **319** (1986) 618.
Glauber R. J., *J. Math. Phys.* **4** (1963) 294.
Goles Chacc E., "Comportement Dynamique de Reseaux d'Automates", Thesis of l'Universite Scientifique et Medicale de Grenoble, (1985).
Goltsev A. V., *J. Phys.* **A16** (1983) 1337.
Goltsev A. V., *J. Phys.* **A17** (1984) 237.
Grossman S., F. Wegner and K. H. Hoffman, *J. Phys. Lett.* **46** (1985) L575.
Hebb D. O., in "The Organization of Behavior" (Wiley, New York, 1949).

Hinton G. F. and Anderson J. A., eds., "Parallel Models of Associative Memory" (Erlbaum Hillsdale N.J., 1981).

Hinton G. F., T. J. Sejnowski and D. H. Ackley, *Cognitive Sci.* **9** (1985) 147.

Hopfield J. J., *Proc of Nac. Acad. Sci. USA* **79** (1982) 2554, reprint 29, this book.

Hopfield J. J. and D. W. Tank, in "Disordered Systems and Biological Organization", eds. E. Bienenstock, F. Fogelman and G. Weisbuch (Springer, Berlin, 1985).

Hopfield J. J. and D. W. Tank, *Science* **233** (1986) 625.

Hörner H., *Z. Phys.* **B57** (1984) 29, 39.

Huberman B. A. and M. Kerszberg, *J. Phys.* **A18** (1985) L331.

Hurlbert A. and T. Poggio, *Nature* **321** (1986) 651.

Janssen H. K., *Z. Phys.* **B23** (1976) 377.

Jerne N. K., *Ann. Immunol. (Inst. Pasteur)* **125C** (1974) 373.

Kanter I. and H. Sompolinsky, preprint, (Bar Ilan University, Tel Aviv).

Kauffman S., in "Disordered Systems and Biological Organization", eds. E. Bienenstock, F. Fogelman and G. Weisbuch, (Springer, Berlin, 1985).

Kinzel W., *Phys. Rev.* **B33** (1986) 5086.

Kirkpatrick S., in *Lecture notes in Physics* **149** (1981) 280, (Springer, 1981).

Kirkpatrick S., C. D. Gelatt Jr. and M. P. Vecchi, *Science* **220** (1983) 671.

Kirkpatrick S. and G. Toulouse, *J. Phys.* **46** (1985) 1277.

Klatzky R. L., "Human Memory" (W. A. Freeman and Company, San Francisco, 1976).

Koch C. and T. Poggio, "Biophysics of Computation: Neuron, Synapses and Membranes", Internal Report MIT Artificial Intelligence Laboratory and Center for Biological Information Processing, Whitaker College, (1984).

Kohonen T., "Self-organization and Associative Memory" (Springer Verlag, Berlin, 1984).

Kondor I., *J. Phys.* **A16** (1983) L217.

Kondor I. and C. De Dominicis, *Europhys. Lett.* **2** (1986) 617.

Lawler E. L., J. K. Lenstra, A. H. G. Rinnooy Kan and D. B. Shmoys, eds., "The traveling salesman problem", (Wiley, Chichester, 1985).

LeCun Y., in "Disordered Systems and Biological Organization", eds. E. Bienenstock, F. Fogelman and G. Weisbuch (Springer, Berlin, 1985).

Lin S., *Bell Syst. Tech. J.* **44** (1965) 2245.

Little W. A., *Math. Biosci.* **19** (1974) 101.

Little W. A. and G. L. Shaw, *Math. Biosci.* **39** (1978) 281.

Maritan A. and A. L. Stella, *J. Phys.* **A19** (1986) L259.

Marr D., "Vision", (Freeman, 1982).

Martin P. C., E. D. Siggia and H. A. Rose, *Phys. Rev.* **A8** (1978) 423.

Mattis D. C., *Phys. Lett.* **56A** (1976) 421.

McCullochs W. S. and W. Pitts, *Bulletin of Math. Biophys.* **5** (1943) 115.

Metropolis N., A. W. Rosenbluth, M. N. Rosenbluth, A. H. Teller and E. J. Teller, *J. Chem. Phys.* **21** (1953) 1087.

Mézard M., Nadal J. P. and G. Toulouse, *J. Phys.* **47** (1986) 1457, reprint 35, this book.

Mézard M. and G. Parisi, *J. Phys. Lett.* **45** (1984) L707.

Mézard M. and G. Parisi, *Europhys. Lett.*, to appear, (1986a).

Mézard M. and G. Parisi, *Europhys. Lett.*, to appear, (1986b).

Mézard M., G. Parisi, N. Sourlas, G. Toulouse and M. A. Virasoro, *Phys. Rev. Lett.* **52** (1984) 1156.

Minsky M. and S. Papert, "Perceptrons" (MIT Press, Cambridge, Mass., 1969).

Nadal J. P., G. Toulose and S. Dehaene, *Europhys. Lett.* **1** (1986) 535.

Nemoto K., preprint (Hokkaido University, 1986).

Nemoto K. and H. Takayama, *J. Phys.* **C18** (1985) L529.

Ogielski A. T. and D. L. Stein, *Phys. Rev. Lett.* **55** (1985) 1634.

Oliverio A., "Storia Naturale della mente, L'evoluzione del comportamento", (Ed. Boringhieri, Torino, 1984).

Onsager L., *J. Am. Chem. Soc.* **58** (1936) 1486.

Orland H., *J. Phys. Lett.* **46** (1985) L763.

Paladin G., M. Mézard and C. De Dominicis, *J. Phys. Lett.* **46** (1985) L985.

Palmer R. G. and C. M. Pond, *J. Phys.* **F9** (1979) 1451.

Palmer R. G., D. L. Stein, E. Abrahams and P. W. Anderson, *Phys. Rev. Lett.* **53** (1984) 1958.

Papadimitriou C. H. and K. Steiglitz, "Combinatorial Optimization", (Prentice Hall, New Jersey, 1982).

Parga N. and G. Parisi, *J. of Phys. A*, in press, (1985).

Parga N., G. Parisi and M. A. Virasoro, *J. Phys. Lett.* **45** (1984) L1063.

Parga N. and M. A. Virasoro, *J. Phys.* **47** (1986) 1857.

Parisi G., *J. Phys. Lett.* **44** (1983). L581.

Parisi G. and G. Toulouse, *J. Phys. Lett.* **41** (1980) L361.

Parisi G., J. of Phys. **A19** (1986a) L617.

Parisi G., J. of Phys. **A19** (1986b) L675, reprint 34, this book.

Plefka T., *J. Phys.* **A15** (1982) 1971.

Purves D. and J. W. Lichtman, "Principles of Neural Development", (Sinauer Associates Inc., Sunderland Mass., 1985).

Rammal R., G. Toulouse and M. A. Virasoro, *Rev. Mod. Phys.* **58** (1986) 765.

Rokhsar D. S., P. W. Anderson and D. L. Stein, *J. of Molec. Evolution* **23** (1986) 119.

Rosenblatt F., "Principles of Neurodynamics", (Spartan. Washington DC., 1961).

Ruelle D., preprint, (IHES, Bures-Sur-Yvettes, 1986).

Rummelhart D. E., G. E. Hinton and R. J. Williams, *Nature* **323** (1986) 533.

Schreckenberg M., *Z. Phys.* **B60** (1985) 483.

Siarry P. and G. Dreyfus, *J. Phys. Lett.* **45** (1984) L39.

Sejnowski T. J. and C. R. Rosenberg, "NetTalk: A parallel network that learns to read aloud", (The John Hopkins University Technical, Report JHU/EECS-86/01, 1986).

Solla S. A., G. B. Sorkin and S. R. White, in "Disordered Systems and Biological Organization", eds. E. Bienenstock, F. Fogelman and G. Weisbuch, (Springer, Berlin, 1985).

Sommers H. J., *Z. Phys.* **B31** (1978) 301.

Sommers H. J., *J. Phys.* **A16** (1983a) 447.

Sommers H. J., *Z. Phys.* **B50** (1983b) 97.

Sommers H. J., contribution to the "Heidelberg colloquium on glassy dynamics and optimization", eds. Van Hemmen and Morgenstern, (Springer, 1986).
Sommers H. J. and W. Dupont, *J. Phys.* **C17** (1984) 5785.
Sompolinsky H., *Phys. Rev.* **B23** (1980) 1371.
Sompolinsky H., *Phys. Rev. Lett.* **47** (1981) 935.
Sompolinsky H. and A. Zippelius, *Phys. Rev. Lett.* **47** (1981) 359.
Soukoulis C. M., K. Levin and G. S. Grest, *Phys. Rev. Lett.* **48** (1982) 1756.
Sourlas N., *J. Phys. Lett.* **45** (1984) L969.
Sourlas N., to appear in *Europhys. Lett.*, (1986).
Tanaka and Edwards, *J. Phys.* **F10** (1980) 2471.
Teitel S. and E. Domany, *Phys. Rev. Lett.* **55** (1985) 2176.
Toulouse G., *Comm. Phys.* **2** (1977) 115.
Toulouse G. and B. Derrida, *Proc VI Brazilian Symposium on Theoretical Physics*, (Ed. du CNPq, 1981).
Toulouse G., S. Dehaene and J. P. Changeux, *Proc. of Nac. Acad. Sci. USA* **83** (1986) 1695.
Young A. P., A. J. Bray and M. A. Moore, *J. Phys.* **C17** (1984) L155.
Van Hemmen J. L. and R. G. Palmer, *J. Phys.* **A12** (1979) 563.
Van Hemmen J. L. and R. G. Palmer (1983).
Vannimenus J. and M. Mézard, *J. Phys. Lett.* **45** (1984) L1145.
Vannimenus J., G. Toulouse and G. Parisi, *J. Phys.* 42 (1981) 565.
Viana L. and A. J. Bray, *J. Phys.* **C18** (1985) 3037.
Wilson A. C., *Sc. Am.* **253**, No. 4 (1985) p. 148.
Zuckerlandl E. and L. Pauling, *J. Theor. Biol.* **8** (1965) 357.